OPEN PIT MINE PLANNING & DESIGN
VOLUME 1 – FUNDAMENTALS

 BALKEMA – Proceedings and Monographs
in Engineering, Water and Earth Sciences

OPEN PIT MINE PLANNING & DESIGN

Volume 1 – Fundamentals

WILLIAM HUSTRULID
Department of Mining Engineering, University of Utah, Salt Lake City, Utah

MARK KUCHTA
Department of Mining Engineering, Colorado School of Mines, Golden, Colorado

LONDON/LEIDEN/NEW YORK/PHILADELPHIA/SINGAPORE

Cover photo credit: Courtesy of Kennecott Utah Copper

Library of Congress Cataloging-in-Publication Data

Open pit mine planning & design/William Hustrulid, Mark Kuchta. Rotterdam;
Brookfield, VT: A.A. Balkema, 1998.
2 v. (xv, 836 p.): ill.; 26 cm. + 1 computer disk (3½ in.)
v. 1. Fundamentals – v. 2. CSMine software package.
Includes bibliographical references and indexes.
    ISBN 90-5410-173-3 (hardbound: set);   ISBN 90-5410-183-0 (pbk.: set);
    ISBN 90-5410-181-4 (hardbound: v. 1);   ISBN 90-5410-184-9 (pbk.: v. 1);
    ISBN 90-5410-182-2 (hardbound: v. 2);   ISBN 90-5410-185-7 (pbk.: v. 2)
1. Strip mining   2. Strip mining–Design–Data processing.
I. Hustrulid, William   II. Kuchta, Mark.

TN291.H87 1998
622/.292–dc21

                                                                                                99226853

1st edition, 1st print: 1995
2nd print: 1998
Revised and extended 2nd edition: 2006

Copyright © 2006 Taylor & Francis plc., London, UK

All rights reserved. No part of this publication or the information contained herein may be reproduced, stored in a retrieval system, or transmitted in any form or by any means, electronic, mechanical, by photocopying, recording or otherwise, without written prior permission from the publishers.

Although all care is taken to ensure the integrity and quality of this publication and the information herein, no responsibility is assumed by the publishers nor the authors for any damage to property or persons as a result of operation or use of this publication and/or the information contained herein.

Published by: Taylor & Francis/Balkema
                  P.O. Box 447, 2300 AK Leiden, The Netherlands
                  e-mail: Pub.NL@tandf.co.uk
                  www.taylorandfrancis.co.uk/engineering, www.crcpress.com

Hardbound edition
Complete set of two volumes plus CD-ROM:
        ISBN 10: 0 415 40737 0   ISBN 13: 978 0 415 40737 3
Volume 1: ISBN 10: 0 415 40738 9   ISBN 13: 978 0 415 40738 0
Volume 2: ISBN 10: 0 415 40739 7   ISBN 13: 978 0 415 40739 7
CD-ROM: ISBN 10: 0 415 40740 0   ISBN 13: 978 0 415 40740 3

Paperback edition
Complete set of two volumes plus CD-ROM:
        ISBN 10: 0 415 40741 9   ISBN 13: 978 0 415 40741 0
Volume 1: ISBN 10: 0 415 40742 7   ISBN 13: 978 0 415 40742 7
Volume 2: ISBN 10: 0 415 40743 5   ISBN 13: 978 0 415 40743 4
CD-ROM: ISBN 10: 0 415 40740 0   ISBN 13: 978 0 415 40740 3

Printed in Great Britain by Antony Rowe Ltd, Chippenham, Wiltshire

Contents

PREFACE   XIII

ABOUT THE AUTHORS   XV

1 MINE PLANNING   1

   1.1   Introduction   1
         1.1.1   The meaning of ore   1
         1.1.2   Some important definitions   2
   1.2   Mine development phases   5
   1.3   An initial data collection checklist   7
   1.4   The planning phase   11
         1.4.1   Introduction   11
         1.4.2   The content of an intermediate valuation report   12
         1.4.3   The content of the feasibility report   12
   1.5   Planning costs   17
   1.6   Accuracy of estimates   17
         1.6.1   Tonnage and grade   17
         1.6.2   Performance   17
         1.6.3   Costs   18
         1.6.4   Price and revenue   18
   1.7   Feasibility study preparation   19
   1.8   Critical path representation   24
   1.9   Mine reclamation   24
         1.9.1   Introduction   24
         1.9.2   Multiple-use management   25
         1.9.3   Reclamation plan purpose   28
         1.9.4   Reclamation plan content   28
         1.9.5   Reclamation standards   29
         1.9.6   Surface and ground water management   31
         1.9.7   Mine waste management   32
         1.9.8   Tailings and slime ponds   33
         1.9.9   Cyanide heap and vat leach systems   33
         1.9.10   Landform reclamation   34

1.10	Environmental planning procedures		35
	1.10.1	Initial project evaluation	35
	1.10.2	The strategic plan	37
	1.10.3	The environmental planning team	38
1.11	A sample list of project permits and approvals		40
	References		40
	Review questions and exercises		42

2 MINING REVENUES AND COSTS — 46

2.1	Introduction		46
2.2	Economic concepts including cash flow		46
	2.2.1	Future worth	46
	2.2.2	Present value	47
	2.2.3	Present value of a series of uniform contributions	47
	2.2.4	Payback period	48
	2.2.5	Rate of return on an investment	48
	2.2.6	Cash flow (CF)	49
	2.2.7	Discounted cash flow (DCF)	50
	2.2.8	Discounted cash flow rate of return (DCFROR)	50
	2.2.9	Cash flows, DCF and DCFROR including depreciation	51
	2.2.10	Depletion	52
	2.2.11	Cash flows, including depletion	54
2.3	Estimating revenues		55
	2.3.1	Current mineral prices	55
	2.3.2	Historical price data	63
	2.3.3	Trend analysis	76
	2.3.4	Econometric models	86
	2.3.5	Net smelter return	86
	2.3.6	Price-cost relationships	92
2.4	Estimating costs		95
	2.4.1	Types of costs	95
	2.4.2	Costs from actual operations	96
	2.4.3	Escalation of older costs	96
	2.4.4	The original O'Hara cost estimator	116
	2.4.5	The updated O'Hara cost estimator	118
	2.4.6	Detailed cost calculations	137
	2.4.7	Quick-and-dirty mining cost estimates	152
	2.4.8	Current equipment, supplies and labor costs	153
	References		159
	Review questions and exercises		164

3 OREBODY DESCRIPTION — 168

3.1	Introduction	168
3.2	Mine maps	168
3.3	Geologic information	183

3.4	Compositing and tonnage factor calculations		187
	3.4.1	Compositing	187
	3.4.2	Tonnage factors	193
3.5	Method of vertical sections		198
	3.5.1	Introduction	198
	3.5.2	Procedures	198
	3.5.3	Construction of a cross-section	199
	3.5.4	Calculation of tonnage and average grade for a pit	203
3.6	Method of vertical sections (grade contours)		212
3.7	The method of horizontal sections		219
	3.7.1	Introduction	219
	3.7.2	Triangles	219
	3.7.3	Polygons	223
3.8	Block models		227
	3.8.1	Introduction	227
	3.8.2	Rule-of-nearest points	230
	3.8.3	Constant distance weighting techniques	231
3.9	Statistical basis for grade assignment		235
	3.9.1	Some statistics on the orebody	238
	3.9.2	Range of sample influence	242
	3.9.3	Illustrative example	243
	3.9.4	Describing variograms by mathematical models	248
	3.9.5	Quantification of a deposit through variograms	250
3.10	Kriging		251
	3.10.1	Introduction	251
	3.10.2	Concept development	252
	3.10.3	Kriging example	254
	3.10.4	Example of estimation for a level	258
	3.10.5	Block kriging	258
	3.10.6	Common problems associated with the use of the kriging technique	259
	3.10.7	Comparison of results using several techniques	260
	References		261
	Review questions and exercises		266

4 GEOMETRICAL CONSIDERATIONS 270

4.1	Introduction		270
4.2	Basic bench geometry		270
4.3	Ore access		277
4.4	The pit expansion process		290
	4.4.1	Introduction	290
	4.4.2	Frontal cuts	290
	4.4.3	Drive-by cuts	293
	4.4.4	Parallel cuts	293
	4.4.5	Minimum required operating room for parallel cuts	296
	4.4.6	Cut sequencing	302

VIII   *Open pit mine planning and design*: Fundamentals

4.5	Pit slope geometry		303
4.6	Final pit slope angles		312
	4.6.1	Introduction	312
	4.6.2	Geomechanical background	313
	4.6.3	Planar failure	314
	4.6.4	Circular failure	320
	4.6.5	Stability of curved wall sections	320
	4.6.6	Slope stability data presentation	322
	4.6.7	Slope analysis example	323
	4.6.8	Economic aspects of final slope angles	324
4.7	Plan representation of bench geometry		326
4.8	Addition of a road		330
	4.8.1	Introduction	330
	4.8.2	Design of a spiral road – inside the wall	336
	4.8.3	Design of a spiral ramp – outside the wall	341
	4.8.4	Design of a switchback	344
	4.8.5	The volume represented by a road	347
4.9	Road construction		352
	4.9.1	Introduction	352
	4.9.2	Road section design	353
	4.9.3	Straight segment design	358
	4.9.4	Curve design	361
	4.9.5	Conventional parallel berm design	364
	4.9.6	Median berm design	364
	4.9.7	Haulage road gradients	365
	4.9.8	Practical road building and maintenance tips	368
4.10	Stripping ratios		369
4.11	Geometric sequencing		374
4.12	Summary		377
	References		377
	Review questions and exercises		383

5 PIT LIMITS   388

5.1	Introduction		388
5.2	Hand methods		389
	5.2.1	The basic concept	389
	5.2.2	The net value calculation	392
	5.2.3	Location of pit limits – pit bottom in waste	398
	5.2.4	Location of pit limits – pit bottom in ore	404
	5.2.5	Location of pit limits – one side plus pit bottom in ore	404
	5.2.6	Radial sections	405
	5.2.7	Generating a final pit outline	411
	5.2.8	Destinations for in-pit materials	416
5.3	Economic block models		418
5.4	The floating cone technique		420
5.5	The Lerchs-Grossmann 2-D algorithm		429

5.6	Modification of the Lerchs-Grossmann 2-D algorithm to a 2½-D algorithm		438
5.7	The Lerchs-Grossmann 3-D algorithm		441
	5.7.1	Introduction	441
	5.7.2	Definition of some important terms and concepts	444
	5.7.3	Two approaches to tree construction	447
	5.7.4	The arbitrary tree approach (Approach 1)	448
	5.7.5	The all root connection approach (Approach 2)	450
	5.7.6	The tree 'cutting' process	454
	5.7.7	A more complicated example	456
5.8	Computer assisted methods		457
	5.8.1	The RTZ open-pit generator	457
	5.8.2	Computer assisted pit design based upon sections	463
	References		475
	Review questions and exercises		479

6 PRODUCTION PLANNING 482

6.1	Introduction		482
6.2	Some basic mine life – plant size concepts		483
6.3	Taylor's mine life rule		493
6.4	Sequencing by nested pits		494
6.5	Cash flow calculations		499
6.6	Mine and mill plant sizing		511
	6.6.1	Ore reserves supporting the plant size decision	511
	6.6.2	Incremental financial analysis principles	515
	6.6.3	Plant sizing example	518
6.7	Lane's algorithm		526
	6.7.1	Introduction	526
	6.7.2	Model definition	527
	6.7.3	The basic equations	528
	6.7.4	An illustrative example	529
	6.7.5	Cutoff grade for maximum profit	530
	6.7.6	Net present value maximization	538
6.8	Material destination considerations		556
	6.8.1	Introduction	556
	6.8.2	The leach dump alternative	557
	6.8.3	The stockpile alternative	562
6.9	Production scheduling		568
	6.9.1	Introduction	568
	6.9.2	Phase scheduling	580
	6.9.3	Block sequencing using set dynamic programming	586
	6.9.4	Some scheduling examples	598
6.10	Push back design		604
	6.10.1	Introduction	604
	6.10.2	The basic manual steps	611
	6.10.3	Manual push back design example	613

	6.10.4	Time period plans	625
	6.10.5	Equipment fleet requirements	627
	6.10.6	Other planning considerations	629
6.11	The mine planning and design process – summary and closing remarks		631
	References		633
	Review questions and exercises		640

7 REPORTING OF MINERAL RESOURCES AND ORE RESERVES — 644

7.1	Introduction		644
7.2	The JORC code – 2004 edition		645
	7.2.1	Preamble	645
	7.2.2	Foreword	645
	7.2.3	Introduction	645
	7.2.4	Scope	649
	7.2.5	Competence and responsibility	650
	7.2.6	Reporting terminology	652
	7.2.7	Reporting – General	653
	7.2.8	Reporting of exploration results	653
	7.2.9	Reporting of mineral resources	654
	7.2.10	Reporting of ore reserves	658
	7.2.11	Reporting of mineralized stope fill, stockpiles, remnants, pillars, low grade mineralization and tailings	661
7.3	The CIM best practice guidelines for the estimation of mineral resources and mineral reserves – general guidelines		662
	7.3.1	Preamble	662
	7.3.2	Foreword	662
	7.3.3	The resource database	664
	7.3.4	Geological interpretation and modeling	666
	7.3.5	Mineral resource estimation	669
	7.3.6	Quantifying elements to convert a Mineral Resource to a Mineral Reserve	672
	7.3.7	Mineral reserve estimation	674
	7.3.8	Reporting	676
	7.3.9	Reconciliation of mineral reserves	680
	7.3.10	Selected references	683
	References		683
	Review questions and exercises		685

8 RESPONSIBLE MINING — 688

8.1	Introduction	688
8.2	The 1972 United Nations Conference on the Human Environment	689
8.3	The World Conservation Strategy (WCS) – 1980	693
8.4	World Commission on Environment and Development (1987)	696

8.5	The 'Earth Summit'		698
	8.5.1	The Rio Declaration	698
	8.5.2	Agenda 21	701
8.6	World Summit on Sustainable Development (WSSD)		703
8.7	Mining industry and mining industry-related initiatives		704
	8.7.1	Introduction	704
	8.7.2	The Global Mining Initiative (GMI)	704
	8.7.3	International Council on Mining and Metals (ICMM)	706
	8.7.4	Mining, Minerals, and Sustainable Development (MMSD)	708
	8.7.5	The U.S. Government and federal land management	709
	8.7.6	The position of the U.S. National Mining Association (NMA)	712
	8.7.7	The view of one mining company executive	714
8.8	'Responsible Mining' – the way forward is good engineering		716
	8.8.1	Introduction	716
	8.8.2	The Milos Statement	716
8.9	Concluding remarks		719
	References		719
	Review questions and exercises		723

Index 727

Preface to the 2nd Edition

The first edition of Open Pit Mine Planning and Design appeared in 1995. We have been very pleased with the response received and have been encouraged to provide an updated and somewhat expanded version. The result is what you now hold in your hands. We hope that you will find some things of value.

The CSMine software included in the first edition was written for the DOS operating system which was current at that time. Although the original program does work in the Windows environment, it is not optimum. Furthermore, with the major advances in computer power that have occurred during the intervening ten-year period, many improvements could be incorporated. Of prime importance, however, was to retain the user friendliness of the original CSMine. We hope that you will agree that we have succeeded in this regard.

Chapters 1, and 3–6 have remained largely the same but the reference lists have been updated. The costs and prices included in Chapter 2 – Mining Costs and Revenues have been changed to reflect today's values. Two new chapters have been added to cover some important new developments regarding resource/reserve definition (Chapter 7) and responsible mining (Chapter 8).

To facilitate the use of this book in the classroom, review questions and exercises have been added at the end of chapters 1 through 9. As will be quickly realized, the "answers" have not been provided. There are several reasons for this. First, most of the answers will be found by the careful reading, and perhaps re-reading, of the text material. Secondly, for practicing mining engineers, the answers to the opportunities offered by their operations are seldom provided in advance. The fact that the answers are not given will help introduce the student to the real world of mining. Finally, for those students using the book under the guidance of a professor, some of the questions will offer discussion possibilities. There is no single "right" answer for some of the included exercises.

A total of eight drill hole data sets involving three iron properties, two gold properties and three copper properties have been included on the distribution CD. Each of these properties is described in some detail in Chapter 11. It is intended that when used in conjunction with the CSMine software these data sets might form the basis for capstone surface mine design courses.

The authors would like to acknowledge the Canadian Institute of Mining, Metallurgy and Petroleum (CIM) for permission to include their 'Estimation of Mineral Resources and Mineral Reserves: Best Practices Guidelines' in Chapter 7. The Australasian Institute of Mining and Metallurgy (AusIMM) was very kind to permit our inclusion of the 'JORC-2004 Code' in Chapter 7. The current commodity prices were kindly supplied by Platt's Metals Week, the Metal Bulletin, Minerals Price Watch, and Skillings Mining Review. The Engineering News-Record graciously allowed the inclusion of their cost indexes. The CMJ Mining Sourcebook, EquipmentWatch/PRIMEDIA, and Western Mine Engineering provided updated costs. The authors drew very heavily on the statistics carefully compiled by the U.S. Department of Labor, U.S. Bureau of Labor Statistics, and the U.S. Geological Survey.

Kennecott Utah Copper generously provided the beautiful photo of their Bingham Canyon mine for use on the cover.

The drill hole sets were kindly supplied by Kennecott Barneys Canyon mine, Newmont Mining Corporation, Minnesota Department of Revenue, Minnesota Division of Minerals (Ironton Office), Geneva Steel and Codelco.

Finally, we would like to thank those of you who bought the first edition and provided useful suggestions for improvement.

William Hustrulid
Mark Kuchta
August 2005

About the Authors

William Hustrulid studied Minerals Engineering at the University of Minnesota. After obtaining his Ph.D. in 1968, his career has included responsible roles in both mining academia and in the mining business itself. He has served as Professor of Mining Engineering at the University of Utah and at the Colorado School of Mines and as a Guest Professor at the Technical University in Luleå, Sweden. In addition, he has held mining R&D positions for companies in the USA, Sweden, and the former Republic of Zaire. Consulting opportunities have taken him to many parts of the world including Chile, Canada, Mexico, Australia and South Africa. He has received several awards,  including the Halliburton Award for Outstanding Professional Achievement from the Colorado School of Mines and the Daniel C. Jackling Award from the Society for Mining, Metallurgy and Exploration (SME). In addition, he has been elected Foreign Member of the Swedish Royal Academy of Engineering Sciences. He currently holds the rank of Professor Emeritus at the University of Utah and manages Hustrulid Mining Services in Bonita Springs, Florida.

Mark Kuchta studied Mining Engineering at the Colorado School of Mines and received his Ph.D. degree from the Technical University in Luleå, Sweden. He has had a wide-ranging career in the mining business. This has included working as a contract miner in the uranium mines of western Colorado and 10 years of experience in various positions with LKAB in northern Sweden. At present, Mark is an Associate Professor of Mining Engineering at the Colorado School of Mines. He is actively involved in the education of future mining engineers at both undergraduate and graduate levels and conducts a very active research program. His professional interests include the use of  high-pressure waterjets for rock scaling applications in underground mines, the development of a deep underground science laboratory in Colorado, strategic mine planning, advanced mine production scheduling and the development of user-friendly mine software.

CHAPTER 1

Mine planning

1.1 INTRODUCTION

1.1.1 *The meaning of ore*

One of the first things discussed in an Introduction to Mining course and one which students must commit to memory is the definition of 'ore'. One of the more common definitions (USBM, 1967) is given below:

Ore: A metalliferous mineral, or an aggregate of metalliferous minerals, more or less mixed with gangue which from the standpoint of the miner can be mined at a profit or, from the standpoint of a metallurgist can be treated at a profit.

This standard definition is consistent with the custom of dividing mineral deposits into two groups: metallic (ore) and non-metallic. Over the years, the usage of the word 'ore' has been expanded by many to include non-metallics as well. The definition of ore suggested by Banfield (1972) would appear to be more in keeping with the general present day usage.

Ore: A natural aggregate of one or more solid minerals which can be mined, or from which one or more mineral products can be extracted, at a profit.

In this book the following, somewhat simplified, definition will be used:

Ore: A natural aggregation of one or more solid minerals that can be mined, processed and sold at a profit.

Although definitions are important to know, it is even more important to know what they mean. To prevent the reader from simply transferring this definition directly to memory without being first processed by the brain, the 'meaning' of ore will be expanded upon.

The key concept is 'extraction leading to a profit'. For engineers, profits can be expressed in simple equation form as

$$\text{Profits} = \text{Revenues} - \text{Costs} \tag{1.1}$$

The revenue portion of the equation can be written as

$$\text{Revenues} = \text{Material sold (units)} \times \text{Price/unit} \tag{1.2}$$

The costs can be similarly expressed as

$$\text{Costs} = \text{Material sold (units)} \times \text{Cost/unit} \tag{1.3}$$

Combining the equations yields

$$\text{Profits} = \text{Material sold (units)} \times (\text{Price/unit} - \text{Cost/unit}) \tag{1.4}$$

As has been the case since the early Phoenician traders, the minerals used by modern man come from deposits scattered around the globe. The price received is more and more being set by world wide supply and demand. Thus, the price component in the equation is largely determined by others. Where the mining engineer can and does enter is in doing something about the unit costs. Although the development of new technology at your property is one answer, new technology easily and quickly spreads around the world and soon all operations have the 'new' technology. Hence to remain profitable over the long term, the mining engineer must continually examine and assess smarter and better site specific ways for reducing costs at the operation. This is done through a better understanding of the deposit itself and the tools/techniques employed or employable in the extraction process. Cost containment/reduction through efficient, safe and environmentally responsive mining practices is serious business today and will be even more important in the future with increasing mining depths and ever more stringent regulations. A failure to keep up is reflected quite simply by the profit equation as

$$\text{Profits} < 0 \tag{1.5}$$

This, needless to say, is unfavorable for all concerned (the employees, the company, and the country or nation). For the mining engineer (student or practicing) reading this book, the personal meaning of ore is

$$\text{Ore} \equiv \text{Profits} \equiv \text{Jobs} \tag{1.6}$$

The use of the mathematical equivalence symbol simply says that 'ore' is equivalent to 'profits' which is equivalent to 'jobs'. Hence one important meaning of 'ore' to us in the minerals business is jobs. Probably this simple practical definition is more easily remembered than those offered earlier. The remainder of the book is intended to provide the engineer with tools to perform even better in an increasingly competitive world.

1.1.2 Some important definitions

The exploration, development, and production stages of a mineral deposit (Banfield & Havard, 1975) are defined as:

Exploration: The search for a mineral deposit (prospecting) and the subsequent investigation of any deposit found until an orebody, if such exists, has been established.

Development: Work done on a mineral deposit, after exploration has disclosed ore in sufficient quantity and quality to justify extraction, in order to make the ore available for mining.

Production: The mining of ores, and as required, the subsequent processing into products ready for marketing.

It is essential that the various terms used to describe the nature, size and tenor of the deposit be very carefully selected and then used within the limits of well recognized and accepted definitions.

Over the years a number of attempts have been made to provide a set of universally accepted definitions for the most important terms. These definitions have evolved somewhat as the technology used to investigate and evaluate orebodies has changed. On February 24, 1991, the report, 'A Guide for Reporting Exploration Information, Resources and Reserves' prepared by Working Party No. 79 – 'Ore Reserves Definition' of the Society of Mining, Metallurgy and Exploration (SME), was delivered to the SME Board of Directors (SME,

1991). This report was subsequently published for discussion. In this section, the 'Definitions' and 'Report Terminology' portions of their report (SME, 1991) are included. The interested reader is encouraged to consult the given reference for the detailed guidelines. The definitions presented are tied closely to the sequential relationship between exploration information, resources and reserves shown in Figure 1.1.

With an increase in geological knowledge, the exploration information may become sufficient to calculate a resource. When economic information increases it may be possible to convert a portion of the resource to a reserve. The double arrows between reserves and resources in Figure 1.1 indicate that changes due to any number of factors may cause material to move from one category to another.

Definitions
Exploration information. Information that results from activities designed to locate economic deposits and to establish the size, composition, shape and grade of these deposits. Exploration methods include geological, geochemical, and geophysical surveys, drill holes, trial pits and surface underground openings.

Resource. A concentration of naturally occurring solid, liquid or gaseous material in or on the Earth's crust in such form and amount that economic extraction of a commodity from the concentration is currently or potentially feasible. Location, grade, quality, and quantity are known or estimated from specific geological evidence. To reflect varying degrees of geological certainty, resources can be subdivided into measured, indicated, and inferred.

– Measured. Quantity is computed from dimensions revealed in outcrops, trenches, workings or drill holes; grade and/or quality are computed from the result of detailed sampling. The sites for inspection, sampling and measurement are spaced so closely and the geological character is so well defined that size, shape, depth and mineral content of the resource are well established.

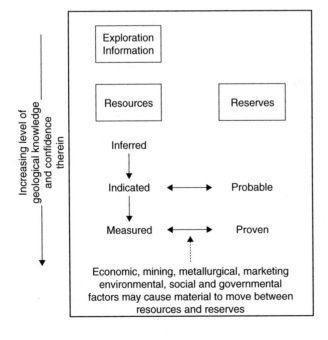

Figure 1.1. The relationship between exploration information, resources and reserves (SME, 1991).

– Indicated. Quantity and grade and/or quality are computed from information similar to that used for measured resources, but the sites for inspection, sampling, and measurements are farther apart or are otherwise less adequately spaced. The degree of assurance, although lower than that for measured resources, is high enough to assume geological continuity between points of observation.

– Inferred. Estimates are based on geological evidence and assumed continuity in which there is less confidence than for measured and/or indicated resources. Inferred resources may or may not be supported by samples or measurements but the inference must be supported by reasonable geo-scientific (geological, geochemical, geophysical, or other) data.

Reserve. A reserve is that part of the resource that meets minimum physical and chemical criteria related to the specified mining and production practices, including those for grade, quality, thickness and depth; and can be reasonably assumed to be economically and legally extracted or produced at the time of determination. The feasibility of the specified mining and production practices must have been demonstrated or can be reasonably assumed on the basis of tests and measurements. The term reserves need not signify that extraction facilities are in place and operative.

The term economic implies that profitable extraction or production under defined investment assumptions has been established or analytically demonstrated. The assumptions made must be reasonable including assumptions concerning the prices and costs that will prevail during the life of the project.

The term 'legally' does not imply that all permits needed for mining and processing have been obtained or that other legal issues have been completely resolved. However, for a reserve to exist, there should not be any significant uncertainty concerning issuance of these permits or resolution of legal issues.

Reserves relate to resources as follows:

– Proven reserve. That part of a measured resource that satisfies the conditions to be classified as a reserve.

– Probable reserve. That part of an indicated resources that satisfies the conditions to be classified as a reserve.

It should be stated whether the reserve estimate is of in-place material or of recoverable material. Any in-place estimate should be qualified to show the anticipated losses resulting from mining methods and beneficiation or preparation.

Reporting terminology

The following terms should be used for reporting exploration information, resources and reserves:

1. Exploration information. Terms such as 'deposit' or 'mineralization' are appropriate for reporting exploration information. Terms such as 'ore,' 'reserve,' and other terms that imply that economic extraction or production has been demonstrated, should not be used.

2. Resource. A resource can be subdivided into three categories:
   (a) Measured resource;
   (b) Indicated resource;
   (c) Inferred resource.

The term 'resource' is recommended over the terms 'mineral resource, identified resource' and 'in situ resource.' 'Resource' as defined herein includes 'identified resource,' but excludes 'undiscovered resource' of the United States Bureau of Mines (USBM) and United

States Geological Survey (USGS) classification scheme. The 'undiscovered resource' classification is used by public planning agencies and is not appropriate for use in commercial ventures.

3. Reserve. A reserve can be subdivided into two categories:
   (a) Probable reserve;
   (b) Proven reserve.

The term 'reserve' is recommended over the terms 'ore reserve,' 'minable reserve' or 'recoverable reserve.'

The terms 'measured reserve' and 'indicated reserve,' generally equivalent to 'proven reserve' and 'probable reserve,' respectively, are not part of this classification scheme and should not be used. The terms 'measured,' 'indicated' and 'inferred' qualify resources and reflect only differences in geological confidence. The terms 'proven' and 'probable' qualify reserves and reflect a high level of economic confidence as well as differences in geological confidence.

The terms 'possible reserve' and 'inferred reserve' are not part of this classification scheme. Material described by these terms lacks the requisite degree of assurance to be reported as a reserve.

The term 'ore' should be used only for material that meets the requirements to be a reserve.

It is recommended that proven and probable reserves be reported separately. Where the term reserve is used without the modifiers proven or probable, it is considered to be the total of proven and probable reserves.

1.2 MINE DEVELOPMENT PHASES

The mineral supply process is shown diagrammatically in Figure 1.2. As can be seen a positive change in the market place creates a new or increased demand for a mineral product.

In response to the demand, financial resources are applied in an exploration phase resulting in the discovery and delineation of deposits. Through increases in price and/or advances in technology, previously located deposits may become interesting. These deposits must then be thoroughly evaluated regarding their economic attractiveness. This evaluation process will be termed the 'planning phase' of a project (Lee, 1984). The conclusion of this phase will be the preparation of a feasibility report. Based upon this, the decision will be made as to whether or not to proceed. If the decision is 'go', then the development of the mine and concentrating facilities is undertaken. This is called the implementation, investment, or design and construction phase. Finally there is the production or operational phase during which the mineral is mined and processed. The result is a product to be sold in the marketplace. The entrance of the mining engineer into this process begins at the planning phase and continues through the production phase. Figure 1.3 is a time line showing the relationship of the different phases and their stages.

The implementation phase consists of two stages (Lee, 1984). The design and construction stage includes the design, procurement and construction activities. Since it is the period of major cash flow for the project, economies generally result by keeping the time frame to a realistic minimum. The second stage is commissioning. This is the trial operation of the individual components to integrate them into an operating system and ensure their readiness

6 *Open pit mine planning and design: Fundamentals*

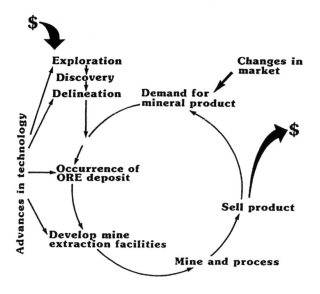

MINERAL SUPPLY PROCESS

Figure 1.2. Diagrammatic representation of the mineral supply process (McKenzie, 1980).

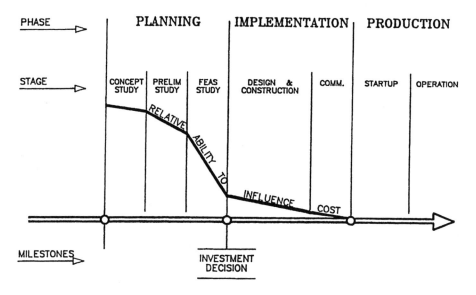

Figure 1.3. Relative ability to influence costs (Lee, 1984).

for startup. It is conducted without feedstock or raw materials. Frequently the demands and costs of the commissioning period are underestimated.

The production phase also has two stages (Lee, 1984). The startup stage commences at the moment that feed is delivered to the plant with the express intention of transforming it into product. Startup normally ends when the quantity and quality of the product is sustainable at the desired level. Operation commences at the end of the startup stage.

As can be seen in Figure 1.3, and as indicated by Lee (1984),

the planning phase offers the greatest opportunity to minimize the capital and operating costs of the ultimate project, while maximizing the operability and profitability of the venture. But the opposite is also true: no phase of the project contains the potential for instilling technical or fiscal disaster into a developing project, that is inherent in the planning phase....

At the start of the conceptual study, there is a relatively unlimited ability to influence the cost of the emerging project. As decisions are made, correctly or otherwise, during the balance of the planning phase, the opportunity to influence the cost of the job diminishes rapidly.

The ability to influence the cost of the project diminishes further as more decisions are made during the design stage. At the end of the construction period there is essentially no opportunity to influence costs.

The remainder of this chapter will focus on the activities conducted within the planning stage.

1.3 AN INITIAL DATA COLLECTION CHECKLIST

In the initial planning stages for any new project there are a great number of factors of rather diverse types requiring consideration. Some of these factors can be easily addressed, whereas others will require in-depth study. To prevent forgetting factors, checklist are often of great value. Included below are the items from a 'Field Work Program Checklist for New Properties' developed by Halls (1975). Student engineers will find many of the items on this checklist of relevance when preparing mine design reports.

Checklist items (Halls, 1975)
1. Topography
   (a) USGS maps
   (b) Special aerial or land survey
       Establish survey control stations
       Contour
2. Climatic conditions
   (a) Altitude
   (b) Temperatures
       Extremes
       Monthly averages
   (c) Precipitation
       Average annual precipitation
       Average monthly rainfall
       Average monthly snowfall
       Run-off
         Normal
         Flood
       Slides – snow and mud
   (d) Wind
       Maximum recorded
       Prevailing direction
       Hurricanes, tornados, cyclones, etc.

(e) Humidity
   Effect on installations, i.e. electrical motors, etc.
(f) Dust
(g) Fog and cloud conditions
3. Water – potable and process
   (a) Sources
      Streams
      Lakes
      Wells
   (b) Availability
      Ownership
      Water rights
      Cost
   (c) Quantities
      Monthly availability
      Flow rates
      Drought or flood conditions
      Possible dam locations
   (d) Quality
      Present sample
      Possibility of quality change in upstream source water
      Effect of contamination on downstream users
   (e) Sewage disposal method
4. Geologic structure
   (a) Within mine area
   (b) Surrounding areas
   (c) Dam locations
   (d) Earthquakes
   (e) Effect on pit slopes
      Maximum predicted slopes
   (f) Estimate on foundation conditions
5. Mine water as determined by prospect holes
   (a) Depth
   (b) Quantity
   (c) Method of drainage
6. Surface
   (a) Vegetation
      Type
      Method of clearing
      Local costs for clearing
   (b) Unusual conditions
      Extra heavy timber growth
      Muskeg
      Lakes
      Stream diversions
      Gravel deposits

7. Rock type – overburden and ore
   (a) Submit sample for drillability test
   (b) Observe fragmentation features
       Hardness
       Degree of weathering
       Cleavage and fracture planes
       Suitability for road surface
8. Locations for concentrator – factors to consider for optimum location
   (a) Mine location
       Haul uphill or downhill
   (b) Site preparation
       Amount of cut and/or fill
   (c) Process water
       Gravity flow or pumping
   (d) Tailings disposal
       Gravity flow or pumping
   (e) Maintenance facilities
       Location
9. Tailings pond area
   (a) Location of pipeline length and discharge elevations
   (b) Enclosing features
       Natural
       Dams or dikes
       Lakes
   (c) Pond overflow
       Effect of water pollution on downstream users
       Possibility for reclaiming water
   (d) Tailings dust
       Its effect on the area
10. Roads
    (a) Obtain area road maps
    (b) Additional road information
        Widths
        Surfacing
        Maximum load limits
        Seasonal load limits
        Seasonal access
        Other limits or restrictions
        Maintained by county, state, etc.
    (c) Access roads to be constructed by company (factors considered)
        Distance
        Profile
        Cut and fill
        Bridges, culverts
        Terrain and soil conditions

11. Power
    (a) Availability
        Kilovolts
        Distance
        Rates and length of contract
    (b) Power lines to site
        Who builds
        Who maintains
        Right-of-way requirements
    (c) Substation location
    (d) Possibility of power generation at or near site
12. Smelting
    (a) Availability
    (b) Method of shipping concentrate
    (c) Rates
    (d) If company on site smelting – effect of smelter gases
    (e) Concentrate freight rates
    (f) Railroads and dock facility
13. Land ownership
    (a) Present owners
    (b) Present usage
    (c) Price of land
    (d) Types of options, leases and royalties expected
14. Government
    (a) Political climate
        Favorable or unfavorable to mining
        Past reactions in the area to mining
    (b) Special mining laws
    (c) Local mining restrictions
15. Economic climate
    (a) Principal industries
    (b) Availability of labor and normal work schedules
    (c) Wage scales
    (d) Tax structure
    (e) Availability of goods and services
        Housing
        Stores
        Recreation
        Medical facilities and unusual local disease
        Hospital
        Schools
    (f) Material costs and/or availability
        Fuel oil
        Concrete
        Gravel
        Borrow material for dams

(g) Purchasing
       Duties
16. Waste dump location
    (a) Haul distance
    (b) Haul profile
    (c) Amenable to future leaching operation
17. Accessibility of principal town to outside
    (a) Methods of transportation available
    (b) Reliability of transportation available
    (c) Communications
18. Methods of obtaining information
    (a) Past records (i.e. government sources)
    (b) Maintain measuring and recording devices
    (c) Collect samples
    (d) Field observations and measurements
    (e) Field surveys
    (f) Make preliminary plant layouts
    (g) Check courthouse records for land information
    (h) Check local laws and ordinances for applicable legislation
    (i) Personal inquiries and observation on economic and political climates
    (j) Maps
    (k) Make cost inquiries
    (l) Make material availability inquiries
    (m) Make utility availability inquiries

1.4 THE PLANNING PHASE

In preparing this section the authors have drawn heavily on material originally presented in papers by Lee (1984) and Taylor (1977). The permission by the authors and their publisher, The Northwest Mining Association, to include this material is gratefully acknowledged.

1.4.1 *Introduction*

The planning phase commonly involves three stages of study (Lee, 1984).

Stage 1: Conceptual study
A conceptual (or preliminary valuation) study represents the transformation of a project idea into a broad investment proposition, by using comparative methods of scope definition and cost estimating techniques to identify a potential investment opportunity. Capital and operating costs are usually approximate ratio estimates using historical data. It is intended primarily to highlight the principal investment aspects of a possible mining proposition. The preparation of such a study is normally the work of one or two engineers. The findings are reported as a preliminary valuation.

Stage 2: Preliminary or pre-feasibility study
A preliminary study is an intermediate-level exercise, normally not suitable for an investment decision. It has the objectives of determining whether the project concept justifies a detailed

analysis by a feasibility study, and whether any aspects of the project are critical to its viability and necessitate in-depth investigation through functional or support studies.

A preliminary study should be viewed as an intermediate stage between a relatively inexpensive conceptual study and a relatively expensive feasibility study. Some are done by a two or three man team who have access to consultants in various fields others may be multi-group efforts.

Stage 3: Feasibility study

The feasibility study provides a definitive technical, environmental and commercial base for an investment decision. It uses iterative processes to optimize all critical elements of the project. It identifies the production capacity, technology, investment and production costs, sales revenues, and return on investment. Normally it defines the scope of work unequivocally, and serves as a base-line document for advancement of the project through subsequent phases.

These latter two stages will now be described in more detail.

1.4.2 *The content of an intermediate valuation report*

The important sections of an intermediate valuation report (Taylor, 1977) are:
- Aim;
- Technical concept;
- Findings;
- Ore tonnage and grade;
- Mining and production schedule;
- Capital cost estimate;
- Operating cost estimate;
- Revenue estimate;
- Taxes and financing;
- Cash flow tables.

The degree of detail depends on the quantity and quality of information. Table 1.1 outlines the contents of the different sections.

1.4.3 *The content of the feasibility report*

The essential functions of the feasibility report are given in Table 1.2.

Due to the great importance of this report it is necessary to include all detailed information that supports a general understanding and appraisal of the project or the reasons for selecting particular processes, equipment or courses of action. The contents of the feasibility report are outlined in Table 1.3.

The two important requirements for both valuation and feasibility reports are:

1. Reports must be easy to read, and their information must be easily accessible.

2. Parts of the reports need to be read and understood by non-technical people.

According to Taylor (1977):

There is much merit in a layered or pyramid presentation in which the entire body of information is assembled and retained in three distinct layers.

Layer 1. Detailed background information neatly assembled in readable form and adequately indexed, but retained in the company's office for reference and not included in the feasibility report.

Layer 2. Factual information about the project, precisely what is proposed to be done about it, and what the technical, physical and financial results are expected to be.

Layer 3. A comprehensive but reasonably short summary report, issued preferably as a separate volume.

The feasibility report itself then comprises only the second and third layers. While everything may legitimately be grouped into a single volume, the use of smaller separated volumes makes for easier reading and for more flexible forms of binding. Feasibility reports always need to be reviewed by experts in various specialities. The use of several smaller volumes makes this easier, and minimizes the total number of copies needed.

Table 1.1. The content of an intermediate valuation report (Taylor, 1977).

Aim: States briefly what knowledge is being sought about the property, and why, for guidance in exploration spending, for joint venture negotiations, for major feasibility study spending, etc. Sources of information are also conveniently listed.

Concept: Describes very briefly where the property is located, what is proposed or assumed to be done in the course of production, how this may be achieved, and what is to be done with the products.

Findings: Comprise a summary, preferably in sequential and mainly tabular forms, of the important figures and observations from all the remaining sections. This section may equally be termed Conclusions, though this title invites a danger of straying into recommendations which should not be offered unless specially requested.

Any cautions or reservations the authors care to make should be incorporated in one of the first three sections. The general aim is that the non-technical or less-technical reader should be adequately informed about the property by the time he has read the end of Findings.

Ore tonnage and grade: Gives brief notes on geology and structure, if applicable, and on the drilling and sampling accomplished. Tonnages and grades, both geological and minable and possibly at various cut-off grades, are given in tabular form with an accompanying statement on their status and reliability.

Mining and production schedule: Tabulates the mining program (including preproduction work), the milling program, any expansions or capacity changes, the recoveries and product qualities (concentrate grades), and outputs of products.

Capital cost estimate: Tabulates the cost to bring the property to production from the time of writing including the costs of further exploration, research and studies. Any prereport costs, being sunk, may be noted separately.
An estimate of postproduction capital expenditures is also needed. This item, because it consists largely of imponderables, tends to be underestimated even in detailed feasibilities studies.

Operating cost estimate: Tabulates the cash costs of mining, milling, other treatment, ancillary services, administration, etc. Depreciation is not a cash cost, and is handled separately in cash flow calculations. Postmine treatment and realization costs are most conveniently regarded as deductions from revenues.

Revenue estimate: Records the metal or product prices used, states the realization terms and costs, and calculates the net smelter return or net price at the deemed point of disposal. The latter is usually taken to be the point at which the product leaves the mine's plant and is handed over to a common carrier. Application of these net prices to the outputs determined in the production schedule yields a schedule of annual revenues.

Financing and tax data: State what financing assumptions have been made, all equity, all debt or some specified mixture, together with the interest and repayment terms of loans. A statement on the tax regime specifies tax holidays (if any), depreciation and tax rates, (actual or assumed) and any special

---

(Continued)

Table 1.1. (Continued).

features. Many countries, particularly those with federal constitutions, impose multiple levels of taxation by various authorities, but a condensation or simplification of formulae may suffice for early studies without involving significant loss of accuracy.

Cash flow schedules: Present (if information permits) one or more year-by-year projections of cash movements in and out of the project. These tabulations are very informative, particularly because their format is almost uniformly standardized. They may be compiled for the indicated life of the project or, in very early studies, for some arbitrary shorter period.

Figures must also be totalled and summarized. Depending on company practice and instructions, investment indicators such as internal rate of return, debt payback time, or cash flow after payback may be displayed.

Table 1.2. The essential functions of the feasibility report (Taylor, 1977).

1. To provide a comprehensive framework of established and detailed facts concerning the mineral project.
2. To present an appropriate scheme of exploitation with designs and equipment lists taken to a degree of detail sufficient for accurate prediction of costs and results.
3. To indicate to the project's owners and other interested parties the likely profitability of investment in the project if equipped and operated as the report specifies.
4. To provide this information in a form intelligible to the owner and suitable for presentation to prospective partners or to sources of finance.

Table 1.3. The content of a feasibility study (Taylor, 1977).

General:
- Topography, climate, population, access, services.
- Suitable sites for plant, dumps, towns, etc.

Geological (field):
- Geological study of structure, mineralization and possibly of genesis.
- Sampling by drilling or tunnelling or both.
- Bulk sampling for checking and for metallurgical testing.
- Extent of leached or oxidized areas (frequently found to be underestimated).
- Assaying and recording of data, including check assaying, rock properties, strength and stability.
- Closer drilling of areas scheduled for the start of mining.
- Geophysics and indication of the likely ultimate limits of mineralization, including proof of non-mineralization of plan and dump areas.
- Sources of water and of construction materials.

Geological and mining (office):
- Checking, correcting and coding of data for computer input.
- Manual calculations of ore tonnages and grades.
- Assay compositing and statistical analysis.
- Computation of mineral inventory (geological reserves) and minable reserves, segregated as needed by orebody, by ore type, by elevation or bench, and by grade categories.
- Computation of associated waste rock.
- Derivation of the economic factors used in the determination of minable reserves.

Mining:
- Open pit layouts and plans.
- Determination of preproduction mining or development requirements.
- Estimation of waste rock dilution and ore losses.

(Continued)

Table 1.3. (Continued).

- Production and stripping schedules, in detail for the first few years but averaged thereafter, and specifying important changes in ore types if these occur.
- Waste mining and waste disposal.
- Labor and equipment requirements and cost, and an appropriate replacement schedule for the major equipment.

Metallurgy (research):
- Bench testing of samples from drill cores.
- Selection of type and stages of the extraction process.
- Small scale pilot plant testing of composited or bulk samples followed by larger scale pilot mill operation over a period of months should this work appear necessary.
- Specification of degree of processing, and nature and quality of products.
- Provision of samples of the product.
- Estimating the effects of ore type or head grade variations upon recovery and product quality.

Metallurgy (design):
- The treatment concept in considerable detail, with flowsheets and calculation of quantities flowing.
- Specification of recovery and of product grade.
- General siting and layout of plant with drawings if necessary.

Ancillary services and requirements:
- Access, transport, power, water, fuel and communications.
- Workshops, offices, changehouse, laboratories, sundry buildings and equipment.
- Labor structure and strength.
- Housing and transport of employees.
- Other social requirements.

Capital cost estimation:
- Develop the mine and plant concepts and make all necessary drawings.
- Calculate or estimate the equipment list and all important quantities (of excavation, concrete, building area and volume, pipework, etc.).
- Determine a provisional construction schedule.
- Obtain quotes of the direct cost of items of machinery, establish the costs of materials and services, and of labor and installation.
- Determine the various and very substantial indirect costs, which include freight and taxes on equipment (may be included in directs), contractors' camps and overheads plus equipment rental, labor punitive and fringe costs, the owner's field office, supervision and travel, purchasing and design costs, licenses, fees, customs duties and sales taxes.
- Warehouse inventories.
- A contingency allowance for unforeseen adverse happenings and for unestimated small requirements that may arise.
- Operating capital sufficient to pay for running the mine until the first revenue is received.
- Financing costs and, if applicable, preproduction interest on borrowed money.

A separate exercise is to forecast the major replacements and the accompanying provisions for postproduction capital spending. Adequate allowance needs to be made for small requirements that, though unforeseeable, always arise in significant amounts.

Operating cost estimation:
- Define the labor strength, basic pay rates, fringe costs.
- Establish the quantities of important measurable supplies to be consumed – power, explosives, fuel, grinding steel, reagents, etc. – and their unit costs.
- Determine the hourly operating and maintenance costs for mobile equipment plus fair performance factors.
- Estimate the fixed administration costs and other overheads plus the irrecoverable elements of townsite and social costs.

(Continued)

Table 1.3. (Continued).

Only cash costs are used thus excluding depreciation charges that must be accounted for elsewhere. As for earlier studies, post-mine costs for further treatment and for selling the product are best regarded as deductions from the gross revenue.

Marketing:
- Product specifications, transport, marketing regulations or restrictions.
- Market analysis and forecast of future prices.
- Likely purchasers.
- Costs for freight, further treatment and sales.
- Draft sales terms, preferably with a letter of intent.
- Merits of direct purchase as against toll treatment.
- Contract duration, provisions for amendment or cost escalation.
- Requirements for sampling, assaying and umpiring.

The existence of a market contract or firm letter of intent is usually an important prerequisite to the loan financing of a new mine.

Rights, ownership and legal matters:
- Mineral rights and tenure.
- Mining rights (if separated from mineral rights).
- Rents and royalties.
- Property acquisitions or securement by option or otherwise.
- Surface rights to land, water, rights-of-way, etc.
- Licenses and permits for construction as well as operation.
- Employment laws for local and expatriate employees separately if applicable.
- Agreements between partners in the enterprise.
- Legal features of tax, currency exchange and financial matters.
- Company incorporation.

Financial and tax matters:
- Suggested organization of the enterprise, as corporation, joint-venture or partnership.
- Financing and obligations, particularly relating to interest and repayment on debt.
- Foreign exchange and reconversion rights, if applicable.
- Study of tax authorities and regimes, whether single or multiple.
- Depreciation allowances and tax rates.
- Tax concessions and the negotiating procedure for them.
- Appropriation and division of distributable profits.

Environmental effects:
- Environmental study and report; the need for pollution or related permits, the requirements during construction and during operation.
- Prescribed reports to government authorities, plans for restoration of the area after mining ceases.

Revenue and profit analysis:
- The mine and mill production schedules and the year-by-year output of products.
- Net revenue at the mine (at various product prices if desired) after deduction of transport, treatment and other realization charges.
- Calculation of annual costs from the production schedules and from unit operating costs derived previously.
- Calculation of complete cash flow schedules with depreciation, taxes, etc. for some appropriate number of years – individually for at least 10 years and grouped thereafter.
- Presentation of totals and summaries of results.
- Derived figures (rate of return, payback, profit split, etc.) as specified by owner or client.
- Assessment of sensitivity to price changes and generally to variation in important input elements.

1.5 PLANNING COSTS

The cost of these studies (Lee, 1984) varies substantially, depending upon the size and nature of the project, the type of study being undertaken, the number of alternatives to be investigated, and numerous other factors. However, the order of magnitude cost of the technical portion of studies, excluding such owner's cost items as exploration drilling, special grinding or metallurgical tests, environmental and permitting studies, or other support studies, is commonly expressed as a percentage of the capital cost of the project:

Conceptual study: 0.1 to 0.3 percent
Preliminary study: 0.2 to 0.8 percent
Feasibility study: 0.5 to 1.5 percent

1.6 ACCURACY OF ESTIMATES

The material presented in this section has been largely extracted from the paper 'Mine Valuation and Feasibility Studies' presented by Taylor (1977).

1.6.1 *Tonnage and grade*

At feasibility, by reason of multiple sampling and numerous checks, the average mining grade of some declared tonnage is likely to be known within acceptable limits, say $\pm 5\%$, and verified by standard statistical methods. Although the ultimate tonnage of ore may be known for open pit mines if exploration drilling from surface penetrates deeper than the practical mining limit, in practice, the ultimate tonnage of many deposits is nebulous because it depends on cost-price relationships late in the project life. By the discount effects in present value theory, late life tonnage is not economically significant at the feasibility stage. Its significance will grow steadily with time once production has begun. It is not critical that the total possible tonnage be known at the outset. What is more important is that the grade and quality factors of the first few years of operation be known with assurance.

Two standards of importance can be defined for most large open pit mines:

1. A minimum ore reserve equal to that required for all the years that the cash flows are projected in the feasibility report must be known with accuracy and confidence.

2. An ultimate tonnage potential, projected generously and optimistically, should be calculated so as to define the area adversely affected by mining and within which dumps and plant buildings must not encroach.

1.6.2 *Performance*

This reduces to two items – throughput and recovery. Open pit mining units have well established performance rates that can usually be achieved if the work is correctly organized and the associated items (i.e. shovels and trucks) are suitably matched. Performance suffers if advance work (waste stripping in a pit) is inadequate. Care must be taken that these tasks are adequately scheduled and provided for in the feasibility study.

The throughput of a concentrator tends to be limited at either the fine crushing stage or the grinding stages. The principles of milling design are well established, but their application requires accurate knowledge of the ore's hardness and grindability. These qualities must therefore receive careful attention in the prefeasibility test work. Concentrator performance is part of a three way relationship involving the fineness of grind, recovery, and the grade of concentrate or product. Very similar relations may exist in metallurgical plants of other types. Again, accuracy can result only from adequate test-work.

1.6.3 *Costs*

Some cost items, notably in the operating cost field, differ little from mine to mine and are reliably known in detail. Others may be unique or otherwise difficult to estimate. Generally, accuracy in capital or operating cost estimating goes back to accuracy in quantities, reliable quotes or unit prices, and adequate provision for indirect or overhead items. The latter tend to form an ever increasing burden. For this reason, they should also be itemized and estimated directly whenever possible, and not be concealed in or allocated into other direct cost items.

Contingency allowance is an allowance for possible over expenditures contingent upon unforeseen happenings such as a strike or time delaying accident during construction, poor plant foundation conditions, or severe weather problems. To some extent the contingency allowance inevitably allows for certain small expenditures always known to arise but not foreseeable nor estimable in detail. Caution is needed here. The contingency allowance is not an allowance for bad or inadequate estimating, and it should never be interpreted in that manner.

The accuracy of capital and operating cost estimates increases as the project advances from conceptual to preliminary to feasibility stage. Normally acceptable ranges of accuracy are considered to be (Lee, 1984):

Conceptual study: ± 30 percent
Preliminary study: ± 20 percent
Feasibility study: ± 10 percent

It was noted earlier that the scope of work in the conceptual and preliminary studies is not optimized. The cost estimate is suitable for decision purposes, to advance the project to the next stage, or to abort and minimize losses.

1.6.4 *Price and revenue*

The revenue over a mine's life is the largest single category of money. It has to pay for everything, including repayment of the original investment money. Because revenue is the biggest base, measures of the mine's economic merit are more sensitive to changes in revenue than to changes of similar ratio in any of the expenditure items.

Revenue is governed by grade, throughput, recovery, and metal or product price. Of these, price is: (a) by far the most difficult to estimate and (b) the one quantity largely outside the estimator's control. Even ignoring inflation, selling prices are widely variable with time. Except for certain controlled commodities, they tend to follow a cyclic pattern.

The market departments of major metal mining corporations are well informed on supply/demand relationships and metal price movements. They can usually provide forecasts of average metal prices in present value dollars, both probable and conservative, the latter being with 80% probability or better. Ideally, even at the conservative product price, the proposed project should still display at least the lowest acceptable level of profitability.

1.7 FEASIBILITY STUDY PREPARATION

The feasibility study is a major undertaking involving many people and a variety of specialized skills. There are two basic ways through which it is accomplished.
 1. The mining company itself organizes the study and assembles the feasibility report. Various parts or tasks are assigned to outside consultants.
 2. The feasibility work is delegated to one or more engineering companies.

Contained on the following pages is an eleven step methodology outlining the planning (Steps 1–4) organizing (Steps 5–10) and execution (Step 11) steps which might be used in conducting a feasibility study. It has been developed by Lee (1984, 1991).

Phase A. Planning
Step 1: Establish a steering committee. A steering committee consisting of managers and other individuals of wide experience and responsibility would be formed to overview and evaluate the direction and viability of the feasibility study team. One such steering committee might be the following:
 – Vice-President (Chairman);
 – General Manager, mining operations;
 – Vice-President, finance;
 – Chief Geologist, exploration;
 – Vice-President, technical services;
 – Consultant(s).

Step 2: Establish a project study team. The criteria for selection of the study team members would emphasize these qualities:
 – Competent in their respective fields.
 – Considerable experience with mining operations.
 – Complementary technical abilities.
 – Compatible personalities – strong interpersonal qualities.
 – Commitment to be available through the implementation phase, should the prospect be viable.

The team members might be:
 – Project Manager;
 – Area Supervisor, mining;
 – Area Supervisor, beneficiation;
 – Area Supervisor, ancillaries.

Step 3: Develop a work breakdown structure. The Work Breakdown Structure (WBS) is defined by the American Association of Cost Engineers (AACE) as:

a product-oriented family tree division of hardware, software, facilities and other items which organizes, defines and displays all of the work to be performed in accomplishing the project objectives.

The WBS is a functional breakdown of all elements of work on a project, on a geographical and/or process basis. It is a hierarchy of work packages, or products, on a work area basis. The WBS is project-unique, reflecting the axiom that every project is a unique event.

A WBS is a simple common-sense procedure which systematically reviews the full scope of a project (or study) and breaks it down into logical packages of work. The primary challenge is normally one of perspective. It is imperative that the entire project be visualized as a sum of many parts, any one of which could be designed, scheduled, constructed, and priced as a single mini-project.

There are a number of categories which can be used to construct a work breakdown structure. These include:
   (1) Components of the product;
   (2) Functions;
   (3) Organizational units;
   (4) Geographical areas;
   (5) Cost accounts;
   (6) Time phases;
   (7) Configuration characteristics;
   (8) Deliverables;
   (9) Responsible persons;
   (10) Subpurposes.

It is not a rigid system. WBS categories can be used in any sequence desired, including using the same category several times. A sample WBS is shown in Figures 1.4 and 1.5.

An alternative to this is the Work Classification Structure (WCS). This commodity-based classification of goods and services is commonly used by construction contractors and consulting engineering firms as the primary cost-collection system. The specific intent of the WCS is to provide a consistent reference system for storage, comparison and evaluation of technical, man-hour and cost data from work area to work area within a project; and from project to project; and from country to country. The WCS may have different names in different organizations, but it is the 'original' costing system. It is the basis for virtually all of the estimating manuals and handbooks which identify unit costs for commodities such as concrete, or piping, or road construction, or equipment installation. The WCS provides a commodity based method to estimate and control costs. The key to the success of the WCS system within an organization is the absolute consistency with which it is used.

The WBS is of primary interest to owners and project managers – both of whom are interested in tracking cost and schedule on a work area basis. The WCS is primarily of interest to construction contractors and engineering consultants, who measure actual performance against forecast performance on a commodity basis.

Professional project managers and cost engineers normally use cost coding systems which encode both the commodity and the work package. This allows them to evaluate job-to-date performance, then forecast cost or productivity trends for the balance of the project.

Step 4: Develop an action plan for the study. An action plan in its simplest form, is just a logical (logic-oriented) time-bar plan listing all of the activities to be studied. Figure 1.6 is one example of such a time-bar graph. A more general action plan would have these characteristics:

1. Purpose: the action plan serves as a control document during the execution of the feasibility study. It functions as a master reference, against which change can be measured and resolved. It provides a visual communication of the logic and progress of the study.

2. Methodology: it may be possible for one person, working in isolation, to develop an action plan. However, it is substantially more desirable to have the project study team develop their plan on a participative, interactive basis. (Texas Instruments' Patrick Haggerty insisted that 'those who implement plans must make the plans'). This interaction

Mine planning 21

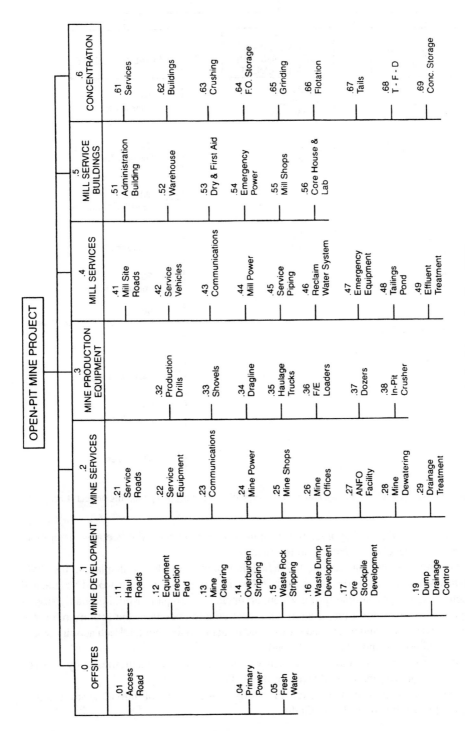

Figure 1.4. Typical work breakdown structure (WBS) directs for an open pit mining project (Lee, 1991).

22 *Open pit mine planning and design: Fundamentals*

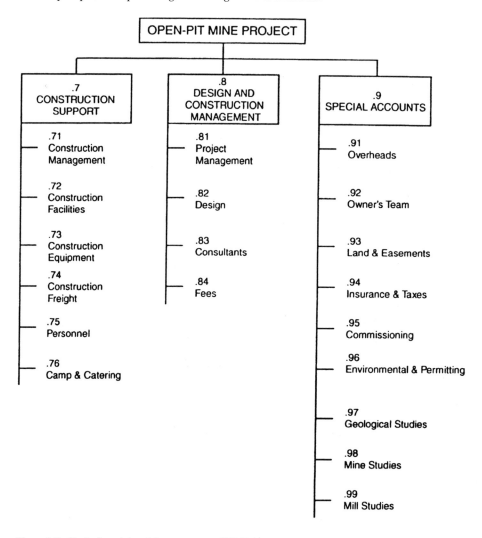

Figure 1.5. Typical work breakdown structure (WBS) indirects for an open pit mining project (Lee, 1991).

fosters understanding and appreciation of mutual requirements and objectives; even more importantly, it develops a shared commitment.

3. Format: a simple master time-bar schedule would be produced, displaying the study activities in a logic-oriented fashion. Brief titles and a reference number would be attached to each activity. For a simple in-house job, an operating company would probably stop at this point. However, for a major study on a new mining operation, the activity reference numbers and titles would be carried into a separate action plan booklet. Each activity would be described briefly, and a budget attached to it.

4. WBS reference: the most convenient way to organize these activities is by referencing them to the first and second levels of the WBS.

5. Number of activities: the practical limitation on the number of individual study activities would be of the order of one hundred.

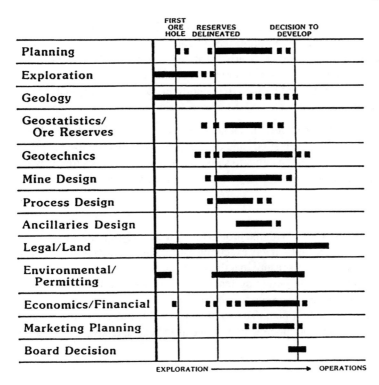

Figure 1.6. Bar chart representation for a mine feasibility/decision-making sequence (McKelvey, 1984).

Phase B. Organizing

Step 5: *Identify additional resource requirements.* While developing a comprehensive action plan, needs for additional resources normally become apparent.

Step 6: *Identify secondary project team members.*

Step 7: *Develop organization chart and responsibilities.* There are a number of ways to organize a project study team for a large study. A separate task force can be established by removing personnel from existing jobs and developing a project-oriented hierarchy, military style. This can work effectively, but can discourage broad participation in the evolution of the project. In a large company, a matrix system can be used very successfully, if used very carefully and with understanding. The management of a matrix organization is based on the management of intentional conflict; it works exceedingly well in a positive environment, and is an unequivocal disaster in an unfavorable environment.

Step 8: *Develop second-level plans and schedules.* Using the master time-bar, the action plan and the WBS as primary references, the enlarged project study team develops second-level plans and schedules, thus establishing their objectives and commitments for the balance of the study. These schedules are oriented on an area-by-area basis, with the primary team members providing the leadership for each area.

Step 9: *Identify special expertise required.* The project study team after reviewing their plan, with the additional information developed during Step 8, may identify a number of areas of the job which require special expertise. Such items may be packaged as separate Requests For Proposals (RFP's), and forwarded to pre-screened consultants on an invitation basis. The scope of work in each RFP should be clearly identified, along with the objectives for the work. A separate section provides explicit comments on the criteria for selection of the successful bidder; this provides the bidder with the opportunity to deliver proposals which can be weighted in the directions indicated by the project team.

Step 10: *Evaluate and select consultants.* Evaluation of the consultant's bids should be thorough, objective, and fair. The evaluations and decisions are made by the use of spread sheets which compare each bidder's capability to satisfy each of the objectives for the work as identified in the RFP. The objectives should be pre-weighted to remove bias from the selection process.

Phase C. Execution
Step 11: *Execute, monitor, control.* With the project study team fully mobilized and with the specialist consultants engaged and actively executing well-defined contracts, the primary challenge to the project manager is to ensure that the study stays on track.

A number of management and reporting systems and forms may be utilized, but the base-line reference for each system and report is the scope of work, schedule and cost for each activity identified in the action plan. The status-line is added to the schedule on a bi-weekly basis, and corrections and modifications made as indicated, to keep the work on track.

1.8 CRITICAL PATH REPRESENTATION

Figure 1.7 is an example of a network chart which has been presented by Taylor (1977) for a medium sized, open pit base metal mine. Each box on the chart contains:
 – activity number,
 – activity title,
 – responsibility (this should be a person/head of section who would carry the responsibility for budget and for progress reports),
 – starting date,
 – completion date,
 – task duration.

The activities, sequential relationships and critical/near critical paths can be easily seen.

Figure 1.8 is the branch showing the basic mining related activities. This progression will be followed through the remainder of the book.

1.9 MINE RECLAMATION

1.9.1 *Introduction*

In the past, reclamation was something to be considered at the end of mine operations and not in the planning stage. Today, in many countries at least, there will be no mine without first thoroughly and satisfactorily addressing the environmental aspects of the proposed project. Although the subject of mine reclamation is much too large to be covered in this brief chapter, some of the factors requiring planning consideration will be discussed. In the western United States, a considerable amount of mineral development takes place on federal and Indian lands. The Bureau of Land Management (BLM) of the U.S. Department of the Interior has developed the *Solid Minerals Reclamation Handbook* (BLM, 1992) with the objective being 'to provide the user with clear guidance which highlights a logical sequence for managing the reclamation process and a summary of key reclamation principles.'

The remaining sections of this chapter have been extracted from the handbook. Although they only pertain directly to those lands under BLM supervision, the concepts have more general application as well. Permission from the BLM to include this material is gratefully acknowledged.

1.9.2 *Multiple-use management*

Multiple-use management is the central concept in the Federal Land Policy and Management Act (FLPMA) of 1976. FLPMA mandates that 'the public lands be managed in a manner that will protect the quality of scientific, scenic, historical, ecological, environmental, air and atmospheric, water resource and archeological values.' Multiple-use management is defined in FLPMA (43 USC 1702(c)) and in regulations (43 CFR 1601.0-5(f)) as, in part, the 'harmonious and coordinated management of the various resources without permanent impairment of the productivity of the lands and the quality of the environment with consideration being given to the relative values of the resources and not necessarily to the combination of uses that will give the greatest economic return or the greatest unit output.' In addition, FLPMA mandates that activities be conducted so as to prevent 'unnecessary or undue degradation of the lands' (43 USC 1732 (b)).

The Mining and Minerals Policy Act of 1970 (30 USC 21(a)) established the policy for the federal government relating to mining and mineral development. The Act states that it is policy to encourage the development of 'economically sound and stable domestic mining, minerals, metal and mineral reclamation industries.' The Act also states, however, that the government should also promote the 'development of methods for the disposal, control, and reclamation of mineral waste products, and the reclamation of mined land, so as to lessen any adverse impact of mineral extraction and processing upon the physical environment that may result from mining or mineral activities.'

In accordance with the National Environmental Policy Act (NEPA), an environmental document will be prepared for those mineral actions which propose surface disturbance. The requirements and mitigation measures recommended in an Environmental Assessment (ERA) or Environmental Impact Statement (EIS) shall be made a part of the reclamation plan.

It is a statutory mandate that BLM ensure that reclamation and closure of mineral operations be completed in an environmentally sound manner. The BLM's long-term reclamation goals are to shape, stabilize, revegetate, or otherwise treat disturbed areas in order to provide a self-sustaining, safe, and stable condition that provides a productive use of the land which conforms to the approved land-use plan for the area. The short-term reclamation goals are to stabilize disturbed areas and to protect both disturbed and adjacent undisturbed areas from unnecessary or undue degradation.

26 *Open pit mine planning and design: Fundamentals*

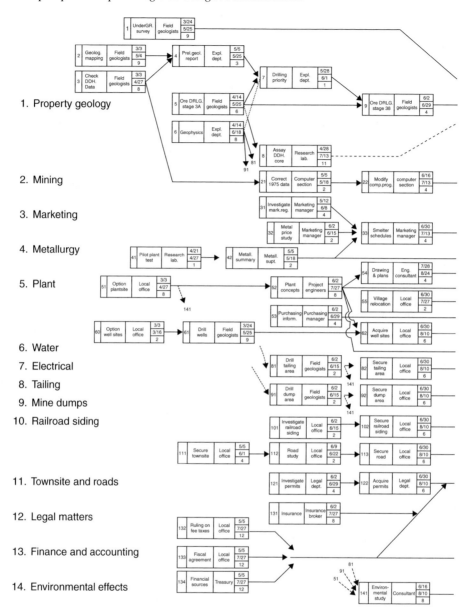

Figure 1.7. Activity network for a feasibility study (Taylor, 1977).

Mine planning

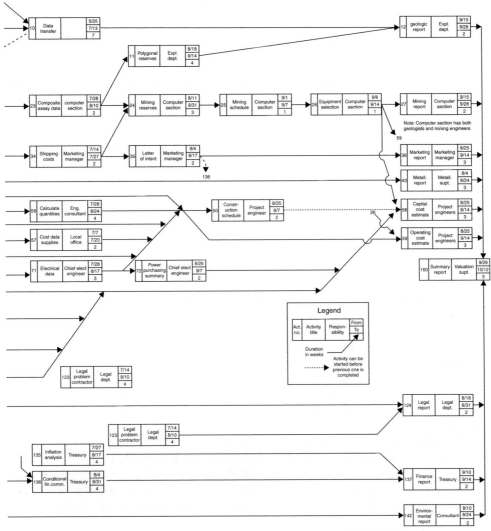

Feasibility report
Volumes:
1. Geology, ore reserves & mining
2. Metallurgy
3. Capital & operating cost estimates
4. Legal, finance & marketing
5. Summary & economic analysis

28 *Open pit mine planning and design: Fundamentals*

Figure 1.8. Simplified flow sheet showing the mining department activities.

1.9.3 *Reclamation plan purpose*

The purposes of the reclamation plan are as follows:

1. Reclamation plans provide detailed guidelines for the reclamation process and fulfill federal, state, county and other local agencies requirements. They can be used by regulatory agencies in their oversight roles to ensure that the reclamation measures are implemented, are appropriate for the site, and are environmentally sound.

2. Reclamation plans will be used by the operator throughout the operational period of the project and subsequent to cessation of exploration, mining, and processing activities. In turn, responsible agencies, including the BLM, will use the reclamation plan as a basis to review and evaluate the success of the reclamation program.

3. Reclamation plans should provide direction and standards to assist in monitoring and compliance evaluations.

1.9.4 *Reclamation plan content*

The reclamation plan should be a comprehensive document submitted with the plan of operations notice, exploration plan, or mining plan. A reclamation plan should provide the following:

1. A logical sequence of steps for completing the reclamation process.
2. The specifics of how reclamation standards will be achieved.
3. An estimate of specific costs of reclamation.
4. Sufficient information for development of a basis of inspection and enforcement of reclamation and criteria to be used to evaluate reclamation success and reclamation bond release.

The reclamation plan shall guide both the operator and the BLM toward a planned future condition of the disturbed area. This requires early coordination with the operator to produce a comprehensive plan. The reclamation plan will serve as a binding agreement between the operator and the regulatory agencies for the reclamation methodology and expected reclamation condition of the disturbed lands and should be periodically reviewed and modified as necessary.

Although the operator will usually develop the reclamation plan, appropriate pre-planning, data inventory, and involvement in the planning process by the regulatory agencies, is essential to determine the optimum reclamation proposal. Most determinations as to what is expected should be made before the reclamation plan is approved and implemented.

It is expected that there will be changes to planned reclamation procedures over the life of the project. Any changes will generally be limited to techniques and methodology needed to attain the goals set forth in the plan. These changes to the plan may result from oversights or omissions from the original reclamation plan, permitted alterations of project activities, procedural changes in planned reclamation as a result of information developed by on-site revegetation research undertaken by the operator and studies performed elsewhere,

Mine planning 29

and/or changes in federal/state regulations. Specific requirements are given in the next section.

In preparing and reviewing reclamation plans, the BLM and the operator must set reasonable, achievable, and measurable reclamation goals which are not inconsistent with the established land-use plans. Achievable goals will ensure reclamation and encourage operators to conduct research on different aspects of reclamation for different environments. These goals should be based on available information and techniques, should offer incentives to both parties, and should, as a result, generate useful information for future use.

1.9.5 *Reclamation standards*

An interdisciplinary approach shall be used to analyze the physical, chemical, biological, climatic, and other site characteristics and make recommendations for the reclamation plan. In order for a disturbed area to be considered properly reclaimed, the following must be complied with:

1. Waste management. All undesirable materials (e.g. toxic subsoil, contaminated soil, drilling fluids, process residue, refuse, etc.) shall be isolated, removed, or buried, or otherwise disposed as appropriate, in a manner providing for long-term stability and in compliance with all applicable state and federal requirements:

(a) The area shall be protected from future contamination resulting from an operator's mining and reclamation activities.

(b) There shall be no contaminated materials remaining at or near the surface.

(c) Toxic substances that may contaminate air, water, soil, or prohibit plan growth shall be isolated, removed, buried or otherwise disposed of in an appropriate manner.

(d) Waste disposal practices and the reclamation of waste disposal facilities shall be conducted in conformance to applicable federal and state requirements.

2. Subsurface. The subsurface shall be properly stabilized, holes and underground workings properly plugged, when required, and subsurface integrity ensured subject to applicable federal and state requirements.

3. Site stability.

(a) The reclaimed area shall be stable and exhibit none of the following characteristics:
 – Large rills or gullies.
 – Perceptible soil movement or head cutting in drainages.
 – Slope instability on or adjacent to the reclaimed area.

(b) The slope shall be stabilized using appropriate reshaping and earthwork measures, including proper placement of soils and other materials.

(c) Appropriate water courses and drainage features shall be established and stabilized.

4. Water management. The quality and integrity of affected ground and surface waters shall be protected as a part of mineral development and reclamation activities in accordance with applicable federal and state requirements:

(a) Appropriate hydrologic practices shall be used to protect and, if practical, enhance both the quality and quantity of impacted waters.

(b) Where appropriate, actions shall be taken to eliminate ground water co-mingling and contamination.

(c) Drill holes shall be plugged and underground openings, such as shafts, slopes, stopes, and adits, shall be closed in a manner which protects and isolates aquifers and prevents infiltration of surface waters, where appropriate.

(d) Waste disposal practices shall be designed and conducted to provide for long-term ground and surface water protection.

5. Soil management. Topsoil, selected subsoils, or other materials suitable as a growth medium shall be salvaged from areas to be disturbed and managed for later use in reclamation.

6. Erosion prevention. The surface area disturbed at any one time during the development of a project shall be kept to the minimum necessary and the disturbed areas reclaimed as soon as is practical (concurrent reclamation) to prevent unnecessary or undue degradation resulting from erosion:

(a) The soil surface must be stable and have adequate surface roughness to reduce run-off, capture rainfall and snow melt, and allow for the capture of windblown plant seeds.

(b) Additional short-term measures, such as the application of mulch or erosion netting, may be necessary to reduce surface soil movement and promote revegetation.

(c) Soil conservation measures, including surface manipulation, reduction in slope angle, revegetation, and water management techniques, shall be used.

(d) Sediment retention structures or devices shall be located as close to the source of sediment generating activities as possible to increase their effectiveness and reduce environmental impacts.

7. Revegetation. When the final landform is achieved, the surface shall be stabilized by vegetation or other means as soon as practical to reduce further soil erosion from wind or water, provide forage and cover, and reduce visual impacts. Specific criteria for evaluating revegetation success must be site-specific and included as a part of the reclamation plan:

(a) Vegetation production, species diversity, and cover (on unforested sites), shall approximate the surrounding undisturbed area.

(b) The vegetation shall stabilize the site and support the planned post-disturbance land use, provide natural plant community succession and development, and be capable of renewing itself. This shall be demonstrated by:

- Successful on-site establishment of the species included in the planting mixture and/or other desirable species.
- Evidence of vegetation reproduction, either spreading by rhizomatous species or seed production.
- Evidence of overall site stability and sustainability.

(c) Where revegetation is to be used, a diversity of vegetation species shall be used to establish a resilient, self-perpetuating ecosystem capable of supporting the postmining land use. Species planted shall includes those that will provide for quick soil stabilization, provide litter and nutrients for soil building, and are self-renewing. Except in extenuating circumstances, native species should be given preference in revegetation efforts.

(d) Species diversity should be selected to accommodate long-term land uses, such as rangeland and wildlife habitat, and to provide for a reduction in visual contrast.

(e) Fertilizers, other soil amendments, and irrigation shall be used only as necessary to provide for establishment and maintenance of a self-sustaining plant community.

(f) Seedlings and other young plants may require protection until they are fully established. Grazing and other intensive uses may be prohibited until the plant community is appropriately mature.

(g) Where revegetation is impractical or inconsistent with the surrounding undisturbed areas, other forms of surface stabilization, such as rock pavement, shall be used.

8. Visual resources. To the extent practicable, the reclaimed landscape should have characteristics that approximate or are compatible with the visual quality of the adjacent area with regard to location, scale, shape, color, and orientation of major landscape features.

9. Site protection. During and following reclamation activities the operator is responsible for monitoring and, if necessary, protecting the reclaimed landscape to help ensure reclamation success until the liability and bond are released.

10. Site-specific standards. All site-specific standards must be met in order for the site to be properly and adequately reclaimed.

1.9.6 *Surface and ground water management*

The hydrologic portion of the reclamation plan shall be designed in accordance with all federal, state, and local water quality standards, especially those under the Clean Water Act National Pollutant Discharge Elimination System (NPDES) point source and non point source programs.

The baseline survey should be conducted to identify the quantity and quality of all surface and subsurface waters which may be at risk from a proposed mineral operation. All aspects of an operation which may cause pollution need to be investigated, so that every phase of the operation can be designed to avoid contamination. It is better to avoid pollution rather than subsequently treat water. The diversion of water around chemically reactive mining areas or waste dumps must be considered during the planning stage. Site selection must be considered during the planning stage. Site selection for waste dumps should be conducted to minimize pollution.

Reclamation plans should be prepared to include a detailed discussion of the proposed surface water run-off and erosion controls including how surface run-off will be controlled during the ongoing operations, during interim shutdowns, and upon final closure.

Reclamation plans should also include a properly designed water monitoring program to ensure operator compliance with the approved plan. The purpose of the monitoring program is to determine the quantities and qualities of all waters which may be affected by mineral operations.

Operators should consider controlling all surface flows (i.e. run-on and run-off) with engineered structures, surface stabilization and early vegetative cover. Where the threat to the downstream water quality is high, the plan should provide for total containment, treatment, or both, if necessary, of the surface run-off on the project site. Sediment retention devices or structures should be located as near as possible to sediment source.

The physical control of water use and routing is a major task for mining projects. The analysis includes the need to:
- Minimize the quantity of water used in mining and processing.
- Prevent contamination and degradation of all water.
- Intercept water so that it does not come in contact with pollutant generating sources.
- Intercept polluted water and divert it to the appropriate treatment facility.

Control may be complicated by the fact that many sources of water pollution are non point sources and the contaminated water is difficult to intercept.

1.9.7 *Mine waste management*

Handling of the waste materials generated during mining has a direct and substantial effect on the success of reclamation. Materials which will comprise the waste should be sampled and characterized for acid generation potential, reactivity, and other parameters of concern. Final waste handling should consider the selective placement of the overburden, spoils, or waste materials, and shaping the waste disposal areas. Creating special subsurface features (rock drains), sealing toxic materials, and grading or leveling the waste dumps are all waste handling techniques for enhancing reclamation. Any problems with the placement of waste discovered after the final handling will be very costly to rectify. Therefore, the selective placement of wastes must be considered during the mine plan review process in order to mitigate potential problems. Waste materials generated during mining are either placed in external waste dumps, used to backfill mined out pits, or used to construct roads, pads, dikes, etc. The design of waste management practices must be conducted in cooperation with the State, the Environmental Protection Agency (EPA), the BLM, other involved federal agencies and the operator.

The most common types of waste dumps include: (1) head of valley fills, (2) cross valley fills, (3) side hill dumps, and (4) flat land pile dumps. In the design and construction of large waste dumps it is important to consider appropriate reclamation performance standards for stability, drainage, and revegetation. Some guidance to consider during the mine plan review process includes the following:

1. Waste dumps should not be located within stream drainages or groundwater discharge areas unless engineered to provide adequate drainage to accommodate the expected maximum flow.

2. Waste dumps will be graded or contoured and designed for mass stability. Design criteria should include a geotechnical failure analysis. It is also recommended that prior to the construction of large waste dumps, a foundation analysis and geophysical testing be conducted on the dump site to ensure basal stability, especially on side hill dump locations. The effects of local groundwater conditions and other geohydrologic factors must be considered in the siting and designing of the dump.

3. Cross valley fills should provide for stream flow through the base of the dump. This is usually done using a rubble drain or french drain. At a minimum, the drain capacity should be capable of handling a design storm flow. To be effective, the drain must extend from the head of the upstream fill to the toe of the downstream face and should be constructed of coarse durable rock which will pass a standard slake test. Toxic or acid-producing materials should not be placed in valley fills.

4. Drainage should be diverted around or through head of valley and sidehill dumps.

5. Drains must be constructed of durable, nonslaking rock or gravel.

6. Topsoil or other suitable growth media should be removed from the proposed dump site and stockpiled for future use in reclamation.

7. Placement of coarse durable materials at the base and toe of the waste dump lowers the dump pore pressure and provides for additional internal hydrologic stability. An exception to this guidance would be in the case where the spoils materials exhibit high phytotoxic properties and the spoils must be sealed to prevent water percolation.

8. The finer textured waste materials which are more adaptable for use as a growing medium should be placed on the outside or mantel of the waste dump.

9. After the waste dump has been shaped, scarified, or otherwise treated to enhance reclamation, available topsoil or other selected subsoils should be spread over the surfaces of the dump as a growing medium. Grading and scarification may be required.

10. The dump should be designed to provide for controlled water flow which minimizes erosion and enhances structural stability.

11. Control erosion on long face slopes by requiring some form of slope-break mitigation, such as benches to intercept the flow of water or rock/brush terraces to slow down the velocity of the run-off.

12. Waste dump benches should be bermed or constructed wide enough to handle the peak design flows and to prevent overflowing onto the face of the dump in the event of freezing conditions. Dump benches should be constructed to allow for mass settling of the dump.

Safety requirements must be calculated for large waste dumps or waste embankments.

1.9.8 *Tailings and slime ponds*

Tailings and slime ponds consist of impounded mill wastes. Slime ponds are tailings ponds with high percentages of silts and clays, which cause very slow sediment drying conditions. Slime ponds are commonly associated with phosphate and bauxite processing operations. Reclamation of slime ponds is complicated by the slow dewatering.

Tailings impoundments are typically placed behind dams. Dams and the impounded wastes may require sealing on a case-by-case basis to avoid seepage below the dam or contamination of the groundwater. This measure only may be done before emplacement of the wastes. Long-term stability of the structure must be assured in order to guarantee ultimate reclamation success.

The nature of the tailings to be impounded should be determined as early as possible during the development of any plan. Tailings exhibiting phytotoxic or other undesirable physical or chemical properties will require a more complex reclamation plan. Analysis should include a thorough review of groundwater flow patterns in the area and a discussion of potential groundwater impacts. An impermeable liner or clay layer may be required to avoid contamination of groundwater. Where tailings include cyanide, final reclamation may include either extensive groundwater monitoring or pumpback wells and water treatment facilities to assure (ensure) groundwater quality is protected. The presence of cyanide in the tailings will not normally complicate reclamation of the surface.

1.9.9 *Cyanide heap and vat leach systems*

Dilute solutions of sodium cyanide (NaCN) or potassium cyanide (KCN) are used to extract precious metals from ores. Concentrations of cyanide solution utilized range from 300 to 500 ppm for heap leach operations to 2000 ppm (0.2%) for vat leach systems.

Low-grade ores can be economically leached in heaps placed on impermeable pads where a cyanide solution is sprinkled onto the ore. The solution preferentially collects the metals as it percolates downward and is recovered at the bottom of the heap through various means. Other metals besides gold and silver are mobilized by cyanide solutions.

Higher grade ores may be crushed, ground and agitated with cyanide solution in vats or tanks. The solids are then separated from the gold or silver-bearing (pregnant) solution. The precious metals are recovered from the pregnant solution and the solids are transferred to a tailings impoundment. The tailings are often deposited in a slurry form and may contain several hundred parts per million of cyanide.

Part of the overall mine reclamation plan includes cyanide detoxification of residual process solutions, ore heaps, tailings impoundments, and processing components.

A key to reclamation of cyanide facilities is planning for the solution neutralization process. The first step is to set a detoxification performance standard. This will have to be site specific dependent on the resources present and their susceptibility to cyanide and metal contamination. A minimum requirement would have to be the specific state standard. BLM may need to require more stringent standards if sensitive resources are present. Other considerations include the health advisory guideline used by EPA of 0.2 mg/l for cyanide in drinking water; and the freshwater chronic standard of 0.0052 mg/l for aquatic organisms. Some species of fish are especially sensitive to cyanide. Likewise metals, and other constituent levels, should be specified for detoxification of cyanide solutions.

There are a variety of methods for achieving detoxification of cyanide solutions. These range from simple natural degradation, to active chemical or physical treatment of process waters. A thorough understanding of the metallurgical process generating the waste, and of the chemistry of the waste stream is necessary to select the most effective cyanide destruction technique.

1.9.10 *Landform reclamation*

Shaping, grading, erosion control, and visual impact mitigation of an affected site are important considerations during review of the reclamation plan. The review process not only ensures that the topography of the reclaimed lands blend in as much as possible with the surrounding landforms, natural drainage patterns, and visual contrasts, but also enhances the success of revegetation.

The final landform should:
– be mechanically stable,
– promote successful revegetation,
– prevent wind and water erosion,
– be hydrologically compatible with the surrounding, landforms, and
– be visually compatible with the surrounding landforms.

Pit backfilling provides an effective means for reclamation of the disturbed lands to a productive post-mining land use. However, development of some commodities and deposit types may not be compatible with pit backfilling.

Open pit mine optimization is achieved by extending the pit to the point where the cost of removing overlying volumes of unmineralized 'waste' rock just equal the revenues (including profit) from the ore being mined in the walls and bottom of the pit. Because there is usually mineralization remaining, favorable changes in an economic factor (such as an increase in the price of the commodity or new technology resulting in a reduced operating cost) can result in a condition where mining can be expanded, or resumed at a future time. This economically determined pit configuration is typical of the open pit metal mining industry and is of critical importance in efforts to maximize the recovery of the mineral resource. To recover all the known ore reserves the entire pit must remain exposed through progressively deeper cuts. Backfilling where technologically and economically feasible, can not begin until the ore reserves within the specific pit are depleted at the conclusion of mining. Additionally, some waste material is not suitable for use as backfilling material.

Depending upon the size of the open pit, backfilling can extend the duration of operations from a few months to several years.

Final highwall configuration, including consideration of overall slope angle, bench width, bench height, etc., should be determined during the review of the plan. The maximum height of the highwall should be determined using site-specific parameters such as rock type and morphology. In most cases, the maximum height is regulated by various state agencies.

The normal procedures are to either leave the exposed highwall or to backfill and bury the highwall either totally or partially. Appropriate fencing or berming at the top of the highwall is necessary to abate some of the hazards to people and animals.

It is important that the backfill requirements be determined during the plan review process and included in the approved plan.

1.10 ENVIRONMENTAL PLANNING PROCEDURES

As described by Gilliland (1977), environmental planning consists of two distinct phases:
- Initial project evaluation,
- The strategic plan.

The components involved in each of these as extracted from the Gilliland paper will be outlined below.

1.10.1 *Initial project evaluation*

1. Prepare a detailed outline of the proposed action. This should include such items as drawings of land status, general arrangement of facilities, emission points and estimates of emission composition and quantities, and reclamation plans. It is also helpful to have information on the scope of possible future development and alternatives that might be available which could be accommodated within the scope of the proposed action.

For example, are there other acceptable locations for tailings disposal if the initial location cannot be environmentally marketed? A schedule for engineering and construction of the proposed action and possible future development should also be available.

2. Identify permit requirements. Certain permits can take many months to process and must be applied for well in advance of construction. Further, some permits will require extensive data, and very long lead times may be encountered in the collection of such data.

For example, biotic studies for environmental impact statements require at least a year, and sometimes longer, to evaluate seasonal changes in organisms. Are there points of conflict between permit requirements and the nature of the proposed action? Can the proposed action be altered to overcome these discrepancies or to avoid the need for permits that could be particularly difficult or significantly time-consuming to obtain? For example, a 'zero' effluent discharge facility could well avoid the Federal Water Pollution Control Action requirement for an Environmental Impact Statement (EIS).

3. Identify major environmental concerns. This includes potential on-site and off-site impacts of the proposed action and from possible future development. Land use and socio-economic issues as well as those of pollutional character must be taken into account. Although there may be little concern about the impacts of an exploratory activity itself, when bulldozers and drill rigs begin to move onto a property, it becomes apparent to the public that there may indeed ultimately be a full development of the property. Public concern may surface from speculation about the possible impacts of full development, and this could result in considerable difficulty in obtaining even the permits necessary to proceed with the proposed activity.

4. Evaluate the opportunity for and likelihood of public participation in the decision-making process. Recent administrative reforms provide for expanded opportunity for public participation in the decision-making process. Projects to be located in areas of minimal environmental sensitivity may stir little public interest and permits will not be delayed beyond their normal course of approval. A project threatening material impact to an area where the environmental resources are significant, however, will probably receive careful public scrutiny and may be challenged every step in the permit process.

5. Consider the amount and effect of delay possibly resulting from public participation during each state of the project. This could also be called intervention forecasting. When can a hearing be requested? When would it be possible for a citizen to bring suit? How long would it take to secure a final court action? Could the plaintiffs enjoin work on the project during the pendency of litigation? Can the project tolerate such delays? Can the project schedule be adjusted to live with such delays?

6. Evaluate the organization and effectiveness of local citizens groups. Attitudes are also part of this evaluation. Local citizen groups can be a powerful ally in positive communications with the public. They can also be effective adversaries. This evaluation should be extended to all groups which could have a significant voice in opinion making within a community. The working relationship of local groups with state or national counterpart groups should also be assessed.

7. Determine the attitudes and experience of governmental agencies. Identify any inter-agency conflicts. New ventures face an intricate web of federal, state, and local laws and regulations which are often complicated by inconsistencies in the policy goals which underlie these laws, and overlapping jurisdiction of the regulatory agencies.

Sometimes you must deal with personnel who have little knowledge of the business world or of the nature of operations being proposed. A company must be prepared, therefore, to dedicate considerable time and effort in promoting and understanding of the project.

Further, it is imperative for a company to recognize that government agency personnel have a public responsibility to see that the various laws and regulations within their jurisdiction are complied with. They may not always agree that the requirements of the law are practical, fair or equitable, but it is their job to ensure their applicability. Sometimes areas of apparent frustration or conflict will resolve themselves by re-evaluating your position with regard to the role that must be performed by regulatory personnel.

8. Consider previous industry experience in the area. This involves a determination of public attitudes toward previous or existing industry in the area and the posture and performance of these industries as a responsible member of the community. It is extremely helpful to the cause of your project if industry enjoys the status of being a good citizen. Where negative attitudes prevail, is there something about your project that could invite similar censure or could it be so designed to change these public attitudes?

9. Consider recent experience of other companies. Have new industries located or tried to locate within the area? Were there any issues involved relative to their success or failure to locate that might also be issues of concern to the proposed activity?

10. Identify possible local consultants and evaluate their ability and experience. Local consultants can be invaluable in assisting the company in many areas of inquiry. Their familiarity with the local scene on environmental, legal, socioeconomic, land use and other matters can enhance the credibility of a company's planning efforts and acceptability within a community.

11. Consider having a local consultant check the conclusions of the initial evaluation.

This initial project evaluation is essentially an identification procedure which in many instances can be largely produced in-house with possibly some modest assistance from outside consultants. Correspondingly, the cost could range from a thousand dollars or less to several thousand dollars depending upon the familiarity of personnel with this type of work and the amount of outside consultant help needed.

1.10.2 *The strategic plan*

Following the initial project evaluation the next step is to prepare a strategy or game plan for dealing with the identified issues. The elements would include:

1. Outline of technical information needed to obtain permits and to address legitimate environmental, land use and socioeconomic concerns. There are good reasons for the reluctance of planners to develop hard data before they are sure that they will be permitted to proceed with a project.

However, if a project is worthwhile, every practical effort must be made to develop information that demonstrates impacts have been carefully assessed, legitimate environmental concerns have been addressed, and controls and mitigation measures will be adequate to meet all existing standards and to protect the environment. In cases where standards are stringent and controls are not demonstrated technology, substantial extra effort may have to be made to develop predictions of performance. In case where better data cannot be developed without delaying construction, plans may have to include a proposal for eventually securing such data and adjusting permit requirements before operations begin.

Where predictive data are not practically obtainable, a plan might provide for operational monitoring with post-startup alteration of permit requirements if problems arise. This plan element, therefore, provides a specific checklist for the information gathering system.

2. Categorically assign responsibilities for the acquisition of the technical information and hire necessary consultants. Coordinate this work with governmental agencies when appropriate.

The primary responsibility for each element of data collection should be clearly designated so that misunderstandings do not arise. Governmental agencies can be an important source of background information on air quality, water quality and other pertinent data. Further, government studies may be intended or in progress which in scope would include the location and environmental concerns of the company's proposed action. Data collection by the company could complement these studies and vice versa.

3. Prepare a schedule for obtaining information and data and for submitting permit applications to the appropriate agencies. Firm target dates must be established for the finalization of reports, permit applications or other necessary authorizations. Interim reporting periods should also be set to ascertain the status of progress and to provide whatever adjustments are necessary to keep on the appropriate schedule. A critical path chart would include a display of this sequence. If a project is properly planned, its proponents require nothing more from government except even handed operation of the approval mechanism.

4. Select local legal, technical and public relations consultants. Sometimes the local consultants may be those who will be directly involved in the data development. In other instances, these consultants would have more of a role in planning, data evaluation and public communications.

5. Avoid hostile confrontations with environmental groups. There is nothing to be gained from a shouting match where both sides become so highly polarized that reason and credibility cannot be maintained. No-growth advocates will probably continue to be unyielding in their opposition no matter how much progress is made in devising effective environmental controls.

Project planners who view citizen opposition as monolithic and implacable miss, however, an opportunity to reduce the risks of intervention and delay. Citizen attitudes are subject to change, and many citizen activists are sincerely, and very properly, seeking to secure for themselves and others the maintenance of a quality environment.

If the proposed activity is demonstrably sound, both industrially and environmentally, and the public has access to all the facts, it is likely that people will make sound judgements and that mineral development will be permitted.

6. Develop a consistent program for the generation of credible factual information. Good factual information needed to refute or substantiate concerns regarding possible impacts of the proposed action or future development is not always available. Such deficiencies are not uncommon or unacceptable if they are honestly faced and a program is designed to acquire the necessary information. Many projects have been seriously delayed or stopped because of a company's failure to admit that a concern exists. This can become a focal point for attacking the credibility of a company's entire program.

1.10.3 *The environmental planning team*

The environmental planning effort, due to the wide diversity of tasks involved requires the participation of many specialists drawn from various functional areas of mining organizations and from outside consulting firms. To coordinate this effort there must be a team leader who has the perspective to understand the requirements of the disciplines involved and the eventual use of the information evolved. This team leader must also have the acknowledged responsibility and authority for the performance of this coordinating role.

Table 1.4. Types of permits and approvals which may be required for the Kensington Gold Project (Forest Service, 1990).

1. *Federal government*
   Forest Service
   1. NEPA compliance and record of decision on EIS
   2. Plan of operations
   3. Special use permits

   Environmental Protection Agency
   1. National Pollutant Discharge Elimination System (NPDES)
   2. Spill Prevention Control and Countermeasure (SPCC) plan
   3. Review of section 404 Permit
   4. Notification of hazardous wages activity
   5. NEPA compliance and record of decision on EIS (cooperating agency)

   Army Corps of Engineers
   1. Section 404 Permit – Clean Water Act (dredge and fill)
   2. Section 10 Permit – Rivers and Harbor Act
   3. NEPA compliance and record of decision on EIS (cooperating agency)

   Coast Guard
   1. Notice of fueling operations
   2. Permit to handle hazardous materials
   3. Application for private aids to navigation

   Federal Aviation Administration
   1. Notice of landing area and certification of operation
   2. Determination of no hazard

   Federal Communications Commission
   1. Radio and microwave station authorizations

   Treasury Department (Dept of Alcohol, Tobacco & Firearms)
   1. Explosives user permit

   Mine Safety and Health Administration
   1. Mine I.D. number
   2. Legal identity report
   3. Miner training plan approval

   U.S. Fish and Wildlife Service
   1. Threatened and endangered species clearance
   2. Bald Eagle Protection Act clearance

   National Marine Fisheries Services
   1. Threatened and endangered species clearance

2. *State of Alaska*
   Alaska Division of Government Coordination
   1. Coastal project questionnaire
   2. Coastal management program certification

   Alaska Department of Environmental Conservation
   1. Air quality permit
   2. Burning permit
   3. Certification of reasonable assurance
   4. Solid Waste Management permit
   5. Oil facilities approval of financial responsibility
   6. Oil facilities discharge contingency plan
   7. Water and sewer plan approval
   8. Food service permit

   Alaska Department of Natural Resources
   1. Water rights permits
   2. Tidelands lease
   3. Right-of-way permit
   4. Permit to construct or modify a dam
   5. Land use permit

   Alaska Department of Fish and Game
   1. Fishway or fish passage permit
   2. Anadromous fish protection permit

   Alaska Department of Public Safety
   1. Life and fire safety plan check

   Alaska Department of Labor
   1. Fired and unfired pressure vessel certificate
   2. Elevator certificate of operation

   Alaska Department of Revenue
   1. Affidavit for non-resident business taxation
   2. Alaska business license
   3. Alaska mining license

   Alaska Department of Health and Social Services
   1. Health care facilities construction license
   2. Certificate of need (townsite with health care facilities)

3. *Local government*
   City and Bureau of Juneau
   1. Mining permit
   2. Grading permit
   3. Building permits
   4. Burning permits
   5. Explosive permits

40   *Open pit mine planning and design: Fundamentals*

The team members include such personnel as the project manager, project engineers, attorneys, environmental specialists, technical and public relations experts.

1.11 A SAMPLE LIST OF PROJECT PERMITS AND APPROVALS

The 'Final Scoping Document, Environmental Impact Statement' for the Kensington Gold Project located near Juneau, Alaska was published by the U.S. Forest Service (Juneau Ranger District) in July 1990 (Forest Service, 1990). To provide the reader with an appreciation for the level of effort involved just in the permitting process, a listing of the various federal, state, and local government permits/approvals which may be required for this underground gold mine/mill, is given in Table 1.4.

REFERENCES

Anonymous. 1976. Principles of the Mineral Resource Classification System of the U.S. Bureau of Mines and U.S. Geological Survey. Geological Survey Bulletin 1450-A. Washington: U.S. Government Printing Office.
Anonymous. 1980. Principles of a Resource/Reserve Classification for Minerals. Geological Survey Circular 831. Arlington, VA: U.S. Geological Survey.
Anonymous. 1998. A heavy metal harvest for the millennium. E/MJ. 199(11): 32YY–32ZZ.
AusIMM. 2004. JORC information. www.jorc.org/main.
AusIMM. 2005. Australasian Code for the reporting of exploration results, mineral resources and ore reserves: The JORC Code – 2004 Edition. www.jorc.org/pdf/jorc2004print.pdf.
AusIMM. 1995. Code and guidelines for technical assessment and/or valuation of mineral and petroleum assets and mineral and petroleum securities for independent expert reports (The VALMIN Code). http://www.ausimm.com.au/codes/valmin/valcode0.asp.
AusIMM. 1998a. The revised and updated VALMIN Code: Code and guidelines for technical assessment and/or valuation of mineral and petroleum assets and mineral and petroleum securities for independent expert reports (The VALMIN Code). http://www.ausimm.com.au/codes/valmin.
AusIMM. 1998b. The revised VALMIN Code and guidelines: An aide memoire to assist its interpretation. http://www.ausimm.com.au/codes/valmin/eexp11.asp.
Banfield, A.F. 1972. Ore reserves, feasibility studies and valuations of mineral properties. Paper presented at the AIME Annual Meeting, San Francisco, CA February 20–24, 1972. Society of Mining Engineers of AIME, Preprint 72-AK-87.
Banfield, A.F. & J.F. Havard 1975. Let's define our terms in mineral valuation. *Mining Engineering* 27(7): 74–78.
BLM (Bureau of Land Management) 1992. *Solid Minerals Reclamation Handbook* (BLM Manual Handbook H-3042-1). U.S. Department of the Interior.
Böde, K. 1999 KUBUS – Estimating and controlling system for project management in the construction and construction-related industry. 28th International Symposium on Application of Computers and Operations Research in the Mineral Industry: 155–164. Colorado School of Mines:CSM.
CIM. 2000. Exploration Best Practice Guidelines. Aug 20. www.cim.org/definitions/explorationBEST PRACTICE.pdf.
CIM. 2003a. Standards and guidelines for valuation of mineral properties. February. www.cim.org/committees/CIMVal_final_standards.pdf.
CIM. 2003b. Guidelines for the reporting of diamond exploration results. Mar 9. www.cim.org/committees/diamond_exploration_final.cfm.
CIM. 2003c. Estimation of mineral resources and mineral reserves: Best practice guidelines. Nov 23. www.cim.org/committees/estimation.cfm.
CIM. 2004. Definition Standards – On mineral resources and mineral reserves. www.cim.org/committees/StdsAppNovpdf.

CIM. 2005. Standards and guidelines. www.cim.org/committees/guidelinesStandards_main.cfm.
Danni, J.W. 1992. Mineral Hill Mine – A case study in corporate environmentalism. *Mining Engineering.* 44(1): 50–53.
Danilkewich, H., Mann, T., and G. Wahl. 2002. Preparing a feasibility study request for proposal in the 21*st* century. Pre-print Paper 02–101. 2002 SME Annual Meeting, Feb 26–28. Phoenix, AZ.
Davis, G. 1994. U.S. share of world mineral markets: Where are we headed. *Mining Engineering.* 46(9): 1067–1069.
Dessureault, S., Scoble, M., and S. Dunbar. 2002. Activity based costing and information engineering: Formulation of an annual budget. 30th International Symposium on Application of Computers and Operations Research in the Mineral Industry: 601–614. Alaska:SME.
De Voto, R.H., and T.P. McNulty. 2000. Banning cyanide use at McDonald – an attack on open-pit mining. *Mining Engineering.* 52(12): 19–27.
Forest Service (Juneau Ranger District) 1990. Kensington Gold Project, Alaska: Final Scoping Document, Environmental Impact Statement. U.S. Department of Agriculture (7).
Francis, D. 1997. Bre-X: The Inside Story. Key Porter Books, Toronto. 240 pp.
Gilliland, J.C. 1977. The environmental requirements of mine planning. *Mineral Industry Costs* Northwest Mining Association: 57–65.
Goold, D., and A. Willis. 1997. Bre-X Fraud. McClelland & Stewart. 272 pp.
Grace, K.A. 1984. Reserves, resources and pie-in-the-sky. *Mining Engineering* 36(10): 1446–1450; 1985. Discussion. *Mining Engineering* 37(8): 1069–1072.
Halls, J.L. 1975. *Personal communication.* Kennecott Copper Company.
Hustrulid, W. 2002. 2002 Jackling Lecture: You can't make a silk purse out of a sow's ear. *Mining Engineering.* 54(9): 41–48.
Jacus, J.R. and T.E. Root 1991. The emerging federal law of mine waste: administrative, judicial and legislative developments. Land and Water Law Review, University of Wyoming, vol. XXVI, n. 2, 1991.
Kelly, T.D. 2002. Raw materials and technology fuel U.S. economic growth. *Mining Engineering.* 54(12): 17–21.
King, B.M. 1998. Impact of rehabilitation and closure costs on production rate and cut-off grade strategy. 27th International Symposium on Application of Computers and Operations Research in the Mineral Industry: 617–630. London:IMM.
Kirk, S.J. & T.J. O'Neil 1983. Preplanning of end land uses for an open pit copper mine. *Mining Engineering* 35(8): 1191–1195.
Kirk, W.S. 1998. Iron ore reserves and mineral resource classification systems. Skillings Mining Review. June 6. pp 4–11.
Kral, S. 1993. Risk assessment/management in the environmental planning of mines. *Mining Engineering* 45(2): 151–154.
Lee, T.D. 1984. Planning and mine feasibility study – An owners perspective. In: *Proceedings of the 1984 NWMA Short Course 'Mine Feasibility – Concept to Completion'* (G.E. McKelvey, compiler). Spokane, WA.
Lee, T.D. 1991. Personal communication. 12/10/91.
Marcus, J. 1990. Mining environment – regulatory control of mining at the federal level, part 1. *E&MJ* 191(6).
McKelvey, G.E. 1984. Mineral Exploration. *Mine Feasibility – Concept to Completion* (G.E. McKelvey, compiler). Spokane, WA: Northwest Mining Association.
McKenzie, B.W. 1980. Looking for the improbable needle in a haystack. *The Economics of Base Metal Exploration in Canada.* Queens University, Ontario: Working Paper No. 19.
Miller, J.D. 1984. Legal considerations in mine planning and feasibility. *Mine Feasibility – Concept to Completion* (G.E. McKelvey, compiler). Spokane, WA: Northwest Mining Association.
Moore, R.T. 1984. Environmental/Permitting. *Mine Feasibility – Concept to Completion* (G.E. McKelvey, compiler). Northwest Mining Association.
Noble, A.C. 1993. Geologic resources vs. ore reserves. *Mining Engineering* 45(2): 173–176.
Paulsen, K.R. 1982. Environmental and regulatory costs of mining projects. *Mineral Industry Costs* (J.R. Hoskins, compiler): 27–32. Northwest Mining Association.
Rendu, J-M. 2003. SME meets with the SEC – resources and reserves reporting discussed. *Mining Engineering.* July. P35.

Rogers, J.P. 1999. ProjectWorks® – Bechtel's automated solution for projects. 28th International Symposium on Application of Computers and Operations Research in the Mineral Industry: 165–172. Colorado School of Mines:CSM.

SAIMM. 2000. South African Code for reporting of mineral resources and mineral reserves (the SAMREC Code). www.saimm.co.za/pages/comppages/samrec_version.pdf.

SEC. 2005. Description of property by users engaged or to be engaged in significant mining operations. Guide 7. www.sec.gov/divisions/corpfin/forms/industry.htm#secguide7/.

SME (Society of Mining, Metallurgy and Exploration) 1991. A guide for reporting exploration information, resources and reserves. *Mining Engineering* 43(4): 379–384.

SME. 1991. A guide for reporting exploration information, resources and reserves. *Mining Engineering* April. pp 379–384.

SME. 1999. A guide for reporting exploration information, mineral resources, and mineral reserves. www.smenet.org.digital_library/index.cfm.

SME. 2003. Reporting mineral resources and reserves: Regulatory, financial, legal, accounting, managerial and other aspects. Conference October 1–3, 2003. SME.

Sykes, T. 1978. The Money Miners: Australia's Mining Boom 1969–1970. Wildcat Press, Sydney. 388 pages.

Sykes, T. 1988. Two Centuries of Panic: A History of Corporate Collapses in Australia. Allen & UnwinLondon. 593 pages.

Sykes, T. 1996. The Official History of Blue Sky Mines. Australian Financial Review. 165 pages.

Taylor, H.K. 1977. Mine valuation and feasibility studies. *Mineral Industry Costs* (J.R. Hoskins & W.R. Green, editors): 1–17. Spokane, WA: Northwest Mining Association.

U.S. Department of the Interior (Office of Surface Mining Reclamation and Enforcement) 1988. Surface Mining Control and Reclamation Act of 1977. (Public Law 95–87).

U.S. Department of Agriculture (Forest Service) 1988. Best minerals management practices – a guide to resource management and reclamation of mined lands in the Black Hills of South Dakota and Wyoming.

USBM 1967. *A Dictionary of Mining, Mineral and Related Terms* (Paul W. Thrush, editor). Washington: U.S. Department of the Interior.

USGS. 1976. Principles of the mineral resource classification system of the U.S. Bureau of Mines and U.S. Geological Survey. Bulletin 1450-A.

USGS. 1980. Principles of a resource/reserve classification for minerals. USGS Circular 831. http://pubs.er.usgs.gov/pubs/cir/cir831/ or http://imcg.wr.usgs.gov/usbmak/c831.html.

Wober, H.H. & P.J. Morgan 1993. Classification of ore reserves based on geostatistical and economic parameters. *CIM Bulletin* 86(966): 73–76.

REVIEW QUESTIONS AND EXERCISES

1. What is meant by ore?
2. Express the meaning of 'profit' in your own words. How does it relate to your future opportunities?
3. Define
   - Exploration
   - Development
   - Production
4. Discuss the changes that have occurred between the SME 1991 and SME 1999 guidelines regarding the 'Reporting of Exploration Information, Resources and Reserves.' See the Reference section of this chapter.
5. Distinguish the meanings of 'Resource' and 'Reserve.'
6. Using Figure 1.1 discuss the basis upon which 'Resources' and 'Reserves' change category.

Mine planning 43

7. The U.S. Securities and Exchange Commission (SEC) have their own guidelines regarding the public reporting of resources and reserves. Referring to the website provided in the References, summarize their requirements. How do they compare to the SME 1991 guidelines? To the 1999 guidelines?
8. The most recent version of the USGS/USBM classification system is published as USGS Circular 831. Download the Circular from their website (see References). What was the main purpose of these guidelines? Who was the intended customer?
9. What is meant by 'Hypothetical Resources'?
10. What is meant by 'Undiscovered Resources'?
11. In the recent Bre-X scandal, the basis for their resource/reserve reporting was indicated to be USGS Circular 831. In which of the classification categories would the Bre-X resources/reserves fall? Explain your answer. See the References for the website.
12. Compare the 1999 SME guidelines with those provided in the JORC code included in Chapter 7.
13. Compare the 1999 SME guidelines with those provided in the CIM guidelines included in Chapter 7.
14. Discuss the relevance of the Mineral Supply Process depicted in Figure 1.2 to iron ore for the period 2002 to 2005.
15. Discuss the relevance of the Mineral Supply Process depicted in Figure 1.2 to molybdenum for the period 2002 to 2005.
16. Discuss the relevance of the Mineral Supply Process depicted in Figure 1.2 to copper for the period 2002 to 2005.
17. Figure 1.3 shows diagrammatically the planning, implementation and production phases for a new mining operation. What are the planning stages? What are the implementation stages? What are the production stages?
18. Does the 'Relative Ability to Influence Cost' curve shown in Figure 1.3 make sense? Why or why not?
19. What is the fourth phase that should be added to Figure 1.3?
20. In the initial planning stages for any new project there are a great number of factors of rather diverse nature that must be considered. The development of a 'checklist' is often a very helpful planning tool. Combine the items included in the checklist given in section 1.3 with those provided by Gentry and O'Neil on pages 395–396 of the SME Mining Engineering Handbook (2nd edition, Volume 1).
21. How might the list compiled in problem 20 be used to guide the preparation of a senior thesis in mining engineering?
22. What is the meaning of a 'bankable' mining study?
23. Summarize the differences between a conceptual study, a pre-feasibility study and a feasibility study.
24. Assume that the capstone senior mine design course extends over two semesters each of which is 16 weeks in duration. Using the information provided in Tables 1.3 and 1.4 regarding the content of an intermediate valuation report (pre-feasibility study) and a feasibility study, respectively, develop a detailed series of deliverables and milestones. It is suggested that you scan the two tables and cut-and-paste/edit to arrive at your final product.
25. Assume that the estimated capital cost for an open pit project is $500 million. How much would you expect the conceptual study, the preliminary study and the feasibility study to cost?

26. Section 1.6 concerns the accuracy of the estimates provided. These are discussed with respect to tonnage and grade, performance, costs, and price and revenue. Summarize each.
27. Discuss what is meant by the contingency allowance. What is it intended to cover? What is it not meant to cover?
28. In section 1.6.4 it is indicated that both probable and average metal prices expressed in present value dollars need to be provided. The 'conservative' price is considered to be that with an 80% probability of applying. Choose a mineral commodity and assign a probable price and a conservative price for use in a pre-feasibility study. Justify your choices.
29. What are the two common ways for accomplishing a feasibility study?
30. Summarize the steps involved in performing a feasibility study. What is the function of the steering committee? Who are the members?
31. Who are the members of the project teams?
32. What is meant by a Work Breakdown Structure? What is its purpose?
33. What is the difference between a Work Breakdown Structure and a Work Classification Structure?
34. Construct a bar chart for the activities listed in the project schedule developed in problem 24. It should be of the type shown in Figure 1.6.
35. What is meant by an RFP? How should they be structured?
36. What is the goal of a Critical Path representation?
37. Section 1.9 deals with mine reclamation. What is the rationale for including this material in Chapter 1 of this book and not later?
38. What is the concept of multiple use management? What is its application to minerals?
39. What is the practical implication of the statement from the Mining and Minerals Policy act of 1970?
40. What is the U.S. National Materials and Mineral Policy, Research and Development Act of 1980 (Public Law 96-479)? Is it being followed today?
41. Define the following acronyms:
    a. BLM
    b. FLPMA
    c. CFR
    d. NEPA
    e. EA
    f. EIS
    g. NPDES
    h. EPA
42. Summarize the purposes of a reclamation plan.
43. What should a reclamation plan contain?
44. For a disturbed area to be properly reclaimed, what must be achieved? Summarize the major concepts.
45. Discuss the most important concepts regarding surface and ground water management.
46. Discuss the important concepts regarding mine waste management.
47. Tailings and slime ponds. What is the difference between them? What are the engineering concerns?
48. What means are available for the detoxification of cyanide heap and vat leach systems?
49. What is meant by landform reclamation?

50. Do open pit mines have to be refilled? Discuss the pro's and con's regarding the backfilling of open pit mines.
51. In section 1.10 'Environmental Planning Procedures' Gilliland divided environmental planning into two distinct phases: (1) Initial project evaluation and (2) The strategic plan. Summarize the most important aspects of each.
52. What members would be part of an environmental planning team?
53. In section 1.11 a list of the project permits and approvals required for the Kensington Gold project in Alaska have been provided. List the permits and approvals required for a new mining project in your state/country.

CHAPTER 2

Mining revenues and costs

2.1 INTRODUCTION

For one to know whether the material under consideration is 'ore' or simply 'mineralized rock', both the revenues and the costs must be examined. It is the main objective of this chapter to explore in some detail each of these topics.

2.2 ECONOMIC CONCEPTS INCLUDING CASH FLOW

In Chapter 6, the production planning portion of this text, an economic basis will be used to select production rate, mine life, etc. This section has been included to support that chapter. It is not intended to be a textbook complete in itself but rather to demonstrate some of the important concepts and terms.

2.2.1 *Future worth*

If someone puts $1 in a savings account today at a bank paying 10% simple interest, at the end of year 1 the depositor would have $1.10 in his account. This can be written as

$$FW = PV(1 + i) \tag{2.1}$$

where FW is the future worth, PV is the present value, i is the interest rate.

If the money is left in the account, the entire amount (principal plus interest) would draw interest. At the end of year 2, the account would contain $1.21. This is calculated using

$$FW = PV(1 + i)(1 + i)$$

At the end of year n, the accumulated amount would be

$$FW = PV(1 + i)^n \tag{2.2}$$

In this case if $n = 5$ years, then

$$FW = \$1(1 + 0.10)^5 = \$1.61$$

2.2.2 Present value

The future worth calculation procedure can now be reversed by asking the question 'What is the present value of $1.61 deposited in the bank 5 years hence assuming an interest rate of 10%?' The formula is rewritten in the form

$$PV = \frac{FW}{(1+i)^n} \quad (2.3)$$

Substituting FW = $1.61, $i = 0.10$, and $n = 5$ one finds, as expected, that the present value is

$$PV = \frac{\$1.61}{(1+0.10)^5} = \$1$$

2.2.3 Present value of a series of uniform contributions

Assume that $1 is to be deposited in the bank at the end of 5 consecutive years. Assuming an interest rate of 10%, one can calculate the present value of each of these payments. These individual present values can then be summed to get the total.

Year 1: Payment

$$PV_1 = \frac{\$1}{(1.10)^1} = \$0.909$$

Year 2: Payment

$$PV_2 = \frac{\$1}{(1.10)^2} = \$0.826$$

Year 3: Payment

$$PV_3 = \frac{\$1}{(1.10)^3} = \$0.751$$

Year 4: Payment

$$PV_4 = \frac{\$1}{(1.10)^4} = \$0.683$$

Year 5: Payment

$$PV_5 = \frac{\$1}{(1.10)^5} = \$0.621$$

The present value of these 5 payments is

$$PV = \$3.790$$

The general formula for calculating the present value of such equal yearly payments is

$$PV = FW \left[\frac{(1+i)^n - 1}{i(1+i)^n} \right] \quad (2.4)$$

Applying the formula in this case yields

$$PV = \$1 \left[\frac{(1.10)^5 - 1}{(0.10)(1.10)^5} \right] = \$3.791$$

The difference in the results is due to roundoff.

2.2.4 Payback period

Assume that $5 is borrowed from the bank today (time $=0$) to purchase a piece of equipment and that a 10% interest rate applies. It is intended to repay the loan in equal yearly payments of $1. The question is 'How long will it take to repay the loan?' This is called the payback period. The present value of the loan is

$$PV \text{ (loan)} = -\$5$$

The present value of the payments is

$$PV \text{ (payments)} = \$1 \left[\frac{(1.10)^n - 1}{0.10(1.10)^n} \right]$$

The loan has been repaid when the net present value

$$\text{Net present value (NPV)} = PV \text{ (loan)} + PV \text{ (payments)}$$

is equal to zero. In this case, one substitutes different values of n into the formula

$$NPV = -\$5 + \$1 \left[\frac{(1.10)^n - 1}{0.10 (1.10)^n} \right]$$

For $n = 5$ years NPV $= -\$1.209$; for $n = 6$ years NPV $= -\$0.645$; for $n = 7$ years NPV $= -\$0.132$; for $n = 8$ years NPV $= \$0.335$.
  Thus the payback period would be slightly more than 7 years ($n \cong 7.25$ years).

2.2.5 Rate of return on an investment

Assume that $1 is invested in a piece of equipment at time $=0$. After tax profits of $1 will be generated through its use for each of the next 10 years. If the $5 had been placed in a bank at an interest rate of i then its value at the end of 10 years would have been using Equation (2.2).

$$FW = PV(1+i)^n = \$5(1+i)^{10}$$

The future worth (at the end of 10 years) of the yearly $1 after tax profits is

$$FW = A_m \left[\frac{(1+i)^n - 1}{i} \right] \tag{2.5}$$

where A_m is the annual amount and $[(1+i)^n-1]/i$ is the uniform series compound amount factor.

The interest rate i which makes the future worths equal is called the rate of return (ROR) on the investment.

In this case

$$\$5(1+i)^{10} = \$1\left[\frac{(1+i)^{10}-1}{i}\right]$$

Solving for i one finds that

$$i \cong 0.15$$

The rate of return is therefore 15%. One can similarly find the interest rate which makes the net present value of the payments and the investment equal to zero at time $t=0$.

$$\text{NPV} = -\$5 + \$1\left[\frac{(1.10)^{10}-1}{i(1+i)^{10}}\right] = 0$$

$$i \cong 0.15$$

The answer is the same.

The process of bringing the future payments back to time zero is called 'discounting'.

2.2.6 Cash flow (CF)

The term 'cash flow' refers to the net inflow or outflow of money that occurs during a specific time period. The representation using the word equation written vertically for an elementary cash flow calculation is

  Gross revenue
  − Operating expense
  = Gross profit (taxable income)
  − Tax
  = Net profit
  − Capital costs
  ─────────────────
  = Cash flow

A simple example (after Stermole & Stermole, 1987) is given in Table 2.1.

In this case there is a capital expense of $200 incurred at time $t=0$ and another $100 at the end of the first year. There are positive cash flows for years 2 through 6.

Table 2.1. Simple cash flow example (Stermole & Stermole, 1987).

Year	0	1	2	3	4	5	6
Revenue			170	200	230	260	290
− Operating cost			−40	−50	−60	−70	−80
− Capital costs	−200	−100					
− Tax costs			−30	−40	−50	−60	−70
Project cash flow	−200	−100	+100	+110	+120	+130	+140

2.2.7 Discounted cash flow (DCF)

To 'discount' is generally used synonymously with 'to find the present value'. In the previous example, one can calculate the present values of each of the individual cash flows. The net present value assuming a minimum acceptable discount rate of 15% is

Year 0 $NPV_0 = -200 \quad = \quad -200.00$

Year 1 $NPV_1 = \dfrac{-100}{1.15} \quad = \quad -86.96$

Year 2 $NPV_2 = \dfrac{100}{(1.15)^2} \quad = \quad 75.61$

Year 3 $NPV_3 = \dfrac{110}{(1.15)^3} \quad = \quad 73.33$

Year 4 $NPV_4 = \dfrac{120}{(1.15)^4} \quad = \quad 68.61$

Year 5 $NPV_5 = \dfrac{130}{(1.15)^5} \quad = \quad 64.63$

Year 6 $NPV_6 = \dfrac{140}{(1.15)^6} \quad = \quad 60.53$

Discounted cash flow = $55.75

The summed cash flows equal $55.75. This represents the additional capital expense that could be incurred in year 0 and still achieve a minimum rate of return of 15% on the invested capital.

2.2.8 Discounted cash flow rate of return (DCFROR)

To calculate the net present value, a discount rate had to be assumed. One can however calculate the discount rate which makes the net present value equal to zero. This is called the discounted cash flow rate of return (DCFROR) or the internal rate of return (ROR). The terms DCFROR or simply ROR will be used interchangeably in this book. For the example given in Subsection 2.2.6, the NPV equation is

$$NPV = -200 - \dfrac{100}{1+i} + \dfrac{100}{(1+i)^2} + \dfrac{110}{(1+i)^3} + \dfrac{120}{(1+i)^4} + \dfrac{130}{(1+i)^5} + \dfrac{140}{(1+i)^6} = 0$$

Solving for i one finds that

$$i \cong 0.208$$

In words, the after tax rate of return on this investment is 20.8%.

2.2.9 Cash flows, DCF and DCFROR including depreciation

The cash flow calculation is modified in the following way when a capital investment is depreciated over a certain time period.

    Gross revenue
    − Operating expense
    − Depreciation
    = Taxable income
    − Tax
    = Profit
    + Depreciation
    − Capital costs
    ─────────────
    = Cash flow

In this book no attempt will be made to discuss the various techniques for depreciating a capital asset. For this example it will be assumed that the investment (Inv) has a Y year life with zero salvage value. Standard straight line depreciation yields a yearly depreciation value (Dep) of

$$\text{Dep} = \frac{\text{Inv}}{Y} \tag{2.6}$$

The procedure will be illustrated using the example adapted from Stermole & Stermole (1987).

Example. A $100 investment cost has been incurred at time $t = 0$ as part of a project having a 5 year lifetime. The salvage value is zero. Project dollar income is estimated to be $80 in year 1, $84 in year 2, $88 in year 3, $92 in year 4, and $96 in year 5. Operating expenses are estimated to be $30 in year 1, $32 in year 2, $34 in year 3, $36 in year 4, and $38 in year 5. The effective income tax rate is 32%.

The cash flows are shown in Table 2.2.

The net present value (NPV) of these cash flows assuming a discount rate of 15% is $43.29

$$\text{NPV} = -100 - \frac{40.4}{1.15} + \frac{41.8}{(1.15)^2} + \frac{43.1}{(1.15)^3} + \frac{44.5}{(1.15)^4} + \frac{45.8}{(1.15)^5}$$
$$= \$43.29$$

Table 2.2. Cash flow example including depreciation (Stermole & Stermole, 1987).

Year	0	1	2	3	4	5	Cumulative
Revenue		80.0	84.0	88.0	92.0	96.0	440.0
− Oper costs		−30.0	−32.0	−34.0	−36.0	−38.0	−170.0
− Depreciation		−20.0	−20.0	−20.0	−20.0	−20.0	−100.0
= Taxable		30.0	32.0	34.0	36.0	38.0	170.0
− Tax @ 32%		−9.6	−10.2	−10.9	−11.5	−12.2	−54.4
= Net income		20.4	21.8	23.1	24.5	25.8	115.6
+ Depreciation		20.0	20.0	20.0	20.0	20.0	100.0
− Capital costs	−100.0	−	−	−	−	−	−100.0
Cash flow	−100.0	40.4	41.8	43.1	44.5	45.8	115.6

The DCFROR is the discount rate which makes the net present value equal to zero. In this case

$$\text{NPV} = -100 - \frac{40.4}{1+i} + \frac{41.8}{(1+i)^2} + \frac{43.1}{(1+i)^3} + \frac{44.5}{(1+i)^4} + \frac{45.8}{(1+i)^5} = 0$$

The value of i is about

$$i \cong 0.315$$

2.2.10 Depletion

In the U.S. special tax consideration is given to the owner of a mineral deposit which is extracted (depleted) over the production life. One might consider the value of the deposit to 'depreciate' much the same way as any other capital investment. Instead of 'depreciation', the process is called 'depletion'. The two methods for computing depletion are:
 (1) cost depletion,
 (2) percentage depletion.

Each year both methods are applied and that which yields the greatest tax deduction is chosen. The method chosen can vary from year to year. For most mining operations, percentage depletion normally results in the greatest deduction.

To apply the cost depletion method, one must first establish the cost depletion basis. The initial cost basis would normally include:
- the cost of acquiring the property including abstract and attorney fees.
- exploration costs, geological and geophysical survey costs.

To illustrate the principle, assume that this is $10. Assume also that there are 100 tons of reserves and the yearly production is 10 tons. The $10 cost must then be written off over the 100 total tons. For the calculation of cost depletion the cost basis at the end of any year (not adjusted by the current years depletion) is divided by the estimated remaining ore reserve units plus the amount of ore removed during the year. This gives the unit depletion. In this simple case, for year 1

$$\text{Unit depletion} = \frac{\$10}{100} = \$0.10$$

The unit depletion is then multiplied by the amount of ore extracted during the year to arrive at the depletion deduction,

$$\text{Depletion deduction} = 10\,\text{tons} \times \$0.10 = \$1$$

The new depletion cost basis is the original cost basis minus the depletion to date. Thus for the year 2 calculation:

$$\text{Depletion cost basis} = \$10 - \$1 = \$9$$

$$\text{Remaining reserves} = 90\,\text{tons}$$

The year 2 unit depletion and depletion deduction are:

$$\text{Unit depletion} = \frac{\$9}{90} = \$0.10$$

$$\text{Depletion deduction} = 10 \times \$0.10 = \$1$$

Table 2.3. Percentage depletion rates for the more common minerals. A complete list of minerals and their percentage depletion rates are given in section 613(b) of the Internal Revenue Code.

Deposits	Rate
Sulphur, uranium, and, if from deposits in the United States, asbestos, lead ore, zinc ore, nickel ore, and mica	22 %
Gold, silver, copper, iron ore, and certain oil shale, if from deposits in the United States	15 %
Borax, granite, limestone, marble, mollusk shells, potash, slate, soapstone, and carbon dioxide produced from a well	14 %
Coal, lignite, and sodium chloride	10 %
Clay and shale used or sold for use in making sewer pipe or bricks or used or sold for use as sintered or burned lightweight aggregates	7½ %
Clay used or sold for use in making drainage and roofing tile, flower pots, and kindred products, and gravel, sand, and stone (other than stone used or sold for use by a mine owner or operator as dimension or ornamental stone)	5 %

IRS 2005. Depletion.
Ref: www.irs.gov/publications/p535/ch10/

Once the initial cost of the property has been recovered, the cost depletion basis is zero. Obviously, the cost depletion deduction will remain zero for all succeeding years.

The percent depletion deduction calculation is a three step process. In the first step, the percent deduction is found by multiplying a specified percentage times the gross mining income (after royalties have been subtracted) resulting from the sale of the minerals extracted from the property during the tax year. According to Stermole & Stermole (1987):

'Mining' includes, in addition to the extraction of minerals from the ground, treatment processes considered as mining applied by the mine owner or operator to the minerals or the ore, and transportation that is not over 50 miles from the point of extraction to the plant or mill in which allowable treatment processes are applied. Treatment processes considered as mining depend upon the ore or mineral mined, and generally include those processes necessary to bring the mineral or ore to the stage at which it first becomes commercially marketable; this usually means to a shipping grade and form. However, in certain cases, additional processes are specified in the Internal Revenue Service regulations, and are considered as mining. Net smelter return or its equivalent is the gross income on which mining percentage depletion commonly is based. Royalty owners get percentage depletion on royalty income so companies get percentage depletion on gross income after royalties.

As shown in Table 2.3, the percentage which is applied varies depending on the type of mineral being mined.

In step 2, the taxable income (including all deductions except depletion and carry forward loss) is calculated for the year in question. Finally in step 3, the allowable percentage depletion deduction is selected as the lesser of the percent depletion (found in step 1) and 50% of the taxable income (found in step 2).

With both the allowable cost depletion and percentage depletion deductions now calculated, they are compared. The larger of the two is the 'allowed depletion deduction'. The overall process is summarized in Figure 2.1.

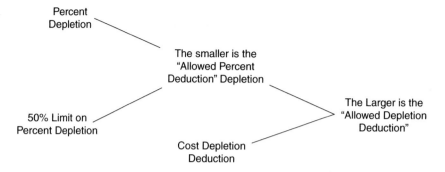

Figure 2.1. Flow sheet for determining the depletion deduction (Stermole & Stermole, 1987).

2.2.11 *Cash flows, including depletion*

As indicated the depletion allowance works exactly the same way in a cash flow calculation as depreciation. With depletion the cash flow becomes:

    Gross revenue
    − Operating expense
    − Depreciation
    − Depletion
    − Taxable income
    − Tax
    = Profit
    + Depreciation
    + Depletion
    − Capital costs
    ─────────────
    = Cash flow

The following simplified example adapted from Stermole & Stermole (1987) illustrates the inclusion of depletion in a cash flow calculation.

Example. A mining operation has an annual sales revenue of $1,500,000 from a silver ore. Operating costs are $700,000, the allowable depreciation is $100,000 and the applicable tax rate is 32%. The cost depletion basis is zero. The cash flow is:

(1) Preliminary. Calculation without depletion.

Gross revenue	$1,500,000
− Operating expense	−700,000
− Depreciation	−100,000
= Taxable income before depletion	$700,000

(2) Depletion calculation. Since the depletion basis is zero, percentage depletion is the only one to be considered. One must then choose the smaller of:
  (a) 50% of the taxable income before depletion and carry-forward-losses
  (b) 15% of the gross revenue

In this case the values are:
    (a) $0.50 \times \$700,000 = \$350,000$
    (b) $0.15 \times \$1,500,000 = \$225,000$
Hence the depletion allowance is $225,000.

(3) Cash flow calculation.

Gross revenue	$1,500,000
− Operating expense	− 700,000
− Depreciation	− 100,000
− Depletion	− 225,000
= Taxable income	$475,000
− Tax @ 32%	− 152,000
= Profit	$323,000
+ Depreciation	+100,000
+ Depletion	+225,000
= Cash flow	$648,000

The cash flow calculation process is expressed in words (Laing, 1976) in Table 2.4.

Laing (1976) has summarized (see Table 2.5) the factors which should be considered when making a cash flow analysis of a mining property.

The distinction between the 'exploration' and 'development' phases of a project is often blurred in actual practice. However from a tax viewpoint a sharp distinction is often made. The distinction made by the U.S. Internal Revenue Service (1988a) is paraphrased below.

1. The exploration stage involves those activities aimed at ascertaining the existence, location, extent or quality of any deposit of ore or other mineral (other than oil or gas). Exploration expenditures paid for or incurred before the beginning of the development stage of the mine or other natural deposit may for tax purposes be deducted from current income. If, however, a producing mine results, these expenditures must be 'recaptured' and capitalized. These are later recovered through either depreciation or cost depletion.

2. The development stage of the mine or other natural deposit will be deemed to begin at the time when, in consideration of all the facts and circumstances, deposits of ore or other mineral are shown to exist in sufficient quantity and quality to reasonably justify exploitation. Expenditures on a mine after the development stage has been reached are treated as operating expenses.

2.3 ESTIMATING REVENUES

2.3.1 *Current mineral prices*

Current mineral prices may be found in a number of different publications. *Metals Week*, *Skillings Mining Review*, *Metal Bulletin* and *Industrial Minerals* are four examples. Spot

Table 2.4. Components of an annual cash flow analysis for a mining property (Laing, 1976).

Calculation	Component
	Revenue
less	Operating costs
equals	Net income before depreciation and depletion
less	Depreciation and amortization allowance
equals	Net income after depreciation and amortization
less	Depletion allowance
equals	Net taxable income
less	State income tax
equals	Net federal taxable income
less	Federal income tax
equals	Net profit after taxes
add	Depreciation and amortization allowances
add	Depletion allowance
equals	Operating cash flow
less	Capital expenditures
less	Working capital
equals	Net annual cash flow

Table 2.5. Factors for consideration in a cash flow analysis of a mining property (Laing, 1976).

Preproduction period

Exploration expenses	Land and mineral rights
Water rights	Environmental costs
Mine and plant capital requirements	Development costs
Sunk costs	Financial structure
Working capital	Administration

Production period

Price	Capital investment-replacement and expansions
Processing costs	Royalty
Recovery	Mining cost
Post concentrate cost	Development cost
Reserves and percent removable	Exploration cost
Grade	General and administration
Investment tax credit	Insurance
State taxes	Production rate in tons per year
Federal taxes	Financial year production begins
Depletion rate	Percent production not sent to processing plant
Depreciation schedule	Operating days per year

Post production period

Salvage value	Contractual and reclamation expenditures

prices for the major metals are listed in the business sections of many daily newspapers together with futures prices. Tables 2.6 through 2.8 contain prices for certain:
- metals
- nonmetallic minerals
- miscellaneous metals
- ferro alloys
- ores and concentrates.

In reviewing the tables it is seen that there is considerable variation in how the prices are quoted. In general, the prices depend upon:
- quality
- quantity
- source
- form
- packaging.

The units in which the prices are expressed also vary. Some examples in this regard are presented below.

1. For many minerals, the 'ton' is unit of sale. There are three different 'tons' which might be used. They are:

$$1 \text{ short ton (st)} = 2000 \text{ lbs} = 0.9072 \text{ metric tons}$$

$$1 \text{ long ton (lt)} = 2240 \text{ lbs} = 1.01605 \text{ metric tons}$$

$$1 \text{ metric ton (mt or tonne)} = 2204.61 \text{ lbs}$$
$$= 1000 \text{ kilograms}$$
$$= 0.9842 \text{ long tons}$$
$$= 1.1023 \text{ short tons}$$

Iron ore, sulfur, and manganese ore are three materials normally sold by the long ton. The prices for iron ore and manganese ore are expressed in X dollars (or cents) per long ton unit (ltu). A 'unit' refers to the unit in which the quality of the mineral is expressed. For iron ore the quality is expressed in $Y\%$ Fe. Therefore one unit means 1%.

If 1 long ton (2240 lbs) of iron ore contained 1% iron (22.40 lbs), then it would contain 1 long ton unit (1 ltu) of iron. If the long ton assayed at 65% iron then it would contain 65 ltu. If the quoted price for pellets is 70 ¢/ltu, then the price of 1 long ton of pellets running 65% iron would be:

$$\text{price/long ton} = 65 \times 70 \text{ ¢} = 4550 \text{ ¢/lt}$$
$$= \$45.50/\text{lt}$$

Metric ton units (mtu) and short ton units (stu) are dealt with in the same way.

The reason for using the 'unit' approach is to take into account varying qualities.

2. For most metals, the unit of weight is the pound (lb) or kilogram (kg).

3. Gold, silver, platinum, palladium, and rhodium are sold by the troy ounce.

Table 2.6. Metal prices (*E/MJ*, 1993).

Aluminum, ¢/lb
Used beverage cans,
   12/24/92 40–42
Comex, 99.7% closing, Mar. 58.00
                            May 58.00

N.Y. merchant, 99.7%,
   1/11/93 .... 57.00–57.25

Antimony, $/lb
Merchant, 3/1/90 .... 0.85–0.92
Antimony oxide, 3/1/90 .... 1.10–1.20

Beryllium copper, 9/17/90
Strip (No. 25) 9.25
Rod, bar, and wire (No. 25) 10.24

Bismuth, $/lb, ton lots
Merchant, 3/1/92 .... 2.40–2.45

Cadmium, $/lb, ton lots
Producers, 6/1/92 .... 1.80–2.00

Chromium, $/lb, 1/1/91
Electrolytic metal, standard 3.75

Cobalt, 99%, $/lb
Afrimet, F.O.B. New York
Cathodes, etc., 1/1/92 .... 13.00
Powder, 1/27/89 .... NQ
Extra fine, 1/27/89 .... NQ
Sherritt Gordon
   'S' power, 12/31/90 .... NQ

Copper, ¢/lb
LME, grade A, closing
   cash bid 100.26
   3 mo 100.58
Comex, high grade, Jan. closing 98.35
U.S. producers, cathode 100.00
Warrenton refining, wirebar 110.35
N.Y. merchant, cathodes, Mar. 101.00

Gold, $/tr oz
Zurich avg., opg 328.00
Paris, p.m. 329.21
London, 3:00 p.m. 327.65
Handy & Harman, N.Y. 327.65
Engelhard bullion 328.86
Engelhard fabricated 345.30

Lead, ¢/lb
U.S. and Canadian producers,
   11/6/92 .... 32.00–35.00
Secondary fabricated,
   11/9/92 .... 36.00–40.00
London fix, $/mt

Rudolf Wolff, spot 434.18
Rudolf Wolff, 3 mo 445.88

Lithium, $/lb
99.9%, 1,000 lb lots,
   11/2/92 .... 31.80–32.45
Carbonate tech., 11/2/92 .... 1.91–1.96

Magnesium, ¢/lb, 5-st lots
Ingots, 99.8%, 11/16/91 143–153
Grinding slab, 1/1/91 143
Sticks, 1.3-in.-dia, 1/2/91 223

Manganese, ¢/lb
Electrolytic, 99.9%, 11/15/90 105

Mercury, 99.9%, $/flask
New York prompt, 8/26/92 205–210
C.I.F. European port, 12/8/92 115–130

Molybdic oxide, $/lb
Producer, 1/6/92 .... 3.35

Nickel, $/lb
Melting briquettes, 10/10/91 .... 3.48–3.52
N.Y. merchant, spot, 1/14/93 .... 2.64–2.68

Platinum, $/tr oz
London p.m. fix 1/14/93 .... 358.75
Comex, 99.9%, Apr. 355.50
Engelhard fabricated 459.00
U.S. merchant, 1/14/93 .... 358.00–359.00

Silver, ¢/tr oz
Engelhard bullion 368.00
Handy & Harman, N.Y. 397.00
London fix, spot 369.15
   3 mo 371.95
   6 mo 374.95
   12 mo 383.00
Zurich fix 369.70

Tin
Kuala Lumpur spot, ringgit/kilo 14.81
Spot exchange, $/ringgit 0.3843
AMM N.Y. ex-dock, $/lb 2.76

Uranium, $/lb U_3O_8
Nuexco, 12/31/92 .... 7.85

Zinc, ¢/lb
U.S. and foreign producers,
slab, delivered in U.S., 1/14/93
   High grade 53.76–54.74
   Special high grade 51.00–55.24
   CGG 51.75–54.49
U.S. producers, die casting alloys
   No. 3, 7/30/91 .... NQ
   No. 5, 7/30/91 .... NQ

Table 2.7. Prices for some common non-metallic minerals (*Industrial Minerals*, December 1992).

Asbestos
All prices quoted are F.O.B. mine
Canadian chrysotile
  Group No. 3 .... C$1,450–1,750
  Group No. 4 .... C$1,080–1,400
  Group No. 5 .... C$645–850
  Group No. 6 .... C$525–575
  Group No. 7 .... C$180–350
South African chrysotile
  Group No. 5 .... $360–410
  Group No. 6 .... $300–390
  Group No. 7 .... $180–220
South African amosite
  Long $660–1,000
  Medium $610–700
  Short $425–625
South African crocidolite
  Long $720–880
  Medium $645–715
  Short $640–695

Bentonite
Wyoming, foundry grade, 85%
  200 mesh, bagged, 10 ton lots,
  del U.K. £120–130
F.O.B. plants, Wyoming rail
  hopper cars, bulk st $18.00–35.00
F.O.B. plants, Wyoming, bagged,
  rail cars, st $33.00–45.00
Fullers' Earth, soda ash-treated,
  del, U.K. foundry grade, bagged £85–95
Civil engineering grade, bulk £60–70
OCMA, bulk del U.K. £65–70
API, F.O.B. plant, Wyoming,
  rail cards, bagged, st $34.50

Feldspar
Ceramic grade, powder, 300 mesh,
  bagged, ex-storage U.K. £140
Sand, 28 mesh, glass grade,
  ex-store U.K. £65
Ceramic grade, bulk, st
  F.O.B. Spruce Pine, NC, 170-250 mesh $50.00
  F.O.B. Monticello, Ga, 200 mesh
    high potash $82.50
  F.O.B. Middleton, Con, 200 mesh ... $67.50
Glass grade, bulk, st
  F.O.B. Spruce Pine, NC, 97.8%
    > 200 mesh $33.50
  F.O.B. Monticello, Ga, 92%
    > 200 mesh, high potash $64.75
  F.O.B. Middleton, Con, 96%
    > 200 mesh $45.50

Fluorspar
Metallurgical, min 70% CaF_2,
  ex-U.K. mine £85–90
Acidspar, dry basis 97% CaF_2
  bagged ex-works £140–150
Acidspar, dry, bulk ex-works
  tankers £125–135

Acidspar Chinese dry bulk, C.I.F.
  Rotterdam $100–110
Mexican, F.O.B. Tampico,
  Acidspar filtercake $122–127
  Metallurgical $90–95
South African acidspar dry basis,
  F.O.B. Durban $110–115
USA, Illinois district, bulk, st
  acidspar $190–195

Iodine
Crude iodine crystal, 50 kg
  drums 99.5% min, per kg del
  U.K. $15–16

Phosphates
Florida, land pebble, run of mine, st
dry basis, unground, bulk, ex-mine, avg.

	Domestic	Export
60–66% BPL	$34.99	$30.36
66–70% BPL	$25.99	$34.67
70–72% BPL	$27.84	$37.38
72–74% BPL	$36.24	$41.80
74% BPL	$35.10	$50.20

Morocco, 75–77% BPL, FAS
  Casablanca $48.50
70–72% BPL, FAS
  Casablanca $46
Tunisia, 65–68% BPL, FAS Sfax $32–38
Nauru, 83% BPL, lt, F.O.B. .... –

Potash
Muriate of potash, bulk, 60% K_2O
  Std, C.I.F. UP port £71–74
  Granular, C.I.F. U.K. port £81–84
  Std, F.O.B. Vancouver $90–100
F.O.B. Saskatchewan, bulk per, st
  Standard $83
  Coarse $87
  Granular $89
F.O.B. Carlsbad, bulk, per ton,
  Coarse $90–100
  Granular $95

Salt
Ground rocksalt, 15–20 tonne lots,
  avg. price del U.K. £20

Soda ash
  US natural, F.O.B. Wyoming, Dense,
  st $80

Sulphur
US Frasch, liquid, dark
  ex-terminal, Tampa, lt $88
Canadian, liquid, bright, F.O.B.
  Rotterdam, tonne $90
French, Polish, liquid, ex-terminal
  Rotterdam, tonne $105.75
Canadian, solid/slate, F.O.B.
  Vancouver, spot $65.75
Canadian, solid/slate, F.O.B.
  Vancouver, contract $65.70

To accord with trade practices, certain prices are quoted in US$ (sterling now floating at around $1.50 = £1). All quotations are © *Metal Bulletin* plc 1992.

Table 2.8. Prices* for some common non-ferrous ores (*Metal Bulletin*, 1993).

Antimony .... Per metric tonne unit Sb. C.I.F.	*Tantalite* .... Per lb Ta_2O_5
Clean sulphide conc., 60% Sb ......$14.00–$15.50 Lump sulphide ore, 60% Sb .... $14.50–$16.00 Chinese conc., 60% Sb, Se typically 60 ppm.	25/40% basis 30% Ta_2O_5 C.I.F., max 0.5% U_3O_8 and ThO_2 combined .... $30.00–$33.00 Greenbushes 40% basis $40.00
Hg 30 ppm max. ....$12.00–$13.00	*Tin conc.* .... T/C per tonne
Beryl ....Per short ton unit of BeO	20/30% Sn (including deduction) ...£400–£530 30/50% Sn (including deduction) ...£350–£500
Cobbed lump min. 10% BeO C.I.F. ...... $75–$80	50/65% Sn (including deduction) ...£300–£600 65/75% Sn (including deduction) ...£400–£525
Chromite .... Per tonne delivered	*Titanium ores* .... Australian per tonne
Transvaal, friably lumpy, basis 40% Cr_2O_3 F.O.B. .... $55–$65 Albanian, hard lumpy, min. 42% F.O.B. . $70–$80 Albanian, conc., 51% F.O.B. .... $100–$110 Turkish, lumpy, 48% 3:1 (scale pro rata) F.O.B. .... $160–$180 Russian, lumpy, 40% min. 36% F.O.B. ... $75–$95	Rutile conc.min.95% TiO_2 bagged, F.O.B./Fid .... A$550–A$600 Rutile bulk conc.min.95% TiO_2 F.O.B./Fid .... A$500–A$560 Limenite bulk conc.min.54% TiO_2 F.O.B. .... A$83–A$90
Columbium ores .... Per lb pentoxide content	*Tungsten ore* .... Per metric tonne unit WO_3
Columbite min. 65% $Cb_2O_5 + Ta_2O_5$, 10:1 C.I.F. .... $2.60–$3.05	Min. 65% WO_3 C.I.F. .... $40–$50
Lead conc.	*Uranium* .... Per lb U_3O_8
70/80% Pb $500–550 basis C.I.F. ...... $170–$180	Nuexco exchange value December ...... $7.85 Nuexco restricted American market penalty .... $2.10 Nukem December restricted spot . $9.90–$10.35 Nukem December unrestricted spot $7.90–$8.00
Lithium ores .... Per tonne	
Petalite, 3.5–4.5% C.I.F. .... £135–£140 Spodumene 4–7% Li_2O C.I.F. .... £178–£183	*Vanadium* .... Per lb V_2O_5
Manganese ore .... Metallurgical per mtu Mn 48/50% Mn Max. 0.1%P C.I.F. .... $3.35–$3.55	Highveld, fused min. 98% V_2O_5 C.I.F. ...$1.95 Other sources .... $1.75–$1.85
Molybdenite ....Per lb Mo in MoS_2	*Zinc conc.* .... T/C per metric dry tonne Sulphide 49/55% Zn basis
Conc. C.I.F. .... $1.95–$2.05 Conc. C.I.F. U.S. .... $2.80–$3.00	$1,000 C.I.F. main port ....$188–$190 Sulphide 56/61% Zn basis $1,000 C.I.F. main port ....$190–$194
Monazite ....Australian per tonne	*Zircon* ....Australian per tonne
Conc. Min.55% REO + Thoria, F.O.B./Fid .... A$300–A$350	Std. min. 65% ZrO_2 F.O.B./Fid . A$230–A$270 Premium max. 0.05% Fe_2O_3 F.O.B./Fid .... A$250–A$325

*Prices expressed C.I.F. Europe unless otherwise indicated.

The relationship between the troy ounce and some other units of weight are given below.

Troy weight (tr)

1 tr oz = 31.1035 grams

= 480 grains

= 20 pennyweights (dwt)

= 1.09714 oz avoird

U.S. Standard (avoirdupois)

1 oz avoird = 28.3495 grams

= 437.5 grains

1 lb = 16 oz

= 14.5833 tr oz

= 453.59 grams

4. Mercury (quicksilver) is sold by the flask. A 'flask' is an iron container which holds 76 lbs of mercury.

5. Molybdenum is sometimes quoted in the oxide (roasted) form (MoO_3) or as the sulfide (MoS_2). The price given is per pound of Mo contained. For the oxide form this is about 67% and for the sulfide 60%.

6. The forms in which minerals are sold include bulk, bags, cathodes, ingots, rods, slabs, etc.

7. The purity of the mineral products often have a substantial effect on price (see for example Feldspar).

8. The fiber length and quality is extremely important to asbestos.

The point of sale also has a considerable effect on the price. Two abbreviations are often used in this regard.

The abbreviation 'F.O.B.' stands for 'free-on-board'. Thus the designation 'F.O.B. mine' means that the product would be loaded into a transport vessel (for example, rail cars) but the buyer must pay all transport charges from the mine to the final destination.

The abbreviation 'C.I.F.' means that cost, insurance and freight are included in the price.

Table 2.9 illustrates the difference in price depending on delivery point. For USX Corp. pellets, the price of 37.344¢/ltu is at the mine (Mountain Iron, Minnesota). The Cleveland-Cliffs price of 59.4¢/ltu (also for Minnesota taconite pellets) is at the hold of the ship at the upper lake port. Hence a rail charge has now been imposed. For the Oglebay Norton Co. (Minnesota) pellets, the price of 72.45¢/ltu is at the lower lake port. Thus it includes rail transport from the mine to the upper lake port plus ship transport to the lower lake port. Table 2.10 provides freight rates for iron ore and pellets. Lake freight rates are given in Table 2.11.

Many mineral products are sold through long term contracts arranged between supplier and customer. The prices will reflect this shared risk taking. There will often be significant differences between the short term (spot) and long term prices.

Recent prices for metals, industrial minerals and iron ore are provided in Tables 2.12, 2.13 and 2.14, respectively.

Table 2.9. Iron ore prices (*Skillings Mining Review*, 1993a).

Lake Superior iron ore prices
(Per gross ton, 51.50% iron natural, at rail of vessel lower lake port)
Mesabi non-Bessemer $30.03–31.53

Cleveland-Cliffs Inc iron ore pellet prices
(Per iron natural unit, at rail of vessel lower lake port)
Cleveland-Cliffs Inc .... 72.45¢
Per gross ton unit at hold of vessel upper lake port .... 53.40¢
Wabush pellets per gross ton unit F.O.B. Pointe Noire .... 63.50¢

Cyprus Northshore Mining Corp. pellet price
(Per gross ton iron unit natural F.O.B. Silver Bay)
Cyprus Northshore pellets .... 48.76¢

IOC Ore Sales Co. pellet price
(Per natural gross ton unit delivered rail of vessel lower lake port)
Carol pellets .... 74.65¢

Inland Steel Mining Co. pellet price
Per gross ton natural iron unit at hold vessel upper lake port .... 46.84¢

Oglebay Norton Co. pellet prices
(Per natural gross ton unit, at rail of vessel lower lake port)
Standard grade .... 72.45¢
Eveleth special .... 74.00¢

U.S. Steel pellet price
Per dry gross ton iron unit at Mtn. Iron .... 37.344¢

Table 2.10. Rail tariff rates on iron ore and pellets (*Skillings Mining Review*, 1993b).

Rail freight rates ($/gross ton) from mines to Upper Lake Port
Marquette Range to Presque Isle
    Pellets $2.50
    Natural ore 2.56
Pellets from Marquette Range to Escanaba delivered into vessel 3.39
Mesabi Range plants on BN to:
    Allouez delivered direct into vessel 6.16
    When consigned to storage subject to storage charges 6.53
Winter ground storage charges on pellets:
    At Allouez per gross ton per month .... 16.0¢
    At Escanaba: storage per gross ton per month .... 2.4¢
        handling to storage .... 21.8¢
        handling from storage .... 21.8¢

Dock charges on iron ore per gross ton
Car to vessel at Duluth and Two Harbors $1.05

All-rail freight rates to consuming districts
Mesabi Range to:
    Chicago district 20.42
    Geneva, Utah *46.97
    Granite City and East St. Louis, Ill. 19.69
    Valley district 40.17
Marquette Range to Detroit 28.60
Marquette & Menominee Ranges to:
    Chicago district **24.09
    Granite City and East St. Louis, Ill. 23.18

*Conditional on tender of not less than 4800 GT nor more than 5200 GT.
**Multiple car rate.

Table 2.11. Lake freight tariff rates on iron ore, pellets and limestone (*Skillings Mining Review*, 1993c).

Lake freight rates from Upper Lake ports to Lower Lake ports

Iron ore ($/gross ton)	Self unloading vessels
Head of Lakes to Lower Lakes	$6.50
Marquette to Lower Lakes	5.40
Escanaba to Lake Erie	4.88
Escanaba to Lake Michigan	3.90

Limestone ($/gross ton)
Calcite, Drummond, Cedarville and Stoneport to

Lower Lake Michigan	3.98
Lake Erie ports	4.10

Note: The above cargo rates apply after April 15 and before December 15, 1993. Winter formulas apply during other periods. Rates are further subject to surcharges, if warranted.

Dock, handling and storage charges ($/gross ton) on iron ore at Lower Lake ports RCCR X-088C

Ex self-unloading vessels at Cleveland, Ohio

Dockage	$0.26
From dock receiving area into cars, via storage	1.60
From dock receiving area to cars	1.05
Rail of vessel receiving area to cars	1.16

At Conneaut, Ohio BLE

Dockage of self-unloading vessel	$0.15
From receiving bin to storage	0.53
From storage to railcars	0.68

Ex bulk vessels at Cleveland C&P

From hold to rail of vessel	$1.03
From rail of vessel into car	1.26
From rail of vessel via storage into car	2.25

2.3.2 Historical price data

Mineral prices as monitored over a time span of many years exhibit a general upward trend. However, this is not a steady increase with time but rather is characterized by cyclic fluctuations. Table 2.15 shows the average yearly prices for 10 common metals from 1900 through 1998 (USGS, 2001). The monthly and average prices for 11 common metals for the period January 1997 through February 2004 are given in Table 2.16 (Metal Bulletin, 2005). To provide an indication of the price unpredictability, consider the case of copper. In 1900 for example the copper price was about 16.2 ¢/lb (Table 2.15). In 2000, 100 years later, the price had risen to 82 ¢/lb. The average rate of price increase per year over this period using the end point values is 1.6 percent. The price dropped to a low of 5.8 ¢/lb (1932) and reached a high of 131 ¢/lb (1989) over this period. Using the average price increase over the period of 1900 to 2000, the predicted price in 1950 should have been 36.4¢/lb. The actual value was 21.2¢/lb.

A mining venture may span a few years or several decades. In some cases mines have produced over several centuries. Normally a considerable capital investment is required to bring a mine into production. This investment is recovered from the revenues generated over the life of the mine. The revenues obviously are strongly dependent upon mineral price. If the actual price over the mine life period is less than that projected, serious revenue shortfalls would be experienced. Capital recovery would be jeopardized to say nothing of profits.

Price trends, for metals in particular, are typically cyclic. Figures 2.2 through 2.6 show the prices for 10 metals over the period 1950 through 2004. The period and amplitude of the cycles varies considerably. For nickel consider the period from 1983 through 1998. In 1983,

Table 2.12. Recent Weekly Metal Prices (Platt's Metals Week, Feb 2, 2004).

Major Metals		
Aluminum		
		cts/lb
MW US Market		77.500/78.500
US Six-Months P1020		4.750
US 6063 Billet Upcharge		7.500/8.750
US UBCs		62.000/63.000
US 6063 press scrap		3.000/3.500
		Eur/mt
Alloy 226 delivered European works		1340.000/1370.000
Copper		
		cts/lb
MW No.1 Burnt Scrap Disc		2.500
MW No.1 Bare Bright Disc		0.000
MW No.2 Scrap Disc		12.500
NY Dealer Premium cathodes		4.250/4.750
US Producer cathodes		115.740/116.800
		$/mt
Grade A Cathode CIF R'dam		2542.000/2548.000
Grade A Premium CIF R'dam		65.000/70.000
Grade A CIF Livorno/Salerno		2547.000/2553.000
Grade A Prem CIF Livorno/ Salerno		70.000/75.000
Russian Standard CIF R'dam		2522.000/2528.000
Russian Standard Prem CIF R'dam		45.000/50.000
Lead		
		cts/lb
North American Market		37.928/40.332
		$/mt
European dealer		887.000/893.000
European 99.985% Prem IW (R'dam)		110.000/115.000
In-Warehouse S'pore Prem		28.000/30.000
Nickel		
		$/lb
NY Dealer/Cathode		6.880/7.000
NY Dealer/Melting		6.880/7.000
NY Dealer/Plating		6.980/7.100
		cts/lb
NY Dealer/Cathode Premium		24.000
NY Dealer/Melting Premium		24.000
NY Dealer/Plating Premium		34.000
		$/mt
Melting Grade (Cut), IW R'dam		15053.000/15116.000
Melting R'dam Prem (Cut) IW R'dam		200.000/250.000
Plating Grade IW R'dam		15062.000/15147.000
Plating Grade Prem IW R'dam		200.000/250.000
Russia Full-Plate		14937.000/14997.000
Russia Full-Plate Prem IW R'dam		75.000/100.000
Briquette Premium IW R'dam		200.000/300.000
In-Warehouse S'pore Prem		200.000/240.000
Tin		
		$/mt
Europe 99.85% IW R'dam		6550.000/6576.000
Europe 99.85% Prem IW R'dam		170.000/190.000
Europe 99.90% IW R'dam		6570.000/6616.000
Europe 99.90% Prem IW R'dam		190.000/230.000
Zinc		
		cts/lb
US Dealer SHG		50.993
MW SHG Premium		4.250
MW SHG Galv. Prem.		4.750
MW SHG Alloyer #3 Prem.		11.500
		$/mt
Europe physical SHG IW R'dam		1093.000/1099.000
Europe physical SHG Prem IW R'dam		80.000/85.000
In-Warehouse S'pore Prem		45.000/50.000
Precious Metals		
	All PGM figures in $/tr oz	
Iridium		
MW NY Dealer		67.000/84.000
Osmium		
MW NY Dealer		350.000/450.000
Palladium		
MW NY Dealer		222.000/245.000
Platinum		
MW NY Dealer		824.000/865.000
Rhodium		
MW NY Dealer		480.000/505.000
Ruthenium		
MW NY Dealer		39.000/43.000
Minor Metals		
Antimony		
		cts/lb
MW NY Dealer		122.000/132.000

(Continued)

Table 2.12. (Continued).

	$/mt	US Primary Ingot/Prod.	180.000/180.000
99.65% HK	2350.000/2450.000	MW US Spot Western	110.000/117.000
Arsenic		MW US Dealer Import	105.000/110.000
	$/lb		**$/mt**
MW Dealer	0.500/0.550	Europe Free Market	1850.000/1950.000
Bismuth			**Eur/kg**
	$/lb	Eur Hydro Alloy Prod	2.500
MW NY Dealer	2.600/2.900	*Titanium*	
Cadmium			**$/st**
	$/lb	Ore/Rutile	800.000/850.000
MW NY Dealer	0.500/0.600		**$/lb**
Free Market HG	0.650/0.750	US SG Ingot Producer	5.750
Indium		MW US 70% Ferrotitanium	3.500/3.650
	$/kg		**$/kg**
Producer: US Prod Indium Corp	375.000	Eur. 70% Ferrotitanium	7.200/7.500
	$/kg		**$/lb**
MW NY Dealer	380.000/420.000	MW US Turning 0.5%	1.750/1.900
Mercury		Eur. Turning .5%	1.850/2.050
	$/fl	*Ferroalloys*	
Free Market International	240.000/260.000	*Cobalt*	
U.S. Domestic	300.000/320.000		**$/lb**
Rhenium		MW 99.8% US Spot Cathode	27.000/28.000
	$/kg	99.8% European	27.000/28.000
MW NY Dealer	1200.000/1450.000	99.3% Russian	25.000/26.000
Selenium		99.6% Zambian	26.000/27.000
	$/lb	*Ferrochrome*	
MW NY Dealer	10.000/11.500		**cts/lb**
Light Metals		Charge 50–55%/Impt.	55.500/58.000
Beryllium		60–65%/Impt.	55.500/58.000
		Low-Carbon 0.05% Imported	82.000/85.000
	$/lb	Low-Carbon 0.10% Imported	75.000/78.000
US BE-CU/Alloy 25	9.950	Low C 0.15% Imported	68.000/73.000
US BE-CU/Cast Ingt	6.400	High C 52% Cr Europe	54.000/57.000
US BE-CU/Mast Ally	160.000	High Carbon 62% Cr Europe	57.000/61.000
Lithium		Low Carbon 0.1% C Europe	66.000/69.000
		High Carbon 60% Hong Kong	50.000/53.000
	cts/lb	50–55% Regular CIF Japan	56.000
US Carbonate	197.000	50–55% Spot CIF Japan	60.000/63.000
US Carbonate Pellet	202.500	*Ferromanganese*	
	$/lb		**$/lt**
US Ingot 01-29-04	41.000	MW 78% Mn/Impt.	675.000/710.000
			cts/lb
US Rod	72.600/72.600	Medium Carbon/Imported	52.000/55.000
Magnesium			**$/mt**
	cts/lb	High C 75% Hong Kong	580.000/590.000
US Die Cast Alloy: Producer	170.000/170.000	*Ferromolybdenum*	
US Die Cast Alloy: Transaction	110.000/115.000		**$/lb**
		MW US FeMo	8.500/9.000

(Continued)

Table 2.12. (Continued).

	$/kg		
MW Europe FeMo	18.700/19.500		
Hong Kong FeMo	17.000/17.300		$/mt
Spot CIF Japan	17.300/17.600	Hong Kong 65% Mn	600.000/620.000
		Regular CIF Japan	480.000/480.000
Ferrosilicon		CIS CIF Japan	520.000/525.000
		Chinese CIF Japan	640.000/650.000
	cts/lb		Yen/mt
MW 75% Si Imported	46.000/47.000	Non-Origin	57000.000/58000.000
	$/mt		Eur/mt
Regular CIF Japan	950.000/950.000	Std 16–20% Si Eur	520.000/540.000
Spot CIF Japan	660.000/670.000		
Chinese CIF Japan	605.000/610.000	*Silicon*	
	Yen/mt		cts/lb
Non-Origin	64000.000/67000.000	MW Dealer Import	66.000/67.500
	$/mt		$/mt
Hong Kong 75% Si	610.000/630.000	Hong Kong 98.5% Si	1020.000/1050.000
	Eur/mt	Spot CIF Japan	1000.000/1200.000
Standard 75% Si Europe	680.000/730.000		Eur/mt
		98% Europe	1100.000/1200.000
Ferrovanadium			
	$/lb	*Stainless Scrap*	
Free Market V2O5	3.500/4.000		$/lt
US Ferrovanadium	10.000/11.000	NA FREE MKT 18-8	1475.000/1500.000
	$/kg	*Tantalum*	
Europe Ferrovanadium	17.000/17.500		
			$/lb
Manganese		Spot Tantalite Ore	30.000/40.000
	$/mt	*Tungsten*	
99.7% Hong Kong	1350.000/1400.000		$/stu
Molybdenum		MW US Spot Ore	40.000/45.000
	$/lb	APT-US	65.000/70.000
MW Dealer Oxide	7.600/8.000		$/mtu
Silicomanganese		APT-Hong Kong	60.000/62.000
	cts/lb		$/kg
MW 2% Carbon Imported	47.000/50.000		

Republished per agreement with Platts, A Division of The McGraw-Hill companies.

the price began at $2.12/lb, dropped to $1.76/lb in 1986, and rose to a high of $6.25/lb in 1988. It then dropped to $2.40/lb in 1993 before increasing once again to $3.73/lb in 1995 before landing in 1998 at about the same price it started at in 1983. In 1975, silver was around $4.50/tr oz. It shot up to nearly $40.00/tr oz (January 1980) due to the buying of the Hunt brothers from Texas. By the end of 1991, the price had dropped back to about $4.50/tr oz. The price performance of iron ore fines over the period 1900–2004 is presented in Table 2.17 (USGS, 2004). Figure 2.7 depicts the price development for LKAB iron ore fines over the period 1950–2004 (LKAB, 2005). Comparative prices for LKAB fines and pellets are presented in Figure 2.8 (LKAB, 2005). The interested reader is encouraged to carefully study the price trends presented for each of these metals and try to explain the fluctuations. In some cases there are clear causes while in others an explanation is difficult to find.

For making the valuation calculations, the first problem is deciding what base price should be used. The second problem is forecasting the future price history.

Table 2.13. Prices for some common industrial minerals (Mineral Price Watch, Indmin, 2005).

Alumina

94% Al_2O_3 CIF

Brown, FEPA 8-220, European/US	$650–850
Brown, FEPA 8-220, Chinese	$400–450
White, 25 kg bags, CIF UK	$900–1100

Hydrated alumina (ATH)

Damp (57–60% Al_2O_3, 5–8% moisture) bulk FOB refinery	$290–320
Dry (65% Al_2O_3) bulk FOB refinery	$315–350

Antimony

Antimony oxide

99.5% Sb_2O_3 (5t lots) FOB Antwerp	$2500–3000

Asbestos

Canadian chrysotile, ex-mine

Group No 3	C$1494–1803
Group No 4	C$1030–1442
Group No 5	C$684–950
Group No 6	C$425–610
Group No 7	C$210–435

Baddeleyite

Contract price, CIF main

European port

Refractory/abrasive grade	$2200–2600
Ceramic grade (98% ZrO_2 + HfO_2)	$2800–3200

Barytes

Paint grade

Micronised, off white <20 microns del UK, per tonne, min 99%	£140–150
ex-works USA, min. 95%, per s.ton	$275–325

Drilling grade

Ground OCMA grade bulk, del Aberdeen	£50–55
API grade, lump, CIF US Gulf Coast,	
Chinese	$64–66
Indian	$70–73
Moroccan	$62–65

Refractory bauxite

Chinese, min 87% Al_2O_3, FOBT (0/50 mm, undried)

Shanxi, shaft lump	$135–145
Shanxi, rotary lump	$150–160
Guizhou, rotary lump	$145–155
Guyanese, FOB barge US Gulf	$160–170
Brazilian, bulk, FOB Brazil	$115–130

Bentonite

Wyoming, ex-works, USA, per s.ton

Rail hopper cars, crude, bulk all grades	$26–63
Foundry grade, bagged (100 lb),	$50–76
API grade, bagged (100 lb),	$43–53

FOB main European port, bulk, per tonne

Cat litter, grade 1–5 mm	€37–55
Foundry, crude, 10,000 t ship	$55–60
API Section 6 grade	$52–57

Indian, FOB Kandla, crushed and dried, loose in bulk

OCMA/API grade	$30–40
Cat litter grade	$32–40
Foundry grade	$40–45

Borates

Paper bags (25 kg), del UK

Anhydrous borax	£840–900
Decahydrate borax, granular, technical	£400–450
Pentahydrate borax, granular, refined	£300–350
Boric acid, granular, technical	£350–400

Bulk, FOB California

Anhydrous borax	$840–900
Decahydrate borax, technical	$340–380
Pentahydrate borax, granular, refined	$400–430
Boric acid, granular, technical	$900–925

Boron Minerals

Turkish

Lump colemanite, 40–42% B_2O_3, FOB USA/Japan	$270–290

Latin American

Ulexite, 40% B_2O_3 FOB Lima	$250–300

Calcium Carbonate

GCC

ex-works UK

chalk, uncoated	£30–52
coated, fine grade	£80–103

FOB USA, per s.ton

5–7μ	$110–160
2–0.5μ	$140–290
High brightness for paper (1.5μ)	$170–180

(Continued)

Table 2.13. (Continued).

PCC			*Fluorspar*	
ex-works UK				
uncoated	£300–390		*Acidspar filtercake*	
coated	£300–417		Chinese, dry basis, CIF US Gulf Port	$200–210
FOB USA, per s.ton			South African, FOB Durban	$134–145
Fine (0.4–1μ)	$250–270		Mexican, FOB Tampico	$130–150
Ultrafine, surface treated (0.02–0.36μ)	$375–750		Mexican, FOB Tampico, As <5 ppm	$168–178
			Graphite	
Celestite			CIF European port, FCL	
Mexican, 94% SrSO$_4$, FOB USA	$80–100		Crystalline medium, 85–87%C, +100–80 mesh	$630–810
Spanish, 96% SrSO$_4$, FOB Motril	$50–60			
Turkish, 96% SrSO$_4$ FOB Iskenderun	$65–80		Crystalline fine, 90%C, −100 mesh	$525–715
Moroccan, 94% SrSO$_4$ FOB Nador	$54–56		Crystalline medium, 90%C, +100–80 mesh	$370–410
			Crystalline large, 90%C, +80 mesh	$800–1100
Chromite			Crystalline fine, 94–97%C, +100 mesh	$710–820
Transvaal, 46% Cr$_2$O$_3$ wet bulk, FOB			Crystalline medium, 94–97%C, +100–80 mesh	$650–840
Chemical grade	$125–150		Crystalline large, 94–97%C, +80 mesh	$900–1150
Foundry grade	$170–190			
Refractory grade	$100–120		*Ilmenite*	
Metallurgical grade, friable lumpy, 40% Cr$_2$O$_3$	$80–90		*Australian, min 54% TiO$_2$, FOB*	
			Bulk concentrates,	US$72–90
Philippine, refractory grade, FOB	$125–145		Spot prices,	US$70–90
			Iodine	
Diatomite			*Crystal, 99.5% min, drums, per kg*	
US calcined filter-aids, del UK	£370–410		Spot & contract	$17.20–20.00
US flux-calcined filter-aids, del UK	£380–420			
			Kaolin	
Feldspar				
Ex-works USA, per s.ton, bulk			*Ex-Georgia plant per s.ton*	
Ceramic grade,			Filler, bulk	$80–100
170–200 mesh, (Na)	$60–75		Paper coating grade	$85–185
200 mesh (K)	$125		Calcined, bulk	$320–375
Glass grade,			Sanitaryware grade, bagged	$65–75
30 mesh (Na)	$40–52		Tableware grade, bagged	$125
80 mesh (K)	$85–90		*Ceramic grade, bulk*	
			Refined, ex-works France	£40–100
Turkish, FOB Gulluk, Na feldspar			Refined, FOB Rotterdam	£60–100
Crude, −10 mm size, bulk	$13–14		*Leuceoxene*	
Ground, −63 microns, bagged	$75–80			
Glass grade, −500 microns, bagged	$54–56		*FOB Western Australia, typically 91% TiO$_2$, max. 1% ZrO$_2$*	
South African, FOB Durban, bagged			Bulk	US$350–380
Ceramic grade	$112–165		Bagged	US$380–400
Micronised (2,5,10 microns)	$205		*Lithium Minerals*	
Indian, FOB India			*Petalite*	
Ceramic grade (K), bulk	$25–27		4.2% Li$_2$O, big bags FOB Durban	$165–260
Powder grade, 200 mesh	$70			

(Continued)

Table 2.13. (Continued).

Spodumene, FOB W. Virginia per s.ton, bulk concentrate >7.25% Li$_2$O	$330–350	FOB Turkey Raw, crushed, graded, bulk/big bags	$32–60
Glass 5% Li$_2$O	$195–200	Raw, bulk	$14–17
Lithium carbonate		Aggregate, expanded, ex-works, UK	£320–650
del continental USA, large contracts, $ per lb	$1.15–1.50	Filter-aids, expanded, ex-works, USA	$210–410
Magnesite		Silica Sand	
Greek, raw, <3.5% SiO$_2$, FOB E.Mediterranean	€50–55	Ex-works, UK Foundry sand, dry, bulk Glass sand, flint, container	£15.50–16.50 £15–17

The prices in Table 2.13 appeared in the February 2005 issue of Mineral Price Watch on pp 12–15. Published by Industrial Minerals Information, a division of Metal Bulletin plc, U.K. © Metal Bulletin plc 2005.

Table 2.14. Iron ore prices announced for year 2005 (Skillings, August 2005).

Europe (cents/mtu)	
CVRD Carajas Sinter Feed FOB Ponta da Maderia	65.00
CVRD Blast Furnace Pellets (FOB Ponta da Maderia)	118.57
CVRD Direct Reduction Pellets (FOB Ponta da Maderia)	130.43
CVRD Standard Sinter Feed (FOB Tubarao)	62.51
CVRD Blast Furnace Pellets (FOB Tubarao)	115.51
CVRD Direct Reduction Pellets (FOB Tubarao)	127.06
IOC Concentrate (FOB Sept-Iles)	66.71
IOC Pellets (FOB Sept-Iles)	120.06
Japan (cents/ltu)	
BHP Billiton Mt. Newman (DMT) Fines	61.72
BHP Billiton (DMT) Lump	78.77
CVRD Carajas Sinter Feed FOB Ponta da Maderia	57.08
CVRD Blast Furnace Pellets (FOB Ponta da Maderia)	116.86
CVRD Standard Sinter Feed (FOB Tubarao)	56.23
CVRD Blast Furnace Pellets (FOB Tubarao)	113.84
Rio Tinto Hamersley Lump	78.77
Rio Tinto Hamersley Fines	61.72
Rio Tinto Yandicoogina Ore	58.02

Note: The European and Japanese customers have negotiated somewhat different contracts.

Figure 2.9 shows a plot of the average price for copper as a function of time over the period of 1935 to 1992. This is based on the data given in Table 2.18. As can be seen, the price exhibits an upward trend but a cyclic variation is observed.

If the year in which the valuation was made was 1980, then the average copper price is 101.42 ¢/lb. If this current price had been selected as the base price, since it was at the peak of a cycle, the average price would never reach this base price again for many years. In this particular case, that would not occur until 1988. The revenue projection would have been very far off. The same would have been true if the base price for 1985 (a local low) had been selected. Here however the revenue projections would be too pessimistic and possibly the

Table 2.15. Average annual metal prices 1900–1998 (USGS, 2001).

Year	Al (¢/lb)	Cu (¢/lb)	Pb (¢/lb)	Zn (¢/lb)	Au ($/tr oz)	Pt ($/tr oz)	Ag ($/tr oz)	Mo ($/lb)	Ni ($/lb)	Sn ($/lb)
1900	32.7	16.2	na	4.4	20.67	6	0.62	na	0.50	0.30
1901	33.0	16.1	na	4.1	20.67	20	0.60	na	0.56	0.17
1902	33.0	11.6	na	4.8	20.67	20	0.53	na	0.45	0.27
1903	33.0	13.2	na	5.4	20.67	19	0.54	na	0.40	0.28
1904	35.0	12.8	na	5.1	20.67	21	0.58	na	0.40	0.28
1905	35.0	15.6	na	5.9	20.67	17	0.61	na	0.40	0.31
1906	35.8	19.3	na	6.1	20.67	28	0.67	na	0.40	0.40
1907	45.0	20	na	5.8	20.67	na	0.66	na	0.45	0.38
1908	28.7	13.2	na	4.6	20.67	21	0.53	na	0.45	0.30
1909	22.0	13.1	4.3	5.4	20.67	25	0.52	na	0.40	0.30
1910	22.3	12.9	4.4	5.4	20.67	33	0.54	na	0.40	0.34
1911	20.1	12.6	4.4	5.7	20.67	43	0.54	na	0.40	0.42
1912	22.0	16.5	4.5	6.9	20.67	45	0.62	0.20	0.40	0.46
1913	23.6	15.5	4.4	5.6	20.67	45	0.61	0.30	0.42	0.44
1914	18.6	13.3	3.9	5.1	20.67	45	0.56	1.02	0.41	0.34
1915	34.0	17.5	4.7	14.2	20.67	47	0.51	1.02	0.41	0.39
1916	60.8	28.4	6.9	13.6	20.67	83	0.67	1.02	0.42	0.44
1917	51.7	29.2	8.8	8.9	20.67	103	0.84	1.43	0.42	0.62
1918	33.5	24.7	7.4	8.0	20.67	106	0.98	1.48	0.41	0.89
1919	32.1	18.2	5.8	7.0	20.67	115	1.12	1.17	0.40	0.63
1920	32.7	17.5	8.0	7.8	20.67	111	1.02	0.51	0.42	0.48
1921	22.1	12.7	4.5	4.7	20.67	75	0.63	0.71	0.42	0.30
1922	18.7	13.6	5.7	5.7	20.67	98	0.68	0.22	0.38	0.33
1923	25.4	14.7	7.3	6.7	20.67	117	0.65	0.77	0.36	0.43
1924	27.0	13.3	8.1	6.3	20.67	119	0.67	0.92	0.30	0.50
1925	27.0	14.3	9.0	7.7	20.67	119	0.69	0.41	0.33	0.58
1926	27.0	14.1	8.4	7.4	20.67	113	0.62	0.71	0.36	0.65
1927	25.4	13.1	6.8	6.3	20.67	85	0.57	0.77	0.35	0.64
1928	24.3	14.8	6.3	6.0	20.67	79	0.58	1.02	0.37	0.50
1929	24.3	18.4	6.8	6.5	20.67	68	0.53	0.51	0.35	0.45
1930	23.8	13.2	5.5	4.6	20.67	44	0.38	0.56	0.35	0.32
1931	23.3	8.4	4.2	3.6	20.67	32	0.29	0.43	0.35	0.25
1932	23.3	5.8	3.2	2.9	20.67	32	0.28	0.51	0.35	0.22
1933	23.3	7.3	3.9	4.0	20.67	31	0.35	0.76	0.35	0.39
1934	23.4	8.7	3.9	4.2	35.00	34	0.48	0.71	0.35	0.52
1935	20.0	8.9	4.1	4.4	35.00	33	0.64	0.71	0.35	0.50
1936	20.5	9.7	4.7	4.9	35.00	42	0.45	0.67	0.35	0.46
1937	19.9	13.4	6.0	6.5	35.00	47	0.45	0.69	0.35	0.54
1938	20.0	10.2	4.7	4.6	35.00	34	0.43	0.71	0.35	0.42
1939	20.0	11.2	5.1	5.1	35.00	36	0.39	0.69	0.35	0.50
1940	18.7	11.5	5.2	6.4	35.00	36	0.35	0.70	0.35	0.50
1941	16.5	12.0	5.8	7.5	35.00	36	0.35	0.69	0.35	0.52
1942	15.0	12.0	6.5	8.3	35.00	36	0.38	0.72	0.32	0.52
1943	15.0	12.0	6.5	8.3	35.00	35	0.45	0.72	0.32	0.52
1944	15.0	12.0	6.5	8.3	35.00	35	0.45	0.72	0.32	0.52
1945	15.0	12.0	6.5	8.3	35.00	35	0.52	0.72	0.32	0.52
1946	15.0	14.1	8.1	8.7	35.00	53	0.80	0.69	0.35	0.55
1947	15.0	21.3	14.7	10.5	35.00	62	0.72	0.69	0.35	0.78
1948	15.7	22.3	18.0	13.6	35.00	92	0.74	0.70	0.36	0.99

(Continued)

Table 2.15. (Continued).

1949	17.0	19.5	15.4	12.2	35.00	75	0.72	0.84	0.40	0.99
1950	17.7	21.6	13.3	13.9	35.00	76	0.74	0.86	0.45	0.96
1951	19.0	24.5	17.5	18.0	35.00	93	0.89	0.97	0.54	1.27
1952	19.4	24.5	16.5	16.2	35.00	93	0.85	0.98	0.57	1.21
1953	20.9	29.0	13.5	10.8	35.00	93	0.85	0.98	0.60	0.96
1954	21.8	29.9	14.1	10.7	35.00	88	0.85	1.02	0.61	0.92
1955	23.7	37.5	15.1	12.3	35.00	94	0.89	1.05	0.66	0.95
1956	24.0	42.0	16.0	13.5	35.00	105	0.91	1.14	0.65	1.01
1957	25.4	30.2	14.7	11.4	35.00	90	0.91	1.20	0.74	0.96
1958	24.8	26.3	12.1	10.3	35.00	66	0.89	1.21	0.74	0.95
1959	24.7	31.0	12.2	11.5	35.00	72	0.91	1.27	0.74	1.02
1960	26.0	32.3	11.9	13.0	35.00	83	0.91	1.27	0.74	1.01
1961	25.5	30.3	10.9	11.6	35.00	83	0.92	1.32	0.78	1.13
1962	23.9	31.0	9.6	11.6	35.00	83	1.09	1.36	0.80	1.15
1963	22.6	31.0	11.1	12.0	35.00	82	1.28	1.36	0.79	1.17
1964	23.7	32.3	13.6	13.6	35.00	90	1.29	1.50	0.79	1.58
1965	24.5	35.4	16.0	14.5	35.00	100	1.29	1.59	0.79	1.78
1966	24.5	36.0	15.1	14.5	35.00	100	1.29	1.59	0.79	1.64
1967	25.0	38.1	14.0	13.8	35.00	111	1.55	1.63	0.88	1.53
1968	25.6	41.2	13.2	13.5	40.12	117	2.14	1.63	0.95	1.48
1969	27.2	47.4	14.9	14.7	41.68	124	1.79	1.68	1.05	1.64
1970	28.7	58.1	15.7	15.3	36.39	133	1.77	1.72	1.29	1.74
1971	29.0	52.1	13.9	16.2	41.37	121	1.55	1.68	1.33	1.67
1972	25.0	51.4	15.0	17.7	58.48	121	1.68	1.68	1.40	1.78
1973	26.4	59.5	16.3	20.7	97.98	150	2.56	1.63	1.53	2.28
1974	43.1	77.3	22.5	36.0	159.87	181	4.71	2.00	1.74	3.96
1975	34.8	64.1	21.5	39.0	161.43	164	4.42	2.50	2.07	3.40
1976	41.2	69.6	23.1	37.0	125.35	162	4.35	2.95	2.25	3.80
1977	47.6	66.8	30.7	34.4	148.36	157	4.62	3.63	2.27	5.35
1978	50.8	65.8	33.7	31.0	193.47	261	5.40	4.72	2.04	6.30
1979	70.8	92.2	52.6	37.3	307.62	445	11.09	6.17	2.66	7.36
1980	76.2	101.3	42.5	37.4	612.74	677	20.63	9.12	2.83	8.46
1981	59.9	84.2	36.5	44.6	460.34	446	10.52	8.08	2.71	7.33
1982	46.7	72.8	25.5	38.5	376.36	327	7.95	6.72	2.18	6.54
1983	68.5	76.5	21.7	41.4	423.01	424	11.44	3.45	2.12	6.55
1984	61.2	66.9	25.6	48.5	360.80	357	8.14	3.22	2.17	6.24
1985	49.0	67.0	19.1	40.4	317.26	291	6.14	3.13	2.26	5.96
1986	55.8	66.0	22.1	38.0	367.02	461	5.47	2.54	1.76	3.83
1987	72.1	82.5	35.9	41.9	479.00	553	7.01	2.59	2.19	4.19
1988	110.2	120.5	37.1	60.3	438.56	523	6.53	2.72	6.25	4.41
1989	88.0	131.0	39.4	82.1	382.58	507	5.50	3.22	6.04	5.20
1990	73.9	123.0	46.0	74.4	385.69	467	4.82	2.59	4.02	3.86
1991	59.4	109.3	33.5	52.6	363.91	371	4.04	2.09	3.70	3.63
1992	57.6	107.4	35.1	58.5	345.25	361	3.94	2.22	3.18	4.02
1993	53.5	91.6	31.7	46.3	360.80	375	4.30	1.72	2.40	3.50
1994	71.2	111.0	37.2	49.4	385.69	411	5.29	1.13	2.88	3.69
1995	85.7	138.3	42.3	55.8	385.69	425	5.15	3.77	3.73	4.16
1996	71.2	109.0	48.8	51.3	388.80	398	5.19	2.27	3.40	4.12
1997	77.1	107.0	46.5	64.4	332.81	397	4.89	2.27	3.14	3.82
1998	65.3	78.7	45.3	51.3	295.17	373	5.10	2.63	2.10	3.73

72 Open pit mine planning and design: Fundamentals

Table 2.16. Metal prices (after Metal Bulletin, 2004).

Year	Month	Al (¢/lb)	Cu (¢/lb)	Pb (¢/lb)	Zn (¢/lb)	Au ($/tr oz)	Pt ($/tr oz)	Ag ($/tr oz)	Mo Oxide ($/lb)	Ni ($/lb)	Sn ($/lb)	Pd ($/tr oz)
1997	Jan.	71	110	31.4	49	355	359	4.77	4.51	3.21	2.67	121
	Feb.	72	109	29.9	53	347	365	5.07	4.77	3.51	2.67	136
	Mar.	74	110	31.5	57	352	380	5.20	4.77	3.58	2.68	149
	Apr.	71	108	29.1	56	344	371	4.77	4.77	3.32	2.59	154
	May	74	114	28.0	59	344	390	4.76	4.71	3.39	2.59	171
	June	71	118	27.9	61	341	431	4.75	4.76	3.20	2.52	204
	July	72	111	28.8	69	324	416	4.37	4.73	3.10	2.47	188
	Aug.	78	102	27.6	75	324	425	4.50	4.73	3.07	2.46	215
	Sept.	73	96	28.8	74	323	425	4.73	4.51	2.95	2.49	191
	Oct.	73	93	27.2	58	325	424	5.03	4.23	2.89	2.52	205
	Nov.	73	87	25.5	53	307	293	5.08	3.99	2.79	2.57	208
	Dec.	69	80	23.9	50	289	367	5.79	3.98	2.70	2.50	199
	Avg.	73	103	28.3	60	331	387	4.90	4.54	3.14	2.56	178
1998	Jan.	67	77	24.1	50	297	375	5.88	3.97	2.49	2.36	226
	Feb.	66	75	23.4	47	289	386	6.83	4.04	2.44	2.38	237
	Mar.	65	79	25.4	47	296	399	6.24	4.49	2.45	2.48	262
	Apr.	64	82	25.9	50	308	414	6.33	4.40	2.45	2.59	321
	May	62	79	24.6	48	299	389	5.56	4.11	2.28	2.66	354
	June	59	75	23.9	46	292	356	5.27	4.09	2.03	2.71	287
	July	59	75	24.8	47	293	378	5.46	3.99	1.96	2.56	307
	Aug.	59	73	24.3	47	284	370	5.18	3.58	1.85	2.58	288
	Sept.	61	75	23.6	45	289	360	5.00	3.11	1.86	2.49	283
	Oct.	59	72	22.3	43	296	343	5.00	2.71	1.76	2.46	277
	Nov.	59	71	22.4	44	294	347	4.97	2.38	1.87	2.48	277
	Dec.	57	67	22.7	43	291	350	4.88	2.75	1.76	2.39	297
	Avg.	62	75	24.0	46	294	372	5.55	3.64	2.10	2.51	285
1999	Jan.	55	65	22.3	42	287	355	5.15	2.81	1.94	2.32	322
	Feb.	54	64	23.3	46	287	365	5.53	2.85	2.10	2.39	352
	Mar.	54	63	23.0	47	286	370	5.19	2.82	2.27	2.43	353
	Apr.	58	66	23.5	46	282	358	5.07	2.64	2.31	2.45	362
	May	60	69	24.5	47	277	356	5.27	2.61	2.45	2.56	330
	June	60	65	22.5	45	261	357	5.03	2.74	2.36	2.39	337
	July	64	74	22.5	49	256	349	5.18	2.69	2.59	2.37	332
	Aug.	65	75	22.8	51	257	350	5.27	2.73	2.93	2.37	340
	Sept.	68	79	23.0	54	265	372	5.23	2.91	3.19	2.42	362
	Oct.	67	78	22.5	52	311	423	5.41	2.81	3.32	2.46	387
	Nov.	67	78	21.7	52	293	435	5.16	2.72	3.61	2.65	401
	Dec.	70	80	21.7	54	284	441	5.16	2.71	3.67	2.60	425
	Avg.	62	71	22.8	49	279	378	5.22	2.75	2.73	2.45	359
2000	Jan.	76	84	21.4	53	284	441	5.19	2.67	3.77	2.69	452
	Feb.	76	82	20.5	50	300	517	5.25	2.65	4.38	2.56	636
	Mar.	72	79	20.0	51	286	481	5.06	2.65	4.66	2.48	667
	Apr.	66	76	19.1	51	280	498	5.06	2.65	4.41	2.44	572
	May	67	81	18.7	52	275	527	4.99	2.69	4.59	2.47	571
	June	68	79	19.0	51	286	560	5.00	2.93	3.82	2.48	647
	July	71	82	20.5	52	281	560	4.97	2.94	3.70	2.42	702
	Aug.	69	84	21.5	53	274	578	4.88	2.79	3.63	2.41	760

(Continued)

Table 2.16. (Continued).

	Sept.	73	89	22.1	56	274	593	4.89	2.77	3.92	2.48	728
	Oct.	68	86	22.0	50	270	579	4.83	2.69	3.48	2.40	739
	Nov.	67	81	21.2	48	266	594	4.68	2.55	3.33	2.39	784
	Dec.	71	84	21.0	48	272	611	4.64	2.51	3.32	2.37	917
	Avg.	70	82	20.6	51	279	545	4.95	2.71	3.92	2.47	681
2001	Jan.	73	81	21.7	47	266	622	4.66	2.33	3.17	2.35	1040
	Feb.	73	80	22.7	46	262	601	4.55	2.38	2.96	2.32	975
	Mar.	68	79	22.6	46	263	586	4.40	2.41	2.78	2.29	782
	Apr.	68	75	21.6	44	261	595	4.37	2.41	2.87	2.24	696
	May	70	76	21.1	43	272	610	4.43	2.44	3.20	2.24	655
	June	66	73	20.1	41	270	580	4.36	2.58	3.01	2.19	614
	July	64	69	20.9	39	268	532	4.25	2.63	2.69	1.97	526
	Aug.	62	66	21.9	38	272	451	4.20	2.45	2.50	1.77	455
	Sept.	61	65	21.0	36	284	458	4.35	2.45	2.28	1.68	445
	Oct.	58	62	21.2	35	283	432	4.40	2.45	2.19	1.70	335
	Nov.	60	65	22.0	35	276	430	4.12	2.44	2.30	1.83	328
	Dec.	61	67	21.9	34	276	462	4.35	2.43	2.39	1.82	400
	Avg.	65	72	21.6	40	271	530	4.37	2.45	2.70	2.03	604
2002	Jan.	62	68	23.3	36	281	473	4.51	2.62	2.74	1.75	411
	Feb.	62	71	21.8	35	295	471	4.42	2.75	2.73	1.69	374
	Mar.	64	73	21.8	37	294	512	4.53	2.86	2.97	1.74	374
	Apr.	62	72	21.4	37	303	541	4.57	2.89	3.16	1.83	370
	May	61	72	20.5	35	314	535	4.71	2.99	3.07	1.88	357
	June	61	75	20.0	35	322	557	4.89	7.05	3.23	1.94	335
	July	61	72	20.2	36	314	526	4.92	6.05	3.24	1.96	323
	Aug.	59	67	19.2	34	310	545	4.55	4.76	3.05	1.74	324
	Sept.	59	67	19.1	34	319	555	4.55	4.78	3.01	1.79	327
	Oct.	59	67	19.0	34	317	581	4.40	4.71	3.09	1.92	317
	Nov.	62	72	20.0	35	319	588	4.51	4.04	3.32	1.92	286
	Dec.	62	72	20.1	36	333	597	4.63	3.44	3.26	1.92	243
	Avg.	61	71	20.5	35	310	540	4.60	4.08	3.07	1.84	337
2003	Jan.	63	75	20.1	35	357	630	4.81	3.62	3.64	2.01	255
	Feb.	65	76	21.5	36	360	682	4.65	3.75	3.91	2.07	253
	Mar.	63	75	20.7	36	341	677	4.49	4.51	3.80	2.09	226
	Apr.	60	72	19.8	34	328	625	4.49	5.21	3.59	2.07	163
	May	63	75	21.0	35	355	650	4.74	5.21	3.78	2.15	167
	June	64	76	21.2	36	356	662	4.53	5.67	4.03	2.13	180
	July	65	78	23.3	38	351	682	4.80	5.87	3.99	2.15	173
	Aug.	66	80	22.5	37	360	693	4.99	5.61	4.24	2.19	182
	Sept.	64	81	23.6	37	379	705	5.17	5.61	4.52	2.23	211
	Oct.	67	87	26.6	41	379	732	5.00	6.11	5.01	2.38	202
	Nov.	68	93	28.2	41	389	760	5.18	6.25	5.48	2.43	197
	Dec.	71	100	31.4	44	407	808	5.62	6.25	6.42	2.75	198
	Avg.	65	81	23.3	38	364	692	4.87	5.31	4.37	2.22	201
2004	Jan.	73	110	34.4	46	414	851	6.32	7.67	6.95	2.94	216
	Feb.	76	125	40.3	49	405	846	6.44	8.15	6.87	3.03	235

74 Open pit mine planning and design: Fundamentals

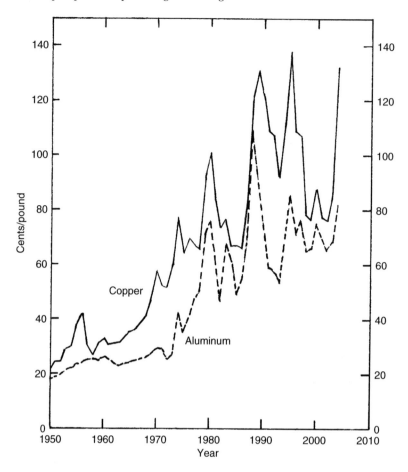

Figure 2.2. Price performance of copper and aluminium over the period 1950–2004.

proposed project would be shelved. The conclusion is that choosing the current price as the base price for the valuation is generally poor due to the cyclic behavior of the prices. The problem is shown diagrammatically in Figure 2.10.

One must decide the base price to be used as well as trend angle and project the results over the depreciation period as a minimum. Another alternative to the selection of the current price as the base price is to use a recent price history over the past two or perhaps five years. For a valuation being done in July 1989 the price was $1.15/lb. Averaging this value with those over the past three years would yield

Years	Base value	% change
1989	1.15	0
1988–89	1.18	−4.8
1987–89	1.06	+28.3
1986–89	0.96	+42.5

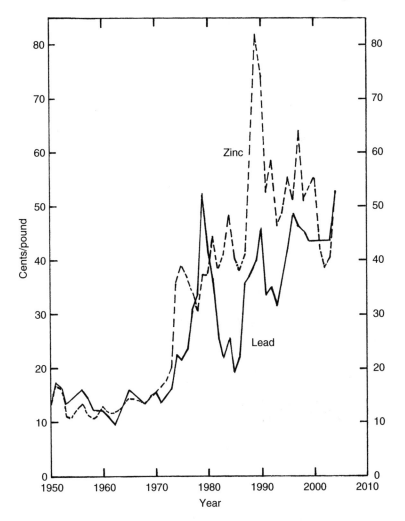

Figure 2.3. Price performance of lead and zinc over the period 1950–2004.

Inflation has not been accounted for in these figures. The point being that a wide range of base values can be calculated. The same is obviously true for determining the 'slope' of the trend line. This can be reflected by the percent change over the period of interest. These values have been added to the above table. They have been calculated by

$$\text{Percent change} = \frac{(\text{Price (1989)} - \text{Price Y})}{\text{Price Y}} 100\%$$

The conclusion is that due to the cyclic nature of the prices, several cycles must be examined in arriving at both a representative base price and a trend.

76 *Open pit mine planning and design: Fundamentals*

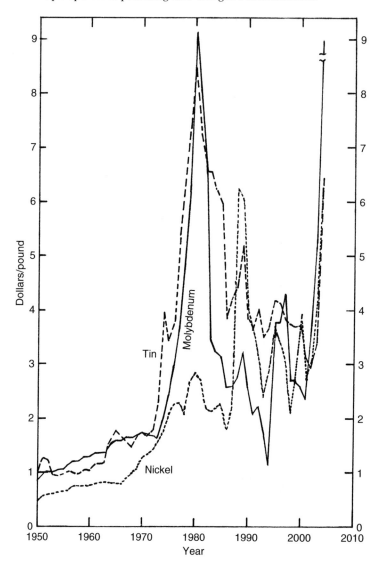

Figure 2.4. Price performance of molybdenum, nickel and tin over the period 1950–2004.

There are two approaches which will be briefly discussed for price forecasting. These are:
- trend analysis
- use of econometric models.

2.3.3 *Trend analysis*

The basic idea in trend analysis is to try and replace the actual price-time history with a mathematical representation which can be used for projection into the future. In examining a 'typical' curve one can see that it is cyclic and the cycles have different amplitudes. One

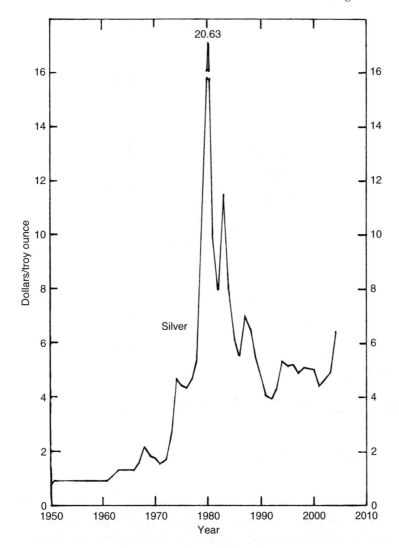

Figure 2.5. Price performance of silver over the period 1950–2004.

could try to fit a function describing the behavior quite closely over a given time period using a type of regression analysis which is commonly available on computers as part of a statistical software package.

The general objective is to determine an equation of the form

$$y = a_0 + a_1 x + a_2 x^2 + a_3 x^3 + a_4 x^4 + \cdots + a_m x^m \qquad (2.7)$$

where a_i are coefficients; y is the price in year x; x is the year relative to the initial year ($x = 0$).

If one has 10 pairs of data (price, year), then the maximum power of the polynomial which could be fitted is $m = 9$. As the power is increased, the actual behavior of the data

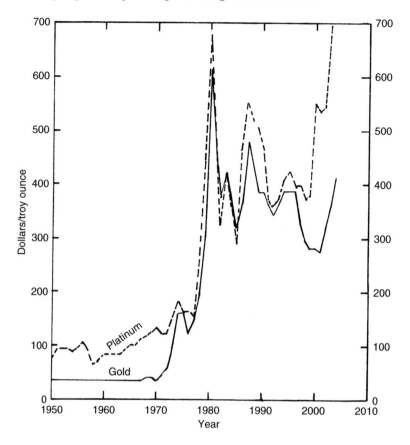

Figure 2.6. Price performance of gold and platinum over the period 1950–2004.

could be more and more closely represented. Unfortunately while this is a good procedure for interpolation, that is, defining values for points within the range of the data, the equation cannot be used for determining values beyond the endpoints (extrapolation). This however is what is desired, that of projecting the historical data past the end points into the future. It can be easily demonstrated that some of the terms of power 2 and higher can vary wildly both in sign and magnitude over only one year. Thus such a general power series representation is not of interest. There are some other possibilities however based upon the fitting of the first two terms of a power series. The simplest representation

$$y = a_0 + a_1 x \tag{2.8}$$

represents a straight line with intercept a_0 $(x=0)$ and slope a_1. For this to apply the data should plot as a straight line on rectangular graph paper. Figure 2.9 shows such a plot for the copper data over the time period 1935–1992. The average trend over the period 1935 through 1970 might be fitted by such a straight line but then there is a rapid change in the rate of price growth. In examining the average trend it appears as if some type of non-linear function

Table 2.17. Annual price of iron ore fines (USGS, Iron Ore Statistics).

Year	$/ton	Year	$/ton	Year	$/ton
1900	2.35	1935	2.66	1970	10.39
1901	1.68	1936	2.64	1971	10.92
1902	1.82	1937	2.82	1972	12.09
1903	1.88	1938	2.56	1973	12.75
1904	1.55	1939	2.99	1974	15.50
1905	1.75	1940	2.52	1975	19.44
1906	2.09	1941	2.65	1976	22.56
1907	2.53	1942	2.61	1977	25.00
1908	2.25	1943	2.62	1978	27.74
1909	2.13	1944	2.69	1979	30.79
1910	2.46	1945	2.72	1980	34.48
1911	1.98	1946	3.01	1981	37.46
1912	1.95	1947	3.44	1982	38.68
1913	2.12	1948	3.88	1983	46.31
1914	1.76	1949	4.46	1984	39.92
1915	1.83	1950	4.92	1985	38.58
1916	2.40	1951	5.40	1986	34.22
1917	3.12	1952	6.21	1987	29.64
1918	3.46	1953	6.81	1988	28.33
1919	3.20	1954	6.76	1989	31.31
1920	4.14	1955	7.21	1990	30.89
1921	3.00	1956	7.68	1991	30.11
1922	3.32	1957	8.10	1992	28.58
1923	3.44	1958	8.27	1993	25.79
1924	2.83	1959	8.48	1994	25.16
1925	2.57	1960	8.35	1995	27.73
1926	2.52	1961	9.13	1996	28.90
1927	2.50	1962	8.82	1997	29.92
1928	2.45	1963	9.22	1998	31.16
1929	2.65	1964	9.46	1999	26.77
1930	2.47	1965	9.25	2000	25.81
1931	2.36	1966	9.50	2001	24.52
1932	1.41	1967	9.64	2002	25.83
1933	3.52	1968	9.78	2003	26.86
1934	2.64	1969	10.15	2004	31.00

U.S. dollars per metric ton contained iron.

is required. The first approach by the engineer might be to try an exponential function such as

$$y = ae^{bx} \tag{2.9}$$

Taking natural logs of both sides one finds that

$$\ln y = \ln a + bx \ln e$$

Since the natural log of e is 1, then

$$\ln y = \ln a + bx \tag{2.10}$$

Letting

$$y^1 = \ln y$$
$$a^1 = \ln a$$

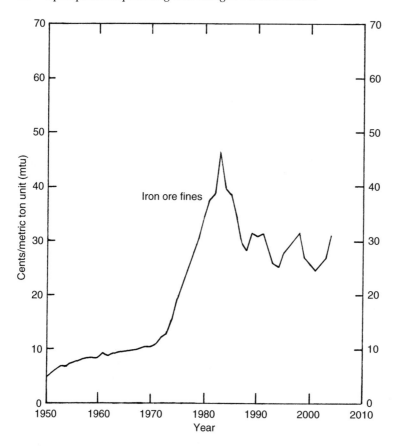

Figure 2.7. Price performance of iron ore fines over the period 1950–2004. LKAB (2005).

Equation (2.10) becomes

$$y^1 = a^1 + bx \qquad (2.11)$$

A straight line should now result when the natural log of the price is plotted versus the year. Such a plot, easily made using semi-log paper, is shown in Figure 2.11. A straight line can be made to fit the data quite well. In 1977, Noble (1979) fitted an equation of the form

$$y = ae^{bx}$$

to the data in Table 2.18 for the period 1935 to 1976. For the least squares approach employed, the constants a and b are given by

$$b = \frac{\sum (x_i \ln y_i) - \frac{1}{n} \sum x_i \sum \ln y_i}{\sum x_i^2 - \frac{1}{n} \left(\sum x_i\right)^2} \qquad (2.12)$$

$$a = \exp\left(\frac{\sum \ln y_i}{n} - b \frac{\sum x_i}{n}\right) \qquad (2.13)$$

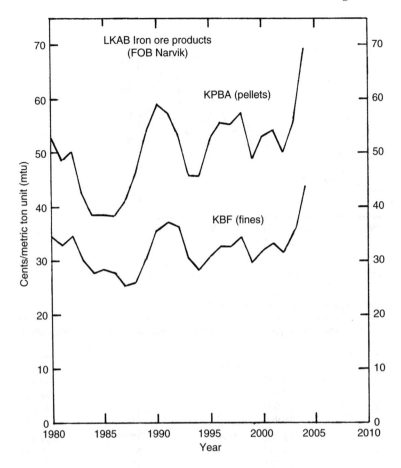

Figure 2.8. Price performance of LKAB iron ore products over the period 1980–2004. LKAB (2005).

For the period of 1935 to 1976

$n = 42$

$\sum (x_i \ln y_i) = 3085.521$

$\sum x_i = 861$

$\sum \ln y_i = 136.039$

$\sum x_i^2 = 23,821$

$\sum (\ln y_i)^2 = 456.0775$

Substituting the appropriate values into Equations (2.12) and (2.13) one finds that

$$b = \frac{3085.521 - \frac{1}{42} 861 \times 136.039}{23821 - \frac{1}{42}(861)^2} = 0.04809$$

82 *Open pit mine planning and design: Fundamentals*

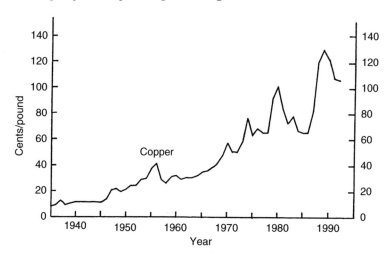

Figure 2.9. Average copper price by year over the time period 1935 to 1992 (Noble, 1979; *E/MJ*, 1992).

Table 2.18. Average annual copper prices 1935–92 (*E/MJ*, 1935–1992).

Calendar year	Relative year	Domestic copper (¢/lb)	Calendar year	Relative year	Domestic copper (¢/lb)
1935	0	8.649	1965	30	35.017
	1	9.474		31	36.170
	2	13.167		32	38.226
	3	10.000		33	41.847
	4	10.965		34	47.534
1940	5	11.296	1970	35	57.700
	6	11.797		36	51.433
	7	11.775		37	50.617
	8	11.775		38	58.852
	9	11.775		39	76.649
1945	10	11.775	1975	40	63.535
	11	13.820		41	68.824
	12	20.958		42	65.808
	13	22.038		43	65.510
	14	19.202		44	92.234
1950	15	21.235	1980	45	101.416
	16	24.200		46	83.744
	17	24.200		47	72.909
	18	28.798		48	77.861
	19	29.694		49	66.757
1955	20	37.491	1985	50	65.566
	21	41.818		51	64.652
	22	29.576		52	81.097
	23	25.764		53	119.106
	24	31.182		54	129.534
1960	25	32.053	1990	55	121.764
	26	29.921		56	107.927
	27	30.600		57	106.023
	28	30.600			
	29	31.960			

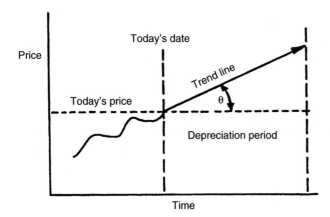

Figure 2.10. Diagrammatic representation of the price projection process.

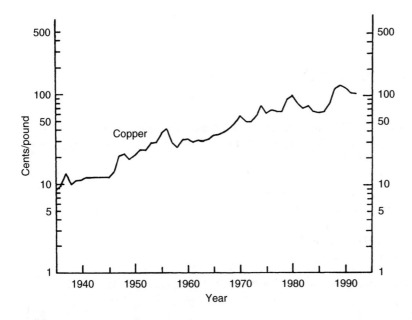

Figure 2.11. Logarithmic plot of copper price versus time for the time period 1935 to 1992.

$$a = \exp\left(\frac{136.039}{42} - \frac{0.04809}{42}861\right) = 9.5185$$

Hence the price predictor equation becomes

$$y = 9.5185e^{0.04809x} \tag{2.14}$$

The correlation coefficient r may be calculated using

$$r = \frac{S_{xy}}{\sqrt{S_{xx}S_{yy}}} \tag{2.15}$$

where

$$S_{xx} = n\sum x_i^2 - \left(\sum x_i\right)^2 = 2,59161 \qquad (2.16)$$

$$S_{yy} = n\sum (\ln y_i)^2 - \left(\sum \ln y_i\right)^2 = 648.65 \qquad (2.17)$$

$$S_{xy} = n\sum (x_i \ln y_i) - \left(\sum x_i\right)\left(\sum \ln y_i\right) = 12,462.30 \qquad (2.18)$$

In this case

$$r = 0.961$$

This high correlation coefficient indicates a strong relationship between price and time. Confidence limits (CL) for the estimates may be calculated using

$$\mathrm{CL}(y) = ae^{bx \pm c} \qquad (2.19)$$

The constant c is given by

$$c = t_{\alpha/2}(1-r^2)^{1/2}\left[S_y^2 + \frac{(n-\bar{x})^2}{(n-2)S_x^2}\right]^{1/2} \qquad (2.20)$$

where:
α is the probability (expressed as a decimal) of y being outside the confidence limits;
$t_{\alpha/2}$ is the value read from a Student 't' table for a cumulative probability (P) of $1-\alpha/2$ and $n-2$ degrees of freedom;
S_y^2 is the population variance of $\ln y$ ($= (\mathrm{SD}_y)^2$);
S_x^2 is the population variance of x ($= (\mathrm{SD}_x)^2$);
$\bar{x}$ is the mean (arithmetic value) of x;
(SD_y) is the standard deviation of $\ln y$;
(SD_x) is the standard deviation of x;
r is the correlation coefficient.

The required mean values, variances and standard deviations are given by

$$\overline{\ln y} = \frac{\sum \ln y_i}{n} = 3.2390 \qquad (2.21)$$

$$S_y^2 = \frac{\sum \left(\overline{\ln y} - \ln y_i\right)^2}{n-1} = 0.3767 \qquad (2.22)$$

$$S_y = \mathrm{SD}_y = \sqrt{S_y^2} = 0.6137 \qquad (2.23)$$

$$\bar{x} = \frac{\sum x_i}{n} = 20.5 \qquad (2.24)$$

$$S_x^2 = \frac{\sum (\bar{x} - x_i)^2}{n-1} = 150.5 \qquad (2.25)$$

$$S_x = \mathrm{SD}_x = \sqrt{S_x^2} = 12.2678 \qquad (2.26)$$

Table 2.19. Predicted and actual copper prices (Noble, 1979).

X	Year	Predicted average price (¢/lb)	Predicted range (¢/lb)		Actual price (¢/lb)
			Low	High	
42	1977	71.74	56.28	91.44	65.81
43	1978	75.27	58.94	96.13	65.51
44	1979	78.98	61.72	101.07	92.33
45	1980	82.87	64.63	106.27	101.42
46	1981	86.96	67.66	111.75	83.74
47	1982	91.24	70.84	117.51	72.91
48	1983	95.73	74.16	123.59	77.86
49	1984	100.45	77.63	129.98	66.76
50	1985	105.40	81.26	136.71	65.57
51	1986	110.59	85.05	143.80	66.10
52	1987	116.04	89.01	151.27	82.50
53	1988	121.76	93.16	159.14	120.51
Average		94.75	73.36	122.39	80.09

Substitution of these values into Equation (2.20) yields

$$c = t_{\alpha/2}(0.2765)\left[0.3767 + \frac{(x-20.5)^2}{6020}\right]^{1/2} \qquad (2.27)$$

If the probability of y being outside of the confidence limits is $\alpha = 0.20$ (80% probability that the price is within the upper and lower limits), then

$$P = 1 - 0.20/2 = 0.90$$

$$DF = 42 - 2 = 40$$

The value of $t_{\alpha/2}$ as read from the Student 't' table is

$$t_{\alpha/2} = 1.303$$

Equation (2.27) becomes

$$c = 0.3603\left[0.3767 + \frac{(x-20.5)^2}{6020}\right]^{1/2} \qquad (2.28)$$

The average trend price value for 1976 ($x = 41$) is

$$y = 68.37 \ \text{¢/lb}$$

This happens to be very close to the actual 1976 price of 68.82 ¢/lb. The predicted average price and price ranges for the time period 1977–1988 ($x = 42$ to 53) are given in Table 2.19.

86 *Open pit mine planning and design: Fundamentals*

The actual average prices for the same years are also given. As can be seen for the time period 1982–1987, the actual value was considerably lower than the predicted lower limit. This corresponded to a very difficult time for producers. The conclusion is that it is very difficult to predict future price trends.

2.3.4 *Econometric models*

A commodity model is a quantitative representation of a commodity market or industry (Labys, 1977). The behavioral relationships included reflect supply and demand aspects of price determination, as well as other related economic, political and social phenomena.

There are a number of different methodologies applied to modelling mineral markets and industries. Each concentrates on different aspects of explaining history, analyzing policy and forecasting. The methodologies chosen for a model depend on the particular economic behavior of interest. It could be price determination, reserve and supply effects, or other aspects.

The market model is the most basic type of micro economic structure and the one from which other commodity methodologies have developed. It includes factors such as:
- Commodity demand, supply and prices;
- Prices of substitute commodities;
- Price lags;
- Commodity inventories;
- Income or activity level;
- Technical factors;
- Geological factors;
- Policy factors influencing the supply.

Market models, which balance supply and demand to produce an equilibrium price, are commonly used in the mineral business for: (a) historical explanation, (b) policy analysis decision making, and (c) prediction.

They are also used to simulate the possible effects of stockpiles and/or supply restrictions over time.

2.3.5 *Net smelter return*

For base metals such as copper, lead, and zinc, prices are not quoted for concentrates rather the refined metal price is given. The payment received by the company from the smelter for their concentrates (called the net smelter return (NSR)) (Lewis et al., 1978; Huss, 1984; Werner & Janakiraman, 1980) depends upon many factors besides metal price. The process by which the net smelter return is calculated is the subject of this section.

Assume that a mill produces a copper concentrate containing G percent of copper metal. The amount of contained metal in one ton of concentrate is

$$\text{CM} = \frac{G}{100} 2000 \tag{2.29}$$

where CM is the contained metal (lbs), G is the concentrate grade (% metal), and the ratio lbs/ton is 2000.

Most smelters and refiners pay for the contained metal based upon prices published in sources such as *Metals Week*. If the current market price ($/lb) is P then the contained copper value is

$$CV = \frac{G}{100} 2000 P \qquad (2.30)$$

where CV is the contained copper value ($/ton), P is the current market price ($/lb).

It is never possible for smelting and refining operations to recover one hundred percent of the contained metal. Some metal is lost in the slag, for example. To account for these losses, the smelter only pays for a portion of the metal content in the concentrate. The deductions may take one of three forms:

(1) Percentage deduction. The smelter pays only for a percentage (C) of the contained metal.

(2) Unit deduction. The concentrate grade is reduced by a certain fixed amount called the unit deduction. For minerals whose grade is expressed in percent one unit is one percentage point. For minerals whose grade is expressed in troy ounces, one unit is one troy ounce.

(3) A combination of percentage and unit deductions.

The 'effective' concentrate grade (G_e) is thus

$$G_e = \frac{C}{100}(G - u) \qquad (2.31)$$

where G_e is the effective concentrate grade (%), u is the fixed unit deduction (%), C is the credited percentage of the metal content (%).

The payable (accountable) metal content in one ton of concentrates is

$$M_e = \frac{C}{100} \frac{G-u}{100} 2000 \qquad (2.32)$$

where M_e is the payable metal content (lbs).

Smelters sometimes pay only a certain percentage of the current market price. The factor relating the price paid to the market price is called the price factor. If 100% of the market price is paid then the price factor is 1.00. The gross value of one ton of concentrates is thus

$$GV = M_e Pf \qquad (2.33)$$

where f is the price factor, GV is the gross value ($/ton of concentrate).

To obtain the basic smelter return (BSR) the charges incurred during treatment, refining and selling must be taken into account. The basic equation is that given below

$$BSR = M_e(Pf - r) - T \qquad (2.34)$$

where r is the refining and selling cost ($/pound of payable metal), T is the treatment charge ($/ton of concentrate).

Often there are other metals/elements in the concentrate. Their presence can be advantageous in the sense that a by-product credit (Y) is received or deleterious resulting in a penalty charge (X).

The net smelter return (NSR) is expressed by

$$\text{NSR} = M_e(Pf - r) - T - X + Y \tag{2.35}$$

where X is the penalty charge due to excessive amounts of certain elements in the concentrate (\$/ton of concentrate) and Y is the credit for valuable by-products recovered from the concentrate (\$/ton of concentrate).

Letting

$$P_e = Pf - r \tag{2.36}$$

where P_e is the effective metal price (after price reductions and refining charges) the NSR expression can be simplified to

$$\text{NSR} = M_e P_e - T - X + Y \tag{2.37}$$

Long-term refining and treatment agreements generally contain cost and price escalation provisions. Escalation of refining charges (e_1) can be grouped into five distinct forms:

(1) No escalation.
(2) Predictive escalation. A specified rate of increase for each year of the contract based upon predictions of cost/price changes.
(3) Cost-indexed escalation. Escalation based upon published cost indices (e.g. wages, fuel and energy).
(4) Price based cost escalation. If the metal price increases above a certain level, the refining cost is increased. This allows the refinery to share in the gain. On the other hand if the price decreases below a certain level, the refining cost may or may not be decreased.
(5) Some combination of (2), (3) and (4).

Escalation of treatment charges (e_2) is generally either by predictive (No. 2) or cost-indexed (No. 3) means. Including escalation, the general NSR equation can be written as

$$\text{NSR} = M_e(Pf - r \pm e_1) - (T \pm e_2) - X + Y \tag{2.38}$$

Smelter contracts must cover all aspects of the sale and purchase of the concentrates from the moment that they leave the mine until final payment is made. Table 2.20 lists the elements of smelter contract and the questions to be addressed. Although the terms of an existing smelter contract are binding upon the contracting parties, supplementary agreements are usually made in respect to problems as they arise.

The net value of the concentrate to the mine is called the 'at-mine-revenue' or AMR. It is the net smelter return (NSR) minus the realization cost (R)

$$\text{AMR} = \text{NSR} - R \tag{2.39}$$

The realization cost covers such items as:
(a) Freight.
(b) Insurance.
(c) Sales agents' commissions.
(d) Representation at the smelter during weighing and sampling.

Table 2.20. The major elements of a smelter contract and the questions addressed (Werner & Janakiraman, 1980).

1. *The parties to the contract.* Who is the seller and who is the buyer?
2. *The material and its quality.* What is the nature of the material; for example, ore or concentrate? What are its assay characteristics and on what basis (wet or dry) are any assays made and reported?
3. *The quantity and duration of the contract.* What are the amounts involved? Are the shipments sent wet or dry? What is the size of a lot? What is the duration of the contract and how are the shipments to be spread out over the period in question?
4. *What is the mode of delivery.* What is the mode of delivery (in bulk or otherwise)? In what manner is the shipment to be made and how is it to be spread over the life of the contract? Is it to be delivered by truck, rail or ship and on what basis; for example, C.I.F. or F.O.B.? Which of the parties to the contract pays the transportation charges?
5. *The price to be paid.* How much of the metal contained in the shipment is to be paid for? Is there any minimum deduction from the reported content of the concentrate? What is the nature of the prices; for example, simple or weighted average? Which particular price or prices are to be used, for example; New York or London Metal Exchange basis, and in what publications are these prices published? What is quotational period involved? How is the price derived averaged over what quotational period?

   In what current is the price to be paid? If it is an average price, what is the basis for the averaging? What forms of weight measurement are to be used: for example, troy, avoirdupois or metric? What premiums, if any, are to be offered? What are the provisions if the quotation ceases to be published longer reflects the full value of the metal?
6. *The smelter's charges.* What is the smelter's base charge for treatment? On what grounds might it be increased or decreased, and if so, how might this be done – in relation to the price of metal or to some other change such as a variation in the cost of labour at the smelter? What are the penalties or premiums for excess moisture and for exceeding minimum or maximum limits in content of other elements of mineral constituents?
   What are the associated refining charges, if any?
7. *The quotational period.* What is the quotational period (for example, the calendar month following the month during which the vessel reports to customs at a designated port) and how is this specified?
8. *Settlement.* Whose weights and samples are to used and what represents a smelter lot?
9. *Mode of payment.* How much is to be paid and when (for example, 80% of the estimated value of each smelter lot shall be paid for at the latest two weeks after the arrival of the last car in each of the lot at buyers' plant)?
   Where, in what form and how are the payments to be made and what are the provisions for the making of the final payment?
10. *Freight allowances.* Which party pays duties, import taxes, etc.?
11. *Insurance.* Which party pays for insurance and to what point? For example, 'alongside wharf'.
12. *Other conditions.* What are the provisions concerning events beyond the control of the buyer or seller? When may 'force majeure' be invoked? What are the conditions under which the contract may be assigned to a third party by the buyer or the seller? What, if any, are the provisions in regard to the presence of specific minor minerals or metals in the concentrate?
13. *Payment conversion rates.* What price quotations and currency conversion rates are to be used and in what publication are they quoted? What are the provisions for the adoption of new bases if those used cease to be published?
14. *Arbitration.* Where are any problems to be arbitrated? How and under what jurisdiction?
15. *Weighing, sampling and moisture determination.* Who does this, where, and who pays for it? Does the seller have the right to be represented at the procedure at his own expense? How are the samples taken to be distributed among buyer, seller and umpire?

(Continued)

Table 2.20. (Continued).

16. *Assaying.* What are the procedures for assaying? What is an acceptable assay for settlement purposes and what are the splitting limits? Which party pays for umpire assaying?
17. *Definitions of terms.* What are the agreed upon definitions of terms? For example: 'Short ton equivalent to 2000 pounds avoirdupois.' Do dollars and cents refer to, for example, Canadian or United States currency?
18. *Insurance.* What insurance coverage is required and which party is to pay for it?
19. *Title.* At what point shall title pass from seller to buyer?
20. *Risk of loss.* At what point shall risk of loss pass from seller to buyer?
21. *Arbitration and jurisdiction.* What are the procedures for settling any disputes that may arise between the parties and which country's law are to govern the parties?

O'Hara (1980) presented the following formula for freight cost F in $ Canadian (1979) per ton of concentrates

$$F = 0.17T_m^{0.9} + \$0.26R_m^{0.7} + \$0.80D_o \qquad (2.40)$$

where F is the freight cost ($/ton); T_m are miles by road (truck); R_m are miles by railroad; D_o are days of loading, ocean travel and unloading on a 15,000-ton freighter.

In 1989 US$, the formula becomes

$$F = 0.26T_m^{0.9} + 0.39R_m^{0.7} + \$1.20D_o \qquad (2.41)$$

The relationship between the AMR and the gross value of the metal contained in the concentrate (CV) is called the percent payment (PP)

$$PP = 100 \times \frac{AMR}{CV} \qquad (2.42)$$

For base-metal concentrates the percent payment can vary from as little as 50% to more than 95%.

Smelter terms are, therefore, a significant factor in the estimation of potential revenue from any new mining venture.

To illustrate the concepts, an example using the model smelter schedule for copper concentrates (Table 2.21) will be presented. It will be assumed that the copper concentrate contains 30% copper and 30 tr oz of silver per ton. It also contains 2% lead. All other deleterious elements are below the allowable limits. Assumed prices are $1/lb for copper, $6/tr oz for silver and $0.50/lb for lead. The payments, deductions and assessments are:

Payments
    Copper: $C = 98\%$
                  $f = 1.0$
                  $u = 1\%$
    Silver: $C = 95\%$
                  $f = 1.0$
                $u = 1.0$ tr oz

Table 2.21. Model 1989 smelter schedule for copper concentrates (Western Mine Engineering, 1989).

Payments	copper	Pay for 95% to 98% of the copper content at market value, minimum deduction of 1.0 unit per dry ton for copper concentrates grading below 30%. Unit deductions and treatment charges may be higher for concentrates above 30% copper.
	gold	Deduct 0.02 to 0.03 troy ounce per dry ton and pay for 90% to 95% of the remaining gold content at market value.
	silver	Deduct 1.0 troy ounce per dry ton and pay for 95% of the remaining silver content at market value.
Deductions	treatment charge	$65 to $100 per dry ton ore or concentrate. Concentrates sold on the spot market are commanding a treatment charge of about $90 to $100 per short ton.
	refining charges	$5.00 to $6.00 per ounce of accountable gold. $0.30 to $0.40 per ounce of accountable silver. $0.075 to $0.105 per pound of accountable copper, March, 1989 spot concentrate deals are carrying refining charges of about $0.11 per pound, long term refining charges are usually $0.09 to $0.095 per pound.
Deleterious element assessments		Excessive amounts of some of the following elements may result in rejection
	lead	Allow 1.0 units free; charge for excess at up to $10.00 per unit. (No penalty at some plants).
	zinc	Allow 1.0 to 3.0 units free; charge for excess at up to $10.00 per unit. (No penalty at some plants).
	arsenic	Allow 0.0 units free; charge for excess at $10.00 per unit.
	antimony	Allow 0.0 units free; charge for excess at $10.00 per unit.
	bismuth	Allow 0.5 units free; charge for excess at $20.00 per unit.
	nickel	Allow 0.3 units free; charge for excess at $10.00 per unit. (No penalty at some plants).
	moisture	Allow 10.0 units free; charge for excess at $1.00 per unit. (No penalty at some plants).
	alumina	Allow 3.0 units free; charge for excess at $1.00 per unit.
	other	Possible charges for fluorine, chlorine, magnesium oxide, and mercury.

Deductions
Copper: $T = \$75/\text{ton}$
         $r = \$0.10/\text{lb}$
Silver: $r = \$0.35/\text{tr oz}$ of accountable silver

Assessments
Lead: 1 unit is free
        additional units charged at $10.00 per unit.

The copper provides the major source of income. Using Equation (2.34) one finds that the basic smelter return is

$$\text{BSR} = M_e(Pf - r) - T$$

$$= \frac{C}{100}\frac{G-u}{100}2000(Pf - r) - T$$

$$= \frac{98}{100}\frac{30-1}{100}2000(\$1 \times 1 - 0.10) - \$75$$

$$= \$436.56/\text{ton of concentrate}$$

The penalty charge for excess lead is

$X =$ is the (Units present − Units allowable) × Charge/unit
$= (2 − 1)\$10.00 = \$10.00$

The by-product credit for silver is

$$Y = \left(\frac{C}{100}\right)(G - u)(Pf - r)$$

$$= \left(\frac{95}{100}\right)(30.0 - 1.0)(\$6 \times 1.0 - 0.35) = \$155.66/\text{ton of concentrate}$$

Hence the net smelter return is

$$\text{NSR} = \$436.56 - \$10.00 + \$155.66 = \$582.22/\text{ton of concentrate}$$

It will be assumed that the concentrates are shipped 500 miles by rail to the smelter. The transport cost (F) per ton of concentrate is

$$F = 0.39 R_m^{0.7} = 0.39(500)^{0.7} = \$30.22$$

The at-mine-revenue becomes

$$\text{AMR} = \$582.22 - \$30.22 = \$552$$

The gross value of the metal contained in one ton of concentrates is

$$\begin{array}{ccc}(\text{copper}) & (\text{silver}) & (\text{lead})\end{array}$$
$$\text{GV} = 2000\frac{30}{100}\$1 + 30 \times \$6 + 2000\frac{2}{100}\$0.50$$
$$= \$600 + \$180 + \$20 = \$800$$

The percent payment is

$$\text{PP} = 100\frac{552}{800} = 69\%$$

Table 2.22 is an example of a current model smelter schedule for copper concentrates (Western Mine Engineering, 2003).

2.3.6 Price-cost relationships

Using the net smelter return formula it is possible to calculate the revenue per ton of concentrate. The revenue to the mine every year depends upon the tons of concentrate produced and the price. The costs to the mine on the other hand depend upon the amount of material mined and processed.

Mining revenues and costs 93

Table 2.22. Model 2003 smelter schedule for copper concentrates (Western Mine Engineering, 2003).

Payments	copper	Pay for 95% to 98% of the copper content at market value. Minimum deduction of 1.0 unit per dry metric ton for copper concentrates grading below 30%. Unit deductions and treatment charges may be higher for concentrates above 40% copper.
	gold	Deduct 0.03 to 0.05 troy ounces per dry metric ton and pay for 90% to 95% of the remaining gold content at market value.
	silver	Deduct 1.0 troy ounce per dry metric ton and pay for 95% of the remaining silver content at market value.
Deductions	treatment charge	Treatment charges in long-term contracts ranged from $45 to $58 per metric ton in 2003. Contracts often include price participation clauses which can reduce the treatment charge if the copper price is below $.90/lb. Typical price participation clauses include up and down escalators of 3% to 5%. Some contracts settled this year contained price participation caps, limiting the up and down escalators to $.0120 to $.0225 per pound refined copper. Treatment charges for "dirty" concentrates are typically $10 to $20 per metric ton higher. Until recently, copper concentrates grading over 40% were charged up to $10 per metric ton more for treatment. Now most high grade producers are escaping this charge. Spot treatment charges over the last 12 months have ranged wildly from below $10 per metric ton to $27 per metric ton, with spot refining charges of $0.0097 to $0.027 per pound. Over the next two years, long-term treatment charges are expected to average in the $55 to $70 per metric ton range as worldwide smelting capacity continues to increase.
	refining charges	$4.00 to $6.00 per ounce of accountable gold $0.30 to $0.50 per ounce of accountable silver $0.045 to $0.058 per pound of accountable copper.
Deleterious element assessments		Copper concentrates containing excessive amounts of the following elements may be penalized or rejected: lead, zinc, arsenic, antimony, bismuth, nickel, alumina, fluorine, chlorine, magnesium oxide, and mercury. Lead, zinc, and arsenic levels above 2% each often result in rejection. For fluxing ores, iron must be less than 3%, and alumina must be kept at low levels so available silica remains high. High moisture content may also be penalized due to material handling difficulties. See individual smelter descriptions for details.

If one assumes that K tons of concentrate are produced every year, then the yearly revenue depends directly on the price received for the product.

A large capital investment is required at the start of the mining. As will be discussed later this must be recovered from the yearly profits. If the yearly profits are not as expected, then the payments cannot be made. Therefore it is important that the price projections or price forecasts be made covering at least the depreciation period (that period in which the investment is being recovered).

By examining the simplified net smelter return formula,

$$\text{NSR} = \frac{C}{100} 20(G - u) \frac{P - r}{100} - T \qquad (2.43)$$

where P is the price, r is the refining and selling cost, T is the treatment cost, G is the percent of metal, u is the fixed unit deduction (%), one can see that the revenue is equal to

$$k(P - r) - T - F$$

where k is a constant.

Assume for definiteness that

$$C = 100\%$$
$$r = 12\cent$$
$$T = \$60$$
$$u = 1.3\%$$
$$G = 28.5\%$$
$$P = 90\cent$$

Thus

$$\text{NSR} = 1 \times 20(28.5 - 1.3)\left(\frac{P - r}{100}\right) - T = \frac{544}{100}(P - r) - T$$
$$= 5.44(90 - 12) - 60 = \$364.32$$

Assume that next year both the price and the costs increase by 5% but that C, G and u remain constant

$$P = 90 \times 1.05 = 94.5 \cent$$
$$r = 12 \times 1.05 = 12.6 \cent$$
$$T = 60 \times 1.05 = \$63$$

Hence

$$\text{NSR} = 5.44 \times 81.9 - 63 = \$382.54$$

The net present value would be

$$\text{NPV} = \frac{382.54}{1.05} = \$364.32$$

If the price however *decreased* by 5% and the costs *increased* by 5%, then

$$\text{NSR} = 5.44(85.5 - 12.6) - 63 = \$333.58$$

If the price increased by 10% and the costs increased by 5%, then

$$\text{NSR} = 5.44(99 - 12.6) - 63 = \$407.02$$

The conclusion is that the net smelter return depends upon the relative changes of the price and the costs. If the prices and costs *escalate* at the same rate then the expected return remains intact. If however, there is a difference then the return may be significantly more or significantly less than expected. Obviously the problem area is if the costs are significantly more than expected or the price significantly less.

2.4 ESTIMATING COSTS

2.4.1 *Types of costs*

There are a number of different types of costs which are incurred in a mining operation (Pfleider & Weaton, 1968). There are also many ways in which they can be reported.

Three cost categories might be:
- Capital cost;
- Operating cost;
- General and administrative cost (G&A).

The capital cost in this case might refer to the investment required for the mine and mill plant. The operating costs would reflect drilling, blasting, etc. costs incurred on a per ton basis. The general and administrative cost might be a yearly charge. The G&A cost could include one or more of the following:
- Area supervision;
- Mine supervision;
- Employee benefits;
- Overtime premium;
- Mine office expense;
- Head office expense;
- Mine surveying;
- Pumping;
- Development drilling;
- Payroll taxes;
- State and local taxes;
- Insurance;
- Assaying;
- Mine plant depreciation.

The capital and G&A costs could be translated into a cost per ton basis just as the operating costs. The cost categories might then become:
- Ownership cost;
- Production cost;
- General and administrative costs.

The operating cost can be reported by the different unit operations:
- Drilling;
- Blasting;
- Loading;
- Hauling;
- Other.

The 'other category' could be broken down to include dozing, grading, road maintenance, dump maintenance, pumping, etc. Some mines include maintenance costs together with the operating costs. Others might include it under G&A. Material cost can be further broken down into components. For blasting this might mean:
- Explosive;
- Caps;
- Primers;
- Downlines.

96 *Open pit mine planning and design: Fundamentals*

The operating cost could just as easily be broken down for example into the categories:
– Labor;
– Materials, expenses and power (MEP);
– Other.

At a given operation, the labor expense may include only the direct labor (driller, and driller helper, for example). At another the indirect labor (supervision, repair, etc.) could be included as well.

There are certain costs which are regarded as 'fixed', or independent of the production level. Other costs are 'variable', depending directly on production level. Still other costs are somewhere in between.

Costs can be charged against the ore, against the waste, or against both.

For equipment the ownership cost is often broken down into depreciation and an average annual investment cost. The average annual investment cost may include for example taxes, insurance and interest (the cost of money).

The bottom line is that when discussing, calculating or presenting costs one must be very careful to define what is meant and included (or not included). This section attempts to present a number of ways in which costs of various types might be estimated.

2.4.2 *Costs from actual operations*

Sometimes it is possible to obtain actual costs from 'similar' operations. However great care must be exercised in using such costs since accounting practices vary widely. For many years the *Canadian Mining Journal* has published its 'Reference Manual and Buyers Guide'. A great deal of useful information is contained regarding both mine and mill. Table 2.23 contains information from the 1986 edition for the Similkameen Mine.

Similar information for eleven open pit operations of different types and sizes as extracted from the 1993 edition of the *Reference Manual* (Southam Mining Group, 1992), is included in Table 2.24. Since Similco Mines Ltd. is the successor of Similkameen Property described in Table 2.23, one can examine changes in the operation and in the costs with time. Information as complete as this is seldom publicly available.

The 2004 edition of the CMJ Mining Source book (CMJ, 2004) has included the detailed cost information for the Huckleberry Mine given in Table 2.25. The authors are grateful to the Canadian Mining Journal for permission to include this valuable set of information.

2.4.3 *Escalation of older costs*

Publications from years past often contain valuable cost information. Is there some simple technique for updating so that these costs could be applied for estimating even today? The answer is a qualified yes. The qualification will be discussed later in this section. The procedure involves the escalation of costs through the application of various published indexes. Table 2.26 is an example of the:
– Construction cost;
– Building cost;
– Skilled labor;
– Common labor;
– Materials.

Table 2.23. Cost information (Canadian $) for the Similkameen Mine, Newmont Mining Company Mines Limited (*CMJ*, 1986).

1. *Location*: Princeton, British Columbia, Canada

2. *Pit geometry*
   (a) Pit size at surface: 3200′ × 1000′
   (b) Pit depth: 330′
   (c) Bench height: 40′
   (d) Bench face angle: 70°
   (e) Berm width: 40′
   (f) Road grade: 10%

3. *Capacity*
   (a) Mining: ore = 22,000 tpd
       waste = 25,500 tpd
       ore and waste = 5,111,500 tpy (actual)
   (b) Milling: capacity = 20,000 tpd
       ore = 2,945,000 tpy (actual)
       mill heads = 0.43% Cu
       minerals recovered = Cu, Au, Ag
       recovery = 85.5%
       concentrate grade = 30% Cu
       principal processes: primary SAG, secondary
          cone crusher, ball milling, Cu flotation

4. *Pit equipment*
   (a) Ore and waste loading = 4 P & H 1900A shovels (10 yd^3)
   (b) Ore and waste haulage = 15 Lectra Haul M100 trucks
   (c) Other = 4 Cat D8K dozers
       1 Cat 14E grader
       3 Cat 824 r.t.d.
       1 Dart 600C f.e.l. (15 yd^3)
       2 Komatsu 705A graders

5. *Blasting in ore and waste*
   (a) Explosives 85% bulk Anfo
         15% packaged slurry
   (b) Powder factor (lb/ton) = 0.46
   (c) Loading factor (lb/yd^3) = 0.70

6. *Drilling in ore and waste*
   (a) Drills = 3 Bucyrus-Erie 60R
   (b) Hole diameter = $9^7/_8''$
   (c) Pattern (burden × spacing) = 18′ × 24′
   (d) Feet drilled/shift = 490′
   (e) Tons/foot = 26
   (f) Bit life = 6000 ft
   (g) Rod life = 320,000 ft

7. *Power requirements*
   (a) Total (all motors) = 53,600 HP
   (b) Peak demand = 35,776 kVA
   (c) Annual mill demand = 255,282,172 kWh
   (d) Total annual demand = 266,573,172 kWh

(Continued)

Table 2.23. (Continued).

8. *Personnel*	
(a) Open pit	
Staff personnel	20
Equipment operators, labor	78
Mechanical, maintenance crew	44
Total open pit workforce	142
(b) Mineral processing plant	
Staff personnel	25
Operators (all classifications)	37
Repair and maintenance crew	45
Total mill workforce	107
(c) Surface plant	
Staff personnel	1
Mechanical and maintenance crew	13
Total surface plant workforce	14
(d) Other	
Office and clerical personnel	22
Warehouse	14
Total other	36
Total employees	299
9. *Mining costs for ore and waste ($/ton)*	
(a) Dozing and grading	0.05
(b) Drilling	0.07
(c) Blasting	0.13
(d) Loading	0.14
(e) Hauling	0.21
(f) Crushing	0.11
(g) Conveying	0.06
(h) Pumping	0.01
(i) Maintenance	0.10
(j) Supervision	0.02
(k) Other	0.02
Total	$0.92
10. *Milling costs ($/ton)*	
(a) Crushing	0.096
(b) Grinding	1.844
(c) Flotation	0.244
(d) Drying	0.075
(e) Assaying	0.015
(f) Conveying	0.091
(g) Power	0.921
(h) Tailings disposal	0.137
(i) Labor	0.437
(j) Supervision	0.094
Total	$3.954

Table 2.24. Operating and cost data from some Canadian open pit mines (*CMJ*, 1993).

1. Open pit mines included

Company, mine	Location	Minerals recovered	Pit size at surface Depth	Ore mined Waste removed	Bench height Slope	Berm road Grade
BHP Minerals Canada Ltd., Island Copper	Port Hardy, BC	Cu, Mo	7500' × 4000' 1300'	57,500 tpd ore 77,500 tpd waste	40' 45°	25' 10%
Equity Silver Mines Ltd.	Houston, BC	Ag, Au, Cu	1100 m × 500 m 240 m	10,000 mtpd ore 8000 mtpd waste	5 m 52°	8 m 12%
Hudson Bay Mining & Smelting Co. Ltd., Chisel Lake	Snow Lake, Man	Zn, Cu, Pb, Ag, Au	805' × 200' 200'	1315 tpd ore 656 tpd waste	20' 90°	11'6" 9%
Iron Ore Co. of Canada, Carol	Labrador City, Nfld	Fe ore	Avg. size, 5 pits: 1500 m × 500 m 90 m	106,000 mtpd ore 18,500 m³ waste	13.7 m 40° and 60°	15 m 8%
Mines Selbaie, A1 zone	Joutel, Que	Cu, Zn, Ag, Au	1090 m × 875 m 80 m	6000 mtpd ore 19,000 mtpd waste	8–10 m 45–54°	8–10 m 9%
Placer Dome Inc., Dome	South Porcupine, Ont	Au, Ag	100' × 500' 130'	350,000 tpy ore 700,000 tpy waste	40' 40°	20' 10%
Similco Mines Ltd. No. 1	Princeton, BC	Cu, Au, Ag	1500' × 1200' 800'	Total 3 pits: 22,500 tpd ore 22,500 tpd waste	40' 55°	40' 8%
Similco Mines Ltd. No. 3	Princeton, BC	Cu, Au, Ag	4000' × 2500' 1200'	Total 3 pits: 22,500 tpd ore 22,500 tpd waste	40' 55°	40' 8%
Similco Mines Ltd., Virginia	Princeton, BC	Cu, Au, Ag	1350' × 1200' 440'	Total 3 pits: 22,500 tpd ore 22,500 tpd waste	40' 55°	40' 8%
Stratmin Graphite Inc.	Lac-des-Iles, Que	Graphite	650 m × 350 m 25 m	1000 mtpd ore 3500 mtpd waste	6 m 50°	4 m 10%
Williams Operating Corp., C zone	Hemlo, Ont	Rockfill and some Au	450 m × 350 m 36 m	330 mtpd ore 4500 mtpd waste	10 m 70°	8 m 10%

(Continued)

Table 2.24. (Continued).
2. Deposit description

Company, mine	Proven and probable reserves	In situ grade	Ore type	Dimensions ($L \times W \times D$)	Host rock
BHP, Island Copper	95 million tonnes	0.355% Cu 0.017% Mo	Porphyry Cu	5000' × 2000' × 1600'	Andesite
Equity Silver Mines	6 million tonnes	72 g/t Ag 0.22% Cu 0.83 g/t Au	Disseminated and brecciated Sulphides	2.5 km × 500 m × 250 m	Pyroclastic
Hudson Bay, Chisel Lake	439,384 tonnes	0.056 g/t Au 1.23 g/t Ag 0.29% Cu 9.2% Zn 1.09% Pb	Mainly massive sulphides	750' × 100' × 150'	Altered felsic volcanics
Iron Ore Co., Carol	3 billion tonnes	39% Fe 19% magnetite	Specular hematite and magnetite in sedimentary iron formation	8 km × 250 m × 300 m	Quartzites and quartz-carbonate sandstones
Mines Selbaie	21.2 million tonnes	0.77% Cu 2.19% Zn 28.15 g/t Ag 0.47 g/t Au	Cu, Zn and pyrite lenses	1090 m × 875 m × 176 m	Welded acitic tuff, rhyodacitic breccia, massive pyrite and pyrite breccia
Placer Dome, Dome	9.3 million tonnes	0.143 Au	Hydrothermal quartz vein	200' × 1800' open at depth	Volcanic porphyry, conglomerate and ultramatic
Similco Mines	Pvn: 24.9 million tonnes Prb: 121.9 million tonnes stockpile: 13 million tonnes	0.45% Cu 0.40% Cu 0.25% Cu	Bornite, pyrite and chalcopyrite		Ore in adesitic volcanics with barren diorite gabbro intrusive
Stratmin Graphite	4 million tonnes	7.4% Cg	Graphite flakes	2700 m × 1700 m × 350 m	Marble and quartzite

3. Pit equipment

Company, mine	Ore or waste	Shovels	Loaders	Trucks	Other
BHP, Island Copper	Ore and waste	3 P&H 2100BL, 15 yd^3 2 Marion 191M, 15 yd^3		16 Euclid R170 3 Unit Rig Mark 36, 170-t 2 Euclid R190	4 Cat D10N dozers 1 Cat D9L dozer 3 Cat 824 dozers 5 Cat 16G graders 1 Cat 988 loader 1 Hitachi UH20 backhoe
Equity Silver Mines	Ore and waste	3 P&H 1600, 7 yd^3	1 Cat 992C, 10 m^3	2 Wabco, 80-t 5 Cat 777B, 80-t 3 Cat 773, 45-t	2 Cat D8L dozers 2 Cat 14G graders 2 Cat 824 dozers 1 Cat D6K dozer 4 Cat 16G graders 2 Cat 824C dozers 2 Cat 235 backhoes
Hudson Bay, Chisel Lake	Ore and waste	1 Komatsu PC640, 4.6 yd^3	1 Cat 988B, 7 yd^3	3 Euclid, 35-t	1 Cat D8N dozer 1 Champion 780A grader
Iron Ore Co., Carol	Ore and waste	5 B-E 295bII, 14 m^3 2 P&H 2300, 14 m^3 3 Kubota 280, 9 m^3	1 Letourneau L1100, 11.5 m^3 1 Cat 992C, 7.5 m^3 2 Cat 938B, 6 m^3	22 Titan T2200, 180-t 8 Terex 33-15, 154-t	5 Cat 16G graders 4 Komatsu D375A dozers 2 Cat D10N dozers 6 Cat 834B dozers
Mines Selbaie, A1 zone	Ore Waste	1 Hitachi VH801, 11 yd^3	1 Cat 992C, 11 yd^3 1 Cat 992C, 11 yd^3	3 Cat 777, 77-t 5 Cat 777, 77-t	3 Cat D8L dozers 2 Cat 16G graders 1 Cat 824 dozer 2 water trucks 1 dewatering truck 1 Anfo truck 2 Cat 235 shovels
Similco Mines	Ore and waste	4 P&H 1900A 1 P&H 1900A1	1 Cat 992 1 Terex 7271, 7 yd^3	4 Cat 785, 120-t 9 Unit Rig, 120-t	2 Cat D8 dozers 2 Cat 16G graders 2 Cat 824 dozers 1 Cat D9N dozer
Stratmin Graphite	Ore and waste	1 Cat 235C	2 Cat 980C, 3.8 m^3 1 Cat 988B, 5.5 m^3	4 Cat 769C, 35-t 2 Cat D25C, 25-t	1 Cat 14G grader
Williams Operating Corp.	Ore and waste	1 P&H 1900A1	1 Cat 992	4 Cat 777	1 Cat 14G grader 1 Cat D7 dozer 1 Cat D9N dozer

(Continued)

Table 2.24. (Continued).
4. Drilling equipment and practices

Company, mine	Ore or waste	Drills	Hole diameter Pattern	Feet per shift	Tons per foot	Feet per bit	Feet per shank	Feet per rod
BHP, Island Copper	Ore and waste	2 Bucyrus-Erie 60R2 2 Bucyrus-Erie 60RIII	$9^{7}/_{8}''$ $25' \times 25'$	828	44	12,600	654,000	292,900
Equity Silver Mines	Ore and waste	3 Bucyrus-Erie 40R	230 mm 5 m × 5 m	625	22	37,400	215,225	269,030
Hudson Bay, Chisel Lake	Ore	1 GD SCH3500BU 1 with HPR 1H hammer 1 Copco ROC 812HCSO	$4^{1}/_{2}''$ $7' \times 8'$	443	7	69	836	5297
	Waste	2 with 1238ME hammer	$4^{1}/_{2}''$ $8' \times 10'$	280	5.8	423	871	7405
Iron Ore Co., Carol	Ore and waste	4 Bucyrus-Erie 49RH 3 Gardner Denver 120	381 mm Ore: 8 m × 8 m Waste: 8.5 m × 8.5 m	Ore: 260 Waste: 22	Ore: 65.5 Waste: 55.5	1660		
Mines Selbaire, A1 zone	Ore and waste	2 Driltech D40KII 1 Driltech D60KII 1 Copco ROC 712H (secd)	200 mm Ore: 5.1 m × 5.8 m Waste: var	435	Ore: 70 Waste: 77	1090	14,925	13,355
Placer Dome, Dome	Ore		$4''$ $10' \times 10'$		8			
Similco Mines	Ore and waste	3 Bucyrus-Erie 60R	$9^{7}/_{8}''$	540	30	4500		
Stratmin Graphite	Ore and waste	2 Atlas Copco 812HC5001	$5''$ 4 m × 4 m	340	13.6	13,000	3000	200,000
Williams Operating Corp.	Waste	1 Gardner Denver 100	$10^{5}/_{8}''$ 5.5 m × 6.5 m	328	23	1804		6000

5. Blasting practices

Company, mine	Ore			Waste		
	Explosives	Loading factor	Powder factor	Explosives	Loading factor	Powder factor
BHP, Island Copper	100% emulsion	0.86 lb/yd^3	0.38 lb/ton	Magnafrac	0.86 lb/yd^3	0.38 lb/ton
Equity Silver Mines	60% Anfo 40% slurry	0.53 kg/m^3	0.18 kg/t	60% Anfo 40% slurry		
Hudson Bay, Chisel Lake	Dry holes: Amex Wet holes: Magnafrac	2.4 lb/yd^3	0.53 lb/ton	Dry holes: Amex Wet holes: Magnafrac	1.66 lb/yd^3	0.63 lb/ton
Iron Ore Co., Carol	Magnafrac B9000	1.71 kg/m^3	0.44 kg/t	Magnafrac B9000	1.65 kg/m^3	0.43 kg/t
Mines Selbaie, A1 zone	67% Anfo (column charge) 33% slurry (bottom charge)	0.80–0.86 kg/m^3	0.23–0.32 kg/t	67% Anfo (column charge) 33% slurry (bottom charge)	0.75–0.80 kg/m^3	0.26–0.28 kg/t
Placer Dome, Dome	90% Anfo 10 slurry	1.125 lb/yd^3	0.5 lb/ton	Amex & Detagel	1.125 lb/yd^3	0.5 lb/ton
Similco Mines	95% Fragmax 5% emulsion		0.59 lb/ton	Fragmax NBL 1019		0.59 lb/ton
Stratmin Graphite	90% Anfo 10% slurry	1.35 lb/yd^3	0.06 lb/ton	90% Anfo 10% slurry	1.12 lb/yd^3	0.5 lb/ton
Williams Operating Corp.	95% Anfo 5% packaged			95% Anfo 5% packaged	0.40 kg/m^3	

(Continued)

104 *Open pit mine planning and design: Fundamentals*

Table 2.24. (Continued).
6. Primary crushing

Company, mine	Crusher site	Transport and distance to crusher	Crusher	Crusher setting	Throughput capacity per hour	Transport and distance to mill
BHP, Island Copper	In pit	170-t trucks, 2000′	A-C gyratory 54″ × 72″	6″	4500 tph	54″ conveyor, 4100′ 54″ conveyor, 800′
	At mill	170-t trucks, 11,200′	A-C gyratory 54″ × 72″	6″	4000 tph	
Equity Silver Mines	At mill	80-t trucks, 1200 m	A-C gyratory 1.1 m × 1.65 m	175 mm	1500 mtph	1.2 m conveyor
Hudson Bay, Chisel Lake	Adjacent to pit	35-t trucks, 2000′	Kue Ken jaw 36″ × 43″	6″	300 tph	Bottom dump rock wagons, 9 miles
Mines Selbaie, A1 zone	Between pit and mill	77-t trucks, 1050 m	A-C gyratory 42″ × 65″	6″	1200 mtph	5 conveyors, 375 m
Placer Dome, Dome	At pit edge	Cat 988 loader, 200′	Piedmont portable jaw 48″ × 60″	6″	250 tph	30″ conveyor, 1000′
Similco Mines		Trucks, 3500′ to 6600′	A-C gyratory 54″ × 72″	8″	120 tph	40″ conveyor, 6600′
Stratmin Graphite	At mill	35-t trucks, 7000 m	915 mm × 1220 mm	150 mm	150 mtph	915 mm conveyor, 150 m
Williams Operating Corp.	Outside pit	4 Cat 777 trucks	A-C gyratory 42″ × 65″	175 mm	100 mtph	1066 mm conveyor (stockpile), 450 m

7. Mineral processing

Company, mill	Location	Daily throughput	Products	Mill heads	Recovery	Concentrate grade	Principal processes
BHP Minerals Canada Ltd., Island Copper	Port Hardy, BC	57,500 tpd	Cu conc Mo conc	0.39% Cu 0.016% Mo	84% Cu 65% Mo	24.0% Cu 45% Mo	Primary SAG, secondary ball milling, flotation, filtering and drying
Equity Silver Mines Ltd.	Houston, BC	9000 mtpd	Cu-Ag conc Ag-Au doré	0.22% Cu 86 g/t Ag 0.88 g/t Au	69% Cu 62% Ag 59% Au	11.1% Cu 3927 g/t Ag 21.9 g/t Au	Rod and ball milling, flotation, filtering, drying and CIL circuit
Hudson Bay Mining and Smelting Co. Ltd., Flin Flon	Flin Flon, Man	7000 tpd	Cu conc Zn conc	1.8% Cu 6.3% Zn 0.06 g/t Au	94.3% Cu 92.3% Zn 83% Au	21.0% Cu 51.5% Zn	Rod and ball milling, roughing and cleaning for both Cu and Zn conc., thickening and filtering
Mines Selbaie	Joutel, Que	A1 ore: 5800 mtpd	Cu conc Zn conc	0.67% Cu 2.28% Zn	87% Cu 82% Zn	24% Cu 56% Zn	SAG and ball milling, differential flotation and pressure filtering
		A2 & B ore: 1650 mtpd	Cu conc Zn conc	2.93% Cu 0.77% Zn	96% Cu 40% Zn	27% Cu 55% Zn	Rod and ball milling, differential flotation and pressure filtering
Placer Dome Inc., Dome	south Porcupine, Ont	4000 tpd	Au	0.125 g/t Au	95.5%		Crushing, rod and ball milling, jig concentrating, NaCN leaching and CIP
Similco Mines Ltd.	Princeton, BC	25,000 tpd	Cu conc	0.50% Cu	80%	28.5% Cu	SAG, ball milling, Cu flotation, thickening, filtering and drying
Stratmin Graphite Inc.	Lac-des-Iles, Que	1000 mtpd	Natural flake graphite	6.2% Cg	95%	96% Cg	Mechanical processes
Williams Operating Corp.	Hemlo, Ont	6000 mtpd	Au bullion	7.77 g/t	95%	873.7 fine Au	SAG, leaching, CIP, carbon stripping, E/W and refining

(Continued)

Table 2.24. (Continued).
8. Personnel numbers and distribution

Company, operation	Underground					Open pit				Mineral processing			Surface plant				Other		Total employees
	Staff personnel	Stopping, production miners	Haulage, hoisting crew	Development, maintenance crew	Total underground workforce	Staff personnel	Equipment operators, labor	Mechanical and maintenance crew	Total open pit workforce	Staff personnel	Operators (all classifications)	Repair and maintenance crew	Total mill workforce	Staff personnel	Mechanical and maintenance crew	Total surface plant workforce	Office and clerical personnel	Others	
BHP, Island Copper						47	155	135	337	41	60	83	184				34	17[1]	572
Equity Silver Mines Ltd.						12	37		49	15	33		48	8	42	50	12		159
Hudson Bay, Chisel Lake						2	11	2	15										
Iron Ore Co., Carol						80	247	235	562										
Mines Selbaie	12	31	46	43	132	6	58	29	93	15	57	21	93	47	99	146	42	32	538
Placer Dome Inc., Dome	18	54	63	41	176				Note 2	10	17	49	27	16	70	86	53		342
Similco Mines Ltd.						22	75	42	139	25	43		117	1	12	13	25		294
Stratmin Graphite						5	22	6	33	4	21	10	35	3	5	8	12		88
Williams Operating Corp.	25	120	125	50	320	2	13	7	18	19	19	10	48	24	130	124	80		619

[1] Warehouse and safety personnel. [2] Employees of contractor.

9. Mine operating costs (Canadian $/ton)

Company, mine	Dozing and grading	Drilling	Blasting	Loading	Haulage	Crushing	Conveying	Pumping	Maintenance	Labour	Power	Other	Total
BHP, Island Copper		0.040	0.088	0.345		0.177[1]		0.020	0.044			0.240[2]	0.993
Equity Silver Mines*		0.161	0.161	0.230	0.440			0.033			0.050	0.116	1.189
Hudson Bay, Chisel Lake													
waste		1.640	0.480		0.870[3]	0.050		0.125	1.360		0.840	0.550[4]	5.915
ore		0.950[5]			0.820[3]	0.510		0.125	1.360		0.840	0.550[4]	5.155
Iron Ore Co., Carol													1.887
Mines Selbaie*													
A zone ore	0.137	0.270	0.285	0.090	0.232	0.070			0.955	0.052		0.668[6]	2.759
overburden	0.137			0.090	0.232				0.488			0.097	1.044
waste	0.137	0.270	0.285	0.090	0.232				0.615			0.097	1.726
Placer Dome, Dome													
ore													2.900
waste													3.100
Similco Mines	0.077	0.063	0.118	0.144	0.328	0.090	0.090			0.082		0.015	1.007
Stratmin Graphite*	0.163	0.254	0.308	0.336	0.336			0.082				0.390	1.869
Williams Operating Corp.*	0.073	0.327	0.218	0.308	0.354	0.744					0.136	0.227	2.387

* Amounts reported in metric units have been converted to imperial equivalents. [1] Includes conveying. [2] Includes reclamation $0.055, mine shipping expense $0.096, and other $0.089. [3] Includes loading. [4] Supervision. [5] Includes blasting. [6] Includes engineering and geology.

(Continued)

108 *Open pit mine planning and design: Fundamentals*

Table 2.24. (Continued).
10. Processing costs (Canadian $/ton)

Company, mill	Crushing	Grinding	Reagents	Flotation	Leaching	Dewatering	Tailings disposal	Assaying	Power	Supervision and labor	Maintenance	Other	Total
BHP Island Copper			0.160						0.740	0.430	0.180	0.6947	2.200
Equity Silver Mines*	0.299	1.887	0.064	0.318	0.916	0.191	0.027	0.127				0.490	4.318
Placer Dome, Dome	1.220	1.690			1.340							0.960	5.210
Williams Operating Corp.*		1.660			0.680		0.218	0.308	1.642		1.769	0.889	7.167

*Amounts reported in metric units have been converted to imperial equivalents. [7] Includes grinding balls $0.480, mill liners $0.160, fuel and lubricants $0.008, and operating supplies $0.046.

Table 2.25. Cost information (Canadian $) for the Huckleberry Mine (*CMJ*, 2004).

1. *Location*: Houston, British Columbia, Canada
2. *Pit geometry*
    (a) Pit size at surface: 1050 m × 600 m
    (b) Pit depth: 320 m
    (c) Bench height: 12 m
    (d) Bench face angle: 70°
    (e) Berm width: 10 m
    (f) Road grade: 10%
    (g) Slope angle: 52°
3. *Capacity*
    (a) Mining: ore = 21,000 mtpd
    waste = 44,000 mtpd
    ore and waste = 19,034,000 tonnes (actual)
    (b) Milling: capacity = 20,333 mtpd
    (c) ore = 7,422,000 tonnes (actual)
    mill heads = 0.534% Cu, 0.014% Mo
    minerals recovered = Cu, Mo
    recovery = 88.38% Cu, 47.54% Mo
    concentrate grade = 27.19% Cu, 48.73% Mo
    principal processes: SAG & ball milling, bulk float,
    regrinding & dewatering, Moly float, Cu-Mo
    separation, float and regrinding.
4. *Pit equipment*
    (a) Ore and waste loading = 1 P & H 1900AL shovel
    1 P & H 2100BL shovel
    1 Cat 992C FEL
    1 Cat 416
    (b) Ore and waste haulage = 5 Cat 777C trucks
    4 Cat 785B trucks
    1 Cat 777B truck
    (c) Other = 2 Cat D9N dozers
    2 Cat D8N dozers
    1 Cat D10N dozer
    1 Cat 824 RTD
    2 Cat 16G graders
    1 Hitachi excavator
    1 Cat 769 water truck
5. *Blasting in ore and waste*
    (a) Emulsion
    (b) Powder factor (kg/t) = 0.20
    (c) Loading factor (kg/m^3) = 0.54
6. *Drilling in ore and waste*
    (a) Drills = 2 Bucyrus-Erie 60R
    (b) Hole diameter = 251 mm
    (c) Pattern (burden × spacing) = 7.5 m × 7.5 m to 9.3 m × 9.3 m offset & wall control
    (d) Drilling per shift = 140 m
    (e) Tons/m = 177
    (f) Bit life = 2227 m
    (g) Rod life = NA

(Continued)

Table 2.25. (Continued).

7. *Power requirements*
   (a) Total (all motors) = 190,900,000 kWh
   (b) Peak demand = 30 kVA
   (c) Annual mill demand = 190,500,000 kWh
   (d) Total annual demand = 195,300,000 kWh

8. *Personnel*
   (a) Open pit
Staff personnel	18
Equipment operators, labor	68
Mechanical, maintenance crew	34
Total open pit workforce	120

   (b) Mineral processing plant
Staff personnel	12
Operators (all classifications)	33
Repair and maintenance crew	12
Total mill workforce	68

   (c) Other
Office and clerical personnel	14
Others	9
Total other	23
Total employees	211

9. *Mining costs for ore and waste ($/tonne)*
(a) Dozing and grading	0.158
(b) Drilling	0.053
(c) Blasting	0.111
(d) Loading	0.221
(e) Hauling	0.210
(f) Crushing	incl
(g) Conveying	incl
(h) Pumping	0.024
(i) Maintenance	incl
(j) Supervision and labor	incl
(k) Power	incl
(l) Other	0.135
Total	$1.123

10. *Milling costs ($/tonne)*
(a) Crushing	0.269
(b) Grinding	1.940
(c) Reagents	0.009
(d) Flotation	0.545
(e) Dewatering	0.122
(f) Tailings disposal	0.056
(g) Assaying	0.047
(h) Conveying	0.091
(i) Power	incl
(j) Supervision & labor	incl
(k) Maintenance	incl
(l) Other	0.359
Total	$3.342

Table 2.26. ENR cost indices (ENR, 1976–2004) Reference year 1967 = 100.

Year	Construction Cost	Building Cost	Skilled Labor	Common Labor	Materials
1976	2499	1425	2136	4700	1055
1977	2577	1545	2264	4977	1159
Jun-78	2754	1664	2376	5241	1229
1978	2776	1674	2405	5303	1289
1979	3003	1819	2564	5676	1427
1980	3237	1941	2767	6168	1488
1981	3535	2097	3025	6802	1527
1982	3825	2234	3358	7545	1548
1983	4066	2384	3591	8020	1651
1984	4146	2417	3721	8269	1621
1985	4182	2425	3778	8396	1617
1986	4295	2483	3867	8616	1634
1987	4406	2541	3986	8869	1659
1988	4519	2598	4085	9120	1694
Jun-89	4593	2626	4166	9336	1686
1989	4615	2634	4174	9381	1693
1990	4732	2702	4310	9646	1720
1991	4835	2751	4457	9935	1709
1992	4985	2834	4589	10243	1761
1993	5210	2996	4703	10525	1953
1994	5408	3111	4818	10856	2068
1995	5471	3112	4923	11146	1993
1996	5620	3203	5085	11444	2046
1997	5826	3364	5229	11697	2226
1998	5920	3391	5374	12024	2179
1999	6059	3456	5537	12383	2184
2000	6221	3539	5740	12790	2195
2001	6334	3574	5965	13242	2113
2002	6538	3623	6208	13871	2044
2003	6694	3693	6496	14386	1981
2004	7115	3984	6747	14978	2296

Reprinted from Engineering News-Record, Copyright The McGraw-Hill Companies, 2005, All rights reserved.

indices published weekly in the *Engineering News Record* (ENR). The average yearly values are given except where noted. To illustrate the application of the index system, assume that the cost of the mine maintenance building was $100,000 in June of 1978. The estimated cost of the same building in June of 1989 would be

$$\text{Cost June 1989} = \text{Cost June 1978} \times \frac{\text{Building cost index (June 1989)}}{\text{Building cost index (June 1978)}}$$

In this case

$$\text{Cost June 1989} = 100,000 \times \frac{2626}{1664} = 100,000 \times 1.58 = \$158,000$$

The escalation factor of 1.58 is the ratio of the index values for the years involved. In a similar way one can compute the escalation factors for the other ENR indexes over this period. They are summarized below:

$$\text{Construction cost factor} = \frac{4593}{2754} = 1.67$$

112   *Open pit mine planning and design: Fundamentals*

$$\text{Building cost factor} = \frac{2626}{1664} = 1.58$$

$$\text{Skilled labor factor} = \frac{4166}{2376} = 1.75$$

$$\text{Common labor factor} = \frac{9336}{5241} = 1.78$$

$$\text{Materials factor} = \frac{1686}{1229} = 1.37$$

Other indexes are also available. Table 2.27 gives the average hourly earnings for mining production/non supervisory workers as published by the U.S. Bureau of Labor Statistics (BLS) over the period 1964 through 2002. These values can also serve as a labor cost escalator. For the period 1978 to 1989 the factor would be

$$\text{Mining hourly wage factor} = \frac{13.25}{7.67} = 1.73$$

Table 2.28 gives average hourly earnings broken down by industry. Contained within the publication *Statistical Abstract of the United States* are values for the producer price index for construction machinery and equipment. The values for the time period 1978 to 2004 are given in Table 2.29. The resulting factor for the 1978 to 1989 time period is

$$\text{Construction machinery and equipment factor} = \frac{117.2}{67.7} = 1.73$$

Considering the five ENR indices plus the two from the BLS, an average escalation factor of 1.71 is selected. The average cost inflation rate r over this 11 year period is computed by

$$(1 + r)^{11} = 1.71$$

Hence

$$r = 0.050$$

The rate is 5%/year.

It was indicated earlier that such escalation has to be done with some care. A major reason for this is the change in labor productivity which has occurred over time.

Productivity is a very important aspect of cost estimation. It deals with the rate at which a certain task can be accomplished. If for example the daily production for a one shift per day mining operation is 20,000 tons with 100 employees, then one way of expressing the productivity is

$$\text{Productivity} = \frac{20,000}{100} = 200 \text{ tons/manshift}$$

Assume that the payroll is $10,000/day or $100/manshift. The labor cost would be $0.50/ton.

If through some type of change, the daily production could be raised to 30,000 tons, with the same employees, then the productivity would be

$$\text{Productivity} = \frac{30,000}{100} = 300 \text{ tons/manshift}$$

Table 2.27. Average hourly wage of production workers broken down by industry using the SIC base (BLS, 2004).

Year	Average hourly earnings ($/hr)							
	Mining*	Metal mining	Iron ores	Copper ores	Coal mining	Bitum. Coal/ lignite	Nonmetallic minerals*	Crushed/ broken stone
1964	$2.81	2.96	3.13	3.04	3.26	3.30	2.48	2.41
1965	2.92	3.06	3.16	3.15	3.45	3.49	2.57	2.47
1966	3.05	3.17	3.28	3.22	3.63	3.66	2.70	2.61
1967	3.19	3.24	3.30	3.26	3.73	3.76	2.85	2.72
1968	3.35	3.42	3.47	3.44	3.83	3.86	3.04	2.93
1969	3.60	3.64	3.70	3.65	4.20	4.24	3.27	3.21
1970	3.85	3.88	3.90	3.93	4.54	4.58	3.47	3.39
1971	4.06	4.12	4.19	4.16	4.78	4.83	3.70	3.64
1972	4.44	4.56	4.60	4.64	5.27	5.31	3.95	3.94
1973	4.75	4.84	4.88	4.90	5.70	5.75	4.22	4.21
1974	5.23	5.44	5.53	5.54	6.22	6.26	4.50	4.49
1975	5.95	6.13	6.29	6.36	7.21	7.24	4.95	4.84
1976	6.46	6.76	7.04	7.04	7.74	7.77	5.36	5.20
1977	6.94	7.28	7.49	7.49	8.25	8.27	5.81	5.67
1978	7.67	8.23	8.48	8.46	9.51	9.55	6.33	6.14
1979	8.49	9.27	9.57	9.53	10.28	10.31	6.90	6.62
1980	9.17	10.26	10.95	10.61	10.86	10.90	7.52	7.16
1981	10.04	11.55	12.16	11.83	11.91	11.95	8.28	7.93
1982	10.77	12.31	12.97	12.53	12.69	12.73	8.90	8.53
1983	11.28	12.58	12.39	13.10	13.73	13.78	9.31	8.70
1984	11.63	13.05	12.74	13.56	14.82	14.87	9.87	9.26
1985	11.98	13.38	13.01	13.62	15.24	15.3	10.18	9.59
1986	12.46	13.19	13.88	12.29	15.40	15.46	10.38	9.80
1987	12.54	12.94	14.36	11.42	15.76	15.81	10.60	10.00
1988	12.80	13.24	14.19	11.62	16.06	16.26	10.94	10.37
1989	13.25	13.58	14.24	11.80	16.26	16.39	11.25	10.69
1990	13.69	14.05	14.59	12.48	16.71	16.85	11.58	11.07
1991	14.21	14.87	16.36	13.36	17.06	17.21	11.93	11.24
1992	14.54	15.26	16.52	13.83	17.15	17.29	12.26	11.55
1993	14.60	15.29	16.67	14.03	17.27	17.46	12.70	12.01
1994	14.88	16.08	17.87	14.31	17.76	17.97	13.11	12.47
1995	15.30	16.77	18.49	14.93	18.45	18.70	13.39	12.66
1996	15.62	17.35	18.70	15.72	18.74	19.03	13.75	13.15
1997	16.15	17.82	18.85	16.32	19.01	19.30	14.19	13.56
1998	16.91	18.24	19.90	16.53	19.17	19.43	14.67	14.03
1999	17.05	18.26	20.43	16.18	19.15	19.33	14.97	14.47
2000	17.22	18.60	21.48	15.65	19.09	19.20	15.28	14.83
2001	17.56	18.74	21.63	15.77	18.94	19.05	15.62	15.02
2002	17.77	18.81	21.93	16.02	19.64	19.77	15.99	15.49

* Except fuels
BLS, 2005. Employment, Hours, and Earnings from the Current Employment Statistics Survey (National) SIC.
Series Reports EEU10000006, EEU10100006, EEU10101006, EEU10102006, EEU10120006, EEU10122006, EEU10140006, EEU10142006
http://www.bls.gov/data/

Table 2.28. Average hourly earnings of production workers using the NAICS base (BLS, 2004).

Year	Average hourly earnings ($/hr)			
	Mining*	Metal mining	Bituminous coal/ lignite surface mining	Nonmetallic mineral mining and quarrying
1990	15.47	15.36	15.19	11.82
1991	15.96	15.58	16.15	12.16
1992	16.11	15.67	16.5	12.52
1993	16.08	16.04	16.53	12.97
1994	16.67	16.33	17.41	13.37
1995	17.15	17.09	18.17	13.66
1996	17.51	17.41	18.77	14.07
1997	17.85	17.54	19.28	14.51
1998	18.08	18.14	19.77	14.99
1999	18.04	18.74	19.81	15.31
2000	18.07	19.02	20.07	15.65
2001	18.22	19.12	19.97	16.09
2002	18.61	19.73	20.53	16.56
2003	19.14	20.34	21.9	17.14
2004	19.85	21.88	22.91	17.74

* Except oil and gas

Table 2.29. Producer price indexes for construction machinery and equipment.

Year	Index (1982 = 100)	Year	Index (1982 = 100)
1978	67.7	1992	128.7
1979	74.5	1993	132
1980	84.2	1994	133.7
1981	93.3	1995	136.7
1982	100	1996	139.8
1983	102.3	1997	142.2
1984	103.8	1998	145.2
1985	105.4	1999	147.2
1986	106.7	2000	148.6
1987	108.9	2001	149.1
1988	111.7	2002	151.1
1989	117.2	2003	153.2
1990	121.6	2004	158.5
1991	125.2		

http://www.bls.gov/data/ then go to Producer Price Commodity index, most requested statistics construction machinery and equipment
Machinery and equipment

and the labor component of the cost would drop to $0.333/ton. If this productivity has come about through the purchase of new, larger equipment then the decrease in unit labor cost will be accompanied by an increase in other costs (ownership, etc.).

A copper mining example will be used to demonstrate the effect of productivity changes on cost escalation.

In 1909, the use of steam shovels was just beginning in the Utah Copper Company Bingham Canyon Mine of Kennecott (Anonymous, 1909a,b; Finlay, 1908; Jackling, 1909). The following data are available from that time.

1. Direct ore mining cost = 15.39 ¢/ton
2. General mining expense = 9.83 ¢/ton
   (includes fixed change per ton to retire prepaid stripping)

   Total mining cost = 25.22 ¢/ton

3. Average stripping cost = 31.43 ¢/yd³
4. Direct milling cost = 47 ¢/ton
5. General milling expense = 5.16 ¢/ton

   Total milling cost = 52.16 ¢/ton

6. Ore grade = 36 lbs/ton (1.8% Cu)
   Milling rate ≅ 7000 tpd
   Recovery ≅ 70%
   Production rate = 63,000,000 lbs Cu/year
7. Total average cost = 8.125 ¢/lb Cu
   (mining, milling, smelting)
8. Labor wages = $2/day
9. Copper price ≅ 12.7 ¢/lb

The index values in 1913 (the closest available year to 1910) were 100. In January 1993 the index values were as follows:

Category	Index	Ratio = $\frac{index 1993}{index 1913}$
Skilled labor	4650	47
Common labor	10,395	104
Materials	1806	18
Building cost	2886	29
Construction cost	5070	51

In 1992 the average mining wages were about $15.00/hour or $120/day. The labor cost ratio (LCR) of 1992 to 1910 is

$$\text{LCR} = \frac{\$120}{\$2} = 60$$

This is similar to the index values for skilled labor. The copper price ratio (CPR) for the same period is about

$$\text{CPR} = \frac{100}{12.7} = 7.9$$

Due to major changes in productivity over these intervening 80 years, the price and overall cost increase has been much less than would be expected due to labor costs alone. Therefore,

Table 2.30. Productivity (relative output per hour) for mine production workers (BLS, 1992a) (1982 = 100).

Year	Iron mining (crude ore)	Copper mining (crude ore)	Crushed and broken stone	Non-metallic minerals except fuels
1967	85.3	58.7	78.5	87.2
1968	92.7	65.4	85.5	94.8
1969	96.7	73.1	87.3	97.2
1970	99.0	79.7	88.2	100.5
1971	97.7	81.4	86.0	99.7
1972	107.0	86.3	90.8	103.7
1973	112.2	86.4	99.9	108.9
1974	106.9	80.5	97.8	104.8
1975	111.6	81.9	97.2	101.5
1976	112.5	93.2	99.5	107.7
1977	99.1	94.0	106.2	112.0
1978	114.4	103.1	115.1	117.3
1979	121.6	102.5	113.6	115.0
1980	123.6	93.4	107.6	108.1
1981	131.5	95.8	102.8	106.1
1982	100.0	100.0	100.0	100.0
1983	136.8	121.0	109.9	110.1
1984	169.4	130.1	111.6	117.7
1985	182.3	153.9	109.7	120.0
1986	192.2	181.9	110.0	120.8
1987	243.0	179.1	125.7	127.8
1988	260.8	190.4	126.9	130.5
1989	251.2	187.7	123.6	131.8
1990	229.7	182.8	125.6	134.8

when using productivity factors, one must bear in mine the effect of productivity changes over the intervening time.

Table 2.30 gives productivity figures for the mining of iron ore, copper ore, crushed and broken stone and non metallic minerals over the period 1967 through 1990. One can see the major productivity increase which has taken place over the period 1978 through 1990 in iron and copper mining. Table 2.31 provides productivity figures for the period 1987 through 2000. Here a major productivity increase in the gold mining sector is shown. In June 2003, the Standard Industrial Classification system (SIC) was replaced by the North American Industry Classification System (NAICS). Table 2.32 provides the NAICS based labor productivity figures for 1987 through 2002.

It is often very difficult to interpret productivity figures since it makes a big difference as to who has or has not been included.

2.4.4 The original O'Hara cost estimator

In 1980 O'Hara (1980) published what has become a classic paper 'Quick guides to the evaluation of orebodies'. He has since produced an updated version which is the subject of the next section. However, one of his original curves (Fig. 2.12), which relates mine/mill capital cost C to daily milling rate T_p will be used to demonstrate cost escalation procedures.

The mill generally has a much larger capital cost per daily ton of ore, and hence dominates the curve. It was assumed that the mining operations run only 5 days/week, but that the mill

Table 2.31. Productivity (relative output per hour) for mine production workers (SIC code) (1987 = 100).

Year	Iron ores	Copper ores	Gold Ores	Crushed and broken stone	Non-metallic minerals*
1987	100	100	100	100	100
1988	103	109.2	99	101.3	101
1989	98.4	106.6	108.9	98.7	99.6
1990	88.5	102.7	119.4	102.2	101.4
1991	85	100.5	118.2	99.8	98.5
1992	83.3	115.2	130.1	105	103
1993	86.9	118.1	144.7	103.6	100.8
1994	85	126	146	108.7	104.4
1995	94.8	117.2	131.9	105.4	104.5
1996	90.7	116.5	128.6	107.2	104.3
1997	89.1	118.9	146.6	112.6	107.3
1998	93	118.3	176.2	110.2	108.6
1999	89.2	110	186.8	105	108.6
2000	103.2	122.6	229.3	101.9	103.3

* Except fuels.
http://ftp.bls.gov/pub/special.requests/opt/dipts/oaehhiin.txt

Table 2.32. Labor productivity output per hour (Index, 1997 = 100) NAICS code.

Year	Mining*	Metal ore mining	Nonmetallic mineral mining & quarrying
1987	69.5	70.9	87.9
1988	74.2	74.7	89.5
1989	77.1	75.1	91.1
1990	79.3	79.9	92.3
1991	80	82.7	89.5
1992	86.8	91.7	96.1
1993	89.9	102.2	93.6
1994	93	104.1	96.9
1995	94	98.5	97.3
1996	96	95.3	97.1
1997	100	100	100
1998	104.6	109.5	101.3
1999	105.9	112.7	101.2
2000	106.8	124.4	96.2
2001	109	131.8	99.3
2002	111.7	143.9	103.8

* Excludes oil and gas.
http://data.bls.gov/PDQ/outside.jsp?/survey=ip

is operated continuously 7 days per week. Thus, the daily ore tonnage mined and crushed T_o will be 40% higher than the milling rate T:

$$T_o = \text{Ore mining rate} = \frac{7}{5}T = 1.4\,T$$

118 *Open pit mine planning and design: Fundamentals*

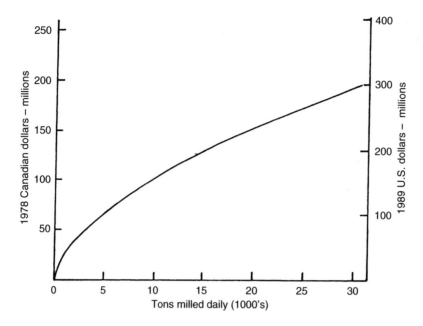

Figure 2.12. Mine/mill project capital cost as a function of milling rate (O'Hara, 1980).

The combined mine/mill capital cost expressed in mid-1978 Canadian dollars is

$$C = \$400,000 \, T^{0.6} \tag{2.44}$$

This must first be converted to U.S. dollars and then escalated to 1989 U.S. dollars. In mid-1978, one Canadian dollar had a value of 0.877 U.S. dollars. The approximate escalation factor from mid-1978 using ENR indices is 1.71. Combining the escalation factor and the exchange rate factor yields an overall multiplying factor of 1.50. Applying this, the expected capital cost in U.S. dollars for mid-1989 is

$$C = \$600,000 T^{0.6} \tag{2.45}$$

These values are reflected by the right hand axis in Figure 2.12. The interested student is encouraged to escalate these costs to the present time.

2.4.5 *The updated O'Hara cost estimator*

Introduction
Included in the 2nd edition of the *SME Mining Engineers Handbook* (Hartman, 1992), is an updated chapter on 'Costs & Cost Estimation', prepared by O'Hara and Suboleski (1992). They cover the costs associated with both open pit and underground mining. This section presents material extracted from their paper. The presentation, however, is organized somewhat differently from theirs. All of the costs are expressed in U.S. dollars appropriate for the third quarter of 1988.

Pits may vary greatly in shape, size, and pit slope, especially in mountainous areas or where the ore and/or waste rock varies greatly in competence. The typical open pit mine in North America produces about 43,000 tpd (39 kt/day) of ore and waste from a pit depth of about 400 to 500 ft (120 to 150 m), with an oval shaped periphery 2200 ft (670 m) wide and 4700 ft (1430 m) long. Pit benches are typically 40 ft (12 m) high, and overall pit slope (excluding roads) is about 57° in pits with competent rock, and 44° in pits with oxidized or altered rock, with in-pit haulage road gradients averaging 9%.

The formulas given for equipment sizing, preproduction stripping, and maintenance facilities presume that the shape and type of open pit is similar, except in daily tonnage, to the 'typical' open pit.

Daily tonnage
The most important factor affecting costs is the size of the mine, primary crusher, and processing plant as expressed in terms of the tons of material handled per day of operation. To simplify the discussion the following terms will be introduced:

T = tons of ore milled/day
T_o = tons of ore mined/day
T_w = tons of waste mined/day
T_c = tons of ore passing the primary crusher/day
$T_p = T_o + T_w$ = total material mined/day

In this estimator it is assumed that the mill operates three 8-hour shifts per day and 7 days/week irregardless of the shifts worked by the open pit. Many open pit mines operate 7 days/week, but others may operate only 5. In the case of a 5 day/week mining operation.

$$T = \frac{5}{7} T_o = 0.71 T_o \qquad (2.46)$$

The cost guides in this section are based upon this assumption that the mill capacity is 71% of the daily mined ore tonnage.

The crushing plant may operate 5, 6, or 7 days/week, depending on the mine schedule and whether or not there is adequate fine ore storage capacity to keep the mill supplied with ore when the crusher is shut down for repairs or regular maintenance.

It is assumed that the crushing plant has the same daily capacity as the mine, but will work 6 days/week to ensure that the mill will be supplied with crushed ore if the fine ore bins have insufficient capacity to keep the mill supplied with ore during the two-day mine shutdown.

Personnel numbers
It may seem somewhat unusual to begin the cost discussion with personnel, but their productivity is extremely important to the profitability of an operation and their compensation is a major cost item.

The number of mine personnel N_{op} required in open pit mines using shovels and trucks for loading and hauling the ore may be estimated from the following formulas:

$$N_{op} = \begin{cases} 0.034 \, T_p^{0.8} \text{ for hard rock} \\ 0.024 \, T_p^{0.8} \text{ for competent soft rock} \end{cases} \qquad (2.47)$$

120 Open pit mine planning and design: Fundamentals

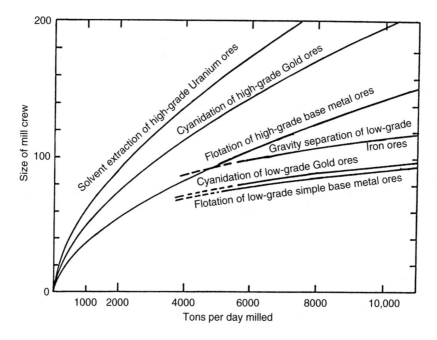

Figure 2.13. Mill crew size versus mill process and size (O'Hara & Suboleski, 1992).

The number personnel N_{ml} required to operate mills treating T tons of low-grade ore may be estimated from the following formulas:

$$N_{ml} = \begin{cases} 5.90\,T^{0.3} \text{ for cyanidation of precious metal ores} \\ 5.70\,T^{0.3} \text{ for flotation of low-grade base metal ores} \\ 7.20\,T^{0.3} \text{ for gravity concentration of iron ores} \end{cases} \quad (2.48)$$

The mill crew size (which includes those involved in crushing and/or grinding as well as beneficiation) as a function of process type and mill rate is shown in Figure 2.13.

The number of service personnel N_{sv} required for open pits mining low grade ore may be estimated as a percentage of the total mine and mill personnel as shown below:

$$N_{sv} = 25.4\% \text{ of } (N_{op} + N_{ml}) \quad (2.49)$$

The number of administrative and technical personnel N_{at} required for a mining and milling plant may be estimated as a percentage of the total required for mining, milling, and services:

$$N_{at} = 11\% \text{ of } (N_{op} + N_{ml} + N_{sv}) \quad (2.50)$$

It should be noted that the formulas do not include the personnel required for smelters, refineries, mine townsite services, concentrate transport, or offsite head offices, since these services may not be required for many mine projects. Whenever these services can be financially justified for the mine project circumstances, the additional personnel should be estimated separately.

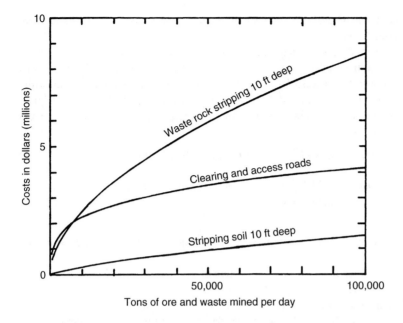

Figure 2.14. Clearing, stripping and road costs for open pit mines (O'Hara & Suboleski, 1992).

Mine associated capital costs
Mine site clearing. Prior to beginning construction, the mine/mill site must be first cleared of trees, plants and topsoil. The soil overburden should be stripped to the limits of the ultimate pit and stockpiled.

The average soil thickness can be found from drilling logs or ultrasonic techniques. By multiplying the average thickness times the pit area, the volume is determined. As an aid to tonnage calculations, an acre of moist soil averaging 10 ft in thickness contains about 23,000 tons of material. For the pit, the required area A_p in acres is

$$A_p = 0.0173 \, T_p^{0.9} \tag{2.51}$$

The clearing costs depend upon the topography, the type of cover, and the total area. They are expressed as

$$\text{Total clearing cost} = \begin{cases} \$1600 \, A_p^{0.9} & \text{for 20\% slopes with light tree growth} \\ \$300 \, A_p^{0.9} & \text{for flat land with shrubs and no trees} \\ \$2000 \, A_p^{0.9} & \text{for 30\% slopes with heavy trees} \end{cases} \tag{2.52}$$

Clearing, initial stripping and access road costs are plotted as a function of T_p in Figure 2.14.

Pre-production waste stripping. The rock overburden above the ore must be stripped to expose a sufficient amount of ore to supply the planned daily ore tonnage for a period of four to six months. If insufficient ore has been exposed by the pre-production stripping of waste, it may become difficult to continue ore mining due to the close proximity of waste benches where blasting, loading, and haulage of waste is taking place.

The location and required area of the ore exposure is determined from ore body mapping. Once this has been done, the average thickness and area of the waste rock overlying this ore can be computed. Each acre of waste rock averaging 10 ft in thickness contains about 40,000 tons of waste.

Because of the inverted conical shape of the ultimate open pit, the waste/ore tonnage ratio at each horizontal bench decreases with each lower bench. Typically, the uppermost ore bench to be exposed has a waste/ore ratio of at least twice the waste/ore ratio of the ultimate pit. If T_s is the tons of soil, and T_{ws} is the tons of waste rock that must be stripped to expose an amount of ore to sustain four to six months ore production, then the estimated costs of waste stripping will be

$$\text{Soil stripping costs} = \$3.20\, T_s^{0.8} \text{ for soil not more than 20 ft deep} \qquad (2.53a)$$

$$\text{Waste stripping costs} = \$340\, T_{ws}^{0.6} \text{ for rock requiring blasting, loading, and haulage} \qquad (2.53b)$$

Mine equipment
(a) Drills. The size, hole diameter, and number of drills required depends on the tons of ore and waste to be drilled off daily.

Typically, drill hole sizes have standard diameters of 4, 5, 6(1/2), 7(7/8), 9, 9(7/8), 10(5/8), 12(1/8), 13(3/8), 15, and 17(1/2) inches (or 102, 125, 165, 200, 229, 250, 270, 310, 350, 380, and 445 mm). Thus drill selection will be limited to one of these sizes.

The tons of ore or waste that are drilled off per day by a drill with a hole diameter of d inches is:

$$\text{tons of medium drillable rock} = 170\, d^2$$
$$\text{tons of easily drillable rock} = 230\, d^2 \qquad (2.54)$$
$$\text{tons of hard drillable rock} = 100\, d^2$$

For the rock defined as 'medium' drillable, the expected production rate is about 500 ft per shift.

The number of drills N_d should never be less than two. For tonnages up to 25,000 tpd, two drills of appropriate hole diameter should be chosen. Three drills should be adequate for up to 60,000 tpd and four or more drills will be required for daily tonnages over 60,000.

The cost of the drilling equipment is given by:

$$\text{Drilling equipment costs} = N_d \times \$20,000\, d^{1.8} \qquad (2.55)$$

This formula includes a 25% allowance for drilling and blasting supplies and accessory equipment.

(b) Shovels. The optimum shovel size S expressed in cubic yards of nominal dipper capacity in relation to daily tonnage of ore and waste T_p to be loaded daily is

$$S = 0.145\, T_p^{0.4} \qquad (2.56)$$

The number of shovels N_s with dipper size S that will be required to load a total of T_p tons of ore and waste daily will be

$$N_s = 0.011 \frac{T_p^{0.8}}{S} \qquad (2.57)$$

In practice, the size of shovel chosen will be one with a standard dipper size close to the size calculated by Equation (2.56). The calculated number of shovels N_s usually is not a whole number. It should be rounded down. The omitted fractional number expresses the need for either a smaller-sized shovel or a front-end loader for supplemental loading service. This smaller shovel or front end loader must, of course, be capable of loading trucks of a size appropriate to the shovels with dipper size S.

The total costs of the fleet of shovels supplemented by auxiliary bulldozers and front end loaders will be

$$\text{Loading equipment cost} = N_s \times \$510{,}000 \, S^{0.8} \tag{2.58}$$

(c) Trucks. The optimum truck size t in tons that is well matched with shovels of bucket size S (cubic yards) is

$$\text{Truck size t (tons)} = 9.0 \, S^{1.1} \tag{2.59}$$

The total number of trucks N_t of t tons capacity required for the open pit truck fleet, plus an allowance for trucks under repair, is approximated by the following formula:

$$N_t \text{ (Number of trucks required)} = 0.25 \frac{T_p^{0.8}}{t} \tag{2.60}$$

The formula for N_t determines the size of the truck fleet under the typical conditions where the average haulage distance and gradient outside the pit periphery is less than the haulage distance and gradient inside the pit periphery. If the waste dump and the ore dump by the primary crusher are well removed from the pit boundaries, or if the haulage road beyond the pit has a steep gradient, it may be necessary to increase the truck fleet size to allow for the longer trip time per load.

The cost of haulage equipment including the accessory road maintenance equipment is given by:

$$\text{Haulage equipment cost} = N_t \times \$20{,}000 \, t^{0.9} \tag{2.61}$$

The capital costs for the production fleet are given in Figure 2.15.

Pit services
(a) Maintenance facilities. The size of maintenance facilities for repair and maintenance of open pit equipment depends primarily on the number and size of the mine haulage trucks, which in turn depends on the daily tonnage of ore and waste to be hauled. Repair and maintenance of the shovels and drills is normally performed on site by mobile repair vehicles.

The area in square feet required by the open pit maintenance shop (which should be located close to the open pit) is as follows:

$$\text{Area of open pit repair shot} = 360 \, T_p^{0.4} \tag{2.62}$$

Thus the areas of repair shops required for open pit mines are:

Mine size, tpd	10,000	20,000	40,000	80,000
Repair shop area, ft^2	14,300	18,900	25,000	33,000

The cost of constructing and equipping the shop is expressed by

$$\text{Cost of pit maintenance facilities} = \$6000 \, A^{0.6} t^{0.1} \tag{2.63}$$

(b) Communication and electrical distribution. This cost includes the installed costs for a surface telephone system with mobile and base radio units and one or more repeaters

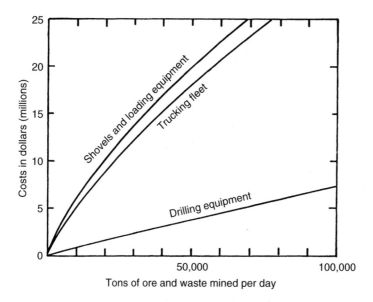

Figure 2.15. Costs for open pit equipment (O'Hara & Suboleski, 1992).

depending on the size of the mine. The electrical distribution includes the installed costs of primary substations, transmission lines, portable skid-mount transformers, and trailing cables, all of which depend on the size of the open pit mine as measured by the daily tons T_p of ore and waste mined.

$$\text{Cost of communications/electrical} = \$250\, T_p^{0.7} \tag{2.64}$$

(c) Fueling system. This cost includes the storage and services for diesel fuel, gasoline, lubricants, and coolants for the truck haulage fleet and mobile service vehicles

$$\text{Cost of refueling system} = \$28\, T_p^{0.7} \tag{2.65}$$

The open pit services costs are shown in Figure 2.16.

Mill associated capital costs

Mill site clearing and foundation preparation costs. The area A_c (in acres) to be cleared for the concentrator building, crusher building, substation, warehouse, and ancillary buildings is given by

$$A_c = 0.05\, T^{0.5} \tag{2.66}$$

In addition to this clearing, roads must be constructed from the nearest existing suitable road to provide access to the concentrator site, the hoisting plant, the proposed tailings basin, and the source of the water supply. Costs for clearing and access roads for the surface plant are estimated to be:

$$\text{Clearing costs} = \$2000\, A_c^{0.9} \text{ for lightly treed area with slopes of not more than 20\% gradient} \tag{2.67a}$$

$$\text{Access roads} = \$280{,}000 \text{ per mile for 30-ft (9-m) wide graveled road in mildly hilly region} \tag{2.67b}$$

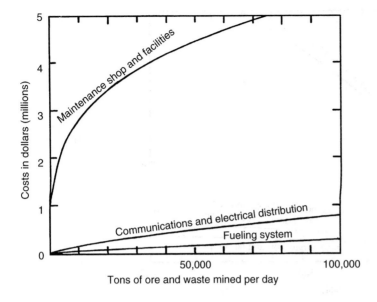

Figure 2.16. Costs for open pit services (O'Hara & Suboleski, 1992).

The formulas should be modified ±30% for more adverse or more favorable slope and tree growth conditions.

Soil overburden must be stripped wherever buildings and facilities are to be sited. The cost of stripping soil overburden D_o feet deep over and area of A acres will be:

$$\text{Cost of soil stripping} = \$1000 A^{0.8} D_o \qquad (2.68)$$

After the soil overburden is removed and the underlying rock or basal strata is exposed, this rock or strata will require localized removal, probably by drilling and blasting, to establish sound foundation conditions over levelled areas for the plant buildings and plant equipment. If there are C_u cubic yards of rock requiring drilling, blasting, and haulage to a dump site, this mass excavation will cost:

$$\text{Cost of mass excavation} = \$200 C_u^{0.7} \qquad (2.69)$$

for excavations of up to 100,000 yd³.

If the mass excavation is in rock that can be broken by ripping, the cost will be only 20% of that indicated.

When the mass excavation has been completed, detailed excavation to tailor the rock surface to the exact levels for pouring concrete foundations can be done. At the same time, suitable fill will be placed and compacted over level areas where deep trenches of soft soil have been removed. If there are C_d cubic yards of rock to be excavated by detailed excavation and F_c cubic yards of compacted fill to be placed, the cost will be:

$$\text{Excavated and fill compaction} = \$850 C_d^{0.6} + \$75 F_c^{0.7} \qquad (2.70)$$

Concrete costs for the foundations of the concentrator building, fine ore bins, and concentrator equipment probably will cost between \$350 and \$900/yd³, depending on whether the

126 *Open pit mine planning and design: Fundamentals*

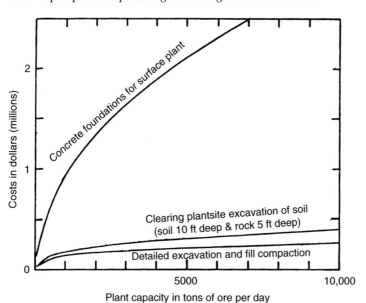

Figure 2.17. Costs for surface plant clearing, excavation and foundations (O'Hara & Suboleski, 1992).

concrete pour is for a simple form with little reinforcing steel or for a complex form that is heavily reinforced. The concrete cost may be significantly higher per cubic yard if concrete is scheduled to be poured in winter months when the temperature is below 40°F (4.4°C) and heating of aggregate and water and heating of concrete forms is required for sound concrete.

It is difficult to estimate the shape and volume of concrete forms before these forms have been designed, and hence concrete costs related to concrete volume are unreliable for preliminary estimation. Assuming no difficulties

$$\text{Approximate concrete foundation costs} = \$30,000 \, T^{0.5} \tag{2.71}$$

These different costs are shown in Figure 2.17 as a function of daily plant capacity.

Concentrator building. The costs of the concentrator building include all costs of constructing the building above the concrete foundations and enclosing the building, plus the cost of internal offices, laboratories, and changerooms. It does not include the cost of process equipment, piping, or electrical wiring, because these items are included in the costs of each functional area. The equipment in operating concentrators generates a substantial amount of heat and comfortable working conditions can be attained with little or no insulation, as long as the concentrator is located in a region with a mild climate. For flotation mills located in a mild climate

$$\text{Cost of building} = \$27,000 \, T^{0.6} \tag{2.72}$$

A 'mild climate' is defined as a region where the degree-days are about 7000 (in °F) or 4000 (in °C) per year. Weather stations usually record the 'degree-days' (°F × D, or °C × D), which represents the average number of days times the degrees that the temperature is below 65°F or 18°C. In hot climates, where freezing temperatures are not experienced, the building costs may be reduced by only partially enclosing the building and by locating thickeners and

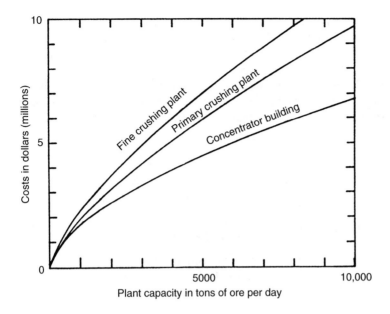

Figure 2.18. Costs for the concentrator building and crushing plant (O'Hara & Suboleski, 1992).

other hydrometallurgical equipment outside the building. In cold climates, the additional cost of insulation, heating, and snow loading is likely to increase the building cost by about 10% for each increase of 1800 (°F × D) above 7000 or 1000 (°C × D) above 4000.

Primary crushing plant with gyratory crusher. Open pit mines generally place the primary crusher on the surface outside the pit, within convenient conveying distance to the coarse or stockpile and the fine ore crushing plant. Open pit trucks normally dump the ore onto a grizzly mounted over the gyratory crusher which discharges crushed ore to a conveyor. Because of the headroom required to operate and discharge the crushed ore from a gyratory crusher, a substantial excavation and volume of concrete is required for the primary crusher plant. The cost of the primary crusher depends on the size and capacity of the gyratory crusher selected for crushing T_c tons of ore daily:

$$\text{Cost of gyratory crusher} = \$63\, T_c^{0.9} \tag{2.73}$$

The cost of excavating and concreting the foundations for the primary crusher, installing the crusher, construction of the truck dump and grizzly, plus the coarse ore conveyor and feeder under the crusher is:

$$\text{Cost of primary crushing plant} = \$15{,}000\, T_c^{0.7} \tag{2.74}$$

The cost of the crusher itself is not included.

Fine ore crushing and conveyors. This cost includes the crushing plant building, installed equipment and conveyors.

$$\text{Cost of fine ore crushing plant} = \$18{,}000\, T_c^{0.7} \tag{2.75}$$

Note: The cost may be 12% higher if the conveyors must be enclosed and heated.

Grinding section and fine ore storage. The fine storage bins must have sufficient live capacity to provide mill feed for at least the number of days that the crushing plant is idle per week. The cost of the fine ore bins will be proportional to the weight of steel used in constructing these bins, and the weight of steel will be proportional to $T^{0.7}$.

The size and cost of the grinding mills depend on the tons of ore to be ground daily by each mill, but they also depend on the hardness of the ore as measured by the work index and the fineness of grind that is required to attain the desired concentration and recovery of valuable minerals.

$$\text{Cost of grinding and bins} = \begin{cases} \$18{,}700\ T^{0.7} \text{ for medium hard ore} \\ \text{with a work index of 15, ground} \\ \text{to 70\% passing 200 mesh} \\ \$12{,}500\ T^{0.7} \text{ for soft ores ground} \\ \text{to 55\% passing 200 mesh} \\ \$22{,}500\ T^{0.7} \text{ for hard ores with a} \\ \text{work index of higher than 17, ground} \\ \text{to 85\% passing 200 mesh} \end{cases} \quad (2.76)$$

These costs are plotted in Figure 2.19 as a function of plant capacity.

Processing and related sections. The capital costs in this section cover the purchase and installation of all equipment required to concentrate or extract valuable minerals from the slurried ground ore, and process the concentrates or extracted minerals into dried solids or impure metals that are directly salable as dry concentrates, ingots of precious metals, uranium yellowcake, or impure metallic gravity concentrates of alloy metals. These capital

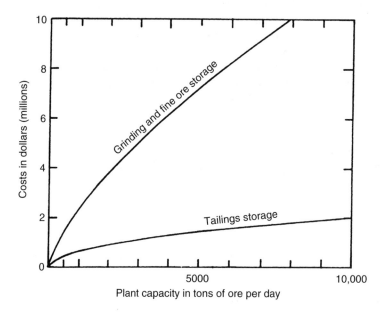

Figure 2.19. Costs for the grinding section, storage bins and tailings storage (O'Hara & Suboleski, 1992).

costs include equipment and tanks for thickening, filtering, precipitation, leaching, solvent extraction, etc., plus all process piping, electrical wiring, and process control.

Process costs for different types of ore by different methods are listed below:

1. High-grade gold ores leached by cyanidation, followed by zinc dust precipitation of gold by Merrill Crowe process, filtering, drying, and gold refining:

$$\text{Process capital costs} = \$60,200\ T^{0.5} \tag{2.77}$$

2. Low-grade ores, cyanide leaching, CIP (carbon-in-pulp) or CIL (carbon-in-leach) adsorption, refining:

$$\text{Process capital costs} = \$47,300\ T^{0.5} \tag{2.78}$$

3. High-grade gold ores with base metal sulfides; cyanide leaching, secondary flotation, carbon adsorption by CIP or CIL process, filtering, thickening, drying, and refining:

$$\text{Process capital costs} = \$103,200\ T^{0.5} \tag{2.79}$$

4. Simple low-grade base metal ores of copper with minor content of gold, which can be recovered as smelter credits. Flotation, thickening, filtering, and drying of auriferous copper concentrates:

$$\text{Process capital costs} = \$13,700\ T^{0.6} \tag{2.80}$$

5. Pyritic gold/silver ores where the precious metals are locked in the pyritic minerals. Differential flotation, selective roasting, recovery of deleterious materials, cyanidation, thickening, precipitation, filtering, and refining.

$$\text{Process capital costs} = \$180,000\ T^{0.5} \tag{2.81}$$

6. High-grade Cu/Pb ores, Cu/Zn ores, Pb/Zn ores, Cu/Ni ores. Recovery by differential flotation, thickening, filtering, and drying of separate concentrates:

$$\text{Process capital costs} = \$20,600\ T^{0.6} \tag{2.82}$$

7. Complex base metal ores containing at least three valuable metals, with recoverable minor amounts of precious metals; Cu/Zn/Pb ores, Pb/Zn/Ag ores, Cu/Pb/Ag ores, Cu/Zn/Au ores. Recovery by differential flotation, separate thickening, filtering, and drying of several concentrates and/or bulk concentrates.

$$\text{Process capital costs} = \$30,100\ T^{0.6} \tag{2.83}$$

8. Non-sulfide ores containing specialty metals such as columbium (niobium), tantalum, tungsten, and tin in minerals that do not respond to flotation, and which are separated by specialized gravity concentration methods:

$$\text{Process capital costs} = \$5000\ T^{0.7} \text{ to } \$13,000\ T^{0.7} \tag{2.84}$$

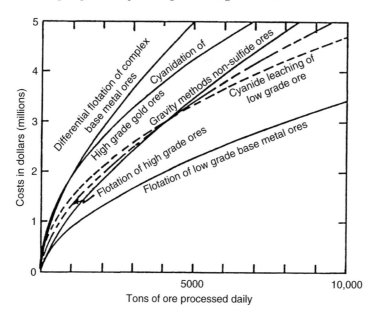

Figure 2.20. Processing section costs (O'Hara & Suboleski, 1992).

9. Uranium ores: acid leaching, countercurrent decantation, clarification, solvent extraction and yellowcake precipitation:

$$\text{Process capital costs} = \$150{,}000\, T^{0.5} \text{ to } \$200{,}000\, T^{0.5} \tag{2.85}$$

Figure 2.20 is a plot of these relationships.

Initial tailings storage. There are many aspects of tailings storage such as topography, distance from mill to tailings site, localized environmental concerns, etc., that could drastically alter the costs of tailings storage. If, however, all adverse aspects are absent, and a suitable tailings site is available within two miles of the mill, and the nature of the tailings does not have adverse environmental effects, the minimum cost of tailings storage may be:

$$\text{Minimum tailings storage cost} = \$20{,}000\, T^{0.5} \tag{2.86}$$

Very few mines have such favorable conditions, and if the area topography is steep or the environmental constraints are stringent, the tailings storage costs could be several times as high as the foregoing cost guide.

General plant capital cost
Water supply system. The cost of fresh water pumping plants, reclaim water plants, and provision for fire protection water supply, plus potable water supply, varies according the local topography and the proximity and nature of nearby sources of year-round supplies of water. If there is a suitable source of water within two miles of the mill, and the intervening topography is moderately level, the water supply system would cost:

$$\text{Cost of water supply system} = \$14{,}000\, T^{0.6} \tag{2.87}$$

The cost of the water supply system for the mine, mill, and plant (but excluding the mine water distribution system) will be much higher if the local topography is steep and rugged or if there are severe constraints on sources of fresh water.

Electrical substation and surface electrical distribution. The capital cost of electrical facilities for a mining/milling plant depends primarily on the size of the electrical peak load in kilowatts.

The peak load (PL) expressed in kilowatts per month and the average daily power consumption in kilowatt hours can be estimated from the following formulas:

$$\text{Peak load (PL)} = 78\, T^{0.6} \text{ for open pit mines milling } T \text{ tons of ore daily} \quad (2.88)$$

$$\text{Power Consumed} = 1400\, T^{0.6} \text{ for open pit mines with shovel and truck haulage to concentrator} \quad (2.89)$$

Typically, the concentrator and related facilities account for about 85% of the total power consumption for open pit mines and concentrators.

The cost of power supply depends on whether the power is generated by an existing electric utility or by a mine diesel-electric plant. Small mines in remote areas may be forced to generate their own electric power, because the cost of a lengthy transmission line from an existing utility may be too high due to the low peak load and low electric power consumption of a small mine.

If the mine is supplied with utility power, the cost of a utility substation with step-down transformers will be

$$\text{Cost of substation} = \$580\,(\text{PL})^{0.8} \quad (2.90)$$

The cost of installing low-voltage power distribution to the surface concentrator, crushing plant, and surface facilities, but excluding the distribution to the surface open pit is likely to be

$$\text{Cost of surface power distribution} = \$1150(\text{PL})^{0.8} \quad (2.91)$$

A diesel-electric generating plant may be required for a small mine in a remote area or by a larger mine supplied with utility power that may require a standby electric power plant for protection of vital equipment.

$$\text{Cost of diesel-electric plant} = \$6000(\text{PL})^{0.8} \quad (2.92)$$

General plant services. These costs include the costs of constructing, furnishing, and equipping the general administrative office, general warehouse, electrical and mechanical repair shop (for smaller mill equipment and services equipment), vehicle garages, changehouses, first aid and mine rescue stations, security stations plus general purpose vehicles, parking lots, and yard fencing.

The size of the buildings tends to depend on the number of employees served by each building. It is necessary to estimate the building size in square feet before estimating building cost, which will vary with the area of each type of building.

(a) Administrative office. The floor space per person tends to increase as the number of administrative and technical staff N_{at} becomes larger. This reflects the more complex records

132 Open pit mine planning and design: Fundamentals

of accounting and technical staff and the consequent requirement of more space for computer facilities, mining plans, and reference file facilities.

$$A = \text{Office are required in ft}^2 = 35 N_{at}^{1.3}$$

$$\text{Cost of office} = \$155 A^{0.9} \tag{2.93}$$

(b) Maintenance shop. Maintenance personnel N_{sv} will require about 85 ft^2/person for maintenance and repair of movable equipment from the mill and service departments.

$$\text{Cost of shop} = \$102(85 N_{sv})^{0.9} \tag{2.94}$$

(c) Mine changehouse. The mine changehouse requires about 24 ft^2/person on the mine payroll and includes the first aid station and mine rescue facilities.

$$\text{Changehouse cost} = \$125(24 N_{op})^{0.9} \tag{2.95}$$

(d) Surface warehouse. This should accommodate all supplies and spare parts for the mine, mill, and service facilities that must be kept indoors. Bulky supplies such as rough lumber, structural steel, etc., can be stored outdoors in most climates.

$$\text{Surface warehouse cost} = \$5,750 \, T^{0.4} \tag{2.96}$$

(e) Miscellaneous surface facilities. This includes general purpose vehicles and garages, security stations and fencing, parking lots, and miscellaneous services.

$$\text{Miscellaneous surface facilities} = \$10,000 \, T^{0.5} \tag{2.97}$$

Those general plant capital costs dependent on plant capacity are shown in Figure 2.21.

Mine project overhead costs
In addition to the direct costs for specific facilities for a mine project, which may total many millions of dollars, there are substantial costs and expenses involved in project design, general site costs, supervision and administration, and provision of working capital. These overhead costs may be estimated as a function of the total direct costs D in dollars.

Engineering. This includes the costs of feasibility studies, environmental impact studies, design engineering, equipment specifications and procurement, and specialized consulting services:

$$\text{Engineering costs} = \$2.30 \, D^{0.8} \tag{2.98}$$

General site costs. This includes construction camp costs, specialized construction equipment, and general construction site costs:

$$\text{General site costs} = \$0.310 \, D^{0.9} \tag{2.99}$$

Project supervision. This includes project supervision, scheduling and budgeting, and construction management:

$$\text{Project supervision costs} = \$1.80 \, D^{0.8} \tag{2.100}$$

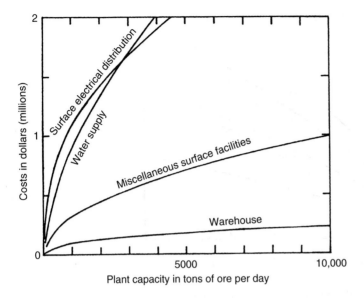

Figure 2.21. Costs of plant service facilities (O'Hara & Suboleski, 1992).

Administration. This includes local office administration by corporate owner's representatives, accounting and payment of general contractor, legal costs, plus preproduction employment of key operating staff:

$$\text{Administration costs} = \$1.50 \, D^{0.8} \qquad (2.101)$$

Project overhead costs as a percentage of direct project costs tend to vary depending on the size and complexity of the project. The lower percentages of 4 to 6% would be typical for $100 million projects and conventional technology, whereas the higher percentages of 8 to 11% would apply to smaller $10 million project that are technically novel or complex.

Working capital. The allowance for working capital for a mining project should be sufficient to cover all operating costs plus purchase of the initial inventory of capital spares and parts until revenue is received from smelters or purchasers of metallic products. The time period elapsing before receipt of revenue sufficient to pay imminent operating costs will vary depending on the smelter terms or marketing terms, but the typical allowance is about 10 weeks after the concentrator is operating at full capacity.

Typical working capital allowance is equal to the operating costs for 10 weeks after commissioning of concentrator plus cost of purchasing initial inventory of capital spares and parts.

Whenever the mine or mill design is based on extensive usage of reconditioned used equipment, there is a higher frequency of equipment downtime that requires additional time allowance of working capital; this will decrease the apparent savings of used equipment.

The total overhead costs as a summation of the different components are shown in Figure 2.22.

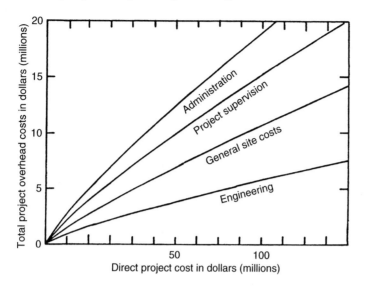

Figure 2.22. Project overhead versus direct costs (O'Hara & Suboleski, 1992).

Daily operating costs
Introduction. In this section the operating costs per day for each activity will be presented in the form

$$\text{Operating cost} = KT_I^x \tag{2.102}$$

The operating cost per ton can be derived from the given formula simply by dividing the operating cost per day by the tons mined (or processed) per day. If, for example,

$$\text{Operating cost per day} = 100\, T_I^{0.7} \tag{2.103}$$

the operating cost per ton is

$$\text{Operating cost per ton} = 100 \frac{T_I^{0.7}}{T_I} = 100\, T_I^{-0.3} \tag{2.104}$$

Pit operating costs. The operating costs of open pit mines depends on the size and numbers of drills, shovels, and trucks, which in turn is dependent on the tons per day of ore and waste. In most open pit mines mining low grade ore, there is little if any difference in the specific gravities, blasting characteristics, and drillabilities of ore or waste, and the haulage distance to the ore dump usually does not differ very much from the waste haulage distance. Consequently, the cost of mining a ton of ore will be virtually the same as the cost of mining a ton of waste.

The daily operating costs are:

$$\text{Drilling cost per day} = \$1.90\, T_p^{0.7} \tag{2.105}$$

$$\text{Blasting cost per day} = \$3.17\, T_p^{0.7} \tag{2.106}$$

$$\text{Loading cost per day} = \$2.67\, T_p^{0.7} \tag{2.107}$$

$$\text{Haulage cost per day} = \$18.07\, T_p^{0.7} \tag{2.108}$$

$$\text{General services cost per day} = \$6.65\, T_p^{0.7} \tag{2.109}$$

The open pit general services cost includes the cost of pit maintenance, road grading, waste dump grading, pumping, and open pit supervision, but it does not include the cost of primary crushing or electric power.

Concentrator operating costs. Although the gyratory crusher may be located at the edge of the open pit, the costs of operating it are grouped under milling costs (rather than as open pit operating costs) since it is the first stage of ore treatment.

The design of the milling flowsheet is usually optimized after extensive testwork on the types of processes tailored to the characteristics of the ore. At the preliminary feasibility stage however, the optimum processing requirements are not known with accuracy, and the costs of processing can only be approximately estimated.

The following cost guides are offered as rough estimates of crushing and concentrating costs per day.

(a) Primary crushing. This cost includes the cost of primary crushing, the cost of conveying the primary crushed ore to the coarse ore stockpile, plus operating costs of the coarse ore stockpile.

$$\text{Crushing costs per day} = \$7.90\, T^{0.6} \tag{2.110}$$

(b) Fine crushing and conveying. This includes fine crushing, conveying from coarse ore storage, and conveying to the fine ore bins.

$$\text{Fine crushing costs per day} = \$12.60\, T^{0.6} \tag{2.111}$$

(c) Grinding. This cost includes the fine ore bin storage and the rod mills, ball mills, and/or SAG (semiautogenous grinding) mills:

$$\text{Grinding section costs per day} = \$4.90\, T^{0.8} \tag{2.112}$$

(d) Process section. This includes the operating costs of all sections that involve concentration of ore by flotation or by gravity, leaching of metals from ore, thickening of slurries, ion exchange, precipitation, filtering, drying, and recovery of metallic concentrations, or deleterious materials that would otherwise penalize smelter revenue.

$$\text{Processing costs per day} = \begin{cases} \$65\, T^{0.6} \text{ for cyanidation of gold/silver ores} & (2.113a) \\ \$54\, T^{0.6} \text{ for flotation of simple base metal ores} & (2.113b) \\ \$34 \text{ to } \$41\, T^{0.7} \text{ for complex base metal ores varying in complexity} & (2.113c) \\ \$65\, T^{0.7} \text{ for uranium ores by leaching, CCD, solvent extraction, and precipitation} & (2.113d) \\ \$45\, T^{0.7} \text{ for nonfloatable nonsulfide ores responding to gravity separation} & (2.113e) \end{cases}$$

$$\text{Tailings costs per day} = \$0.92\, T^{0.8} \text{ for all concentrators} \tag{2.113f}$$

$$\text{Assaying costs per day} = \$1.27\, T^{0.8} \text{ for all concentrators} \tag{2.113g}$$

136 *Open pit mine planning and design: Fundamentals*

Supervision, maintenance, = $40.80T^{0.8}$ for all concentrators
and general costs per day (2.113h)

Processing costs would be decreased to 55% of those shown by the foregoing formulas when low-grade ore, typically mined by open pit mining, is being treated by a concentrator that rejects tailings at an early stage.

Other operating costs
(a) Electrical power. Expressions for the peak load and daily power requirements for the open pit, crushing plant and concentrator, etc. have been given earlier. The power cost for open pit mines and plants processing T tons of ore per day is

Cost of electric power = $145 T^{0.56}$ (2.114)

(b) Surface services. The daily cost of each person in the surface maintenance and general services departments is estimated to be $141 in wages and fringe benefits, plus an average cost of $16 in supplies consumed. If the number of maintenance and general services personnel is N_{sv}, then the daily costs of maintenance and general services departments is

Services cost per day = $157 N_{sv}$ (2.115)

The daily costs of the administrative and technical staff, including supplies and services required by them, plus fixed costs for local property taxes and legal fees paid by administrative services, are proportional to the number of staff N_{at}.

Each staff person is estimated to cost on the average $185 in salary per day, and to consume $37.60 in supplies and services per day.

Total cost per day for administrative and = $222.60 N_{at}$
technical staff salaries and supplies (2.116)

(c) Additional assistance in cost estimation. O'Hara and Suboleski (1992) suggest that the following sources/publications may be of assistance to those making cost estimates.
  1. *General Construction Estimation Standards*, 6 volumes, revised annually, published by Richardson Engineering Services, Inc., P.O. Box 1055, San Marcos, CA 92069.
  2. *Means Construction Costs*, revised annually, and published by Robert Snow Means Co., Inc., 100 Construction Pl., Kingston, MA 02364.
  3. *US Bureau of Mines Cost Estimating System Handbook*, 2 volumes, Information Circular 9142 (surface and underground mining), and Information Circular 9143 (mineral processing). Mining and milling costs are as of January 1984. The two volumes, IC 9143, are available from the Superintendent of Documents, U.S. Government Printing Office, Washington, DC 20402.
  4. *Canadian Construction Costs: Yardsticks for Costing*, revised annually; available from Southam Business Publications, 1450 Don Mills Rd., Don Mills, ON, Canada, M3B 2X7.
  5. *Mining and Mineral Processing Equipment Costs and Preliminary Capital Cost Estimations*, Special Vol. 25, 1982; published by The Canadian Institute of Mining and Metallurgy, 1 Place Alexis Nihon, 1210-3400 de Maisonneuve Blvd. W., Montreal, PQ, Canada H3Z 3B8.

A recent addition to this list of useful publications is "CAPCOSTS: A Handbook for Estimating Mining and Mineral Processing Equipment Costs and Capital Expenditures and Aiding Mineral Project Evaluations" by Andrew L. Mular and Richard Powlin. Special Vol. 47, 1998; published by The Canadian Institute of Mining, Metallurgy and Petroleum, Xerox Tower, 1210-3400 de Maisonneure Blud. W., Montreal, PQ, Canada H3Z 3B8. Since the passing of T. A. O'Hara, no one, unfortunately, has continued to update his useful curves.

2.4.6 Detailed cost calculations

Before discussing some techniques for estimating mining costs, a more detailed cost examination will be presented. With this as a basis, the costs will be grouped in several ways to show the dependence on accounting practice. The overall process is as follows:

Step 1. Given the annual production requirements for ore and waste plus the operating schedule, determine the daily production rate.

Step 2. Select a basic equipment fleet.

Step 3. Calculate the expected production rate for each type of equipment. Calculate the number of machines required. Determine the amount of support equipment needed.

Step 4. Determine the number of production employees required. Determine the number of support employees.

Step 5. Calculate the owning and operating costs for the equipment.

Step 6. Calculate the other costs.

Step 7. Calculate the overall cost per ton.

This procedure will be demonstrated using an example presented by Cherrier (1968). Although the costs are old, the process remains the same.

The cross-sections through the molybdenum orebody used in this example will be presented in Chapter 3. The initial mine design has indicated a pit for which
- Rock type is granite porphyry,
- 32,300,000 tons of waste,
- 53,000,000 tons of ore,
- Stripping ratio SR equals 0.6:1, and
- Average ore grade is 0.28% MoS_2.

The waste will be hauled by trucks to a dump area. The ore will be hauled by trucks to one of two ore passes. The ore is crushed and then transported by underground conveyor to the mill. The step-by-step process will be developed.

Step 1: Daily production rate determination. It has been decided that the annual production rate will be
- 3,000,000 tons ore and
- 2,000,000 tons waste.

The mine will operate 2 shifts/day, 5 days/week, 52 weeks per year, with 10 holidays.
 The ore and waste production per shift becomes:
- Ore: 6000 tons/shift,
- Waste: 4000 tons/shift.

138 *Open pit mine planning and design: Fundamentals*

Step 2: Selection of a consistent set of pit equipment. The major types of production equipment to be selected are:
- Drills,
- Shovels,
- Trucks.

The bench height has been chosen to be 30 ft. The basic equipment fleet selected consists of:
- 6 yd^3 electric shovels,
- 35 ton capacity rear dump trucks,
- Rotary drills capable of drilling 9 7/8″ diameter hole.

Step 3: Production capacity/Number of machines. Based upon a detailed examination of each unit operation, the following production rates were determined:
- Drills: 35 ft of hole per hour,
- Shovels (waste): 630 tons/hour,
- Shovels (ore): 750 tons/hour,
- Trucks (waste): 175 tons/hour,
- Trucks (ore): 280 tons/hour.

From this, of the number production units and required scheduling were determined for ore and waste:
- Ore
    1 drill (1 shift/day),
    1 shovel (2 shifts/day),
    3 trucks (2 shifts/day).
- Waste
    1 drill (1 shift/day),
    1 shovel (2 shifts/day),
    4 trucks (2 shifts/day).

The support equipment includes:
- 4 dozers (2 shifts/day),
- 2–5 yd^3 rubber tired front end loaders (2 shifts/day),
- 2 road graders (2 shifts/day),
- 1 water truck (2 shifts/day),
- 1 explosives truck (1 shift/day).

The reserve production equipment to be purchased is:
- 2–35 ton trucks,
- 1–5 yd^3 front end loader.

In case of shovel breakdown, a front end loader will substitute.

Step 4: Determine the number of production employees. A manpower scheduling chart is prepared such as is shown in Table 2.33. The overall numbers are summarized below.
- 1 assistant superintendent,
- 4 shift foreman,
- 4 shovel operators,
- 4 oilers,
- 14 truck drivers,
- 2 drillers,
- 2 driller helpers,

Table 2.33. Production employees manpower estimate.

(a) Salaried manpower estimate

Classification	Shift requirements																					Manshifts per week			No. of req'd men
	Sun.			Mon.			Tue.			Wed.			Thur.			Fri.			Sat.						
	A	B	C	A	B	C	A	B	C	A	B	C	A	B	C	A	B	C	A	B	C	A	B	C	Total	
Assist. supt.				1			1			1			1			1						5			5	1
Shift				2			2			2			2			2						10				
foreman					2			2			2			2			2						10		20	4
																								Total 5	

(b) Day pay manpower estimate

Classification	Shift requirements																					Manshifts per week			No. of req'd men
	Sun.			Mon.			Tue.			Wed.			Thur.			Fri.			Sat.						
	A	B	C	A	B	C	A	B	C	A	B	C	A	B	C	A	B	C	A	B	C	A	B	C	Total	
Shovel				2			2			2			2			2						10				
operator					2			2			2			2			2						10		20	4
Shovel				2			2			2			2			2						10				
oiler					2			2			2			2			2						10		20	4
Truck				7			7			7			7			7						35				
driver					7			7			7			7			7						35		70	14
Doser				4			4			4			4			4						20				
operator					4			4			4			4			4						20		40	8
Drill				2			2			2			2			2						10			10	2
operator																										
Drill				2			2			2			2			2						10			10	2
helper																										
Loader				2			2			2			2			2						10				
operator					2			2			2			2			2						10		20	4
Grader				2			2			2			2			2						10				
operator					2			2			2			2			2						10		20	4
Water truck				1			1			1			1			1						5				
driver					1			1			1			1			1						5		10	2
Truck				2			2			2			2			2						10				
spotters					2			2			2			2			2						10		20	4
Blaster				1			1			1			1			1						5			5	1
Blaster				1			1			1			1			1						5			5	1
helper																										
Crusher	1			1			1			1			1			1			1			7				
operator		1			1			1			1			1			1			1			7			
			1			1			1			1			1			1			1		7	21	4
Conveyer	1			1			1			1			1			1			1			7				
operator		1			1			1			1			1			1			1			7			
			1			1			1			1			1			1			1		7	21	4
Laborers				5			5			5			5			5						25				
					5			5			5			5			5						25	50	10
																								Total 68	

140 *Open pit mine planning and design: Fundamentals*

ADMINISTRATION	ENGINEERING	MINE	MAINTENANCE
1 Office Manager	1 Chief Engineer	1 Assistant Supt.	1 Shop Foreman
1 Clerk	2 Engineers	4 Shovel Operators	2 Shift Foreman
10 General and Administrative Personnel	2 Surveyors	4 Oilers	1 Master Mech.
	4 Survey Helpers	14 Truck Drivers	6 Mechanics
	1 Geologist	2 Drillers	2 Electricians
—	2 Samplers	2 Drillers Helpers	2 Geasers
13*	3 Draftsmen	1 Blaster	2 Oilers
	1 Safety Engineer	1 Blaster Helper	2 Welders
	1 Industrial Eng.	8 Dozer Operators	4 Helpers
	—	4 Grader Operators	2 Janitors
	17	2 Water Truck Drivers	—
		4 Shift Foreman	24
		4 Truck Spotters	
		4 Crusher Oper's	
* Includes Mine Superintendent		4 Crusher Oper's	
		10 Laborers	
		—	
		73	

Figure 2.23. Mine personnel requirements (Cherrier, 1968).

- 1 blaster,
- 1 blaster helper,
- 8 dozer operators,
- 4 loader operators,
- 4 grader operators,
- 2 water truck drivers,
- 4 truck spotters,
- 4 crusher operators,
- 4 conveyor operators,
- 10 laborers.

As can be seen, there are 73 production employees.

The crusher/conveyor part of the production system operates 7 days/week and 3 shifts per day. There is one crusher operator and one conveyor operator per shift. The ore passes contain enough storage capacity so that the mill can run 7 days/week even though the mine runs 5.

Step 5: Determine the number of other employees. There are four basic departments at the mine:
- Administration,
- Engineering,
- Mine,
- Maintenance.

The overall structure is shown in Figure 2.23. The administration and engineering departments work straight day shift. The maintenance department works two shifts/day, 5 days per week as shown in the manpower chart (Table 2.34).

Step 6: Determine the payroll cost. Table 2.35 summarizes the basic annual wage figures for the various job classifications. Fringe benefits amounting to 25% are not included in this table.

Table 2.34. Maintenance employees manpower estimate.

(a) Salaried manpower estimate

Classification	Shift requirements																						Manshifts per week			No. of req'd men
	Sun.			Mon.			Tue.			Wed.			Thur.			Fri.			Sat.							
	A	B	C	A	B	C	A	B	C	A	B	C	A	B	C	A	B	C	A	B	C		A	B	C	Total	
Shop foreman	1			1			1			1			1										5			5	1
Shift foreman	1	1		1	1		1	1		1	1		1	1			1						5	5		10	2
Master mechanic	1			1			1			1			1										5			5	1
																										Total	4

(b) Day pay manpower estimate

Classification	Shift requirements																						Manshifts per week			No. of req'd men
	Sun.			Mon.			Tue.			Wed.			Thur.			Fri.			Sat.							
	A	B	C	A	B	C	A	B	C	A	B	C	A	B	C	A	B	C	A	B	C		A	B	C	Total	
Mechanic	3			3	3		3	3		3	3		3	3			3						15	15		30	6
Electrician	1			1	1		1	1		1	1		1	1			1						5	5		10	2
Machinist	1			1	1		1	1		1	1		1	1			1						5	5		10	2
Greaser	1			1	1		1	1		1	1		1	1			1						5	5		10	2
Oiler	1			1	1		1	1		1	1		1	1			1						5	5		10	2
Welder	1			1	1		1	1		1	1		1	1			1						5	5		10	2
Helper	2			2	2		2	2		2	2		2	2			2						10	10		20	4
Janitor	1			1	1		1	1		1	1		1	1			1						5	5		10	2
																										Total	22

Step 7: Determine the operating costs. The total operating costs for the various unit operations include materials, supplies, power and labor. These are summarized in Table 2.36. The labor cost includes the 25% fringe benefits.

Note that there are three different units which have been used to express the costs ($/ft, $/hr and $/ton). Some conversion factors are required to obtain the desired common values of $/ton. These are:
 – Drilling: 27.2 tons/ft,
 – Loading (ore): 750 tons/hr,
 – Loading (waste): 630 tons/hr,

142 *Open pit mine planning and design: Fundamentals*

Table 2.35. Annual payroll (Cherrier, 1968).

Classification	Number required	Unit annual age	Total annual age
Administration			
Mine superintendent	1	$25,000	$25,000
Office manager	1	13,000	13,000
Clerk	1	7,000	7,000
Pro-rated G&A personnel			
(average salary $8,000)	10	8,000	80,000
Total	13		$125,000
Engineering			
Chief engineer	1	$16,000	$16,000
Mining engineers	2	9,500	19,000
Safety engineer	1	8,000	8,000
Industrial engineer	1	8,500	8,500
Geologist	1	7,500	7,500
Surveyors	2	7,500	15,000
Surveyor helpers	4	5,500	22,000
Draftsmen	3	6,000	18,000
Samplers	2	5,500	11,000
Total	17		$125,000
Mine			
Ass't mine superintendent	1	$18,000	$18,000
Shift foreman	2	11,000	22,000
Shovel operators	4	8,500	34,000
Shovel oilers	4	6,000	24,000
Truck drivers	14	7,000	84,000
Drillers	3	8,000	24,000
Driller helpers	3	6,500	19,500
Blaster	1	8,500	8,500
Blaster helper	1	6,500	6,500
Dozer operators	8	8,000	64,000
Loader operators	6	8,000	48,000
Grader operators	4	8,000	32,000
Water truck driver	2	7,000	14,000
Crusher operators	4	7,500	30,000
Conveyor operators	4	7,500	30,000
Truck spotters	4	5,500	22,000
Laborers	10	5,000	50,000
Total	73		$526,500
Maintenance			
Shop foreman	1	$14,000	$14,000
Shift foreman	2	10,500	21,000
Master mechanic	1	11,000	11,000
Mechanics	6	8,500	51,000
Electricians	2	8,000	16,000
Machinist	2	7,000	14,000
Greaser	2	6,000	12,000
Oiler	2	6,000	12,000
Welder	2	6,500	13,000
Helpers	4	5,500	22,000
Janitor	2	5,000	10,000
Total	26		$196,000

Table 2.36. Mine operating costs (Cherrier, 1968).

Unit operation	Cost category		$/ft	$/hr	$/ton
Drilling	Bits		0.100		
	Maintenance, lub, repairs		0.200		
	Supplies		0.015		
	Fuel		0.250		
	Labor		0.133		
Blasting	Ammonium nitrate				0.0125
	Fuel oil				0.0010
	Primers				0.0020
	Primacord, caps, fuse				0.0030
	Labor				0.0030
Loading	Repairs, maintenance, supplies	cost		7.20	
	Power	per		1.65	
	Lubrication	shovel		0.10	
	Labor			8.75	
Hauling	Tires cost	cost		1.65	
	Tire repairs	per		0.25	
	Repairs, maintenance	truck		2.70	
	Fuel			0.96	
	Oil, grease			0.40	
	Labor			3.60	
Auxiliary	Tracked dozer	materials			0.015
equipment	Rubber tired dozer	supplies			0.013
	Front end loader	repairs			0.010
	Graders	fuel			0.020
	Water truck				0.005
	Labor				0.040
Secondary drilling and blasting	MEP				0.005
	Labor				0.005
Crushing	MEP				0.020
	Labor				0.010
Conveying	MEP				0.018
	Labor				0.013
Snow removal	MEP				0.005
	Labor				0.005

- Hauling (ore): 750 tons/hr (3 trucks),
- Hauling (waste): 630 tons/hr (4 trucks).

To simplify the presentation, only two cost categories will be carried further. These are operating labor and MEP (material, expenses and power) which includes all of the rest. The ore-waste separation will be made. The results are summarized in Tables 2.37 and 2.38.

It is noted that the labor cost is strictly for operating labor and does not include repair labor nor does it include supervision.

Step 8: *Determine the capital cost and the ownership cost for the equipment.* The capital cost for the pit equipment is given in Table 2.39.

During the life of the mine, some of the equipment will have to be replaced. This is indicated in Table 2.40. As can be seen, the original equipment falls into three lifetime

Table 2.37. Summary of direct operating expenses for ore (Cherrier, 1968).

Unit operation	Cost ($/ton) MEP	Cost ($/ton) Labor	Total
Drilling	0.021	0.005	0.026
Blasting	0.019	0.003	0.022
Loading	0.012	0.012	0.024
Hauling	0.024	0.014	0.038
Auxilliary	0.063	0.040	0.103
Secondary D&B	0.005	0.005	0.010
Crushing	0.020	0.010	0.030
Conveying	0.018	0.013	0.031
Snow removal	0.005	0.005	0.010
Total	0.187	0.107	0.294

Table 2.38. Summary of direct operating expenses for waste (Cherrier, 1968).

Unit operation	Cost ($/ton) MEP	Cost ($/ton) Labor	Total
Drilling	0.021	0.005	0.026
Blasting	0.019	0.003	0.022
Loading	0.014	0.018	0.032
Hauling	0.038	0.029	0.067
Auxilliary	0.063	0.040	0.103
Secondary D&B	0.005	0.005	0.010
Snow removal	0.005	0.005	0.010
Total	0.165	0.105	0.270

groups (5, 10 and 20 years). These are summarized below:

Life (yrs)	Total original cost ($)
5	861,000
10	396,000
20	1,401,000

The equipment ownership cost consists of two parts:
(1) Depreciation,
(2) Average annual investment cost.

Straight line depreciation with zero salvage value is assumed. Thus the average equipment depreciation per year is $281,850. The average annual investment (AAI) is calculated using the following formula.

$$\text{AAI} = \frac{n+1}{2n} \text{CC} \qquad (2.117)$$

where n is the life (yrs) and CC is the capital cost ($).

The depreciation categories of 5, 10 and 20 years will be used. Thus the AAI for $n = 5$ yrs is

$$\text{AAI (5 yrs)} = \frac{6}{10} \times \$861{,}000 = \$516{,}600$$

Table 2.39. Capital costs and annual depreciation for the mining equipment (Cherrier, 1968).

Operation and Item	No.	Unit cost	Total cost	Life (yrs)	Annual deprec.
Drilling					
Bucyrus Erie 40-R					
Rotary drill	2	$75,000	$150,000	10	$15,000
Blasting					
Powder truck	1	8,000	8,000	10	800
Excavating					
P&H shovel model 1400	2	288,000	576,000	20	28,800
Haulage and Transportation					
Haulpak 35 ton trucks					
rear dump	9	58,000	522,000	5	104,400
Crusher	2	150,000	300,000	20	15,000
Conveyor	1	395,000	395,000	20	19,750
Chloride spray system	2	15,000	30,000	20	1,500
Pit maintenance					
Bulldozers	3	50,000	150,000	5	30,000
Rubber tired dozers	3	40,000	120,000	5	24,000
Front end loaders	2	45,000	90,000	10	9,000
Road graders	3	25,000	75,000	10	7,500
Sprinkler truck	2	7,000	14,000	10	1,400
Fuel truck	1	9,000	9,000	10	900
Mobile maintenance	1	15,000	15,000	10	1,500
Miscellaneous					
Pick up trucks	4	3,000	12,000	5	2,400
9 passenger trucks	4	4,000	16,000	5	3,200
Truck crane	1	35,000	35,000	10	3,500
Radio system	1	7,000	7,000	5	1,400
Power	1	100,000	100,000	20	5,000
Contingencies			34,000	5	6,800
(20%)(170,000)					
Total			$2,658,000		$281,850

The total AAI is

AAI = $1,469,930

To obtain the average annual investment cost AAIC a percent P (expressed as a ratio) is applied. Included in P are interest, taxes and insurance.

$$\text{AAIC} = P \times \text{AAI} \qquad (2.118)$$

In this case 10 percent will be used. Hence

AAIC = 0.10 × $1,469,930 = $146,993

The average annual equipment ownership cost becomes

Ownership cost = Depreciation + AAIC

$$= \$281{,}850 + 146{,}993 \qquad (2.119)$$

= $428,843

Step 9: Calculation of other capital expenditures (mine). The other capital expenditures at the mine include those for the required mine buildings and the costs associated with the mine development period.

Table 2.40. Equipment replacement schedule and cost (Cherrier, 1968).

Item	No.	Year of replacement	Total cost
Haulpak truck	9	6th, 11th, 16th	$522,000
Bulldozers	3	6th, 11th, 16th	150,000
Rubber tired dozers	3	6th, 11th, 16th	120,000
Pick up trucks	4	6th, 11th, 16th	12,000
9 Passenger trucks	4	6th, 11th, 16th	16,000
Radio system	1	6th, 11th, 16th	7,000
Contingencies			34,000
Total			$861,000
Bucyrus Erie drill	2	11th	$150,000
Powder truck	1	11th	8,000
Front end loader	2	11th	90,000
Road grader	3	11th	75,000
Sprinkler truck	2	11th	14,000
Fuel truck	1	11th	9,000
Mobile maintenance	1	11th	15,000
Truck crane	1	11th	35,000
Total			$396,000

The following list includes all required mine buildings, and their required capital outlays.

Office building (Admin. and eng.)	
Building (brick)	$180,000
Equipment	
Office (desks, files, typewriters, etc.)	40,000
Engineering (calculators, survey equip. etc.)	25,000
Total	$245,000
Change house	
Building (brick)	$30,000
Equipment (lockers, showers, fixtures, etc.)	$5,000
Total	$35,000
Warehouse	
Building (sheet metal)	$40,000
Fuel station	
Building (sheet metal)	$6,000
Equipment (tanks, pumps)	$10,000
Total	$16,000
Maintenance shops	
Building (sheet metal)	$105,000
Equipment (electrical, cranes, tools, etc.)	$130,000
Total	$235,000

Powder hopper
    Purchase and erection $12,000
Total for all buildings $583,000
Contingencies (10%) $58,300

Total estimated capital expenditures for buildings $641,300

This will be depreciated over the 20 year mine life. The development expenses are as follows:

Schedule of capital expenditures
1969
    Development
        Extended exploration $205,000
        Access road and site preparation 200,000
        Preliminary stripping 400,000
        Tunnel excavation 537,000
    Equipment
        Surface installations 641,000
        Equipment 1,232,000

    Total $3,215,000

1970
    Development
        Preliminary stripping $675,000
        Ore pass excavation 415,000
    Equipment
        Crusher 300,000

    Total $1,390,000

1971
    Equipment
        Mill 12,000,000
        Equipment 1,126,000

    Total $13,126,000

Development drilling 205,000
Access road 200,000
Development stripping 1,075,000
Tunnel excavation 537,000
Ore pass excavation 415,000

    Total $2,432,000

Both the buildings and the development costs are considered as sunk. They will be recovered against the ore produced.

Step 10: *Calculation of milling costs.* In this case the open pit ore tonnage (averaged over 7 days) of 8,600 tpd will go to a mill having a capacity of 34,400 tpd. The additional 25,800 tpd will be supplied by an underground mine. Since the total mill investment is $48,000,000 the share for the open pit operation is $12,000,000. This will be depreciated over the 20 year mine life. The ownership cost (depreciation plus average annual investment) per year is

$$\text{Ownership cost} = \$600,000 + \$630,000$$
$$= \$1,230,000/\text{year}$$

The primary crushing will be done at the mine, therefore the first step in milling will be the secondary crushing. Operating costs for milling are estimated to be the following (includes administration and overhead):

Labor	$0.13/ton
Material	0.39/ton
Power	0.26/ton
Total	$0.78/ton

Step 11: Expression of the mining costs. There are a variety of ways by which the mining costs can be expressed. A series of cases will be presented to illustrate this.

Case 1. Direct operating costs. The simplest way is to examine the direct operating mining costs for ore and waste. These are:

Ore
$\quad$ MEP = $0.187/ton
$\quad$ Labor = $0.107/ton
$\quad$ Total = $0.294/ton

Waste
$\quad$ MEP = $0.165/ton
$\quad$ Labor = $0.105/ton
$\quad$ Total = $0.270/ton

The weighted average costs are:
$\quad$ MEP = $0.178/ton
$\quad$ Labor = $0.106/ton
$\quad$ Total = $0.284/ton

Using the average costs, the percent breakdown is:
$\quad$ MEP = 63%
$\quad$ Labor = 37%

Case 2. Total operating cost. The labor costs used in Case 1 did not include all of the people involved under the mine category. The total labor expense is equal to $530,500. Including fringes of 25%, this figure increases to $663,125. The labor, MEP (ore), and MEP (waste) costs incurred in the mining of 5,000,000 tons are

Labor (with fringes) = $663,125
MEP (ore) = $0.187 \times 3,000,000 = \$561,000$
MEP (waste) = $0.165 \times 2,000,000 = \$330,000$

The total MEP cost is $891,000. The costs per ton of material moved are

Labor = $0.133/ton
MEP = $0.178/ton
Total = $0.311/ton

In terms of percentages one finds

Labor = 43%
MEP = 57%

Case 3. Direct operating cost plus maintenance. The basic annual maintenance labor cost is $196,000. With the 25% fringe this becomes

Maintenance labor = $245,000

The overall mine plus maintenance labor cost is therefore

Labor (Mine + Maintenance) = $663,125 + $245,000
$$= \$908,125$$

The average operating cost per ton of material moved is

$$\text{Operating cost} = \frac{\$891,000 + \$908,125}{5,000,000}$$
$$= \$0.360/\text{ton}$$

The percent distribution is now

MEP = 49.8%
Labor = 50.2%

Case 4. All mine related costs. The costs for the engineering and administration departments can now be added. The annual wages (including fringes) are $312,500. The associated MEP is $150,000. Thus the average operating cost per ton of material moved is

$$\text{Operating cost} = \frac{\$1,799,125 + \$312,500 + \$150,000}{5,000,000}$$
$$= \frac{\$2,261,625}{5,000,000} = \$0.452/\text{ton}$$

The cost breakdown is:

$$MEP = \$1,041,000$$
$$Labor = 1,220,625$$

The percent distribution is:

$$MEP = 46\%$$
$$Labor = 54\%$$

Step 11: *Productivity calculations.* The productivity in terms of tons per manshift can now be calculated. It will be assumed that each employee works 250 shifts per year in producing the 5,000,000 tons of total material. The productivity will change depending on the number of departments included:

$$\text{Mine productivity} = \frac{5,000,000}{73 \times 250} = 274 \, t/ms$$

$$(\text{Mine} + \text{Maintenance}) \text{ productivity} = \frac{5,000,000}{99 \times 250} = 202 \, t/ms$$

$(\text{Mine} + \text{Maintenance} + \text{Engineering} + \text{Administration})$ productivity

$$= \frac{5,000,000}{129 \times 250} = 155 \, t/ms$$

Step 12: *Mine ownership costs/ton.* As indicated earlier, in addition to the operating costs, there are a number of capital costs to be charged against the material moved. These are:
– Equipment ownership costs;
– Development costs;
– Mine buildings.

The development and mine buildings will be amortized over the total amount of material moved. In this case

$$\text{Cost/ton} = \frac{\$641,300 + \$2,432,000}{84,800,000} = \$0.036$$

The equipment ownership cost is

$$\text{Cost/ton} = \frac{\$428,843}{5,000,000} = \$0.086$$

Hence the ownership cost/ton is

$$\text{Ownership cost/ton} = \$0.036 + \$0.086 = \$0.122$$

Step 13: *Total mining cost.* The total mining cost is equal to the total operating cost plus the ownership cost. In this case it is

$$\text{Total mining cost} = \$0.452/\text{ton} + \$0.122/\text{ton} = \$0.574/\text{ton}$$

Note that the ownership cost here is about 27% of the operating cost and 21% of the total mining cost.

Step 14: Milling cost. As was indicated earlier, the mills operating costs per ton is

Mill operating cost = $0.78/ton

Mill recovery = 90%

The mill ownership cost/year is the depreciation/year + AAIC/year.
In this case

Depreciation/year = $600,000

AAIC/year = $630,000

Total = $1,230,000

The ownership cost per ton milled is

Ownership cost/ton = $0.41

This is about 53% of the mill operating cost and 34% of the total milling cost.

Step 15: Profitability estimate. The revenues are attributable to the ore and all the costs must now be charged against the ore as well.

Revenue per ton of ore

Average grade	= 0.28% MoS_2
Recovery	= 90%
Mo contained	= 60% of recovered MoS_2
Price per lb of contained Mo	= $1.62

Revenue = $2000 \times 0.0028 \times 0.90 \times 0.60 \times 1.62 = \$4.90$/ton

Cost per ton of ore

Mining of ore	= $0.452
Stripping of waste	= 0.452 × 0.6 = $0.271
Mine operating (Mining + Stripping)	= $0.72
Mine overhead (27% of mine operating)	= $0.20
Mill operating	= $0.78
Milling overhead (53% of mill operating)	= $0.41
Total cost	= $2.11/ton
Profitability per ton	= $4.90 − $2.11 = $2.79/ton

2.4.7 Quick-and-dirty mining cost estimates

After going through the detailed cost estimate, one might ask about ways in which ballpark estimates could be made. Even if existing operations are not willing to release cost data they sometimes will provide data concerning cost distributions, production rates (ore and waste), total personnel and stripping ratio. If so then the following formulae may be used for making preliminary cost estimates (Pfleider & Wheaton, 1968):

$$\text{Labor cost per ton} = \frac{\text{Estimated average labor cost per man shift (including fringe benefits)}}{A} \quad (2.120)$$

$$\text{Total operating cost per ton} = \frac{\text{Labor cost per ton}}{B} \quad (2.121)$$

$$\text{Total mining cost per ton} = \text{Total operating cost}(1 + C) \quad (2.122)$$

where A is the estimated average tons per manshift (t/ms),

$$B = \frac{\text{Labor cost}}{\text{Total operating cost}}, \quad \text{as an estimated value}$$

$$C = \frac{\text{Ownership costs}}{\text{Total operating costs}}, \quad \text{as an estimated value}$$

The information contained in the previous section will be used in this example.
Given:

Productivity	= 155 t/ms
Operating cost percentage: Labor	= 54%
MEP	= 46%
Overhead cost	= 27% of operating cost
Stripping ratio	= 0.6:1
Estimate: base labor cost/hour	= $4
Average fringe benefits	= 25%

Calculation:

Labor cost/shift $= 8 \text{ hr} \times \$4/\text{hr} \times 1.25 = \$40/\text{shift}$

Labor cost/ton $= \dfrac{\$40}{155} = \$0.26/\text{ton}$

Total operating cost/ton $= \dfrac{0.26}{0.54} = \$0.48/\text{ton}$

Total mining cost/ton $= 0.48\,(1+0.27) = \$0.61$/ton

Total mining cost/ton of ore $= 0.61(1+SR)$
$\phantom{\text{Total mining cost/ton of ore }}= 0.61(1.6) = \$0.97$/ton ore

As can be seen, this compares well with the value of $0.92 obtained in the long calculation.

2.4.8 *Current equipment, supplies and labor costs*

Current prices for mining equipment and supplies can easily be obtained from the individual manufacturer or supplier. Lists of such suppliers are published yearly in special publications of the *Engineering/Mining Journal (E/MJ), Canadian Mining Journal* and *Coal Age*. Several companies publish guidebooks which are extremely useful for making preliminary cost estimates. The *Cost Reference Guide For Construction Equipment* published by Equipment Guide-Book Company, Palo Alto, California (PRIMEDIA, 2004), contains ownership and operating costs for a wide range of mining equipment. The table of contents is given below:

Section	Description
1	Introduction
2	Index
3	Air tools
4	Crushing and conveying
5	Asphalt and bituminous
6	Compaction
7	Concrete
8	Drilling
9	Tractors and earthmoving
10	Excavating
11	Motors and generators
12	Hoists and derricks
13	Lifting
14	Marine
15	Pile driving
16	Pumping
17	Road maintenance
18	Shop tools
19	Trailers
20	Trucks
21	Tunnel
22	Miscellaneous
23	Cost and production formulas

Table 2.41 extracted from the section on electric powered rotary blasthole drills provides an indication of the detailed information provided. Operating costs for selected mining equipment are provided in Table 2.42. The costs are updated on a regular basis.

154 *Open pit mine planning and design: Fundamentals*

Table 2.41. Specifications and costs associated with Bucyrus Erie crawler mounted rotary blasthole drills (Cost Reference Guide for Construction Equipment, 2003, PRIMEDIA, 2004).

Equipment specifications

Model	Power	Maximum pull down capacity	Maximum rotary hole size	Compressor CFM	CWT
35-R	Diesel	50,000 lbs	9″	900	–
39-R	Diesel	90,000 lbs	12-¼″	1645	1900
49-R	Electric	120,000 lbs	16″	2600	3400
59-R	Electric	140,000 lbs	17-½″	3450	4050

Hourly ownership and overhaul expenses

		Ownership			Overhaul	
Model	Econ. hours	Depreciation ($)	CFC ($)	Overhead ($)	Labor ($)	Parts ($)
35-R	12,800	17.46	5.37	1.89	13.37	9.43
39-R	18,000	63.08	25.50	23.67	33.58	34.80
49-R	18,000	73.33	29.18	27.23	25.25	35.04
59-R	18,000	87.08	34.65	32.67	25.25	41.61

Hourly field repair and fuel expenses

Model	Labor ($)	Parts ($)	Elec./Fuel ($)	Lube ($)	Tires ($)	GEC ($)	Total operating costs/hr ($)	Total hourly costs/hr ($)
35-R	25.58	16.48	18.02	4.08	0.00	1.34	65.60	113.12
39-R	62.70	60.65	0.00	11.03	0.00	6.07	140.45	321.08
49-R	50.33	56.25	0.00	12.60	0.00	5.63	125.69	304.35
59-R	50.33	66.80	0.00	14.96	0.00	6.68	139.65	346.14

Notes:
1. All models include water injection, cab heater, second pipe rack and automatic lubrication for upper and lower works.
2. Rates do not include drill bits.
3. Operating costs do not include the cost of electricity for electric powered models.
4. Cost of money rate = 5.250%.
5. Mechanic's wage = $34.85 (including fringe benefits).

Western Mine Engineering of Spokane, Washington provides their *Mining Cost Service* to subscribers. The *Service* contains the following cost and related information:
– Electric power;
– Natural gas;
– Transportation;
– Labor;
– Cost indexes;
– Supplies and miscellaneous items;
– Equipment;
– Smelting;

Table 2.42. Operating costs for selected mining equipment (Cost Reference Guide, 2003, PRIMEDIA, 2004).

Equipment	Make	Model	Operating cost ($/hr)
Standard crawler dozer	Caterpillar	D8R	41
		D9R	57
		D10R	76
		D11R	114
4-WD articulated wheel loader	Caterpillar	972G	30
		980G	35
		988G	50
		990	74
		992G	97
Articulated frame grader	Caterpillar	12H	18
		14H	26
		16H	35
Crawler mounted rotary blasthole drills, diesel powered*	DrilTech	D25KS	56
		D40KS	69
		D45KS	70
		D50KS	71
		D60KS	125
		D75KS	126
		D90KS	174
Mechanical drive rear dump trucks	Caterpillar	773E	52
		775E	55
		777D	71
Electric drive rear dump trucks	Euclid	R170	94
		R190	110
	Hitachi	EH3000	108
		EH3500	125
		EH4000	84
		EH4500	96
	Komatsu	630E	115
		730E	112
		830E	67
		930E(2000)	120
		930 E-2	89
	LeTourneau	T2190	98
		T2200	106
		T2240	55
	Unit Rig	MT3300	45

* Rates do not include drill bits.
None of the operating costs include the cost of the operator.
Diesel cost = $1.50/gallon.
Mechanics wage = $34.85/hr (including fringe benefits).

– Cost models;
– Taxes;
– Miscellaneous.

As an example of the content Table 2.43 presents the 2003 labor rates for a large (555 employees, 42,900,000 tons of ore/year) copper mine in Arizona. Similar information is provided on mines in other states involving a range of minerals. Table 2.44 also taken from the Service, provides cost data on rotary bits.

156 *Open pit mine planning and design: Fundamentals*

Table 2.43. Typical labor rates (2003) for an open pit copper mine in Arizona. Mining Cost Service (2004).

Job classification	Hourly wage base	Job classification	Hourly wage base
Tailings Operations Team Lead	$20.30	Mill Electrician	18.05
SXEW Team Leader	19.83	Crusher Operator	18.06
Senior/Lead Mill Operator	19.82	Tailings Mechanic	17.84
Senior/Lead Mine Mechanic	19.87	Mill Process Operator	17.95
Senior/Lead Mine Electrician	19.77	Pwr Plant Eng	16.58
Utilities Mechanic	18.94	SXEW Mechanic	17.69
Utility Plant Electrician	18.85	Mill Mechanic	17.58
Senior/Lead Mill Mechanic	19.16	Leach Utilityman	16.00
Blast Operator	18.72	Mine Mechanic	18.03
SXEW Electrician	18.98	Tailings Header Operator	15.95
Utility Plant Operator	18.39	Mill Electrical Assistant	15.60
Mine Electrician	18.61	Mill Production Operator	16.95
Heavy Equipment Mechanic	18.55	Mine Mechanical Assistant	15.49
Equipment Operator	18.33	Mill Mechanical Assistant	15.24
Shovel Operator	18.35	EW Production Operator	15.64
Tailings Process Operator	18.54	Haul Truck Driver	16.12
Drill Operator	18.07	Production Operator	14.50
SXEW Plant Operator	18.00	SX Utility	14.45
Concentrate Leach Plant Operator	17.88	Mill Operations Utility	12.60

Wages increased 3%, effective 3/1/04.

Benefits			Vacation	
		% paid	Years Service	Days Vacation
Life Insurance	yes	100 1 × annual		
Medical Insurance	yes	100	1	10
Dental Insurance	yes	100	3	10
Vision Insurance	yes	100	5	15
Retirement Benefits	yes	100	10	15
401K plan – company matches 100% of the first 3% of			15	20
employee's contribution and matches the next 2% of			20	25
employee's contribution at 50%.			25/plus	25
Sick Leave	no			
Disability Income Ins.	yes			
ESOP	no			

10 days paid holidays per year.

Shift differential: evening – $.45/hour; night – $.45/hour.

Cost of benefits: Mandated – 13.5% of wages; voluntary – 35% of wages.

Other benefits: 401K plan, tuition aid, employee assistance program, additional life and dependent life available at minimal cost to employee. Optional health care and day care spending accounts.

Incentive bonus plan: Company pays a bonus based on meeting financial goals. Target payout is 6% of wages – minimum is 0% and maximum payout is 12%.

Typical prices for explosives and blasting accessories are provided in Table 2.45. In performing an economic analysis, capital costs for mining equipment are needed. Such costs are provided by Western Mine Engineering in the convenient form shown in Table 2.46.

The costs are updated on a regular basis.

Table 2.44. Price of large diameter rotary blast hole bits (Western Mine Engineering, 2003).

Cutter type	Diameter (ins)	Unit	Price($)
Tungsten carbide	7-7/8″	ea.	2716
	9″	ea.	3225
	9-7/8″	ea.	4301
	10-5/8″	ea.	4952
	12-1/4″	ea.	5932
	15″	ea.	9314
	16″	ea.	9897
Steel-tooth	9-7/8″	ea.	2272
	10-5/8″	ea.	2567
	12-1/4″	ea.	3098
	15″	ea.	4786
	17-1/2″	ea.	5895

Table 2.45. Price of explosives and blasting accessories (Western Mine Engineering, 2003).

1. Bulk explosives	Strength (cal/cm^3)	Unit	Price ($)
*ANFO	739	100 lb.	36.00
*Site mixed emulsion	820–945	100 lb.	34.00
*Repumpable emulsion	770–900	100 lb.	48.00–71.00
*High density emulsion	815–975	100 lb.	48.00–54.00
2. Non-electric shock tube, milliseconds delays	Length (ft.)	Unit	Price ($)
	8	100	195
	12	100	203
	16	100	219
	20	100	243
	30	100	303
	40	100	365
	50	100	425
	60	100	515
	80	100	663
	100	100	780
	120	100	916
3. Detonating cord	Strength (grains/ft)	Unit	Price ($)
	7.5	1000 ft.	101
	15	1000 ft.	117
	18	1000 ft.	117
	25	1000 ft.	149
	40	1000 ft.	171
	50	1000 ft.	192
4. Cast primers	Weight (lbs)	Unit	Price($)
	1/3	ea.	2.11
	1/2	ea.	2.69
	3/4	ea.	2.97
	1	ea.	3.54
	2	ea.	6.38
	3	ea.	9.43
	5	ea.	15.47

Table 2.46. Capital cost of mining equipment (Western Mine Engineering, 2003).

1. Rotary drills, electric	Hole size (ins)	Price ($1000)
	9″ to 12-¼″	2100
	9-⅞″ to 16″	2700
	10-¾″ to 17-½″	3100
2. Graders	Blade width (ft)	Price ($1000)
	12	250
	14	360
	16	512
	24	1280
3. Wheel loaders	Bucket capacity (yd^3)	Price ($1000)
	11	973
	12.5	1450
	14	1385
	17	1433
	22	1990
	26	2678
	33	3470
	53	5000
4. Electric cable shovels (rock)	Bucket capacity (yd^3)	Price ($1000)
	11	2400
	16	3250
	20	3800
	27	6250
	34	7129
	55	9150
5. Trucks, rear dump (mechanical drive)	Capacity (tons)	Price ($1000)
	85	900
	100	1000
	150	1540
	170	1560
	220	1910
	260	2240
	360	3400
6. Trucks, rear dump (electric drive)	Capacity (tons)	Price ($1000)
	120	1641
	140	1725
	170	1964
	255	2000
	320	2800
	360	3270
7. Crawler tractors (dozers) (with ripper)	Horsepower	Price ($1000)
	260	492
	285	514
	330	451
	350	581
	460	805
	520	929
	770	1478

REFERENCES

Anonymous 1909a. Monthly Average Metals Price. *The Engineering and Mining Journal* 88(11): 1054.
Anonymous 1909b. Utah Copper Quarterly Report. *The Engineering and Mining Journal* 88(11): 1028.
Benning, I. 2000. Bankers' perspective of mining project finance. JSAIMM. 100(3): 145–152.
Berry, C.W. 1984. Economic Evaluation Phases of a Mining Prospect. *Mine Feasibility – Concept to Completion.* (G.E. McKelvey, compiler). Northwest Mining Association.
Bhappu, R.R., & J. Guzman. 1995. Mineral investment decision making. E/MJ. 196(7): 36–38.
Bigler, L.M., & J.L. Dobra. 1994. Economic analysis of a hypothetical Mexican gold mine. E/MJ. 195(7): 16EE–16FF.
Blossom, J.S. 1991. Molybdenum. *1989 Minerals Yearbook.* (Volume 1): 719–728. Bureau of Mines, U.S. Government Printing Office.
BLS (Bureau of Labor Statistics) 1991. Employment, Hours, and Earnings, United States 1989–90. Bulletin 2370, vol. 1(3). U.S. Department of Labor.
BLS 1992a. Productivity Measures for Selected Industries and Government Services. Bulletin 2406 (4). U.S. Department of Labor.
BLS 1992b. *Producer Price.* Indexes – Data for 1991 (9).
BLS 1993a. *Producer Price.* Indexes – Data for October 1992 (2).
BLS 1993b. *Employment and Earnings.* (6). U.S. Department of Labor.
BLS 2003. Productivity for mine production and nonsupervisory workers. ftp://ftp.bls.gov/pub/special. requests/opt/dipts/oaehhiin.txt.
BLS 2004a. Average hourly earnings of mine production workers(SIC base). http:/www.bls.gov/data/. Series Report. EEU10000006, EEU10100006, EEU10101006, EEU10102006, EEU10120006, EEU10122006, EEU10140006, EEU10142006.
BLS 2004b. Average hourly earnings of production workers (NAICS base). http://www.bls.gov/data/. Employment, Hours and Earnings from the Current Employment Statistics Survey (National). Create customized tables (one screen). Average hourly earnings of production workers, natural resources and mining, and mining (except oil and gas).
BLS 2004c. Producer price indexes for construction machinery and equipment. http://www.bls.gov/data/. Producer price index commodity data. Most requested statistics. Construction machinery and equipment. Machinery and equipment.
BLS 2004d. Labor productivity expressed in output per hour. http://data.bls.gov/PDQ/outside.jsp?/survey=ip/.
BLS 2005a. What is NAICS? http://www.bls.gov/sae/saewhatis.htm
BLS 2005b. NAICS 11 & 21: Natural resources and mining. http://www.bls.gov/iag/natresmining.htm
BLS 2005c. CES Series for Natural Resources and Mining Under NAICS. http://www.bls.gov/ces/naturalres/htm.
Bourne, H.L. 1994. What it's worth: A review of mineral royalty information. Mining Engineering. 46(7): 641–644.
Bourne, H.L. 1995. What it's worth: A review of mineral royalty information. Mining Engineering. 47(7): 654–657.
Bourne, H.L. 1996. What its worth: A review of mineral royalty information. Mining Engineering. 48(7): 35–38.
Camm, T.W. 1991. Simplified cost models for prefeasibility mineral evaluations. USBM. *Information Circular* 9298.
Camm, T.W. 1994. Simplified cost models for prefeasibility mineral evaluations. Mining Engineering. 46(6): 559–562.
Camus, J. 2002. Management of mineral resources. Mining Engineering. 54(1): 17–25.
Camus, J.P. 2002. Management of Mineral Resources: Creating Value in the Mining Business. SME. Littleton, Colorado. 107 pages.
Cavender, B. 2001. Leadership vs management: Observations on the successes and failures of cost-reduction programs. Mining Engineering. 53(8): 45–50.
Cherrier, T.E. 1968. A Report on the Ceresco Ridge Extension of the Climax Molybdenite Deposit. MSc. Thesis. University of Minnesota.

CMJ (Canadian Mining Journal) 1986. *Reference Manual and Buyers Guide.* Southam Business Publications.
CMJ 2004. Mining Source book. Canadian Mining Journal. www.canadian mining journal.com
Coal Age 1992. Buyers Guide. *Maclean Hunter Publication* (6): 57–107.
Dataquest Inc. *Cost Reference Guide for Construction Equipment.* Palo Alto, Calif.: Equipment Guide-Book Company.
Davis, G. 1996. Using commodity price projections in mineral project valuation. Mining Engineering. 48(4): 67–70.
Davis, G. 1998. One project, two discount rates. Mining Engineering. 50(4): 70–74.
Davis, G.A. 1995. (Mis)use of Monte Carlo simulations in NPV analysis. Mining Engineering. 47(1): 75–79.
Dixon, S. 2004. Definition of economic optimum for leaching of high acid consuming copper ores. Pre-print Paper 04–04. 2004 SME Annual Meeting, Feb 23–25. Denver, CO.
Dran, J.J. Jr. & H.N. McCarl 1977. An examination of interest rates and their effect on valuation of mineral deposits. *Mining Engineering* 29(6): 44–47.
Dysinger, S.M. 1997. Capital budgeting: forecasting the future. Mining Engineering. 49(9): 35–38.
E/MJ 1986. *Average Annual Metal Prices.* 1925–1985 (3) 27.
E/MJ 1992. Buyers Guide 1992. *E&MJ* 193(10): BG1-45.
E/MJ 1993. *Prices from the American Metal Market.* Friday January 15, 1993: 74WW.
ENR (Engineering News Record) 1976–1992. *Annual Summary.*
ENR 1993. *ENR Market Trends.* (2).
ENR 2005. ENR's Cost Indexes. www.enr.com.
Equipment Quest. 2003. Cost Reference Guide for Construction Equipment. Primedia Publications.
Finlay, J.R. 1908. Cost of producing the world's supply of copper. *The Engineering and Mining Journal* 86(7): 165–168.
Finlay, J.R. 1909. Cost of pig iron made from Lake Superior ores. *The Engineering and Mining Journal* 87(15): 739–745.
Frimpong, S. & J.M. Whiting. 1996. The pricing of mineral investment options in competitive markets. In Surface Mining 1996 (H.W. Glen, editor): 37–42. SAIMM.
Gallagher, T.C. 1982. Developing a construction cost escalation index for the U.S. mining and metals industry. *Cost Engineering* 24(3): 147–152.
Gentry, D.W. 1988. Minerals project evaluation – An overview. *Trans. Instn. Min. Metall. (Sect. A: Min. Industry)* 97(1): A25–A35.
Gentry, D.W. 1998. 1998 Jackling Lecture: Constrained prospects of mining's future. Mining Engineering. 50(6): 85–94.
Hajdasinski, M.M. 1993. A generalized true rate of return of a project. 24th International Symposium on Application of Computers and Operations Research in the Mineral Industry: 2(280–287). Montreal: CIMM.
Hajdasinski, M.M. 2000. Internal rate of return (IRR) – a flawed and dysfunctional project evaluation criterion. Mining Engineering. 52(8): 49–56.
Hajdasinski, M.M. 2000. Internal rate of return (IRR) as a project-ranking tool. Mining Engineering. 52(11):60–64.
Hansen, B.D. 1999. Reducing Newmont's operating costs and increasing reserves and production at a mid-sized international mining company. Pre-print Paper 99-031. 1999 SME Annual Meeting, Mar 1–3. Denver, CO.
Heath, K.C.G. 1988. Mine costing. *Trans. Inst. Min. Metall. (Sect. A: Min. industry)* 97(1): A1–A8.
Hrebar, M.J. 1977. Financial analysis *Mineral Industry Costs.* (J.R. Hoskins & W.R. Green, editors): 205–224. Northwest Mining Association.
Hummel, R. 2001. The road to NPV+. JSAIMM. 101(8): 411–420.
Humphreys, D. 1989. The metal markets since 1983 – and into 1989. *International Mining* 6(1): 14–22.
Huss, M.P. 1984. Marketing planning. *Mine Feasibility Concept to Completion* (G.E. McKelvey, compiler). Northwest Mining Association.
Industrial Minerals 1992. *Prices.* 303(12): 66.
IRS (U.S. Internal Revenue Service) 1988a. Business Expenses. Publication 535 (Rev. Nov. 88).
IRS 1988b. Depreciation. Publication 534. (Rev. Nov. 88).
IRS 2004. Publication 535 (2004) Business Expenses. http://www.irs.gov/publications/p535/ch10.html

IRS 2004. Publication 946 (2004): How to Depreciate Property. http://www.irs.gov/publications/p946/index.html
IRC 2004C. Publication 535 (2004): Depletion. http:/www.irs.gov/publications/P535/ch10.html
Jackling, D.C. 1909. Operations of Utah Copper Co. During 1908. *The Engineering and Mining Journal* 88(6): 1185–1188.
Ji, C. & Y. Zhang. 1999. Gray-correlation analysis of production cost. 28th International Symposium on Application of Computers and Operations Research in the Mineral Industry: 173–176. Colorado School of Mines: CSM.
Jolly, J.L.W. & D.L. Edelstein 1991. Copper. *1989 Minerals Yearbook.* (Volume 1): 333–386. Bureau of Mines. U.S. Government Printing Office.
Kirkman, R.C. 1979. Discussion of the H.M. Wells paper. *Mining Engineering* 31(4): 406–407.
Kirk, W.S. 1998. Iron ore reserves and mineral resource classification systems. Skillings Mining Review. June 6. Pp 4–11.
Kitco 2005. Charts and Data. http://www.kitco.com/scripts/hist_charts/yearly_graphs.cgi
Kovisars, L. 1983. Mine and mill labor productivity study shows wide range of values. *Mining Engineering* 36(11): 1536–1537.
Kuck, P.H. & C.M. Cvetic 1991. Iron ore. *1989 Minerals Yearbook.* (Volume 1): 521–564. Bureau of Mines, U.S. Government Printing Office.
Labys, W.C. 1977. Minerals commodity modeling – the state of the art. *Mineral Policies in Transition.* (J.H. DeYoung, Jr., editor). The Council of Economics of AIME (11): 80–106.
Laing, G.J.S. 1976. *An Analysis of the Effects of State Taxation on the Mining Industry In the Rocky Mountain States.* MSc. Thesis T-1900. Colorado School of Mines.
Lawrence, M.J. 2000. DCF/NPV modeling: Valuation practice or financial engineering? Pre-print Paper 00-58. 2000 SME Annual Meeting, Feb 28–Mar 1. SLC, UT.
Lawrence, R.D. 2000. Should discounted cash flow projections for the determination of fair market value be based solely on proven and probable reserves? Pre-print Paper 00-64. 2000 SME Annual Meeting, Feb 28–Mar 1. SLC, UT.
Lawrence, R.D. 2001. Should discounted cash flow projections for the determination of fair market value be based solely on proven and probable reserves. Mining Engineering. 53(4): 51–56.
Lesemann, R.H. 1977. Commercial and price considerations in mine feasibility and cost studies. *Mineral Industry Costs.* (J.R. Hoskins & W.R. Green, editors): 195–204. Northwest Mining Association.
Lewis, F.M., C.K. Chase & R.B. Bhappu 1978. Copper production costs update. *Mining Engineering* 30(4): 371–373.
Lewis, P.J. & C.G. Streets 1978. An analysis of base-metal smelter terms. *Proceedings of the 11th Commonwealth Mining and Metallurgical Congress:* 753–767. Hong Kong: IMM.
Lillico, T.M. 1973. How to maximize return on capital when planning open pit mines. *World Mining* 9(6): 22–27.
LKAB. 2005. Personal communication. www.lkab.com
Manssen, L.B. 1983. Financial evaluation of mining projects: Is 'common practice' enough? *Mining Engineering* 35(6): 601–606.
Mason, P.M. 1984. Capital and operational planning for open pit in a modern economy. *18th International Symposium on Application of Computers and Mathematics in the Mineral Industry*: 791–802. London: IMM.
Metal Bulletin 1993. Metal Prices (1). Trans-world Metals (USA) Ltd.
Metal Bulletin 2004. www.metalbulletin.com. Personal communication.
Metals Week 1993. Weekly Prices (2).
Metals Week 2005. July. www.plats.com.
Miner, R.B. 2003. Using long-term international concentrate agreements in major project financing. CIM Bulletin 96(1074):56–64.
Mining Cost Service 1989. *Model Smelter Schedules.* Spokane, WA: Western Mine Engineering.
Mining Cost Service 1992. *Prices for Large Diameter Blast Hole Bits.* Spokane, WA: Western Mine Engineering.
Mining Sourcebook 2004. Canadian Mining Journal. www.canadianminingjournal.com
Mireku-Gyimah, D. & J.O. Darko 1996. Economic risk analysis of mineral projects in Ghana – A simulation approach. In Surface Mining 1996 (H.W. Glen, editor): 71–76. SAIMM.

Mular, A.L. 1978. Mineral Processing Equipment Costs and Preliminary Capital Cost Estimates. Special Volume 18. CIM.
Mular, A.L. 1982. Mining and Mineral Processing Equipment Costs and Preliminary Capital Cost Estimates. Special Volume 25. CIM.
Mular, A.L. & R. Poulin 1998. CAPCOSTS: A Handbook for Estimating Mining and Mineral Processing Equipment Costs and Capital Expenditures and Aiding Mineral Project Evaluations. Special Volume 47. CIM.
Murdy, W.W., and R.R. Bhappu. 1997. Risk mitigation in global mining development. Mining Engineering. 49(9): 30–34.
Noble, A.C. 1979. Price forecasting and sensitivity analysis for economic analysis of final pit limit. *Open Pit Mine Planning and Design* (J.T. Crawford and W.A. Hustrulid, eds): 127–133. New York: SME-AIME.
O'Hara, T.A. 1980. Quick guides to the evaluation of orebodies. *CIM Bulletin (no. 814)* 73(2): 87–99.
O'Hara, T.A. 1981. Chapter 6: Mine evaluation. *Mineral Industry Costs* (J.R. Hoskins, compiler): 89–99. Northwest Mining Association.
O'Hara, T.A. & S.C. Suboleski 1992. Chapter 6.3: Costs and cost estimation. *SME Mining Engineering Handbook* 2nd Edition (Volume 1) (H.L. Hartman, Senior Editor): 405–424. SME.
O'Neil, T.J. 1982. Mine evaluation in a changing investment climate (Parts 1 and 2). *Mining Engineering* Part 1 vol. 34(11): 1563–1566, Part 2 vol. 34(12): 1669–1672.
Osborne, T.C. 1985. What is the cost of producing copper? *International Mining* 2(3): 14–15.
Paulsen, K.R. 1981. Environmental and regulatory costs of mining projects. *Mineral Industry Costs* (J.R. Hoskins, compiler): 27–32. Northwest Mining Association.
Petersen, U. & R.S. Maxwell 1979. Historical mineral production and price trends. *Mining Engineering* 31(1): 25–34.
Pfleider, E.P. & G.F. Weaton 1968. Chapter 13.4: Iron ore mining. *Surface Mining* (E.P. Pfleider, editor): 917–921. New York: AIME.
Pierson, G. 1997. Commodity prices from a monetary perspective. Mining Engineering. 49(10): 75–80.
PRIMEDIA 2004. Cost Reference Guide for Construction Equipment. Primedia Publications.
Randal, R.L. 1979. Economic modeling and market forecasting *Computer Methods for the '80's*: 938–949. Society of Mining Engineers of AIME.
Randall, R.L. 1980. Sources of cyclical variation in the mineral industries. *Proceedings of the Council of Economics* AIME: 11–118.
Rendu, J-M. 2003. SME meets with the SEC – resources and reserves reporting discussed. Mining Engineering. July. P35.
Roberts, M.C. 2000. Cycles in mineral prices. Pre-print Paper 00-73. 2000 SME Annual Meeting, Feb 28–Mar 1. SLC, UT.
Roberts, M.C. 2002. Cycles in metal prices. Mining Engineering. 54(2): 40–48.
Rosta, J. 1992. Gold, silver. *Mining Engineering* 44(5): 449.
Runge, I.C. 1998. Mining Economics and Strategy. SME. Littleton, Colorado. 295 pages.
Simonsen, H. & J. Perry 1999. Risk identification, assessment and management in the mining and metallurgical industries. JSAIMM. 99(6): 321–328.
Skillings Mining Review 1993a. *Iron Ore Prices.* 82(2).
Skillings Mining Review 1993b. *Rail Freight Rates on Iron Ore and Pellets Per Gross Ton.* 82(2).
Skillings Mining Review 1993c. *Rail and Lake Tariff Rates on Iron Ore and Pellets Per Gross Ton.* vol. 82(2).
Skillings 2005. Metal/ore prices. August www.skillings.net.
Smith, L.D. 1999. The argument for a "bare bones" base case. CIM Bulletin 92(1031): 143–150.
Soderberg, A. 1930. *Mining method and costs at the Utah Copper Co., Bingham Canyon, Utah. US Bureau of Mines Information Circular 6234.*
Soma, J.L. 1977. Estimating the cost of development and operating of surface mines. *Mineral Industry Costs* (J.R. Hoskins & W.R. Green, editors): 67–90. Northwest Mining Association.
Southam Mining Group 1993. *Mining Source Book.* Southam Mining Group. Don Mills, Ontario.
Staff 1987. *Bureau of Mines Cost Estimating System Handbook. Part 1. Surface and Underground Mining.* U.S. Bureau of Mines Information Circular 9142: 631 pages.
Staff 1987. *Bureau of Mines Cost Estimating System Handbook. Part 2. Mineral Processing.* U.S. Bureau of Mines Information Circular 9143: 566 pages.

Stebbins, S. 2002. Evaluating the economics of development alternatives early in the project feasibility process. 30th International Symposium on Application of Computers and Operations Research in the Mineral Industry: 587–600. Alaska: SME.

Stermole, F.J. 1979. Economic evaluation of open pit mines. *Open Pit Mine Planning and Design* (J.T. Crawford & W.A. Hustrulid, editors): 136–144. New York: Society of Mining Engineers of the AIME.

Stermole, F.J. & J.M. Stermole 1987. *Economic Evaluation and Investment Decision Methods.* (6th Edition). Investment Evaluations Corporation.

Stinnett, L.A., and J.A. Wehinger. 1995. Labor productivity and its impact on product price. Mining Engineering. 47(7): 658–659.

Taylor, H.K. 1977. Mine valuation and feasibility studies. *Mineral Industry Costs*: 1–17. Spokane, WA: Northwest Mining Association.

Taylor, H.K. 1988. Discussion of the article 'Mine Costing' by K.G.G. Heath. *Trans. Inst. Min. Metall. (Sect A: Min industry)* 97(10): A196–197.

Taylor, H.K. 1992. Private communication.

Tilton, J.E. 1997. Surviving in the competitive and global mining industry. Mining Engineering. 49(9): 27–29.

Torries, T.F. 1998. NPV or IRR? Why not both? Mining Engineering. 50(10): 69–73.

Ulatowski, T. 1982. Inflationary impacts on project evaluation; real and nominal hurdle rates. *Mineral Industry Costs* (J.R. Hoskins, compiler): 245–248. Northwest Mining Association.

U.S. Bureau of the Census 1992. *Statistical Abstract of the United States.* (112th edition) Washington, D.C.

U.S. Census Bureau. 2005. 1997 NAICS Matched to 1987 SIC – Mining. http://www.census.gov/epcd/naics/NAICS21.htm

USGS. 2005a. Aluminum Statistics and Information. http://minerals.usgs.gov/minerals/pubs/commodity/aluminum/.

USGS. 2005b. Copper Statistics and Information. http://minerals.usgs.gov/minerals/pubs/commodity/copper/.

USGS. 2005c. Lead Statistics and Information. http://minerals.usgs.gov/minerals/pubs/commodity/lead/.

USGS. 2005d. Zinc Statistics and Information. http://minerals.usgs.gov/minerals/pubs/commodity/zinc/.

USGS. 2005e. Gold Statistics and Information. http://minerals.usgs.gov/minerals/pubs/commodity/gold/.

USGS. 2005f. Platinum-Group Metals Statistics and Information. http://minerals.usgs.gov/minerals/pubs/commodity/platinum/.

USGS. 2005g. Silver Statistics and Information. http://minerals.usgs.gov/minerals/pubs/commodity/silver/.

USGS. 2005h. Molybdenum Statistics and Information. http://minerals.usgs.gov/minerals/pubs/commodity/molybdenum/.

USGS. 2005i. Nickel Statistics and Information. http://minerals.usgs.gov/minerals/pubs/commodity/nickel/

USGS. 2005j. Tin Statistics and Information. http://minerals.usgs.gov/minerals/pubs/commodity/tin/.

USGS. 2005k. Iron Ore Statistics and Information. http://minerals.usgs.gov/minerals/pubs/commodity/iron_ore/.

USGS. 2005l. Historical statistics for mineral and material commodities in the United States. http://minerals.usgs.gov/minerals/pubs/of01-006/

USGS. 2005m. Metal prices in the United States through 1998. http://minerals.usgs.gov/minerals/pubs/metal_prices/.

Viera, F.M. 1993. How to choose projects by means of risk analysis. 24th International Symposium on Application of Computers and Operations Research in the Mineral Industry: 2(324–331). Montreal: CIMM.

Warnken, D.E. 2005. A case study in valuation of non-producing mineral rights. Pre-print Paper 05-113. 2005 SME Annual Meeting, Feb 28–Mar 2. SLC, UT.

Wells, H.M. 1978. Optimization of mining engineering design in mineral valuation. *Mining Engineering* 30(12): 1676–1684.

Werner, A.B.T. & C. Janakiraman 1980. Smelter contracts and their place in mineral reserve evaluation. *Proceedings of the Council of Economics* AIME: 85–92.

Western Mine Engineering 2003. Mine and Mill Equipment Costs – Estimator's Guide.

Western Mine Engineering 2004. Mining Cost Service.

Williamson, D. 1997. Into the next millennium – The long bull market in minerals. Mining Engineering. 49(6): 61–64.

Wright, E.A. 1993. A simulation model for assessing mining investment incentives. 24th International Symposium on Application of Computers and Operations Research in the Mineral Industry: 2(332–342). Montreal: CIMM.
Yernberg, W.R. 2000. Copper 99: Mine closures help copper prices improve. Mining Engineering. 52(1): 36–36.

REVIEW QUESTIONS AND EXERCISES

1. What is meant by the term "future worth"?
2. What is meant by the term "present value"?
3. Assume that the interest rate is 5% compounded daily. What would be the equivalent simple interest rate?
4. If $1000 is placed in a bank savings account earning 5% annually, what is the value after 10 years?
5. Would you rather have $10,000 today or $20,000 in 5 years? Assume an interest rate of 8%.
6. Would you rather have $10,000 today or receive payments of $1000/year for the next 20 years?
7. What is meant by the term 'payback period'?
8. If you borrowed $10,000 from the bank today, how long would it require to repay the loan at $1000/year? Interest rate of 5%.
9. What is meant by the term 'rate of return'?
10. You invest $1000 today. It will be repaid in 15 equal payments of $200 over a period of 15 years. What is your rate of return? If you could put the money in a certificate of deposit paying 5%, what should you do?
11. What is meant by the term 'cash flow'?
12. Redo the cash flow example in Table 2.1 assuming that the capital cost in year 0 was −$300 and in year 1 was −$200. What is the total cash flow?
13. For problem 12, using a discounting rate of 15%, what is the net present value?
14. What is meant by the term 'discounted cash flow'?
15. What is meant by the term 'discounted cash flow rate of return'?
16. For problem 12, what is the DCFROR?
17. Redo the cash flow example in Table 2.2 assuming that the capital cost in year 0 is $150 rather than $100.
18. What is meant by depreciation? How is it included in a cash flow calculation?
19. Redo the example in section 2.2.9 assuming that the salvage value is $50.00.
20. What is meant by the 'depletion allowance'?
21. What are the two types of depletion? Describe how they differ.
22. How is depletion included in a cash flow calculation?
23. Redo the cash flow including depletion example given in section 2.2.1 assuming a nickel ore rather than silver.
24. What are the different factors that must be considered when performing a cash flow analysis of a mining property?
25. What is the weight of a
    - short ton
    - metric ton
    - long ton

Mining revenues and costs 165

26. What is meant by the expression 'long ton unit'? Short ton unit? Metric ton unit? Provide an example of a mineral which is sold in this way.
27. Assume that the price for iron ore fines is 45¢ U.S./mtu. What is the value of a concentrate running 63.45% Fe?
28. How much is a troy ounce? What metals are sold by the troy ounce?
29. How is mercury sold?
30. In what form is molybdenum normally sold?
31. What is meant by F.O.B.? What is meant by C.I.F.? Why is it important that you understand the meaning of these terms?
32. Select a metal from Table 2.6 and determine the price today.
33. Select a non-metal from Table 2.7 and determine the price today.
34. Select an ore from Table 2.8 and determine the price today.
35. Compare the price of the Cleveland-Cliffs, Inc. iron ore pellets given in Table 2.9 with the price today.
36. Table 2.10 gives the rail tariff rates for iron ore and pellets. Try to find similar rail tariff data which are applicable today.
37. Table 2.11 gives the lake freight tariff rates for iron ore and pellets. Try to find similar data which are applicable today.
38. Select a metal from Table 2.15 and plot the price versus time. Describe the price pattern observed. Try to attribute reasons for the major fluctuations observed.
39. Using the endpoint data in problem 38, what would be the average annual percent price increase?
40. Using the data in problem 38, predict what the price might be in 5 years, 10 years and 20 years in the future.
41. Plot the data from Table 2.16 for the same metal selected in problem 38. Describe the price pattern with time. What reasons can you provide, if any, for the major fluctuations?
42. Using the data from problem 41, what price might you expect in 1 year, 5 years, 10 years and 20 years?
43. The price of iron ore fines is given in Table 2.17 for the period 1900 through 2004. Plot the data from the year that you were born until today. Any explanation for the pattern?
44. Using the data provided in Figure 2.8 as an example, what is meant by value-added at the mine site? How could the value-added be further increased? Why might this be interesting to a company or to a country?
45. Trend analysis is one way to predict future prices based upon historical data but it must be carefully done. To illustrate the types of problems that can occur, select one of the metal price – time data sets for analysis. Fit a second order polynomial to the data. Fit a third order polynomial to the data. Fit an arbitrary higher order polynomial to the data. Compare the results obtained using each equation if you extrapolate 5 years, 10 years and 20 years into the future? What are your conclusions?
46. What is the difference between extrapolation and interpolation?
47. What is meant by an econometric model? When applied in mining, what factors might be included?
48. What is meant by 'net smelter return'? How is it calculated?
49. What are the major elements of a smelter contract?
50. What is meant by 'at-mine-revenue'? Why is it important?
51. Redo the example in section 2.3.5 using the Model Smelter Schedule given in Table 2.22. Make the necessary assumptions.

166 *Open pit mine planning and design: Fundamentals*

52. What are some of the different cost categories used at a mining operation? Describe the costs included in each.
53. What is meant by 'general and administrative' cost? What does it include?
54. How might the operating cost be broken down?
55. What are meant by fixed costs? Should they be charged to ore, waste, or both?
56. Discuss the value of information such as contained in Table 2.23 to a student performing a senior pre-feasibility type analysis.
57. Table 2.23 presents the mining costs at the Similkameen operation. Determine a percentage distribution. Present these costs in a pie-type plot.
58. Compare the information presented in Tables 2.23 and 2.24 for the Similkameen/Similco operation.
59. What is the value of data such as contained in Table 2.24?
60. Compare the Similkameen and Huckleberry mines based upon the data provided in Tables 2.23 and 2.25.
61. In section 2.4.3, the estimation of costs based upon the escalation of older costs is described. Estimate the cost of the building in year 2004 based upon the different ENR indices. Which value do you think might be the most appropriate? Why?
62. Select one of the categories in Table 2.27. What has been the average hourly wage increase over the last 20 years?
63. What was the reason to change from the SIC to the NAICS base?
64. Using Table 2.29, how much would you estimate the price of a mining truck has increased over the past 10 years?
65. Why is it important to include the change in productivity when escalating older costs? How can you do this?
66. In 1980, O'Hara published the curve relating mine/mill project capital cost to the milling rate. Discuss how you might try to produce a similar curve today.
67. Actual data for a number of open pit mining operations have been provided in Tables 2.23, 2.24, and 2.25. Compare some of the estimates from section 2.4.5 with these actual data.
68. Discuss how the curves included in section 2.4.5 might be updated.
69. Compare the mine and mill operating costs predicted using the O'Hara estimators with the actual costs given in Tables 2.23, 2.24 and 2.25.
70. A detailed cost calculation has been provided in section 2.4.6. Repeat the example but assume that the ore and waste mining rates are twice those given.
71. In section 2.4.7 a 'quick-and-dirty' mining cost estimation procedure has been provided. Sometimes it is possible to obtain a percentage cost breakdown for a mining operation even if the actual costs are unavailable. How might such a cost distribution be beneficial?
72. Apply the quick-and-dirty approach to the Similkameen data.
73. Apply the quick-and-dirty approach to the Huckleberry data.
74. Some detailed cost information is provided in section 2.4.8. Using the data in Table 2.41, what might be the expected purchase price of a Bucyrus-Erie 59-R drill?
75. When using published costs and/or cost estimators it is very important to know what is included. In Table 2.41, describe the basis/meaning for the different cost components.
76. What are typical depreciation lives for mining equipment (trucks, shovels, drills, front-end loaders, graders, dozers, pickup trucks, etc.)?

77. What items are included in equipment ownership costs? What is the approximate split between ownership and operating cost for a piece of equipment?
78. Assume that you have a small equipment fleet consisting of
    – Caterpillar D9R dozer (1)
    – Caterpillar 992G wheel loader (1)
    – Caterpillar 777D trucks (3)
    – Caterpillar 16H grader (1)
    – DrilTech D25KS drill (1)

    What would be the expected total hourly operating cost? Does this include the operator?
79. Using the values given in Table 2.43, what would be the cost of the operators to run the equipment in problem 78?
80. The DrilTech D25KS drill will complete a 7-7/8" diameter hole, 40 ft long in 40 minutes. The life of the tungsten carbide cutter is 5000 ft. What is the cost of each hole?
81. Except for 10 ft of stemming, the hole in problem 80 is filled with ANFO. One 1/3 lb primer is used and the downline is detonating cord. What would be the cost of explosive per hole.
82. For the fleet described in problem 78, what would be the approximate capital cost based upon the information contained in Table 2.45?

CHAPTER 3

Orebody description

3.1 INTRODUCTION

Today, most potential orebodies are explored using diamond core drilling. The small diameter core collected from each hole provides a continuous 'line' of geologic information. Each of the recovered cores is studied in detail and the contained information recorded. The process is called 'logging'. Each 'line' is subsequently subdivided into a series of segments representing a particular rock type, structural feature, type of mineralization, grade, etc. By drilling a pattern of such holes, a series of similarly segmented lines are located in space. Using this information, together with a knowledge of the geologic setting and other factors, the mining geologist proceeds to construct a 3-dimensional representation of the mineralized body. The objective is to quantify, as best possible, the size, shape and distribution of the observable geologic features. The distribution of ore grades are correlated to lithology, alteration, structure, etc. The result is a mineral inventory or geological reserve. At this point in the evaluation process, economics have not been introduced so that terms such as 'ore' or 'ore reserve' are not involved.

The development of a mineral inventory involves substantial judgement, assumptions being made regarding sample and assay quality, and the interpretation and projection of geologic features based upon very limited data. The geologic data base, properly gathered and interpreted, should remain useful for many years. It forms the basis for current and future feasibility studies, mine planning and financial analyses. The success or failure of a project can thus be directly linked to the quality of its recorded data base, the drill logs and the maps. This chapter covers some of the basic techniques involved in the development and presentation of a mineral inventory.

3.2 MINE MAPS

The fundamental documents in all stages of mine planning and design are the maps. Maps are essential for the purpose of:
 – collecting,
 – outlining, and
 – correlating

a large portion of the data required for a surface mining feasibility study. These maps are drawn to various *scales*. The 'scale' is the ratio between the linear distances on the map and

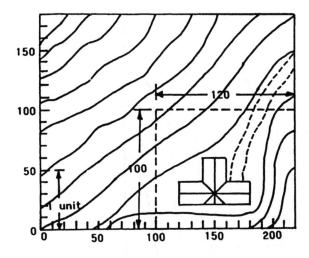

Figure 3.1. Plan map drawn to scale 1:50.

the corresponding distances at the site. In the English system this scale relates 'inches' as measured on the map to 'feet' in the field. This may be expressed as

$$\text{Map distance (in)} = K_E \times \text{Actual distance (ft)} \qquad (3.1)$$

where K_E is the English map scale. A typical map scale might be

$$K_E = \frac{1}{200} = 1{:}200$$

This means that an actual distance of 200 ft would be represented by a length of 1 inch on the map. In the metric system, the map scale relates similar map and actual distance units:

$$\text{Map distance (m)} = K_M \times \text{Actual distance (m)}$$

or $\qquad (3.2)$

$$\text{Map distance (cm)} = K_M \times \text{Actual distance (cm)}$$

where K_M is the metric map scale. A scale of 1:1000 means that a length of 1 meter on the map represents 1000 meters in the field. Similarly a length of 1 cm represents a distance of 1000 cm. A metric scale of 1:1250 is very close to the English scale of 1 in = 100 ft.

One speaks of a map being of larger or smaller scale than another. Figure 3.1 shows a particular area drawn to a scale of 1:50. In Figure 3.2 the region within the dashed lines of Figure 3.1 has been drawn to a scale of 1:20. In this figure the building appears *larger*. Thus the scale of the 1:20 map is larger than that of the 1:50 map. A map of scale 1:40 would be of larger scale than one drawn to 1:200.

The general rule is 'the greater the ratio (50 is greater than 20), the smaller is the scale.'
The selection of the most appropriate scale for any map depends upon:
1. The size of the area to be represented.
2. The intended uses for the map.

As more detail and accuracy is required, the scale should be increased.

Mine planning, for example, should be done at a scale that keeps the whole pit on one sheet and yet permits sufficient detail to be shown. For medium to large size metal mines, common planning scales are:

1 in = 100 ft
1 in = 200 ft

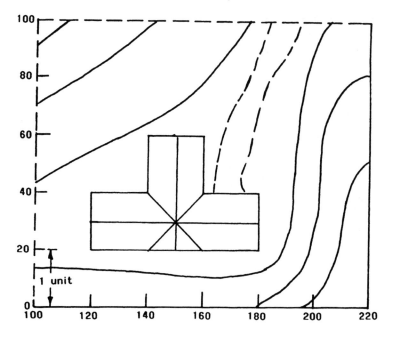

Figure 3.2. A portion of the plan map in Figure 3.1 drawn to scale 1:20.

In the metric system, common scales are:
  1:1000
  1:1250
  1:2000

Geologic mapping is commonly done on a larger scale such as 1 in = 40 ft (the corresponding metric scale is 1:500). For planning purposes, the geologic features (outlines) are replotted onto the smaller scale maps.

The types of maps prepared and used depends upon the stage in the life of the property. At the exploration stage, satellite maps may provide important information regarding structural regimes and potential exploration sites. These can be complemented with infra-red photos, etc. For certain types of information, for example the location of smelters, a small scale map, such as a map of the U.S., may be the most appropriate. Certain materials, such as crushed rock, are highly dependent on transportation costs. Regional maps overlain with circles corresponding to different freight tariffs are useful for displaying potential markets.

A state map Figure 3.3 can provide a considerable amount of basic information:
– nearest highways,
– closest towns,
– property location,
– railroad lines, and
– gross property ownership.

A typical scale for a state map is 1 inch equal to 15 miles.

Very quickly, however, one needs maps of larger scale for the more detailed planning. In the U.S., these often are the topographic maps prepared by the U.S. Geological Survey. These 'quadrangle' maps are prepared in two series. The 7½ minute-series (covers an area

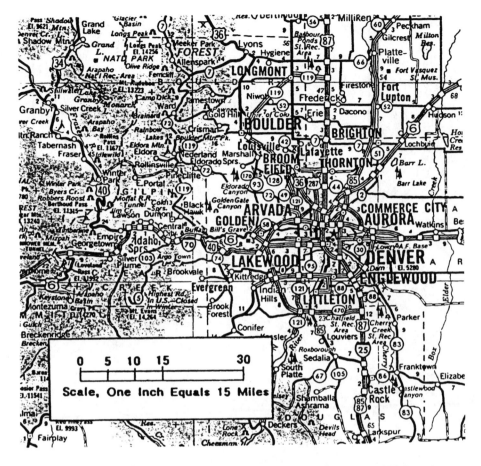

Figure 3.3. A section of the Colorado state highway map (Colorado Department of Highways).

of 7½ minutes latitude (high) by 7½ minutes longitude (wide)) with a unitless (metric) scale of 1:24000. This corresponds to a scale of 1 inch equal to 2000 ft. The 15-minute series includes an area 15 minutes in latitude by 15 minutes in longitude. One minute, it should be noted, represents one sixtieth of a degree.

Quadrangle maps show the topographic features, roads, rivers and drainage regions (Fig. 3.4).

Such maps can be enlarged to any desired scale (Figs 3.5 and 3.6), to serve as base maps until more detailed surveying is done. Aerial photos of the area are sometimes available through the state or county engineer's office or a federal agency. Many sections of the U.S. have been mapped on relief maps by the U.S. Corps of Engineers.

Very early in the life of a prospective mining area it is necessary to develop an ownership map. In the U.S., the best available ownership map can be obtained from the office of the county surveyor or the county clerk.

In the western U.S., much of the land and minerals are owned by either the state or by the federal government. A four-step process is followed (Parr & Ely, 1973) to determine the current status of the land.

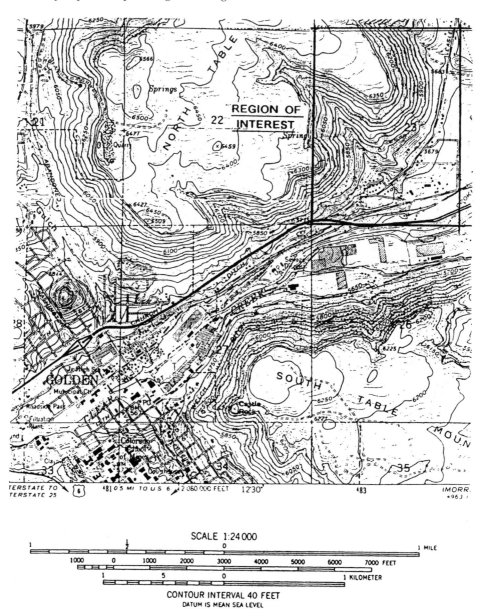

Figure 3.4. The USGS quadrangle map including Golden, Colorado (USGS, 1976).

Step 1. Consult the appropriate State land office, U.S. Geological Survey office or state or regional land office of the U.S. Bureau of Land Management. These offices can determine whether the land in question is available or if there are prospecting permits/leases in effect.

Step 2. If State lands are involved, the appropriate State land office should be consulted regarding the status.

Orebody description 173

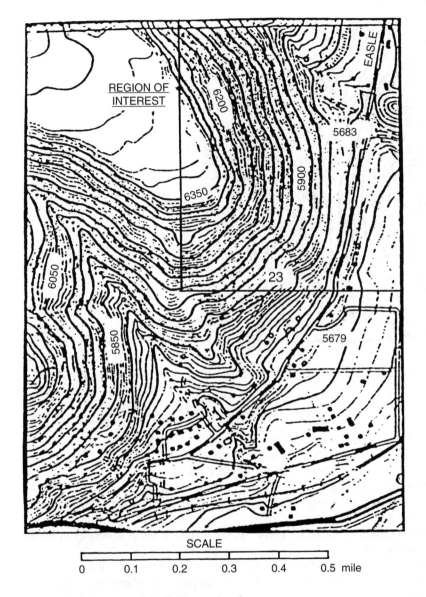

Figure 3.5. Enlargement of the Golden quadrangle map.

Step 3. The records of the appropriate County Recorder or other office should be checked regarding mining claims.

Step 4. A visit to the site should be conducted to determine whether there is anything (mining claim location markers, mine workings, etc.) not disclosed in the other check out steps.

For mine planning and design there are three map types (Phelps, 1968) of different scales:
1. General area map,
2. General mine map,
3. Detailed mine map (plans and cross sections).

174 *Open pit mine planning and design: Fundamentals*

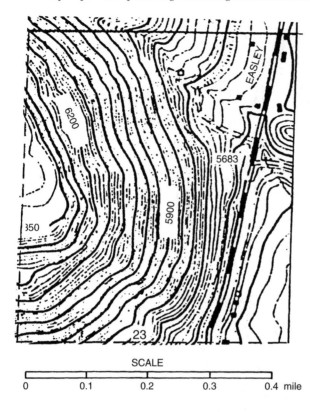

Figure 3.6. Further enlargement of the map.

The objective of the general area map is to show many pertinent features:
- geology (extent of orebodies, mineralized zones),
- transportation routes (highways, railroad, water routes),
- property ownership and control,
- distances to market, processing or transfer points (applicable freight rates),
- available access,
- location of transmission lines for power supply (capacity and construction distances required for connections),
- location of both present and future potential water supply/reservoir areas,
- areas suitable for tailings, slurry and refuse disposal in relation to mining and processing.

As such it is considered to be a small scale map. One can superimpose the data on these maps either directly or through the use of transparent overlays. Figure 3.7 is one example of a general area map. Figure 3.8 is an example of a general area geologic map.

The general mine map is a map of 'medium' scale. It covers a particular region within the general area map. Because the scale is larger, greater detail may be examined. Figure 3.9 is one example of a general mine map. The types of things which might be shown on such a map include:
- processing plant location,
- mine structures,
- power lines,

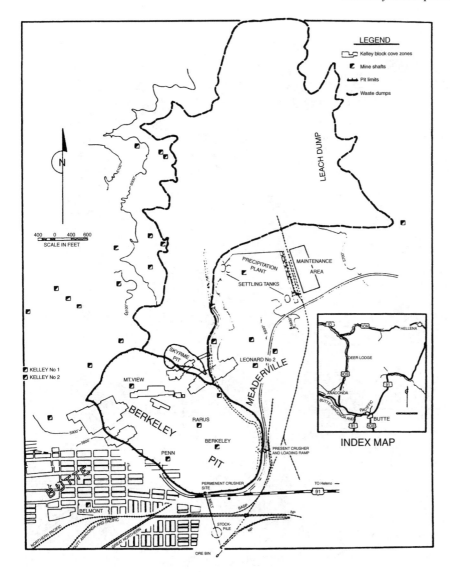

Figure 3.7. An example of a general area map (McWilliams, 1959).

- water supply,
- access roads,
- railroad lines,
- conveyor lines,
- pipelines,
- location of the orebody,
- location of a few drillholes,
- dump/tailing pond locations,
- property ownership and control,
- proposed timing of mining development.

176 *Open pit mine planning and design: Fundamentals*

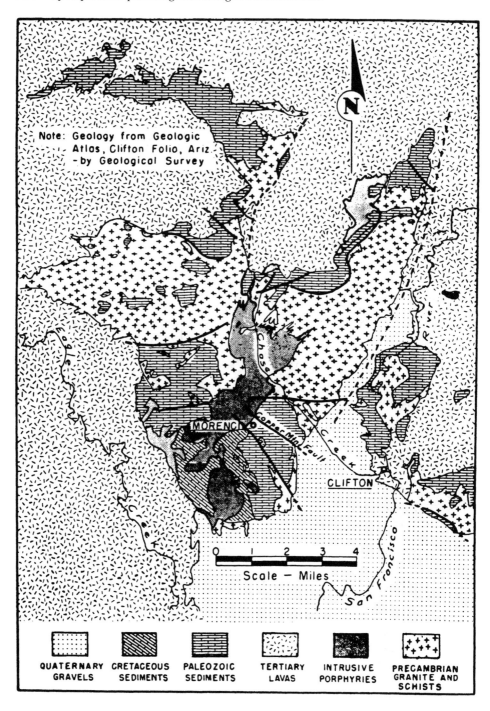

Figure 3.8. An example of a general area geologic map (Hardwick, 1959).

Orebody description 177

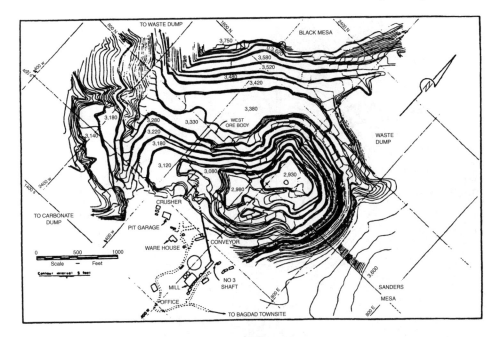

Figure 3.9. An example of a general area geologic map (Hardwick & Jones, 1959).

A general mine geologic map is shown in Figure 3.10.

The detailed mine maps are those used for the actual pit planning. The basic planning package consists of both plan maps and cross sections. Drill hole locations are laid out on plan maps at a scale appropriate to the requirements. In the Northern hemisphere the map is normally laid out in the northeast quadrant (Fig. 3.11).

The active portion of the map is selected with initial coordinates sufficiently large so that there is no danger that other quadrants will be involved. When this occurs the awkward use of +, − coordinates or West and South nomenclature is required. Both lead to confusion and possible errors. In the Southern hemisphere, the southwestern quadrant is often used.

The selection of the map 'north' direction is not universal. Some mines select geographic north as their north. Others use magnetic north. One large mine reputedly selected mine north with respect to the direction that the mine manager looked out over the mine from his office window. For elongated deposits, such as shown in Figure 3.12, the deposit lends itself to primary sections running N-S.

Local mine North is then conveniently selected as running perpendicular to the long axis of the deposit. For massive, more or less circular deposits (Fig. 3.13), the choice is less obvious.

Sometimes there is a different coordinate system for the pit as dictated by orebody geometry and that for the mine as a whole.

The exploratory and development diamond drill holes are generally laid out in a more or less regular pattern. When orebody evaluation is to be done using sections, it is convenient, although not imperative, if the drilling and mine grids are aligned. When block models are used to represent the orebodies, this is not as important.

Table 3.1 contains some guidelines for the preparation of mine maps. The line widths for the borders and coordinate systems should be carefully selected so that they can be

178 *Open pit mine planning and design: Fundamentals*

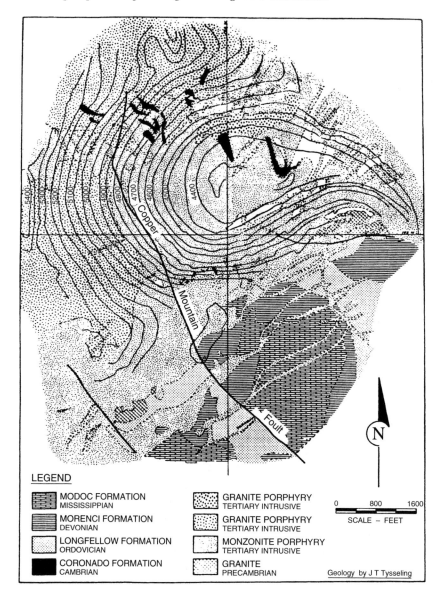

Figure 3.10. An example of a general mine geologic map (Hardwick, 1959).

easily seen, yet do not interfere with the major purpose of the map – that of presenting the graphical information. Modern CAD (computer-aided design) drafting systems have greatly simplified this previously, very tedious and time consuming job.

It is very important that revision/versions of the different maps be maintained.

The maps coordinates are labelled as 1600N, 1400E, etc. These are shown in Figure 3.14.

Vertical sections are made based upon these plan maps. As seen in Figure 3.14, there are two ways of constructing the N-S running sections.

Orebody description 179

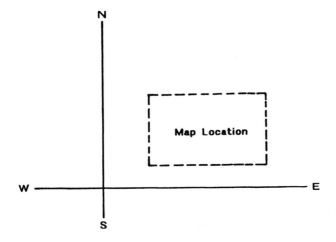

Figure 3.11. Typical NE quadrant plan map location.

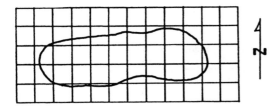

Figure 3.12. Grid system superimposed on an elongated deposit.

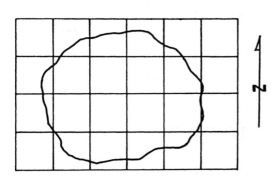

Figure 3.13. Grid system superimposed on a circular deposit.

In constructing section 1050E, one would get the results shown in Figures 3.15a,b depending on whether one is looking from east to west or west to east. Since one is used to numbers increasing from left to right, the east to west choice is made.

The location of the drill holes have been added to the plan map in Figure 3.16. The nomenclature DDH is often used to identify diamond drill holes. These holes are also added to the vertical sections (Fig. 3.17). Normally, the vertical scale for the sections is chosen to be the same as the horizontal. If this is not done, then pit slopes and other features become distorted.

180 *Open pit mine planning and design: Fundamentals*

Table 3.1. Guidelines for preparing mine maps (source unknown).

1. Title or subject, location of the area, and an extra identification or indexing notation on the outside or in an upper corner, and a reference to the associated report.
2. Compiler's name, the names of the field mapper, and an index diagram identifying sources of data.
3. Date of field work and date of compilation.
4. Scale; graphic and numerical.
5. Orientation of maps and sections, with magnetic declination shown on maps.
6. Isoline intervals and datum, with sufficient line labels and an explanation of heavier lines, dashed lines, and changes in interval.
7. Legend or explanation, with all units, symbols, and patterns explained.
8. Sheet identification and key, where more than one sheet or a series of overlays is involved.
9. Lines of cross section on the maps and map coordinates or key locations on the sections.
10. Reference grid and reference points.
11. Clarity: All areas enclosed by boundary lines should be labeled even in several places if necessary, so that they can still be identified if photo-reproduction in black-and-white changes the color pattern into shades of gray. Lines and letters should still be distinguishable after intended reductions in size. A trial reproduction can help in selecting colors and lines weights.
12. Size: The size should be compatible with reproduction equipment.

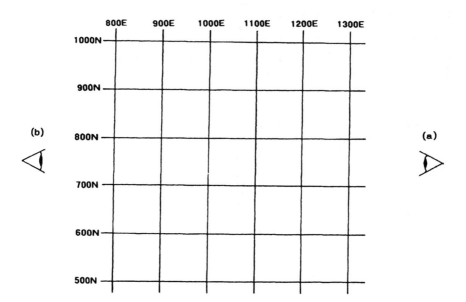

Figure 3.14. The two viewing directions for sections.

By examining the rock types and grades present in the holes on a given section, the geologist connects up similar features (Fig. 3.18). In this way a preliminary view of the size, shape and extent of the orebody is achieved. Such sections and their related plan maps form the basic elements used in mine planning and design. Sometimes, however, it is very helpful for visualization purposes to use isometric projections. Figure 3.19 is such a projection for the Bingham Canyon Mine.

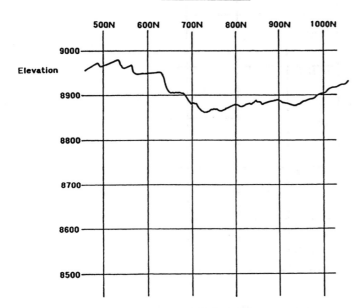

Figure 3.15a. The sections created looking from east to west.

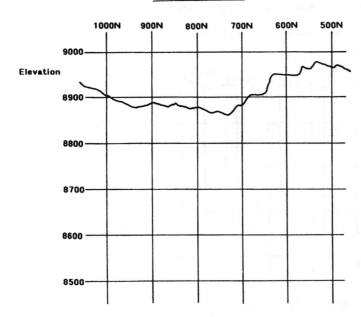

Figure 3.15b. The sections created looking from west to east.

182 Open pit mine planning and design: Fundamentals

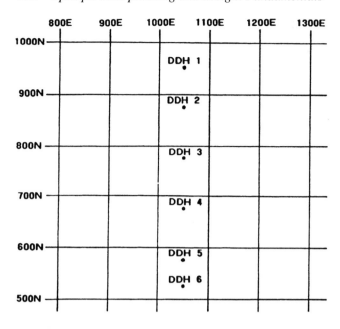

Figure 3.16. Additions of the diamond drill hole (DDH) locations to the plan map.

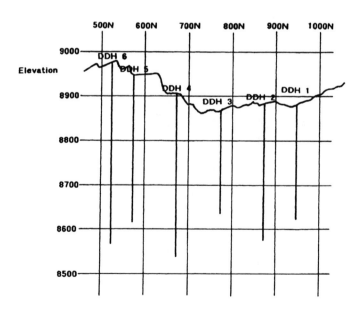

Figure 3.17. Typical sections with the drill holes added.

Orebody description 183

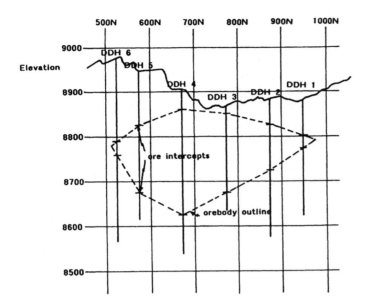

Figure 3.18. Addition of the ore zone to a section.

Figure 3.19. An isometric view of the Bingham Pit (Kennecott, 1966).

3.3 GEOLOGIC INFORMATION

Mining operations at any given mineral deposit may be divided into four stages. Prospecting (Stage 1), is the act of searching for valuable minerals. With the discovery of such minerals, the property becomes a mineral prospect. The property is then explored to gain some initial information regarding the size, shape, position, characteristics and value of the deposit. If this exploration stage (Stage 2) is successful, then the decision to proceed into Stage 3 (development) may be made. Detailed (Waterman & Hazen, 1968) geologic information

184 *Open pit mine planning and design: Fundamentals*

must be collected and made available early in this stage to facilitate planning and design. The following points should be included:
1. Geology of the mineralized zone;
2. Physical size and shape of the deposit;
3. Quantitative data on grade and tons of material within pertinent cut-off limits;
4. Mineralogical and metallurgical characteristics of the ore;
5. Physical characteristics of the ore and waste; and
6. Data on ground conditions, groundwater and other factors that affect mine design and operation.

Stage 4 is the actual mining of the deposit.

Although trenching, the sinking of shafts, and the driving of tunnels are all sometimes used in the evaluation of a surface mining prospect, most often the initial and continuing collection of geologic information is accomplished through a drilling program. The various drilling methods and their characteristics are summarized in Table 3.2. Here the focus will be on diamond drilling which produces a core for logging and assaying. The two basic types of diamond core drilling are: (a) conventional and (b) wireline. In conventional drilling the core is retained by a core spring or core lifter in the core barrel which is located just behind the bit. When the core barrel is full or after drilling a certain length, the entire drill string must be removed in order to extract the core.

Table 3.2. Characteristics of exploratory/development drilling techniques (Peters, 1978).

	Diamond core	Rotary	Reverse circulation	Downhole rotary	Downhole hammer	Percussion	Churn
Geologic information	Good	Poor	Fair	Poor	Poor	Poor	Poor
Sample volume	Small	Large	Large	Large	Large	Small	Large
Minimum hole diameter	30 mm	50 mm	120 mm	50 mm	100 mm	40 mm	130 mm
Depth limit	3000 m	3000 m	1000 m	3000 m	300 m	100 m	1500 m
Speed	Low	High	High	High	High	High	Low
Wall contamination	Variable	Variable	Low	Variable	Variable	Variable	Variable
Penetration-broken or irregular ground	Poor	Fair	Fair	Fair	Good	Good	Good
Site, surface, and underground	S+U	S	S	S+U	S+U	S+U	S
Collar inclination, range from vertical and down	180°	30°	0° ↑	30°	180°	180°	0°
Deflection capability	Moderate	Moderate	None	High	None	None	None
Deviation from course	High	High	Little	Little	Little	High	Little
Drilling medium, air or liquid	L	A+L	L	A+L	A	A+L	L
Cost per unit depth	High	Low	Moderate	Low	Low	Low	High
Mobilization cost	Low	Variable	Variable	Variable	Variable	Low	Variable
Site preparation cost	Low	Variable	Variable	Variable	Variable	Low	High

↑ Reverse circulation has recently been used at inclinations up to 40°.

The wireline method uses a core barrel removable through the inside of the drill stem with a latching device on the end of a cable. With this method, core can be retrieved at any desired point. Due to the space taken up by the inner barrel, wireline cores are smaller than those obtained in conventional drilling for the same hole size. The most common core size is NX/NQ which means that for a nominal 3-inch diameter hole, a $1\frac{7}{8}$ inch diameter core is recovered. The cores are typically recovered over 5 to 10 ft intervals. Sometimes the core recovery is poor and the grade is obtained by analyzing the cuttings/sludge. The recovered core is placed in order in core boxes for study, transport and storage. Table 3.3 is an example of a drill hole log. The amount of core recovered has been noted and a description of the material provided for each interval. Representative samples of the core are selected and sent out for assaying. Table 3.4 shows the results for the core described in Table 3.3. Information of this type may be plotted directly on cross sections or entered into a data file for computer processing. Table 3.5 is an example of one such computer file.

Table 3.3. Drill hole log from the Comstock Mine (Blais, 1985).

Mine: Comstock		Driller			Machine	Hole No. 144	
Level: Surface		Location: Iron County, Utah			Elev. 6246.50	Date 6-11-53	
Lat. 890.60 S		Dep. 2099.69 E				Angle Vert.	Bearing
Date	Drilled		Core		Material/Remarks		
	From	To	ft	ft	in		

Date	From	To	ft	ft	in	Material/Remarks
6-11-53D	0	5	5	2	0	Porphyry, plag. & K-feldspars about 1 to 1.5 mm long in gray, siliceous looking groundmass. Some euhedral biotite present. Rock somewhat altered.
	5	10	5	2	6	Porphyry, same as 0–5 except for 4″ hard, black, magnetite veinlet at 10′.
	10	16	6	1	0	Porphyry, same as 0–5
	16	26	10	8	0	Porphyry, highly argililized & bleached, very friable, some limonite staining.
	26	36	10	4	6	Porphyry, same as 16–26 except no limonite staining.
	36	46	10	10	0	Porphyry, same as 26–36
	46	56	10	1	6	Porphyry, same as 16–26
	56	66	10	5	0	Porphyry, same as 26–36
	66	71	5	1	6	2′ porphyry, same as 26–36 3′ siltstone, fine grained, light gray.
	71	75	4	1	6	Siltstone, very dense, fine grained, med gray color, occasional small cherty nodules.
	75	77	2	1	6	Siltstone, same as 71–75
	77	80	3	1	0	Siltstone, same as 71–75
	80	86	6	2	0	Siltstone, same as 71–75
	86	90	4	2	6	Siltstone, med grained, sandy, gray to purplish gray color.
	90	96	6	2	0	Siltstone, same as 71–75
	96	100	4	1	6	Siltstone, very dense, fine grained, dark gray with well developed joints dipping 45°, usually filled with calcite.

(Continued)

186 Open pit mine planning and design: Fundamentals

Table 3.3. (Continued).

100	106	6	1	4	Siltstone, same as 96–100
106	116	10	4	6	Siltstone, fine grained, slightly sandy, somewhat banded with bands of light gray, purplish gray, and greenish gray about $1/16''$ to $1/4''$ thick joints are filled with calcite.
116	126	10	7	6	Siltstone, fine grained, dense, gray with many wide calcite filled joints
126	136	10	2	6	Siltstone, same as 116–126 except $1/4''$ magnetite vein as 126–127.
136	146	10	6	0	Siltstone, med grained, slightly sandy, gray with many calcite and magnetite veinlets and stringers at 136–140.
146	150	4	0	6	$2'$ siltstone, fine grained, greenish-gray, slightly altered. $2'$ magnetite, black, soft, with much admixed silty material.
150	153	3	2	6	Magnetite, soft, sooty, black with many silty bands and streaks.
153	156	3	2	7	Magnetite, soft, sooty, black, with some streaks and bands of shiny black, med crystalline magnetite. Silty material present as streaks and tiny blobs.
156	161	5	4	4	Magnetite, same as 153–156
161	166	5	3	10	Magnetite, same as 153–156 except locally quite vuggy.
166	171	5	3	4	Magnetite, same as 153–156
171	176	5	4	6	Magnetite, same as 153–156 except silty material is green.
176	181	5	5	0	Magnetite, same as 153–156 except many calcite streaks and blobs from 177–178.
181	183	2	1	6	Magnetite, fine grained, soft to med hard, dark gray, very little silt.
183	188	5	4	9	Magnetite, same as 181–183.
188	189½	1½	1	6	Magnetite, same as 181–183.
189½	195	5½	3	4	Magnetite, same as 181–183 except some med crystalline, shiny black magnetite.
195	201	6	6	0	Magnetite, same as 189½–195 except much pyrite.
201	203	2	0	11	Magnetite, fine grained, soft, sooty, black with some med grained shiny black magnetite.
203	206	3	2	6	Limestone, very fine grained, dense, bluish gray with some magnetite as disseminated fine grains and veinlets.
206	211	5	4	10	Limestone, same as 202–206 except no magnetite.
					Bottom of hole

In addition to the type of information needed to compute the grade and tonnage, rock structural data are important for pit slope design. A simplified and a more comprehensive data sheet used in the logging of structural information are given in Tables 3.6 and 3.7 respectively. These data are the foundations for the planning and design steps. The information represented by a single hole is extended to a rather large region including the hole. Thus mistakes in evaluation, poor drilling practices, poor core recovery, sloppy record keeping, etc. may have very serious consequences. As soon as these data are entered into the computer or onto

Table 3.4. Drill sample analysis from the drill hole in Table 3.3 (Blais, 1985).

Mine: Comstock		Driller		Machine		Hole No. 144
Depth 211′		Core identified by P. Kalish				Date 6/11/53
Core recovery %		Analysis by				Date 2/23/53

Interval		Sample No.		Fe	Mn	SiO$_2$	Al$_2$O$_3$	P	S	CaO	MgO	Insol.	R$_2$O$_3$
From	To	Core	Sludge										
150	153	6177		50.7		12.0	3.8	0.270	0.03	2.5		1.7	
153	156	6178		51.8		8.1	2.7	0.280	0.03	3.7		2.0	
156	161	6179		55.3		9.1	2.8	0.280	0.04	1.5		1.2	
161	166	6180		56.0		10.2	2.5	0.216	0.05	1.8		1.0	
166	171	6181		58.5		8.8	1.9	0.058	0.03	1.2		1.0	
171	176	6182		53.6		9.6	3.0	0.148	0.03	2.0		1.6	
176	181	6183		56.6		8.8	2.7	0.206	0.02	1.5		1.0	
181	183	6184		51.2		12.0	2.5	0.020	0.02	2.6		2.1	
183	188	6185		54.7		8.0	2.3	0.170	0.03	1.8		1.6	
183	189½	6186		55.0		5.9	1.8	0.025	0.03	1.3		2.3	
189½	195	6187		56.3		4.9	1.7	0.057	0.03	1.3		2.3	
195	201	6188		54.6		6.9	1.9	0.033	0.67	1.1		2.5	
201	203	6189		54.2		7.6	2.1	0.027	0.61	1.3		1.7	
153	156		6190	46.8		13.2	4.4	0.075	0.03	1.6		1.0	
156	161		6191	52.4		9.9	3.1	0.130	0.02	2.5		1.3	
161	166		6192	54.4		8.9	2.0	0.318	0.02	2.0		1.1	
166	171		6193	57.0		7.2	2.4	0.116	0.02	1.9		1.5	
189½	195		6194	53.8		10.2	2.9	0.138	0.06	1.7		1.5	

sections and plans, the level or uncertainty associated with them vanishes. Good numbers and less good numbers all carry the same weight at that stage. Hence it is imperative that utmost care be exercised at this early stage to provide a thorough and accurate evaluation of all information. Each hole is quite expensive and there is pressure to keep them to a minimum. On the other hand, poor decisions based on inadequate data are also expensive. This weighing of real costs versus project benefits is not easy.

3.4 COMPOSITING AND TONNAGE FACTOR CALCULATIONS

3.4.1 *Compositing*

As discussed in the previous section, after the diamond core has been extracted it is logged by the geologist and representative samples are sent out for assaying. Upon receipt, the assays are added to the other collected information. These individual assay values may represent core lengths of a few inches up to many feet. Compositing is a technique by which these assay data are combined to form weighted average or composite grades representative of intervals longer than their own. The drill log shown diagrammatically in Figure 3.20 contains a series of ore lengths l_i and corresponding grades g_i.

In this case the boundaries between ore and waste are assumed sharp. The first question which might be asked is 'What is the average grade for this ore intersection?' The weighted average is found by first tabulating the individual lengths l_i and their corresponding grades g_i.

188 *Open pit mine planning and design: Fundamentals*

Table 3.5. Typical computerized drill hole data file, after Stanley (1979).

Report No: 01															Page No.	1
Site	Hole	Type	Collar coordinates		Elevation		Azimuth	Inclination			Interval				Run date:	03/03/77
HU	0002	D	East	North	05509		+0	090			010				Run time:	14:30:06
			08054	05796											Hole depth	Remarks
															00500	none
Seq.	Coordinates		Elevation	Distance				Percent	Max		PC	RK	R	RK	Alteration	Mineral
No	East	North			AZ1	INC	INT	MOS2	WO3	LNG	CR	QD	C	TP	PSSACFT	MFPFU
001	8054	05796	05504	5	+0	90	10	0.382	NA	0.4	37	54	6	U1	332405	0000
002	8054	05796	05494	15	+0	90	10	0.305	0.004	0.3	48	4	6	U1	332405	000R
003	0854	05796	05484	25	+0	90	10	0.246	0.002	0.2	59	0	5	U1	342406	0000
004	8054	05796	05474	35	+0	90	10	0.257	0.002	0.2	69	4	4	U1	232505	0000
005	8054	05796	05464	45	+0	90	10	0.229	0.002	0.4	60	4	5	U1	232505	0000
006	8054	05796	05454	55	+0	90	10	0.411	0.001	0.7	48	25	6	U1	132505	000R
007	8054	05796	05444	65	+0	90	10	0.277	0.004	0.7	38	18	7	U1	132304	P000
008	8054	05796	05434	75	+0	90	10	0.400	0.003	0.7	35	42	7	U1	132304	000R
009	8054	05796	05424	85	+0	90	10	0.287	0.001	0.5	42	12	6	U1	131304	000F
010	8053	05796	05414	95	−1	90	10	0.283	0.002	0.9	32	61	7	U1	132303	000R
011	8053	05796	05404	105	−1	90	10	0.290	NL	0.6	60	18	5	U1	122403	000F
012	8053	05796	05394	115	−1	90	10	0.504	NL	0.6	38	7	7	U1	144304	0000
013	8053	05796	05384	125	−1	90	10	0.286	0.002	0.9	26	76	7	U1	123306	0000
014	8053	05796	05374	135	−1	90	10	0.390	0.002	0.5	65	30	5	U1	122303	0000
015	8052	05796	05364	145	−1	90	10	0.545	0.002	0.8	44	42	6	U1	232204	000B
016	8052	05796	05354	155	−2	90	10	0.429	0.001	0.4	57	26	5	U1	232303	000F
017	8052	05796	05344	165	−2	90	10	0.346	NL	0.9	55	27	6	U1	123203	000R
018	8051	05796	05334	175	−2	90	10	0.253	NL	0.6	68	5	6	U1	122203	0000
019	8051	05796	05324	185	−2	90	10	0.374	NL	0.7	50	34	7	U1	122303	0000
020	8051	05796	05314	195	−2	90	10	0.248	NL	0.9	65	21	5	U1	332303	0000
021	8050	05796	05304	205	−2	90	10	0.483	NL	0.7	63	13	6	U1	253303	0000

Orebody description 189

Table 3.6. Example data collection form for core recovery and RQD (Call, 1979).

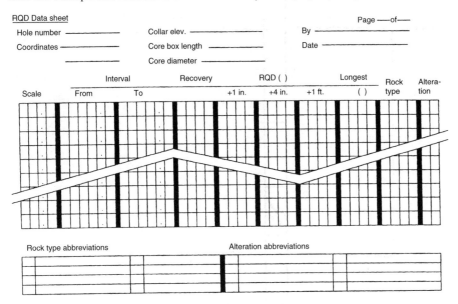

Table 3.7. Example data collection form for oriented core (Call, 1979).

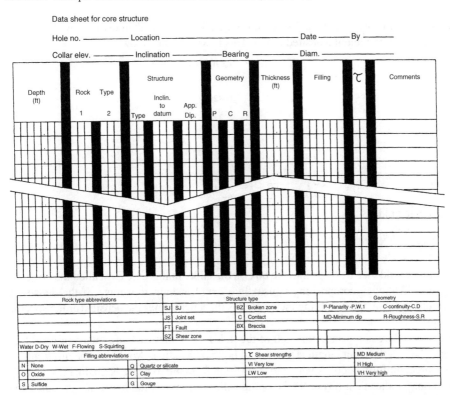

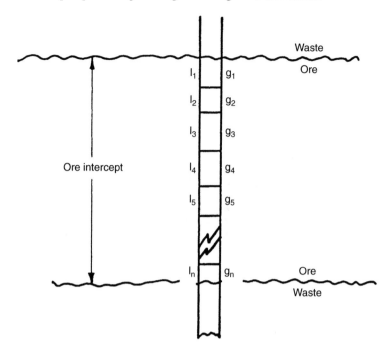

Figure 3.20. Ore intercept compositing.

The products $l_i g_i$ are formed and summed $\sum l_i g_i$. This sum is then divided by the sum of the lengths $\sum l_i$ to yield the desired grade. This is written out below:

Length	Grade	Length × Grade
l_1	g_1	$l_1 g_1$
l_2	g_2	$l_2 g_2$
l_3	g_3	$l_3 g_3$
.	.	.
.	.	.
.	.	.
l_n	g_n	$l_n g_n$
$\sum l_i$	$\boxed{\bar{g}}$	$\sum l_i g_i$

The average grade is

$$\bar{g} = \frac{\sum l_i g_i}{\sum l_i} \qquad (3.3)$$

This value would then be filled into the box on the table. In this case $\bar{g}$ is called the *ore-zone composite*. Although compositing is usually a length-weighted average, if the density is extremely variable, the weighting factor used is the length times the density (or the specific gravity).

This procedure is repeated for each of the holes. Note that each ore intercept would, in general, be of a different length. The top and bottom elevations would also be different.

For large, uniform deposits where the transition from ore to waste is gradual (the cut-off is economic rather than physical) the compositing interval is the bench height and fixed

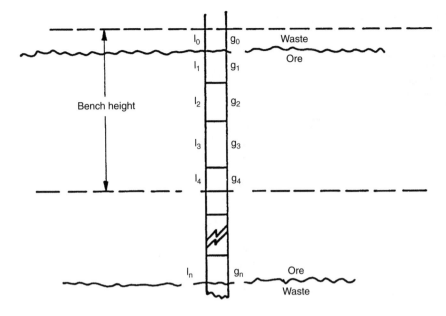

Figure 3.21. Bench height compositing.

elevations are selected. This *bench compositing* is the method most often used for resource modeling in open pit mining today. In Figure 3.21, a bench is shown by the horizontal dashed lines.

In this particular case, the upper portion of the bench lies in 'waste.' The bench composite grade is

Length	Grade	Length × Grade
l_0	g_0	$l_0 g_0$
l_1	g_1	$l_1 g_1$
l_2	g_2	$l_2 g_2$
l_3	g_3	$l_3 g_3$
l_4	g_4	$l_4^* g_4$
$\sum l_i$	$\bar{g}$	$\sum l_i g_i$
$\sum l_i = H$		

where H is bench height. Hence

$$\bar{g} = \frac{\sum l_i g_i}{H}$$

Compositing with fixed intervals and elevations makes it very easy to present and analyze the results for a deposit containing a number of drill holes.

Some of the reasons for and the benefits of compositing include:

1. Irregular length assay samples must be composited to provide representative data for analysis.

2. Compositing incorporates dilution such as that from mining constant height benches in an open pit mine.

3. Compositing reduces erratic variation due to very high or very low assay values.

4. By compositing, the number of data, and hence the required computational times, are reduced.

Table 3.8. Example of drill-hole log, after (Davey, 1979).

Drill-hole ID = C-22
Collar location = 1800.0 N 800.0 E
Elevation = 45198.0
Azimuth = 0.0 Attitude = −90

Depth	Assay
5	0.400
10	0.560
15	0.440
20	0.480
25	0.400
30	0.380
35	0.330
40	0.590
45	0.480
50	0.600
55	0.560
60	0.320
65	0.700
70	0.210
75	0.180
80	0.080
85	0.200
90	0.070

To illustrate the principles presented, consider the simplified drill hole log (Davey, 1979) given in Table 3.8.

It has been decided that 40 ft high benches and a 5200 ft reference elevation will be used. This means that bench crest elevations would be at 5200 ft, 5160 ft, 5120 ft, etc. The upper 38 ft of hole C-22 would lie in bench 1. The next 40 ft would be in bench 2 and the hole would terminate in bench 3. Using the procedure outlined above, the composite grade at this hole location for bench 2 is determined as:

Length (ft)	Grade (%)	Length × Grade (ft%)
2	0.590	1.18
5	0.480	2.40
5	0.600	3.00
5	0.560	2.80
5	0.320	1.60
5	0.700	3.50
5	0.210	1.05
5	0.180	0.90
3	0.080	0.24
40	0.417	16.67

$$\bar{g} = \frac{16.67}{40} = 0.417$$

Orebody description 193

The mid-elevation of the bench 2 is 5140.0 ft. Composites of the remaining portions of the drill hole lying above and below this bench may be found in the same way. The results are given below.

Bench	Center coordinates			Grade
	E	N	Elevation	
1	800.00	1800.00	5179.00	0.440
2	800.00	1800.00	5140.00	0.417
3	800.00	1800.00	5114.00	0.126

If material running 0.3% and higher is understood to be ore, then the ore-zone at this hole extends from the surface to a depth of 65 ft. The ore-zone composite would be:

Length (ft)	Grade (%)	Length × Grade (ft%)
5	0.40	2.00
5	0.56	2.80
5	0.44	2.20
5	0.48	2.40
5	0.40	2.00
5	0.38	1.90
5	0.33	1.65
5	0.59	2.95
5	0.48	2.40
5	0.60	3.00
5	0.56	2.80
5	0.32	1.60
5	0.70	3.50
65	0.48	31.20

$$\bar{g} = \frac{31.2}{65} = 0.48$$

In this case when the lengths are all equal, the average grade is just the simple average of the grades.

$$\bar{g} = \frac{6.24}{13} = 0.48$$

The same compositing technique can be used when dealing with grades representing different areas or volumes. This will be demonstrated in Section 3.5.

3.4.2 *Tonnage factors*

In mining, although *volumes* of material are removed, payment is normally received on the basis of the *weight* of the valuable material contained. This is in contrast to civil construction projects where normally payment is received based simply upon the material volume removed or emplaced. Even here, however, the conversion from volume to weight must often be made due to the lifting and carrying limitations of the loading and hauling equipment used. The conversion from volume V to weight W and vice versa is done in the English system of units with the help of a tonnage factor TF (volume/weight):

$$V = \text{TF} \times W \tag{3.4}$$

where TF is the tonnage factor (volume/weight), V is the volume, and W is the weight. The determination of a representative factor(s) is quite important to mining operations.

In the English system of measurement, the basic unit for describing the weight of materials is the weight of a cubic foot of water. The density W_D of water is

$$W_D(H_2O) = 62.4 \, \text{lb/ft}^3 \tag{3.5}$$

and its specific gravity SG is 1. If the mined material has a specific gravity of 2.5, its weight density is

$$W_D = \text{SG} \times W_D(H_2O) = 2.5 \times 62.4 \, \text{lb/ft}^3 = 156 \, \text{lb/ft}^3 \tag{3.6}$$

The tonnage factor TF for the material (assuming that the short ton (st) applies) is

$$\text{TF} = \frac{2000 \, \text{lb/st}}{156 \, \text{lb/ft}^3} = 12.82 \, \text{ft}^3/\text{st} \tag{3.7}$$

In the metric system, the density of water is

$$W_D(H_2O) = 1 \, \text{g/cm}^3 = 1000 \, \text{kg/m}^3 = 1 \, \text{t/m}^3 \tag{3.8}$$

Since the specific gravity of the mined material is 2.5, the density is $2.5 \, \text{t/m}^3$. The tonnage factor is

$$\text{TF} = \frac{1}{2.5} = 0.4 \, \text{m}^3/\text{t} \tag{3.9}$$

Although the tonnage factor as defined here with units of volume per weight is probably the most commonly used, the inverse (TF*) is also used:

$$\text{TF}^* = \frac{W}{V}$$

Other units such as yd^3 instead of ft^3 are sometimes used for convenience.

Although simple in principle, it is not as easy in practice to determine the appropriate material densities to be used in the calculations. There can be many different materials involved in an open pit mine and each 'material' can vary in density from point to point.

Three techniques are available for determining material density:

1. Density testing of small samples in the laboratory.
2. Careful excavation and weighing of a large volume.
3. Calculation based upon composition (mineralogy) using published densities such as given in Tables 3.9 through 3.11.

Depending upon the requirements, all three are sometimes used. For Technique 1, there are two primary tests which are done. In the first, the sample is first weighed (W) in air. The sample volume V is then determined by water displacement (the water level in a graduated cylinder is, for example, compared before immersion and after immersion of the sample).

Table 3.9. Average density of minerals (Westerfelt, 1961).

Material	Mineral	Density (g/cm³)	Material	Mineral	Density (g/cm³)
Antimony	Native	6.7	Iron	Arsenopyrite	6.0
	Stibnite	4.6	(continued)	Hematite	5.0
Arsenic	Orpiment	3.5		Magnetite	5.0
	Realgar	3.5		Limonite	3.8
Barium	Barite	4.5		Siderite	3.8
	Witherite	4.3	Lead	Galena	7.5
Calcium	Calcite	2.7		Cerussite	6.5
	Aragonite	3.0		Anglesite	6.3
	Gypsum	2.3		Crocoite	6.0
	Fluorspar	3.2		Pyromorphite	7.0
	Apatite	3.2	Manganese	Pyrolusite	4.8
Coal	Anthracite	1.5		Psilomelane	4.2
	Bituminous	1.3		Rhodochrosite	3.6
Cobalt	Linnaite	4.9		Rhodonite	3.6
	Smaltite	6.5	Mercury	Native	14.4
	Cobaltite	6.2		Cinnabar	8.1
	Erythrite	3.0	Molybdenum	Molybdenite	4.7
Copper	Native	8.9	Nickel	Millenite	5.6
	Chalcocite	5.7		Niccolite	7.5
	Chalcopyrite	4.2	Platinum	Native	17.5
	Bornite	5.0	Silver	Native	10.5
	Enargite	4.4		Argentite	7.3
	Tetrahedrite	4.9		Sylvanite	8.0
	Atacamite	3.8		Pyrargyrite	5.8
	Cuprite	6.0		Cerargyrite	5.4
	Chalcanthite	2.2	Sulphur	Native	2.1
	Malachite	3.9	Tin	Cassiterite	6.8
	Azurite	3.7		Stannite	4.5
	Chrysocolla	2.2	Tungsten	Wolframite	7.3
	Dioptase	3.3		Scheelite	6.0
Gold	Native	19.0	Zinc	Blende	4.0
Iron	Pyrite	5.1		Zincite	5.7
	Marcasite	4.8		Smithsonite	4.4
	Pyrrhotite	4.6			

The density d is then calculated:

$$d = \frac{W}{V} \tag{3.10}$$

In the second type of test, the sample is first weighed (W) in air and then weighed (S) when suspended in water. The specific gravity is

$$SG = \frac{W}{W - S} \tag{3.11}$$

Care must be taken to correct both for porosity and moisture.

Technique 2 is the most expensive and time consuming, but provides the best site specific results. Such tests would have to be made for different locations in the mine.

196 *Open pit mine planning and design: Fundamentals*

Table 3.10. Average density of some common rock types (Reich, 1961).

Origin	Rock type	Density (g/cm³)	Origin	Rock type	Density (g/cm³)
1. Igneous (plutonic)	Nepheline syenite	2.62	3. Metamorphic (continued)	Phyllite	2.74
	Granite	2.65		Marble	2.78
	Quartz	2.65		Chlorite schist	2.87
	Anorthosite	2.73		Serpentine	2.95
	Syenite	2.74	4. Sedimentary (consolidated)	Greywacke	2.69
	Quartz diorite	2.79		Sandstone	2.65
	Diorite	2.93		Limestone	2.73
	Gabbro	3.00		Argillaceous shale	2.78
	Peridotite	3.06		Calcareous shale	2.67
	Pyroxene	3.22		Chert	2.76
2. Igneous (hypabasal/ volcanic)	Quartz porphyry	2.63	5. Sedimentary (unconsolidated)	Humus soil	1.45
	Porphyry	2.67		Surface soil	1.73
	Diabase	2.94		Clayey sand/sandy	1.93
	Rhyolite	2.50		Gravel, very damp	2.00
	Phonolite	2.56		Dry, loose soil	1.13
	Trachyte	2.58		Very fine, sandy alluvium	1.33
	Dacite	2.59			1.51
	Andesite	2.62		Carbonaceous loam	1.65
	Basalt	2.90		Glayey, sandy soil	2.25
3. Metamorphic	Orthoclase gneiss	2.70		Very wet, quartz sand	
	Plagioclase gneiss	2.84			
	Quartz schist	2.68		Loess	2.64
	Mica schist	2.73		Clay	2.58

Table 3.11. Density of some metals (CRC Handbook of Chemistry and Physics, 1991–1992).

Metal	Density (g/cm³)
Chromium	6.92
Copper	8.89
Gold	19.3
Iron	7.86
Lead	11.34
Molybdenum	9.0
Nickel	8.6
Platinum	21.37
Silver	10.5
Zinc	7.13

To illustrate the use of Technique 3, consider a gold ore made up of 94% quartz and 6% iron pyrite by weight. From Tables 3.9 and 3.10 one finds that the respective specific gravities are:
- Quartz: 2.65
- Iron pyrite: 5.1

The overall specific gravity for the ore is

$$SG = 2.65 \times 0.94 + 5.1 \times 0.06 = 2.80$$

and the tonnage factor (English system) is

$$TF = \frac{2000}{2.80 \times 62.4} = 11.45 \, \text{ft}^3/\text{st}$$

A margin of safety is introduced when applying the results of any of the techniques. In this case a value of 12 or even greater might be used. This is the in-situ or in-place tonnage factor.

To illustrate the principles involved in the conversion from volumes to weight and vice versa assume that a mining company has a contract to sell 5000 tons of metal X per year. The mined material contains 1% of the contained metal and the processing plant recovers 50%. The total tonnage T_A which must be mined and processed each year is given by

$$T_A = \frac{5000 \, \text{st}}{0.01 \times 0.50} = 1,000,000 \, \text{st}$$

Assuming that the layer being mined has a thickness t of 20 ft, the question becomes how large a plan area A must be exposed to produce the required tonnage.

The annual volume V_A is

$$V_A = tA$$

To solve the problem, the relationship between the volume V_A and the weight T_A must be known. Assuming that the specific gravity of the mined material is 2.5, the tonnage factor is $12.82 \, \text{ft}^3/\text{st}$.

Hence, the volume removed per year is

$$V_A = 12.82 \, \text{ft}^3/\text{st} \times 1,000,000 \, \text{st} = 12,820,000 \, \text{ft}^3$$

Hence, the area to be exposed is

$$A = \frac{12,820,000}{20} = 641,000 \, \text{ft}^2$$

The acre is commonly used to describe land area:

$$1 \, \text{acre} = 43,560 \, \text{ft}^2$$

Thus, a total of 14.72 acres would be mined each year.

The same problem will now be worked using the metric system. It is assumed that 4537 t of mineral are produced from a seam 6.1 m thick. The numerical value of the density and the specific gravity are the same in this system, which simplifies the calculations.

Since the specific gravity of the mined material is 2.5, the density is $2.5 \, \text{t/m}^3$. The tonnage factor is

$$TF = \frac{1}{2.5} = 0.4 \, \text{m}^3/\text{t}$$

Therefore

$$V_A = \frac{4,537}{0.5 \times 0.01} \times 0.4 = 362,960 \, \text{m}^3$$

$$A = 59,502 \, \text{m}^2$$

198 *Open pit mine planning and design: Fundamentals*

In the metric system, land area is expressed in terms of the hectare:

$$1 \text{ hectare} = 100 \text{ m} \times 100 \text{ m} = 10{,}000 \text{ m}^2$$

Thus a total of 5.95 hectares would be mined each year.

3.5 METHOD OF VERTICAL SECTIONS

3.5.1 *Introduction*

The traditional method for estimating ore reserves has been through the use of sections. The method has a number of advantages, the primary one is that it can be done by hand. Other advantages are that it can be easily depicted, understood and checked. It will be assumed that the method is done by hand. However, a number of computer techniques are available to allow designer input/flexibility while doing the calculations by machine. Some computer programs have been designed to essentially reproduce the interpretation logic currently done by engineers and geologists by hand.

3.5.2 *Procedures*

The general procedures described below have been used by the Office of Ore Estimation-University of Minnesota (Weaton, 1972, 1973) for preparing and/or reviewing iron ore reserve estimates for the State of Minnesota. They can easily be adapted to other types of mineralization and deposits.

Planning materials
1. A current, up-to-date plan map. This is made to a convenient scale (usually 1 in = 100 ft) and shows the following:
    (a) Pit surface conditions, existing banks and details of the immediate vicinity.
    (b) Location of all drill holes.
    (c) Location of all quarter section lines and property lines.
    (d) Location of pit cross sections.
2. A complete set of cross sections. These are drawn to any convenient scale (usually 40 feet to the inch) and contain the following:
    (a) All exploration drill holes which fall on or close to the section (half way to the next section), with the detailed analysis of each sample taken. Results of hand wash or heavy density tests if they were made. Also, the location and analysis of any bank samples that have been taken.
    (b) A line showing the current top of the material remaining in the ground undisturbed.
    (c) Geologic structure lines showing an interpretation of the limits of the ore areas, and the various lean ore (low grade) or waste formations.

Planning procedures
1. Drill samples are evaluated on the cross sections and zones of different types of material are color coded for convenience. If the pit has been operating, any pit operations or observations which may disprove drilling samples in any way are taken into consideration in outlining zones of the various types of materials.

Orebody description 199

2. Limits of ore materials are transposed to the plan map as a general outline for the pit area.

3. The pit plan layout is developed to recover all of the ore that is economically minable with the necessary removal of the waste materials. Many factors enter into this plan, and govern the amount of material which must be removed. Some of these are:

(a) The nature of the surface capping; i.e., sand, clay, gravel, muskeg, etc., and the angle at which this material will remain stable in the bank.

(b) The nature of the rock and waste material and the angle at which it will remain stable when exposed.

(c) The local terrain and the location of the mine facilities, plant and dump areas in relation to the pit.

(d) The grade or steepness of the haulage road and the width required by the haulage trucks.

(e) The number of berms or protective benches that will be required to insure pit safety and bank stability.

4. After the pit plan is laid out, and the bank slopes are drawn on the cross sections, tonnages can be computed.

5. Unless disproven by other drilling or samplings, the material on each cross section is assumed to extend to a point one half the distance to the section on each side or 100 feet beyond the end section.

6. Computation of volumes in cubic feet are made by measuring the area of each type of material as shown on the cross section, and multiplying this by the distance represented by the section (one-half the distance to each adjacent section). By experience, the factor of cubic feet per ton has been established, for both ore and other materials. Concentration tests on drill samples of materials requiring plant treatment establish the recovery figure or how much concentrate will remain after being run through a concentration plant.

7. Tonnages of each section are totalled to give the final reserve tonnage figures.

8. A weighted average of the chemical analyses of each type of material is computed to produce the final estimated grade of the products included in the estimate.

3.5.3 *Construction of a cross-section*

An E-W section (640 N) taken through an iron deposit is shown in Figure 3.22. The objective is to begin with the drill hole data and proceed through to the determination of the areas of the different materials which would be included in the final pit. The symbols which have been used to denote the layers are:

SU = surface (overburden) material (soil, glacial till, etc.) which can be removed without drilling and blasting.
DT = decomposed taconite.
OP = ore and paint rock.
OT = ore and taconite.
SWT = sandy wash ore and taconite.

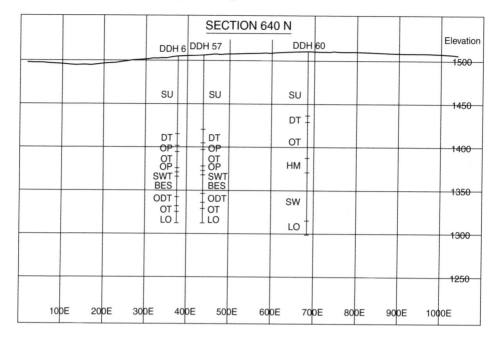

Figure 3.22. Drill holes and topography added to the section (Step 1).

BES = Bessemer ore.
ODT = ore and decomposed taconite.
LO = lean (low grade) ore.
SW = sandy wash ore.
HM = heavy media ore.

Step 1. The drill holes and surface topography are plotted on the section.

Step 2. The bisector between DDH57 and DDH60 is constructed. The surface-rock interface is drawn. Points common to all 3 holes are connected (Fig. 3.23).

Step 3. Starting from the surface, connect the remaining common points in holes DDH57 and DDH6 and extend them to the left of DDH6. To the right of DDH57 extend the layers over to the bisector line. These are drawn parallel to the known overlying surfaces. Fill in the region between DDH57 and the bisector by extending the layers parallel to the known overlying trend lines. (Fig. 3.24).

Step 4. The remaining layers intersected by DDH60 are extended left to the bisector and to the right. (Fig. 3.25).

Step 5. The pit outline is superimposed on the section. In this case the following rules have been used:
 – the lean ore intercept forms the pit bottom,
 – an extension of 50 ft outside of the drill holes at the pit bottom is assumed,
 – the allowable pit slope angle in the surface material is 27° whereas in the rock layers near the pit bottom it is 54°. A transition of 41° is used between these. (Fig. 3.26).

Step 6. An access road 50 ft in width crosses this section at the position indicated (Fig. 3.27).

Orebody description 201

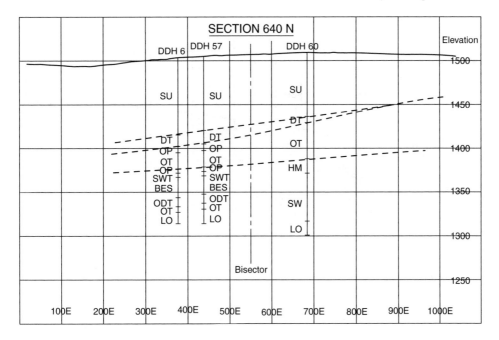

Figure 3.23. The Step 2 section.

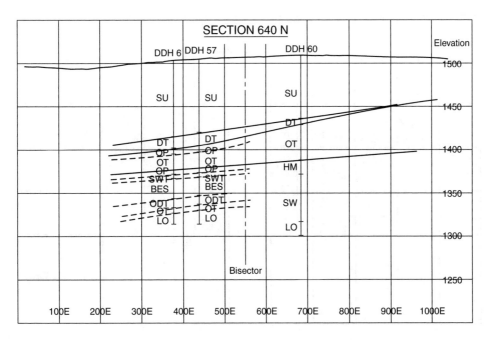

Figure 3.24. The Step 3 section.

202 *Open pit mine planning and design: Fundamentals*

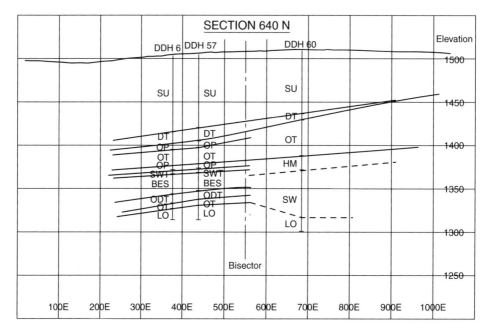

Figure 3.25. The Step 4 section.

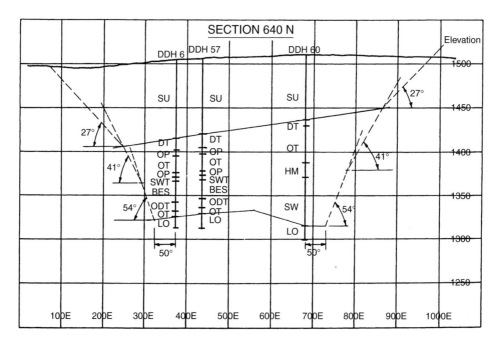

Figure 3.26. The Step 5 section.

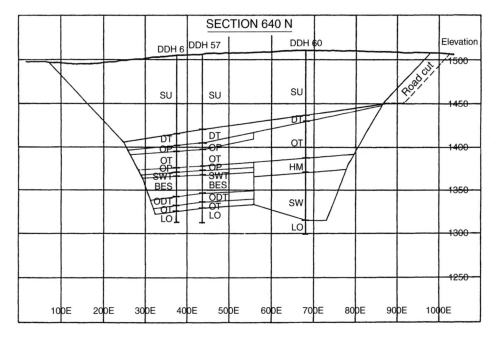

Figure 3.27. The Step 6 section.

Step 7. The areas of the different materials on the section are determined using a planimeter (Fig. 3.28).

Even in this relatively simple case, it is clear that a good knowledge of the structure of the orebody is required (Fig. 3.29) as well as some judgement in order to create such sections. As will be discussed in more detail later, final pit outlines can only be determined by considering all sections together.

3.5.4 *Calculation of tonnage and average grade for a pit*

This simplified example has also been taken from iron mining practice. The following concepts will be illustrated:
1. Side completion for sections.
2. Development of a final pit outline including pit ends.
3. Determination of tons and average grade for a section.
4. Determination of tons and average grade for the pit.

Although most of the discussion will revolve around section $1+00$, the same approach would be used on all sections.

Side completion

As described in the previous example, the section $1+00$ (Fig. 3.30) has been extended 50 ft past the positions of the outermost drill holes. On the left side of the section, the ore appears to pinch out within this zone. The pit slope of 27° has been drawn to pass through the mid-height of this extension. The width associated with hole 6 would be 50 ft plus half the distance between holes 6 and 1. On the right-hand side, the ore is quite thick (25 ft) and

204 *Open pit mine planning and design: Fundamentals*

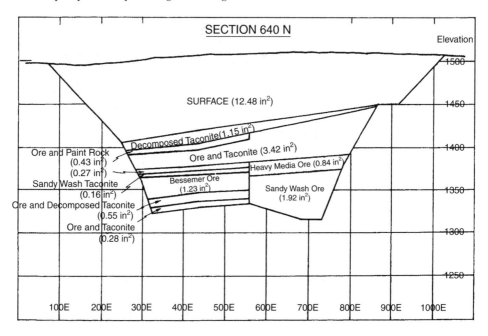

Figure 3.28. The Step 7 (final) section.

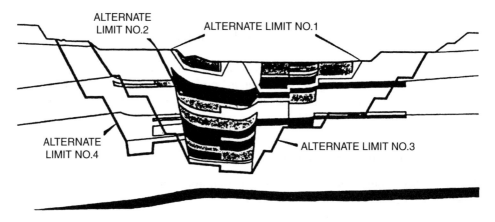

Figure 3.29. A typical cross section showing the structure of a Minnesota Mesabi range iron orebody (Axelson, 1963).

would appear to continue. The slope has been drawn at a point measured 50 ft along the pit bottom. The ore width associated with hole 5 becomes 75 ft.

Final pit outline
The surface is assumed to be flat and at 0 elevation. A bench height of 25 will be used. Through an examination of all sections, it is seen that the overburden-rock interface lies at an average elevation of about 100 ft. In the plan view shown in Figure 3.31, the surface ($\times$) and 100 ft intercepts ($\circ$) read from the 6 sections have been marked.

Orebody description 205

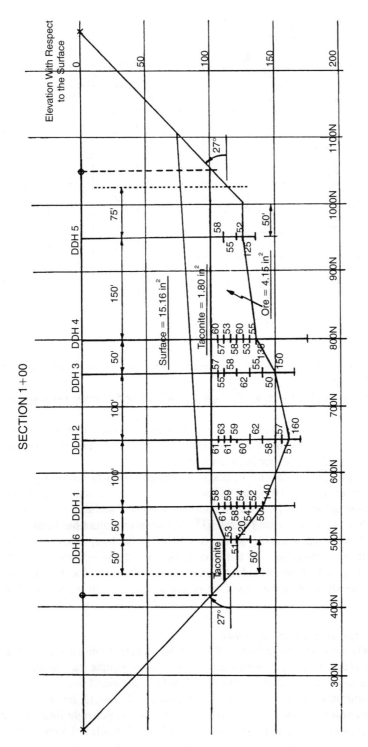

Figure 3.30. Section used for the average sectional grade and tonnage calculation.

206 *Open pit mine planning and design: Fundamentals*

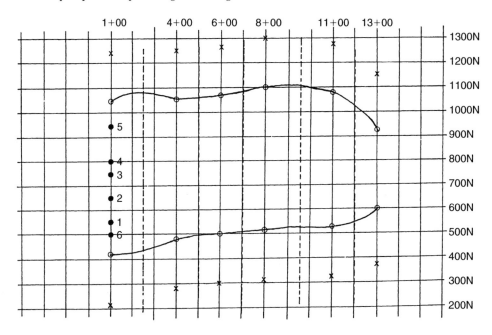

Figure 3.31. Plan smoothing of the pit limits.

The ends of the pit must now be completed. There are several possible constructions which might be used.

1. Construct longitudinal sections through the orebody and estimate ore continuation by projection.
2. Extend the ore a distance equal to the ore thickness observed in the end section.
3. Extend the ends some fixed distance past the last section. For example 100 ft or a distance equal to half the distance between sections.

In this particular case, the second method was selected. As can be seen in Figure 3.30, the maximum ore thickness of 60 ft occurs in hole 2. This thickness extends in plan from holes 1 to 4. Projection of the lines from the pit bottom to the 100 ft and surface elevations yields the points a, b, c and d shown on Figure 3.32. The same procedure has been followed on the east end of the pit where the ore thickness is 40 ft. The final step is to connect the points by smooth curves. In this case, the transition between sections is smooth and no further adjustment is required. If the individual sections do not fit together as nicely as in Figure 3.32, then obviously an iterative procedure of going from plan to section to plan, etc. is required.

Tonnage and average grade for a section
Section 1 + 00 is defined by 6 drillholes. The first step in determining average grade for the section is to find the average grade for each drill hole. If the sample interval was always the same then a simple average of the assays would suffice. In general this is not the case and compositing must be done. The results of this are given in Table 3.12. An influence area for each hole must now be calculated. This area is the ore intercept height times a width. For interior holes, the width extends halfway to adjacent holes. For side holes the width extends

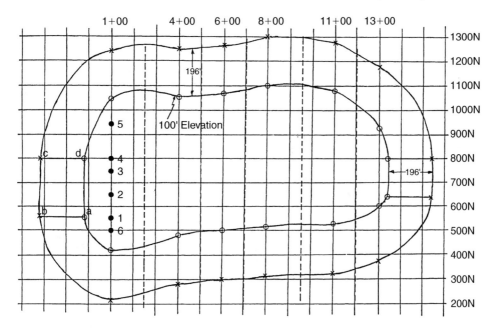

Figure 3.32. Addition of pit ends.

from the side line halfway to the adjacent hole. For this section, the influence areas are as shown in Figure 3.33 and summarized in Table 3.13. Using the grades and influence areas, an average grade for the section is determined by compositing. In this case it is 56.8% Fe.

Tons and average grade for the pit
The ore volume attributed to the section is obtained by multiplying the section area (determined by calculation or planimeter) by the section interval which for section 1 + 00 is 210 ft (Table 3.14). For this section the ore volume is 4,357,500 ft^3. Applying a tonnage factor of 14 ft^3/lt yields 311,000 tons.

Similar figures are developed for each of the other sections. The average grade for the pit is obtained by compositing. The total ore tonnage is just the sum of the tons (A).

The associated waste consists of two types:
– surface material, and
– taconite.

On each section, the areas of each material type are determined by planimetering. The results for section 1 + 00 are given in Tables 3.15 and 3.16.

Pit end tonnage
A somewhat troublesome problem is deciding how to include the material making up the pit ends. The west end of the pit will be examined in this example assuming that all of the material above the 100 ft contour is surface material. In Figure 3.33, the end has been divided into five sectors which will be approximated by the following two shapes.
– prism (A_1), and
– frustum of a right cone ($A_2 \rightarrow A_5$).

Table 3.12. Calculating average hole analyses (Pfleider, 1962).

Cross-sect.	DH	From	To	Length (ft)	Avg Analysis-dry (Fe%)	Length × Analysis (ft × %Fe)
1+00	1	100	105	5	58	290
		105	110	5	61	305
		110	115	5	59	295
		115	120	5	58	290
		120	125	5	54	270
		125	130	5	54	270
		130	135	5	52	260
		135	140	5	50	250
	Avg	100	140	40	55.8	2230
	2	100	105	5	61	305
		105	110	5	63	315
		110	115	5	61	305
		115	120	5	59	295
		120	130	10	60	600
		130	140	10	62	620
		140	150	10	58	580
		150	155	5	57	285
		155	160	5	51	255
	Avg	100	160	60	59.3	3560
	3	100	105	5	57	285
		105	110	5	55	275
		110	120	10	58	580
		120	130	10	62	620
		130	140	10	55	550
		140	150	10	50	500
	Avg	100	150	50	56.2	2810
	4	100	105	5	60	300
		105	110	5	57	285
		110	115	5	53	265
		115	120	5	58	290
		120	125	5	60	300
		125	130	5	53	265
		130	135	5	55	275
	Avg	100	135	35	56.6	1980
	5	100	110	10	58	580
		110	120	10	55	550
		120	125	5	52	260
	Avg	100	125	25	55.6	1390
	6	110	115	5	53	265
		115	120	5	51	255
	Avg	110	120	10	52.0	520

An isometric drawing of sectors A_1, A_2 and A_3 is shown in Figure 3.34. The individual parts are shown in Figure 3.35. The general formula for the volume of a prism is

$$V_p = \frac{1}{2}(S_1 + S_2)h \tag{3.12}$$

Orebody description 209

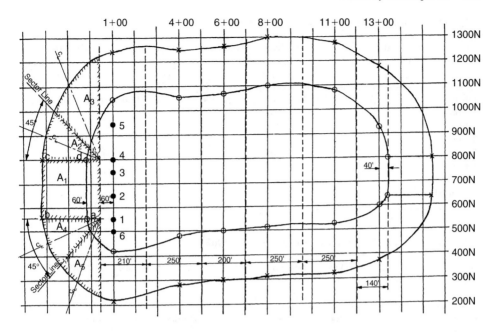

Figure 3.33. Pit end volume calculations.

Table 3.13. Calculating average section analyses (Pfleider, 1962).

Cross-sect.	Hole No.	From	To	Length	Dist. bet holes	Area (ft²)	Avg Analysis (%Fe)	Area × % (ft² × %Fe)
1+00	1	100	140	40	75	3,000	55.8	167,400
	2	100	160	60	100	6,000	59.3	355,800
	3	100	150	50	75	3,750	56.2	210,750
	4	100	135	35	100	3,500	56.6	198,100
	5	100	125	25	150	3,750	55.6	208,500
	6	110	120	100	75	750	52.0	39,000
	Total					20,750	56.8	1,180,000

where S_1, S_2 are the areas of the top and bottom surfaces, respectively, and h is the altitude. The formula for the volume of a right cone is

$$V_c = \frac{1}{3}\pi r^2 h \tag{3.13}$$

where r is the radius of the base.
The formula for the volume of the frustum of a right cone is

$$V_{fc} = \frac{\pi h}{3}(r_1^2 + r_1 r_2 + r_2^2) \tag{3.14}$$

where r_1 is the radius of the base and r_2 is the radius of the top.

210 Open pit mine planning and design: Fundamentals

Table 3.14. Summary sheet for ore tons and grade (Pfleider, 1962).

Cross-sect.	Planimetered area (in²)	Area factor (ft²/in²)	Section area (ft²)	Section interval (ft)	Volume (ft³)	Tonnage factor (ft³/lt)	Tons (lt)	Grade (%) (%Fe)	Tons × % (lt × %Fe)
1+00	4.15	50 × 100	20,750	210	4,357,500	14	311,000	56.8	17,664,800
4+00				250					
6+00				200					
8+00				250					
11+00				250					
13+00				140					
Grand total							A	B/A	B

Table 3.15. Summary sheet for surface material (Pfleider, 1962).

Cross-sect.	Planimetered area (in²)	Area factor (ft²/in²)	Section area (ft²)	Section interval (ft)	Volume (ft³)	Volume (yd³)	Tonnage factor (ft³/st)	Tons (st)
1+00	15.16	50 × 100	75,800	210	15,918,000	590,000	19	838,000
4+00								
6+00								
8+00								
11+00								
13+00								
Total						A		B

Table 3.16. Summary sheet for taconite (Pfleider, 1962).

Cross-sect.	Plan. area (in²)	Area factor (ft²/in²)	Section area (ft²)	Section interval (ft)	Volume (ft³)	Tonnage factor (ft³/st)	Tons (st)
1+00							
4+00							
6+00							
8+00							
11+00							
13+00							
Total							C

Applying formula (3.12) to sector A_1, one finds

$$V_{A_1} = \frac{1}{2}(60 \times 240 + 256 \times 240)100 = 3{,}792{,}000 \text{ ft}^3$$

For sector A_2, base and top radii are determined along the sector centerlines. These become:

$r_1 = 293$ ft

$r_2 = 60$ ft

Orebody description   211

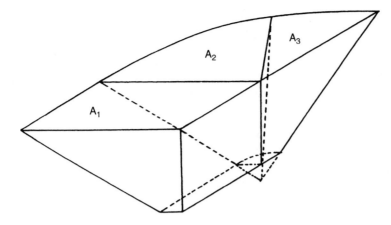

Figure 3.34. Isometric view of sections A_1, A_2 and A_3.

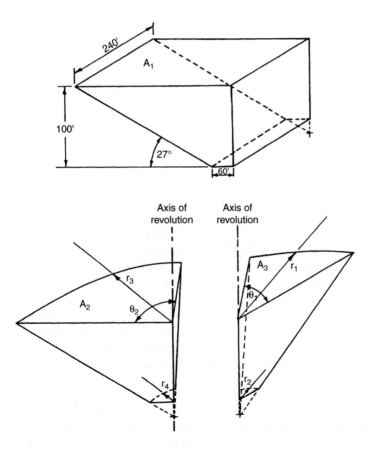

Figure 3.35. Exploded view of the pit end volumes.

212 *Open pit mine planning and design: Fundamentals*

The included angle (θ) of the sector is 45°. Hence

$$V_{A_2} = \frac{\pi}{3}100(293^2 + 293 \times 60 + 60^2)\frac{45}{360} = 1,401,010\,\text{ft}^3$$

For sector A_3:

$\theta_2 = 45°$

$r_3 = 387\,\text{ft}$

$r_4 = 93\,\text{ft}$

$$V_{A_3} = \frac{\pi}{3}100(387^2 + 387 \times 93 + 93^2)\frac{45}{360} = 2,544,814\,\text{ft}^3$$

Similarly for sectors A_4 and A_5:

$$V_{A_4} = \frac{\pi}{3}100(293^2 + 293 \times 53 + 53^2)\frac{45}{360} = 1,363,808\,\text{ft}^3$$

$$V_{A_5} = \frac{\pi}{3}100(346^2 + 346 \times 93 + 93^2)\frac{45}{360} = 2,101,508\,\text{ft}^3$$

The total volume of the west end then becomes

$$V_{we} = 11,204,000\,\text{ft}^3$$

Applying a tonnage factor of 19 ft³/st, yields 590,000 st. The split between taconite and rock can be found by including the interface in the drawings.

In estimating actual grade and tonnage from the pit one must take into account:
– ore losses in pit,
– dilution, and
– mill recovery.

3.6 METHOD OF VERTICAL SECTIONS (GRADE CONTOURS)

A less commonly used technique to represent grades on sections is through iso-grade contours. Although there are several reasons for adopting such an approach, the primary one in the example to be considered is the fact that the directions of the exploratory drill holes were highly variable. This example, using the data and approach of Cherrier (1968), will demonstrate the application of the technique for determining tons and average grade and also the application of longitudinal sections for completing pit ends. Figure 3.36 is a plan map for a molybdenum orebody showing the location of the drill holes and the surface topography. The grid system has been superimposed. Sections 4, 8, 12, 16 and 20 are given in Figures 3.37 through 3.41. The drill holes on Sections 2 and 22 revealed no ore present. Using the drill hole data and a knowledge of the deposit form, the grade contours have been created. A pit outline has been developed using procedures which will be described in Chapter 5. In viewing the 5 transverse sections, it can be seen that longitudinal section 3200 N runs approximately along the proposed pit axis. This section is shown in Figure 3.42. Rock mechanics studies have suggested a 45° slope angle at the east end and 50° at the west. From the drilling results, the orebody must terminate between Sections 2 to 4 on the west and 20 to 22 on the east. The grade contour lines have been drawn in to represent this (Fig. 3.43) condition. In examining the plan representation (Fig. 3.44) of the final pit,

Orebody description 213

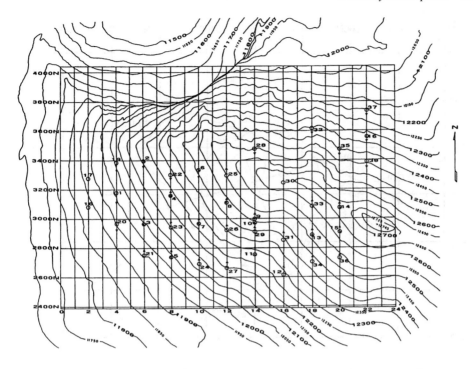

Figure 3.36. Plan map showing drill hole locations (Cherrier, 1968).

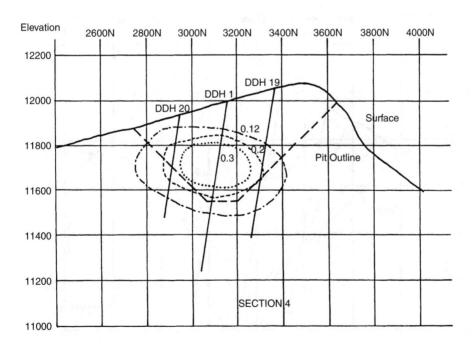

Figure 3.37. Section 4 showing the grade contours and pit outline (Cherrier, 1968).

214 *Open pit mine planning and design: Fundamentals*

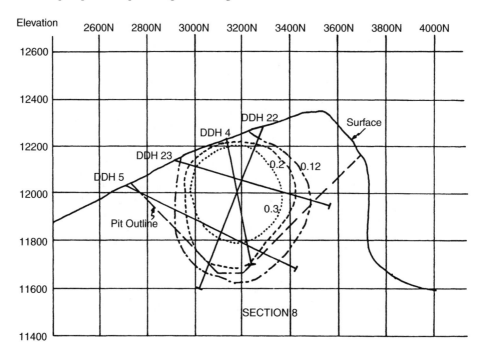

Figure 3.38. Section 8 showing the grade contours and pit outline (Cherrier, 1968).

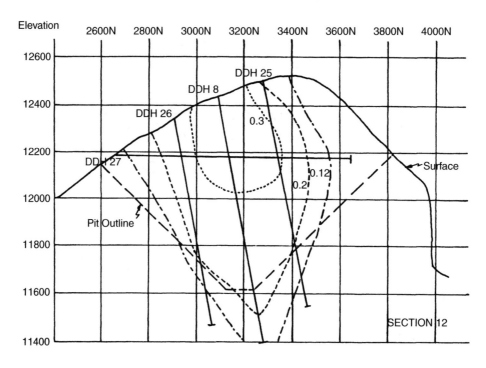

Figure 3.39. Section 12 showing the grade contours and pit outline (Cherrier, 1968).

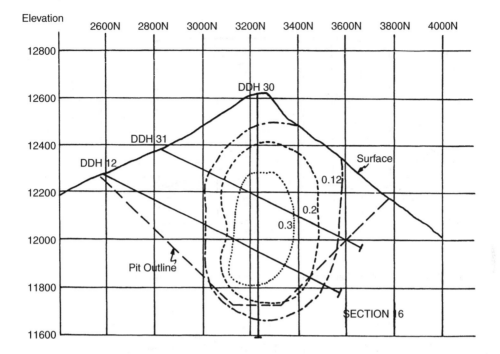

Figure 3.40. Section 16 showing the grade contours and pit outline (Cherrier, 1968).

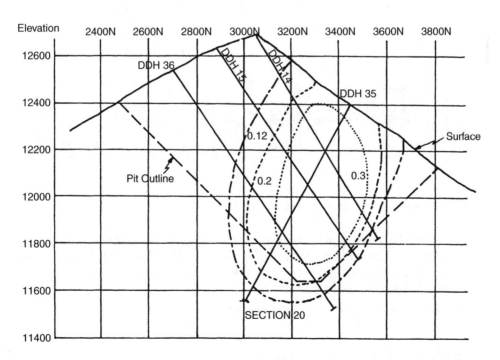

Figure 3.41. Section 20 showing the grade contours and pit outline (Cherrier, 1968).

216 *Open pit mine planning and design: Fundamentals*

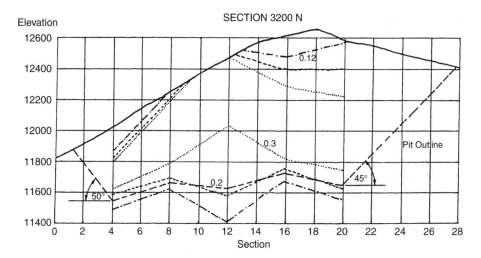

Figure 3.42. Longitudinal Section 3200 N showing the grade contours, surface topography and pit outline (Cherrier, 1968).

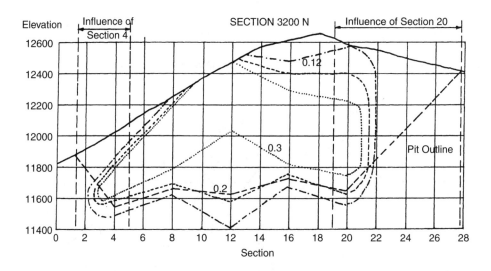

Figure 3.43. Grade contours completed at the pit ends.

and the longitudinal section, it has been decided to incorporate the end volumes of the pit into Sections 4 and 20 rather than to treat them separately. This is accomplished by varying the longitudinal length of influence so that the resulting volumes are correct. For the east and west ends the following have been used:

Section 4	Grade zone	Influence distance (ft)
	+0.3	200
	0.2–0.3	225
	0.12–0.2	250
	Overburden	300

Orebody description 217

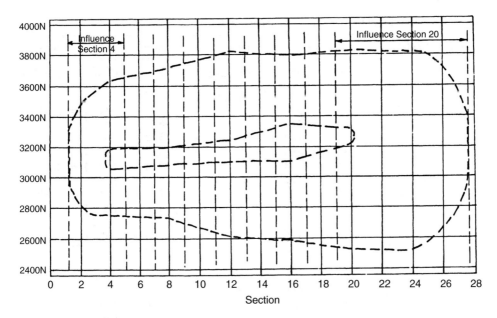

Figure 3.44. Plan view showing outline of pit crest and toe.

20	+0.3	200
	0.2–0.3	250
	0.12–0.2	300
	Overburden	400

The reader is encouraged to check these approximations. For each section the areas corresponding to each grade zone within the pit outline are determined using a planimeter. These are then converted into volumes and tons as summarized in Table 3.17. A summary of ore and waste tons as well as the ratio of the two (average stripping ratio) by section is given in Table 3.18. The overall stripping ratio for the pit is also calculated.

One assay was performed for each 10 ft of core in every drill hole. For each section, the assays lying within a particular grade zone were added together. This sum was then divided by the number of assays to obtain an average. Each average assay was weighted by the volume of influence for the respective zone. These weighted averages were totalled and divided by the total volume for the three zones. This figure is then the average assay for the entire ore zone on that section. The results of this calculation are given in Table 3.19. The overall average grade for the pit is found in Table 3.20.

In summary:

Tonnage of ore = 53,288,000 st
Average grade = 0.281 percent (MoS_2)
Tonnage of overburden = 32,311,000 st
Average stripping ratio = 0.61

Some dilution and ore loss can be expected from this orebody due to the difficulty of defining the ore cut-off grade and mining to this grade in the upper half of the orebody. The following calculation illustrates how one might account for these effects and determine an average grade for the ore actually mined.

Table 3.17. Calculation of tonnage of ore and overburden (Cherrier, 1968).

Section	Grade zone	Area[1] (in²)	Area[1] (10⁴ ft²)	Influence distance (ft)	Volume (10⁶ ft³)	Tons[2] (×10³)	Total tons (×10³)
4	+0.3	1.32	5.28	200	10.56	845	
	0.2–0.3	0.84	3.36	225	7.56	605	= 2,266
	0.12–0.2	1.02	4.08	250	10.20	816	
	Overburden	2.81	11.24	300	33.72	2,698	2,698
6	+0.3	2.52	10.08	200	20.16	1,613	
	0.2–0.3	3.46	13.84	200	27.68	2,214	= 5,075
	0.12–0.2	1.95	7.80	200	15.60	1,248	
	Overburden	2.17	8.68	200	17.36	1,389	1,389
8	+0.3	2.97	11.88	200	23.76	1,901	
	0.12–0.3	1.66	6.64	200	13.28	1,062	= 3,763
	0.12–0.2	1.25	5.00	200	10.00	800	
	Overburden	2.62	10.48	200	20.96	1,677	1,677
10	+0.3	2.17	8.68	200	17.36	1,389	
	0.2–0.3	3.00	12.00	200	24.00	1,920	= 4,423
	0.12–0.2	1.74	6.96	200	13.92	1,114	
	Overburden	5.66	22.64	200	45.28	3,622	3,622
12	+0.3	3.11	12.44	200	24.88	1,990	
	0.2–0.3	6.64	26.56	200	53.12	4,250	= 7,872
	0.12–0.3	2.55	10.20	200	20.40	1,632	
	Overburden	3.17	12.68	200	25.36	2,029	2,029
14	+0.3	2.46	9.84	200	19.68	1,574	
	0.2–0.3	4.30	17.20	200	34.40	2,752	= 5,657
	0.12–0.2	2.08	8.32	200	16.64	1,331	
	Overburden	6.90	27.60	200	55.20	4,416	4,416
16	+0.3	2.57	10.28	200	20.56	1,645	
	0.2–0.3	3.18	12.72	200	25.44	2,035	= 5,760
	0.12–0.2	3.25	13.00	200	26.00	2,080	
	Overburden	5.51	22.40	200	44.08	3,526	3,526
18	+0.3	4.09	16.36	200	32.72	2,618	
	0.2–0.3	5.29	37.52	200	75.04	6,003	= 10,445
	0.12–0.2	2.85	11.40	200	22.80	1,824	
	Overburden	6.10	24.40	200	48.80	3,904	3,904
20	+0.3	4.99	19.96	200	39.92	3,194	
	0.2–0.3	3.27	13.08	250	32.70	2,616	= 8,181
	0.12–0.2	2.47	9.88	300	29.64	2,371	
	Overburden	7.07	28.28	400	113.12	9,050	9,050

[1] Planimetered: $1 \text{ in}^2 = 200' \times 200' = 40,000 \text{ ft}^2$.
[2] Based on a tonnage factor of $12.5 \text{ ft}^3/\text{st}$.

	Tons (st)	Grade (%)	Tons × Grade (st × %)
Ore in place	53.0×10^6	0.281	14.893×10^6
Dilution (est. 5%)	$+2.65 \times 10^6$	0.100	$+0.265 \times 10^6$
Ore loss (est. 5%)	-2.65×10^6	0.140	-0.371×10^6
Total	53.0×10^6		14.787×10^6

$$\text{Average grade of ore mined} = \frac{14.787}{53} = 0.279 \text{ percent.}$$

Table 3.18. Calculation of the average stripping ratio.

Section	Tons of ore (st × 10³)	Tons of overburden (st × 10³)	Avg stripping ratio
4	2,266	2,698	1.191
6	5,075	1,389	0.274
8	3,763	1,677	0.446
10	4,423	3,622	0.819
12	7,872	2,029	0.258
14	5,657	4,416	0.781
16	5,760	3,526	0.612
18	10,445	3,904	0.374
20	8,181	9,050	1.106
Total	53,442	32,311	

Overall average stripping ratio: $\dfrac{32,311}{53,442} = 0.605$.

3.7 THE METHOD OF HORIZONTAL SECTIONS

3.7.1 *Introduction*

Although vertical sections have played a dominant role in ore reserve estimation in the past, today, for many, if not most, deposits this function is rapidly being replaced by techniques based upon the use of horizontal sections. The primary reason being the widespread availability of computers for doing the tedious, time consuming calculations involved and the development of new techniques for estimating the grades between drill holes. Sections taken in the plane of the orebody have generally been used for evaluating relatively thin, flat lying deposits such as uranium, coal, sand, gravel, placer gold, etc. They may be of relatively uniform or varying thickness. Thick deposits are mined in a series of horizontal slices (benches) of uniform thickness. For extraction planning, bench plans showing tons and grade are of utmost importance. Hence even if vertical sections are used for initial evaluation, bench (horizontal) sections are eventually required. In this section hand methods for calculating tons and grade based on triangles and polygons will be discussed. These discussions will use as a basis the drillholes shown in Figure 3.45. The corresponding grades and location coordinates are given in Table 3.21.

3.7.2 *Triangles*

In the *triangular* method, diagrammatically illustrated in Figure 3.46, each hole is taken to be at one corner of a triangle. If the triangular solid formed is of constant thickness t, its volume is just equal to the plan area A times this thickness. To obtain tons, the appropriate tonnage factor is applied. The average grade $\bar{g}$ is given by

$$\bar{g} = \frac{g_1 + g_2 + g_3}{3} \tag{3.15}$$

where g_i are the grades at the three corners.

The area of the triangle can be found using a planimeter or through calculation if the coordinates (x_i, y_i) of the corners are known. This method, shown in Figure 3.47, readily adapts to calculator/computer application. The average grade for the block is strictly correct

Table 3.19. Calculation of average assay for each section (Cherrier, 1968).

Section	Grade zone		Avg assay (%)	Volume ($\times 10^6$ ft^3)	Assay × Volume (% $\times 10^6$ ft^3)	Avg assay for section
4	+0.3		0.452	10.56	4.7731	
	0.2–0.3		0.270	7.56	2.0412	
	0.12–0.2		0.135	10.20	1.3770	
		Total		28.32	8.1913	0.289
6	+0.3		0.435	20.16	8.7696	
	0.2–0.3		0.242	27.68	6.6986	
	0.12–0.2		0.155	15.60	2.4180	
		Total		63.44	17.8862	0.282
8	+0.3		0.485	23.76	11.5236	
	0.2–0.3		0.246	13.28	3.2669	
	0.12–0.2		0.153	10.00	1.5300	
		Total		47.04	16.3205	0.347
10	+0.3		0.379	17.36	6.5794	
	0.2–0.3		0.261	24.00	6.2640	
	0.12–0.2		0.151	13.92	2.1019	
		Total		55.28	14.9453	0.270
12	+0.3		0.411	24.88	10.2257	
	0.2–0.3		0.243	43.12	12.9082	
	0.12–0.2		0.142	20.40	2.8968	
		Total		98.40	26.0307	0.265
14	+0.3		0.403	19.68	7.9310	
	0.02–0.3		0.242	34.40	8.3248	
	0.12–0.2		0.168	16.64	2.7955	
		Total		70.72	19.0513	0.269
16	+0.3		0.398	20.56	8.1829	
	0.2–0.3		0.247	25.44	6.2837	
	0.12–0.2		0.155	26.00	4.0300	
		Total		72.00	18.4966	0.257
18	+0.3		0.393	32.72	12.8590	
	0.2–0.3		0.274	75.04	20.5610	
	0.12–0.2		0.158	22.80	3.6024	
		Total		130.56	37.0224	0.284
20	+0.3		0.405	39.92	16.1676	
	0.2–0.3		0.261	32.70	8.5347	
	0.12–0.2		0.164	29.64	4.8610	
		Total		102.26	29.5633	0.289

only for equilateral triangles. For other triangle shapes the area associated with each grade is not equal as assumed in the formula. For triangular solids which are not of constant thickness, then some additional calculations are required. The average thickness $\bar{t}$ is given by

$$\bar{t} = \frac{t_1 + t_2 + t_3}{3} \tag{3.16}$$

and the average grade $\bar{g}$ is

$$\bar{g} = \frac{g_1 t_1 + g_2 t_2 + g_3 t_3}{3\bar{t}} \tag{3.17}$$

Table 3.20. Calculation of overall average grade (Cherrier, 1968).

Section	Average assay	Volume ($\times 10^6$ ft^3)	Assay × Volume (% × 10^6 ft^3)
4	0.289	28.32	8.1845
6	0.282	63.44	17.8901
8	0.347	47.04	16.3229
10	0.270	55.28	14.9256
12	0.265	98.40	26.0760
14	0.269	70.72	19.0237
16	0.257	72.00	18.5040
18	0.284	130.56	37.0790
20	0.289	102.26	29.5531
Total		668.02	187.5589

Overall average grade of orebody: $\dfrac{187.5589}{668.02} = 0.281$ percent.

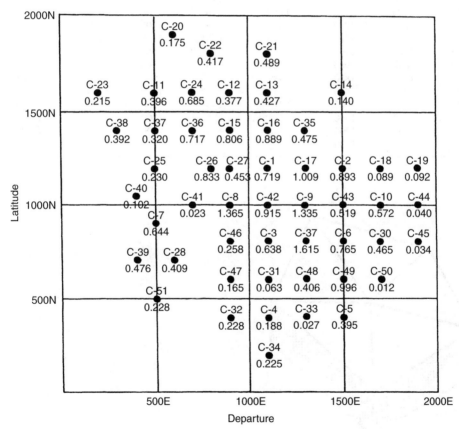

Figure 3.45. Borehole locations and grades for Bench 5140 (Hughes & Davey, 1979).

222 Open pit mine planning and design: Fundamentals

Table 3.21. Coordinates and grades for the holes of Figure 3.45. The composite grades are for Bench 5140 (Hughes & Davey, 1979).

Hole	Coordinates		Grade (% Cu)	Hole	Coordinates		Grade (% Cu)
	East	North			East	North	
C-1	1100	1200	0.719	C-27	900	1200	0.453
C-2	1500	1200	0.893	C-28	600	700	0.409
C-3	1100	800	0.638	C-29	1300	800	1.615
C-4	1100	400	0.188	C-30	1700	800	0.465
C-5	1500	400	0.395	C-31	1100	600	0.063
C-6	1500	800	0.765	C-32	900	400	0.224
C-7	500	900	0.644	C-33	1300	1400	0.027
C-8	900	1000	1.365	C-34	1100	200	0.225
C-9	1300	1000	1.335	C-35	1300	1400	0.475
C-10	1700	1000	0.072	C-36	700	1400	0.717
C-11	500	1600	0.396	C-37	500	1400	0.320
C-12	900	1600	0.377	C-38	300	1400	0.392
C-13	1100	1600	0.427	C-39	400	700	0.476
C-14	1500	1600	0.140	C-40	400	1050	0.102
C-15	900	1400	0.806	C-41	700	1000	0.023
C-16	1100	1400	0.889	C-42	1100	1000	0.915
C-17	1300	1200	1.009	C-43	1500	1000	0.519
C-18	1700	1200	0.089	C-44	1900	1000	0.040
C-19	1900	1200	0.092	C-45	1900	800	0.034
C-20	600	1900	0.175	C-46	900	800	0.258
C-21	1100	1800	0.489	C-47	900	600	0.165
C-22	800	1800	0.417	C-48	1300	600	0.406
C-23	200	1600	0.215	C-49	1500	600	0.996
C-24	700	1600	0.685	C-50	1700	600	0.012
C-25	500	1200	0.230	C-51	500	500	0.228
C-26	800	1200	0.833				

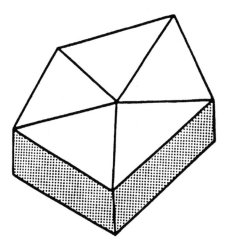

Figure 3.46. Triangular prisms (Barnes, 1979b).

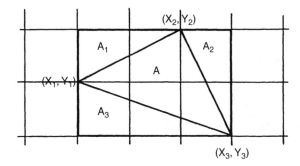

$$A = (X_3 - X_1)(Y_2 - Y_1) - 1/2 (X_3 - X_1)(Y_1 - Y_3) - 1/2 (X_3 - X_2)(Y_2 - Y_3)$$

$$- 1/2 (X_2 - X_1)(Y_2 - Y_1)$$

$$A = \text{Area of Rectangle} - A_1 - A_2 - A_3$$

$$A = (X_{max} - X_{min})(Y_{max} - Y_{min}) - 1/2 \sum_{\substack{i=1,3 \\ j=1,3}} (X_i - X_j)(Y_i - Y_j)$$

Figure 3.47. Calculation of triangular area based upon corner coordinates.

Using the average thickness, the plan area and tonnage factor, the tons can be found. The triangular element formed by holes C-30, C-40, and C-50 (assuming a constant bench thickness of 40 ft and a tonnage factor of 12.5 ft^3/ton), contains 64,000 tons with an average grade 0.17.

3.7.3 Polygons

In the *polygonal* method, each drill hole lies within the center of a polygon. The polygon is constructed such that its boundary is always equidistant from the nearest neighboring hole. Within the polygon, the grade is assumed constant and equal to that of the hole it includes. The thickness of the polygon is also constant and equal to the ore intercept/bench thickness. The steps followed in forming a polygon around hole C-41 are illustrated in Figure 3.48. In step 1, radial lines (similar to the spokes of a wheel) are drawn from the drill hole to its nearest neighbors. The perpendicular bisectors to these lines are constructed and extended until they meet those from adjacent holes (step 2). The area of the polygon is then determined and the tonnage calculated (step 3). At the drilling boundary, since there are holes on only one side, some special procedures are required. Here it will be assumed that an appropriate radius of influence R is known. This concept will be discussed in detail in the following section. Figure 3.49 illustrates the steps necessary to construct the polygon around hole C-14. Step 1 proceeds as before with radial lines drawn to the surrounding holes. To supply the missing sides a circle of radius R is drawn (Step 2). In this case $R = 250$ ft. Chords are drawn parallel to the property boundary (grid) lines along the top and side (Step 3). The

224 *Open pit mine planning and design: Fundamentals*

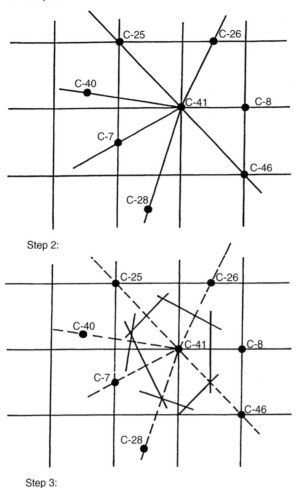

Figure 3.48. The practice of forming polygons for a hole internal to the array.

remaining chords are drawn at angles of 45°, tangent to the circle. In the final step (Step 4) the area is determined, the tonnage calculated and the grade assigned.

Rules developed by Hughes & Davey (1979) which can be followed when constructing polygons are given in Table 3.22. Figure 3.50 shows the hand generated polygons for the drill hole data of Figure 3.45.

Having gone through these two examples it is perhaps of value to list the general steps (after Hughes & Davey, 1979) which are followed:

1) Locations of drill holes and other samples are established for a specified level using available drill-hole survey data. Usually, the drill-hole location and assays of interest are depicted on a horizontal section.

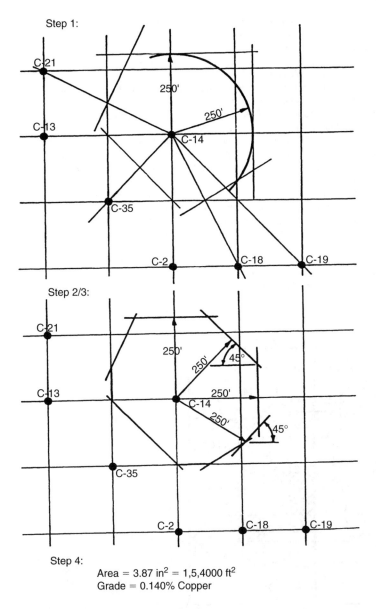

Figure 3.49. The practice of forming polygons at the boundary of the hole array.

2) Drill-hole interval assay data are composited to intervals consistent with bench height. The elevation of the sample is typically determined at the midpoint of the bench.

3) Area of influence or radius of influence is established by geologic and mining experience.

4) Lines are drawn between drill holes that are within two times the radius of influence of each other. This step may be altered by rules such as those in Table 3.22.

5) Perpendicular bisectors are constructed on each of these connecting lines.

226 Open pit mine planning and design: Fundamentals

Table 3.22. Example of polygon interpolation rules (Hughes & Davey, 1979).

1. Ultimate polygon shape is octagonal (eight-sided).
2. Radius of influence is R ft.
3. No polygon exceeds $2R$ ft from a sample point.
4. If drill holes are in excess of $5R$ ft apart, use a radius of R ft to show trend into undrilled area.
5. If holes are in excess of $4R$ ft, but less than $5R$ ft apart:
    a) Construct an R ft radius circle if assays are of unlike character, i.e., rock types, different mineralization, or one ore and the other waste.
    b) Use a $2R$ ft radius if assays are of like character to locate a point on a line between the holes; a line is then drawn to the point and tangent to the R ft diameter circle.
6. For holes less than $4R$ ft apart, construct a perpendicular bisector between the holes.
    a) If the holes are between $3R$ and $4R$ ft apart, use an R ft radius circle and connect the circles by drawing wings at a 30° angle from the center of the R ft radius circle to a $2R$ ft radius circle.
    The perpendicular bisector constructed above becomes the dividing line between $2R$ ft arc intersections.
    b) If the holes are less than $3R$ ft apart and a polygon cannot be constructed entirely from perpendicular lines from adjacent holes, then use a R ft radius circle and connect the circles with tangent lines.
    The dividing line is the perpendicular bisector between the holes.
7. After one pair of holes has been analyzed using these rules, another pair is evaluated, and this procedure is repeated until all combinations have been evaluated.
8. The assay value of the polygon will be the composited assay value of the drill hole that the polygon was constructed around.

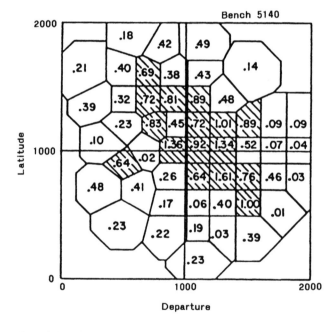

Figure 3.50. Hand-generated polygons from samples in Figure 3.45 (Hughes & Davey, 1979).

6) Bisectors are extended until they intersect. If two lines run parallel or approximately parallel, and it is obvious that they will not intersect before the line closest to the drill hole intersects another line, the bisector that is closest to the drill hole is accepted as the polygon boundary.

7) In areas where drill holes are separated by distances greater than two times the radius of influence, an eight-sided polygon (octagon) form is drawn around the hole location, representing the maximum area of influence. This step may also be altered by rules such as those in Table 3.22.

8) Drill holes along the periphery of the ore body are extrapolated to the radius of influence and the octagonal form is drawn around the drill hole.

If ore is defined as that material for which the grade is

$$g \geq 0.6\%$$

then for this bench the projected tonnage is 1,990,000 st at an average grade of 0.93. A tonnage factor of 13 ft^3/st (specific gravity SG = 2.47) and a 40 ft bench height has been assumed.

It is obvious that the zone/radius of sample influence is over ridden under a number of special conditions. Some examples of such special conditions are:
– Grades should not be projected from one type of formation, mineralization, rock type, etc. to another.
– Sample grades should not be projected from one side of a post mineralization structure such as a fault to the other.

Rules should be determined to deal with assigning a metal grade to in-place material which is less than a full bench height thick, such as near the surface of the deposit.

Using a computer, lists like Table 3.22 can automatically be considered as well as rules concerning mineralization controls for the specific deposit. However, it is very complicated to have a computer draw lines representing polygon boundaries and to assign area grades according to the procedures described previously.

3.8 BLOCK MODELS

3.8.1 *Introduction*

Basic to application of computer techniques for grade and tonnage estimation is the visualization of the deposit as a collection of blocks. Such a block model is shown in Figure 3.51.

Some guidance for the size of the blocks chosen has been provided by David (1977).

Typically in the profession, people like to know as much as possible about their deposit and consequently they ask for detailed estimation on the basis of the smallest possible blocks. This tendency, besides being possibly unnecessarily expensive will also bring disappointing results. One will find that small neighboring block are given very similar grades. One should remember that as the size of a block diminishes, the error of estimation of that block increases. Also, dividing the linear dimensions of a block by 2, multiplies the number of blocks to be estimated and probably the system of equations to be solved by 8! As a rule of thumb, the minimum size of a block should not be less than ¼ of the average drill hole interval, say 50 ft blocks for a 200 ft drilling grid and 200 ft for an 800 ft drilling grid.

The height of the block is often that of the bench which will be used in mining. Furthermore the location of the blocks depends on a variety of factors. For example a key elevation

228 *Open pit mine planning and design: Fundamentals*

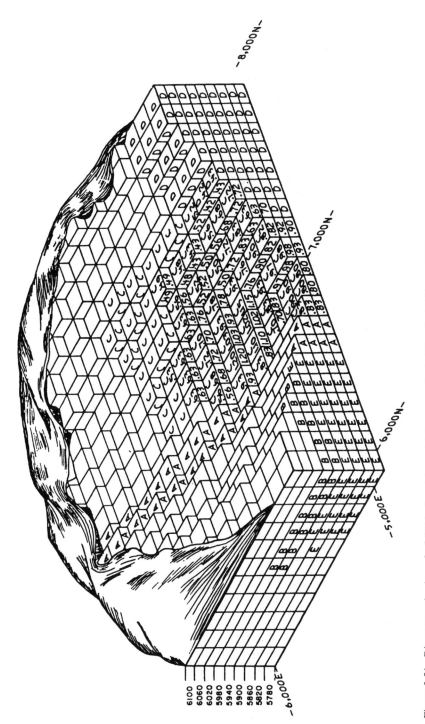

Figure 3.51. Diagrammatic view of a 3-D block matrix containing can orebody (Crawford & Davey, 1979).

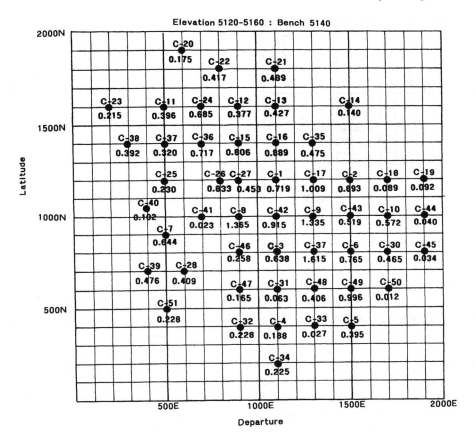

Figure 3.52. Block grid superimposed on Figure 3.45 (Hughes & Davey, 1979).

might be based upon overburden ore contact, the interface between types of mineralization (oxides-sulfides), high grade-low grade zones, etc.

Superposition of a 100 ft × 100 ft block grid on the drill hole data from Figure 3.45 is shown in Figure 3.52. As can be seen, some of the blocks have drill holes in them but most do not.

Some technique must be used to assign grades to these blocks. The tonnage of each block can be easily found from the block volume (the same for all blocks) and the tonnage factor (which may vary). Two techniques will be discussed in this section and an additional one in Section 3.10. They are all based upon the application of the 'sphere of influence' concept in which grades are assigned to blocks by 'weighting' the grades of nearby blocks. Variations in how the weighting factors are selected distinguish the three methods. A simplification which will be made in this discussion is to consider blocks as *point* values rather than as *volumes*. This distinction is illustrated in Figure 3.53. By treating the block as a point one would make one calculation of average block grade based upon the distance from the block center to the surrounding points. If the block is divided into a mesh of smaller blocks, the calculation would be made for each sub-block and the results summed. In the literature this volumetric integration is denoted by integral or summation symbols. Hughes & Davey (1979) has

230 *Open pit mine planning and design: Fundamentals*

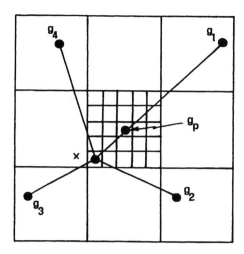

Figure 3.53. Block versus point representation.

indicated that the difference between the point and volume approach is small. We have chosen to take the least complicated approach in presenting the principles. Furthermore, a two-dimensional approach will be focussed upon with only passing reference to extensions into 3 dimensions. The examples used will focus on assignment of grades for a bench using composite grades for that bench alone. Grades lying above or below the bench in question will not be included in the calculations. Finally unless specifically mentioned, all of the grades will be assumed to belong to the same mineralization type and are all useable in assigning grades to the blocks, i.e. there are no characteristics which eliminate certain values (change in mineralization, formation, rock type, structural features). The reader will see how these can be considered.

3.8.2 *Rule-of-nearest points*

The polygon approach described in the previous chapter is an example of the rule-of-nearest points. The area surrounding a drill hole is defined in such a way that the boundary is always equidistant from nearest points. Although computer programs now do exist for doing this procedure, Hughes & Davey (1979) suggests that little accuracy is lost using a regular grid. The computer calculates the distances from the block centers to the surrounding known grade locations, and assigns the grade to the block of the closest grade. If the closest distance is greater than R, no value is assigned. In some cases, the block center may be equidistant from two or more known grades. A procedure must be established to handle this. Sometimes an average value is assigned.

Figure 3.54 shows the application of a computerized polygonal interpolation to the composited values shown as level 5140 in Figure 3.45. If the block contains a hole, it is assigned that value. Blocks without holes are assigned the value of the nearest hole within a 250 ft radius. For blocks having centers outside of this radius a value of 0 has been assigned. The shaded area has been interpolated as mineralization $\geq 0.6\%$ Cu. Because the distance from block to composite is computed from the block center, results vary slightly from the polygons defined in Figure 3.50. Accumulation of blocks with projected grades $\geq 0.6\%$ Cu is calculated as 2,033,778 st at an average grade of 0.92%.

Orebody description 231

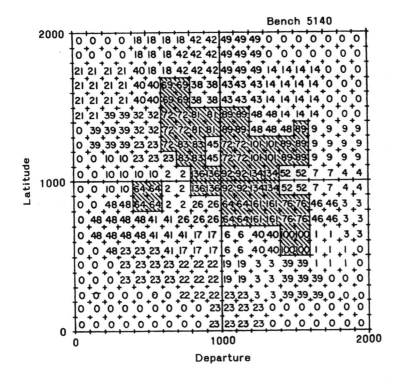

Figure 3.54. Computer generated polygons for Figure 3.45 (Hughes & Davey, 1979).

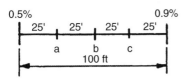

Figure 3.55. Simple example for grade calculation.

3.8.3 *Constant distance weighting techniques*

In the previous technique, the grade was assumed to remain constant over a region extending halfway to the adjacent hole. As the boundary between blocks is crossed, the grade drops to that in the adjacent region. The grade at a point was determined only by the closest grade and none other. A more sophisticated approach would be to allow all of the surrounding grades to influence grade estimation at a point. Figure 3.55 illustrates the assignment of grades along a line between two known grades. Assuming a linear change in grade between the two known grades (Fig. 3.56) one can calculate the expected grades at points a, b, and c. The formula used to calculate this can be written as

$$g = \frac{\sum_{i=1}^{n} \frac{g_i}{d_i}}{\sum_{i=1}^{n} \frac{1}{d_i}} \tag{3.18}$$

where g_i is the given grade at distance d_i away from the desired point.

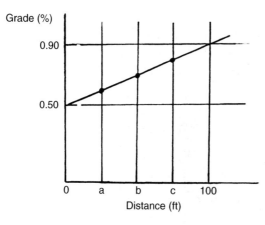

Figure 3.56. Linear variation of grade with separation distance.

For point a

$$g_a = \frac{\frac{0.5}{25} + \frac{0.9}{75}}{\frac{1}{25} + \frac{1}{75}} = \frac{1.5 + 0.9}{3 + 1} = \frac{0.24}{4} = 0.6\%$$

For point b

$$g_b = \frac{\frac{0.5}{25} + \frac{0.9}{75}}{\frac{1}{25} + \frac{1}{75}} = \frac{1.4}{2} = 0.7\%$$

For point c

$$g_c = \frac{\frac{0.5}{75} + \frac{0.9}{25}}{\frac{1}{75} + \frac{1}{25}} = \frac{3.2}{4} = 0.8\%$$

The two-dimensional application of this to the Hughes & Davey (1979) data is shown in Figure 3.57. The calculated grade at the point is given by 0.45%.

This method is called the *inverse distance* weighting technique. The influence of surrounding grades varies inversely with the distance separating the grade and the block center.

It is obvious that the grade of the block should be more similar to nearer points than those far away. To emphasize this dependence, the weighting with distance can be increased. This is done by changing the power of d_i in Equation (3.18). If the dependence varies inversely with the square of the distance rather than linearly, Equation (3.18) becomes

$$g = \frac{\sum_{i=1}^{n} \frac{g_i}{d_i^2}}{\sum_{i=1}^{n} \frac{1}{d_i^2}} \qquad (3.19)$$

This is the commonly used inverse distance squared (IDS) weighting formula.

Applying it to the calculation of grades at points a, b, and c along the line (Fig. 3.55) as before one finds that

$$g_a = \frac{\frac{0.5}{(25)^2} + \frac{0.9}{(75)^2}}{\frac{1}{(25)^2} + \frac{1}{(75)^2}} = \frac{4.5 + 0.9}{10} = 0.54\%$$

Orebody description 233

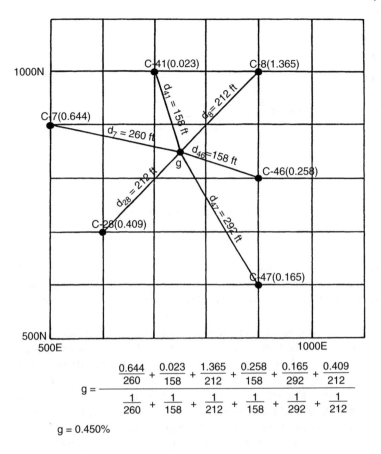

$$g = \frac{\dfrac{0.644}{260} + \dfrac{0.023}{158} + \dfrac{1.365}{212} + \dfrac{0.258}{158} + \dfrac{0.165}{292} + \dfrac{0.409}{212}}{\dfrac{1}{260} + \dfrac{1}{158} + \dfrac{1}{212} + \dfrac{1}{158} + \dfrac{1}{292} + \dfrac{1}{212}}$$

$g = 0.450\%$

Figure 3.57. Application of the inverse distance technique.

$$g_b = \frac{\dfrac{0.5}{(60)^2} + \dfrac{0.9}{(50)^2}}{\dfrac{1}{(50)^2} + \dfrac{1}{(50)^2}} = \frac{0.5 + 0.9}{2} = 0.70\%$$

$$g_c = \frac{\dfrac{0.5}{(75)^2} + \dfrac{0.9}{(25)^2}}{\dfrac{1}{(75)^2} + \dfrac{1}{(25)^2}} = \frac{0.5 + 8.1}{10} = 0.86\%$$

It is obvious that the results are quite different from before. Applying the technique to the 2-D example from Hughes & Davey (Fig. 3.57), one finds that

$$g = \frac{\dfrac{0.644}{(260)^2} + \dfrac{0.023}{(158)^2} + \dfrac{1.365}{(212)^2} + \dfrac{0.258}{(158)^2} + \dfrac{0.165}{(292)^2} + \dfrac{0.409}{(212)^2}}{\dfrac{1}{(260)^2} + \dfrac{1}{(158)^2} + \dfrac{1}{(212)^2} + \dfrac{1}{(158)^2} + \dfrac{1}{(292)^2} + \dfrac{1}{(212)^2}} = 0.412\%$$

If one were to select a different power for d, the results would change. The general formula is

$$g = \frac{\sum_{i=1}^{n} \dfrac{g_i}{d_i^m}}{\sum_{i=1}^{n} \dfrac{1}{d_i^m}} \qquad (3.20)$$

234 *Open pit mine planning and design: Fundamentals*

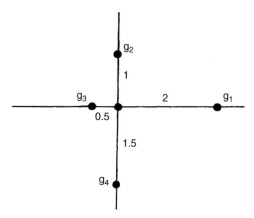

Figure 3.58. Simple example for applying the inverse distance squared (IDS) technique.

For linear dependence $m = 1$, quadratic (squared) dependence, $m = 2$, etc. The value $m = 2$ has been found to be suitable for a number of different kinds of deposits and is widely used. In practice a distribution of m values will be obtained from which a mean can be determined and a best value selected.

To this point in the discussion it has been assumed that the weighting function is independent of the angular position of the known grades with respect to the unknown. Such a function is called isotropic (independent of orientation). This is true for many deposits. For others however the variation of grade with distance does depend upon direction. Thus in one direction, say N-S, the best value for m is m_1 whereas in the E-W direction it would be m_2. Such a deposit would be termed anisotropic. Procedures are available for including these effects. Their discussion is beyond the scope of this book.

Before proceeding, let's take a closer look at the inverse distance squared formula.

$$g = \frac{\sum_{i=1}^{n} \frac{g_i}{d_i^2}}{\sum_{i=1}^{n} \frac{1}{d_i^2}}$$

It will be applied to the simple case shown in Figure 3.58.

Expanding the formula yields

$$g = \frac{\frac{g_1}{4} + \frac{g_2}{1} + \frac{g_3}{0.25} + \frac{g_4}{2.25}}{\frac{1}{4} + \frac{1}{1} + \frac{1}{0.25} + \frac{1}{2.25}}$$

The denominator becomes 5.694 and the equation can be written as

$$g = 0.044 g_1 + 0.176 g_2 + 0.702 g_3 + 0.078 g_4$$

This can be rewritten as

$$g = a_1 g_1 + a_2 g_2 + a_3 g_3 + a_4 g_4 \tag{3.21}$$

where

$$a_i = \frac{\frac{1}{d_i^2}}{\sum_{i=1}^{4} \frac{1}{d_i^2}} \qquad (3.22)$$

The coefficients are:

$a_1 = 0.044$

$a_2 = 0.176$

$a_3 = 0.702$

$a_4 = 0.078$

The sum of the coefficients

$$\sum_{i=1}^{4} a_i = 0.044 + 0.176 + 0.703 + 0.078 = 1 \qquad (3.23)$$

will always equal 1. Furthermore $0 \leq a_i \leq 1$.

In the following section, it will be shown that the geostatistical approach to grade estimation yields the same equation form. The coefficients a_i are simply determined in a different fashion. The constraints on a_i are the same.

To this point, it has been attempted to simply demonstrate how the method works. Some further words are required about the application in practice. Some rules in this regard are given by Hughes & Davey (1979) in Table 3.23. An example showing the application of the following rules:

(1) an angular exclusion of 18° (excludes G_3 and G_5),
(2) maximum of seven nearest holes (excludes G_1 and G_8), and
(3) power $m=2$

is given in Figure 3.59.

Figure 3.60 shows an inverse distance squared ($m=2$) computer evaluation of level 5140. The rules used for the simulation are:

Radius of influence = 250 ft

$m = 2$

Angular exclusion angle = 18°

The accumulation of blocks $\geq 0.6\%$ Cu is calculated to be 2,003,000 st at an average grade of 0.91% Cu.

3.9 STATISTICAL BASIS FOR GRADE ASSIGNMENT

In the previous section one technique for assigning grades to blocks, based upon distance dependent weighting coefficients, was discussed. The application depended upon selecting the power m and a radius of influence for the samples. In some cases a value for m is just picked (often 2), and in others the data set is scanned. Little was mentioned as how to select a value for R. If $m=2$ is used, the decrease of influence with distance is quite rapid and

Table 3.23. Example of inverse distance squared interpolation rules (Hughes & Davey, 1979).

1. Develop rock type distance factors. These factors are sets of A, B, and C coefficients for equations of the form $AX^2 + BX + C = Y$, where Y is the average standard deviation between grades and X is the distance between the sample points. This is done for all combinations of formations plus within each formation.
2. Develop geologic model with rock type code for each block being evaluated. Rock type codes are assigned to each composite value.
3. Radius of influence equals R ft and angle of exclusion equals $\alpha°$.
4. The block must pass one of the following in order to be assigned a grade:
   a) The block must be within R ft of a composite.
   b) The block is within R ft of a line connecting two composites which are within $3R$ ft of each other.
   c) The block is within R ft of a line connecting two composites that are within $3R$ ft of a third composite.
   d) The block is inside a triangle formed by three composites, any two legs of which are equal to or less than $3R$ ft long.
5. Collect all assay composites for the level that are within $5R$ ft of the center of the block.
6. Count the number of composites having the same rock type as the block. A rock type the same as the block is defined as:
   a) The rock type of the composite matches the rock type of the block.
   b) The rock type of the block is unknown or undefined.
7. If no composites are found to match the rock type of the block, extend the radius of search outward by R ft increments until one or more composites are found within an increment. Add these composites to the list of ones affecting the block grade assignment.
8. Compute distances from the block to each composite having a different rock type than the block, such that the new distance would be equivalent to the two points being in the same rock type. If the equivalent distance is less than the original distance, use the original distance. The original distance rather than the equivalent distance will be used by the minimum angle screening.
9. Compute the azimuths from the block to each composite influencing the assay assignment.
10. For each assay of the mineralization model, compute the angle between each pair of composites having data for the assay. Check to see if the angle is less than $\alpha°$. If the angle is less than $\alpha°$ and:
    a) the rock type of the closer composite matches the rock type of the block, the more distant composite is rejected.
    b) both composites match the rock type of the block and only two composites match the rock type of the block, both composites are retained.
    c) the rock type of the closer composite matches the rock type of the block, the more distant composite is rejected.
    d) the rock type of neither composite matches the rock type of the block, the more distant composite is rejected.
11. The grade assignment for the block is computed as:
$$G = \sum_i (G_i/D_i^2) / \sum_i (1/D_i^2)$$
    where G_i is the sample assay value and D_i is the equivalent distance to the i-th composite.
    a) Unless there is a nonzero composite value within or on the boundary of the block, in which case that composite will be used directly.
    b) Unless there is only one composite, in which case the closest composite from the reject list having the same rock type is included. If no second composite can be found with the same rock type, the closest composite from the reject list is included.
12. If the resulting grade assignment is zero, it will be increased to the smallest nonzero number which can be represented in the model.

the use of a large value is not so serious. The minimum value of R is determined by the need to include a sufficient number of points for the calculations. This obviously varies with the drilling pattern. The field of geostatistics has contributed a number of techniques which can be used. Of particular importance is a way to evaluate the radius of influence R and

Orebody description 237

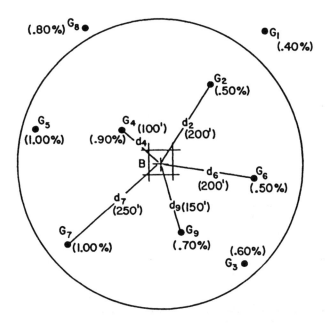

Figure 3.59. A more realistic IDS application (Hughes & Davey, 1979).

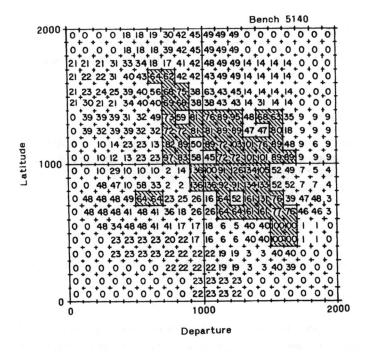

Figure 3.60. Inverse distance squared interpolation for the composited samples in Figure 3.45 (Hughes & Davey, 1979).

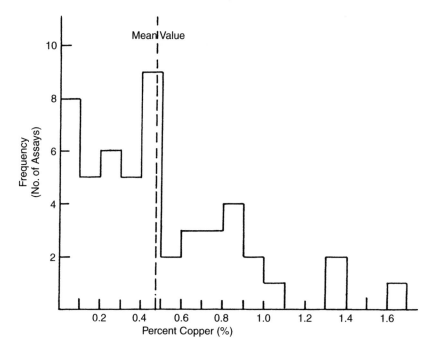

Figure 3.61. Histogram representation of the 5140 bench grades.

assigning grades to blocks. Unfortunately, due to the highly mathematical nature of their presentation, the level of understanding and appreciation of the technique within much of the mining industry is poor. In this section the authors have attempted to clarify some of these concepts.

3.9.1 *Some statistics on the orebody*

One of the first things which can and should be done is to see how the grades are distributed. It is most easily done by plotting a histogram of the data. This has been done in Figure 3.61 using the data from bench 5140 (Table 3.21).

The *average* grade $\bar{g}$ is calculated using

$$\bar{g} = \frac{1}{n} \sum_{i=1}^{n} g_i \qquad (3.24)$$

where n is the number of samples and g_i is the individual grades. In this case the average grade is

$$\bar{g} = 0.477\%$$

It has been superimposed on Figure 3.61. If the grade distribution had been truly normal then a bell shaped curve centered around the average value would be expected. Here, this is not the case. There are many values clustered below the mean and a long tail into the high values. This is termed a positive skew and is quite common for low grade deposits.

Table 3.24. Grades from bench 5140 arranged in increasing order.

i	Grade (% Cu)	C_f (%)	i	Grade (% Cu)	C_f (%)
1	0.012	1.0	27	0.417	52.0
2	0.023	2.9	28	0.427	53.9
3	0.027	4.9	29	0.453	55.9
4	0.034	6.9	30	0.465	57.8
5	0.040	8.8	31	0.475	59.8
6	0.089	10.8	32	0.476	61.8
7	0.092	12.7	33	0.489	63.7
8	0.099	14.7	34	0.519	65.7
9	0.102	16.7	35	0.572	67.6
10	0.140	18.6	36	0.638	69.6
11	0.165	20.5	37	0.644	71.6
12	0.175	22.5	38	0.685	73.5
13	0.180	24.5	39	0.717	75.5
14	0.215	26.5	40	0.719	77.5
15	0.224	28.4	41	0.765	79.4
16	0.225	30.4	42	0.806	81.4
17	0.228	32.4	43	0.833	83.3
18	0.230	34.3	44	0.889	85.3
19	0.252	36.3	45	0.893	87.3
20	0.320	38.2	46	0.915	89.2
21	0.377	40.2	47	0.996	91.2
22	0.392	42.2	48	1.009	93.1
23	0.395	44.1	49	1.335	95.1
24	0.396	46.1	50	1.365	97.1
25	0.406	48.0	51	1.615	99.0
26	0.409	50.0			

The degree of departure from normality can be checked by plotting the values on standard probability paper. First one arranges the grades in order as in Table 3.24. Next the corresponding cumulative frequency of the grades are calculated using

$$C_f = \frac{100(i - \frac{1}{2})}{n} \qquad (3.25)$$

where i is the i-th observation, n is the total number of observations, and C_f is the cumulative frequency.

If n is large it is not necessary to plot every point (every 5-th or 10-th point may be enough). The results are plotted in Figure 3.62. As can be seen, there are departures from a straight line particularly at the lower grades. For grades above 0.3% Cu, the fit is fairly good. If the entire distribution is to be represented, then measures must be taken to convert it into a normal distribution. Two types of logarithmic transformations may be applied to such skewed (whether negatively or positively) distributions. In the simplest case, one plots the natural logarithm of the grade (ln g_i) versus cumulative frequency on log probability paper (Fig. 3.63).

It is observed for grades greater than about 0.3% Cu, a straight line can be fitted. However for lower grades, the points fall below the curve. Hence the simple transformation of ln g_i

240 Open pit mine planning and design: Fundamentals

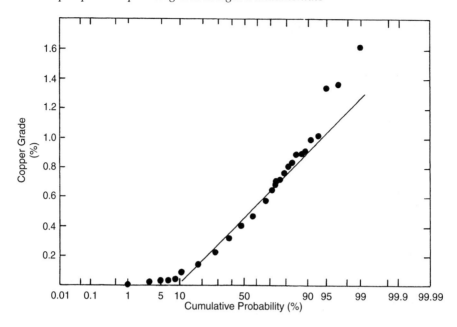

Figure 3.62. Copper grade versus cumulative probability for the 5140 bench grades.

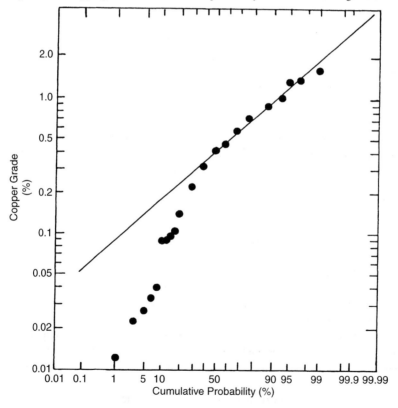

Figure 3.63. The logarithm of copper grade versus cumulative probability.

doesn't yield the desired normal distribution. The next step is to take the natural log of the grade plus an additive constant β and plot $\ln(g_i + \beta)$ on probability paper.

If the number of samples is large enough, one can estimate β using the following formula and values from Figure 3.62:

$$\beta = \frac{m^2 - f_1 f_2}{f_1 + f_2 - 2m} \quad (3.26)$$

where m is the grade at 50% cumulative frequency, f_1 is the sample value corresponding to 15% cumulative frequency, and f_2 is the sample value corresponding to 85% cumulative frequency.

In general f_1 corresponds to frequency P and f_2 to frequency $1 - P$. In theory any value of P can be used but one between 5–20% gives the best results.

Applying this rule, one finds that

$$\beta = \frac{(0.409)^2 - 0.10 \times 0.81}{0.10 + 0.81 - 2 \times 0.409} = \frac{0.086}{0.092} = 0.935\%$$

The resulting value $\ln(g_i + 0.935)$ provides a high degree of normalization to the grade distribution (Fig. 3.64).

The use of log-normal distributions introduces complexities which are beyond the scope of this book. For instance one should be aware that the grade at 50% probability on a log-normal distribution graph, would represent the median – also called the geometric mean – and not the true (arithmetic) mean of the distribution. It will be assumed that the grades from bench 5140 can be adequately represented by a normal distribution.

As can be seen in Table 3.24, there is a large spread or range in the grades. The *range* is from 0.012% to 1.615%. The *variance* s^2, obtained using

$$s^2 = \frac{1}{n} \sum_{i=1}^{n} (g_i - \bar{g})^2 \quad (3.27)$$

is found equal to

$$s^2 = 0.1351(\%)^2$$

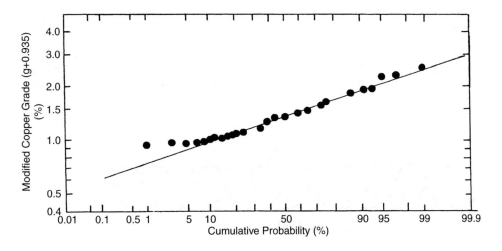

Figure 3.64. Plot of the logarithm $(g_i + \beta)$ versus cumulative probability for the grades of bench 5140.

The standard deviation (s), which is just the square root of the variance, becomes

$$s = 0.368\%$$

It will be recalled that about 68% of the grades should be contained within $\bar{g} \pm s$ and 95% within $\bar{g} \pm 2s$ given a normal distribution.

This traditional statistical approach has treated all the samples as a large group with no special notice being paid to their relative positions within the group. Such attention will be paid in the next section.

3.9.2 Range of sample influence

When using inverse distance weighting techniques, the range of influence of a sample is, in theory, infinite. In practice some finite range is assigned. The question arises as to whether a more quantitative way of determining the effective sample range could be devised? The geostatistical approach described in this section provides one way. The basic logic involved will first be described followed by an example.

If when sampling an orebody, the samples are collected close together, one might expect the resulting assay values to be similar. On the other hand, if they are collected far apart little similarity would be expected. In between these two extremes, one would expect some sort of functional relationship between grade difference and separation distance to apply. If the function could be determined, then the distance (influence range) at which samples first became independent of one another could be found. The basic procedure (Barnes, 1980) would be:

1. Decide on separation distances h into which sample pairs would be grouped. These distances are often called lags. For example, separation distances of 100 ft, 200 ft, 300 ft, etc.

Although each lag is thought of as a specific distance, in practice, the lag distance usually represents the mean of a distance class interval. In other words, the lag distance of 15 m (50 ft) may represent all pairs of samples falling between 11½ and 19½ m (37½ and 62½ ft) apart. Such a practice is necessitated by the uneven spacing of most samples, especially when computing directional variograms that are not parallel or normal to a roughly rectangular sampling pattern.

2. Identify the pairs falling within a particular group. Figure 3.65 illustrates a simple case of n samples separated by a constant lag distance h. Various pairs can be formed. There are $n-1$ pairs of distance h apart, $n-2$ pairs at separation distance $2h$, $n-3$ pairs separated by $3h$, etc.

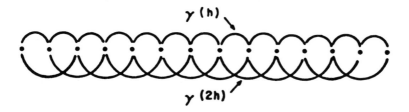

Figure 3.65. Variance computational scheme using sample pairs a given distance apart (Barnes, 1979).

3. Calculate the grade difference $g_i - g_j$ for each of the pairs within each group. It will be found that a distribution of differences exists. As was done before, the average, the variance and the standard deviation could all be calculated. A particular type of variance defined as

$$s^2(h) = \frac{1}{n(h)} \sum_{}^{n} (g_i - g_j)^2 \qquad (3.28)$$

where $n(h)$ is the number of pairs in the group of lag h, $s^2(h)$ is the variance for pairs with lag h, and g_i is the grade at point i of the pair, will be used. For mathematical convenience, one-half of $s^2(h)$, denoted by the symbol $\gamma(h)$ will be used:

$$\gamma(h) = \frac{1}{2n(h)} \sum_{}^{n} (g_i - g_j)^2 \qquad (3.29)$$

This is called the geostatistical variance or the semi-variance (half of the variance).

4. Once values of γ have been found for each of the different groups (called cells), the next step is to plot the results. The plot of γ versus average lag h is called a *variogram* or more properly a *semi-variogram*. In this book the term variogram is retained.

5. The final step is to express the relationship between γ and h in some type of useable form. The value of h beyond which little or no change in γ is observed is called the range of influence 'a'.

3.9.3 *Illustrative example*

To illustrate these concepts an example using the N-S data pairs in Figure 3.45 will be worked. In viewing the plan map it is clear that most of the holes are spaced 200 ft apart. Table 3.25 summarizes the lags and the number of corresponding pairs.

It is found convenient to consider the data in thirteen cells (groups) incremented from each other by 100 ft. The cells are summarized in Table 3.26.

The location of the 19 pairs at a separation distance of 600 ft are shown in Figure 3.66. The calculation of $\gamma(600)$ is shown in Table 3.27.

Table 3.25. Pairs and distances used for computing a N-S variogram.

Distance (ft)	No. of pairs
200	31
300	1
350	1
400	26
500	1
600	19
700	2
800	12
900	1
1000	8
1100	1
1200	6

244 *Open pit mine planning and design: Fundamentals*

Table 3.26. Cells used in the example.

Cell	Separation distance (ft)	No. of pairs	Average separation distance (ft)
1	0 → 99	0	–
2	100 → 199	0	–
3	200 → 299	31	200
4	300 → 399	2	325
5	400 → 499	26	400
6	500 → 599	1	500
7	600 → 699	19	600
8	700 → 799	2	700
9	800 → 899	12	800
10	900 → 999	1	900
11	1000 → 1099	8	1000
12	1100 → 1199	1	1100
13	1200 → 1299	6	1200

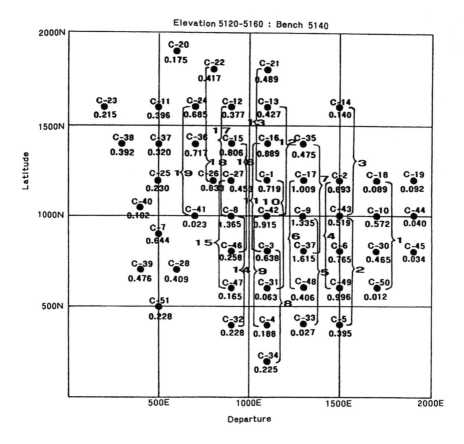

Figure 3.66. Sample calculation using the N-S pairs 600 feet apart.

Orebody description 245

Table 3.27. Steps in the determination of $\gamma(600)$ for use in the N-S variogram.

Pair	Grades $g_i - g_j$ (% × 10^3)	Grade difference (% × 10^3)	(Grade difference)^2 (% × 10^3)^2
1	89–12	77	5929
2	519–395	124	15,376
3	990–893	97	9409
4	519–140	379	143,641
5	1335–27	1308	1,710,864
6	1009–406	603	363,609
7	1615–475	1140	1,299,600
8	638–225	413	170,569
9	915–188	727	528,529
10	719–63	656	430,336
11	889–638	251	63,001
12	915–427	488	238,144
13	719–489	230	52,900
14	1365–224	1141	1,301,881
15	453–165	288	82,944
16	806–258	548	300,304
17	1365–377	988	976,144
18	833–417	416	173,056
19	685–23	662	438,244
		Avg GD = 555	$\Sigma(GD)^2 = 8,304,480$

$$\gamma(600) = \frac{8,304,480}{23 \times 19 \times 10^6} = 0.2185 \text{ percent}$$

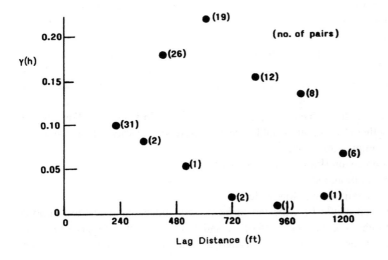

Figure 3.67. The resulting N-S variogram.

In Figure 3.67, the number of data pairs represented by each point is plotted. It is important that sufficient pairs are found at each lag in any direction to assure statistical significance. Ideally at least 30 such pairs are necessary to compute the variance for each lag in any given direction. Sometimes in the early sampling stage, it is difficult to find enough pairs

246 *Open pit mine planning and design: Fundamentals*

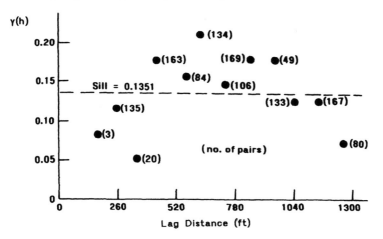

Figure 3.68. The variogram for bench 5140.

at certain lags to produce a viable variogram point, and lesser numbers may be used. A variogram program that will output a different symbol when plotting the $\gamma(h)$ value for all lags having less than 30 pairs is useful for quick recognition of less reliable points (Barnes, 1980b).

As indicated earlier, it is desired to find the value of h at which γ ceases to vary with distance. This is called the range of influence. Due to the small number of sample pairs at the higher separation distances, no particular plateau value is observed. If all directions are included (not just N-S), then a much larger number of pairs is obtained. The resulting figure is shown in Figure 3.68. A definite 'leveling off' is observed with distance although the magnitude of the plateau (termed the sill) is difficult to discern from these data. The variance for the entire set of samples was determined earlier to be 0.1351. It can be shown that this should equal the sill. Hence this line has been superimposed on the figure.

To complete the curve, the behavior in the region of the origin is needed. For samples taken very close together ($h \cong 0$), one would expect a difference in assay values due to
 - lack of care in sample collection.
 - poor analytical precision (limits of analytical precision),
 - poor sample preparation, and
 - highly erratic mineralization at low scale.

This type of variance would be expected to be present independent of the sampling distance. Its magnitude is given the symbol c_o and it is called the nugget effect. In some texts this part of the total variance is called the chaotic or unstructured variance. (Fig. 3.69).

That portion of the variogram lying between the nugget effect and the sill represents the true variability within the deposit for the given mineralization. It is called the structured variance.

A straight line extending from the Y axis to the sill has been drawn through the first few points in Figure 3.70. The value of c_o (the nugget effect) as read from the curve is 0.02. For the spherical model (of which this is an example), it has been found that the straight line

Orebody description 247

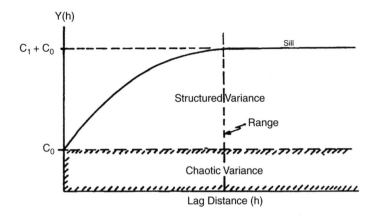

Figure 3.69. Diagrammatic representation of a spherical variogram.

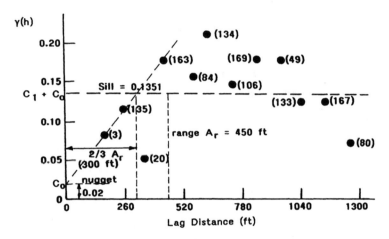

Figure 3.70. Determination of range, sill and nugget.

will intersect the sill at $h = 2/3\,a$, where a is the range. In this case, $a \cong 450$ ft. In summary:

$$c_0 = \text{nugget effect} = 0.02$$
$$c_1 + c_0 = \text{sill} = 0.135$$
$$c_1 = \text{structured variance} = 0.135 - 0.02 = 0.115$$
$$A_r = R = \text{range} = 450\,\text{ft}$$

It is clear that this process can be applied in particular directions to evaluate anisotropic behavior (or variation in range, with direction). The lag h would have an associated direction and thus become a vector quantity denoted by $\vec{h}$. Variograms are prepared for each type of mineralization within a deposit.

The evaluation process can be stopped at this point. The required value of $R = a$ can be used in polygon, inverse distance or other schemes for tonnage-grade calculation. On the other hand the variogram which reflects grade variability with distance can be used to

248 *Open pit mine planning and design: Fundamentals*

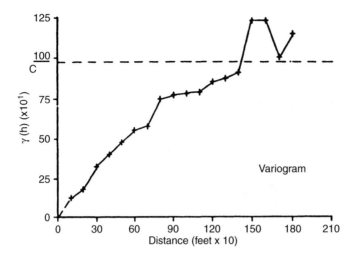

Figure 3.71. Typical variogram for a stratabound deposit (Barnes, 1979b).

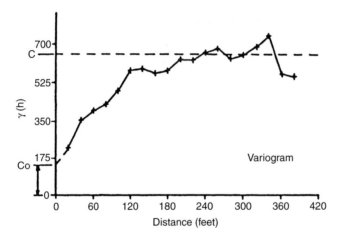

Figure 3.72. Typical variogram for a porphyry copper deposit (Barnes, 1979b).

develop weighting coefficients similar to the a_i's described earlier with respect to the inverse distance method. This process which is called kriging will be described in Section 3.10.

3.9.4 *Describing variograms by mathematical models*

Figures 3.71, 3.72, and 3.73 are real experimental variograms generated from three different types of deposits (Barnes, 1980). The slow steady growth of γ from zero in Figure 3.71 is characteristic of many stratigraphic and stratiform deposits with fairly uniform mineralization having a high degree of continuity. Figure 3.72 was generated from porphyry-copper deposit data where mineral veinlets, changes in structural intensity, and other discontinuous features created a significant nugget effect due to changes over very short distances. Beyond the short range effects, however, $\gamma(h)$ shows a fairly uniform growth curve and reaches a plateau at the sill of the variogram that is the overall variance of all the samples.

Orebody description 249

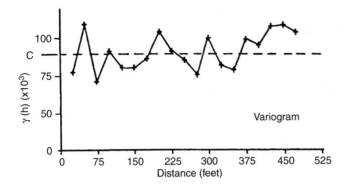

Figure 3.73. Typical variogram for a gold deposit (Barnes, 1979b).

(a) The Linear Model

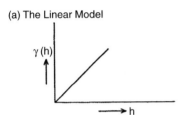

(b) The De wijsian Model

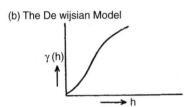

(c) The Spherical Model

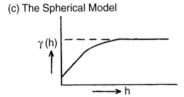

Figure 3.74. Diagrammatic representation of different variogram models (Barnes, 1979b).

The experimental variogram of Figure 3.73 is illustrative of a total random effect found in some gold deposits. The mineral continuity is nonexistent, and the samples appear to be completely independent no matter what the distance between them. Geostatistical ore reserve estimation techniques cannot make any contribution toward evaluating the deposit having a pure nugget effect since no regionalized element is present.

The change of variance with distance between samples can be read directly from the hand-drawn curve fitted through the experimental points. For computer calculations however, it is necessary to have an equation which describes the curve. The three models given in Figure 3.74, (as well as others) have been used to approximate the actual variograms.

Both the linear model and the De Wijsian model, which will produce a straight line when the lag h is plotted to log scale, imply that $\gamma(h)$ increases infinitely with increasing distances. Experience has shown that both models often accurately fit experimental variogram data near the origin, but break down when h becomes large.

The spherical model or Matheron model, as it is sometimes called, is one in which the variogram reaches a finite value as h increases indefinitely. This finite value, referred to as the sill of the spherical variogram, is the overall variance of the deposit and is reached when the grades are far enough apart to become independent of each other and act in a random manner. The spherical model has become the most important one and many practicing geostatisticians have adopted it as an almost universal model. The model has been found to adequately represent such diverse deposits as iron ore bodies, porphyry-copper deposits, stratibound lead-zinc deposits, bauxite and lateritic nickel, as well as uranium and phosphate deposits. This model will be the only one discussed further in this text.

The spherical scheme is defined by the formula

$$\gamma(h) = \begin{cases} c_1 \left(\dfrac{3\,h}{2\,a} - \dfrac{1}{2} \dfrac{h^3}{a^3} \right) + c_0 & \text{when } h \leq a \\ c_1 + c_0 & \text{when } h > a \end{cases} \quad (3.30)$$

where $c_1 + c_0 = \gamma(\infty)$ and is called the sill value, c_0 is the nugget effect (usually present), and a is the range ($a = A_r = R$) or maximum zone of influence.

3.9.5 Quantification of a deposit through variograms

Barnes (1979b, 1980) has summarized very nicely the types of quantitative information provided by variograms.

(a) A measure of continuity of the mineralization: A rate of increase of $\gamma(h)$ near the origin and for small values of h reflects the rate at which the influence of a sample decreases with increasing distance from the sample site. The growth curve demonstrates the regionalized element of the sample, and its smooth steady increase is indicative of the degree of continuity of mineralization.

The intersection of the curve with the origin provides a positive measure of the nugget effect of the samples from which the variogram has been generated and indicates the magnitude of the random element of the samples.

(b) A measure of the area of influence of a sample: The zone of influence of a sample is the distance or range in any direction over which the regionalized element is in effect. When samples reach a point far enough apart so as to have no influence upon each other, we have established the range or zone of influence of the sample. The quantification of the range or zone of influence in various directions has important applications in the design and spacing of development drill holes within a deposit. The total zone of influence is indicated by the point at which the $\gamma(h)$ growth curve reaches a plateau, referred to in the spherical scheme as the sill.

(c) A measure of mineral trend or mineral anisotropy of the deposit: The fact of mineral anisotropism in various types of deposits has long been recognized. The range of influence of a sample is greater along the strike or trend of the deposit than it is normal to trend. Most of the time, another anisotropism is evident in the vertical dimension. Prior to the variogram, there was no satisfactory way of determining the three-dimensional influence of a sample. With the simple process of computing variograms in different directions as well as vertically, one can readily determine not only the mineralogical trend but the magnitude of the directional changes in the zone of influence. Knowing quantitatively the mineralogical range in three dimensions, it is relatively simple to assign directional anisotropic factors that will give proper weighing to samples relative to their location from the point or block being evaluated. For example, if the range of influence along the trend is twice as great as the range normal to trend, one can multiply the distance in the normal direction by a factor of two to restore geometric isotropy in terms of the major trend direction.

3.10 KRIGING

3.10.1 *Introduction*

Prior to going into detail, it is perhaps worthwhile to review the objective and to summarize the approach to be taken. The overall problem (shown in Fig. 3.75) is that of assigning a grade g_0 to the point x_0 knowing the grades g_i at surrounding points x_i.

The objective can be simply expressed as that of determining coefficients a_i's which when multiplied by the known grades g_i and the resulting products summed will yield a best estimate of the grade g_0. The equation developed is called a linear estimator and has the form

$$g_0 = a_1 g_1 + a_2 g_2 + \cdots + a_n g_n \tag{3.21}$$

where g_0 is the grade to be estimated, g_i are the known grades, and a_i are the weighting functions.

As was discussed earlier, the inverse distance method is also of this form. The coefficients would be

$$a_i = \frac{\frac{1}{d_i^m}}{\sum_{i=1}^{n} \frac{1}{d_i^m}} \tag{3.20}$$

The distance weighting factor m is often chosen equal to 2.

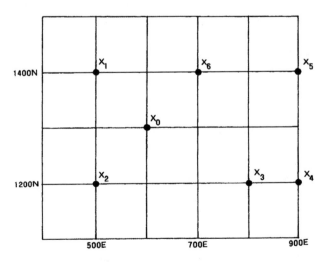

Figure 3.75. Sample locations for the point kriging example.

252 Open pit mine planning and design: Fundamentals

In the present case, we have a curve (a variogram) which expresses the variance as a function of distance. That curve can be used to calculate the total estimated variance of the grade g_0 for different combinations of the a_i coefficients. The best estimate of g_0 is that for which the variance is a minimum. A powerful advantage of this technique over that of other techniques, is that the variance is calculated as well as the estimated grade. The problem therefore boils down to finding the a_i coefficients. Such a set of coefficients must be calculated for each point requiring a grade assignment in the region under consideration. Therefore the use of a high speed computer is a definite requirement.

3.10.2 Concept development

For the example shown in Figure 3.75 there are six surrounding grades, hence an equation of the form

$$g_0 = a_1 g_1 + a_2 g_2 + a_3 g_3 + a_4 g_4 + a_5 g_5 + a_6 g_6 \tag{3.31}$$

is being sought.

Since six coefficients a_1, a_2, a_3, a_4, a_5, and a_6 must be found, at least six equations containing these six unknowns must be developed and solved. Although values of variances γ read directly from the variogram can be used directly in this process, it is found more convenient to use *covariances* σ. The covariance is related to the variance as shown in Figure 3.76.

Whereas γ (the variance) is the distance between the X axis and the curve for a given lag h, the covariance at h is the distance between the curve and the sill $c_0 + c_1$.

At a lag distance of h_0, $\gamma(h_0) = \gamma_0$ and the covariance is

$$\sigma(h_0) = c_1 + c_0 - \gamma_0 = \sigma_0 \tag{3.32}$$

For $h = 0$ (just at the location of the sample itself) the variance of the sample with itself γ is obviously equal to zero, $\gamma(0) = 0$. The corresponding covariance (σ) of the sample with itself is found either from the curve or using the following equation

$$\sigma(0) = c_0 + c_1 - \gamma(0) = c_0 + c_1 \tag{3.33}$$

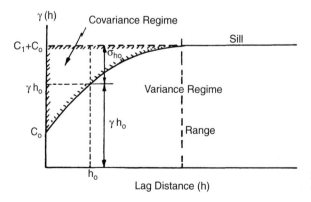

Figure 3.76. The relationship between variance and covariance.

Thus $\gamma(0)$ is just the sill value. For samples taken very close ($h=0+$) but not at the sample, the value of the variance $\gamma(0+)$ jumps to the nugget level c_0 and similarly the covariance becomes

$$\sigma(0+) = c_1 + c_0 - c_0 = c_1 \qquad (3.34)$$

which is the true deposit variability. At the range of sample influence (a), $\gamma(a) = c_1 + c_0$ and the covariance becomes $\sigma(a) = 0$. The reason for using covariances in Formula (3.35) rather than the equivalent gamma values is due primarily to linear programming complexities (David, 1977). In any case, this is a very simple substitution and as such should not present understanding difficulties to the reader.

The equation for the total estimation variance σ_e^2 written in terms of the covariance is given below

$$\sigma_e^2 = \sigma_{x_0 x_0} - 2\sum_{i=1}^{n} a_i \sigma_{x_0 x_i} + \sum_{i=1}^{n}\sum_{j=1}^{n} a_i a_j \sigma_{x_i x_j} \qquad (3.35)$$

where $\sigma_{x_0 x_0}$ is the covariance between the grade at the point and itself, $\sigma_{x_0 x_i}$ is the covariance between the point being considered (x_0) and the sample point x_i, a_i, a_j are the weighting coefficients, $\sigma_{x_i x_j}$ is the covariance between samples x_i and x_j, and σ_e^2 is the total variance.

As can be seen, the total variance consists of three parts:
- The covariance of the unknown grade with itself. As we have just seen, this is equal to ($c_0 + c_1$) which is the sill value and is constant.
- The weighted covariance between the unknown grade and each of the other samples. The covariance can be computed between the point and each of the samples from the variogram since the distance is known. The weighting coefficients a_i are unknown.
- The weighted covariances between each of the known samples. These can be computed from the variogram since the distances are known. The weighting coefficients a_i which apply in this region are unknown.

As stated earlier, the objective is to minimize σ_e^2 by a proper choice of the coefficients a_i. Similar to finding a minimum in many other types of engineering problems, to do this one takes a derivative, sets the resulting equation equal to zero, and solves for the unknown. For a system of equations such as this, partial derivatives with respect to each of the unknown coefficients are taken, the resulting linear equations set equal to zero, and solved for the coefficients.

Taking partial derivatives of Equation (3.35) with respect to a_i one finds that

$$-\sum_{i=1}^{n} \sigma_{x_0 x_i} + \sum_{j=1}^{n} a_j \sigma_{x_i x_j} = 0 \qquad (3.36)$$

In this case, a constraint is imposed on the a_i's to ensure that the grade estimation is unbiased. This means that on the average, the computed grade should be equal to the real grade and not systematically higher or lower. This constraint is written as

$$\sum_{i=1}^{n} a_i = 1 \qquad (3.37)$$

It says simply, that the sum of the weighting factors should equal one. In this new problem of minimizing σ_e^2 in the light of a constraint, a special mathematical procedure involving

254 Open pit mine planning and design: Fundamentals

Lagrange multipliers is used. A treatment of this is beyond the scope of this book, and only the two resulting equations will be given:

$$\sum_{j=1}^{n} a_j \sigma_{x_i x_j} + \lambda = \sigma_{x_0 x_i}, \quad i = 1, \ldots, n \quad (3.38a)$$

$$\sum_{i=1}^{n} a_i = 1 \quad (3.38b)$$

where λ is the Lagrange multiplier. For n grades, there are $n+1$ unknowns ($a_1, \ldots a_n, \lambda$). Equations (3.38) supply the needed $n+1$ equations. Once the a_i's have been found, the estimated grade is

$$g_0 = \sum_{i=1}^{n} a_i g_i \quad (3.39)$$

The estimated variance can then be found substituting the value of λ and a_i's into

$$\sigma_e^2 = \sigma_{x_0 x_0} - \sum_{i=1}^{n} a_i \sigma_{x_0 x_i} - \lambda \quad (3.40)$$

Thus this process provides what the inverse distance squared and other estimation procedures do not, a measure of the confidence associated with the assigned grade. It will be recalled that the actual grade would be expected to fall within the range of the average ± 1 standard deviation 68% of the time (and within ± 2 standard deviations 95% of the time) if the sample distribution is symmetric.

3.10.3 *Kriging example*

The concepts just described will be illustrated by an example. The grades corresponding to points $x_1, \ldots, x_6$ in Figure 3.75 are given in Table 3.28. For the sake of this example, only points x_1, x_2 and x_3 will be used. Hence it will be desired to find the equation

$$g_0 = a_1 g_1 + a_2 g_2 + a_3 g_3 \quad (3.41)$$

where a_1, a_2, a_3 are the weighting coefficients. A spherical variogram having the following values:

$$c_0 = 0.02 \quad c_0 + c_1 = 0.18 = \text{sill}$$
$$c_1 = 0.16$$
$$a = 450 \text{ ft}$$

has been found to describe the deposit (Hughes & Davey, 1979). This variogram is shown in Figure 3.77.

The required distances are first found (Table 3.29). Next the corresponding values of the variance γ are found using the general formulas

$$\gamma = \begin{cases} 0 & \text{if } h = 0 \\ c_1 \left[\dfrac{3h}{2a} - \dfrac{1}{2} \dfrac{h^3}{a^3} \right] + c_0 & \text{if } 0 < h \leq a \\ c_1 + c_0 & \text{if } h > a \end{cases} \quad (3.42)$$

Table 3.28. Grade data for the kriging example. Hole designation after (Hughes & Davey, 1979).

Sample designation	Hole	Grade
x_1	C-37	$g_1 = 0.320$
x_2	C-25	$g_2 = 0.230$
x_3	C-26	$g_3 = 0.833$
x_4	C-27	$g_4 = 0.453$
x_5	C-15	$g_5 = 0.806$
x_6	C-36	$g_6 = 0.717$
x_0		$g_0 =$ (to be determined)

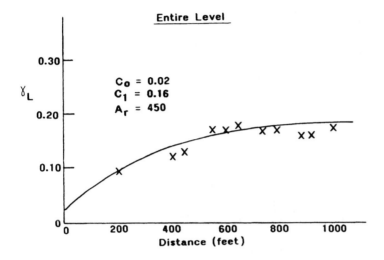

Figure 3.77. The experimental variogram (Hughes & Davey, 1979).

Table 3.29. Separation distance calculation.

From	To	Separation distance (ft)
x_0	x_1	$h_{01} = \sqrt{100^2 + 100^2} = 141$
x_0	x_2	$h_{02} = \sqrt{100^2 + 100^2} = 141$
x_0	x_3	$h_{03} = \sqrt{100^2 + 200^2} = 224$
x_1	x_2	$h_{12} = \sqrt{0^2 + 200^2} = 200$
x_1	x_3	$h_{13} = \sqrt{200^2 + 300^2} = 361$
x_2	x_3	$h_{23} = \sqrt{0^2 + 300^2} = 300$

Substituting the given values into the above equations yields

$$\gamma = \begin{cases} 0 & \text{if } h = 0 \\ 0.16\left[\frac{3}{2}\frac{h}{450} - \frac{1}{2}\left(\frac{h}{450}\right)^3\right] + 0.02 & \text{if } 0 \leq h \leq a \\ 0.18 & \text{if } h > a \end{cases}$$

Table 3.30. Variances for the example.

Lag	Distance (ft)	γ (% × %)
$h_{01} = h_{10}$	141	0.0927
$h_{02} = h_{20}$	141	0.0927
$h_{03} = h_{30}$	224	0.1296
$h_{12} = h_{21}$	200	0.1196
$h_{13} = h_{31}$	361	0.1712
$h_{23} = h_{32}$	300	0.1563
$h_{00} = h_{11} = h_{22} = h_{33}$	0	0.0

Table 3.31. Covariances for the example.

Lag	Distance (ft)	σ (% × %)
$h_{10} = h_{01}$	141	$\sigma_{01} = 0.0873 = \sigma_{10}$
$h_{20} = h_{02}$	141	$\sigma_{02} = 0.0873 = \sigma_{20}$
$h_{30} = h_{03}$	224	$\sigma_{03} = 0.0504 = \sigma_{30}$
$h_{12} = h_{21}$	200	$\sigma_{12} = 0.0604 = \sigma_{21}$
$h_{13} = h_{31}$	361	$\sigma_{13} = 0.0088 = \sigma_{31}$
$h_{23} = h_{32}$	300	$\sigma_{23} = 0.0237 = \sigma_{32}$
$h_{00} = h_{11} = h_{22} = h_{33}$	0	$\sigma_{11} = \sigma_{22} = \sigma_{33} = 0.18 = \sigma_{00}$

For $h_{01} = 141$, one finds that

$$\gamma = 0.16 \left[\frac{3}{2}\left(\frac{141}{450}\right) - \frac{1}{2}\left(\frac{141}{450}\right)^3 \right] + 0.02 = 0.0927$$

The resulting values are summarized in Table 3.30.

In the analysis, the covariances σ defined by

$$\sigma = \begin{cases} c_1 + c_0 & \text{if } h = 0 \\ c_1 + c_0 - \gamma & \text{if } 0 < h \leq a \\ 0 & \text{if } h > a \end{cases} \qquad (3.43)$$

are used.

The covariance σ_{01} corresponding to a lag h_{01} of 141 ft ($\gamma_{01} = 0.0927$) is

$$\sigma_{01} = 0.16 + 0.02 - 0.0927 = 0.0873$$

The covariances are summarized in Table 3.31.

The basic kriging equations are:

$$\sum_{j=1}^{n} a_j \sigma_{x_i x_j} + \lambda = \sigma_{x_0 x_i}, \qquad i = 1, \ldots, n$$

$$\sum_{i=1}^{n} a_i = 1$$

For this example, they become

$$\sum_{j=1}^{3} a_j \sigma_{x_i x_j} + \lambda = \sigma_{x_0 x_i}, \qquad i = 1, 3$$

$$\sum_{i=1}^{n} a_i = 1$$

Expanding one finds that

$$a_1\sigma_{11} + a_2\sigma_{12} + a_3\sigma_{13} + \lambda = \sigma_{01}$$
$$a_1\sigma_{21} + a_2\sigma_{22} + a_3\sigma_{23} + \lambda = \sigma_{02}$$
$$a_1\sigma_{31} + a_2\sigma_{32} + a_3\sigma_{33} + \lambda = \sigma_{03}$$
$$a_1 + a_2 + a_3 = 1$$

The values of the covariances are now substituted from Table 3.31 into the above equations

$$0.18a_1 + 0.0604a_2 + 0.0088a_3 + \lambda = 0.0873$$
$$0.0604a_1 + 0.18a_2 + 0.0237a_3 + \lambda = 0.0873$$
$$0.0088a_1 + 0.0237a_2 + 0.18a_3 + \lambda = 0.0504$$
$$a_1 + a_2 + a_3 = 1$$

One is now faced with solving 4 equations with 4 unknowns. The answers are:

$$a_1 = 0.390$$
$$a_2 = 0.359$$
$$a_3 = 0.251$$
$$\lambda = -0.00677$$

The estimated grade is

$$g_0 = 0.390 \times 0.320 + 0.359 \times 0.230 + 0.251 \times 0.833 = 0.416\% \text{ Cu}$$

The estimation variance is given by

$$\sigma_e^2 = \sigma_{x_0 x_0} - \sum_{i=1}^{n} a_i \sigma_{x_0 x_i} - \lambda$$

Since

$$\sigma_{x_0 x_0} = c_0 + c_1 = 0.18$$
$$\lambda = -0.00677$$
$$a_1\sigma_{01} + a_2\sigma_{02} + a_3\sigma_{03} = 0.390 \times 0.0873 + 0.359 \times 0.0873 + 0.251 \times 0.0504$$
$$= 0.07804$$

Then

$$\sigma_e^2 = 0.18 - 0.07804 + 0.00677 = 0.10873$$

The standard deviation (SD) is the square root of the estimation variance or

$$SD = \sqrt{0.10873} = 0.330\% \text{ Cu}$$

258 *Open pit mine planning and design: Fundamentals*

The interested reader is asked to estimate the grade at x_0 using all 6 surrounding points. The answers are:

$a_1 = 0.263$

$a_2 = 0.326$

$a_3 = 0.167$

$a_4 = -0.0355$

$a_5 = -0.0372$

$a_6 = 0.318$

$\lambda = -0.00023$

$g_{x_0} = 0.480\%$

$SD = 0.307\%$

3.10.4 *Example of estimation for a level*

A kriging evaluation of level 5140 was performed by Hughes & Davey (1979). The variograms along strike (N 45° W to S 45° E) and perpendicular to strike (S 45° W to N 45° E) are shown in (Fig. 3.78).

The rules used by Hughes & Davey in the interpolation are:
- 250 ft radius of influence
- along strike

$c_0 = 0.015 \quad a = 450\,\text{ft}$

$c_1 = 0.20$

- perpendicular to strike

$c_0 = 0.01 \quad a = 400\,\text{ft}$

$c_1 = 0.16$

The results are shown in Figure 3.79.

3.10.5 *Block kriging*

In the preceding sections the discussion has been on *point* kriging. *Block* values can be estimated by considering the blocks to be made up of several points. The values at the points are calculated as before and by averaging them, a block value is obtained. Parker & Sandefur (1976) have however shown that only a small error results when representing the grade of an entire block by a point. Hence the added effort required in block kriging may or may not be warranted.

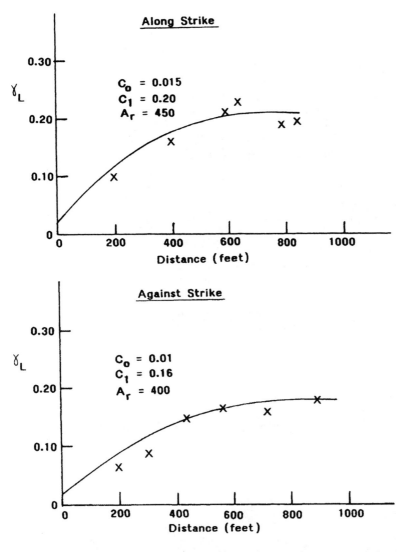

Figure 3.78. The experimental variograms for bench 5140 along strike and normal to strike (Hughes & Davey, 1979).

3.10.6 *Common problems associated with the use of the kriging technique*

Hughes & Davey (1979) have suggested the following problems are commonly associated with the use of the kriging technique.

1. Variograms do not accurately represent the mineralized zone because of inadequate data.
2. Mathematical models do not accurately fit the variogram data, or variograms have been improperly interpreted.
3. Kriging is insensitive to variogram coefficients.
4. Computational problems and expense are associated with repeatedly inverting large matrices.

260 *Open pit mine planning and design: Fundamentals*

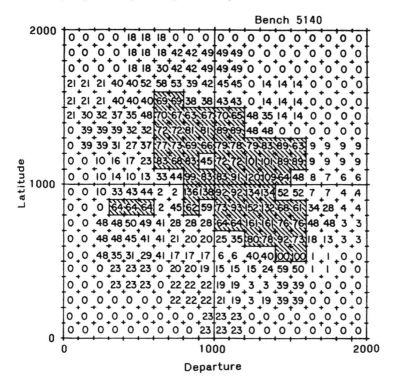

Figure 3.79. The kriged results for bench 5140 (Hughes & Davey, 1979).

Table 3.32. Comparison of the methods applied to the bench 5140 data (Hughes & Davey, 1979).

Method	Block size = 100 × 100 × 40 ft			
	Ore blocks	Tons ore	Avg ore grade (%)	Computer CPU sec.
Hand polygon		1,989,896	0.93	
Computer polygon	66	2,033,778	0.92	6.61
Inverse distance	65	2,002,963	0.91	7.08
Kriging	68	2,095,407	0.86	19.55

5. The matrix involved in finding the coefficients sometimes tends to be ill-conditioned. This means that for certain geometries, kriging doesn't work well.

6. There are problems associated with weighting coefficients.

3.10.7 *Comparison of results using several techniques*

Table 3.32 shows a comparison of some of the methods described in this chapter (Hughes & Davey, 1979). The decision as to which method to apply to any particular deposit evaluation is left to the user, and would certainly depend on the deposit, the data available, sample density, the type of results required, the required accuracy, and the amount of time, money, and energy that one is willing to expend on the evaluation of a specific deposit.

As indicated by Hughes & Davey (1979), there are pros and cons with each method. Many innovative designers have combined what they consider the best of various techniques to their particular application. The inverse distance weighting rules described by Hughes & Davey (1979) in Table 3.23 actually are a combination of inverse distance and geostatistical methods. Although the discussion herein has been restricted to two-dimensional examples, three-dimensional applications, especially with the aid of computers, are quite practical and generally result in a better interpolation of the deposit. There also is no question that the incorporation of geologic data (varying rock types, faults, etc.) into the grade assignment process is essential for generation of the most accurate model of the deposit.

REFERENCES

Adisoma, G.S., and Y.C. Kim. 1993. Jackknife kriging – A simple and robust new estimator. 24th International Symposium on Application of Computers and Operations Research in the Mineral Industry: 2(343–349). Montreal:CIMM.

Adisoma, G.S., and M.G. Hester. 1996. Grade estimation and its precision in mineral resources: the jackknife approach. Mining Engineering. 48(2): 84–88.

Altman, K.A., Taylor, D.L., and R.W. Smith. 2000. How much testing should I do? Are the samples really representative? Pre-print Paper 00-71. 2000 SME Annual Meeting, Feb 28–Mar 1. SLC, UT.

Axelson, A.H. 1963. A practical approach to computer utilization in mine planning. *Twenty-fourth annual mining symposium.* University of Minnesota: 151–165.

Barnes, M.P. 1979a. Case study – ore-body modeling at Sacaton Mine, Arizona. In: *Computer Methods in the 80's* (A. Weiss, editor): 268–275. New York: SME-AIME.

Barnes, M.P. 1979b. Drill-hole interpolation: Estimating mineral inventory. In: *Open Pit Mine Planning and Design* (J.T. Crawford & W.A. Hustrulid, editors): 65–80. SME-AIME.

Barnes, M.P. 1980. *Computer-Assisted Mineral Appraisal and Feasibility.* New York: Society of Mining Engineers of the AIME.

Barnes, N.W. 1989. *The Limitations of Popular Techniques for Preproduction Reserve Estimation in Mining.* Salt Lake City, UT: M. Barnes and Associates, Inc.

Behn, R.G., Beniscelli, J., Carrasco, C.P., Cuadra, P., and G.A. Ferguson. 1998. Estimation of resources and engineering of reserves, Radomiro Tomic Project. 27th International Symposium on Application of Computers and Operations Research in the Mineral Industry: 241–264. London:IMM.

Best, J., and K. Gallant. 2004. Reconciliation wars – Who's data are right? Pre-print Paper 04-183. 2004 SME Annual Meeting, Feb 23–25. Denver, CO.

Blais, J. 1985. Personal Communication. Pueblo, Colorado: CF&I Steel Corporation.

Blackwell, G.H., and A.J. Sinclair. 1993. QMIN, geostatistical software for mineral inventory using personal computers. 24th International Symposium on Application of Computers and Operations Research in the Mineral Industry: 2(350–357). Montreal:CIMM.

Blackwell, G. 1998. Effective modeling of grades of mineral resources. 27th International Symposium on Application of Computers and Operations Research in the Mineral Industry: 265–280. London:IMM.

Call, R.D. 1979. Development drilling. In: *Open Pit Mine Planning and Design* (J.T. Crawford & W.A. Hustrulid, editor): 29–40. New York: SME-AIME.

Carew, T.J. 2002. Unfolding – Getting the geometry right. 30th International Symposium on Application of Computers and Operations Research in the Mineral Industry: 103–112. Alaska:SME.

Case Histories and Methods in Mineral Resource Evaluation (A.E. Annels, editor). 1992. London: The Geological Society.

Cherrier, T.E. 1968. A report on the Ceresco Ridge extension of the Climax Molybdenite Deposit. M.Sc. Thesis. University of Minnesota.

Clark, I. 1979a. Does geostatistics work? In: 16th *APCOM* (T.J. O'Neil, editor): 213–225. SME.

Clark, I. 1979b. *Practical Geostatistics.* London: Applied Science Publishers.

Clark, I. 1999. A case study in the application of geostatistics to lognormal and quasi-lognormal problems. 28th International Symposium on Application of Computers and Operations Research in the Mineral Industry: 407–416. Colorado School of Mines: CSM.

Clark, I., and W. Harper, 2000. Practical geostatistics 2000. Ecosse North American Llc. Columbus, Ohio, USA, 342 pp.

Clark, I. 2000. Erratic highs – A perennial problem in resource estimation. Pre-print Paper 00-110. 2000 SME Annual Meeting, Feb 28–Mar 1. SLC, UT.

Clark, J.W., T.R. Couzens, W.C. Dixon and K.D. Wiley. 1984. Geological and geostatistical applications to mine production at Inspiration. Presented at the SME-AIME Fall Meeting Denver, Colorado Oct. 24–26, 1984. Preprint 84–381.

Cortez, L.P., Sousa, A.J., and F.O. Durão. 1998. Mineral resources estimation using neural networks and geostatistical techniques. 27th International Symposium on Application of Computers and Operations Research in the Mineral Industry: 305–314. London:IMM.

Cortez, L.P., Durão, F.O., and A.J. Sousa. 1999. Mineral resources estimation methods: A comparative study. 28th International Symposium on Application of Computers and Operations Research in the Mineral Industry: 425–434. Colorado School of Mines:CSM.

Colorado Department of Highways. Colorful Colorado Road Map.

Crawford, J.T. and R.K. Davey 1979. Case study in open pit limit analysis. In: *Computer Methods for the 80's in the Mineral Industry* (A. Weiss, editor): 310–318. SME-AIME.

CRC Handbook of Chemistry and Physics, 1991–1992, 72nd Edition (D.R. Lide, editor-in-Chief): 4.1–4.34, 4.150–4.155. Boston: CRC Press.

Dagbert, M. 1982. Orebody modelling. Presented at the 1982 American Mining Congress International Mining Show, Las Vegas, NV, Oct. 11–14, 1982.

Dagbert, M. and M. David 1977. Predicting vanishing tons before production starts or small blocks are no good for planning in porphyry type deposits. Presented at the 1977 AIME Annual Meeting, Atlanta, GA. Society of Mining Engineers of AIME: preprint number 77-AO-83.

David, M. and E. Toh 1989. Grade control problems, dilution and geostatistics: Choosing the required quality and number of samples for grade control. *CIM Bulletin* (931)82(11): 53–60.

David, M. 1971. Geostatistical ore reserve calculations, a step by step case study. *Decision Making in the Mineral Industry,* CIM Special Volume 12: 185–191.

David, M. 1972. Grade-tonnage curve use and misuse in ore reserve estimation. *Trans. Inst. Min. Metall.* (Min. Sect. A): 129–132.

David, M. 1977. *Geostatistical Ore Reserve Estimation*: 283. Amsterdam: Elsevier.

David, M. 1988. *Handbook of Applied Advanced Geostatistical Ore Reserve Estimation.* Amsterdam: Elsevier.

Davey, R.K. 1979. Mineral block evaluation criteria. In: *Open Pit Mine Planning and Design* (J.T. Crawford and W.A. Hustrulid, editors): 83–96. New York: SME-AIME.

Deraimse, J., and A.M. Thwaites. 1998. Use of non-linear geostatistical techniques for recoverable reserves estimation – a practical case study. 27th International Symposium on Application of Computers and Operations Research in the Mineral Industry: 315–326. London:IMM.

De Souza, L.E., Costa, J.F., and J.C. Koppe. 2002. A geostatistical contribution to the use of resource/reserves classification systems. 30th International Symposium on Application of Computers and Operations Research in the Mineral Industry: 73–84. Alaska:SME.

Deutsch, C.V., and R.W. Lewis. 1992. Advances in the practical implementation of indicator geostatistics. 23rd International Symposium on Application of Computers and Operations Research in the Mineral Industry: 169–180. Tucson:SME.

Deutsch, C.V., Magri, V., and K. Norrena. 1999. Optimal grade control using geostatistics and economics: Methodology and examples. Pre-print Paper 99-018. 1999 SME Annual Meeting, Mar 1–3. Denver, CO.

Dowd, P.A. 1998. Geostatistics in the United Kingdom – past, present and future. 27th International Symposium on Application of Computers and Operations Research in the Mineral Industry: 327–336. London:IMM.

Erarslan, K. 2001. Three dimensional variogram modeling and Kriging. 29th International Symposium on Application of Computers and Operations Research in the Mineral Industry: 51–56. Beijing:Balkema.

Francis, H.L., et al. 2000. Intelligent ore delineation: Integrating geosensing and geographical information systems. CIM Bulletin 93(1042): 96–99.

Frempong, P.K., and I. Clark. 1996. An indicator-based geostatistical assessment of the global vermiculite reserves at Palabora Mining Company. In Surface Mining 1996 (H.W. Glen, editor): 1–12. SAIMM.

Gertsch, L., Eloranta, J., and R. Gertsch. 2005. Ore characterization for process control: An update of a Mesabi Range project. Pre-print Paper 05-007. 2005 SME Annual Meeting, Feb 28–Mar 2. SLC, UT.

Hardwick, W.R. 1959. *Open-Pit Copper Mining Methods, Morenci Branch, Phelps Dodge Corp., Greenlee County, Ariz.* U.S. Bureau of Mines Information Circular 7911.

Hardwick, W.R. and E.L. Jones, III. 1959. *Open-Pit Copper Mining Methods and Costs at the Bagdad Mine, Bagdad Copper Corp., Yavapai County, Ariz.* USBM IC 7929.

Hardwick, W.R. and M.M. Stover. 1960. *Open-Pit Copper Mining Methods and Practices, Copper Cities Division, Miami Copper Co., Gila County, Ariz.* Bureau of Mines IC 7985.

Hatch, A.L. 1966. Factors affecting ore grade values. *Mining Engineering* 18(1): 72–75.

Hazen, S.W., Jr. 1967. *Assigning an Area of Influence for an Assay Obtained in Mine Sampling.* U.S. Bureau of Mines RI 6955.

Henley, S., and D.F. Watson. 1998. Possible alternatives to geostatistics. 27th International Symposium on Application of Computers and Operations Research in the Mineral Industry: 337–354. London:IMM.

Hewlett, R.F. 1961. Application of computers to open pit ore reserve calculation. *Short Course on Computers and Computer Applications in the Mineral Industry, volume I.* University of Arizona.

Hewlett, R.F. 1962. *Computing Ore Reserves by the Polygonal Method Using a Medium-Size Digital Computer* RI 5952: 31 pp.

Hewlett, R.F. 1963. Computer methods in evaluation development, and operations of an ore deposit. Annual Meeting of the AIME, Dallas, Texas, Feb 24–28, 1963.

Hewlett, R.F. 1963. *Computing Ore Reserves by the Triangular Method Using a Medium-Size Digital Computer* RI 6176: 26 pp.

Hewlett, R.F. 1965. *Design of drill-hole grid spacings for evaluating low-grade copper deposits.* USBM R.I. 6634.

Hewlett, R.F. 1970. *Comparison of the Triangular, Polygonal, and a Statistical Method of Computing Grade and Tonnage of Ore for the Silver Bell Oxide Porphyry Copper Deposit* U.S. Bureau of Mines RI 7331: 33 pp.

Hughes, W.E. and R.K. Davey 1979. Drill hole interpolation: Mineralized interpolation techniques. In: *Open Pit Mine Planning and Design* (J. Crawford and W. Hustrulid, editors): 51–64. New York: Society of Mining Engineers of the AIME.

Hulse, D.E. 1992. The consequences of block size decisions in ore body modeling. 23rd International Symposium on Application of Computers and Operations Research in the Mineral Industry: 225–232. Tucson:SME.

Hulse, D.E., and F Davies. 1993. Using indicator kriging to analyze the impact of mining dilution. 24th International Symposium on Application of Computers and Operations Research in the Mineral Industry: 2(381–385). Montreal:CIMM.

Inoue, T., and K. Okaya. 1993. Modeling of ore grade variation and its application. 24th International Symposium on Application of Computers and Operations Research in the Mineral Industry: 2(386–393). Montreal:CIMM.

Jackson, C.F. and J.H. Hedges 1939. Metal mining practice. *USBM Bulletin* 419: 58–79.

Jackson, C.F. and J.B. Knaebel 1932. Sampling and estimation of ore deposits. *USBM Bulletin 356*: 114–149.

Jalkanen, G., Greuer, R., and H.S. Chen. 2002. Screening Kriging. 30th International Symposium on Application of Computers and Operations Research in the Mineral Industry: 95–102. Alaska:SME.

Johnson, K.R. 1995. Geological modeling – The way ahead (Concept modelling for exploration and mine planning). 25th International Symposium on Application of Computers and Operations Research in the Mineral Industry: 1–4. Brisbane:AusIMM.

KCC (Scientific and Engineering Computer Center) 1967. *Isometric View of Bingham Pit (5).* Kennecott Copper Company.

Knudsen, H.P., Y.C. Kim and E. Mueller 1978. Comparative study of the geostatistical ore reserve estimation method over the conventional methods. *Mining Engineering* 30(1): 54–58.

Krige, D.G. 1984. Geostatistics and the definition of uncertainty. *Trans. Inst. Min. Metall (Sect A: Mining Ind.)* 93(4): A41–A47.

Krige, D.G., and P.G. Dunn. 1995. Some practical aspects of ore reserve estimation at Chuquicamata Copper Mine, Chile. 25th International Symposium on Application of Computers and Operations Research in the Mineral Industry: 125–134. Brisbane:AusIMM.

Krige, D.G. 1999. Conditional bias and uncertainty of estimation in geostatistics. 28th International Symposium on Application of Computers and Operations Research in the Mineral Industry: 3–14. Colorado School of Mines:CSM.

Krige, D.G., and W. Assibey-Bonsu. 2001. Valuation of recoverable resources by Kriging, direct condition or simulation. 29th International Symposium on Application of Computers and Operations Research in the Mineral Industry: 3–8. Beijing:Balkema.

Kuchta, M.E. 1986. Conceptual mine planning with interactive computer graphics. MSc Thesis T-3194. Colorado School of Mines.

Lemieux, M. 2000. Parametric analysis in surface-mine reserve definition: the inherent error and its correction. Mining Engineering. 52(8): 57–64.

Lemieux, M. 2000. Surface mine reserve definition and the high grade fallacy. Mining Engineering. 52(2): 48–50.

Leonardson, R.W., Weakly, C.G., Lander, A.M., and P.B. Zohar. 2005. Exploring between drill holes yields new ounces at Goldstrike. Pre-print Paper 05-110. 2005 SME Annual Meeting, Feb 28–Mar 2. SLC, UT.

Leuangthong, O., Lyall, G., and C.V. Deutsch. 2002. Multivariate geostatistical simulation of a nickel laterite deposit. 30th International Symposium on Application of Computers and Operations Research in the Mineral Industry: 261–274. Alaska:SME.

Lopez, W.A. 1990. Cokriging: A multivariate geostatistical analysis to enhance ore reserve estimation. MS Thesis T-3577, Colorado School of Mines.

McWilliams, J.R. 1959. *Mining Methods and Costs at the Anaconda Company Berkeley Pit, Butte, Mont.* U.S. Bureau of Mines Information Circular 7888.

Merks, J.W. 1992. Geostatistics or voodoo statistics. *E&MJ* 193(9): 45–49.

Mohanta, H.K., and A.K. Sen. 1995. Computerized orebody modeling and mine design for the Chira Iron Ore deposit. 25th International Symposium on Application of Computers and Operations Research in the Mineral Industry: 21–32. Brisbane:AusIMM.

Murthy, P.S.N., and M. Makki. 1999. Possible alternatives to geostatistics - some views. 28th International Symposium on Application of Computers and Operations Research in the Mineral Industry: 435–444. Colorado School of Mines:CSM.

Nelson, M.G., and N.K. Riddle. 2003. Sampling and analysis for optimal ore-grade control. Pre-print Paper 03-128. 2003 SME Annual Meeting, Feb 24–26. Cincinnati, OH.

O'Donnell, N. 1950. Mine and surface maps. *Mining Engineering* 2(9): 944–946.

Parker, H.M. and R.L. Sandefur 1976. A review of recent developments in geostatistics. AIME Annual Meeting, Las Vegas, NV.

Parker, H.M., Stearley, R.D., Swenson, A.J., Harrison, E.D., Maughan, J.R., Switzer, J.E., Mudge, K.L., and T.W. Smith. 2001. Evolution of computerized resource estimation at Bingham Canyon, Utah. 29th International Symposium on Application of Computers and Operations Research in the Mineral Industry: 25–30. Beijing:Balkema.

Parr, C.J. and N. Ely 1973. Chapter 2. Mining law. *SME Mining Engineering Handbook, Volume 1* (A.B. Cummins and I.A. Given, editors): 2-2–2-4. SME-AIME.

Parrish, I.S. 1993. Tonnage factor – a matter of some gravity. Mining Engineering. 45(10):1268–1271.

Patterson, J.A. 1959. Estimating ore reserves follows logical steps. *E&MJ* 160(9): 111–115.

Pelley, C.W., and O.A. Satti. 1993. Geostatistics for risk analysis in orebody evaluation and ore related mine decisions. 24th International Symposium on Application of Computers and Operations Research in the Mineral Industry: 2(302–309). Montreal:CIMM.

Peters, W.C. 1978. *Exploration and Mining Geology*: 434. New York: John Wiley & Sons.

Pfleider, E.P. 1962. Mine valuation problem. University of Minnesota.

Phelps, E.R. 1968. Correlation of development data and preliminary evaluation. *Surface Mining* (E.P. Pfleider, editor): 122–137. New York: AIME.

Pilger, G.G., Costa, J.F., and J.C. Koppe. 2002. Optimizing the value of a sample. 30th International Symposium on Application of Computers and Operations Research in the Mineral Industry: 85–94. Alaska:SME.

Popoff, C.C. 1966. *Computing Reserves of Mineral Deposits: Principles and Conventional Methods.* Bureau of Mines IC 8283.

Ramsey, R.H. 1944. How Cananea develops newest porphyry copper. *E&MJ* 145(12): 74–83.

Ranta, D.E., A.C. Noble and M.W. Ganster 1984. Geology and geostatistics in ore reserve estimation and mine evaluation. *Mine Feasibility – Concepts to Completion* (G.E. McKelvey, compiler). Northwest Mining Association.

Reich, H. 1961. Density of rocks. *Mining Engineers' Handbook, Third Edition* (R. Peele, editor) 1: 10-A-30. New York: Wiley.

Rendu, J.-M. 1980. A case study: Kriging for ore valuation and mine planning. *E&MJ* 181(1): 114–120.

Rendu, J.-M. 1984. Pocket geostatistics for field geologists. Presented at the SME-AIME Fall Meeting, Denver, Colorado, Oct. 24–26, 1984. Preprint 84-408.

Rendu, J.M. 1994. 1994 Jackling Lecture: Mining geostatistics – forty years passed. What lies ahead. Mining Engineering. 46(6): 557–558.

Rendu, J.M. 1998. Practical geostatistics at Newmont Gold: A story of adaptation. Mining Engineering. 50(2): 40–45.

Rossi, M.E. 1999. Improving the estimates of recoverable reserves. Mining Engineering. 51(1): 50–54.

Rossi, M.E., and J. Camacho Vidakovich. 1999. Using meaningful reconciliation information to evaluate predictive models. Pre-print Paper 99-020. 1999 SME Annual Meeting, Mar 1–3. Denver, CO.

Rossi, M. 2004. Comparing simulated and interpreted geologic models. Pre-print Paper 04-29. 2004 SME Annual Meeting, Feb 23–25. Denver, CO.

Roy, I., and B.C. Sarkar. 2002. Geostatistical orebody modeling and inventory of Gua Iron Ore Deposit, Jharkhand, India. 30th International Symposium on Application of Computers and Operations Research in the Mineral Industry: 243–250. Alaska:SME.

Schofield, D., and B. Denby. 1993. Genetic algorithms: A new approach to pit optimization. 24th International Symposium on Application of Computers and Operations Research in the Mineral Industry: 2(126–133). Montreal:CIMM.

Schurtz, R.F. 1999. A rigorous integral relation between kriging and inverse distance with certain of its implications. 28th International Symposium on Application of Computers and Operations Research in the Mineral Industry: 417–424. Colorado School of Mines:CSM.

Seigel, H.O., Gingerich, J.C., and E.O. Köstlin. 2002. Explore or acquire? The dilemma. CIM Bulletin 95(1058): 62–69.

Sikka, D.B. and R.B. Bhappu. 1994. Economic potential of the Malanjkhand Proterozoic porphyry copper deposit, M.P. India. Mining Engineering. 46(3): 221–229.

Stanley, D. 1993. Geologic modeling at Newmont Gold Company. 24th International Symposium on Application of Computers and Operations Research in the Mineral Industry: 2(445–452). Montreal:CIMM.

Stanley, B.T. 1979. Mineral model construction: Principles of ore-body modelling. *Open Pit Mine Planning and Design* (J.T. Crawford and W.A. Hustrulid, editors): 43–50. New York: SME-AIME.

Tang, Y., and Y. Lan. 2001. A new calculation method of mineral reserves and examination of mineral reserves. 29th International Symposium on Application of Computers and Operations Research in the Mineral Industry: 35–38. Beijing:Balkema.

Thwaites, A.M. 1998. Assessment of geological uncertainty for a mining project. 27th International Symposium on Application of Computers and Operations Research in the Mineral Industry: 391–406. London:IMM.

Thompson, I.S. 2002. A critique of valuation methods for exploration properties and undeveloped mineral resources. CIM Bulletin 95(1061): 57–62.

USGS (United States Geological Survey). 1976. Jefferson County, Colorado – County Map Series (Topographic). Scale 1:50000.

Vallée, M. 2000. Mineral resources + engineering, economy and legal feasibility = ore reserves. CIM Bulletin 93(1033): 53–61.

Vorster, A.P., and W.L. Schöning. 1995. A revision of ore evaluation techniques at Sischen Iron Ore Mine, South Africa. 25th International Symposium on Application of Computers and Operations Research in the Mineral Industry: 139–148. Brisbane:AusIMM.

Vorster, A.P., and J. Smith. 1996. Developments in ore evaluation and grade control at Sishen Iron Ore mine. In Surface Mining 1996 (H.W. Glen, editor): 105–115. SAIMM.

Wade, E.J. 1967. Computerization of ore reserve calculations at Tasu. *Canadian Mining Journal* 88(3).

Williams, W.R. 1983. *Mine Mapping & Layout*. New Jersey: Prentice-Hall.

Waterman, G.C. and S. Hazen 1968. Chapter 3.1. Development drilling and bulk sampling. In: *Surface Mining* (E.P. Pfleider, editor): 69. New York: AIME.

Weaton, G.F. 1972. Valuation of a mineral property. Personal Communication. Minneapolis, Minn.

Weaton, G.F. 1973. Valuation of a mineral property. *Mining Engineering* 25(5).

Westerfelt, W.Y. 1961. Weights of minerals and rocks. Section 25. Mine Examinations. In: *Mining Engineers Handbook – Third Edition* (R. Peele, editor) II: 25–21.

Yalcin, E. and A. Ünal 1992. The effect of variogram estimation on pit-limit design. *CIM Bulletin* (961)85(6): 47–50.

Yamamoto, J.K. 1996. Ore reserve estimation: A new method of block calculation using the inverse of weighted distance. 197(9): 69–72.

REVIEW QUESTIONS AND EXERCISES

1. Briefly discuss the steps in the development of a geologic data base.
2. What is meant by core 'logging'?
3. In a surface mine feasibility study, what role do maps serve?
4. What is meant by the 'scale' of a map?
5. In the English system of units, what is meant by a scale of 1:50?
6. In the metric system, what is meant by a scale of 1:1000?
7. Is a scale of 1:50 larger or smaller than a scale of 1:20? Explain your answer.
8. What different types of maps might be involved in a feasibility study? What type(s) of information might be included/shown on each?
9. In the U.S., what process might be followed to determine the ownership status of a certain parcel of land?
10. What are the three common map types used for mine planning and design? What would be shown on each?
11. In plan, in which quadrant is the map normally positioned?
12. For sections, which is the normal viewing direction chosen?
13. What are some of the different possibilities for the choice of the north direction?
14. What are some of the possible choices for the coordinate system?
15. Practical guidelines for the preparation of maps are provided in Table 3.1. What is meant by 'scale: graphic and numerical'? Why is this important?
16. What does the nomenclature DDH stand for?
17. Mining operations are commonly divided into four stages. What are they? Today, what is an important fifth stage?
18. What types of geologic information should be developed in the exploration stage?
19. What are the two basic types of diamond core drilling? Describe each.
20. What types of information might be collected during the logging of core?
21. Summarize the characteristics of each of the different exploratory techniques.
22. Why is it important that the core and the information gathered there from be treated with utmost respect?
23. In Table 3.4, assay values are given for both core and sludge samples. What are sludge samples? How do the values compare? What is a possible reason for the differences?
24. Where is the ore located based upon the drill hole log from the Comstock mine? What is the mineral? What is the nature of the overburden?
25. What is meant by oriented core? What is the purpose?
26. What is meant by the RQD? How is it calculated? What information does it provide?
27. What is the purpose(s) of compositing?
28. What is the difference between a simple average and a weighted average?
29. What is the difference between an ore zone composite and a bench composite?
30. What are some of the benefits of compositing?
31. Using the drill-hole log given in Table 3.8 and assuming a bench height of 18 ft, what would be the composite grades assigned to the first three benches?
32. How does the choice of bench height affect the compositing?

Orebody description 267

33. How does the choice of bench elevation affect the compositing?
34. What is the purpose of a tonnage factor?
35. Assuming that an ore has a specific gravity of 2.9, what would be the tonnage factor expressed in the English and metric systems?
36. Assuming that the ore density is 3.1 g/cm3, what is the tonnage factor in the English system?
37. Assuming that the ore density is 190 lbs/ft3, what is the tonnage factor in the metric system?
38. What techniques are available to determine the density of the material to be mined? Why is this determination important?
39. In the example given in section 3.4.2, assume that the mining company has a contract to sell 7000 tons of metal X per year. The mined material contains 1.5% of the contained metal and the processing plant recovery is 75%. How large a plan area must be exposed per year? Assume the other factors in the example are the same.
40. Begin with Figure 3.22 and complete the steps required to arrive at the final section shown in Figure 3.27.
41. Repeat Problem 40 but first draw Figure 3.22 in AutoCad. Then use the technique to complete the steps. Compare the areas with those shown in Figure 3.28.
42. What geologic features are evident on the Mesabi iron range of Minnesota? Why is it helpful to know this?
43. In Figure 3.30 the pit slope angles are indicated to be 27°. On the drawing, however, they appear to be at 45°. What is the reason?
44. Describe the process by which the 2D sections are combined to form the 3D pit.
45. In Table 3.13 the ore area and the average grade for a section have been determined. Repeat the process assuming that Hole 6 is missing.
46. How is the overall pit ore tonnage and grade determined?
47. How is the pit end tonnage determined?
48. Discuss the method of vertical sections based upon the use of grade contours.
49. In section 3.6 equivalent influence distances have been chosen for sections 4 and 20 to incorporate end volumes. Show how the numbers were chosen.
50. Discuss the process used to obtain the ore tonnage, the average in-place ore grade, and the average grade of the mined ore in the example given in section 3.6.
51. What is the significance of the 0.12% grade used in the example?
52. In Table 3.18 an average stripping ratio has been determined. What definition of stripping ratio has been used here?
53. Today, horizontal rather than vertical sections are commonly used when describing ore bodies. What has been the primary reason for the change?
54. Determine the average grade for the triangular solid defined by holes C-51, C-28 and C-47 in Figure 3.45? If the thickness of the slice is 40 ft, what is the volume of the solid? If the density is 2.6 g/cm3, how many tons are involved?
55. Repeat problem 54 assuming that the ore intercept is 40 ft for hole C-51, 20 ft for hole C-28, and 30 ft for hole C-47.
56. In section 3.7.3 an example involving a polygon formed around hole C-41 has been presented. Apply the process to hole C-35.
57. In Figure 3.49 the process involving the construction of a polygon at the boundary of a hole array is demonstrated. Apply the technique to hole C-51. Assume $R = 250$ ft.
58. Summarize the steps in developing hand-generated polygons.

268 Open pit mine planning and design: Fundamentals

59. What are some special conditions where the radius of sample influence does not apply?
60. Today, typically block models are used to represent ore bodies. Summarize the important guidelines regarding block size.
61. What 'special case' situation will be used in this section to assign grades to the blocks?
62. What is meant by the rule of nearest points? Assign grades in the region defined by 500N–1000N and 500E–1000E using this rule. Compare your results to what is shown in Figure 3.54.
63. Describe the inverse distance weighting technique.
64. Assign a grade to the block having center coordinates (650E, 850N) in Figure 3.57 using the inverse distance weighting technique.
65. Describe the inverse squared weighting technique. Assign a grade to the block in problem 64 using this technique.
66. How might you determine the most appropriate distance weighting power to be used?
67. What is meant by the angle of exclusion?
68. How do you select the radius of influence?
69. What is meant by the term 'isotropic'? What is meant by the term 'anisotropic'? How would this property be included?
70. Carefully read the rules provided in Table 3.23. How does one take into account the presence of different rock types?
71. Using the grades for bench 5140 as given in Table 3.21, check the given average grade value.
72. Plot a histogram of the values using an interval of 0.2% Cu. Superimpose the mean value.
73. From viewing the histogram developed in problem 72, what type of distribution is shown?
74. Discuss the procedure used to calculate the % cumulative frequency.
75. What is the finesse provided by standard probability plots? Locate this paper on the Web and download an example. Plot the data provided in Table 3.24. Compare your result with that given in Figure 3.62.
76. Repeat the process described in the book to determine the additive constant ß. Develop the plot shown in Figure 3.64.
77. What is meant by the following terms:
    – range
    – variance
    – standard deviation
78. Determine the variance for the data set given in Table 3.24.
79. Describe the logic for determining the range of sample influence using geostatistical techniques.
80. What is meant by:
    – the lag
    – the geostatistical variance
    – the semi-variance
    – the semi-variogram
81. In section 3.9.3 an example is provided of the calculation of $\gamma(600)$ using N-S pairs. Repeat the procedure for $\gamma(800)$ using N-S pairs.
82. In figure 3.68 all of the pairs have been used independent of direction. To be able to do this, what must be true?

Orebody description 269

83. What is meant by
    – the sill
    – the nugget effect
    – the range
84. How does the 'sill' relate to the usual statistical values as determined in problem 77?
85. What is the reason for the occurrence of the 'nugget effect'?
86. What is meant by
    – the chaotic variance
    – the structured variance
    – the total variance
87. What might an experimental variogram look like for
    – a strataform deposit
    – a porphyry copper deposit
    – a placer gold deposit
88. Why do you need a mathematical expression to represent the experimental variogram?
89. What are the characteristics of the spherical model?
90. What is the equation for the spherical model? What parameters must be extracted from the experimental variogram?
91. What type of information concerning the deposit is provided by a variogram?
92. Discuss the basic idea behind the use of kriging.
93. What is meant by covariance?
94. An example of the application of kriging has been presented in section 3.10.3. Follow through the example. Check the answers provided for a_1, a_2, a_3 and γ. Determine the estimated grade value.
95. Assign a grade to point x_0 using the inverse distance technique and the same holes as were used in problem 93.
96. Repeat the kriging example in section 3.10.3 but now use all six of the surrounding holes.
97. Assign the grade to point x_0 using the inverse distance technique and the same holes as in problem 95.
98. Figure 3.78 shows the variograms determined along strike and against strike. What is the conclusion?
99. What is meant by block kriging?
100. What are some of the common problems associated with the use of the kriging technique?
101. What are the advantages of kriging? How is the estimation variance used?

CHAPTER 4

Geometrical considerations

4.1 INTRODUCTION

The ore deposits being mined by open pit techniques today vary considerably in size, shape, orientation and depth below the surface. The initial surface topographies can vary from mountain tops to valley floors. In spite of this, there are a number of geometry based design and planning considerations fundamental to them all. These are the focus of this chapter. By way of introduction consider Figure 4.1 which is a diagrammatic representation of a volume at the earth's surface prior to and after the development of an open pit mine.

The orebody is mined from the top down in a series of horizontal layers of uniform thickness called benches. Mining starts with the top bench and after a sufficient floor area has been exposed, mining of the next layer can begin. The process continues until the bottom bench elevation is reached and the final pit outline achieved. To access the different benches a road or ramp must be created. The width and steepness of this ramp depends upon the type of equipment to be accommodated. Stable slopes must be created and maintained during the creation and operation of the pit. Slope angle is an important geometric parameter which has significant economic impact. Open pit mining is very highly mechanized. Each piece of mining machinery has an associated geometry both related to its own physical size, but also with the space it requires to operate efficiently. There is a complementary set of drilling, loading and hauling equipment which requires a certain amount of working space. This space requirement is taken into account when dimensioning the so-called working benches. From both operating and economic viewpoints certain volumes must or should, at least, be removed before others. These volumes have a certain minimum size and an optimum size.

It is not possible in this short chapter to try and fully cover all of the different geometrical aspects involved in open pit mine planning and design. However, the general principles associated with the primary design components will be presented and whenever possible illustrated by examples.

4.2 BASIC BENCH GEOMETRY

The basic extraction component in an open pit mine is the bench. Bench nomenclature is shown in Figure 4.2.

Geometrical considerations 271

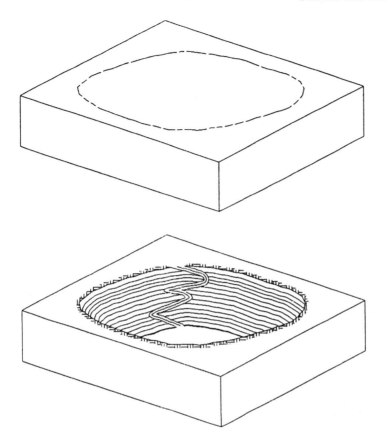

Figure 4.1. Geometry change in pit creation.

Each bench has an upper and lower surface separated by a distance H equal to the bench height. The exposed subvertical surfaces are called the bench faces. They are described by the toe, the crest and the face angle α (the average angle the face makes with the horizontal). The bench face angle can vary considerably with rock characteristics, face orientation and blasting practices. In most hard rock pits it varies from about 55° to 80°. A typical initial design value might be 65°. This should be used with care since the bench face angle can have a major effect on the overall slope angle.

Normally bench faces are mined as steeply as possible. However, due to a variety of causes there is a certain amount of back break. This is defined as the distance the actual bench crest is back of the designed crest. A cumulative frequency distribution plot of measured average bench face angles is shown in Figure 4.3.

The exposed bench lower surface is called the bench floor. The bench width is the distance between the crest and the toe measured along the upper surface. The bank width is the horizontal projection of the bench face.

There are several types of benches. A working bench is one that is in the process of being mined. The width being extracted from the working bench is called the cut. The width of the working bench W_B is defined as the distance from the crest of the bench floor to the new

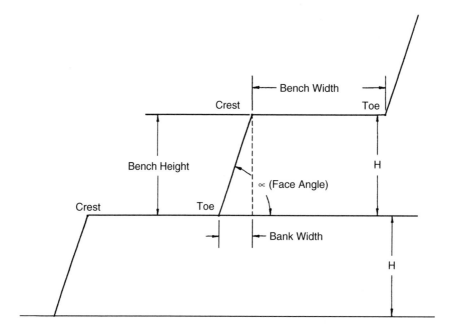

Figure 4.2. Parts of a bench.

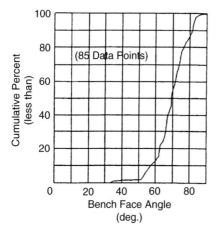

Figure 4.3. Cumulative frequency distribution of measured bench face angles (Call, 1986).

toe position after the cut has been extracted (see Fig. 4.4). A detailed calculation of cut and working bench dimensions is found in Subsection 4.4.5. After the cut has been removed, a safety bench or catch bench of width S_B remains.

The purpose of these benches is to:
(a) collect the material which slides down from benches above,
(b) stop the downward progress of boulders.

During primary extraction, a safety bench is generally left on every level. The width varies with the bench height. Generally the width of the safety bench is of the order of $2/3$ of the

Geometrical considerations 273

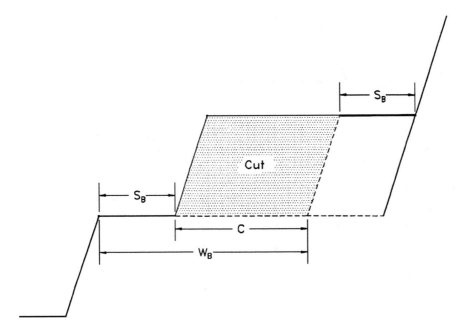

Figure 4.4. Section through a working bench.

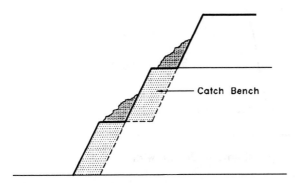

Figure 4.5. Functioning of catch benches.

bench height. At the end of mine life, the safety benches are sometimes reduced in width to about $1/3$ of the bench height.

Sometimes double benches are left along the final pit wall (Fig. 4.6). These are benches of double height which consequently permit, at a given overall slope angle, a single catch bench of double width (and hence greater catching capability). Along the final pit contour careful blasting is done to maintain the rock mass strength characteristics.

In addition to leaving the safety benches, berms (piles) of broken materials are often constructed along the crest. These serve the function of forming a 'ditch' between the berm and the toe of the slope to catch falling rocks. Based upon studies of rock falls made by Ritchie (1963), Call (1986) has made the design catch bench geometry recommendations given in Table 4.1 and illustrated in Figure 4.7.

274 Open pit mine planning and design: Fundamentals

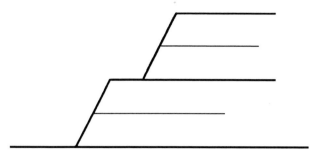

Figure 4.6. Double benches at final pit limits.

Table 4.1. Typical catch bench design dimensions (Call, 1986).

Bench height (m)	Impact zone (m)	Berm height (m)	Berm width (m)	Minimum bench width (m)
15	3.5	1.5	4	7.5
30	4.5	2	5.5	10
45	5	3	8	13

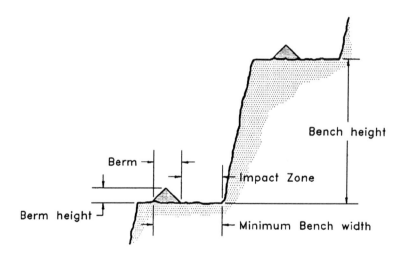

Figure 4.7. Catch bench geometry (Call, 1986).

A safety berm is also left (Fig. 4.8) along the outer edge of a bench to prevent trucks and other machines from backing over. It serves much the same function as a guard rail on bridges and elevated highways. Normally the pile has a height greater than or equal to the tire radius. The berm slope is taken to be about 35° (the angle of repose).

In some large open pits today median berms are also created in the center of haulage roads. In this book the word 'berm' is used to refer to the piles of rock materials used to improve mine safety. Others have used the word 'berm' as being synonymous with bench.

In the extraction of a cut, the drills operate on the upper bench surface. The loaders and trucks work off of the bench floor level.

Geometrical considerations

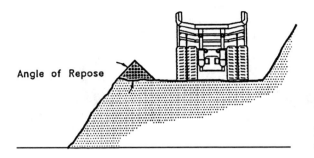

Figure 4.8. Safety berms at bench edge.

A number of different factors influence the selection of bench dimensions. Bench height becomes the basic decision since once this is fixed the rest of the dimensions follow directly. A common bench height in today's large open pits is 50 ft (15 m). For smaller pits the value might be 40 ft (12 m). For small gold deposits a typical value could be 25 ft (7.5 m). A general guideline is that the bench height should be matched to the loading equipment. When using shovels, the bench height should be well within the maximum digging height. For the 9 yd capacity shovel shown in Figure 4.9, it is seen that the maximum cutting height is 43'6". Hence it could be used with 40 ft benches. A general rule of thumb is that the bench height should not be greater than that of the sheave wheel. Operating in benches with heights greater than this sometimes result in overhangs which endanger the loading and other operations.

Figure 4.10 shows typical reach heights for shovels and front end loaders as a function of bucket size.

At one time, bench heights were limited by drilling depth. Modern drills have largely removed such restrictions. However, in large open pit mines, at least, it is desirable to drill the holes in one pass. This means that the drill must have a mast height sufficient to accommodate the bench height plus the required subdrill.

A deposit of thickness T can be extracted in many ways. Two possibilities are shown in Figure 4.11:

(a) 3 benches of height 50 ft,
(b) 6 benches of height 25 ft.

Higher and wider benches yield:
– less selectivity (mixing of high and low grade and ores of different types);
– more dilution (mixing of waste and ore);
– fewer working places hence less flexibility;
– flatter working slopes; large machines require significant working space to operate efficiently.

On the other hand, such benches provide:
– fewer equipment setups, thus a lower proportion of fixed set up time;
– improved supervision possibilities;
– higher mining momentum; larger blasts mean that more material can be handled at a given time;
– efficiencies and high productivities associated with larger machines.

The steps which are followed when considering bench geometry are:

(1) Deposit characteristics (total tonnage, grade distribution, value, etc.) dictate a certain geometrical approach and production strategy.

276 Open pit mine planning and design: Fundamentals

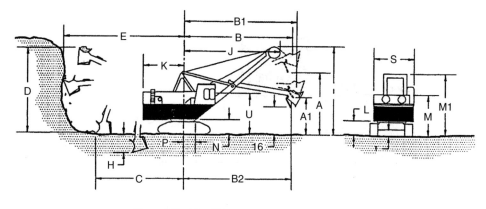

Figure 4.9. Diagrammatic representation of a 9 yd³ shovel (Riese, 1993).

Shovel Working Range

Dipper Capacity (Nominal) cu.yds		9
Dipper Capacities (Range) cu.yds		6 ½–6
Length of Boom		41'–6"
Effective length of dipper handle		25'–6"
Overall length of dipper handle		30'–9"

These dimensions will vary slightly depending upon dipper selection.

	Angle of boom	45°	
A	Dumping height – maximum	28'–0"	A
A1	Dumping height at maximum radius – B1	20'–6"	A1
B	Dumping radius at maximum height – A	45'–6"	B
B1	Dumping radius – maximum	47'–6"	B1
B2	Dumping radius at 16'0" dumping height	47'–0"	B2
D	Cutting height – maximum	43'–6"	D
E	Cutting radius – maximum	54'–6"	E
G	Radius of level floor	35'–3"	G
H	Digging depth below ground level – maximum	8'–6"	H
I	Clearance height – boom point sheaves	42'–3"	I
J	Clearance radius – boom point sheaves	40'–0"	J
K	Clearance radius – revolving frame	19'–9"	K
L	Clearance Under frame – to ground	6'–2"	L
M	Clearance height top of house	18'–10"	M
M1	Height of A-frame	31'–2"	M1
N	Height of boom foot above ground level	9'–11"	N
P	Distance – boom foot to center of rotation	7'–9"	P
S	Overall width of machinery house & operating cab	22'–6"	S
T	Clearance under lowest point in truck frame	14"	T
U	Operator's eye level	18'–0"	U

(2) The production strategy yields daily ore-waste production rates, selective mining and blending requirements, numbers of working places.

(3) The production requirements lead to a certain equipment set (fleet type and size).

(4) Each equipment set has a certain optimum associated geometry.

(5) Each piece of equipment in the set has an associated operating geometry.

(6) A range of suitable bench geometries results.

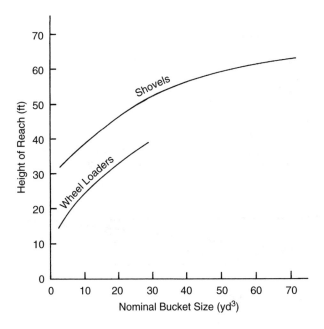

Figure 4.10. Height of reach as a function of bucket size.

Figure 4.11. Two different bench height scenarios.

(7) Consequences regarding stripping ratios, operating vs. capital costs, slope stability aspects, etc. are evaluated.

(8) The 'best' of the various alternatives is selected.

In the past when rail bound equipment was being extensively used, great attention was paid to bench geometry. Today highly mobile rubber tired/ crawler mounted equipment has reduced the detailed evaluation requirements somewhat.

4.3 ORE ACCESS

One of the topics which is little written about in the mining literature is gaining initial physical access to the orebody. How does one actually begin the process of mining? Obviously the approach depends on the topography of the surrounding ground. To introduce the topic it will be assumed that the ground surface is flat. The overlying vegetation has been removed

278 Open pit mine planning and design: Fundamentals

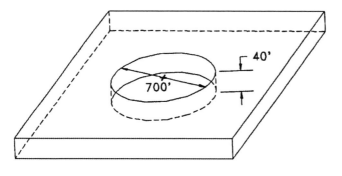

Figure 4.12. Example orebody geometry.

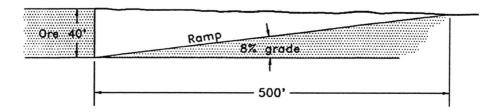

Figure 4.13. Ramp access for the example orebody.

as has the soil/sand/gravel overburden. In this case it will be assumed that the orebody is 700 ft in diameter, 40 ft thick, flat dipping and is exposed by removing the soil overburden. The ore is hard so that drilling and blasting is required. The bench mining situation is shown in Figure 4.12.

A vertical digging face must be established in the orebody before major production can begin. Furthermore a ramp must be created to allow truck and loader access. A drop cut is used to create the vertical breaking face and the ramp access at the same time. Because vertical blastholes are being fired without a vertical free face, the blast conditions are highly constrained. Rock movement is primarily vertically upwards with only very limited sideways motion. To create satisfactory digging conditions the blastholes are normally rather closely spaced. Here only the geometry aspects will be emphasized. To access the orebody, the ramp shown in Figure 4.13 will be driven. It has an 8% grade and a width of 65 ft. Although not generally the case, the walls will be assumed vertical. To reach the 40 ft desired depth the ramp in horizontal projection will be 500 ft in length. There is no general agreement on how the drop cut should be drilled and blasted. Some companies drill the entire cut with holes of the same length. The early part of the ramp then overlies blasted rock while the final portion is at grade. In the design shown in Figure 4.14 the drop cut has been split into three portions. Each is blasted and loaded out before the succeeding one is shot. Rotary drilled holes $9 7/8''$ in diameter are used. The minimum hole depth is 15 ft. This is maintained over the first 90 ft of the ramp. The hole depth is then maintained at 7 ft below the desired final cut bottom. A staggered pattern of holes is used.

The minimum width of the notch is controlled largely by the dimensions of the loading machine being used. In this example, it will be assumed that the loading machine is the 9 yd^3 capacity shovel shown diagrammatically in Figure 4.9.

Geometrical considerations 279

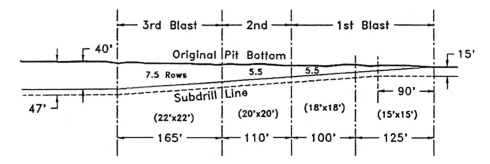

Figure 4.14. Blast design for the ramp excavation.

In the generally tight confines of the drop cut the following shovel dimensions:
K, the clearance radius of the revolving frame,
J, the clearance radius of the boom point sheaves,
G, the maximum digging radius of the level floor, and
E, the maximum cutting radius
are of importance. As can be seen from Figure 4.9, these are:

$K = 19'9''$

$J = 40'0''$

$G = 35'3''$

$E = 54'6''$

The minimum width of the drop cut is given by

Minimum width $= K + J$

In this case it is

Minimum width $= 19'9'' + 40'0'' = 59'9''$

This is such that both the front and rear portions of the machine can clear the banks on the two sides as it revolves in the digging and dumping modes.

The maximum digging radius of the level floor is used to indicate the maximum drop cut width for the shovel working along one cutting path. The maximum value is that which the shovel dipper (bucket) can be moved horizontally outward, thereby accomplishing floor cleanup.

The maximum width of the cut at floor level would be

Maximum cut width (floor) $= 2 \times 35'3'' = 70'6''$

The maximum width of the cut at crest level would be

Maximum cut width (crest) $= 2 \times 54'6'' = 109'$

In practice the cutting width for the shovel moving along one path is relatively tightly constrained by the shovel dimensions. In this case:

Minimum cut width (crest) $\cong 60$ ft

Maximum cut width (floor) $\cong 71$ ft

Maximum cut width (crest) $= 109$ ft

For typical cut slope angles of 60 to 80°, the maximum cut width (floor) is the controlling dimension. When the cutting path is down the center of the cut and the shovel is digging to both sides the maximum floor and minimum crest radii would be

Maximum floor radius = 35'3"

Minimum crest radius = 40'0"

In any case, for laying out the blasting round and evaluating minimum pit bottom dimensions one wants to exceed the minimum working space requirements.

Figures 4.15A through 4.15D show the minimum floor bottom geometry when the shovel moves along the two cutting paths. The loading would first be from one bank. The shovel would then move over and load from the other. This would be considered very tight operating conditions and would be used to create a final cut at the pit bottom.

The usual drop cut is shown in Figures 4.16A through 4.16C where the shovel moves along the cut centerline and can dig to both sides. It will be noted that the shovel must swing through large angles in order to reach the truck.

In both cases the working bench geometry at this stage is characterized by cramped operating conditions.

Two locations for the drop cut/ramp will be considered. The first (case A) is entirely in the waste surrounding the pit. It is desired to have the floor of the ramp at the bottom of ore just as it reaches the ore-waste contact. This is shown diagrammatically in Figure 4.17. The volume of waste rock mined in excavating the ramp is

$$\text{Ramp volume} = \frac{1}{2} H \frac{100H}{g} R_W$$

where R_W is the average ramp width, H is the bench height, and g is the road grade (%). In this case it becomes

$$\text{Ramp volume} = \frac{1}{2} \frac{(40)^2 \times 100 \times 65}{8} = 650,000 \text{ ft}^3$$

This waste must be excavated and paid for before any ore can be removed. However in this arrangement all of the ore can be removed. If it is assumed that the orebody can be extracted with vertical walls, then the ore volume extracted is

$$\text{Ore volume} = \frac{\pi D^2 H}{4} = \frac{\pi}{4}(700)^2 \times 40 = 15,400,000 \text{ ft}^3$$

Upon entering the orebody mining proceeds on an ever expanding front (Fig. 4.18).

As the front expands the number of loading machines which can effectively operate at the same time increases. Hence the production capacity for the level varies with time.

In summary for this ramp placement (case A):

Waste removed (road) = 650,000 ft³

Ore extracted = 15,400,000 ft³

% ore extracted = 100%

Another possibility (case B) as is shown in Figure 4.19 is to place the ramp in ore rather than to place the ramp in waste rock. This would be driven as a drop cut in the same way as discussed earlier. The volume excavated is obviously the same as before but now it is ore. Since the material is ore it can be processed and thereby profits are realized earlier.

Geometrical considerations 281

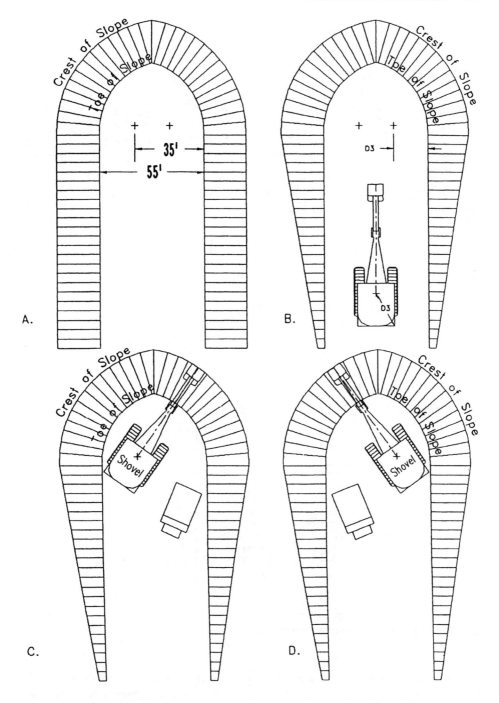

Figure 4.15. Minimum width drop cut geometry with shovel alternating from side to side.

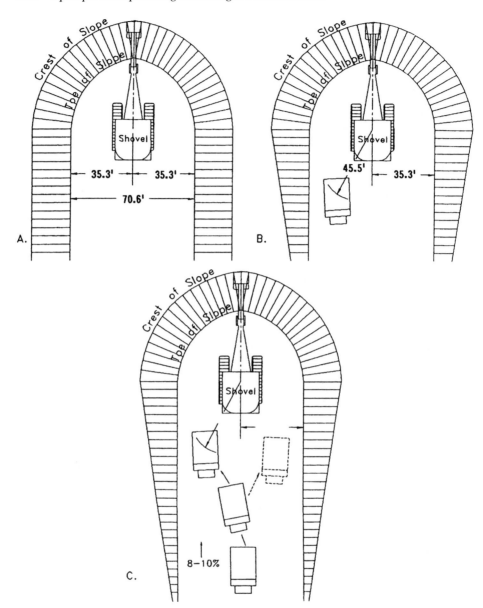

Figure 4.16. Minimum width drop cut geometry with shovel moving along centerline.

From the ramp bottom, the extraction front is gradually increased in length (Fig. 4.20). Obviously the disadvantage is that when mining is completed a quantity of ore remains locked up in the ramp. This quantity is equal to the amount of waste extracted in case A.

Thus the two important points to be made are:

– If the haul road is added external to the planned pit boundaries, then an additional quantity of material equal to the volume of the road must be extracted.

Geometrical considerations 283

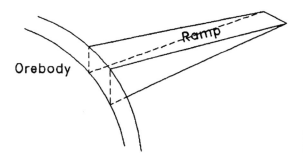

Figure 4.17. Isometric view of the ramp in waste approaching the orebody.

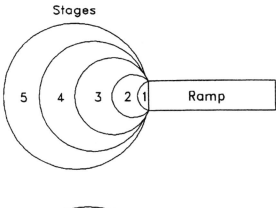

Figure 4.18. Diagrammatic representation of the expanding mining front.

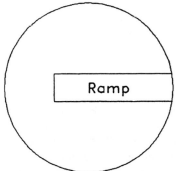 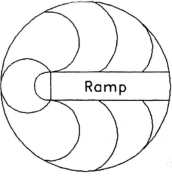

Figure 4.19. Dropcut/ramp placement in ore.   Figure 4.20. Expansion of the mining front.

– If the haul road is added internal to the original planned boundaries, then a quantity of material equal to the road volume must be left in place.

Rather than a straight road such as shown in case A, one might have considered a curved road such as shown in plan in Figure 4.21. With the exception of the final portion, the road is entirely driven in waste. The road could be placed so that the 'ore' left is in the poorest grade.

Assume that the pit is not 1 bench high but instead consists of 2 benches such as is shown in Figure 4.22. The idea is obviously to drive the ramp down to the ore level and establish

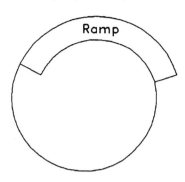

Figure 4.21. Ramp starting in waste and ending in ore.

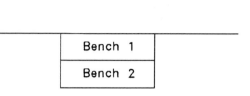

Figure 4.22. Section through a two bench mine.

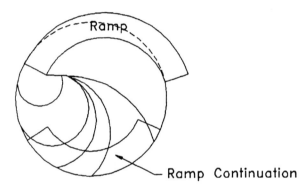

Figure 4.23. Two ramp sections with pit expansion.

the desired production rate. Then while mining is underway on level 1, the ramp would be extended in ore to the lower level as shown in Figure 4.23 through the use of a drop cut. All of the ore lying below the ramp is obviously sterilized. For a multi-bench operation, the procedure continues as shown in Figure 4.24. Note that a flat section having a length of 200 ft has been left in this example between the decline segments. The ramp has a corkscrew shape and the coils get tighter and tighter as the pit is deepened. Rather soon in this example, the pit would reach a final depth simply because the ramp absorbed all of the available working space.

A vertical section taken through the final pit with the orebody superimposed is shown in Figure 4.25. For this particular design where only the initial segment of the ramp is in waste, a large portion of the orebody is sterilized. The amount of waste removed is minimized, however.

An alternative design is one where the ramp is underlain by waste and all of the ore is removed. To make this construction one starts the road design at the lowest bench and works back out. This exercise is left to the reader.

The actual design will generally be somewhere in between these two alternatives with the upper part of the ramp underlain by waste and the lower part by 'ore'.

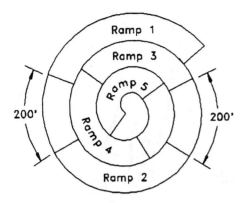

Figure 4.24. Plan view showing ramp locations for a five bench mine.

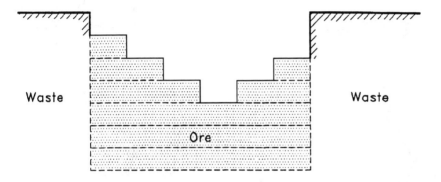

Figure 4.25. Section view showing the sterilization of reserves by ramp.

The excavation may start with attacking the ore first so that the cash flow is improved. Later during the mine life, the waste will be stripped as the main access is gradually moved outward.

In summary:
– there can be considerable volumes associated with the main ramp system;
– the location of the ramp changes with time;
– in the upper levels of the pit, the ramp is underlain by waste; in the lower ranges it is underlain by mineral;
– cash flow considerations are significantly affected by ramp timing;
– the stripping ratio, the percent extraction and the overall extraction are strongly affected simply by the haul road geometry (road width and road grade).

Drop cuts are used on every level to create a new bench. Figures 4.26A through 4.26D show the steps going from the current pit bottom through the mining out of the level. Often the ramp is extended directly off of the current ramp and close to the existing pit wall. This is shown in Figure 4.27. A two level loading operation is shown isometrically in Figure 4.28. The ramp access to both levels in this relatively simple example is easily seen.

286 *Open pit mine planning and design: Fundamentals*

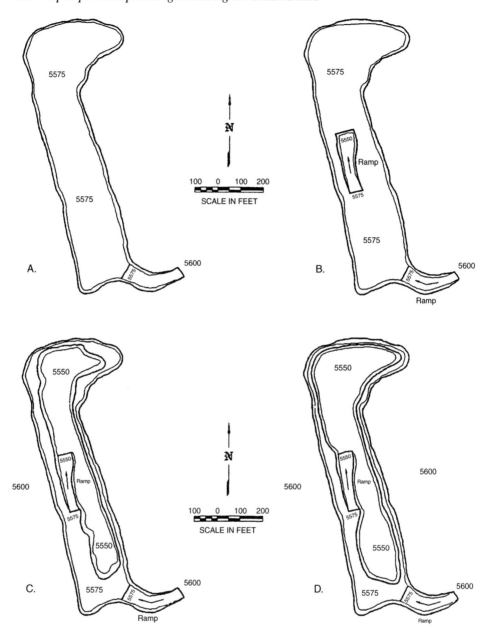

Figure 4.26. Plan view of an actual pit bottom showing drop cut and mining expansion (McWilliams, 1959).

There are many examples where the orebody lies in very rugged terrain. Figure 4.29 shows diagrammatically one possible case. Here the entry to the orebody is made by pushing back the hillside. Bench elevations are first established as shown on the figure. In this case the bench height is 50 ft. Initial benches are established by making pioneering cuts along the surface at convenient bench elevations.

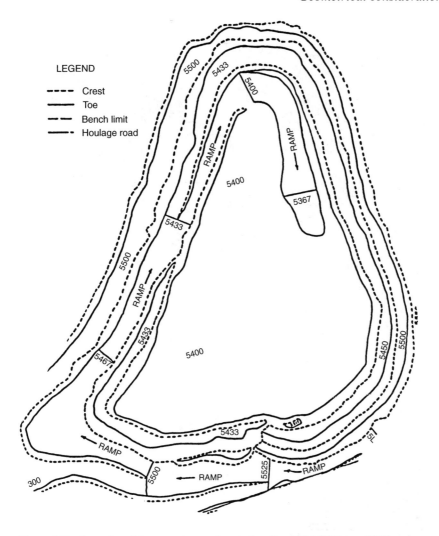

Figure 4.27. Extension of the current ramp close to the pit wall (McWilliams, 1959).

If the slope is composed of softer material, then a dozer can notch it without further assistance (Fig. 4.30). For harder rock types, ripping prior to dozing may be enough. However if the rock is hard or the slope is steep, drilling and blasting will probably be necessary for the pioneer cut. Generally air track types of drills are used. They can reach and drill in very difficult places and can tow their own air compressors/generators.

As shown in Figure 4.31 a shovel can be used instead of a dozer for notching a slope. The notch is enlarged by taking successive cuts until the full bench height is achieved.

Once these initial benches are established, mining of the full faces with vertical blast holes proceeds. Obviously the upper benches have to be advanced before the lower ones.

The final pit outline for this section is shown in Figure 4.33. The reader is encouraged to consider the pit development sequence and the point where drop cuts would be used.

288 *Open pit mine planning and design: Fundamentals*

Figure 4.28. Isometric view of simultaneous mining on several levels (Tamrock, 1978).

Figure 4.29. Deposit located in mountainous terrain.

Geometrical considerations 289

Figure 4.30. Creating initial access/benches (Nichols, 1956).

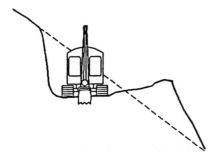

Figure 4.31. Sidehill cut with a shovel (Nichols, 1956).

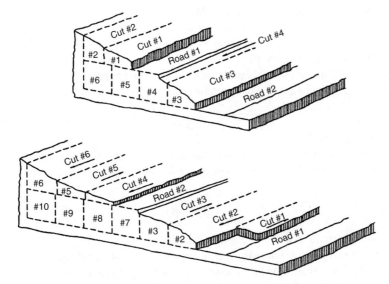

Figure 4.32. Shovel cut sequence when initiating benching in a hilly terrain (Nichols, 1956).

290 Open pit mine planning and design: Fundamentals

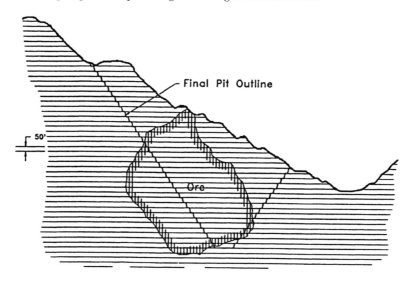

Figure 4.33. Final pit outline superimposed on a section.

4.4 THE PIT EXPANSION PROCESS

4.4.1 Introduction

When the drop cut has reached the desired grade, the cut is expanded laterally. Figure 4.34 shows the steps. Initially (Fig. 4.34A) the operating space is very limited. The trucks must turn and stop at the top of the ramp and then back down the ramp towards the loader. When the pit bottom has been expanded sufficiently (Fig. 4.34B), the truck can turn around on the pit bottom. Later as the working area becomes quite large (Fig. 4.34C) several loaders can be used at the same time. The optimum face length assigned to a machine varies with the size and type. It is of the range 200 to 500 ft.

Once access has been established the cut is widened until the entire bench/level has been extended to the bench limits. There are three approaches which will be discussed here:
 1. Frontal cuts.
 2. Parallel cuts – drive by.
 3. Parallel cuts – turn and back.

The first two apply when there is a great deal of working area available, for example at the pit bottom. The mining of more narrow benches on the sides of the pit is covered under number three.

4.4.2 Frontal cuts

The frontal cut is shown diagrammatically in Figure 4.35.
　The shovel faces the bench face and begins digging forward (straight ahead) and to the side. A niche is cut in the bank wall. For the case shown, double spotting of the trucks is used. The shovel first loads to the left and when the truck is full, he proceeds with the truck on the right. The swing angle varies from a maximum of about 110° to a minimum of 10°. The average swing angle is about 60° hence the loading operation is quite efficient. There

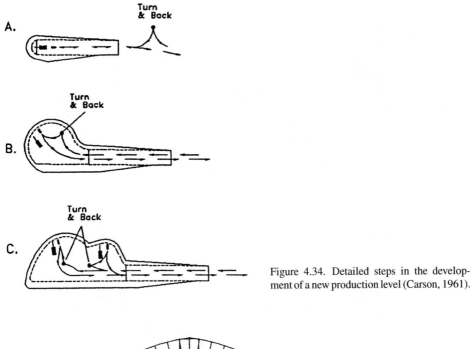

Figure 4.34. Detailed steps in the development of a new production level (Carson, 1961).

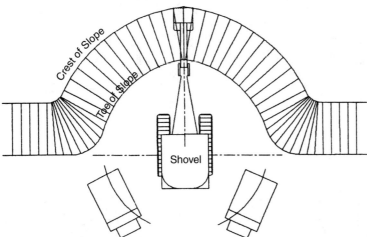

Figure 4.35. Diagrammatic representation of a frontal cutting operation.

must be room for the trucks to position themselves around the shovel. The shovel penetrates to the point that the center of swing is in line with the face. It then moves parallel to itself and takes another frontal cut (Fig. 4.36).

With a long face and sufficient bench width, more than one shovel can work the same face (Fig. 4.37). A fill-in cut is taken between the individual face positions (Fig. 4.38). From the shovels view point this is a highly efficient loading operation. The trucks must however stop and back into position.

292 *Open pit mine planning and design: Fundamentals*

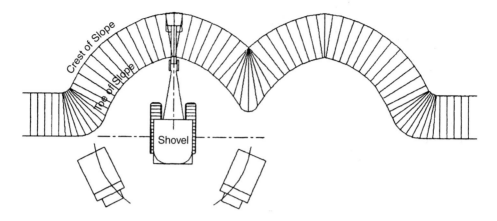

Figure 4.36. Shovel move to adjacent cutting position.

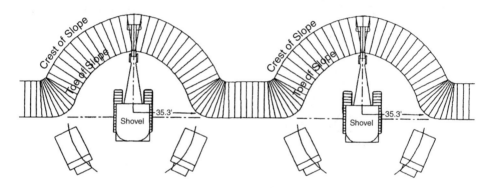

Figure 4.37. Two shovels working the same face.

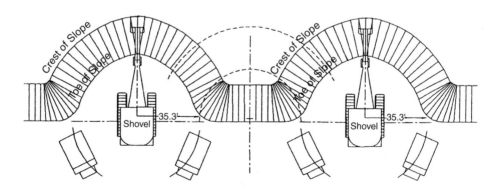

Figure 4.38. Fill-in cutting to complete the face.

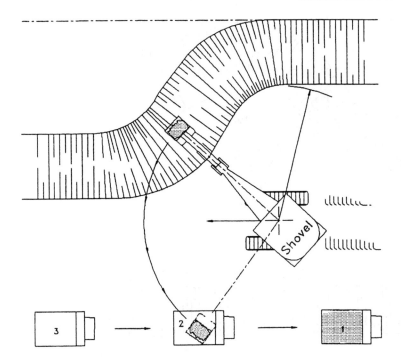

Figure 4.39. Parallel cut with drive-by.

4.4.3 Drive-by cuts

Another possibility when the mine geometry allows is the parallel cut with drive-by. This is shown diagrammatically in Figure 4.39. The shovel moves across and parallel to the digging face. For this case, bench access for the haul units must be available from both directions. It is highly efficient for both the trucks and the loader. Although the average swing angle is greater than for the frontal cut, the trucks do not have to back up to the shovel and spotting is simplified.

4.4.4 Parallel cuts

The expansion of the pit at the upper levels is generally accomplished using parallel cuts. Due to space limitations there is only access to the ramp from one side of the shovel. This means that the trucks approach the shovel from the rear. They then stop, turn and back into load position. Sometimes there is room for the double spotting of trucks (Fig. 4.40) and sometimes for only single spotting (Fig. 4.41).

Pit geometry is made up of a series of trade-offs. Steeper slopes result in a savings of stripping costs. On the other hand they can, by reducing operating space, produce an increase in operating costs.

Figure 4.42 shows the single spotting sequence. Truck 2 (Fig. 4.42B) waits while the shovel completes the loading of truck 1. After truck 1 has departed (Fig. 4.42C), truck 2 turns and stops (Fig. 4.42D) and backs into position (Fig. 4.42E). While truck 2 is being

294 *Open pit mine planning and design: Fundamentals*

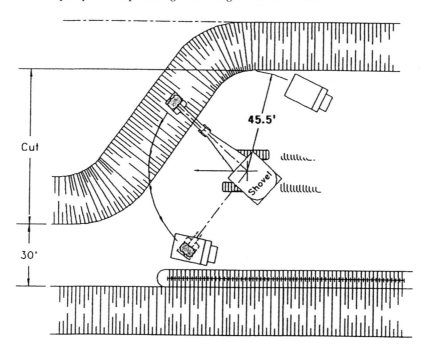

Figure 4.40. Parallel cut with the double spotting of trucks.

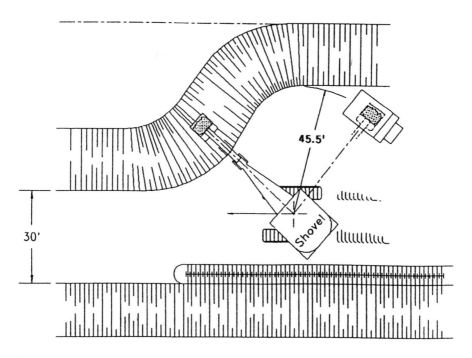

Figure 4.41. Parallel cut with the single spotting of trucks.

Geometrical considerations 295

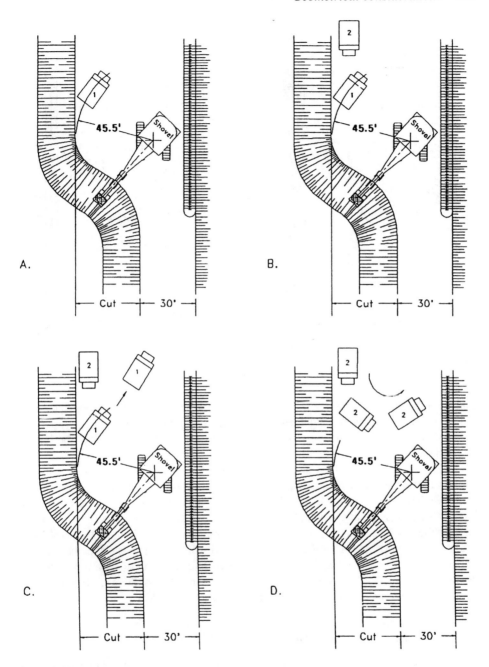

Figure 4.42. Time sequence showing shovel loading with single spotting.

296 *Open pit mine planning and design: Fundamentals*

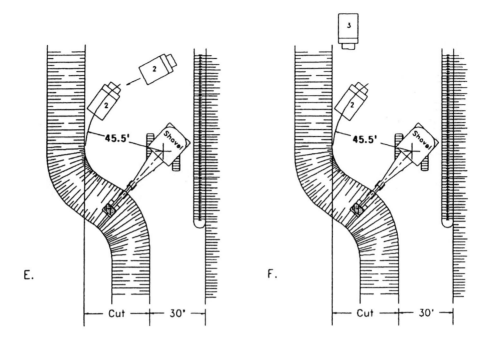

Figure 4.42. (Continued).

loaded truck 3 arrives (Fig. 4.42F). The process then repeats. In this situation both the trucks and the shovel must wait causing a reduction in the overall productivity.

The double spotting situation is shown in Figure 4.43. Truck 1 is first to be loaded (Fig. 4.43A).

Truck 2 arrives (Fig. 4.43B) and backs into position (Fig. 4.43C). When it is just in position the shovel has completed the loading of truck 1. As truck 1 departs (Fig. 4.43D) the shovel begins the loading of truck 2. As truck 2 is being loaded truck 3 arrives. It turns (Fig. 4.43E) and backs into position (Fig. 4.43F). As truck 2 leaves the shovel begins loading truck 3 (Fig. 4.43G). With this type of arrangement there is no waiting by the shovel and less waiting by the trucks. Thus the overall productivity of this system is higher than that for single spotting. The sequencing is unfortunately quite often not as the theory would suggest. Figures 4.43H and 4.43I show two rather typical situations. Both of these can be minimized through the use of an effective communications/dispatching system.

4.4.5 *Minimum required operating room for parallel cuts*

In the previous section the physical process by which a pit is expanded using parallel cuts was described. In this section, the focus will be on determining the amount of operating room required to accommodate the large trucks and shovels involved in the loading operation.

The dimension being sought is the width of the working bench. The working bench is that bench in the process of being mined. This width (which is synonymous with the term 'operating room') is defined as the distance from the crest of the bench providing the floor

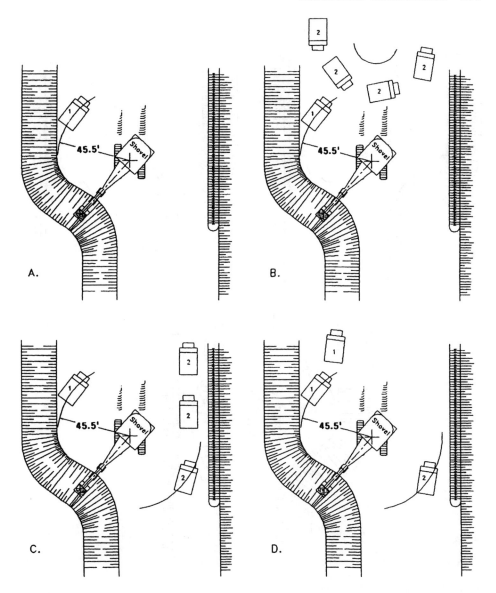

Figure 4.43. Time sequence showing shovel loading with double spotting.

for the loading operations to the bench toe being created as the parallel cut is being advanced. The minimum amount of operating room varies depending upon whether single or double spotting of trucks is used, with the latter obviously requiring somewhat more. The minimum width (W_B) is equal to the width of the minimum required safety bench (S_B) plus the width of the cut (W_C) being taken. This is expressed as

$$W_B = S_B + W_C$$

298 *Open pit mine planning and design: Fundamentals*

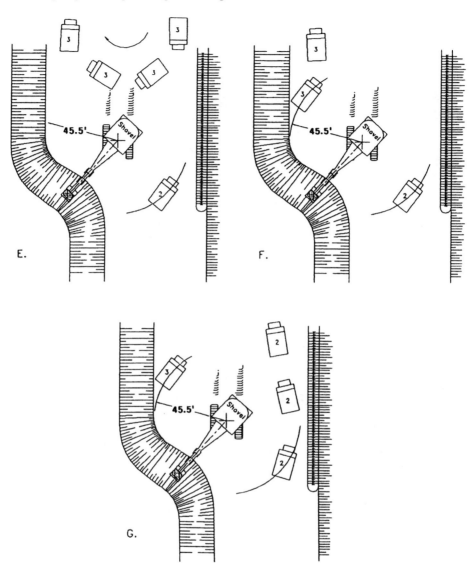

Figure 4.43. (Continued).

The easiest way of demonstrating the principles involved is by way of example. For this, the following assumptions will be made:
– Bench height = 40 ft.
– A safety berm is required.
– The minimum clearance between the outer truck tire and the safety berm = 5 ft.
– Single spotting is used.
– Bench face angle = 70°.
– Loading is done with a 9 yd^3 BE 155 shovel (specifications given in Fig. 4.9).
– Haulage is by 85 ton capacity trucks.

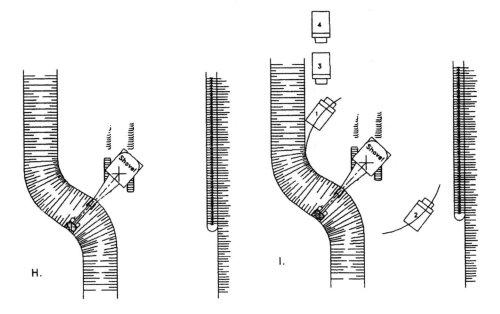

Figure 4.43. (Continued).

- Truck width = 16 ft.
- Tire rolling radius = 4 ft.

The general arrangement in plan and section is shown in Figure 4.44. The design shows that:

$$\text{Working bench width} = 102 \text{ ft}$$
$$\text{Cut width} = 60 \text{ ft}$$
$$\text{Safety bench width} = 42 \text{ ft}$$

The basic calculations (justification) behind these numbers will now be presented.

Step 1. A safety berm is required along the edge of this bench. As will be discussed in Subsection 4.9.5, the height of the berm should be of the order of the tire rolling radius. For this truck, the berm height would be about 4 ft. Assuming that the material has an angle of repose of 45°, the width of the safety berm is 8 ft (see Fig. 4.45). It is assumed that this berm is located with the outer edge at the crest.

Step 2. The distance from the crest to the truck centerline is determined assuming parallel alignment. A 5′ clearance distance between the safety berm and the wheels has been used. Since the truck is 16 ft wide, the centerline to crest distance (T_C) is 21 ft.

Step 3. The appropriate shovel dimensions are read from the specification sheet (Fig. 4.9):

(a) Shovel centerline to truck centerline. This is assumed to be the dumping radius (B) at maximum height,

$$B = 45'6'' = 45.5 \text{ ft}$$

300 Open pit mine planning and design: Fundamentals

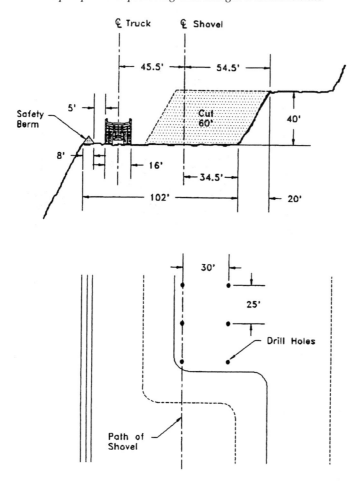

Figure 4.44. Section and plan views through a working bench.

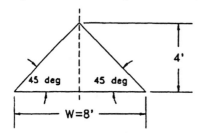

Figure 4.45. Simplified representation of a safety berm.

(b) The maximum dumping height (A) is more than sufficient to clear the truck,

$$A = 28 \text{ ft}$$

(c) The level floor radius dimension (G) is the maximum distance from the shovel centerline which the floor can be cleaned. In this case

$$G = 35'3'' = 35.25 \text{ ft}$$

This will be used as the maximum shovel centerline to toe distance.

Geometrical considerations 301

Step 4. The desired working bench dimension becomes

$$W_B = T_C + B + G = 21 + 45.5 + 35.25 \cong 102 \text{ ft}$$

Step 5. The corresponding width of cut is now calculated. In this case it has been assumed that the shovel moves along a single path parallel to the crest. Information from the shovel manufacturer suggests that the maximum cutting width (W_C) may be estimated by

$$W_C = 0.90 \times 2 \times G = 0.90 \times 2 \times 35.25 \cong 63.5 \text{ ft}$$

This applies to the width of the pile of broken material. Therefore, to allow for swell and throw of the material during blasting, the design cut width should be less than this value. Here a value of 60 ft has been assumed.

Step 6. Knowing the width of the working bench and the cut width, the resulting safety bench has a width

$$S_B = 102 - 60 = 42 \text{ ft}$$

This is of the order of the bench height (40 ft) which is a rule of thumb sometimes employed.

Step 7. Some check calculations are made with regard to other dimensions.

a) The maximum cutting height of the shovel

$$D = 43'6'' = 43.5 \text{ ft}$$

is greater than the 40′ bench height. Thus the shovel can reach to the top of the bench face for scaling.

b) The maximum shovel cutting radius (E) is

$$E = 54'6'' = 54.5 \text{ ft}$$

Since the maximum radius of the level floor (G) is

$$G = 35'3'' = 35.25 \text{ ft}$$

the flattest bench face angle which could be scaled (Fig. 4.46) is

$$\text{Slope} = \tan^{-1} \frac{40}{54.50 - 35.25} = 64.3°$$

Thus the shovel can easily scale the 70° bench face.

Step 8. The cut dimension should be compared to the drilling and blasting pattern being used. In this particular case the holes are $12\frac{1}{4}$ ins. in diameter (D_e) and ANFO is the explosive. Using a common rule of thumb, the burden (B) is given by

$$B = 25 \frac{D_e}{12} \cong 25 \text{ ft}$$

The hole spacing (S) is equal to the burden

$$S = 25 \text{ ft}$$

Thus two rows of holes are appropriate for this cut width.

302 *Open pit mine planning and design: Fundamentals*

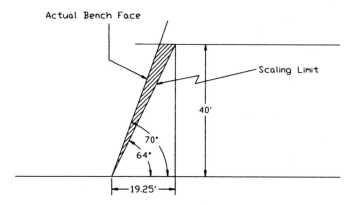

Figure 4.46. The bench/bench face geometry for the example.

A somewhat simplified approach has been applied to the matter of determining working bench width. The complications arise when one examines the best width from an overall economic viewpoint.

As will be discussed in Section 4.5, the working bench is generally one of a set of 3 to 5 benches being mined as a group. The others in the set each have a width equal to that of a safety bench. As the cut is extracted, the remaining portion of the working bench is reduced to a safety bench width. Since the width of the working bench is approximately equal to the combined widths of the others in the set, it has a major impact on the overall slope angle. A wider working bench means that the slope angle is flatter with the extra costs related to earlier/more stripping, but the equipment operating efficiency is higher (with lower related costs). On the other hand a more narrow working bench would provide a steeper overall slope at the cost of operating efficiency. Thus, there are other factors, beside those related to equipment geometries, which must be considered.

4.4.6 *Cut sequencing*

In the previous section the terms 'working bench,' 'cut' and 'safety bench' were introduced. These will now be applied to a simple example in which a 90 ft wide cut 1000 ft long will be taken from the right hand wall of the pit shown in Figure 4.47. As can be seen the wall consists of 4 benches. The entire bench 1 (B1) is exposed at the surface. Benches B2, B3 and B4 are safety benches, 35 ft wide. The process begins with the drill working off the upper surface of B1. The holes forming the cut to be taken from B1 are drilled and blasted (Fig. 4.48). The shovel then moves along the floor of bench B1 (upper surface of B2) and loads the trucks which also travel on this surface. The working bench has a width of 125 ft. When the cut is completed the geometry is as shown in Figure 4.49. The cut to be taken from bench 2 is now drilled and blasted. The shovel moves along the top of bench 3 taking a cut width (W_C). A portion of bench 2 remains as a safety bench. The process is repeated until the bottom of the pit is reached. The shovel then moves back up to bench 1 and the process is repeated. If it is assumed that the shovel can produce 10,000 tons/day, then the overall production from these 4 benches is 10,000 tons/day. The four benches associated with this shovel are referred to as a mine production unit.

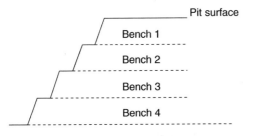

Figure 4.47. Initial geometry for the push back example.

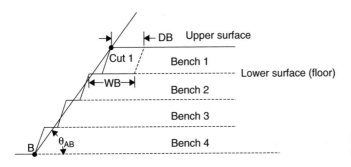

Figure 4.48. Cut mining from bench 1.

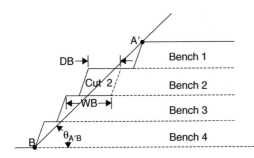

Figure 4.49. Cut mining from bench 2.

4.5 PIT SLOPE GEOMETRY

There are a number of 'slopes' which enter into pit design. Care is needed so that there is no confusion as to how they are calculated and what they mean. One slope has already been introduced. That is the bench face angle (Fig. 4.50). It is defined as the angle made with the horizontal of the line connecting the toe to the crest. This definition of the slope going from the toe to the crest will be maintained throughout this book.

Now consider the slope consisting of 5 such benches (Fig. 4.51). The angle made with the horizontal of the line connecting the lowest most toe to the upper most crest is defined as the overall pit slope,

$$\Theta(\text{overall}) = \tan^{-1} \frac{5 \times 50}{4 \times 35 + \frac{5 \times 50}{\tan 75°}} = 50.4°$$

304 Open pit mine planning and design: Fundamentals

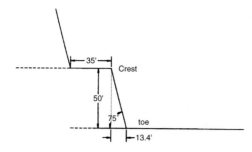

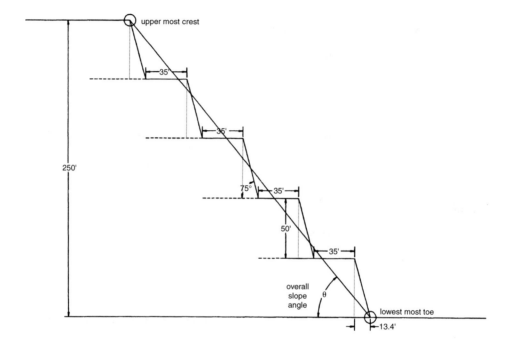

Figure 4.50. Safety bench geometry showing bench face angle.

Figure 4.51. Overall slope angle.

If as is shown in Figure 4.52 an access ramp with a width of 100 ft is located half way up bench 3, the overall pit slope becomes

$$\Theta(\text{overall}) = \tan^{-1} \frac{5 \times 50}{4 \times 35 + \frac{5 \times 50}{\tan 75°} + 100} = 39.2°$$

As can be seen, the presence of the ramp on a given section has an enormous impact on the overall slope angle.

The ramp breaks the overall slope into two portions (Fig. 4.53) which can each be described by slope angles. These angles are called interramp angles (between-the-ramp angles). In this case

$$IR_1 = IR_2 = \tan^{-1} \frac{125}{2 \times 35 + \frac{2 \times 50}{\tan 75°} + \frac{25}{\tan 75°}} = 50.4°$$

Geometrical considerations 305

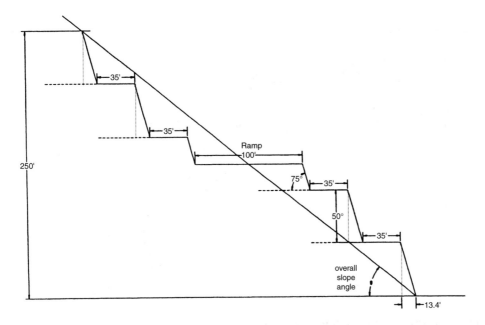

Figure 4.52. Overall slope angle with ramp included.

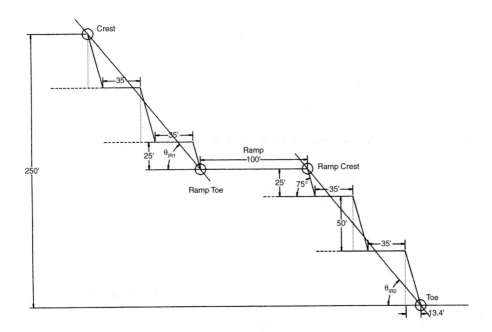

Figure 4.53. Interramp slope angles for Figure 4.52.

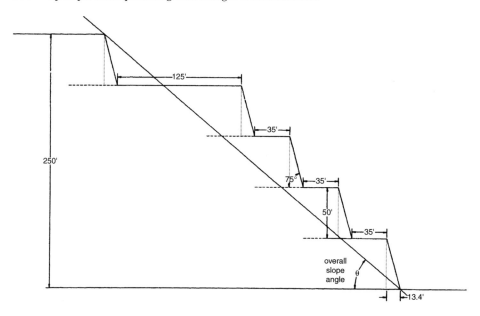

Figure 4.54. Overall slope angle with working bench included.

The interramp wall height is 125 ft for each segment. Generally the interramp wall heights and angles for the different slope segments would not be the same. From a slope stability viewpoint each interramp segment would be examined separately.

While active mining is underway, some working benches would be included in the overall slope. Figure 4.54 shows a working bench 125 ft in width included as bench 2. The overall slope angle is now

$$\Theta = \tan^{-1} \frac{5 \times 50}{125 + 4 \times 35 + \frac{5 \times 50}{\tan 75°}} = 37.0°$$

The working bench is treated in the same way as a ramp in terms of interrupting the slope. The two interramp angles are shown in Figure 4.55. In this case

$$\Theta_{IR_1} = 75°$$

$$\Theta_{IR_2} = \tan^{-1} \frac{200}{3 \times 35 + \frac{4 \times 50}{\tan 75°}} = 51.6°$$

The interramp wall heights are

$$H_1 = 50'$$

$$H_2 = 200'$$

For this section, it is possible that the ramp cuts bench 3 as before. This situation is shown in Figure 4.56.

The overall slope angle has now decreased to

$$\Theta = \tan^{-1} \frac{250}{125 + 3 \times 35 + 100 + \frac{5 \times 50}{\tan 75°}} = 32.2°$$

Geometrical considerations 307

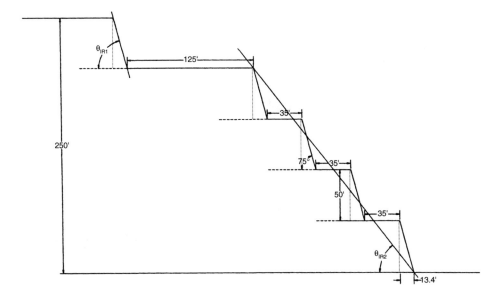

Figure 4.55. Interramp angles associated with the working bench.

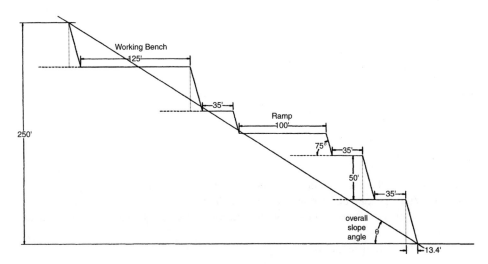

Figure 4.56. Overall slope angle with one working bench and a ramp section.

As shown in Figure 4.57, there are now three interramp portions of the slope. The interramp wall heights and angles are:

Segment 1:

$$\Theta_{IR_1} = 75°$$
$$H = 50'$$

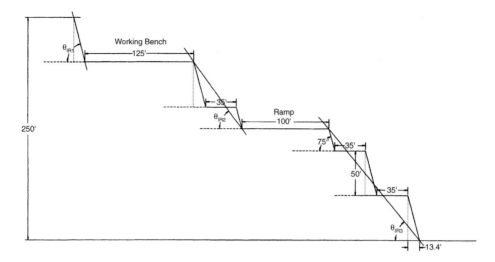

Figure 4.57. Interramp slope angles for a slope containing a working bench and a ramp.

Segment 2:

$$\Theta_{IR_2} = \tan^{-1} \frac{75}{35 + \frac{75}{\tan 75°}} = 35.7°$$

$$H = 75'$$

Segment 3:

$$\Theta_{IR_3} = \tan^{-1} \frac{125}{2 \times 35 + \frac{125}{\tan 75°}} = 50.4°$$

$$H = 125'$$

In Figure 4.57, the overall slope is shown to contain one working bench. Under some circumstances there may be several working benches involved in the mining of the slope. Figure 4.58 shows the case of a slope with 6 benches of which two are working benches 125 ft in width.

The overall (working slope) is given by

$$\Theta = \tan^{-1} \frac{300}{3 \times 35 + 2 \times 125 + \frac{300}{\tan 75°}} = 34.6°$$

The slope associated with each shovel working group is shown in Figure 4.59. In this case it is

$$\Theta = \tan^{-1} \frac{150}{125 + 35 + \frac{150}{\tan 75°}} = 36.8°$$

If the number of working benches is increased to 3 for the slope containing 6 benches, the overall slope would be further reduced. Thus to maintain reasonable slope angles, most mines have one working bench for a group of 4 to 5 benches.

Geometrical considerations 309

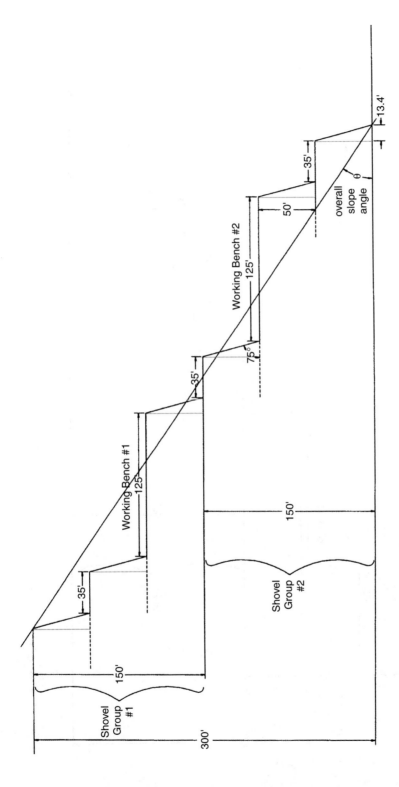

Figure 4.58. Overall slope angle for a slope containing two working benches.

310 *Open pit mine planning and design: Fundamentals*

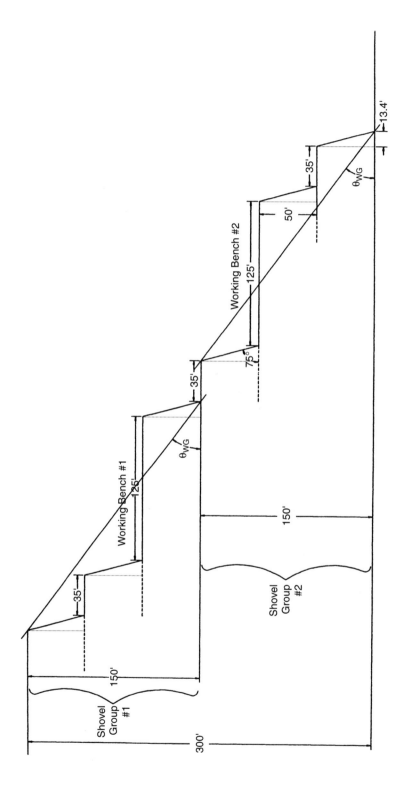

Figure 4.59. Slopes for each working group.

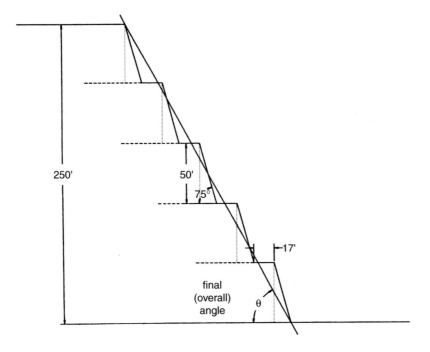

Figure 4.60. Final overall pit slope.

At the end of mining it is desired to leave the final slope as steep as possible. Some of the safety benches will be reduced in width while others may be eliminated entirely. For final walls, a bench width of approximately ⅓ of the bench height is commonly used. For this example with a bench height of 50 ft, the bench width becomes 17 ft. The final pit slope angle, assuming no ramp is needed on this wall (Fig. 4.60), becomes

$$\Theta(\text{final}) = \tan^{-1} \frac{250}{4 \times 17 + \frac{250}{\tan 75°}} = 61.6°$$

If the final bench faces could have been cut at 90° instead of 75°, then the final overall pit slope angle would be

$$\Theta(\text{final}) = \tan^{-1} \frac{250}{4 \times 17} = 74.8°$$

It is much more likely that the final face angles are 60° and the safety benches 20 ft wide. This gives

$$\Theta(\text{final}) = \tan^{-1} \frac{250}{4 \times 20 + \frac{250}{\tan 60°}} = 48°$$

Although much regarding final slope angles has to do with rock structure, care in blasting can make a major impact.

Table 4.2. Classification of open pit slope problems (Hoek, 1970b).

	Category	Conditions	Method of solution
A.	Unimportant slopes	Mining a shallow high grade orebody in favorable geological and climatic conditions. Slope angles unimportant economically and flat slopes can be used.	No consideration of slope stability required.
B.	Average slopes	Mining a variable grade orebody in reasonable geological and climatic conditions. Slope angles important but not critical in determining economics of mining.	Approximate analysis of slope stability normally adequate.
C.	Critical slopes	Mining a low grade orebody in unfavorable geological and climatic conditions. Slope angles critical in terms of both economics of mining and safety of operation.	Detailed geological and groundwater studies followed by comprehensive stability analysis usually required.

4.6 FINAL PIT SLOPE ANGLES

4.6.1 *Introduction*

During the early feasibility studies for a proposed open pit mine, an estimate of the safe slope angles is required for the calculation of ore to waste ratios and for the preliminary pit layout. At this stage generally the only structural information available upon which to base such an estimate is that obtained from diamond drill cores collected for mineral evaluation purposes. Sometimes data from surface outcrops are also available. How well these final slope angles must be known and the techniques used to estimate them depends upon the conditions (Table 4.2) applicable. During the evaluation stage for categories B and C, the best engineering estimate of the steepest safe slope at the pit limits in each pit segment is used. Since the information is so limited, they are hedged with a contingency factor. If the property is large and has a reasonably long lifetime, initially the exact slope angles are of relatively minor importance. The effect of steeper slopes at the pit limits is to increase the amount of ore that can be mined and therefore increase the life of the mine. The effect of profits far in the future has practically no impact on the net present value of the property.

During the pre-production period and the first few years of production, the operating slopes should however be as steep as possible while still providing ample bench room for optimum operating efficiency. The minimization of stripping at this stage has a significant effect on the overall economics of the operation. The working slopes can then be flattened until they reach the outer surface intercepts. Steepening operations then commence to achieve the final pit slopes (Halls, 1970). Cases do occur where the viability of an orebody is highly dependent on the safe slope angle that can be maintained. Special measures, including the collection of drillhole data simply for making slope determinations are then taken.

There are a number of excellent references which deal in great detail with the design of pit slopes. In particular *Rock Slope Engineering* by Hoek & Bray (1977), and the series of publications developed within the *Pit Slope Manual* series produced by CANMET should be mentioned. This brief section focusses on a few of the underlying concepts, and presents some curves extracted largely from the work of Hoek (1970a, 1970b) which may be used for making very preliminary estimates.

Geometrical considerations 313

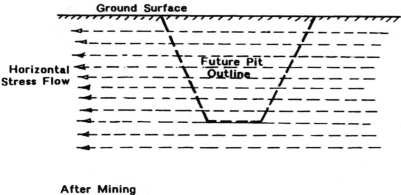

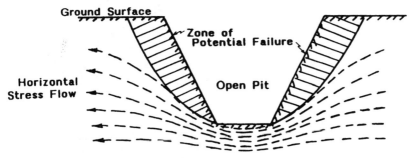

Figure 4.61. Horizontal stress redistribution due to the creation of a pit.

4.6.2 *Geomechanical background*

Figure 4.61 shows diagrammatically the horizontal flow of stress through a particular vertical section both with and without the presence of the final pit. With the excavation of the pit, the pre-existing horizontal stresses are forced to flow beneath the pit bottom (and around the pit ends).

The vertical stresses are also reduced through the removal of the rock overlying the final slopes. This means that the rock lying between the pit outline and these flow lines is largely distressed. As a result of stress removal, cracks/joints can open with a subsequent reduction in the cohesive and friction forces restraining the rock in place. Furthermore, ground water can more easily flow through these zones, reducing the effective normal force on potential failure planes. As the pit is deepened, the extent of this destressed zone increases, and the consequences of a failure becomes more severe. The chances of encountering adverse structures (faults, dykes, weakness zones, etc.) within these zones increase as well. Finally, with increasing pit depth, the relative sizes of the individual structural blocks making up the slopes become small compared to the overall volume involved. Thus the failure mechanism may change from one of structural control to one controlled by the characteristics of a granular mass. Figure 4.62 shows the four major types of failure which occur in an open

314 *Open pit mine planning and design: Fundamentals*

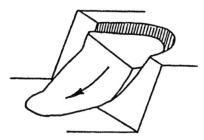
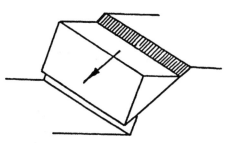

a. Circular failure in overburden soil, waste rock or highly fractured rock with no identifiable structural pattern.

b. Plane failure in rock with highly ordered structure such as slate.

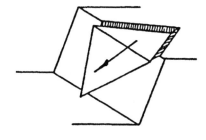
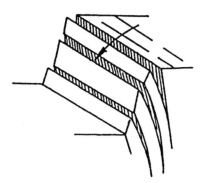

C. Wedge failure on two intersecting discontinuities.

d. Toppling failure in hard rock which can form columnar structure separated by steeply dipping discontinuities.

Figure 4.62. The most common slope failure types (Hoek & Bray, 1977).

pit. In this section the discussion will concentrate on planar failure along major structures and circular failure.

4.6.3 *Planar failure*

Planar failure along various types of discontinuities can occur on the bench scale, interramp scale and pit wall scale (major fault, for example). Bench face instabilities due to the daylighting of major joint planes means that the overall slope must be flattened to provide the space required for adequate safety berms. The final slope is made up of flattened bench faces, coupled with the safety berm steps. The design slope angle may be calculated once an average stable bench face angle is determined. Since one is concerned with final pit wall stability, the analysis in this section applies to a major structure occurring in the pit wall, although the same type of analysis applies on the smaller scale as well. Figure 4.63 shows the dimensions and forces in a rock slope with a potential failure plane. The Mohr-Coulomb failure criterion has been used.

The following definitions apply:

i is the average slope angle from horizontal (degrees),

β is the angle of the discontinuity from the horizontal (degrees),

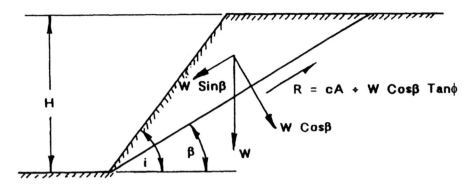

Figure 4.63. Dimensions and forces in a rock slope with a potential failure plane (Hoek, 1970a).

W is the block weight,
R is the resisting force,
c is the cohesion,
ϕ is the friction angle,
$W \cos \beta$ is the normal force,
$W \sin \beta$ is the driving force,
A is the area of the failure plane.

The factor of safety (F) is defined by

$$F = \frac{\text{Total force available to resist sliding}}{\text{Force tending to induce sliding}} \qquad (4.1)$$

For the case shown in Figure 4.63 (drained slope) Equation (4.1) becomes

$$F = \frac{cA + W \cos \beta \tan \phi}{W \sin \beta} \qquad (4.2)$$

If there is water present, then the factor of safety is expressed as

$$F = \frac{cA + (W \cos \beta - U) \tan \phi_a}{W \sin \beta + V} \qquad (4.3)$$

where U is the uplift force along the base of the block due to water pressure, and V is the horizontal force along the face of the block due to water in the tension crack, ϕ_a is the friction angle (as affected by the water). Typical values for the cohesive strength and friction angles of soils and rock are given in Tables 4.3 and 4.4. As the height H of the slope increases the relative contribution of the cohesion to the total resistance decreases. For very high slopes, the stable slope angle approaches the friction angle ϕ. Hoek (1970a) has presented the relationship between slope height and slope angle functions for plane failure in a drained slope given in Figure 4.64.

Assume for example that the average planned slope angle i is 70°, the orientation of the potential failure plane β is 50° and the friction angle ϕ is 30°. Thus

$$X = 2\sqrt{(i - \beta)(\beta - \phi)} = 2\sqrt{20 \times 20} = 40°$$

From Figure 4.64 the slope height function Y is read as

$$Y = 14$$

316 Open pit mine planning and design: Fundamentals

Table 4.3. Cohesive strengths for 'intact' soil and rock (Robertson, 1971).

Material description	c (lb/ft^2)	c (kg/m^2)
Very soft soil	35	170
Soft soil	70	340
Firm soil	180	880
Stiff soil	450	2200
Very stiff soil	1600	7800
Very soft rock	3500	17,000
Soft rock	11,500	56,000
Hard rock	35,000	170,000
Very hard rock	115,000	560,000
Very very hard rock	230,000	1,000,000

Table 4.4. Friction angles (degrees) for typical rock materials (Hoek, 1970a).

Rock	Intact rock ϕ	Joint ϕ	Residual ϕ
Andesite	45	31–35	28–30
Basalt	48–50	47	
Chalk	35–41		
Diorite	53–55		
Granite	50–64		31–33
Graywacke	45–50		
Limestone	30–60		33–37
Monzonite	48–65		28–32
Porphyry		40	30–34
Quartzite	64	44	26–34
Sandstone	45–50	27–38	25–34
Schist	26–70		
Shale	45–64	37	27–32
Siltstone	50	43	
Slate	45–60		24–34

Other materials	Approximate ϕ
Clay gouge (remoulded)	10–20
Calcitic shear zone material	20–27
Shale fault material	14–22
Hard rock breccia	22–30
Compacted hard rock aggregate	40
Hard rock fill	38

Knowing that

$$c = 1600 \text{ lb/ft}^2$$
$$\gamma = 160 \text{ lb/ft}^3$$

the limiting ($F=1$) slope height H with such a structure passing through the toe is found using

$$Y = 14 = \frac{\gamma H}{c} = \frac{160}{1600}H$$

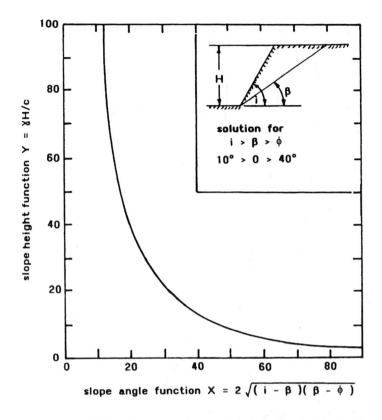

Figure 4.64. Relationship between slope height and slope angle functions for plane failure in a drained slope (Hoek, 1970a).

Thus

$$H = 140 \text{ ft}$$

If the planned pit depth is 500 ft, one could determine the limiting ($F = 1$) pit slope angle. The slope height function is

$$Y = \frac{\gamma H}{c} = \frac{160 \times 500}{1600} = 50$$

From Figure 4.64 one finds that

$$X = 17.5$$

Solving for i yields

$$i = 57.7°$$

The general family of curves corresponding to various safety factors is given in Figure 4.65.

318 Open pit mine planning and design: Fundamentals

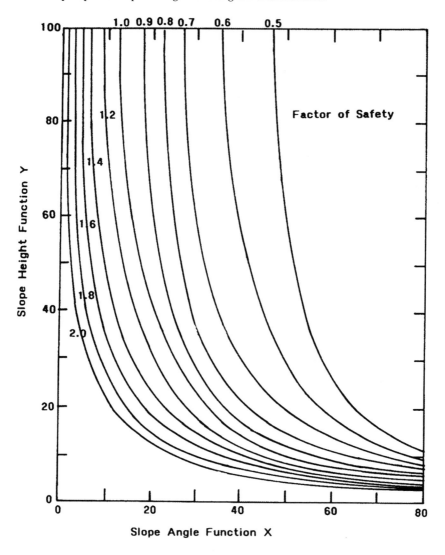

Figure 4.65. Slope design chart for plane failure including various safety factors (Hoek, 1970a).

The question naturally arises as to what an appropriate safety factor might be? This depends on the confidence one has in the 'goodness' of the input data and also on the function of the structure. Jennings & Black (1963) have provided the following advice:

For permanent structures, such as earth dams, F should not be less than 1.5 for the most critical potential failure surface, but for temporary constructions, where engineers are in continual attendance, a lower factor may be accepted. In civil engineering work, construction factors of safety are seldom allowed to be less than 1.30. An open pit is a 'construction' of a very particular type and it is possible that a factor of 1.20–1.30 may be acceptable in this case.

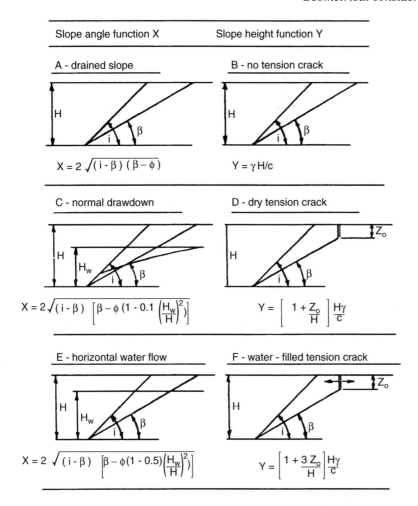

Figure 4.66. Slope angle and slope height functions for different water and tension crack conditions (Hoek, 1970a).

The confidence placed in any value calculated as the factor of safety of a slope depends upon the accuracy with which the various factors involved can be estimated. The critical items are the selection of the most adverse surface for potential failure, the measurement of the shear strength of the materials on this surface and the estimation of the water pressures in the soil pores and in any fissures along the surface.

If one were to select a safety factor of 1.2 for the previous example, one finds that for $Y = 50$, $X = 13.5$. The slope angle becomes

$$i = 54.6°$$

The example applies for the very special case of a drained slope without a tension crack. Often a tension crack will be present and there can be a variety of different slope water conditions. Hoek (1970a) has developed a simple way of handling these. Figure 4.66 provides three different expressions for X corresponding to different slope water conditions

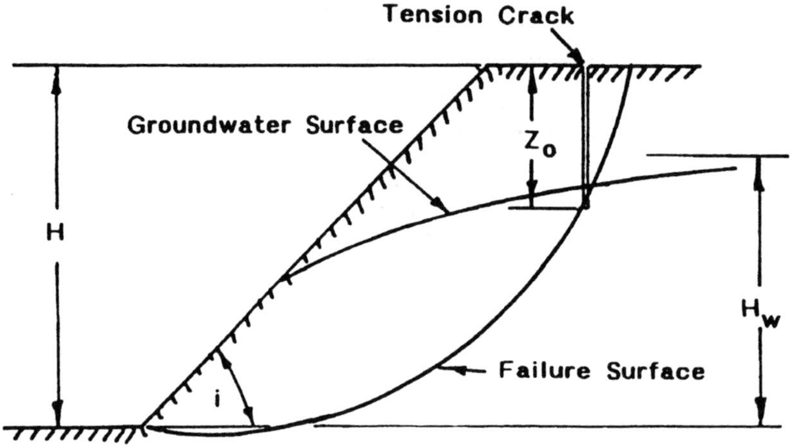

Figure 4.67. Diagrammatic representation of circular failure in a slope (Hoek, 1970a).

and three different expression for Y relating to the tension crack. Thus nine different $X-Y$ combinations are possible. The one used in the earlier examples was combination $A-B$. From Figure 4.66 one finds the $X-Y$ combination most appropriate to the problem at hand.

The known values are substituted and Figure 4.65 is used to determine the desired missing value. The interested reader is encouraged to evaluate the effect of different slope water conditions on the slope angle.

4.6.4 Circular failure

Hoek (1970a) has applied the same approach to the analysis of circular failures (Fig. 4.67).

Such deep seated failures occur when a slope is excavated in soil or soft rock in which the mechanical properties are not dominated by clearly defined structural features. This type of failure is important when considering the stability of:
- very high slopes in rock in which the structural features are assumed to be randomly oriented,
- benches or haul road cuts in soil,
- slimes dams,
- waste dumps.

Figure 4.68 gives the relationship between the slope height function and slope angle function for circular failure in drained slopes without a tension crack ($F = 1$). The corresponding chart, including different safety factor values, is given in Figure 4.69. To accommodate different tension crack and slope water conditions, Figure 4.70 has been developed. This set of curves is used in exactly the same way as described earlier.

4.6.5 Stability of curved wall sections

The approaches discussed to this point have applied to pit wall sections which can be approximated by two-dimensional slices. Open pits often take the form of inverted cones or have portions containing both convex and concave wall portions (Figure 4.71).

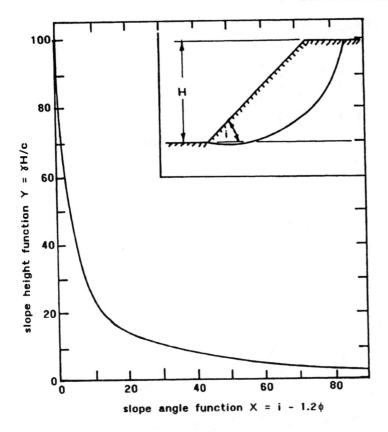

Figure 4.68. Plot of slope height versus slope angle functions for circular failure analysis (Hoek, 1970a).

Very little quantitative information on the effect of pit wall curvature on stability is available from the literature. Convex portions of a pit wall (noses which stick out into the pit) frequently suffer from unstable slopes. The relaxation of lateral stresses give rise to a reduction in the normal stress across potential failure planes and vertical joint systems can open. For concave portions of the pit, the arch shape of the slope tends to induce compressive lateral stresses which increase the normal stress across potential failure planes. The slopes are more stable due to the increased frictional resistance.

Hoek (1970a) suggests that curvature of the slope in plan can result in critical slope differences of approximately 5° from that suggested by the planar analyses. A concave slope, where the horizontal radius of curvature is of the same order of magnitude as the slope height, may have a stable slope angle 5° steeper than for a straight wall (infinite radius of curvature). On the other hand, a convex slope may require flattening by about 5° in order to improve its stability.

However, improved drainage in the convex slopes over that available with the pinched concave shape may provide a stability advantage. Thus, there may be some cancelling of advantages/disadvantages. Hence, each pit curvature situation must be carefully examined.

322 Open pit mine planning and design: Fundamentals

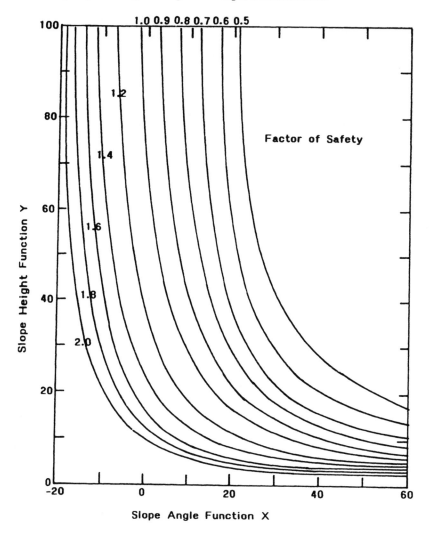

Figure 4.69. Slope design chart for circular failure including various safety factors (Hoek, 1970a).

4.6.6 Slope stability data presentation

Figure 4.72 developed by Hoek & Bray (1977) is a good example of how structural geology information and preliminary evaluation of slope stability of a proposed open pit mine can be presented. A contour plan of the proposed open pit mine is developed and contoured stereoplots of available structural data are superimposed. In this particular case two distinct structural regions denoted by A and B have been identified and marked on the plan. Based simply on geometry (of the pit slopes and structures), the potential failure types are identified. Each of these would then be examined using appropriate material properties and ground water conditions. Required design changes, additional data collection, etc. will emerge.

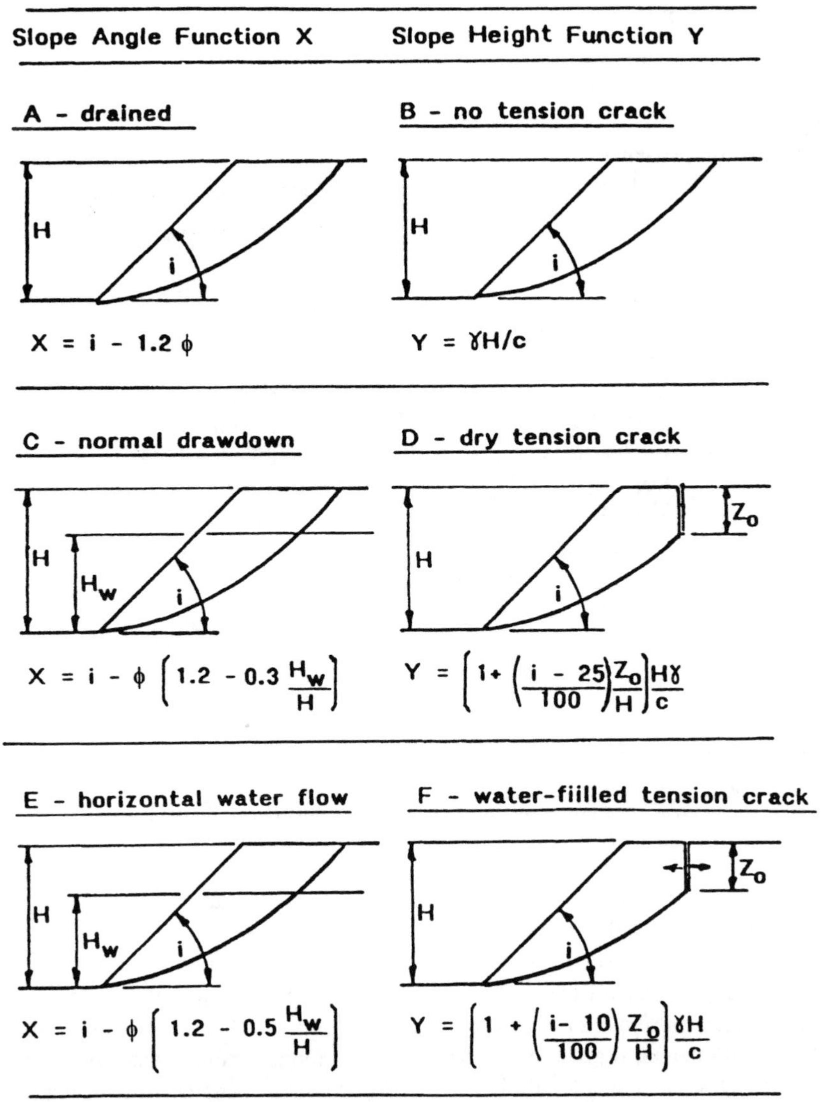

Slope Angle Function X	Slope Height Function Y
A – drained	**B – no tension crack**
$X = i - 1.2\phi$	$Y = \gamma H/c$
C – normal drawdown	**D – dry tension crack**
$X = i - \phi\left(1.2 - 0.3\dfrac{H_w}{H}\right)$	$Y = \left[1 + \left(\dfrac{i-25}{100}\right)\dfrac{Z_o}{H}\right]\dfrac{H\gamma}{c}$
E – horizontal water flow	**F – water-filled tension crack**
$X = i - \phi\left(1.2 - 0.5\dfrac{H_w}{H}\right)$	$Y = \left[1 + \left(\dfrac{i-10}{100}\right)\dfrac{Z_o}{H}\right]\dfrac{\gamma H}{c}$

Figure 4.70. Slope angle and slope height functions for different water and tension crack conditions (Hoek, 1970a).

4.6.7 *Slope analysis example*

Reed (1983) has reported the results of applying the Hoek & Bray (1977) approach to the Afton copper-gold mine located in the southern interior of British Columbia. For the purpose of analyzing the stability of the walls of the open pit, it was divided into 9 structural domains (Fig. 4.73).

For each structural domain, a stability analysis was made of:
- the relative frequency of the various fault and bedding plane orientations, and
- the orientation of the pit wall in that particular domain.

324 *Open pit mine planning and design: Fundamentals*

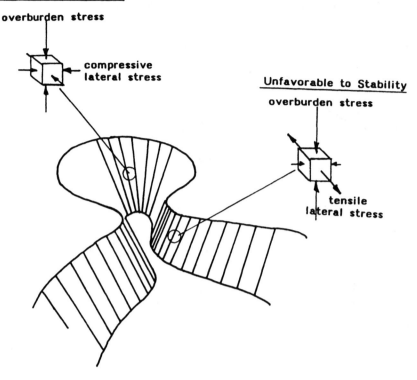

Figure 4.71. Influence of three-dimensional pit shape upon slope stability (Hoek, 1970a).

The safety factors were calculated for plane failure, wedge failure and circular failure in each domain. Table 4.5 shows the results of these stability analyses.

The 'maximum safe slope angle' for the pit wall in each domain corresponds to a calculated safety factor of 1.2. The results in Table 4.5 predict wall failure in all domains if the slopes are wet. The mine, however, lies in a semi-arid area and expected ground water quantities were small. In addition, horizontal drainholes would be used to reduce ground water pressures in domains 3 and 6. Problems would still be expected in domains 3 and 6. Since domain 3 is a relatively narrow domain and the probability of a major slide occurring was small, the design slope of the wall in that area was not flattened. At the time the paper was written (1983), the pit had reached a depth of 480 ft. Two failures had been experienced in domain 3 and several berm failures in domain 6. There was no indication of impending major failures. Final pit depth was planned to be 800 ft.

4.6.8 *Economic aspects of final slope angles*

Figure 4.74 illustrates the volume contained in a conical pit as a function of final slope angle and depth.

For a depth of 500 ft and an overall final pit angle of 45°, 1.4×10^7 tons of rock must be moved. Within the range of possible slopes (20° to 70°) at this depth the volume to be

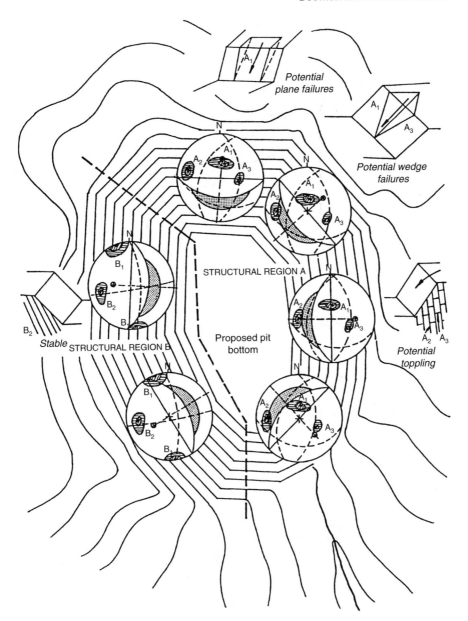

Figure 4.72. Presentation of structural geology information and preliminary evaluation of slope stability of a proposed open pit mine (Hoek & Bray, 1977).

moved approximately doubles for every 10° flattening of the slope. Flattening the slope of the 500 ft deep conical pit from 50° to 40° increases the mass of rock from 1.0×10^7 to 2.0×10^7 tons. This simple example shows that the selection of a particular slope can have a significant impact on the scale of operations and depending upon the shape, size and grade of the ore contained within the pit, on the overall economics.

326 *Open pit mine planning and design: Fundamentals*

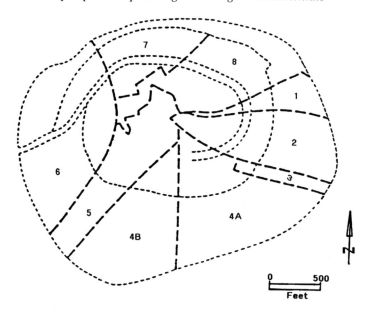

Figure 4.73. Division of the Afton open pit into 9 structural domains (Reed, 1983).

Table 4.5. Calculated and design slope angles for the Afton mine (Reed, 1983).

Domain	Maximum safe slope angle		Design slope angle
	Wet	Dry	
1	24°	54°	45°
2	52°	52°	45°
3	24°	41°	45°
4A	27°	49°	45°
4B	45°	42°	45°
5	22°	42°	45°
6	28°	39°	40°
7	33°	42°	40°
8	32°	43°	40°

4.7 PLAN REPRESENTATION OF BENCH GEOMETRY

Figure 4.75 is a cross-sectional representation of an open pit mine. Figure 4.76 is a 'birds-eye' (plan) view of the same pit. No attempt has been made to distinguish between the toes and crests (which are marked in Fig. 4.77) and hence the figure is difficult to interpret.

Several different techniques are used by the various mines to assist in plan representation and visualization. In Figure 4.78 the bank slopes have been shaded and the benches labelled with their elevations.

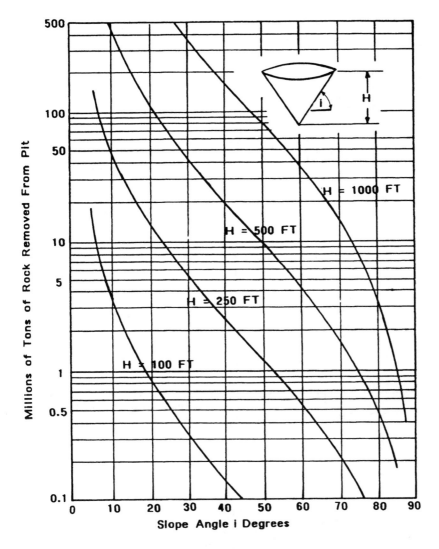

Figure 4.74. Influence of pit depth and slope angle on the amount of rock removed in mining a conical open pit (Hoek & Pentz, 1970).

Figure 4.79 is an example of this type of representation for an actual mine. An alternative is to draw the crests with solid lines and the toes with dashed lines. The result is shown in diagrammatic form in Figure 4.80 and for an actual property in Figure 4.81. Note that the banks have also been shaded. This is however seldom done. This system of identifying toes and crests is recommended by the authors.

Some companies use the opposite system labelling the crests with dashed lines and the toes by solid lines (Fig. 4.82). The Berkeley pit shown in Figure 4.83 is one such example where this system has been applied.

If there are a great number of benches and the scale is large, there can be difficulties in representing both the toes and the crests. Knowing the bench height and the bench face angle

328 *Open pit mine planning and design: Fundamentals*

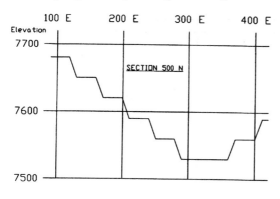

Figure 4.75. Cross-section through an open pit mine.

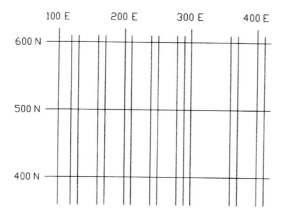

Figure 4.76. Plan view through the portion of the pit shown in cross-section in Figure 4.75 (toes and crests depicted by solid lines).

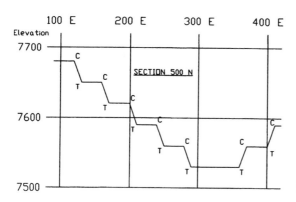

Figure 4.77. Cross-section through a portion of an open pit with toes and crests labelled.

it is a simple matter to construct, if needed, the toes presuming that the crests are given or vice versa. Hence only one set of lines (crests or toes) is actually needed. When only one line is used to represent a bench, the most common technique is to draw the median (mid bench) elevation line at its plan location on the bench face. This is shown in section and plan in Figures 4.84 and 4.85, respectively.

Geometrical considerations 329

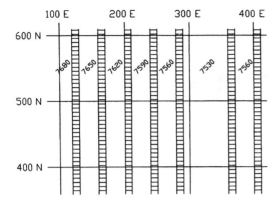

Figure 4.78. Plan view with the bench faces shaded and the flat segment elevations labelled.

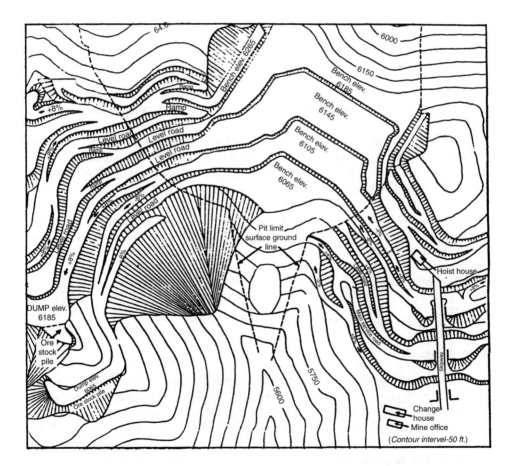

Figure 4.79. Example of slope surface shading described in Figure 4.78 (Ramsey, 1944).

330 *Open pit mine planning and design: Fundamentals*

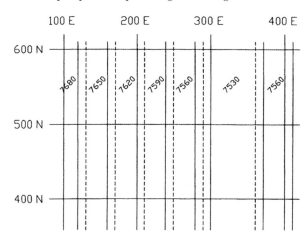

Figure 4.80. Plan view of a portion of the open pit (crests are denoted by solid lines and toes by dashed lines).

An actual example of its use is given in Figure 4.86. An enlarged view of a section of the pit is shown in Figure 4.87. The elevation label is located half way between the median contour lines. This is the actual location for this elevation and corresponds to the bench elevation at that point.

It is a relatively simple matter to go from median lines to actual bench representation (toes and crests) and vice versa. This process is depicted in Figure 4.88. The median contour line in the center will be replaced by the toe-crest equivalent. The road is 100 ft wide and has a grade of 10%. The bench height is 40 ft, the bank width is 30 ft and the width of the safety bench is 50 ft. The process begins by adding the center lines halfway to the next contour lines (Fig. 4.88b). Toe and crest lines are added (Fig. 4.88c) and the edge of road is drawn (Fig. 4.88d). Finally the construction lines are removed (Fig. 4.88e). The reader is encouraged to try this construction going back and forth from toes and crests to median lines.

4.8 ADDITION OF A ROAD

4.8.1 *Introduction*

Roads are one of the more important aspects of open pit planning. Their presence should be included early in the planning process since they can significantly affect the slope angles and the slope angles chosen have a significant effect on the reserves. Most of the currently available computerized pit generating techniques discussed in the following chapter do not easily accommodate the inclusion of roads. The overall slope angles without the roads may be used in the preliminary designs. Their later introduction can mean large amounts of unplanned stripping or the sterilization of some planned reserves. On the other hand a flatter slope angle can be used which includes the road. This may be overly conservative and include more waste than necessary.

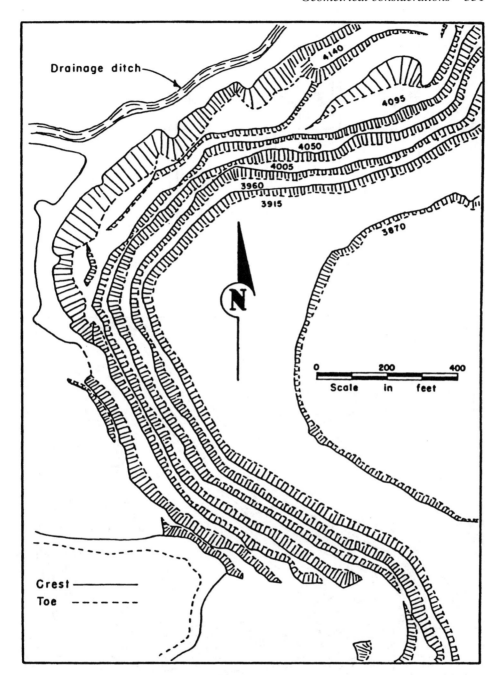

Figure 4.81. Example of the mapping procedure described in Figure 4.80 (Hardwick & Stover, 1960).

332 Open pit mine planning and design: Fundamentals

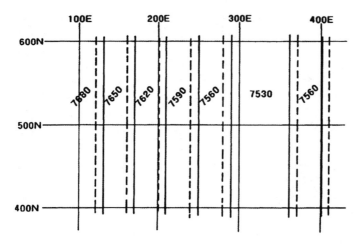

Figure 4.82. Plan view of a portion of the open pit (crests denoted by dashed lines and toes by solid lines).

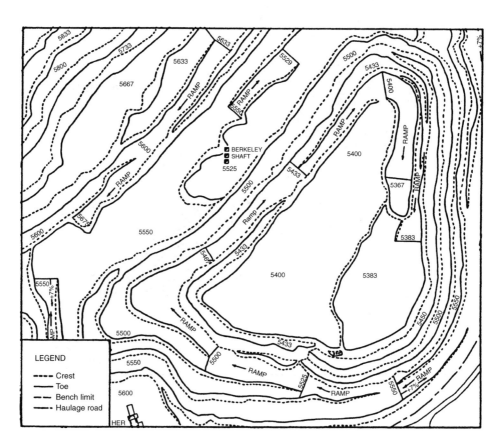

Figure 4.83. Example of the mapping procedure described in Figure 4.82 (McWilliams, 1959).

Geometrical considerations   333

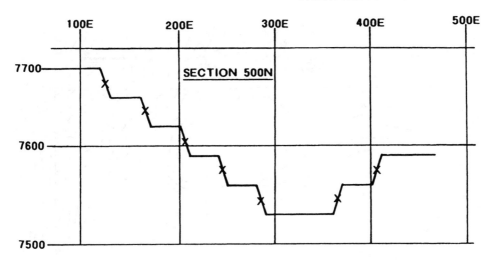

Figure 4.84. Procedure of denoting the median midbench elevation line on the bench face.

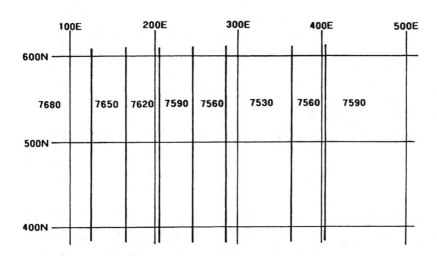

Figure 4.85. Resulting plan view corresponding to the midbench representation of Figure 4.84. The given elevations are bench toe elevations.

Until rather recently, rail haulage was a major factor in open pit operations. Because of the difficulties with sharp turns and steep grades, a great deal of time was spent by mine planners in dealing with track layout and design. Rubber tired haulage equipment has presented great flexibility and ability to overcome many difficulties resulting from inadequate or poor planning in today's pits. However as pits become deeper and the pressure for cost cutting continues, this often neglected area will once again be in focus.

There are a number of important questions which must be answered when siting the roads (Couzens, 1979).

1. The first decision to be taken is where the road exit or exits from the pit wall will be. This is dependent upon the crusher location and the dump points.

334 *Open pit mine planning and design: Fundamentals*

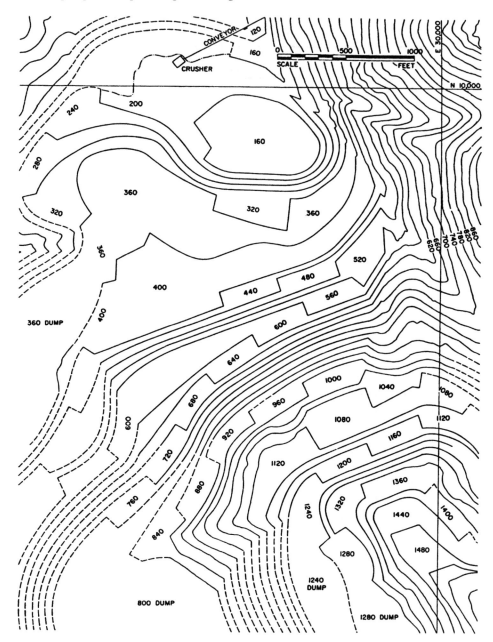

Figure 4.86. Example mining plan composite map based on midbench contours (Couzens, 1979).

2. Should there be more than one means of access? This allows certain flexibility of operation but the cost of added stripping can be high.

3. Should the roads be external or internal to the pit? Should they be temporary or semi-permanent?

Geometrical considerations 335

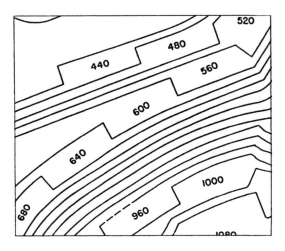

Figure 4.87. An enlarged portion of Figure 4.86 (Couzens, 1979).

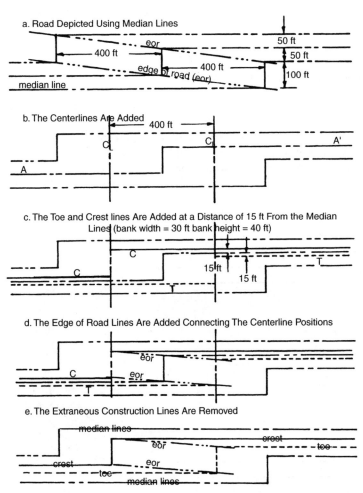

Figure 4.88. Going from midbench contours to a toe and crest representation.

336 *Open pit mine planning and design: Fundamentals*

4. Should the road spiral around the pit? Have switchbacks on one side? Or a combination?

5. How many lanes should the road have? The general rule of thumb for 2 way traffic is: road width $\geq 4 \times$ truck width. Adding an extra lane to allow passing may speed up the traffic and therefore productivity but at an increased stripping cost.

6. What should the road grade be? A number of pits operate at 10% both favorable and unfavorable to the haul. A grade of 8% is preferable since it provides more latitude in building the road and fitting bench entries. That is, providing it does not cause too much extra stripping or unduly complicate the layout.

7. What should be the direction of the traffic flow? Right hand or left hand traffic in the pit?

8. Is trolley assist for the trucks a viable consideration? How does this influence the layout?

This section will not try to answer these questions. The focus will be on the procedure through which haulroad segments can be added to pit designs. The procedures can be done by hand or with computer assist. Once the roads have been added then various equipment performance simulators can be applied to the design for evaluating various options.

4.8.2 *Design of a spiral road – inside the wall*

As has been discussed in Section 4.3, the addition of a road to the pit involves moving the wall either into the pit and therefore losing some material (generally ore) or outward and thereby adding some material (generally waste). This design example considers the first case (inside the original pit wall). The second case will be discussed in the following section. This pit consists of the four benches whose crests are shown in Figure 4.89. Both toes and crests are shown in Figure 4.90. The crest-crest dimension is 60 ft, the bench height is 30 ft

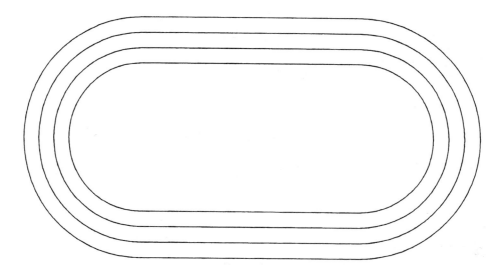

Figure 4.89. The four bench pit with crests shown.

Geometrical considerations 337

and a road having a width of 90 ft and a grade of 10% is to be added to the north wall. The bench face has an angle of 56°.

Step 1. The design of this type of road begins at the pit bottom. For reasons to be discussed later, the point where the ramp meets the first crest line is selected with some care. In this case, the ramp will continue down to lower mining levels along the north and east walls, thus point A in Figure 4.91 has been selected.

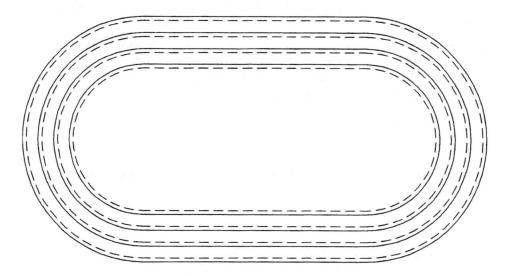

Figure 4.90. The four bench pit with toes added.

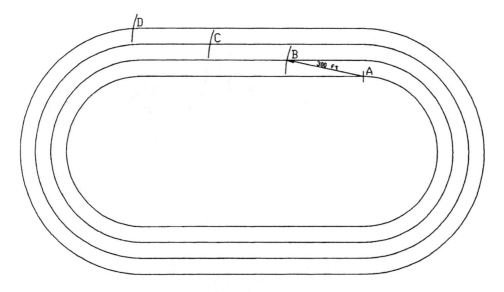

Figure 4.91. Point of ramp initiation and crest intercepts.

338 *Open pit mine planning and design: Fundamentals*

Step 2. The locations where the ramp meets the succeeding crests are now determined. Since the bench height H is 30 ft and the road grade G is 10%, the horizontal distance D travelled by a truck going up to the next level is

$$D = \frac{100H}{G(\%)} = \frac{100 \times 30}{10} = 300 \text{ ft}$$

Point B on the crest of the next bench is located by measuring the 300 ft distance with a ruler or by swinging the appropriate arc with a compass. Points C and D are located in a similar way.

Step 3. The crest line segments indicating the road location will be added at right angles to the crest lines rather than at right angles to the line of the road. Hence they have a length (W_a) which is longer than the true road width (W_t). As can be seen in Figure 4.92, the angle (Θ) that the road makes with the crest lines is

$$\Theta = \sin^{-1} \frac{600}{300} = 11.5°$$

Hence the apparent road width W_a (that which is laid out), is related to the true road width by

$$W_a = \frac{W_t}{\cos \Theta} = 1.02 \, W_t = 1.02 \times 90 = 92 \text{ ft}$$

For most practical purposes, little error results from using

$$W_a \cong W_t = W$$

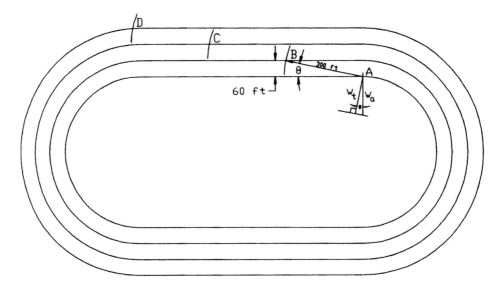

Figure 4.92. Addition of ramp width (Step 3).

Geometrical considerations 339

Lines of length W drawn perpendicular to the crest lines from points A, B, C and D have been added to Figure 4.93a. In addition short lines running parallel to the crest starting at the ends of these lines have been added. Line *a-a'* is one such line.

Step 4. Line *a-a'* is extended towards the west end of the pit. It first runs parallel to the previous crest line but as the pit end approaches it is curved to make a smooth transition with the original crest line. This is shown in Figure 4.93b. The designer has some flexibility

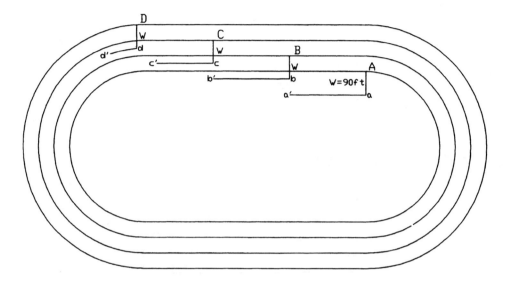

Figure 4.93a. Completing the new crest lines (Step 4).

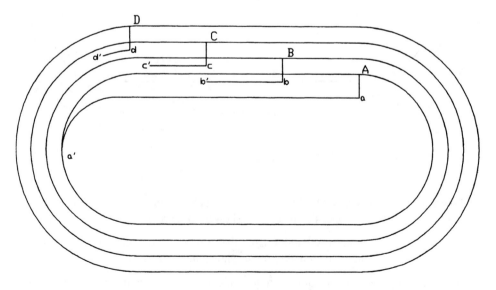

Figure 4.93b. Completing the new crest lines (Step 4).

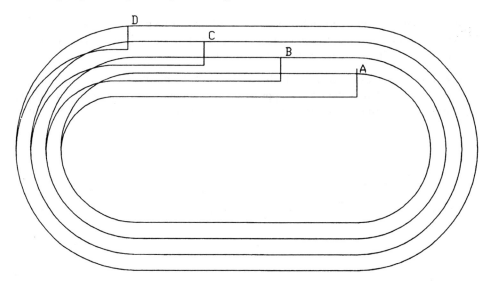

Figure 4.93c. Completing the new crest lines (Step 4).

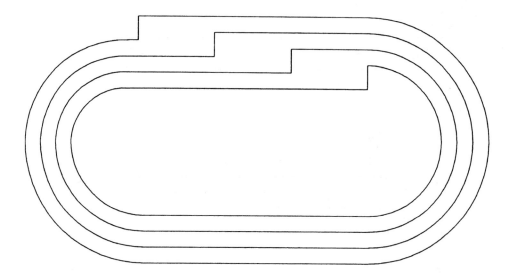

Figure 4.94. The pit as modified by the ramp (Step 5).

on how this transition occurs. Once this decision is made then the remaining crest lines are drawn parallel to this first one. The results are shown in Figure 4.93c.

Step 5. The extraneous lines remaining from the original design are now removed. The resulting crest lines with the included ramp are shown in Figure 4.94.

Step 6. The ramp is extended from the crest of the lowest bench to the pit bottom. This is shown in Figure 4.95. The toe lines have been added to assist in this process. In Figure 4.95,

Geometrical considerations 341

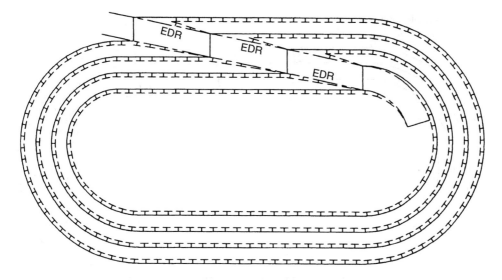

Figure 4.95. Addition of entrance ramp and toe lines (Step 6).

the slopes have been shaded to help in the visualization. The edge of road (EOR) lines shown are also crest lines.

4.8.3 *Design of a spiral ramp – outside the wall*

In the previous section the addition of a spiral ramp lying inside the original pit contours was described. It's addition meant that some material initially scheduled for mining would be left in the pit. For the case described in this section where the ramp is added outside the initial pit shell design, additional material must be removed. The same four bench mine as described earlier will be used:

> Bench height = 30 ft
> Crest-crest distance = 60 ft
> Road width = 90 ft
> Road grade = 10%
> Bench slope angle = 56°

Step 1. The design process begins with the crest of the uppermost bench. A decision must be made regarding the entrance point for the ramp as well as direction. As shown in Figure 4.96, the entrance should be at point A in the direction shown. Mill and dump locations are prime factors in selecting the ramp entrance point. From this point an arc of length L equal to the plan projection of the ramp length between benches is struck. This locates point B. From point B an arc of length L is struck locating point C, etc.

Step 2. From each of the intersection points A, B, C and D, lines of length W_a (apparent road width) are constructed normal to their respective crest lines. This is shown in

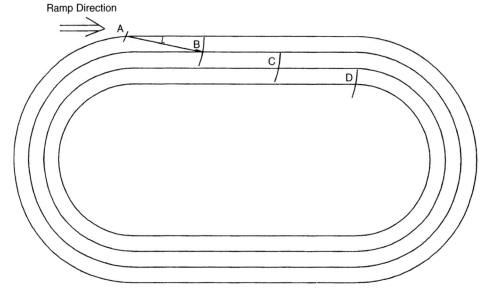

Figure 4.96. Point of ramp initiation and crest intercepts (Step 1).

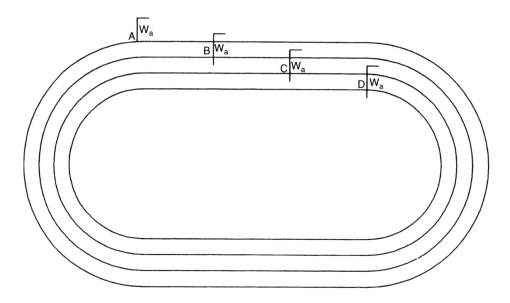

Figure 4.97. Addition of ramp width (Step 2).

Figure 4.97. A short length of line is drawn parallel to the crest line from the end in the ramp direction.

Step 3. Beginning with the lowermost crest, a smooth curve is drawn connecting the new crest with the old. This is shown in Figure 4.98.

Geometrical considerations 343

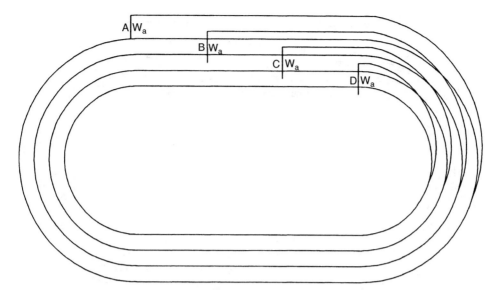

Figure 4.98. Drawing the new crest lines (Steps 3 and 4).

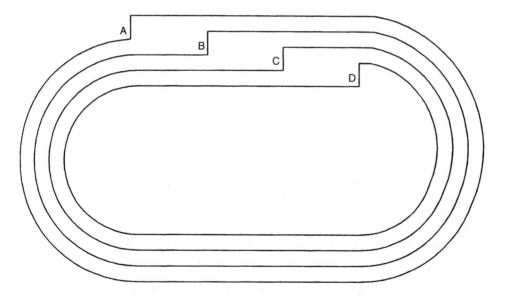

Figure 4.99. The pit as modified by the ramp (Step 5).

Step 4. The remaining new crest line portions are drawn parallel to the first crest working upwards from the lowest bench.

Step 5. The extraneous lines are removed from the design (Fig. 4.99).

Step 6. The toe lines at least for the lowest bench are added and the ramp to the pit bottom added. In Figure 4.100, the slopes have been shaded to assist in viewing the ramp.

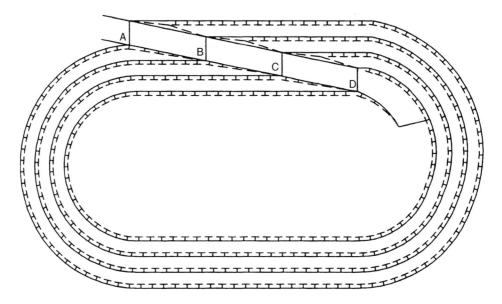

Figure 4.100. Addition of entrance ramp and toe lines (Step 6).

4.8.4 Design of a switchback

In laying out roads the question as to whether to:
(a) spiral the road around the pit,
(b) have a number of switchbacks on one side of the pit, or
(c) use some combination.

Generally (Couzens, 1979) it is desirable to avoid the use of switchbacks in a pit. Switchbacks:
– tend to slow down traffic,
– cause greater tire wear,
– cause various maintenance problems,
– probably pose more of a safety hazard than do spiral roads (vision problems, machinery handling, etc.).

Sometimes the conditions are such that switchbacks become interesting:
– when there is a gently sloping ore contact which provides room to work in switchbacks at little stripping cost;
– it may be better to have some switchbacks on the low side of the pit rather than to accept a lot of stripping on the high side.

The planner must take advantage of such things. The general axiom should be to design the pit to fit the shape of the deposit rather than vice versa. If switchbacks are necessary the planner should:
– leave enough length at the switchbacks for a flat area at the turns so that trucks don't have to operate on extremely steep grades at the inside of curves,
– consider the direction of traffic,
– consider problems the drivers may have with visibility,
– consider the effect of weather conditions on the design (ice, heavy rain, etc.).

Geometrical considerations 345

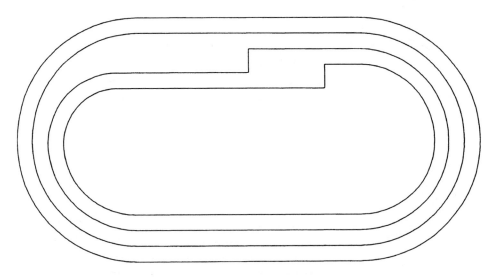

Figure 4.101. The starter pit for switchback addition to the north wall (Step 1).

In this section the steps required to add a switchback to the pit shown in Figure 4.89 will be described. The switchback will occur between the second and third benches on the north pit wall.

Step 1. The design will begin from the pit bottom. In this case the ramp moves into the as-designed pit wall. Figure 4.101 shows the modified pit with the crest lines drawn for benches 4 (lowermost) and 3. This is the same procedure as with the spiral ramp. The bench height has been selected as 30 ft and the road gradient is 10%. Hence the plan distance R is 300 ft.

Step 2. The center C used to construct the switchback is now located as shown in Figure 4.102. There are three distances involved L_1, L_2 and L_3. L_2 is the given crest-crest distance. Distances L_1 and L_3 must now be selected so that

$$L_1 + L_3 = R - L_2$$

In this particular case $L_1 = 0.5R = 150$ ft. Since $L_2 = 60$ ft, then $L_3 = 90$ ft. The center C is located at $L_2/2 = 30$ ft from the 3 construction lines. A vertical line corresponding to road width W is drawn at the end of L_3.

Step 3. In Figure 4.103 the curve with radius $R_2 = L_2/2$ is drawn from C. This becomes the inner road radius. It should be compared with the turning radius for the trucks being used. A second radius $R_3 = 2W$ is also drawn from C. The intersection of this curve with the horizontal line drawn from C becomes a point on the bench 2 crest. It is noted that actual designs may use values of R_3 different from that recommended here. This is a typical value. Portions of the bench 2 crest lines have been added at the appropriate distances.

Step 4. A smooth curve is now added going from line $a - b$ through crest point CP to line $c - d$. The designer can use some judgement regarding the shape of this transition line.

346 *Open pit mine planning and design: Fundamentals*

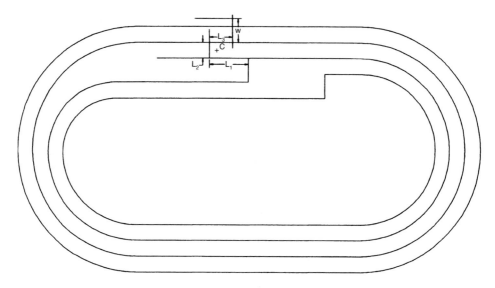

Figure 4.102. Construction lines for drawing the switchback (Step 2).

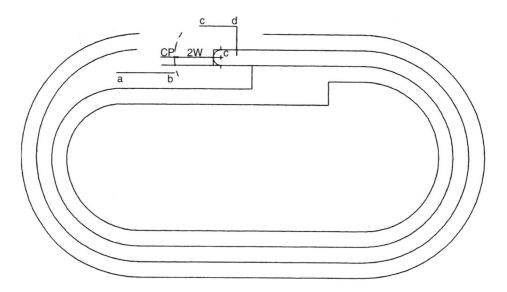

Figure 4.103. Crest lines and the crest point for bench 2 (Step 3).

Figure 4.104 shows the results. The lines surrounding point C simply represent edge of road (EOR).

Step 5. The crest line for bench 1 is then added parallel to that drawn for bench 2 (Fig. 4.105).

Step 6. The final crest line representation of the pit is drawn (Fig. 4.106). As can be seen the switchback occupies a broad region over a relatively short length. Thus it can be logically placed in a flatter portion of the overall pit slope.

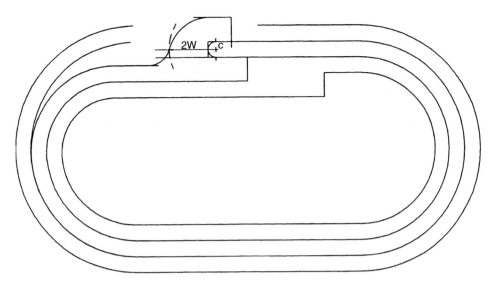

Figure 4.104. The transition curve has been added (Step 4).

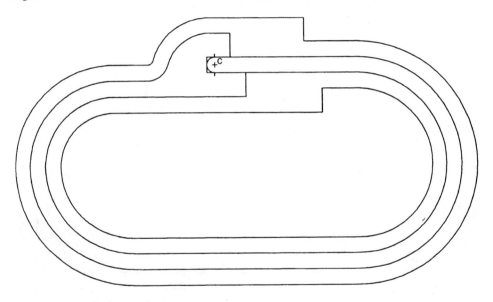

Figure 4.105. The crest line for bench 1 is added (Step 5).

Step 7. The toes are drawn and the lower section of the ramp (between bench 4 crest and the pit floor) added (Fig. 4.107).

Two examples of switchbacks are shown in Figure 4.108.

4.8.5 *The volume represented by a road*

The addition of a haulroad to a pit results in a large volume of extra material which must be removed or a similar volume in the pit which is sterilized (covered by the road). Thus even

348 *Open pit mine planning and design: Fundamentals*

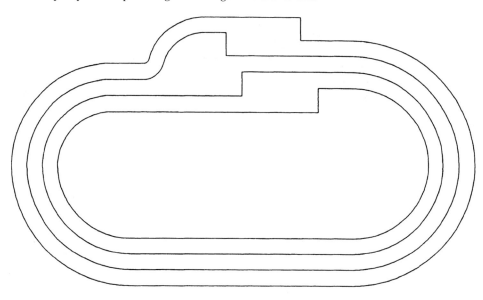

Figure 4.106. The final pit crest lines with the switchback (Step 6).

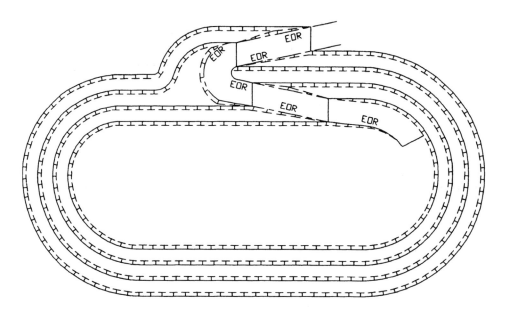

Figure 4.107. The lower entrance ramp and the toes are added (Step 7).

though production flexibility can be improved and the security of having several accesses to the pit can lead to other savings such as steeper interramp slopes, the additional haul roads are associated with significant expense. To demonstrate this, consider the pit shown in Figure 4.109 which contains no haulroad.

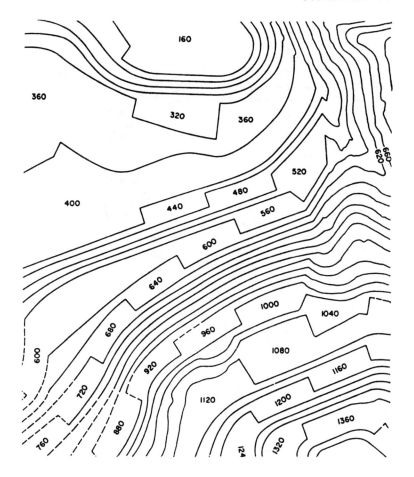

Figure 4.108. An example showing two different switchback (520 region and 1040 region) situations (Couzens, 1979).

The same pit with the road added, is shown in Figure 4.110. The shaded regions show the differences between sections A, B, C, D and E with and without the road.

In plan the length L of the road is

$$L = \frac{(\text{No. of benches} \times \text{Bench height}) \, 100}{\text{Road grade (\%)}} = \frac{4 \times 30 \times 100}{10} = 1200 \, \text{ft} \qquad (4.4)$$

Because the road is oriented at angle Θ to the pit axis, the length projected along the axis is

$$L_2 = L \cos \Theta = 1176 \, \text{ft} \qquad (4.5)$$

The sections are made normal to this axis. They are spaced every 294 ft.

The road areas for each section are shown in Figure 4.111. The shaded boxes are of the same area

$$A = W_A \times \text{Bench height}$$

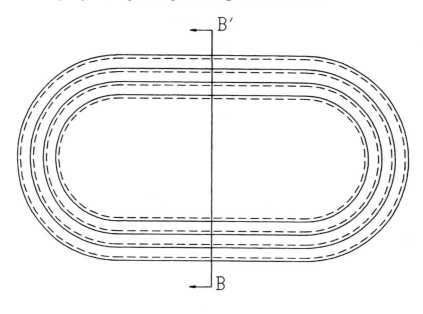

Figure 4.109. Plan and section views of a four bench pit without ramp.

They can be lined up as shown in Figure 4.111a. These in turn can be plotted such as shown in Figure 4.111b.

The volume contained in the ramp is that of a triangular solid of width W_A, length L_2 and height varying linearly from 0 to the pit depth (Fig. 4.112). This can be expressed as

$$V = \frac{1}{2} W_A L_2 \times \text{Pit depth} = \frac{1}{2} W_A L \cos \Theta \times \text{Pit depth} \tag{4.6}$$

which can be simplified to

$$V = \frac{1}{2} W_A \frac{(\text{Pit depth})^2}{\text{Grade }(\%)} 100 \cos \Theta \tag{4.7}$$

Since the apparent road width W_A is equal to

$$W_A = \frac{W_t}{\cos \Theta}$$

The simplified road volume formula becomes

$$V = \frac{1}{2} \frac{100 \times (\text{Pit depth})^2}{\text{Grade }(\%)} W_T \tag{4.8}$$

Geometrical considerations 351

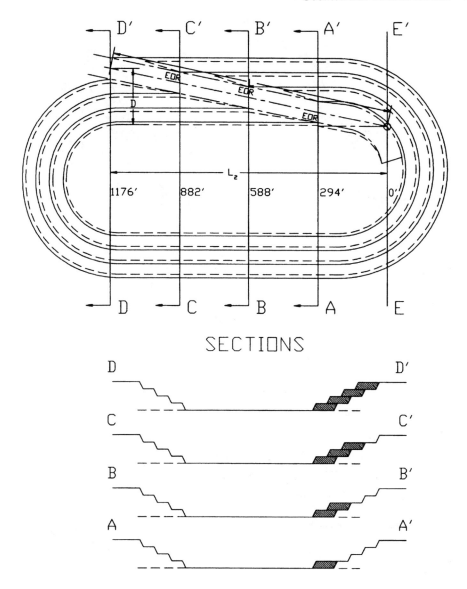

Figure 4.110. Plan and section views of a four bench pit with ramp.

In the present case the volume is

$$V = \frac{1}{2}\frac{100}{10}(120)^2 \times 90 = 6,480,00\text{ ft}^3 = 240,000\text{ yd}^3$$

For a tonnage factor of 12.5 ft³/st, there are 518,400 st involved in the road.
The overall length of the road (L_{ov}) is given by

$$L_{ov} = \sqrt{L^2 + (\text{Pit depth})^2} \qquad (4.9)$$

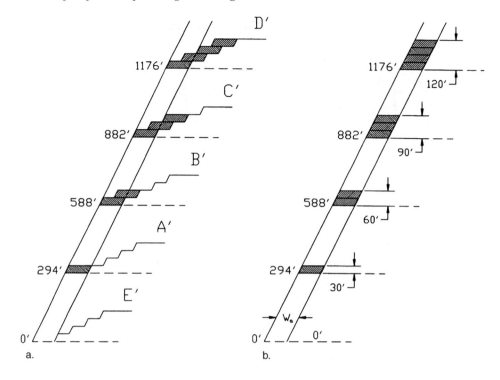

Figure 4.111. Construction to show road volume on each section.

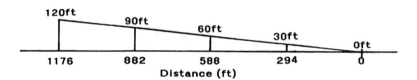

Figure 4.112. The volume involved in the ramp.

In this case it is

$$L_{ov} = \sqrt{(1200)^2 + (120)^2} = 1206 \text{ ft}$$

4.9 ROAD CONSTRUCTION

4.9.1 Introduction

Good haulroads are a key to successful surface mining operations. Poorly designed, constructed and maintained roads are major contributors to high haulage costs and pose safety hazards. In this section some of the basic design aspects will be discussed. Figure 4.113 shows a typical cross section through a road.

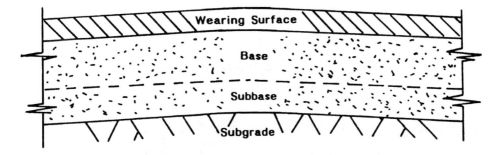

Figure 4.113. Simplified flexible pavement structure (Seelye, 1945).

Generally there are four different layers involved:
- subgrade,
- subbase,
- base,
- wearing surface.

The subgrade is the foundation layer. It is the structure which must eventually support all the loads which come onto the wearing surface. In some cases this layer will simply be the natural earth surface. In other and more usual instances, it will be the compacted rock or soil existing in a cut section or the upper layer of an embankment section.

The wearing surface provides traction, reduces tractive resistance, resists abrasion, ravelling and shear, transmits tire load to the base and seals the base against penetration of surface water. Although this surface may be asphalt or concrete, most typically it is crushed rock.

The base is a layer of very high stability and density. It's principal purpose is to distribute or 'spread' the stresses created by wheel loads acting on the wearing surface, so that they will not result in excessive deformation or the displacement of the subgrade. In addition it insulates the subgrade from frost penetration and protects the working surface from any volume change, expansion and softening of the subgrade.

The subbase which lies between the base and subgrade, may or may not be present. It is used over extremely weak subgrade soils or in areas subject to severe frost action. They may also be used in the interest of economy when suitable subbase materials are cheaper than base materials of a higher quality. Generally the subbase consists of a clean, granular material. The subbase provides drainage, resists frost heave, resists shrinkage and swelling of the subgrade, increases the structural support and distributes the load.

4.9.2 Road section design

In designing the road section, one begins with the maximum weight of the haulage equipment which will use the road. To be as specific as possible, assume that the haulage trucks have a maximum gross vehicle weight of 200,000 lbs including their 58 st payload. The load is distributed as follows:
- 33% on the front tires, and
- 67% on the dual rear tires.

The load on each of the front tires is 33,000 lbs. For each of the four rear tires (2 sets of duals) the load is 33,500 lbs. Thus the maximum loading to the wear surface is applied by

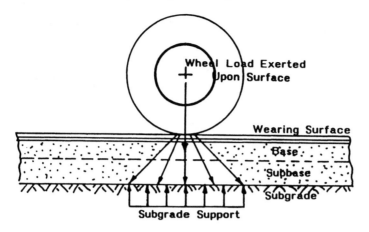

Figure 4.114. Load distribution beneath a tire (Seelye, 1945).

the rear tires. Although the contact pressure between the wheel and the road depends on the tire inflation pressure and the stiffness of the tire side walls, for practical purposes, the contact pressure is assumed to be equal to the tire pressure. Since for this truck, the inflation pressure is about 90 psi, the bearing pressure on the road surface is 90 psi or 12,960 psf. In lieu of knowing or assuming an inflation pressure, Kaufman & Ault (1977), suggest that a value of 16,000 psf (110 psi), will rarely be exceeded. The tire contact area is

$$\text{Contact area (in}^2) = \frac{\text{Tire load (lbs)}}{\text{Tire inflation pressure (psi)}} \qquad (4.10)$$

For the rear tires

$$\text{Contact area (in}^2) = \frac{33,500}{90} = 372 \text{ inch}^2$$

Although the true contact area is approximately elliptical, often for simplicity the contact area is considered to be circular in shape. The contact pressure is usually assumed to be uniformly distributed. Because

$$\pi r^2 = 372 \text{ inch}^2$$

the radius of the tire contact area is

$$r \cong 11 \text{ inch}$$

and the average applied pressure is 90 psi (12,960 psf). As one moves down, away from the road surface, the force of the tire is spread over an ever increasing area and the bearing pressure is reduced. For simplicity, this load 'spreading' is assumed to occur at 45°. This is shown in Figure 4.114. Thus at a depth of 10 inches beneath the tire, the pressure radius would have increased to 21 inches and the pressure has dropped to 24.7 psi (3560 psf). However, for this truck there are dual rear wheels. Tire width is about 22 inches and the centerline spacing for the tires in each set is about 27 inches. This is shown diagrammatically in Figure 4.115.

Geometrical considerations 355

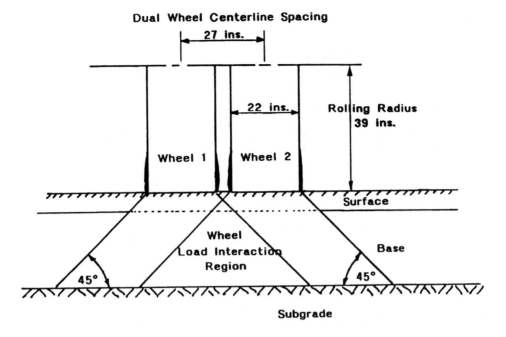

Figure 4.115. Load interaction with dual wheels.

As can be seen, the bearing pressure bulbs from each tire overlap. The greatest effect is observed along the line separating the tires. This interaction changes with tire width, tire separation and depth below the wear surface. To take this into account, Kaufman & Ault (1977) suggest using an equivalent single tire wheel load (L_E) which is 20% higher than the single tire load (L_T). Thus,

$$L_E = 1.20 \times L_T \tag{4.11}$$

In the case of the 58 st capacity truck

$$L_E = 1.20 \times 33,500 \cong 40,000 \text{ lb}$$

The combined subbase, base and wearing surface thickness must be sufficiently large so that the stresses occurring in the subgrade will not cause excessive distortion or displacement of the subgrade soil layer.

As a first guide, one can compare the required wear surface pressure to the bearing capacity of various subgrade materials. These are given in Table 4.6.

As can be seen, any subgrade that is less consolidated than soft rock will require additional material in order to establish a stable base. If, for example, the subgrade is a compact sand-clay soil with a bearing capacity of 6000 psf, then base/subbase materials of suitable strength would have to be placed down to increase the distance between the wear surface and the subgrade. Using the approach described earlier

$$\pi(11+t)^2 \times 6000 = \pi(11)^2 \times 12,960$$

356   *Open pit mine planning and design: Fundamentals*

Table 4.6. Bearing capacities of subgrade materials (Kaufman & Ault, 1977).

Material	1000 psf
Hard, sound rock	120
Medium hard rock	80
Hard pan overlying rock	24
Compact gravel and boulder-gravel formations; very compact sandy gravel	20
Soft rock	16
Loose gravel and sandy gravel; compact sand and gravelly sand; very compact sand – inorganic silt soils	12
Hard dry consolidated clay	10
Loose coarse to medium sand; medium compact fine sand	8
Compact sand-clay soils	6
Loose fine sand; medium compact sand – inorganic silt soils	4
Firm or stiff clay	3
Loose saturated sand clay soils, medium soft clay	2

the minimal required thickness (t) would be

$$t \cong 5 \text{ inch}$$

The technique often applied to determine the working surface, base and subbase thicknesses involves the use of California bearing ratio (CBR) curves. The CBR test is an empirical technique for determining the relative bearing capacity of the aggregate materials involved in road construction. In this test the aggregate material with a maximum size of $3/4$ inch is placed in a 6 in diameter metal mold. The material is compacted by repeatedly dropping a 10 lb weight through a height of 18 in. After compaction, a cylindrical piston having an end area of 3 inch^2 is pushed into the surface at a rate of 0.05 inch/minutes. The CBR is calculated by dividing the piston pressure at 0.1 or 0.2 inch penetration by reference values of 1000 psi for 0.1 inch and 1500 psi for 0.2 inch. These standard values represent the pressures observed for a high quality, well graded crushed stone reference material. The calculated pressure ratios are multiplied by 100 to give the CBR value expressed as a percent. Figure 4.116 shows design curves based upon the use of CBR values. The subbase thickness has been plotted against CBR/soil type for various wheel loads.

To demonstrate the use of these curves, consider the 58-st capacity truck travelling over a haulroad which the subgrade material is a silty clay of medium plasticity (CBR = 5). One finds the intersection of CBR = 5 and the 40,000 lb equivalent single wheel load. Moving horizontally it is found that the required distance between the wear surface and the subgrade must be a minimum of 28 inches.

Fairly clean sand with a CBR of 15 is available to serve as subbase material. Repeating the process, one finds that this must be kept 14 inches away from the wear surface. The base material is well graded, crushed rock with a CBR rating of 80. The intersection of the 40,000 lb curve and CBR = 80 occurs at 6 inches. This 6 inch gap between the top of the base and the wear surface is intended to accommodate the wear surface thickness. If the actual wear surface is thinner than this, the remaining space is simply added to the base thickness (CBR equal to at least 80). Figure 4.117 shows the final results (Kaufman & Ault, 1977).

In most open pit mines, the wear surface is formed by well graded, crushed rock with a maximum dimension smaller than that used as base. Since traffic loading is directly applied

Geometrical considerations 357

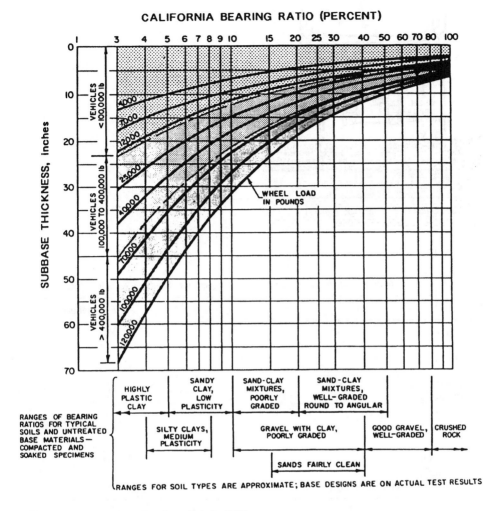

Figure 4.116. CBR curves (Kaufman & Ault, 1977).

to the aggregate layer, the upper most aggregate layer must possess sufficient strength and rutting resistance to minimize both
- bearing capacity failure, and
- rutting failure

within the layer.

The aggregate layer must also possess good wear resistance to minimize attrition under traffic. Table 4.7 indicates an acceptable aggregate size distribution (gradation) for this wearing surface.

Particle gradation is the distribution of the various particle size fractions in the aggregate. A well graded aggregate has a good representation of all particle size fractions from the maximum size through the smaller sizes. This is needed so that particles lock together forming a dense, compact surface. The opposite of a well graded aggregate is one which is poorly graded. Here the particles are all about the same size. Such a distribution might

358 *Open pit mine planning and design: Fundamentals*

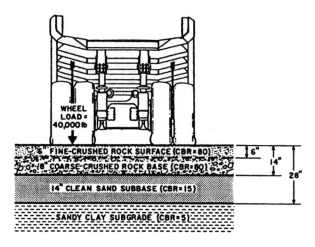

Figure 4.117. Example of mine road construction (Kaufman & Ault, 1977).

Table 4.7. Desired characteristics for a crushed stone running surface (Kaufman & Ault, 1977).

Screen size	Material passing (%)
1½ inches	100
1 inch	98
¾ inch	92
⅜ inch	82
No. 4 mesh	65
No. 10 mesh	53
No. 40 mesh	33
No. 200 mesh	16
Liquid limit	25.2
Plasticity limit	15.8
Plasticity index	9.4
Optimum moisture content during placing	12.2

be used as part of a runaway ramp with the objective being that of creating a high rolling resistance.

The use of CBR curves requires laboratory tests or the assumption of CBR values of subgrade, and available base or subbase materials. The most economical combination is used. The CBR curves show directly the total thickness needed over any subgrade soil. The total subbase and the base thickness is created by putting down a series of relatively thin layers of the correct moisture content. Compaction is done between layers.

4.9.3 *Straight segment design*

Figure 4.118 shows a typical cross-section through a mine haul road carrying two way traffic. As can be seen there are three major components to be considered:
  a) travel lane width,
  b) a safety berm,
  c) a drainage ditch.

Geometrical considerations 359

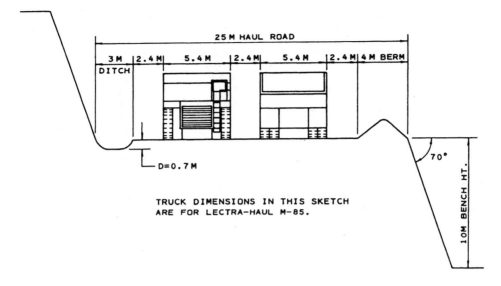

Figure 4.118. Typical design haulroad width for two-way traffic using 85 st capacity trucks (Couzens, 1979).

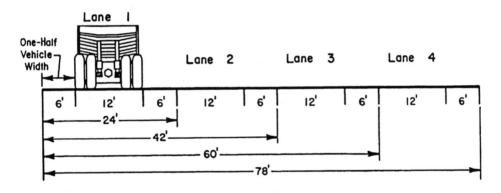

Figure 4.119. Multi-lane road design widths (Kaufman & Ault, 1977).

The width of each is added together to obtain the total roadway width.

The width criteria for the traveled lane of a straight haul segment should be based on the widest vehicle in use.

The 1965 *AASHO Manual for Rural Highway Design* recommends that each lane of travel should provide clearance to the left and right equal to one-half of the vehicle width. This is shown in Figure 4.126 for a 12-ft wide truck.

Values for other truck widths are given in Table 4.8. Typical widths of haulage trucks used in open pit mines are listed in Table 4.9.

For the two-way traffic which is most common in open pit mines, the rule of thumb is that roadway width should be no less than four times the truck width (Couzens, 1979):

$$\text{Roadway width} \geq 4 \times \text{Truck width} \tag{4.12}$$

Table 4.8. Recommended lane widths for tangent sections (Kaufman & Ault, 1977).

Vehicle width (ft)	1 lane	2 lanes	3 lanes
8	16	28.0	40
9	18	31.5	45
10	20	35.0	50
11	22	38.15	55
12	24	42.0	60
13	26	45.5	65
14	28	49.0	70
15	30	52.5	75
16	32	56.0	80
17	34	59.5	85
18	36	63.0	90
19	38	66.5	95
20	40	70.0	100
21	42	73.5	105
22	44	77.0	110
23	46	80.5	115
24	48	84.0	120
25	50	87.5	125
26	52	91.0	130
27	54	94.5	135
28	56	98.0	140

Table 4.9. Widths for various size rear dump trucks (this width includes the safety berm).

Truck size	Approx. width (m)	4 × width (m)	Design width	
			m	ft
35 st	3.7	14.8	15	50
85 st	5.4	21.6	23	75
120 st	5.9	23.6	25	85
170 st	6.4	25.6	30	100

Some mines have two lanes of traffic in one direction to allow passing for loaded uphill traffic. The downhill empty traffic travels in a single lane. A rule of thumb for the width of such a three lane road is 5 times the truck width.

The steps to be followed in selecting a design width are (Kaufman & Ault, 1977):

1. Define the width of all equipment that may have to travel the haulage road.
2. Solicit dimensional data for any anticipated new machines.
3. Determine the overall width of any equipment combinations that may be involved in a passing situation.
4. Delineate the location of road segments requiring a greater than normal width.

There may be wider stretches of road where there is merging of traffic streams such as near a crusher. Curves and switchbacks require special consideration. These will be discussed later.

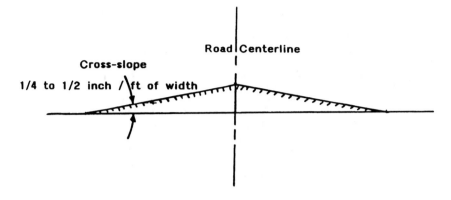

Figure 4.120. Cross slope design.

Table 4.10. Design widths (ft) for curves – single unit vehicles (Kaufman & Ault, 1977).

Radius (R) on inner edge of pavement (ft)	One-lane haulageway, vehicle category				Two-lane haulageway, vehicle category				Three-lane haulageway, vehicle category			
	1	2	3	4	1	2	3	4	1	2	3	4
Minimum	29	34	45	70	51	60	79	123	73	86	113	176
25	27	34	44	68	48	60	76	119	68	86	109	170
50	25	31	41	63	44	54	72	110	63	77	103	158
100	24	29	39	59	42	51	69	103	60	73	99	147
150	24	29	39	58	41	50	68	101	59	72	97	145
200	23	29	38	57	41	50	67	101	59	712	96	144
Tangent	23	28	37	56	40	48	65	98	57	69	93	140

The road surface is often slightly crowned such as shown in Figure 4.120, to facilitate water runoff. The cross slope is expressed in inches per foot of width. Most mine roads are constructed of gravel and crushed rock. In this case, except where ice/mud is a problem, the cross slope should be $\frac{1}{2}$ inch per foot (0.04 ft/ft). For relatively smooth road surfaces such as asphaltic concrete which can rapidly shed water or roads which have ice/mud problems, a cross slope of $\frac{1}{4}$ inch per foot (0.02 ft/ft) is appropriate.

For single lanes, it is necessary to decide whether the left edge should be higher than the right or vice-versa. For three-lane surfaces, there should be a continuous cross slope for the two lanes having traffic in the same direction. It should be noted that the use of a cross-slope increases the steering effort by the driver. Thus there must be a balance between steerability and water drainage.

4.9.4 Curve design

For straight sections it was recommended that the left and right vehicle clearances should be half of the vehicle width. In the case of curves this distance must be increased both due to vehicle overhang and increased driving difficulty.

Tables 4.10 and 4.11 provide the design widths as a function of the inner pavement radius for various combinations of vehicle size, vehicle type and roadway types. For

Table 4.11. Design widths (ft) for curves – articulated vehicles (Kaufman & Ault, 1977).

Radius (R) on inner edge of pavement (ft)	One-lane haulageway, vehicle category			Two-lane haulageway, vehicle category			Three-lane haulageway, vehicle category		
	2	3	4	2	3	4	2	3	4
25	38	68	86	66	119	151	95	170	215
50	32	57	71	56	99	124	80	142	177
100	28	48	58	50	83	101	71	119	144
150	27	44	52	47	76	91	68	109	130
200	26	42	49	46	73	85	66	104	122
Tangent	25	41	41	44	71	72	63	102	103

Table 4.12. Minimum single unit haulage truck turning radius (Kaufman & Ault, 1977).

Vehicle weight classification	Gross vehicle weight (GVW) (lb)	Minimum turning radius (ft)
1	< 100,000	19
2	100–200,000	24
3	200–400,000	31
4	> 400,000	39

reference approximate turning radii are indicated by gross vehicle weight categories in Table 4.12.

For example, if a single unit haulage truck of weight classification 3 is to traverse a 100 ft minimum radius curve, the two lane width should be 69 ft. For a straight road segment the corresponding width is 65 ft. Hence the effect of the curve is to add 4 ft to the width.

Vehicles negotiating curves are forced outward by centrifugal force. For a flat surface this is counteracted by the product of the vehicle weight and the side friction between the roadway and the tires (Fig. 4.121).

For certain combinations of velocity and radius the centrifugal force will equal or exceed the resisting force. In such cases, the vehicle skids sideways. To assist the vehicles around the curves, the roadways are often banked. This banking of curves is called superelevation. The amount of superelevation (cross slope) can be selected to cancel out the centrifugal force. The basic equation is

$$e + f = \frac{V^2}{15R} \qquad (4.13)$$

where e is the superelevation rate (ft/ft); f is the side friction factor; V is the vehicle speed (mph); R is the curve radius (ft). If $f = 0$, then the vehicle would round the curve without steering effort on the part of the operator. If however the operator would maintain a speed different from that used in the design, then he would have to steer upslope (in the case of too low a speed) of downslope (too high a speed) to maintain the desired path. Under ice and snow conditions, too slow a speed on such super elevated curves could lead to sliding down the slope.

Table 4.13 gives recommended superelevation rates as a function of curve radius and vehicle speed. The table can also be used to suggest a safe speed for a given radius and superelevation rate.

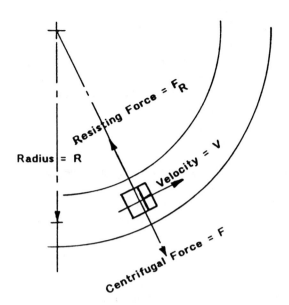

Figure 4.121. Centrifugal force effects on curves.

Table 4.13. Recommended superelevation rates (feet per foot of width) (Kaufman & Ault, 1977).

Radius of curve (ft)	Speed of vehicle (mph)					
	10	15	20	25	30	35 and over
50	0.04	0.04				
100	0.04	0.04	0.04			
150	0.04	0.04	0.04	0.05		
250	0.04	0.04	0.04	0.04	0.06	
300	0.04	0.04	0.04	0.04	0.05	0.06
600	0.04	0.04	0.04	0.04	0.04	0.05
1000	0.04	0.04	0.04	0.04	0.04	0.04

Table 4.14. Recommended rate of cross-slope change (Kaufman & Ault, 1977).

Vehicle speed (mph)	10	15	20	25	30	35 and above
Cross-slope change in 100-ft length of haulageway (ft/ft)	0.08	0.08	0.08	0.07	0.06	0.05

There is a certain distance required to make the transition from the normal cross-slope section to the superelevated portion and back again. This is called the superelevation runout. The purpose is to help ease the operator into and out of the curve. Part of the transition can be placed in the straight (tangent) portion and part in the curve. The design criteria of ⅓ inch curve and ⅔ inch the tangent is used here. The recommended rate of cross-slope change as a function speed is given in Table 4.14.

To illustrate the use of this table, assume a vehicle is traveling at 35 mph on tangent with normal cross slope 0.04 ft/ft to the right. It encounters a curve to the left necessitating a superelevation rate of 0.06 ft/ft to the left. The total cross-slope change required is 0.10 ft/ft (0.04 + 0.06). The table recommends a 0.05 ft/ft cross-slope change in 100 ft. Thus the total runout length is computed as 200 ft [(0.10/0.05) × 100 = 200]. One-third of this length should be placed in the curve and two-thirds on the tangent.

4.9.5 Conventional parallel berm design

U.S. federal law (MSHA, 1992) contains the following guidance regarding the need for berms/guardrails in open pit mines (Section 57.9300):

(a) *Berms or guardrails shall be provided and maintained on the banks of roadways where a drop-off exists of sufficient grade or depth to cause a vehicle to overturn or endanger persons in equipment.*

(b) *Berms or guardrails shall be at least mid-axle height of the largest self-propelled mobile equipment which usually travels the roadway.*

(c) *Berms may have openings to the extent necessary for roadway drainage.*

(d) *Where elevated roadways are infrequently traveled and used only by service or maintenance vehicles, berms or guardrails are not required (when certain very specific conditions are met).*

The principal purpose of these berms is to redirect the vehicle back onto the roadway and away from the edge. Their effectiveness in this regard is controlled by berm face angle, berm facing, the angle of incidence, and primarily by berm height. The stopping of runaway vehicles is accomplished by median berms (Subsection 4.9.6) or special escapeways. One negative effect of berms is the possibility of the vehicles overturning due to climbing the sides.

There are two principal berm designs in common use today. The triangular or trapezoidal shaped berm is generally formed from blasted materials. The sides stand at the angle of repose of the material. The second type is the boulder-faced berm. Here, large boulders, lined up along the haulage road, are backed with earthen material or blasted rock.

For the triangular berms, the design rule of thumb is that the height must be equal to or greater than the static rolling radius (SRR) of the vehicle's tire. For boulder-faced berms, the height of the berm should be approximately equal to the tire height. Figure 4.122 shows the relationship between the static rolling radius and haulage vehicle carrying capacity. Tire height (TH) is about equal to:

$$TH = 1.05 \times 2 \times SRR \qquad (4.14)$$

4.9.6 Median berm design

Some means should be provided on haulroads to reduce truck speed or handle the truck that loses its brakes. This is particularly true when long, downhill loaded hauls are involved. Currently the most successful technique is through the use of median berms, also known as 'straddle berms' or 'whopper stoppers' (Winkle, 1976a,b). These are constructed of sand or some other fine grained material. The height of these berms is designed to impinge on the under-carriage of the truck. Since the typical distance between the road surface to the undercarriage is of the order of 2 to 3 ft for the range of available haulers, it is not necessary to build a big barrier providing just another crash hazard.

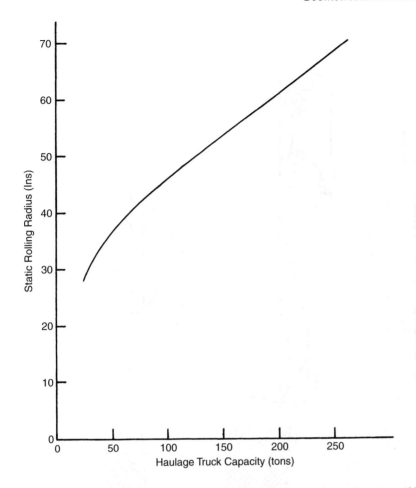

Figure 4.122. Static rolling radius as a function of haulage truck capacity (Goodyear, 1992).

Guidance in median berm design provided by Kaufman & Ault (1977) is given in Figure 4.123. The dimensions corresponding to the letters in the figure are given in Tables 4.15 and 4.16. The vehicle categories are based upon gross vehicle weights.

Training the driver to get onto the berm or into the bank just as soon as they start to lose control of their truck and before they build up speed is as important, or more important, than the berm design itself (Couzens, 1979).

4.9.7 *Haulage road gradients*

A number of rules of thumb regarding haulage road gradients have been provided by Couzens (1979). These are given below:

1. In a pit where there is a considerable vertical component to the haulage requirement, the grade will have to be fairly steep to reduce the length of the road and the extra material necessary to provide the road length. The practical maximum grade is considered to be 10%.

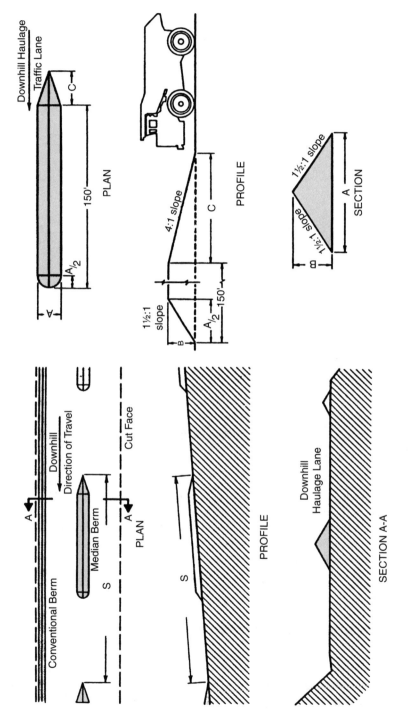

Figure 4.123. Runaway-vehicle collision berms (Kaufman & Ault, 1977)

Table 4.15. Typical median berm dimensions (see Figure 4.123) (Kaufman & Ault, 1977).

	A	B	C
Category 1 13 to 25 st <100,000 lb	11'–12'	3.5'–4'	14'–16'
Category 2 28 to 50 st 100,000–200,000 lb	12'–15'	4'–5'	16'–20'
Category 3 55 to 120 st >200,000–400,000 lb	15'–18'	5'–6'	20'–24'
Category 4 120 to 250 st > 400,000 lb	18'–32'	6'–11'	24'–44'

Table 4.16. Berm spacing (S) expressed in feet assuming that the initial speed at brake failure is 10 mph.

Equivalent downgrade, (%)	Maximum permissible vehicle speed or terminal speed at entrance to safety provision (mph)							
	15	20	25	30	35	40	45	50
1	418	1003	1755	2674	3760	5013	6433	8021
3	140	335	585	892	1254	1671	2145	2674
5	84	201	351	535	752	1003	1287	1604
7	60	144	251	382	537	716	919	1146
9	47	112	195	297	418	557	715	892
11	38	92	160	243	342	456	585	730
13	33	78	135	206	290	386	495	617
15	28	67	117	179	251	335	429	535

A number of pits operate quite well at 10% grades both favorable and unfavorable to the loads.

2. An 8% road grade is probably preferred providing that it does not cause too much extra stripping or unduly complicate the road layout. This grade provides more latitude in: (a) building the road and (b) fitting in bench entries without creating some locally over-steep places, than do steeper grades.

3. There is normally nothing to be gained by flattening the road below 8%, unless there is a long distance to travel without requiring much lift. The extra length on the grade and the complications of fitting the road into the available room or doing extra stripping would probably offset any increase in uphill haul speed.

4. Pit geometry is the prime consideration and roads are designed to fit the particular situation. Thus there often will be a number of different grade segments in haul roads.

4.9.8 Practical road building and maintenance tips

The preceding parts of this section have dealt with some of the general road design principles. Winkle (1976a,b) has provided a number of practical tips based upon many years of practical experience. Some of these have been included below.

1. The size of the orebody and the nature of the overlying topography will have considerable impact upon road design. When the orebody is small it will likely be advantageous to strip immediately to the projected pit limits, since some mining inefficiencies occur when mining areas overlap. This dictates an immediate final road layout on the backslope which is planned to avoid expensive modification. For very large orebodies, particularly where an outcrop of ore is exposed, it is highly unlikely that initial stripping will extend to the final planned perimeter. Careful study of the topography is required to ensure proper rapid access. The cost of rehandling material dumped within eventual pit limits must be weighed against increased haul distances, sharp curves, etc.

2. Change in equipment size frequently is a cause of road modification, particularly width. Pit design should incorporate allowance for reasonable future equipment size increases.

3. When mixed haulage fleets with varying speeds are used or where trucks hauling from two or more shovels are using the same haul roads, passing lanes on long grades should be considered.

4. Short radius curves result in reduced productivity, high tire cost, high maintenance cost (particularly electric wheels) and introduce additional safety hazards into the operation. Switchbacks are to be avoided unless a tradeoff of reduced stripping dictates their construction.

5. When curves are necessary in haulroads, superelevation must be designed into the curves. Excessive superelevation is to be avoided since trucks rounding a curve slippery from rain, ice or overwetting can slide inward and possibly overturn. Overly 'supered' curves result in excessive weight and wear on the inside tires.

6. Often curves are constructed to provide an access road into a mining bench from a steeply inclined haul road. To prevent the inside (and lower) side of this superelevated access curve from being at a steeper gradient than the main haul road, it is necessary to reduce (flatten) the center line grade of the curve. The inside grade should not be allowed to exceed the main road grade.

If enough room is available, the inside gradient of the curve should in fact be flatter than the main road grade to compensate for the increased rolling resistance. To accomplish this the design of a transition spiral is necessary.

7. Curves in the flat haul portion just as the trucks are leaving the shovel are quite critical. Due to the centrifugal forces induced by the curve, spill rock is thrown to the outside. Where possible, the return lane should be on the inside of the curve to avoid spill falling in the path of returning trucks and damaging tires. This can be accomplished by the use of crossovers to change traffic from right hand to left hand or vice versa in the necessary area. Adequate warning lights must be used at night to insured the safety at the crossovers. The costs of the warnings are small compared to savings in tire costs.

Geometrical considerations 369

8. Waste dumps should be designed for placement at a two percent upgrade. This is done for the following reasons:

(a) The increase in dump height and volume occurs with little increase in haul speed or fuel cost. Because of the rapid dump volume increase, the haul distance is reduced.
(b) Better drainage on dumps.
(c) Some additional safety is afforded drivers backing up for dumping.
(d) If eventual dump leaching is planned, the water distribution is less expensive.

9. Within the mining areas, roads are built of the country rock at hand and are surfaced with the best material available within a reasonable haul distance. In the case of using something other than environmental rock to surface roads within the ore zone, double handling costs as well as ore dilution must be considered.

10. Main roads into the pit are usually planned for extended time of use and will justify more expenditure for subbase compaction and surfacing than temporary access roads.

11. If intended for use as a haul road, engineering layout should precede construction of even the shortest road or ramp. Mine survey crews should place desired stakes for initial cuts and fills and grade stakes including finish grade stakes.

12. When shovels are working in coarse, sharp rock faces, loading should be stopped periodically to allow fine material to be brought in and used to cap the surface of the loading area. Similar activity should be performed on waste dumps.

13. Constant attention to haul road surfaces is necessary. Soft spots, holes, 'washboard' areas, etc. should be repaired as soon as possible. Repairs usually consist of digging out the incompetent road material and replacing it with more desirable rock.

14. Grading of roads often results in a buildup of windrows on road edges. These narrow the roads and place sharp rocks in a position to damage tire sidewalls. Windrow buildup should be removed by loader or careful grader application.

15. Balding or grading of roads and dumps should be done when possible at a time when traffic can be moved to other areas. Many tires have been damaged by trucks driving through windrows created by graders assigned to improve roads and thereby reduce tire costs.

16. Maintenance of haul roads is equally important to good haulage costs as are design and construction. As more tires are damaged in shovel pits and dump areas than on actual haul roads, cleanup around an operating shovel is often assigned to the haulroad rather than the loading function. Road maintenance, to be successful, must have responsible supervision assigned to this task alone.

4.10 STRIPPING RATIOS

Consider the orebody shown in Figure 4.124 which has the shape of a right circular cylinder.
It outcrops at the surface and extends to a depth h. The volume of the contained ore is expressed by

$$V_o = \pi r^2 h \tag{4.15}$$

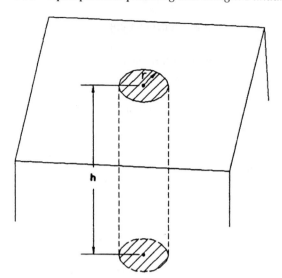

Figure 4.124. Cylindrical orebody.

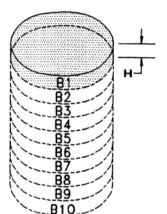

Figure 4.125. Cylindrical orebody mined as a sequence of constant diameter and thickness benches.

where r is the ore radius and h is the ore thickness. In concept, at least, one could remove the ore as a single plug and just leave the remaining hole. In practice, however, the orebody is first divided up into a series of benches of thickness H (Fig. 4.125). The volume of each ore bench B_i is

$$V_b = \pi r^2 H \tag{4.16}$$

In this case it will be assumed that each bench exactly satisfies the required annual production. Hence the pit would increase in depth by one bench per year. The surrounding waste rock has been assumed to have high strength so that these 90° pit walls can be safely achieved and maintained. In this mining scheme no waste is removed.

In reality, vertical rock slopes are seldom achieved except over very limited vertical heights. It is much more common to design using an overall slope angle Θ. As can be seen in Figure 4.126 the shape of the mined space changes from a right circular cylinder to a truncated right circular cone. The height of the truncated portion of the cone is

$$\Delta h = r \tan \Theta \tag{4.17}$$

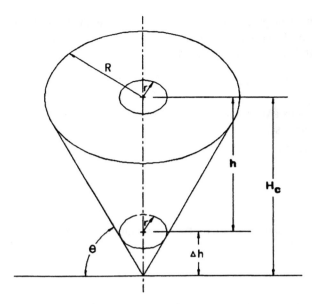

Figure 4.126. The cylindrical orebody mined via a conical pit.

where Θ is the overall slope angle. The height H_c of the cone which includes the orebody is then

$$H_c = h + \Delta h = h + r \tan \Theta \qquad (4.18)$$

The base radius R of the circumscribed cone is

$$R = \frac{H_c}{\tan \Theta} = \frac{h}{\tan \Theta} + r \qquad (4.19)$$

Using the volume formula for a right circular cone

$$V_{rcc} = \frac{1}{3} A_{bc} H_c \qquad (4.20)$$

where A_{bc} is the base area of the cone, H_c is the height of the cone, and V_{rcc} volume of the cone, one can find the following volumes:

Truncated tip

$$V_{tip} = \frac{1}{3} \pi r^2 \Delta h \qquad (4.21)$$

Fully circumscribed cone

$$V = \frac{1}{3} \pi R^2 H_c \qquad (4.22)$$

Mined volume (ore + waste)

$$V_m = V - V_{tip} = \frac{1}{3} \pi R^2 H_c - \frac{1}{3} \pi r^2 \Delta h \qquad (4.23)$$

Volume of waste

$$V_w = V_m - \pi r^2 h \qquad (4.24)$$

One of the ways of describing the geometrical efficiency of a mining operation is through the use of the term 'stripping ratio'. It refers to the amount of waste that must be removed to release a given ore quantity. The ratio is most commonly expressed as

$$\text{SR} = \frac{\text{Waste (tons)}}{\text{Ore (tons)}} \qquad (4.25)$$

however a wide variety of other units are used as well. In strip coal mining operations for example the following are sometimes seen

$$\text{SR} = \frac{\text{Overburden thickness (ft)}}{\text{Coal thickness (ft)}}$$

$$\text{SR} = \frac{\text{Overburden (yd}^3\text{)}}{\text{Coal (tons)}}$$

The ratio of waste to ore is expressed in units useful for the design purpose at hand. For this example, the ratio will be defined as

$$\text{SR} = \frac{\text{Waste (volume)}}{\text{Ore (volume)}} \qquad (4.26)$$

Note that if the waste and ore have the same density, then Equation (4.25) and Equation (4.26) are identical.

If the volumes (or tons) used in the SR calculation correspond to those (cumulatively) removed from the start of mining up to the moment of the present calculation then the overall stripping ratio is being calculated. For this example the overall stripping ratio at the time mining ceases is

$$\text{SR (overall)} = \frac{V_w}{V_o} = \frac{V_m - \pi r^2 h}{\pi r^2 h} \qquad (4.27)$$

On the other hand a stripping ratio can also be calculated over a much shorter time span. Assume that during year 5, X_o tons of ore and X_w tons of waste were mined. The stripping ratio for year 5 is then

$$\text{SR (year 5)} = \frac{X_w}{X_o}$$

This can be referred to as the instantaneous stripping ratio where the 'instant' in this case is 1 year.

If at the end of year 4, X_{o4} tons of ore and X_{w4} tons of waste had been mined then the overall stripping ratio up to the end of year 5 is

$$\text{SR (overall to end of year 5)} = \frac{X_{w4} + X_w}{X_{o4} + X_o}$$

Obviously the 'instant' could be defined as a longer or shorter time period. If during a given day the mine moves 5000 tons of waste and 2000 tons of ore, the instantaneous stripping ratio (for that day) is

$$\text{SR (instantaneous)} = \frac{5000}{2000} = 2.5$$

The determination of final pit limits as will be described in detail in Chapter 5, involves the calculation of a pit limit stripping ratio to be applied to a narrow strip at the pit periphery.

Geometrical considerations 373

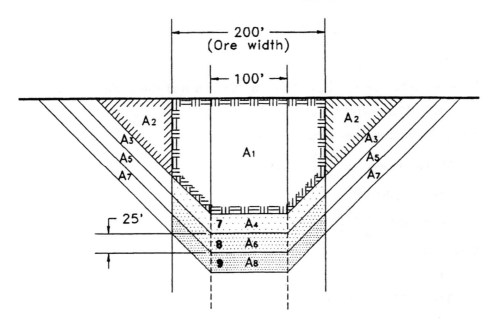

Figure 4.127. Section for stripping ratio calculations.

To illustrate this concept consider the simple cross-section shown in Figure 4.127. It will be assumed that:
 – the pit is deepened in bench height increments of 25 ft;
 – the minimum pit width is 100 ft;
 – overall slope angle is 45°
 – the density of the ore and waste is the same;
 – the ore is of constant grade.
The original pit on this section (Fig. 4.127), consists of 6 benches and has a depth of 150 ft. The area of ore A_o is

$$A_o = A_1 = 200 \times 100 + 50 \times 150 = 27,500 \, \text{ft}^2$$

The area of waste A_w is

$$A_w = 2A_2 = 100 \times 100 = 10,000 \, \text{ft}^2$$

The overall stripping ratio SR (overall) is

$$\text{SR (overall)} = \frac{A_w}{A_o} = \frac{10,000}{27,500} = 0.36$$

Deepening of the pit by one bench (bench 7) requires the removal of $2A_3$ of waste. The amount of ore uncovered is A_4

$$A_4 = 100 \times 25 + 100 \times 25 = 5000 \, \text{ft}^2$$

$$2A_3 = 125 \times 125 - 100 \times 100 = 5625 \, \text{ft}^2$$

The instantaneous stripping ratio is

$$\text{SR (instantaneous)} = \frac{5625}{5000} = 1.125$$

The overall stripping ratio with bench 7 removed is

$$SR \text{ (overall)} = \frac{15,625}{32,500} = 0.48$$

With mining of bench 8, another 5000 ft² of ore (A_6) is removed. This requires the stripping of

$$2A_5 = (150)^2 - (125)^2 = 6875 \text{ ft}^2$$

of waste. The instantaneous stripping ratio is

$$SR \text{ (instantaneous)} = \frac{6875}{5000} = 1.375$$

The overall stripping ratio is

$$SR \text{ (overall)} = \frac{22,500}{37,500} = 0.60$$

For bench 9:

$$A_8 = 5000 \text{ ft}^2$$
$$2A_7 = (175)^2 - (150)^2 = 8125 \text{ ft}^2$$

$$SR \text{ (instantaneous)} = \frac{8125}{5000} = 1.625$$

$$SR \text{ (overall)} = \frac{30,625}{42,500} = 0.72$$

As can be seen in this simple example, with each cut, the same amount of ore 5000 ft² must pay for an increasing amount of waste. The overall stripping ratio is less than the instantaneous value. There becomes a point where the value of the ore uncovered is just equal to the associated costs with the slice. This would yield the maximum pit on this section. Assume that in this case the breakeven stripping ratio is 1.625. Then the final pit would stop with the mining of bench 9. Through pit deepening, the walls of the pit are moved away or 'pushed back' from their original positions. The term 'push-back' is used to describe the process by which the pit is deepened by one bench.

4.11 GEOMETRIC SEQUENCING

There are several ways in which the volume of Figure 4.126 can be mined. As before, the first step in the process is to divide the volume into a series of benches (see Fig. 4.128). If a single bench is mined per year then the ore production would remain constant while both the total production and the stripping ratio would decrease. This would lead to a particular cash flow and net present value.

For most mining projects, a large amount of waste mining in the early years of a project is not of interest.

An alternative mining geometry is shown in Figure 4.129 in which a number of levels are mined at the same time. The overall geometry looks much like that of an onion.

An initial 'starter-pit' is first mined. In this example, the pit bottom extends to the edge of the orebody and the slope angle of the starter pit is the same as that of the final pit (Θ).

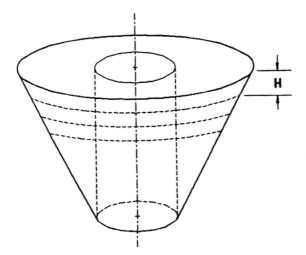

Figure 4.128. Sequential geometry 1 (Fourie, 1992).

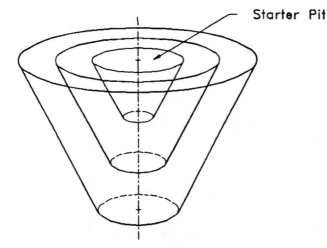

Figure 4.129. Sequential geometry 2 (Fourie, 1992).

In theory one could then slowly 'eat-away' at the sides and bottom of this starter pit until the final pit geometry is achieved. There are practical limits however on the minimum size 'bites' which can be considered both for planning and execution. The 'bites' in surface mining terms are called *push-backs* or phases. For modern large pits the minimum push-back distance (thickness of the bite) is of the order of 200 to 300 ft. For smaller pits it can be of the order of 100 to 200 ft. In this particular example the push-backs result in the pit being extracted in a series of concentric shells. The amount of material (ore and waste) contained in each shell is different. Hence for a constant production rate there might be x years of ore production in shell 1, y years of production in shell 2, etc. Eventually there will be a transition in which mining is conducted in more than one shell at a given time.

Sequencing within a pit shell and between shells becomes important. To this point simple concentric shells have been considered. The next level of complication is to split the overall pit into a number of sectors such as shown in Figure 4.130. Each sector (I → V) can be

376 *Open pit mine planning and design: Fundamentals*

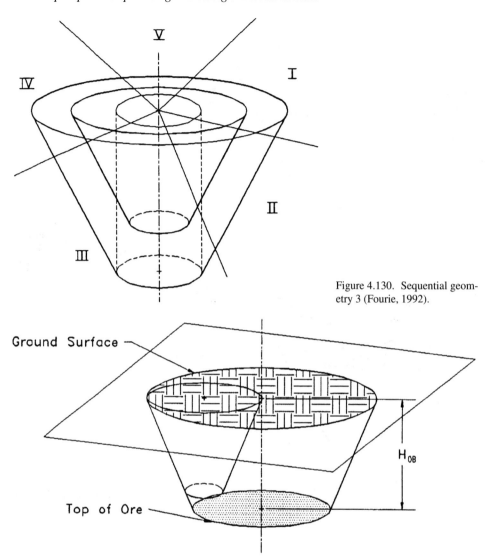

Figure 4.130. Sequential geometry 3 (Fourie, 1992).

Figure 4.131. Sequential geometry 4 (Fourie, 1992).

considered as a separate production or planning unit. A natural basis for dividing the pit this way is due to slope stability/design considerations.

It has been assumed that the orebody outcrops (is exposed) at the surface. If this is not the case, such as is shown in Figure 4.131, then a preproduction or stripping phase must be first considered.

Due to cash flow considerations a variety of aspects enter:
 – desire to reach the ore as quickly as possible,
 – requirement to expose enough ore to maintain the desired plant production,
 – combination of higher grade ore at greater depth versus lower grade at shallower depth.
The geometry-sequencing decisions then become even more complex.

It has been assumed for the sake of simplicity that the 'ore' is of one quality. Generally the values will vary in $X-Y-Z$ space. The same is true for rock quality. Hence new dimensions are added to an already complex overall mine geometry-sequencing problem. Furthermore, the 'simple' addition of a haulage road to provide additional access can have a major effect on mine geometry and economics.

4.12 SUMMARY

In summary, pit geometry at any given time is influenced by many factors. Obviously the overlying material must be removed prior to removing that underlying. A certain operating space is needed by the equipment for efficiently removing the rock. The slope materials largely dictate the slope angles which can be safely used. In addition the sequencing of these geometries is extremely important so that the desired economic result (revenue and costs) is realized. Production rates, ore reserves and mine life are often highly price dependent. Hence mining geometry is a dynamic rather than static concept. To evaluate the many individual possibilities and combinations of possibilities, the computer has become invaluable.

The planning engineer must fully understand the basic geometric components which are combined to yield the overall pit geometry at any time in the life of the mine.

REFERENCES

Anonymous 1978. Road design halts runaway vehicles. *Coal Age* 83(7): 144–147.
Anonymous 1982. Better haul roads speed operations. *Coal Age* 83(3): 96–99.
Armbrust, J.C. 1988. Morenci mine-blasting and dewatering of the 4050 drop cut. *SME Annual Meeting Jan. 25–28, 1988*. Soc. of Mining Engineers: preprint No. 88–146.
Atkinson, T. 1992. Design and layout of haul roads. *SME Mining Engineering Handbook*. 2nd Edition (H.L. Hartman, editor) 2: 1334–1342. Littleton, CO: SME.
Atkinson, T. & G. Walton 1983. Design and layout of haul roads for surface mines. In: *Surface Mining and Quarrying, papers presented at the 2nd Int. Surface Mining and Quarrying Symposium, Oct. 4–6, 1983*. London: Instn. Min. Met.
Barksdale, R.D. (editor) 1991. *The Aggregate Handbook*: 11-1 to 11-23. Washington, D.C.: National Stone Association.
Bauer, A. 1984 Wall control blasting in open pits. In: *Rock Breaking and Mechanical Excavation* (P. Baumgartner, editor) CIM Special Volume 30: 3–10. The Canadian Institute of Mining and Metallurgy.
Beale, G., Luther, A., & J. Foster. 1997. Depressuring the pit wall at Sleeper and at the Mag pit. *Mining Engineering*. 49(11): 40–46.
Brawner, C.O. 1969. Three big factors in stable slope design. *Mining Engineering* 21(8): 73–77.
Caccetta, L., L.M. Giannini & S. Carras 1986. The optimum design of large open pit mines. In: *Aus. IMM/IR Aust. Newman Combined Group Large Open Pit Mining Conference* (J.R. Davidson, editor): 195–200. Australasian Inst. of Mining and Metallurgy.
Call, R.D. 1986. Cost-benefit design of open pit slopes. In: *1st Open Pit Mining Symposium, Antofagasta, Chile, October 1986*: 1–18.
Call, R.D., J.P. Savely & D.E. Nicholas 1977. Preliminary data collection for pit slope design. *Mining Engineering* 29(4): 45–47.
*CANMET. *Pit Slope Manual*. CANMET Publications, 562 Booth Street, Ottawa, Ontario, K1A 0G1, Canada.
Carson, A.B. 1961. *General Excavation Methods*: 42–66. New York: McGraw-Hill.
Caterpillar 1991. *Caterpillar Performance Handbook*. 22nd Edition.

---

* For more information see the end of the references.

Chironis, N.P. 1978. How to build better haul roads. *Coal Age* 83(1): 122–128.

Coates, D., K. McRorie & J. Stubbins 1963. Analysis of pit slides in some incompetent rocks. *Trans AIME* 226(3): 94–101.

Coates, D.F., M. Gyenge & J.B. Stubbins 1965. Slope stability studies at Knob Lake. In: *Proceedings 3rd Canadian Symp. on Rock Mech., Univ. of Toronto, Mines Branch*: 35–46. Queens Printer.

Couzens, T.R. 1979. Aspects of production planning: Operating layout and phase plans. In: *Open Pit Mine Planning and Design* (J.T. Crawford and W.A. Hustrulid, editors): 217–232. SME.

Decision Making in the Mineral Industry. 1971. CIM Special Volume 12: 339–343.

Denby, B., McClarnon, D., & D. Schofield. 1996. The application of virtual reality in the modeling of surface mining operations. In Surface Mining 1996 (H.W. Glen, editor): 85–92. SAIMM.

Dight, P.M. & C. Windsor 1986. The current state of the art of pit design and performance. In: *Aus. IMM/IE Aust. Newman Combined Group Large Open Pit Mining Conference, October 1986*: 285–294. Australasian Institute of Mining and Metallurgy.

Dillon, U., & G. Blackwell. 2003. The use of a geographic information system for open-pit mine development – Technical Note. CIM Bulletin 96(1069): 119–121.

East, D.R. 2000. Tailings dam failures – Why do they continue to occur? Mining Engineering. 52(12): 57–61.

Fishler, S.V. 1991. Personal communication. Harnischfeger Corporation.

Fourie, G.A. 1992. Open pit planning and design – basic concepts. In: *SME Mining Engineering Handbook*. 2nd Edition (H.L. Hartman, editor): 1274–1278. Littleton, CO: SME.

Gill, T. 1999. Planning optimal haul road routes using *Express*. 28th International Symposium on Application of Computers and Operations Research in the Mineral Industry: 375–384. Colorado School of Mines:CSM.

Goodyear 1992. Personal communication.

Halls, J.L. 1970. The basic economics of open pit mining. In: *Planning Open Pit Mines* (P.W.J. Van Rensburg, editor). Johannesburg: SAIMM.

Hanna, T.M., Azrag, E.A., & L.C. Atkinson. 1994. Use of an analytical solution for preliminary estimates of ground water inflow to a pit. Mining Engineering. 46(2): 149–152.

Hardwick, W.R. 1958. *Open-Pit Mining Methods and Practices at Chino Mines Division, Kennecott Copper Corp., Grant County, N. Mex.* U.S. Bureau of Mines IC 7837.

Hardwick, W.R. & M.M. Stover 1960. *Open-Pit Copper Mining Methods and Practices, Copper Cities Division, Miami Copper Co., Gila Count, Arizona*. U.S. Bureau of Mines IC 7985.

Hoek, E. 1970a. Estimating the stability of excavated slopes in open cast mines. *Trans IMM* 79(10): A109–A132.

Hoek, E. 1970b. Design charts for excavated slopes. Contributions to the discussion of Section 6 – Planning Open Pit Mines. In: *Planning of Open Pit Mines* (P.W.J. Van Rensburg, editor): 295–302. Johannesburg: SAIMM.

Hoek, E. and J.S. Bray 1977. *Rock Slope Engineering* (Revised Second Edition): 402 pp. London: Inst. of Mining and Metallurgy.

Hoek, E. & D.L. Pentz 1970. Review of the role of rock mechanics research in the design of opencast mines. In: *Proceedings of the 9th Commonwealth Mining and Metallurgical Congress, Vol. 1, Mining and Petroleum Technology*: 389–412. London: IMM.

Hustrulid, W. A., McCarter, M. K., and D. Vanzyl. (editors) 2000. Slope Stability in Surface Mining, SME, Littleton, CO, USA.

Jennings, J.E. & R.A.L. Black 1963. Factors affecting the angle of slope in open-cast mines. Paper presented at the Annual Meeting of AIME, Dallas, Texas, Feb. 24–28, 1963: preprint 63 A011, and *Trans. of the AIME* 226(3): 42–53.

Kaufman, W.W. & J.C. Ault 1977. *Design of Surface Mine Haulage Roads – A Manual*. USBM IC 8758.

Kim, Y.C., D.F. Coates & T.J. O'Neil 1977. A formal approach to economic analysis of pit slope design. In: *15th APCOM, Brisbane, Aust.*: 405–413.

Kliche, C.A. 1999. Rock Slope Stability. SME. Littleton, Colorado. 253 pp.

Lane, K.S. 1961. Field slope charts for stability studies. In: *5th Intern. Conf. Soil Mech and Foundation Eng.* 2: 651–655.

Long, A.E. 1964. Problems in designing stable open-pit mine slopes. *CIM Bulletin* 57(7): 741–746.

Martin, D.C. & D.R. Piteau 1977. Select berm width to contain local failures. *E/MJ* 178(6): 161–164.

McCarter, M.K. 1976. Monitoring stability of high waste dumps. Presented at the 1976 SME-AIME Meeting, Denver, Sept. 1–3, 1976: preprint no. 76-AO-328: 23 pp.

McWilliams, J.R. 1959. *Mining Methods and Costs at the Anaconda Company Berkeley Pit, Butte, Mont.* USBM IC 7888.
Miller, G.G., G.L. Stecklin & J.J. Labra 1983. *Improved haul road berm design*. U.S. Bureau of Mines IC 8947.
Mines Branch 1972. *Tentative Design Guide for Mine Waste Embankments in Canada*. Technical Bulletin TB 145, March 1972.
Moss, A.S.E. & O.K.H. Steffen 1978. Geotechnology and probability in open-pit mine planning. In: *Proceedings of the 11th Commonwealth Mining and Met. Congress*: 543–550.
MSHA (Mine Safety and Health Administration) 1992. Safety and health standards applicable to surface metal and nonmetal mining and milling operations, Part 56. Code of Federal Regulations (CFR 30), Revised as of July 1, 1992.
Nichols, H.L., Jr 1956. *Modern Techniques of Excavation*: 8–8 to 8–23. Princeton, NJ: Van Norstrand.
Ramsey, R.H. 1944. How Cananea develops newest porphyry copper. *E/MJ* 145(12): 74–87.
Ramsey, R.J. 1945. New Cananea operation now in high gear. *E/MJ* 146(9): 72–78.
Reed, A. 1983. Structural geology and geostatistical parameters of the Afton Copper-Gold Mine, Kamloops, B.C. *CIM Bulletin* (856)76(8): 45–55.
Riese, M. 1993. *Specification Sheet for the Bucyrus-Erie Model 155 Shovel*.
Ritchie, A.M. 1963. Evaluation of rockfall and its control. *Highway Research Record* 17: 13–18.
Robertson, A.M. 1971. The interpretation of geological factors for use in slope theory. In: *Proc. Symposium on Planning Open Pit Mines*. Johannesburg. 55–70. Amsterdam: A.A. Balkema.
Ross-Brown, D.M. 1979. Pit limit slope design – analytical design. In: *Open Pit Mine Planning and Design* (J.T. Crawford and W.A. Hustrulid, editors): 161–184. SME-AIME.
Savely, J.P. 1986. Designing a final wall blast to improve stability. Presented at the SME Annual Meeting, New Orleans, March 2–6, 1986. SME: preprint no. 86–50.
Seegmiller, B.L. 1976. Optimum slopes for future open pit mines: How to obtain them using a rock mechanics approach. Presented at 1976 Fall SME-AIME Meeting, Denver, CO, Sept. 1–3, 1976: preprint 76-F-326.
Seegmiller, B.L. 1978. How to cut risk of slope failure in designing optimum pit slopes. *E/MJ Operating Handbook of Surface Mining* 2: 92–98.
Seegmiller, B.L. 1979. Pit limit slope design – general comments, data collection remedial stability measures. In: *Open Pit Mine Planning and Design* (J.T. Crawford and W.A. Hustrulid, editors): 161–184. SME-AIME.
Seelye, E.E. 1945. *Design Data Book for Civil Engineers*. Volume I. New York: John Wiley.
Sharon, R., Rose, N., & M. Rantapaa. 2005. Design and development of the northeast layback of the Betze-Post open pit. Pre-print Paper 05–009. 2005 SME Annual Meeting, Feb 28–Mar 2. SLC, UT.
Soderberg, R.L. & R.A. Busch 1977. Design guide for metal and nonmetal tailings disposal. Bureau of Mines Information Circular 8755.
Spang, R.M. 1987. Protection against rockfall – stepchild in the design of rock slopes. In: *Proceedings 6th Int. Conference on Rock Mechanics*: 551–557. Rotterdam: A.A. Balkema.
Steffen, O.K.H., W. Holt & V.R. Symons 1970. Optimizing open pit geometry and operational procedure. In: *Planning Open Pit Mines* (P.W.J. Van Rensburg, editor): 9–31. Johannesburg: SAIMM/A.A. Balkema.
Stewart, R.M. & B.A. Kennedy 1971. The role of slope stability in the economics, design and operation of open pit mines. In: *Stability in Open Pit Mining*. SME.
Strachan, C. 2001. Tailings dam performance from USCOLD incident-survey data. Mining Engineering. 53(3): 49–53.
Tamrock 1978. *Handbook of Surface Drilling and Blasting*: 14, J.F. Olán Oy.
Taylor, D.W. 1948. *Fundamental of Soil Mechanics*. New York: John Wiley.
Taylor, J.B. 1971. Incorporation of access roads into computer-generated open pits.
Taylor, P.F. & P.A. Hurry 1986. Design, construction and maintenance of haul roads. In: *The Planning and Operation of Open-Pit and Strip Mines* (J.P. Deetlefs, editor): 137–150. Johannesburg: SAIMM.
Thompson, R.J., & A.T. Visser. 2000. The functional design of surface mine haul roads. JSAIMM. 100(3): 169–180.
Williamson, O.C. 1987. Haul road design for off-highway mining equipment. *World Mining Equipment* 12(3/4): 24–26.
Winkle, R.F. 1976a. Development and maintenance of haulroads in openpit mines. Paper presented at the AIME Annual Meeting Las Vegas, Nevada, Feb. 22–26, 1976. Preprint No. 76-AO-5.
Winkle, R.F. 1976b. Guides to design and control of efficient truck and shovel operations in open-pit mines. M.S. Thesis, Univ. of Arizona.

CANMET Pit Slope Manual – Table of Contents.

*Chapter	Title	Catalogue number
Chapter 1	Summary	M38-14/1-1976
Chapter 2	Structural geology	M38-14/2-1981E
Supplement 2.1	DISCODAT program package	M38-14/2-1981-1E
Supplement 2.2	Domain analysis programs	M38-14/2-1977-2
Supplement 2.3	Geophysics for open pit sites	M38-14/2-1981-3E
Supplement 2.4	Joint mapping by terrestrial photogrammetry	M38-14/2-1977-4
Supplement 2.5	Structural geology case history	M38-14/2-1977-5
Chapter 3	Mechanical properties	M38-14/3-1977
Supplement 3.1	Laboratory classification tests	M38-14/3-1977-1
Supplement 3.2	Laboratory tests for design parameters	M38-14/3-1977-2
Supplement 3.3	In situ field tests	M38-14/3-1977-3
Supplement 3.4	Selected soil tests	M38-14/3-1977-4
Supplement 3.5	Sampling and specimen preparation	M38-14/3-1977-5
Chapter 4	Groundwater	M38-14/4-1977
Supplement 4.1	Computer manual for seepage analysis	M38-14/4-1977-1
Chapter 5	Design	M38-14/5-1979
Supplement 5.1	Plane shear analysis	M38-14/5-1977-1
Supplement 5.2	Rotational shear sliding: analyses and computer programs	M38-14/5-1981-2E
Supplement 5.3	Financial computer programs	M38-14/5-1977-3
Chapter 6	Mechanical support	M38-14/6-1977
Supplement 6.1	Buttresses and retaining walls	M38-14/6-1977-1
Chapter 7	Perimeter blasting	M38-14/7-1977
Chapter 8	Monitoring	M38-14/8-1977
Chapter 9	Waste embankments	M38-14/9-1977
Chapter 10	Environmental planning	M38-14/10-1977
Supplement 10.1	Reclamation by vegetation Vol. 1 – mine waste description and case histories	M38-14/10-1977-1
Supplement 10.1	Reclamation by vegetation Vol. 2 – mine waste inventory by satellite imagery	M38-14/10-1977-1-2

Slope Stability in Surface Mining (W.A. Hustrulid, M.K. McCarter and D. Van Zyl, editors) 2000 SME, Littleton, CO, USA.

Included:

ROCK SLOPE DESIGN CONSIDERATIONS

Hoek, E., Rippere, K.H., and P.F. Stacey. Large-scale slope designs – A review of the state of the art. Pp 3–10.
Nicholas, D.F., and D.B. Sims. Collecting and using geologic structure data for slope design. Pp 11–26.
Ryan, T.M., and P.R. Pryor. Designing catch benches and interramp slopes. Pp 27–38.
Call, R.D., Cicchini, P.F., Ryan, T.M., and R.C. Barkley. Managing and analyzing overall pit slopes. Pp 39–46.
Sjöberg, J. A slope height versus slope angle database. Pp 47–58.
Hoek, E., and A. Karzulovic. Rock-mass properties for surface mines. Pp 59–70.
Sjöberg, J. Failure mechanisms for high slopes in hard rock. Pp 71–80.
Zavodni, Z.M. Time-dependent movements of open-pit slopes. Pp 81–88.
Atkinson, L.C. The role and mitigation of groundwater in slope stability. Pp 89–96.
Glass, C.F. The influence of seismic events on slope stability. Pp 97–106.
Pariseau, W.G. Coupled geomechanic-hydrologic approach to slope stability based on finite elements. Pp 107–114.
Lorig, L., and P. Varona. Practical slope-stability analysis using finite-difference codes. Pp 115–124.
Hagan, T.N., and B. Bulow. Blast designs to protect pit walls. Pp 125–130.
Cunningham, C. Use of blast timing to improve slope stability. Pp 131–134.
Burke, R. Large-diameter and deep-hole presplitting techniques for safe wall stability. Pp 135–138.

CASE STUDIES IN ROCK SLOPE STABILITY

Flores, G., and A. Karzulovic. The role of the geotechnical group in an open pit: Chuquicamata Mine, Chile. Pp 141–152.
Valdivia, C., and L. Lorig. Slope stability at Escondida Mine. Pp 153–162.
Swan, G., and R.S. Sepulveda. Slope stability at Collahuasi. Pp 163–170.
Apablaza, R., Farías, E., Morales, R., Diaz, J., and A. Karzulovic. The Sur Sur Mine of Codelco's Andina Division. Pp 171–176.
Stewart, A., Wessels, F., and S. Bird. Design, implementation, and assessment of open-pit slopes at Palabora over the last 20 years. Pp 177–182.
Sjöberg, J., Sharp, J.C., and D.J. Malorey. Slope stability at Aznalcóllar. Pp 183–202.
Sjöberg, J., and U. Norström. Slope stability at Aitik. Pp 203–212.
Rose, N.D., and R.P. Sharon. Practical rock-slope engineering designs at Barrick Goldstrike. Pp 213–218.
Sharon, R. Slope stability and operational control at Barrick Goldstrike. Pp 219–226.
Jakubec, J., Terbrugge, T.J., Guest, A.R., and F. Ramsden. Pit slope design at Orapa Mine. Pp 227–238.
Pierce, M., Brandshaug, T., and M. Ward. Slope stability assessment at the Main Cresson Mine. Pp 239–250.

Kozyrev, A.A., Reshetnyak, S.P., Maltsev, V.A., and V.V. Rybin. Analysis of stability loss in open-pit slopes and assessment principles for hard, tectonically stressed rock masses. Pp 251–256.

Seegmiller, B. Coal mine highwall stability. Pp 257–264.

STABILITY OF WASTE ROCK EMBANKMENTS

Hawley, P.M. Site selection, characterization, and assessment. Pp 267–274.

Williams, D.J. Assessment of embankment parameters. Pp 275–284.

Campbell, D.B. The mechanism controlling angle-of-repose stability in waste rock embankments. Pp 285–292.

Beckstead, G.R.F., Slate, J., von der Gugten, N., and A. Slawinski. Embankment hydrology storage water controls. Pp 293–304.

Wilson, G.W. Embankment hydrology and unsaturated flow in waste rock. Pp 305–310.

Eaton, T. Operation and monitoring considerations from a British Columbia mountain terrain perspective. Pp 311–322.

Renteria, R.A. Reclamation and surface stabilization. Pp 323–328.

Walker, W.K., and M.J. Johnson. Observational engineering for open-pit geotechnics: A case study of predictions versus performance for the stability of a high overburden embankment over a soft/deep soil foundation at PT Freeport Indonesia's Grasberg open-pit mine. Pp 329–344.

Zeitz, B.K. Construction and operation of a major mined-rock disposal facility at Elkview Coal Corporation, British Columbia. Pp 345–350.

Gerhard, W.L. Steepened spoil slopes at Bridger Coal Company. Pp 351–360.

Buck, B. Design objectives for mine waste rock disposal facilities at phosphate mines in southeastern Idaho. Pp 361–363.

TAILINGS AND HEAP LEACHING

Davies, M., Martin, T., and P. Lighthall. Tailings dam stability: Essential ingredients for success. Pp 365–378.

Oboni, F., and I. Bruce. A database of quantitative risks in tailing management. Pp 379–382.

Blight, G. Management and operational background to three tailings dam failures in South Africa. Pp 383–390.

Welch, D.F. Tailings basin water management. Pp 391–398.

Gowan, M., and G. Fergus. The Gold Ridge mine tailings storage facility: An Australian case history. Pp 399–404.

Verduga, R., Andrade, C., Barrera, S., and J. Lara. Stability analysis of a waste rock dump of great height founded over a tailings impoundment in a high seismicity area. Pp 405–410.

East, D.R., and J.F. Valera. Stability issues related to tailing storage and heap leach facilities. Pp 411–418.

Lupo, J.F., and Terry Mandziak. Case study: Stability analysis of the Cresson Valley leach facility (Cripple Creek and Victor Gold Mining Company). Pp 419–426.

Andrade, C., Bard E., Garrido, H., and J. Campaña. Radomiro Tomic secondary heap leach facility. Pp 427–434.

Smith, M.E., and J.P. Giroud. Influence of the direction of ore placement on the stability of ore heaps on geomembrane-lined pads. Pp 435–438.

REVIEW QUESTIONS AND EXERCISES

1. Summarize the steps in the development of an open pit mine.
2. What "geometries" are involved in pit development?
3. Define or describe the following terms:
   - bench height
   - crest
   - toe
   - bench face angle
   - back break
   - bench floor
   - bench width
   - working bench
   - cut
   - safety bench/catch bench
   - double benches
   - berms
   - angle of repose
4. What are the purposes of safety benches?
5. What is the width of a safety bench?
6. What is the function of a safety berm?
7. What are some typical guidelines for a safety berm?
8. Discuss some of the aspects that enter into bench height selection?
9. Draw a sequence of three benches. Label the crest, toe, bench face angle, bench width and bank width.
10. If the bench face angle is 69°, the bench width is 30 ft and the bench height is 45 ft, determine the overall slope angle.
11. What happens if due to poor excavation practices the actual bench face angle is 66° instead? There are two possibilities to be considered.
12. Discuss the significance of figure 4.3.
13. What are the purposes of safety benches?
14. Discuss the pro's and con's of double benching. Discuss the practical actions required to create double benches.
15. In actual surface mining operations, what happens to the "catch" benches created during the general operations? How does this affect their function? What actions might need to be taken?
16. Call has suggested the catch bench geometries shown in Figure 4.7 and in Table 4.1. Draw the geometry suggested for a 30 m (double bench) height. What would be the corresponding final slope?
17. Discuss the pro's and con's of higher versus lower bench heights.
18. The dimensions for a Bucyrus Erie (BE)9 yd3 shovel are given in Figure 4.9. Some comparable dimensions for larger BE shovels are given below:

Dipper capacity (yd3)	Dimension						
	A	B	D	E	G	H	I
12	22'–3"	40'–9"	37'–6"	50'–0"	34'–0"	7'–1"	39'–3"
20	30'–0"	55'–0"	50'–6"	65'–5"	44'–9"	9'–3"	50'–1"
27	30'–6"	57'–0"	48'–6"	67'–6"	44'–3"	6'–3"	52'–0"

Based upon the shovel geometries, what would be an appropriate maximum bench height for each?

384 Open pit mine planning and design: Fundamentals

19. In the largest open pit operations today, the BE 495 or P&H 4100 shovels are being used. Obtain dimensions similar to those given in the table in problem 18 for them.
20. Using Figure 4.10, what would be the reach height for a shovel with a 56 yd3 dipper capacity?
21. The dipper capacities which are provided for a particular shovel model, normally are based upon material with a density of 3000 lbs/yd3. What is done if the particular shovel model is used to dig coal? To dig magnetite?
22. Summarize the steps which would be followed in considering the appropriate bench geometry.
23. Compare the digging profile for the shovel shown in Figure 4.9 with a bench face drawn at a 65° angle. What is your conclusion?
24. Summarize the discussion of ore access as presented in section 4.3.
25. A drop cut example has been presented based upon the 9 yd3 shovel. Rework the example assuming that the 27 yd3 capacity is used instead. Find the minimum and maximum cut widths. Select an appropriate Caterpillar truck to be used with this shovel (use the information on their website).
26. What would be the volume of the ramp/drop cut created in problem 25?
27. Discuss the different aspects which must be considered when selecting the ore access location.
28. Figure 4.23 shows the situation where the ramp construction is largely in ore. Assume that the diameter of the orebody is 600 ft, the ramp width is 100 ft, the road grade is 10% and the bench height is 40 ft. Determine the approximate amount of ore removed in Figure 4.24.
29. Determine the amount of waste that would be removed if the ramp in Figure 4.24 was entirely constructed in waste.
30. Summarize the factors associated with the ramp location decision.
31. Figures 4.21 through 4.25 show the addition of a ramp to a pit. In this new case the orebody is assumed to be 600 ft in diameter, the road is 100 ft wide and the grade is 10%. Using AutoCad redo the example. What is the final ramp length? As shown, a flat portion 200 ft in length has been left between certain ramp segments.
32. Once the access to the new pit bottom has been established, discuss the three approaches used to widen the cut.
33. Summarize the steps used to determine the minimum required operating room when making parallel cuts.
34. What is the difference between the single and double spotting of trucks? Advantages? Disadvantages?
35. Redo the cut sequencing example described in section 4.4.6 assuming a cut width of 150 ft. If the shovel production rate is 50,000 tpd, how long would it take to exhaust the pushback?
36. A pit is enlarged using a series of pushbacks/laybacks/expansions. What would be the minimum and maximum cut widths using the 27 yd3 shovel and Caterpillar model 789 trucks, assuming single pass mining? Work the problem assuming both single and double spotting of trucks.
37. Discuss the advantages/disadvantages of double spotting.
38. Redo the example described in section 4.4.5 assuming Caterpillar 993 trucks and the P&H 4100 shovel. The bench height is 50 ft and the bench face angle is 70°.

39. Assume that five 50 ft high benches are being worked as a group (Figure 4.51). Using the data from Problem 38, what would the working slope angles be when the working bench is at level 2?
40. For the slopes identified as "critical" in Table 4.2, what types of actions should be taken?
41. How might the design slope angles change during the life of a mine?
42. How do the expected stress conditions in the walls and floor of the pit change as the pit is deepened. Make sketches to illustrate your ideas.
43. What might happen at the pit bottom for the situation shown in Figure 4.61?
44. What are the four most common types of slope failure? Provide a sketch of each.
45. Assume that a 50 ft high bench is as shown in Figure 4.63. The layering goes through the toe. Assume that the following apply:
    $\phi = 32°$
    $c = 100 \, \text{kPa}$
    $\rho = 2.45 \, \text{g/cm3}$
    Bench face angle $= 60°$
    Bedding angle $= 20°$
    What is the safety factor?
46. In section 4.6.3 it was determined that for the given conditions the safety factor was 1 ($F = 1$) for a slope of height 140 ft and a slope angle of 70°. Determine the minimum slope angle if the slope height is 200 ft instead.
47. Redo problem 46 if the required safety factor is 1.2. (See Figure 4.65).
48. Redo problem 46 assuming the presence of a tension crack of length 20 ft. You should consider the two extreme cases: (a) Crack dry; (b) Crack filled with water.
49. What is the safety factor for a slope assuming that the following apply:
    $H = 200 \, \text{ft}$
    Density $= 165 \, \text{lb/ft3}$
    $c = 825 \, \text{lb/ft2}$
    $Z_o = 50 \, \text{ft}$
    $H_w = 100 \, \text{ft}$
    $i = 40°$
    $\beta = 30°$
    $\phi = 30°$
50. A waste dump 500 ft high is planned. It is expected that the face angle will be 50°, the density is assumed to be 1.4 g/cm3, the cohesion $= 0$ and the friction angle is 30°. Will it be stable?
51. What techniques might be used to obtain values for ϕ and c appropriate for the materials making up slopes?
52. Discuss the effect of slope wall curvature (in plan) on stability.
53. A discussion of the Afton copper/gold mine has been presented. Check the literature/Internet to see what eventually happened to the slopes.
54. In Figure 4.72, if the entire pit consisted of Structural Region B, what would happen to the North and East walls?
55. Assume that the conical pit shown in Figure 4.74 has a bottom radius of 100 ft. Redraw the figure showing the volume-slope dependence.
56. Discuss the pros' and con's of the different ways of representing bench positions on a plan map.
57. Redo the example in Figure 4.88 using AutoCad.

386 *Open pit mine planning and design: Fundamentals*

58. List some of the important questions that must be answered when siting a road.
59. The steps in the design of a spiral road inside the wall have been presented in section 4.8.2. Redo the example assuming a grade of 8%, a bench height of 40 ft, a bench face angle of 65° and a crest to crest dimension of 80 ft. The road width is 100 ft.
60. Redo problem 58 using AutoCad.
61. Redo the example in section 4.8.3 using AutoCad.
62. Redo the example in section 4.8.4 using AutoCad.
63. Why is it generally desirable to avoid switchbacks in a pit? Under what conditions might it become of interest?
64. If the conditions are such that a switchback becomes interesting, what should the planner do?
65. Assume that a conical pit has a depth of 500 m and the overall slope angle is 38°. Consideration is being given to adding a second access. The road would have a width of 40 m. How much material would have to be mined? Follow the approach described in section 4.8.5.
66. Using the data for the Caterpillar 797 haulage truck, answer the following questions:
    a. Load distribution of the front and rear tires.
    b. Contact area for an inflation pressure of 80 psi.
    c. What wheel loading should be used in the road design?
67. In the design and construction of a mine haulage road, what layers are involved? Describe each one starting at the lowest layer.
68. Describe the considerations in determining layer thickness.
69. How do you include the effect of dual-wheel loading?
70. What is meant by the California Bearing Ratio? How is it determined?
71. Suggest a haulage road width for two-way traffic involving Cat 797 trucks.
72. The following road cross-section dimensions apply at a particular mining operation which uses Komatsu 930 trucks:
    - safety berm width = 3.5 m
    - truck width = 7.3 m
    - space between trucks = 5.0 m
    - width of drainage ditch = 2.0 m
    - overall road width = 25 m
    - bench face angle = 75°

    Draw the section. How well does this design correspond to the "rules"?
73. Why are well-designed roads important?
74. How is a runaway ramp constructed?
75. What would happen to the road section shown in Figure 4.117 with the passage of a fully-loaded Cat 793 haulage truck? Be as specific as possible.
76. Assume that the sub-grade is a compact sand-clay soil and that you have the following construction materials: well-graded crushed rock, sand, well-graded gravel. What thicknesses would you recommend when using the Cat 793 trucks? Assume all of the materials have the same cost.
77. What is meant by poorly graded material? Well graded material?
78. In the design of a straight haulage road segment, what major factors should be considered?

79. Would the road design shown in Figure 4.118 apply for the Cat 793 truck? Why or why not?
80. What is the design rule for roadway width assuming two-way traffic?
81. List the steps to be followed in selecting a road design width.
82. What is meant by cross-slope? What are the rules involved? Why is it used?
83. If you were a haulage truck driver, what effect would cross-slope have on you?
84. What is meant by centrifugal force? How does it apply to road design?
85. What special design procedures must be applied to curves? What is meant by super-elevation?
86. What is the effect of rain and snow on super-elevated road segments?
87. Assume a curve radius of 50 ft (inner edge of pavement) and two-lane traffic. The truck is a Cat 793. What should be the minimum width? What should be the super-elevation assuming a curve speed of 15 mph?
88. What are transition zones?
89. Summarize the rules regarding the need for berms/guardrails.
90. What are the two principal berm designs in common use today?
91. On an elevated road involving Cat 793 haulage trucks, suggest a parallel berm design.
92. Summarize the rules presented by Couzens regarding haulage road gradients.
93. Summarize the practical road building and maintenance tips offered by Winkle.
94. Define what is meant by:
    a. Stripping ratio
    b. Overall stripping ratio
    c. Instantaneous stripping ratio
95. Indicate some of the different units used to express stripping ratio for different mined materials.
96. Redo the example shown in Figure 4.127 assuming that:
    – ore width = 400 ft
    – bench height = 40 ft
    – minimum pit width = 200 ft
    – original pit depth = 280 ft (7 benches)
    Calculate the appropriate values for SR(instantaneous) and SR(overall) for the mining of benches 8, 9, and 10.
97. What is meant by the term push-back?
98. Discuss the concepts of geometric sequencing. What is meant by a "phase"?
99. Discuss the different possibilities involved in pit sequencing.

CHAPTER 5

Pit limits

5.1 INTRODUCTION

The time has now come to combine the economics introduced in Chapter 2 with the mineral inventory developed in Chapter 3 under the geometric constraints discussed in Chapter 4 to define the mineable portion of the overall inventory. The process involves the development and superposition of a geometric surface called a pit onto the mineral inventory. The mineable material becomes that lying within the pit boundaries. A vertical section taken through such a pit is shown in Figure 5.1. The size and shape of the pit depends upon economic factors and design/production constraints. With an increase in price the pit would expand in size assuming all other factors remained constant. The inverse is obviously also true. The pit existing at the end of mining is called the 'final' or the 'ultimate' pit. In between the birth and the death of an open-pit mine, there are a series of 'intermediate' pits. This chapter will present a series of procedures based upon:
 (1) hand methods,
 (2) computer methods, and
 (3) computer assisted hand methods

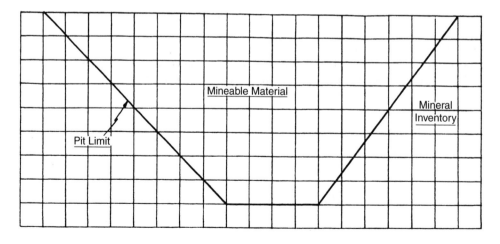

Figure 5.1. Superposition of a pit onto a mineral inventory.

Pit limits 389

for developing pit limits. Within the pit are found materials of differing value. Economic criteria are applied to assign destinations for these materials based on their value (i.e. mill, waste dump, leach dump, stock pile, etc.). These criteria will be discussed. Once the pit limits have been determined and rules established for classifying the in-pit materials, then the ore reserves (tonnage and grade) can be calculated. In Chapter 6, the steps required to go from the ore reserve to production rate, mine life, etc. will be presented.

5.2 HAND METHODS

5.2.1 *The basic concept*

Figure 5.2 shows an idealized cross-section through an orebody which outcrops at the surface and dips to the left at 45°. There are distinct physical boundaries separating the ore from the over- and under-lying waste. The known ore extends to considerable depth down dip and this will be recovered later by underground techniques. It is desired to know how large the open-pit will be. The final pit in this greatly simplified case will appear as in Figure 5.3. The slope angle of the left wall is 45°. As can be seen a wedge of waste (area A) has been removed to uncover the ore (area B). The location of the final pit wall is determined by examining a series of slices such as shown in Figure 5.4.

For this example the width of the slice has been selected as 1.4 units (u) and the thickness of the section (into the page) as 1 unit. Beginning with strip 1 the volumes of waste (V_w) and ore (V_o) are calculated. The volumes are:

Strip 1:

$$V_{w1} = 7.5u^3$$

$$V_{o1} = 5.0u^3$$

The instantaneous stripping ratio (ISR) is defined as

$$\text{ISR}_1 \text{ (instantaneous)} = \frac{V_{w1}}{V_{o1}} \tag{5.1}$$

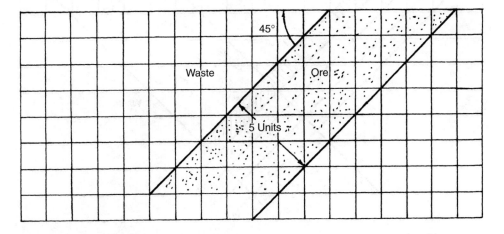

Figure 5.2. Cross-section through an idealized orebody.

390 Open pit mine planning and design: Fundamentals

Hence

$$\text{ISR}_1 = 1.50$$

Assuming that the net value from selling one unit volume of ore (that money remaining after all expenses have been paid) is $1.90 and the cost for mining and disposing of the waste is $1/unit volume, the net value for strip 1 is

$$\text{NV}_1 = 5.0 \times \$1.90 - 7.5 \times \$1 = \$2.00$$

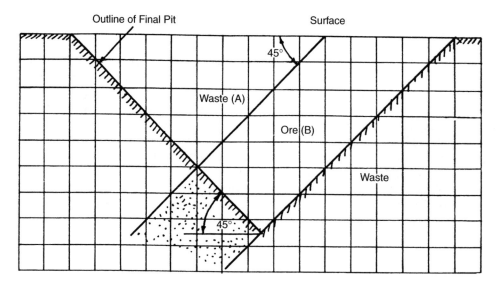

Figure 5.3. Diagrammatic representation of the final pit outline on this section.

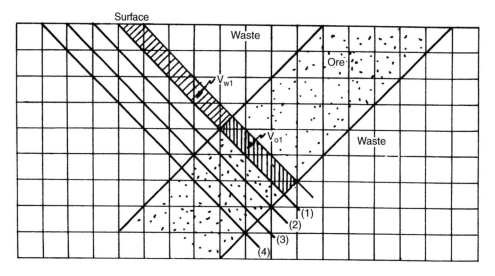

Figure 5.4. Slices used to determine final pit limits.

If the process is now repeated for strips 2, 3 and 4, the results are as given below:

Strip 2:
$$V_{w2} = 8.4u^3$$
$$V_{o2} = 5.0u^3$$
$$ISR_2 = 1.68$$
$$NV_2 = 5.0 \times \$1.90 - 8.4 \times \$1 = \$1.10$$

Strip 3:
$$V_{w3} = 9.45u^3$$
$$V_{o3} = 5.0u^3$$
$$ISR_3 = 1.89$$
$$NV_3 = 5.0 \times \$1.90 - 9.45 \times \$1 = \$0.05 \cong \$0$$

Strip 4:
$$V_{w4} = 10.5u^3$$
$$V_{o4} = 5.0u^3$$
$$ISR_4 = 2.10$$
$$NV_4 = 5.0 \times \$1.90 - 10.5 \times \$1 = -\$1.00$$

As can be seen, the net value changes from (+) to (−) as the pit is expanded. For strip 3, the net value is just about zero. This pit position is termed 'breakeven' since the costs involved in mining the strip just equal the revenues. It is the location of the final pit wall. The breakeven stripping ratio which is strictly applied at the wall is

$$SR_3 = 1.9$$

Since the net value of 1 unit of ore is $1.90 and the cost for 1 unit of waste is $1, one can mine 1.9 units of waste to recover 1 unit of ore (Fig. 5.5).

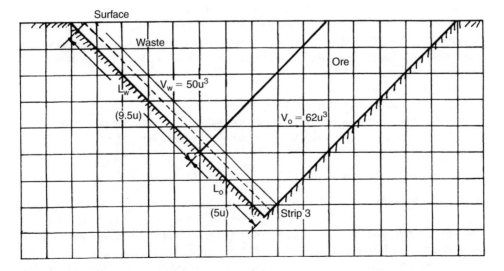

Figure 5.5. Final pit outline showing ore-waste distribution.

The overall stripping ratio (OSR) for this section is calculated as

$$\text{OSR} = \frac{\text{Waste area}}{\text{Ore area}} = \frac{A}{B} \tag{5.2}$$

In this case

$$\text{Waste area} = A = 50u^3$$
$$\text{Ore area} = B = 62u^3$$

hence

$$\text{OSR} \cong 0.8$$

This is compared to the instantaneous stripping ratio at the pit boundary

$$\text{ISR (pit limit)} = 1.9$$

The OSR must always be less than the ISR (pit limit).
The net value for the section (assuming unit thickness) is

$$\text{NV} = \text{Ore area} \times \text{Net ore value} - \text{Waste area} \times \text{Waste removal cost}$$
$$= B \times \$1.90 - A \times \$1 = 62 \times \$1.90 - 50 \times \$1 = \$68$$

Whereas the net value is zero at the pit limit, it is positive for the overall section.

In this example the quantities, costs and revenues were all expressed in terms of volumes. Since the strip width and thickness is the same in both ore and waste, the final pit limit in this situation is that position where the *length* of waste (L_w) is just equal to 1.9 times the *length* of ore (L_o) as measured along the midline of the mined strip.

Often the costs/revenues are expressed as a function of weight (\$/ton). If the density of the ore and waste is the same then the ratio of lengths can still be used. If they are not, then the different densities must be included in the calculations.

As this chapter proceeds, more realistic geometries both for the pit and the orebody will be introduced. A gradual rather than sharp ore-waste transition will be included. As will be seen, even with these changes the following basic steps involved in determining pit limits remain the same:

1. A slice is selected.
2. The contained value is compared with the costs.
3. If the net value is positive, the pit can be expanded. If negative, the pit contracts.
4. The final pit position is where the net value of the slice is zero.

5.2.2 *The net value calculation*

In the previous section, the nature of the deposit was such that there was no ambiguity regarding what was meant by ore and waste. For many deposits however the distinction is much more subtle. The term 'cutoff grades' refers to grades for which the destination of pit materials changes. It should be noted that 'grades' were used rather than 'grade' since there may be several possible destinations. The simplest case would be that in which there are two destinations: the mill or the waste dump. One cutoff grade is needed. For many operations

Pit limits 393

today there are three possible destinations: the mill, the leach dump and the waste dump. Each of the decisions
- mill or leach?
- leach or waste?

requires a cutoff grade. A definition of cutoff grade which is often used (Davey, 1979):

Cutoff grade = the grade at which the mineral resource can no longer be processed at a profit.

applies to the simple ore-waste decision. This will be used in developing the preliminary pit limits. The only destinations allowed are the waste dump or further processing. With this definition the net value of material as a function of grade must be determined. That grade for which the net value is zero is called the breakeven cutoff grade. This calculation will be illustrated using the example provided by Davey (1979) for copper. The copper ore is milled thereby producing a copper concentrate. This mill concentrate is shipped to a smelter and the resulting blister copper is eventually refined.

In this example the following will be assumed:

Mill recovery rate = 80%

Mill concentrate grade = 20%

Smelting loss = 10 lbs/st of concentrate

Refining loss = 5 lbs/st of blister copper

The steps of the net value computation are outlined below for an ore containing 0.55% copper. All of the costs and revenues will be calculated with respect to one ton of ore.

Step 1. Compute the amount of saleable copper (lbs/st of ore).

(a) Contained copper (CC) is

$$CC = 2000 \text{ lbs/st} \times \frac{0.55}{100} = 11.0 \text{ lb}$$

(b) Copper recovered by the mill (RM) is

$$RM = 11.0 \times \frac{80}{100} = 8.8 \text{ lb}$$

(c) Concentration ratio (r). The ratio of concentration is defined as

$$r = \frac{\text{lbs Cu/st of concentrate}}{\text{lbs Cu recovered/st of ore}} \tag{5.3}$$

Since the mill product runs 20% copper there are 400 lb of copper contained in one ton of concentrate. One ton of ore contains 8.8 lb of recoverable copper. Hence

$$r = \frac{400}{8.8} = 45.45$$

This means that 45.45 tons of ore running 0.55% copper are required to produce 1 ton of concentrate running 20%.

(d) Copper recovered by the smelter (RS). The mill concentrate is sent to a smelter. Since the smelting loss is 10 lb/st of concentrate, the smelting loss (SL) per ton of ore is

$$SL = \frac{10 \text{ lb/st of concentrate}}{45.45 \text{ tons of ore/st of concentrate}} = 0.22 \text{ lb}$$

Thus the recovered copper is

$$RS = 8.8 - 0.22 = 8.58 \text{ lb}$$

(e) Copper recovered by the refinery (RR). The number of tons of ore required to produce one ton of blister copper is

$$\frac{2,000 \text{ lb/st of blister copper}}{8.58 \text{ lb of copper/st of ore}} = 233.1$$

Since refining losses are 5 lb/st of blister copper, the refining loss (RL) per ton of ore is

$$RL = \frac{5 \text{ lb of copper/st of blister copper}}{233 \text{ tons of ore/st of blister copper}} = 0.02 \text{ lb}$$

Thus the recovered copper is

$$RR = 8.58 - 0.02 = 8.56 \text{ lb}$$

Step 2. Compute the gross value (GV) for the ore ($/st). The copper price assumed for this calculation is $1.00/lb. Furthermore there is a by-product credit for gold, molybdenum, etc. of $1.77/st of ore. Thus the gross value is

$$GV = 8.56 \times \$1 + \$1.77 = \$10.33$$

Step 3. Compute the associated total costs (TC) ($/st).

(a) Production (operating) costs (PC) excluding stripping are:

Mining	$1.00
Milling	$2.80
General and administration	$0.57
(15% of mining and milling)	
PC =	$4.37

(b) Amortization and depreciation (A&D). This amount is charged against each ton of ore to account for the capital investment in mine and mill plant. If the total A&D is $10,000,000 and overall ore tonnage is 50,000,000 tons, then this value would be $0.20. In this particular case, 20% of the total production costs will be used.

$$A\&D = 0.20 \times \$4.37 = \$0.87$$

(c) Treatment, refining and selling (TRS) cost.

– Shipment of mill concentrate to the smelter. Since transport costs $1.40 per ton of concentrate, the cost per ton of ore is

$$\text{Concentrate transport} = \frac{\$1.40}{45.45} = \$0.03$$

– Smelting cost. Smelting costs $50.00/st of concentrate. The smelting cost per ton of ore is

$$\text{Smelting} = \frac{\$50.00}{45.45} = \$1.10$$

– Shipment of the blister copper to the refinery. There is a transport cost of $50.00/st of blister copper involved. The cost per ton of ore becomes

$$\text{Blister transport} = \$50.00 \frac{8.58}{2000} = \$0.21$$

– Refining cost. Refining costs $130.00/st of blister copper. The refining cost per ton of ore is

$$\text{Refining} = \$130.00 \frac{8.58}{2000} = \$0.56$$

– Selling and delivery cost (S&D). The selling and delivery cost is $0.01/lb of copper. Since 8.56 lb are available for sale

$$\text{S\&D} = \$0.09$$

– General plant (GP) cost. These costs amount to $0.07/lb of copper. Hence the GP cost per ton of ore is

$$\text{GP} = \$0.07 \times 8.56 = \$0.60$$

– Total treatment cost is

$$\text{TRS} = \$2.59$$

(d) Total cost per ton of ore is

$$\text{TC} = \$7.83$$

Step 4. Compute net value per ton of ore. The net value is the gross value minus the total costs. Thus for an initial copper content of 0.55%, the net value becomes

$$\text{NV} = \text{GV} - \text{TC} = \$10.33 - \$7.83 = \$2.50$$

Step 5. Compute the net value for another ore grade. Steps 1 through 4 are now repeated for another copper content. In this case 0.35% Cu has been chosen. The by-product credit will be assumed to vary directly with the copper grade. Hence

$$\text{By-product credit} = \$1.77 \frac{0.35}{0.55} = \$1.13$$

Assuming the recoveries and unit costs remain the same, the net value is −$0.30.

Step 6. Construct a net value – grade curve. The two points on a net value – grade curve which have been determined by the process outlined above

Point	Net value ($/st)	Grade (% Cu)
1	$2.50	0.55
2	−$0.30	0.35

are plotted in Figure 5.6. Assuming that the net value – % Cu relationship is linear it is possible to find an equation of the form $y = a + bx$ relating net value (y) to grade (x). The result is

$$y = -\$5.20 + \$14.0x$$

where y is the net value ($/st of ore) and x is the percent copper.

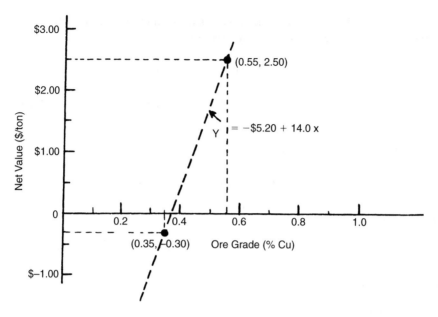

Figure 5.6. Net value – ore grade curve.

Step 7. Determine the breakeven cutoff grade (for application at the pit limit). The breakeven cutoff grade is defined as that grade for which the net value is zero. One can determine that point by inspecting Figure 5.6 or by solving the equation found in Step 6 for $y = 0$. One finds that

$$x\,(\text{breakeven}) = 0.37\%\ \text{Cu}$$

Step 8. Developing a stripping ratio – grade curve. The cutoff grade distinguishes that material which can be mined and processed with a net value greater than or equal to zero. Material with a zero net value cannot pay for any stripping. Thus it must be exposed at the surface or be overlying richer blocks which can pay for the required stripping. Assume that the cost for stripping 1 ton of waste is $1.00. Ore with a net value of $1.00 can pay for the stripping of 1 ton of waste. Ore with a net value of $2.00/ton can pay for the stripping of 2 tons of waste, etc. The stripping ratio axis has been added in Figure 5.7 to show this. The net value – grade equation

$$\text{NV} = -\$5.20 + \$14.00 \times (\%\ \text{Cu}) \tag{5.4}$$

can be modified to yield the stripping ratio (SR) – grade relationship

$$\text{SR} = -\$5.20 + \$14.00 \times (\%\ \text{Cu}) \tag{5.5}$$

For a grade of 0.55% Cu, the breakeven stripping ratio is

$$\text{SR}\,(0.55\%) = 2.5$$

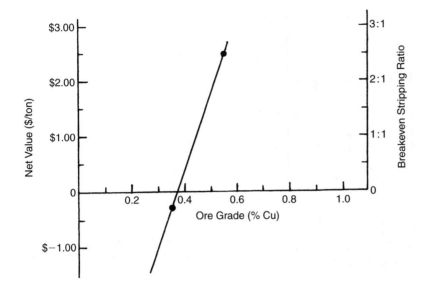

Figure 5.7. Net value and breakeven stripping ratio versus ore grade.

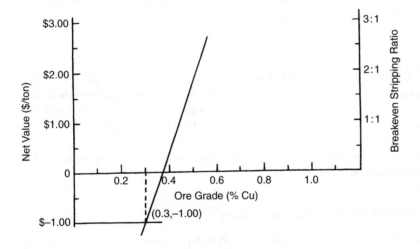

Figure 5.8. Addition of the minimum value portion.

This was as expected since the net value is $2.50 and the stripping cost is $1.00

$$SR = \frac{\$2.50}{\$1.00} = 2.50$$

Step 9. Presenting the final curves. The net value – grade curve should be completed by the addition of the cost of stripping line (SC). This is shown in Figure 5.8. It should be noted that no material can ever have a value less than that of waste. In this case

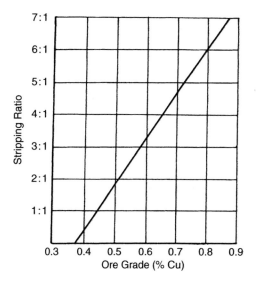

Figure 5.9. Stripping ratio – ore grade curve used for pit limit determinations (Koskiniemi, 1979).

the value of waste is $-\$1.00$. The horizontal line (NV $= -\$1.00$) and the NV-grade line (NV $= -5.20 + 14.00 \times$ (% Cu)) intersect at a grade of

$$\% \text{ Cu} = 0.30$$

For grades less than 0.30%, the material is considered as waste with respect to milling. Depending upon the economics, some other treatment process such as dump leaching may be possible. When using hand methods all material having grades less than the breakeven cutoff (0.37% in this case) is considered as waste. The final stripping ratio-grade curve is shown in Figure 5.9.

For the computer techniques discussed later, the portion of the curve lying between 0.30 and 0.37% Cu is also included.

5.2.3 Location of pit limits – pit bottom in waste

The application of this curve to locating the final pit wall positions will be illustrated using the vertical section (Fig. 5.10) taken through the block model. The basic process used has been presented by Koskiniemi (1979). Locating the pit limit on each vertical section is a trial and error process. It will be assumed that
- Pit slopes:
  Left hand side $= 50°$;
  Right hand side $= 40°$;
- Minimum width of the pit bottom $= 100$ ft;
- Material densities:
  Ore $= 165$ lb/ft^3;
  Waste rock $= 165$ lb/ft^3;
  Overburden $= 165$ lb/ft^3;

Pit limits 399

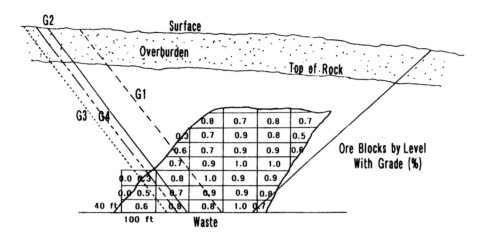

Figure 5.10. Pit limit determination with bottom in waste (Koskiniemi, 1979).

Table 5.1. Bottom in waste: pit limit guess #1 (Line G1).

Length (ft)			Ore grade (g_{oi})	Ore length × ore grade $(l_{oi}g_{oi})$
Overburden (l_{ob})	Waste (l_w)	Ore (l_{oi})		
130				
	296			
		30	0.6	18.0
		52	0.7	36.4
		52	1.0	52.0
		52	0.9	46.8
		52	0.8	41.6
130	296	238	$\bar{g}=0.82$	194.8

SR (actual) = $\frac{130+296}{238} \cong 1.79:1$;
SR (allowable) $\cong 6.2:1$;
Conclusion: move to the left.

- Relative mining characteristics:
  Waste rock = 1;
  Overburden = 1;
- The stripping ratio – ore grade curve of Figure 5.9 applies;
- The pit bottom is at the ore-waste contact.

Overburden as defined here means soil, glacial till, gravel, highly weathered rock, etc. not requiring drilling and blasting prior to removal. Waste rock, on the other hand, does require drilling and blasting. The general procedure will be demonstrated with respect to the left-hand slope.

Step 1. A trial slope (guess #1) is drawn through the section. The lengths and grades are entered into a table such as Table 5.1. The purpose will be to obtain the average ore grade and stripping ratio along this line. The lengths can simply be scaled off the section with enough accuracy. The cutoff grade is 0.37% Cu.

400 Open pit mine planning and design: Fundamentals

Table 5.2. Bottom in waste: pit limit guess #2 (Line G2).

Length (ft)			Ore grade (g_{oi})	Ore length × ore grade $(l_{oi}g_{oi})$
Overburden (l_{ob})	Waste (l_w)	Ore (l_{oi})		
130				
	385			
		52	0.8	41.6
		52	0.7	36.4
		52	0.8	41.6
130	385	156	$\bar{g}=0.77$	119.6

SR (actual) = $\frac{130+385}{156} \cong 33:1$;
SR (allowable) $\cong 5.6:1$;
Conclusion: move to the left.

Table 5.3. Bottom in waste: pit limit guess #3 (Line G3).

Length (ft)			Ore grade (g_{oi})	Ore length × ore grade $(l_{oi}g_{oi})$
Overburden (l_{ob})	Waste (l_w)	Ore (l_{oi})		
130				
	443			
		52	0.5	26.0
		52	0.8	41.6
130	443	104	$\bar{g}=0.65$	67.6

SR (actual) = $\frac{130+443}{104} \cong 5.51:1$;
SR (allowable) $\cong 3.9:1$;
Conclusion: move to the right.

Table 5.4. Bottom in waste: pit limit guess #4 (Line G4).

Length (ft)			Ore grade (g_{oi})	Ore length × ore grade $(l_{oi}g_{oi})$
Overburden (l_{ob})	Waste (l_w)	Ore (l_{oi})		
130				
	435			
		52	0.7	36.4
		52	0.8	41.6
130	435	104	$\bar{g}=0.75$	78.0

SR (actual) = $\frac{130+435}{104} \cong 5.43:1$;
SR (allowable) $\cong 5.4:1$;
Conclusion: final limit.

Step 2. The average ore grade is determined. The products of ore grade × ore length are determined ($l_{oi}g_{oi}$) and summed ($\sum(l_{oi}g_{oi})$). The sum of the ore lengths is found ($\sum l_{oi}$). The average ore grade $\bar{g}$ is found from

$$\bar{g} = \frac{\sum l_{oi}g_{oi}}{\sum l_{oi}} = 0.82\% \text{ Cu} \tag{5.6}$$

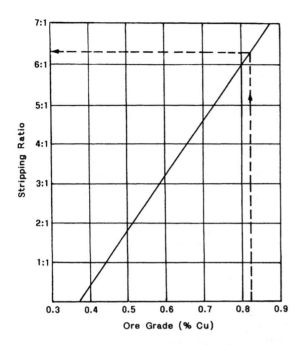

Figure 5.11. Allowable stripping ratio for a grade of 0.82%.

Step 3. The stripping ratio for this line is determined. This must be expressed in the same form as the stripping ratio – ore grade curve. In this case it is required to have tons of waste per ton of ore. Since the densities are all equal and the relative diggability of the overburden and the waste rock is the same, the stripping ratio is simply the ratio of the lengths.

$$SR = \frac{\text{Length overburden} + \text{Length waste rock}}{\text{Length ore}} = \frac{l_{ob} + l_w}{l_o}$$

$$SR = \frac{130 + 296}{238} \cong 1.79 \tag{5.7}$$

Step 4. Determination of the allowable stripping ratio for average ore grade using the SR – ore grade curve (Fig. 5.11). In this case one finds that

$$\text{SR (allowable)} \cong 6.2:1$$

Step 5. Comparison of the actual and allowable SR. Since the actual stripping ratio

$$\text{SR (actual)} = 1.79:1$$

is much less than that allowable (6.2 : 1), the pit slope can be moved to the left.

Step 6. A new guess of the final pit slope location is made and the process repeated. This iteration process is continued until the actual and allowable stripping ratios are close.

Conclusion: Guess #4 is the final position of the left hand slope. Note that the block having grade 0.3 is considered waste since it is below cutoff.

Step 7. Determination of right-hand slope position. The same process is repeated for the right hand wall of the pit. The results are shown in Figure 5.10. The final pit bottom has a width of about 215 ft.

Table 5.5. Bottom in ore: pit limit guess #1 (Line G1).

(a) Left hand side

Length (ft)			Ore grade (g_{oi})	Ore length × ore grade $(l_{oi}g_{oi})$
Overburden (l_{ob})	Waste (l_w)	Ore (l_{oi})		
130				
	330			
		48	0.7	33.6
		52	0.8	41.6
		13	0.7	9.1
		39	0.9	35.1
		22	0.8	17.6
		50	0.8	40.0
130	330	224	$\bar{g}=0.79$	177.0

SR (actual) = $\frac{130+330}{224} \cong 2.05$;
SR (allowable) $\cong 6:1$.

(b) Right hand side

Length (ft)			Ore grade (g_{oi})	Ore length × ore grade $(l_{oi}g_{oi})$
Overburden (l_{ob})	Waste (l_w)	Ore (l_{oi})		
143				
	283			
		48	0.6	28.8
	13			
		48	1.0	48.0
		70	0.9	63.0
		11	0.8	8.8
		57	0.9	51.3
		26	1.0	26.0
		39	1.0	39.0
		11	0.8	8.8
143	296	310	$\bar{g}=0.88$	273.7

SR (actual) = $\frac{143+296}{310} \cong 1.42:1$;
SR (allowable) $\cong 7:1$;
Conclusion: the pit can be 'floated' considerable deeper.

If the waste/overburden have densities different from the ore, then the calculation of stripping ratio using simple length ratios does not work. The generalized stripping ratio calculation becomes

$$SR = \frac{l_{ob}\rho_{ob} + l_w\rho_w}{l_o\rho_o} \qquad (5.8)$$

where ρ_{ob} is the overburden density, ρ_w is the waste density, and ρ_o is the ore density. If the mineability characteristics of the overburden and waste rock are different, then the costs involved in their removal will also be different. It will be recalled that a single waste mining

Table 5.6. Bottom in ore: pit limit guess #2 (Line G2).

(a) Left hand side

Length (ft)			Ore grade (g_{oi})	Ore length × ore grade $(l_{oi} g_{oi})$
Overburden (l_{ob})	Waste (l_w)	Ore (l_{oi})		
139				
	626			
		52	0.5	26.0
		52	0.6	31.2
		52	0.6	31.2
		52	0.6	31.2
		52	0.6	31.2
		52	0.6	31.2
139	626	258	$\bar{g}=0.58$	150.8

SR (actual) $= \frac{139+626}{258} \cong 2.97:1$;
SR (allowable) $\cong 2.95:1$.

(b) Right hand side

Length (ft)			Ore grade (g_{oi})	Ore length × ore grade $(l_{oi} g_{oi})$
Overburden (l_{ob})	Waste (l_w)	Ore (l_{oi})		
157				
	617			
		65	0.6	39.0
		61	0.6	36.6
		65	0.5	32.5
		61	0.5	30.5
		50	0.6	30.0
157	617	302	$\bar{g}=0.56$	168.6

SR (actual) $= \frac{157+617}{302} \cong 2.56:1$;
SR (allowable) $\cong 2.6:1$;
Conclusion: guess #2 is close to the pit slope location.

cost was used in the development of the SR – grade curves. Assume that the given cost (C_w/ton) applies to waste rock and that the overburden removal cost is αC_w/ton. The factor α is the relative mining cost of the overburden to that of the waste rock. The stripping ratio formula can be further modified to

$$SR = \frac{\alpha l_{ob} \rho_{ob} + l_w \rho_w}{l_o \rho_o} \tag{5.9}$$

If the cost/ton to remove the overburden is only half that of the waste rock then $\alpha = 0.5$. This factor changes the overburden into an equivalent waste rock. If the cost used to develop the SR – grade curve had been based on overburden, then one needs to convert waste rock into equivalent overburden.

404 *Open pit mine planning and design: Fundamentals*

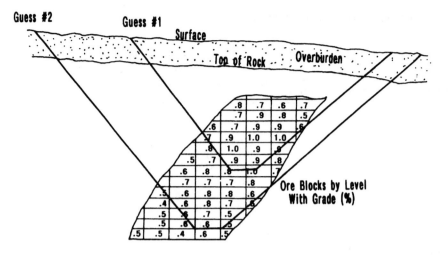

Figure 5.12. Final pit limits with bottom in ore.

5.2.4 *Location of pit limits – pit bottom in ore*

Figure 5.12 shows the case when the orebody continues to depth. Assume as before that

    Left hand slope $= 50°$
    Right hand slope $= 40°$
    Minimum pit bottom width $= 100\,\text{ft}$

Since the pit bottom is now in ore, the costs for stripping can be paid for by the ore in the strip along the pit bottom as well as that along the pit sides. A 50 ft wide allocation is made to both the right and left hand sides. The procedure is similar to that described with the pit bottom in waste. However, now both sides and the bottom must be examined at the same time. The procedure is as follows:

Step 1. Draw to scale a final pit profile using the appropriate left and right hand slopes as well as the minimum pit bottom on a piece of tracing paper. Superimpose this on the section. Guess an initial position.

Step 2. Calculate the average ore grades and stripping ratios for the left and right hand sides. Compare these to the allowable values. For simplicity it will be assumed that the densities and mineabilities of the materials involved are the same.

    Because of the freedom of the pit to 'float' both vertically and horizontally, the iterative procedure can be quite time consuming.

5.2.5 *Location of pit limits – one side plus pit bottom in ore*

A third possible situation is one in which one of the pit slopes is in ore. In Figure 5.13 it will be assumed that the right hand slope follows the ore-waste contact, the left hand slope is at 50°, and a minimum pit bottom width is 100 ft. In this case the ore along the pit bottom

Pit limits 405

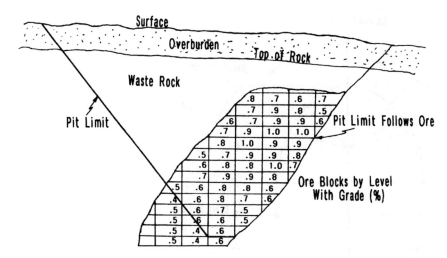

Figure 5.13. Final pit limits with right hand side in ore.

Table 5.7. Final pit limit location for one side plus pit bottom in ore (Fig. 5.13).

Length (ft)			Ore grade (g_{oi})	Ore length × ore grade ($l_{oi}g_{oi}$)
Overburden (l_{ob})	Waste (l_w)	Ore (l_{oi})		
130				
	626			
		43	0.5	12.5
		52	0.4	20.8
		52	0.6	31.2
		52	0.6	31.2
		52	0.4	20.8
		100	0.6	60.0
130	626	351	$\bar{g} = 0.53$	185.5

SR (actual) = $\frac{130 + 626}{351} \cong 2.15$;
SR (allowable) $\cong 2.21$;
Conclusion: this is the final pit location.

contributes to the cost of stripping the left wall. The approximate position of the final pit is shown superimposed on the figure. The corresponding calculation is given in Table 5.7.

5.2.6 *Radial sections*

The types of sections used depends upon the shape of the orebody. For the elongated orebody shown in Figure 5.14, transverse sections yield the best representation in the central portion. Along the axis of the orebody, a longitudinal section may be taken.

Transverse sections such as 1-1', 2-2', etc. in Figure 5.14 have been constructed parallel to one another and normal to the orebody axis. The influence of these sections is assumed to extend halfway to the neighboring sections. They are of constant thickness by construction

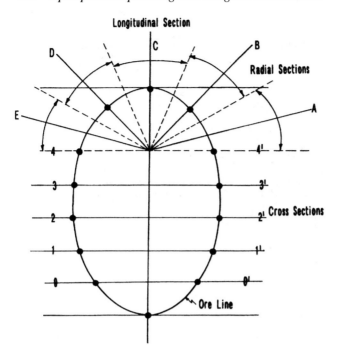

Figure 5.14. Plan view of orebody illustrating the different section types (Koskiniemi, 1979).

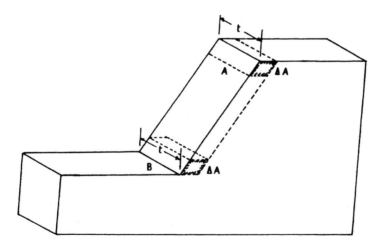

Figure 5.15. Isometric view of a parallel section.

(Fig. 5.15). A small face area (ΔA) at location A at the crest of the pit represents the same volume as the same area located at the toe (location B).

The pit location procedures described in Subsections 5.2.3 through 5.2.5 apply without modification to parallel cross-sections and longitudinal sections. As can be seen in Figure 5.14, radial sections are often needed to describe pit ends. For radial sections such

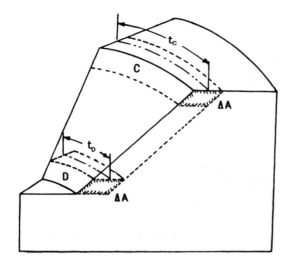

Figure 5.16. Isometric view of a radial section.

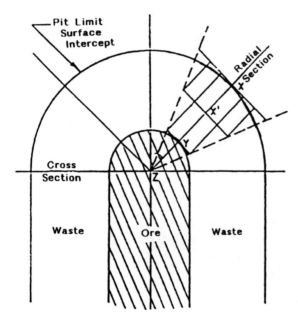

Figure 5.17. Plan representation showing the area of influence for a radial section (Koskiniemi, 1979).

as shown in Figure 5.16, the volume represented by an area ΔA at the crest (location C) is much greater than one at D due to the varying section thickness.

A modification in the procedure used to locate the final pit limit is required. This is accomplished through the development of a curve relating the apparent stripping ratio as measured on the radial section to the true stripping ratio.

Figure 5.17 is a plan view showing the region at the end of the pit in which radial sections are being used.

408 Open pit mine planning and design: Fundamentals

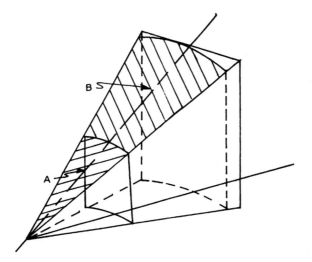

Figure 5.18. Isometric view of the radial sector.

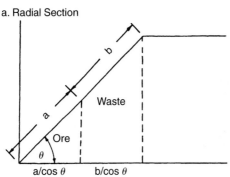

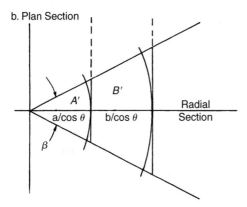

Figure 5.19. Plan and section views of the radial sector shown in Figure 5.18.

Figure 5.18 is an isometric view of the sector in question. The exposed ore area is identified as A and that of waste as B. The apparent (measured) stripping ratio for the radial section as shown on Figure 5.19a would be

$$\text{SR (measured)} = \frac{b}{a} \qquad (5.10)$$

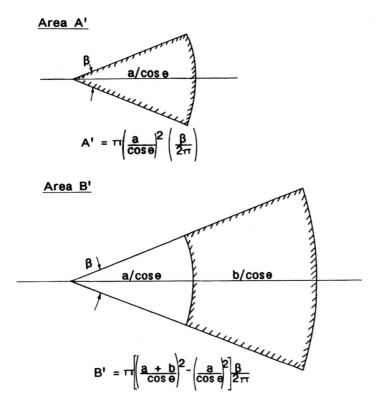

Figure 5.20. The ore and waste areas involved.

From the plan projection of the wall surface for this sector as shown in Figure 5.19b, the area of ore is A' and that of waste B'. Since the included angle for the sector is β, the ore radius in this projection is $a/\cos\beta$ and the thickness of the waste zone is $b/\cos\theta$.

The calculation of the plan projected areas is shown in Figure 5.20. The true stripping ratio can be expressed as

$$\text{SR (true)} = \frac{B'}{A'} = \frac{(a+b)^2 - a^2}{a^2} = \left(1 + \frac{b}{a}\right)^2 - 1$$

Since the measured stripping ratio on the section is as shown in Equation (5.10) then

$$\text{SR (true)} = (1 + \text{SR (measured)})^2 - 1 \tag{5.11}$$

As can be seen, this relationship is independent of both the slope angle and the included angle.

Values of SR (true) and SR (measured) are presented in Table 5.8 and plotted in Figure 5.21. The steps in the location of the final pit limit on a radial section are outlined below:

Step 1. As with parallel sections, one guesses the location of the final pit slope and calculates the average ore grade $\bar{g}$ and the stripping ratio. Assume that $\bar{g} = 0.8$ and the measured stripping ratio is $2:1$.

Table 5.8. Comparison of true versus measured stripping ratios for radial sections.

Stripping ratio	
Measured	True
0	0
0.25	0.56
0.50	1.25
0.75	2.06
1.0	3.0
1.25	4.06
1.5	5.25
2.0	8.0
2.5	11.25
3.0	15.0

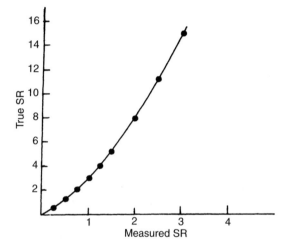

Figure 5.21. True versus measured stripping ratios (Koskiniemi, 1979).

Step 2. With parallel sections one would go directly to the SR – %Cu curve to see what stripping ratio this grade would support. In the case shown in Figure 5.22, it is about 6 : 1 indicating that the limit could be moved outward. For radial sections one must proceed to Step 3.

Step 3. The measured stripping ratio must first be converted to a true stripping ratio through the use of the conversion curve (Fig. 5.21). This indicates that for a measured stripping ratio of 2 : 1, the true stripping ratio is 8 : 1.

Step 4. Returning to Figure 5.22, one finds that the grade required to support a stripping ratio of 8 : 1 is 0.94. This is higher than the 0.8 present and hence the next guess of the final pit limit should be moved toward the pit.

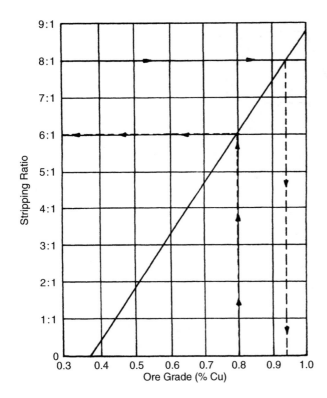

Figure 5.22. Example of stripping ratio calculation for a radial section.

In summary, the treatment of radial sections is done in exactly the same way as with parallel sections except that the use of an intermediate curve to convert measured SR to true SR is required. It should be pointed out that the ore region is pie-shaped and strictly speaking a weighted average approach should be used to calculate an average ore grade.

5.2.7 Generating a final pit outline

Once final pit profiles have been located on the individual sections, they must be evaluated as a group to examine how they fit together relative to one another. One section may suggest a very narrow pit, for example, whereas the adjacent ones yield a wide one (see Fig. 5.23). This smoothing requires adjustments to be made to the various sections. The easiest way of visualizing the pit and performing this task is through the use of level plans (horizontal sections). The final ore reserve estimation will also be done using plans. Thus to proceed requires the development of a composite mine plan map from the vertical sections.

The steps are outlined below (Koskiniemi, 1979):

Step 1. Transfer of intercepts to plan map. The locations of the pit bottom and the surface intercepts of the pit limits are transferred from the vertical sections to the plan map.

Step 2. Handling of discontinuous slopes. If a vertical section does not have a continuous slope line from the pit bottom to the surface intercept this is also shown.

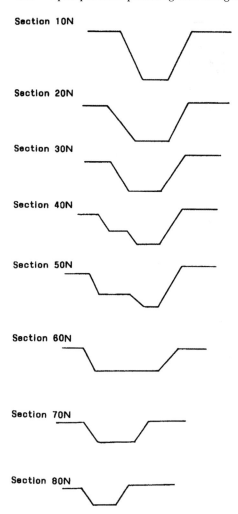

Figure 5.23. A series of initial pit outline sections for an orebody.

Step 3. Deciding on initial bench design location. The actual design of the composite plan generally begins with the pit bottom working upward. (Some designers prefer however to begin the design with a middle bench and work both upwards and downwards.)

Step 4. Smoothing. The points from the sections usually present a very irregular pattern, both vertically and horizontally. In smoothing these and designing the bottom bench several things should be kept in mind (Koskiniemi, 1979):

– Averaging the break-even stripping ratios for adjoining sections. If one section is moved significantly inward or outward, major changes in stripping/ore reserves may result. Thus the sections are carefully evaluated with each change.

– Use of simple geometric patterns for ease of design. The simpler the geometric shape, the easier it is to design the remainder of the pit.

– Location of ramp to pit bottom.

– Watching for patterns that might lead to slope stability problems. For example bulges or noses in the pit often are sources of problems.

Figure 5.24. Plan view of a pit with constant slope angle.

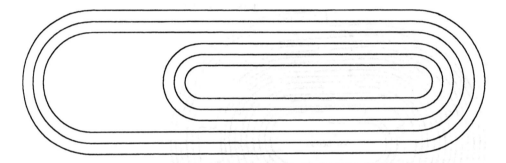

Figure 5.25. Plan view of a pit containing a wide bench.

Step 5. The bottom bench is then drawn. As has been discussed in Chapter 4, there are several possibilities for depicting the geometry. One way is to show both the toe and crest lines. Depending upon map scale and the number of benches, this may or may not be good. A second possibility is to display just one (either the toe or the crest) since the position of the other can be easily obtained knowing the bench height and the bench face angle.

The third possibility is to show the median elevation line (half the distance up the bench face). This is the representation most commonly used and that employed here.

Step 6. Addition of the median lines for the overlying benches. In preliminary designs the roads are often not shown. In such cases, the plan distance between the median lines is:

$$M_d = \frac{H}{\tan \theta} \quad (5.12)$$

where M_d is the plan distance between median lines, H is the bench height and θ is the overall pit slope angle. An overall pit slope angle is chosen so that the space required for the road is included. For the case shown in Figure 5.24, the slope angle is the same throughout the pit and the slope is continuous from toe to crest and hence the median lines are parallel and equally spaced. Figure 5.25 shows the case when there is a wide bench part way up the pit. Figure 5.26 shows the case when the north and south side slope angles are different.

414 Open pit mine planning and design: Fundamentals

Figure 5.26. Plan view of a pit with different north and south wall slopes.

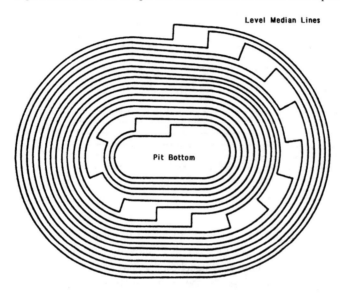

Figure 5.27. Composite ultimate pit plan (Koskiniemi, 1979).

If in the final designs the actual roads are shown, then the horizontal distance M_d between the median lines is equal to

$$M_d = b_w + \frac{H}{\tan \theta_f} \tag{5.13}$$

where b_w is the level berm width, H is the bench height, and θ_f is the bench face angle. A composite ultimate pit plan with a road included is shown in Figure 5.27. The procedure through which a road is added to a pit has been discussed in Chapter 4.

Step 7. Transfer of the pit limits to the individual level plans. When the composite pit plan has been completed, the pit limits are transferred to the individual level plans (Fig. 5.28). Through this process, the design engineer will first check the stripping ratios at the final pit limit. The pit is split into sectors such as shown in Figure 5.29, and the ore-waste relationships checked.

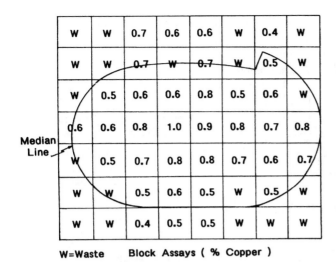

W=Waste   Block Assays (% Copper)

Figure 5.28. Level plan of bench a with the ultimate pit limit superimposed (Koskiniemi, 1979).

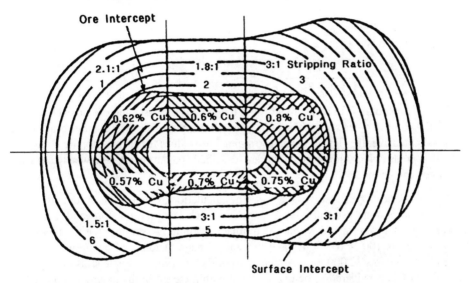

Figure 5.29. Plan of the ultimate pit showing the sectors (Soderberg & Rausch, 1968).

Figure 5.30 shows the median lines and ore grades for a sector taken in the center of the pit. If the pit bottom is in ore then this effect is included simply by adding artificial median lines to this region.

In this case the average grade is 0.7 and the stripping ratio is

$$\text{SR} = \frac{3}{4} = 0.75$$

A radial section is shown in Figure 5.31. In this case

$$\bar{g} = \frac{\sum r_i g_i}{\sum r_i} = \frac{0.9 \times 1 + 0.8 \times 2 + 0.6 \times 3 + 0.5 \times 4}{1 + 2 + 3 + 4} = 0.63$$

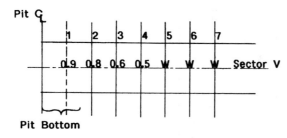

Figure 5.30. Average grade calculation for a parallel sector.

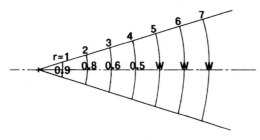

Figure 5.31. Average grade calculation for a radial sector.

The stripping ratio is

$$\text{SR} = \frac{\sum r_{wi}}{\sum r_{oi}} = \frac{5+6+7}{1+2+3+4} = 1.8$$

The general procedure is to measure and sum the lengths of waste and ore on each level at the pit limit within a given sector. The ratio gives the stripping ratio. The average grade is the length weighted average. The grade and stripping ratio values for each sector are compared to the SR – grade curve. Anomalous areas are identified and the design reviewed to see what corrections can be made.

Step 8. Calculation of ore reserves by level. The calculation of final ore reserves and overall stripping ratio is done using the level maps. For each level the ore tons and grade and the waste tons are determined within the ultimate pit limits. Assuming that the block size in Figure 5.28 is $100' \times 100' \times 40'$ and that the tonnage factor is $12.5 \, \text{ft}^3/\text{ton}$, then each whole block contains 32,000 tons. A tabulation of the reserves on this level is presented in Table 5.9. Ore is considered to be that material within the pit limits having a grade equal to or greater than the grade from the stripping curve at a break-even stripping ratio of zero. This is generally called the ore reserve cut off grade. Using a cutoff grade of 0.37%, the total tons of ore and waste are computed. The average ore grade is also obtained.

Step 9. Calculation of final ore reserve and overall stripping ratio. The ore reserves for each level are tabulated such as shown in Table 5.10. From this the ore tons, overall average ore grade, total waste tons and overall stripping ratio can be found. In addition tonnage-grade distribution curves are often plotted both overall and by level to help in mill design and setting of production levels.

5.2.8 *Destinations for in-pit materials*

Once the pit limits have been defined, all of the material within the pit outline will be mined irrespective of its value. Given mined material the decision must be made regarding its

Table 5.9. Reserve summary for bench A.

Grade (%)	Tons ore	Tons waste
<0.30		125,300
0.30–0.36		0
0.37–0.39	0	
0.40–0.49	0	
0.50–0.59	193,340	
0.60–0.69	192,000	
0.70–0.79	144,400	
0.80–0.89	189,000	
0.90–0.99	32,000	
1.00–1.09	32,000	

Table 5.10. Total reserves within the final pit outline.

Bench	Grade									
	<0.30	0.3–0.36	0.37–0.39	0.4–0.49	0.5–0.59	0.6–0.69	0.7–0.79	0.8–0.89	0.9–0.99	1.0–1.09
A	125,300	0	0	0	193,340	192,000	144,400	189,000	32,000	32,000
B										
C										
D										
E										
F										
G										
H										
I										
J										
Totals										

destination. A new cutoff grade based simply upon meeting the processing costs can be calculated. For the example presented in Section 5.2.2, the direct mining cost was given as $1 per ton. If this is subtracted from the total cost then the net value – grade equation becomes

$$y = -4.2 + 14.0x$$

Solving for the breakeven cutoff grade (sending to the mill or the dump) one finds that

$$x = 0.30\% \text{ Cu}$$

Hence for the pit material everything having a grade greater than or equal to 0.30% Cu could be sent to the mill. This assumes, of course, that the transport costs to the dumps and the mill are the same. If not, then the cost differences can be included. Such considerations can be made only if mill capacity is available. Often this is not the case. The economics of this marginal/submarginal material must be evaluated very closely.

Table 5.11. Typical mineral block model data items (Crawford & Davey, 1979).

Copper (Cu)	%
Molybdenite (MoS_2)	%
Gold (Au)	oz/st
Silver (Ag)	oz/st
Copper concentrate recovery	%
Copper concentrate grade	%
Copper smelting recovery	%
Copper in blister	%
Copper refining recovery	%
MoS_2 concentrate recovery	%
MoS_2 conversion recovery	%
Gold concentrate recovery	%
Gold refining recovery	%
Silver concentrate recovery	%
Silver refining recovery	%

5.3 ECONOMIC BLOCK MODELS

The block model representation of orebodies rather than section representation and the storage of the information on high speed computers has offered some new possibilities in open-pit planning. The use of computers allows the rapid updating of plans as well as exploring a wide number of parameters through sensitivity analysis. Although there is still a great deal of opportunity for engineering interaction, much of the tedious work can be done by computer. Both of the major computerized techniques to be described in this chapter: Floating cone and Lerchs-Grossmann, require an initial economic evaluation of the blocks.

Mineral, metallurgical and economic data are combined to assign a net dollar value to each mineral model block. Table 5.11 shows typical mineral block data items for an orebody containing copper, molybdenite, gold and silver. The block size is $50 \times 50 \times 40$ ft and the tonnage factor is 13.5 ft^3/st. The economics format is shown in Table 5.12. All of the costs with the exception of truck haulage and roads are expressed as fixed unit costs for the indicated production quantity unit. All mining and processing costs include operating, maintenance and depreciation costs.

The costs for truck haulage and roads may vary because of haulage profile, length and lift. The costs per hour are first estimated and then converted to unit costs per ton based upon an estimated hourly haulage productivity. The projected haulage productivity may be obtained using haulage simulators. Due to this variation in the haulage and road components, the mining costs for both ore and waste contain a fixed component plus a mining level (bench location) dependent component. This is shown in Table 5.13.

Using Tables 5.11, 5.12, and 5.13, the net value for the block can be determined. In this particular example, the material is considered as either mill feed or waste. Thus the net value calculations are done for these two possibilities. The net value as stripping (NV_{st}) is found by multiplying the block tonnage times the stripping cost.

$$NV_{st} = \text{Block tonnage} \times \text{Stripping cost}_{st} \ (\$/st)$$

As can be seen in Table 5.13, the stripping cost is made up of two items:
- Mining cost (\$/st waste), and
- General plant cost (\$/st waste).

Table 5.12. Block model mining and processing cost items (Crawford & Davey, 1979).

Drilling	$/st ore and waste
Blasting	$/st ore and waste
Loading	$/st ore and waste
Hauling	$/truck hour
Haul roads	$/truck hour
Waste dumps	$/st waste
Pit pumping	$/st ore
Mine general	$/st ore
	$/st waste
Ore reloading	$/st ore
Ore haulage	$/st ore
Concentrating	$/st ore
Concentrate delivery	$/st concentrate
Smelting	$/st concentrate
General plant	$/st ore
	$/st waste
	$/st blister
Blister casting loading and freight	$/st blister
Refining	$/st blister
Selling and delivery	$/lb refined copper
Metal prices	
Copper	$/lb
MoS$_2$	$/lb
Gold	$/oz
Silver	$/oz

Table 5.13. Pit limit analysis cost summary (Crawford & Davey, 1979).

	Stripping (waste)	Mill feed (ore)
Mining ($/st)	$aL^* + b$	$cL^* + d$
Processing and other:		
Ore haulage ($/st ore)		X
Concentrating ($/st ore.)		X
Concentrate delivery ($/st conc.)		X
Smelting ($/st conc.)		X
Blister casting loading, and freight ($/st blister)		X
General plant		
$/st ore		X
$/st waste	X	X
$/st blister		X
Refining ($/st blister)		X
Selling and delivery($/lb Cu)		X

* Level elevation.

With regard to the second cost item (general plant), there are many 'overhead' types of costs which are independent of whether the material being moved is 'ore' or 'waste'.

Although, in the end, the 'ore' must pay for all of the costs, there is some logic for allocating general plant costs to the waste as well (see Subsection 2.4.1).

Table 5.14. Block evaluation calculation format.

Block data	
Block location	Data for block
Block tonnage	from Table 5.11 and 5.13
As mill feed*	
Revenue from metal sales ($)	
Copper: Block tonnage × lb saleable Cu/st × Price ($/lb)	= A
MoS$_2$	= B
Gold	= C
Silver	= D
Total revenue	$\overline{R}$
Cost of copper ($):	
Block tonnage × lb saleable Cu/st × Cost ($/lb)	= C'
Net value as mill feed ($):	
$NV_{mf}(\$) = R - C'$	
Net value as stripping ($):	
$NV_{st}(\$)$ = Block tonnage × Stripping cost ($/st)	

* The block is designated as mill feed and assigned NV_{mf} if net value is greater than NV_{st}, even if negative.

The net value as mill feed (NV_{mf}) requires the calculation of the revenues and costs (excluding any stripping) for the block. Table 5.14 summarizes the calculations. The mill feed and stripping net values are compared for each block and the most positive value is assigned. This net value then becomes the only piece of block data used directly in the mining simulators.

5.4 THE FLOATING CONE TECHNIQUE

In the section on manual techniques for determining the final pit limits, it was shown that for a single pit wall, the line having the slope angle defining the wall was moved back and forth until the actual grade and stripping ratio matched a point on the SR-grade curve. For the case when both walls and the pit bottom were in ore, a geometric figure in the shape of the minimum pit was constructed and moved around on the section (vertically and horizontally) until the stripping ratios and grades measured along the periphery matched that from the curve. For the case of 45° slopes and a 100 ft minimum pit bottom width, the figure to be 'floated' is as shown in Figure 5.32. If this figure is revolved around the axis, the solid generated is called the frustum of a cone. Had the minimum pit width been zero, then the figure would simply be a cone. With sections it was most convenient to use two-dimensional representation and not consider the effect of adjacent sections. This was at least partially done later through the smoothing on plan.

Although there are other techniques for determining the final pit configuration, currently the most popular is through the use of this 'floating cone'. Several changes in the manual process are necessary to accommodate the computer. In the manual process, a net value – grade curve was developed. Knowing the cost of waste removal, this curve was changed into a SR – grade curve. The user then simply evaluated stripping ratio (normally by measuring lengths of ore and waste) and calculated the weighted average grade. Through the use of the SR-grade 'nomograph' one could examine pit expansion.

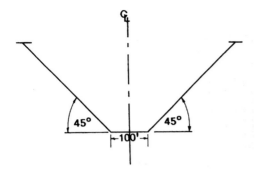

Figure 5.32. Pit profile to be 'floated'.

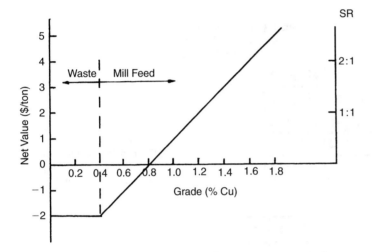

Figure 5.33. Net value – grade curve showing the waste/mill feed split.

When using the computer, it is more convenient to use the net values of the blocks directly. Figure 5.33 shows the type of net value-grade curve which is actually used. Consider the simple example shown in Figure 5.34. This case will be examined using both the manual method based upon grades and stripping ratios, and that in which net values are assigned to the blocks. Three potential pit limits have been superimposed on the grade block model of Figure 5.35. A pit slope of 45° is assumed. The average ore grades and stripping ratios are given. The final pit is represented by case 3.

Using Figure 5.33, the grade block model of Figure 5.34 can be converted into an economic block model. The result is shown in Figure 5.36. By examining the net values of the blocks involved in a particular mining sequence, final pit limits can be determined. Mining is stopped when the net value is negative. The net values for the three cases examined in Figure 5.35 are given in Figure 5.37.

Slice 3 has a NV = 0 and would define the final pit on this section. As can be seen the manual and economic block approaches give equivalent results. The net value approach is by far the easiest to program, particularly when considering a three dimensional array of blocks.

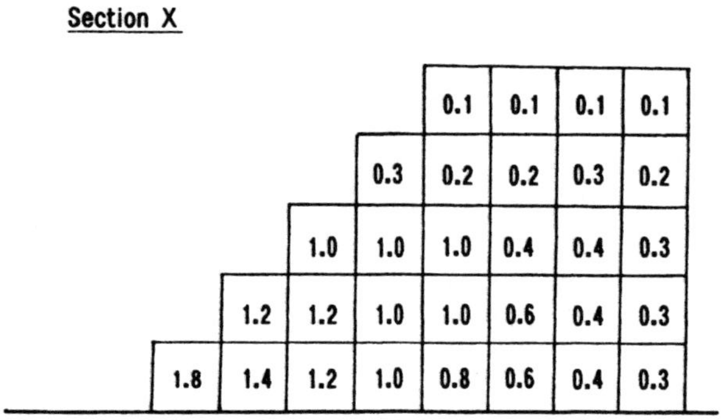

Section X

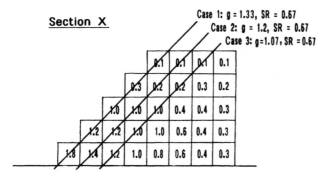

Figure 5.34. Grade block model for pit limit example.

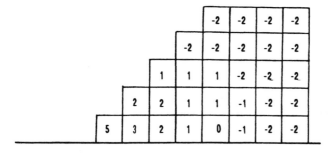

Figure 5.35. Trial final pit limits based on the manual procedure. Case 3 is the final limit.

Figure 5.36. Corresponding economic block model.

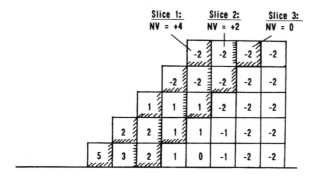

Figure 5.37. Trial final pit limits.

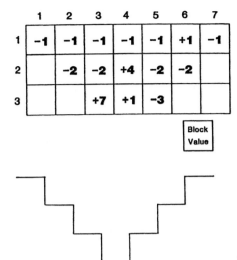

	1	2	3	4	5	6	7
1	−1	−1	−1	−1	−1	+1	−1
2		−2	−2	+4	−2	−2	
3			+7	+1	−3		

Block Value

Figure 5.38. Block model for Example 1 (Barnes, 1982).

Figure 5.39. Stepped cone pit model.

The principles involved in determining the pit outline, given an entire 2-dimensional array of blocks will now be demonstrated. The examples were originally presented by Barnes, 1982. Figure 5.38 shows the sample section to be considered with the net values given. The blocks are equidimensional and the slope angle will be 45°. This means that slope is formed by going up one and over one block. Figure 5.39 shows the stepped cone which will be used for determining the final pit.

The following steps are used:

Step 1. The cone is 'floated' from left to right along the top row of blocks in the section. If there is a positive block it is removed.

Step 2. After traversing the first row, the apex of the cone is moved to the second row. Starting from the left hand side it 'floats' from left to right stopping when it encounters the first positive block. If the sum of all the blocks falling within the cone is positive (or zero), these blocks are removed (mined). If the sum is negative the blocks are left, and the cone floats to the next positive block on this row. The summing and mining or leaving process is repeated.

Step 3. This floating cone process moving from left to right and top to bottom of the section continues until no more blocks can be removed.

Step 4. The profitability for this section is found by summing the values of the blocks removed.

Step 5. The overall stripping ratio can be determined from the numbers of positive (+) and negative (−) blocks.

These rules can now be applied to the section shown in Figure 5.38. There are four positive blocks in the model hence there are four corresponding cones which must be evaluated. Using a top–down rule, the block at row 1/column 6 would initiate the search. Since there are no overlying blocks, the value of the cone is the value of the block: 1. The value is positive, so the block is mined (Fig. 5.40).

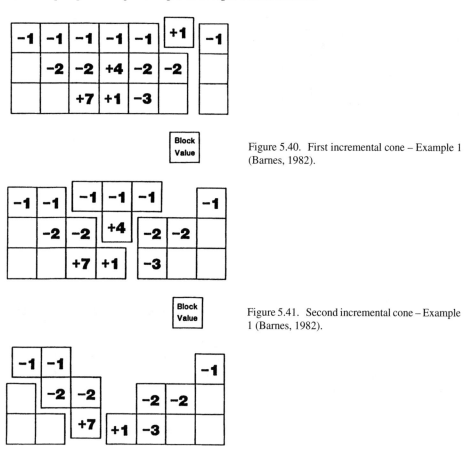

Figure 5.40. First incremental cone – Example 1 (Barnes, 1982).

Figure 5.41. Second incremental cone – Example 1 (Barnes, 1982).

Figure 5.42. Third incremental cone – Example 1 (Barnes, 1982).

The next incremental cone is that defined by the block at row 2/column 4. The value of this cone is

$$-1-1-1+4 = +1$$

Since this value is positive, the cone is mined (Fig. 5.41). For the incremental cone defined by the block at row 3/column 3, its value is

$$-1-1-2-2+7 = +1$$

Again, since the value is positive, this cone is mined (Fig. 5.42). Finally, the value of the incremental cone defined by the block at row 3/column 4 is

$$-2+1 = -1$$

The value of this cone is negative; therefore, the cone is not mined (Fig. 5.43). Figure 5.44 depicts the overall final ultimate pit. The total value of this pit is

$$-1-1-1-1+1-2-2+4+7 = +3$$

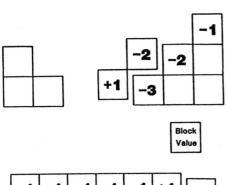

Figure 5.43. Fourth incremental cone – Example 1 (Barnes, 1982).

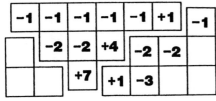

Figure 5.44. Ultimate pit – Example 2 (Barnes, 1982).

The overall stripping ratio is

$$\text{SR (overall)} = \frac{7}{3}$$

In this simple situation, the final pit is 'optimal'. Over the years, considerable effort has been spent in trying to develop procedures which when applied would yield the 'optimal' pit. The 'optimum' might be defined for example, as the one yielding one of the following,
 – maximum profit,
 – maximum net present value,
 – maximum extraction.
Although many contend that the determination of such an 'optimum' pit is largely of academic interest due to changing prices, costs and poorly defined grades, none-the-less efforts continue in this direction. Barnes (1982) has presented three problems representing floating cone situations not leading to an optimum pit.

Problem 1. Missing combinations of profitable blocks
This problem occurs when positive (ore) blocks are investigated individually. A single ore block may not justify the removal of the necessary overburden while combinations of these blocks with overlapping cones are profitable. Johnson (1973) has labelled this 'the mutual support problem'. This is demonstrated in Figures 5.45 to 5.48. The cone defined by the block at row 3/column 3 has a value of

$$-1 - 1 - 1 - 1 - 1 - 2 - 2 - 2 + 10 = -1$$

Since the value is negative, the cone would not be mined by the simple moving cones method (Fig. 5.46). Similarly, for the cone defined by the block at row 3/column 5, the value is

$$-1 - 1 - 1 - 1 - 1 - 2 - 2 - 2 + 10 = -1$$

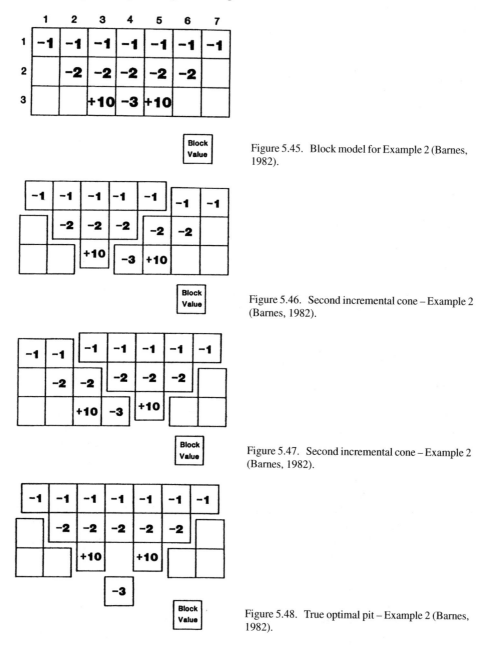

Figure 5.45. Block model for Example 2 (Barnes, 1982).

Figure 5.46. Second incremental cone – Example 2 (Barnes, 1982).

Figure 5.47. Second incremental cone – Example 2 (Barnes, 1982).

Figure 5.48. True optimal pit – Example 2 (Barnes, 1982).

Again, this cone would not be mined (Fig. 5.47). Therefore, using the simple cone analysis, nothing would be mined. However, due to the overlapping (mutual support) portion of the overburden cones, the value of the composite union is positive

$$-1-1-1-1-1-1-1-2-2-2-2-2+10+10 = +3$$

This situation (Fig. 5.48) occurs often in real-world mineral deposits, and a simple moving cones approach misses it.

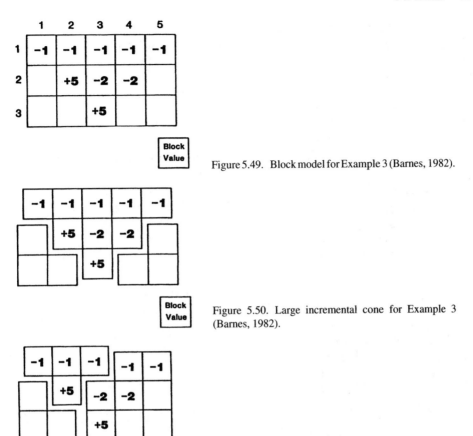

Figure 5.49. Block model for Example 3 (Barnes, 1982).

Figure 5.50. Large incremental cone for Example 3 (Barnes, 1982).

Figure 5.51. Small incremental cone for Example 4 (Barnes, 1982).

Problem 2. Extending the ultimate pit beyond the optimal pit limits
This is the situation where the moving cones algorithm can and often will include non-profitable blocks in the pit design. The inclusion of non-profitable blocks will reduce the net value of the pit. This situation occurs when profitable ore blocks, or profitable combinations of ore blocks, cause a cone defined by an underlying apex to be positive; i.e., the positive values are being extended downward to carry waste below their cones. The two-dimensional block model shown in Figure 5.49 assumes the maximum pit slope to be 45 degrees. The value of the cone defined by the block at row 3/column 3 (Fig. 5.50) is

$$-1 - 1 - 1 - 1 - 1 + 5 - 2 - 2 + 5 = +1$$

The fact that the value of this cone is positive does not imply that the cone should be mined. As shown on Figure 5.51, the block at row 2/column 2 is carrying this cone. The proper design includes only the block at row 2/column 2 and its three overlying blocks, row 1/columns 1, 2, and 3. The value of the optimal design is

$$-1 - 1 - 1 + 5 = +2$$

The value of the small cone is greater than the value of the large cone.

428 *Open pit mine planning and design: Fundamentals*

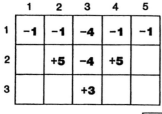

Figure 5.52. Block model for Example 4 (Barnes, 1982).

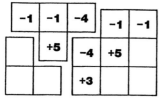

Figure 5.53. First incremental cone for Example 4 (Barnes, 1982).

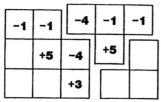

Figure 5.54. Second incremental cone for Example 4 (Barnes, 1982).

Problem 3. Combination of problems 1 and 2

The most common and most difficult situation involving these two problems is their simultaneous occurrence. The two-dimensional block model shown on Figure 5.52 assumes a 45° pit slop. There are three positive blocks, and therefore three possible incremental cones.

The value of the cone defined by the block at row 2/column 2 (Fig. 5.53) is

$$-1 - 1 - 4 + 5 = -1$$

The value of the cone defined by the block at row 2/column 4 (Fig. 5.54) is

$$-4 - 1 - 1 + 5 = -1$$

Yet, the value of the cone defined by the block at row 3/column 3 (Fig. 5.55) is

$$-1 - 1 - 4 - 1 - 1 + 5 - 4 + 5 + 3 = +1$$

This would appear to imply that the pit design shown on Figure 5.55 is optimal; however, this is not the case. The optimal design is shown on Figure 5.56. The value of this pit is

$$-1 - 1 - 4 - 1 - 1 + 5 + 5 = +2$$

This value is one more than the 'initially apparent' pit.

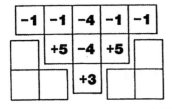

 Figure 5.55. Third incremental cone for Example 4 (Barnes, 1982).

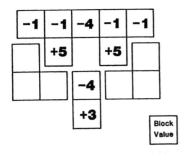

Figure 5.56. Optimal pit design for Example 4 (Barnes, 1982).

In spite of these problems, there are however a number of very positive aspects of the technique which account of its widespread use and popularity (Barnes, 1982):

1. Since the method is a computerization of manual techniques, mining engineers can use the method, understand what they are using, and feel comfortable with the results.

2. Computationally, the algorithm is quite simple. Development and implementation of a moving cones computer program does not require sophisticated knowledge in operations research or computer science. The computer code could be developed in-house, rather than purchased from a software company; thus, a more custom-fitted product can be provided at an operating mine site.

3. The moving cones technique can be used with generalized pit slopes. The single requirement is an unambiguous rule for determining which blocks overlie individual ore blocks.

4. It provides highly useable and sufficiently accurate results for engineering planning.

5.5 THE LERCHS-GROSSMANN 2-D ALGORITHM

In 1965, Lerchs and Grossmann published a paper entitled 'Optimum design of open-pit mines'. In what has become a classic paper they described two numeric methods:
 – a simple dynamic programming algorithm for the two-dimensional pit (or a single vertical section of a mine),
 – a more elaborate graph algorithm for the general three-dimensional pit.
This section will discuss the first method with the second being described in Section 5.7. The easiest way of presenting the technique is through the use of an example. The mathematics which accompany the actual mechanics will be given at the same time. This example was originally presented by Lerchs & Grossmann (1965) and elaborated upon by Sainsbury (1970). The orebody is as shown in Figure 5.57.

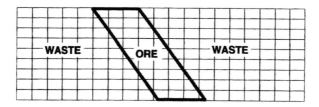

Figure 5.57. Orebody geometry for the Lerchs-Grossmann 2-D example (Sainsbury, 1970; Lerchs & Grossmann, 1965).

The following apply:

$NV_{st} = -\$4 \times 10^3/\text{block}$

$NV_{mf} = \$12 \times 10^3/\text{block}$

Slope angle $= 35.5°$

Bench height $= 40\text{ ft}$

Tonnage factor $= 12.5\text{ ft}^3/\text{st}$

For ease in calculation the values $12 and −$4 will be used. At the conclusion, the factor of 1000 will be reintroduced. To apply this technique the grid (block geometry) is selected based upon the slope angle. The slope is formed by moving up one block and over one block. Hence for a bench height of 40 ft and a slope angle of 35.5° one finds using Equation (5.14)

$$\alpha = \frac{H}{B} = \tan\theta \tag{5.14}$$

where α is the ratio of block height/block width, H is the block height, B is the block width, and θ is the slope angle, that

$$B = \frac{H}{\tan\theta} = \frac{40}{\tan 35.5°} = 56\text{ ft}$$

Hence

$$\alpha = \frac{5}{7}$$

This grid system has been superimposed on the section in Figure 5.57. In Figure 5.58 the net values have been added. As can be seen, the boundary blocks contain both ore and waste elements. A weighted averaging procedure has been used to obtain the block model of Figure 5.59. The block positions will be denoted using an i, j numeration system. In keeping with the nomenclature used by Lerchs and Grossmann, i refers to the rows and j to the columns. The first step in this procedure is to calculate cumulative profits for each column of blocks starting from the top and moving downward. Each vertical column of blocks is independent of the others. This process is shown in Figure 5.60 for column $j = 6$.

The equation which describes this process is

$$M_{ij} = \sum_{k=1}^{i} m_{kj} \tag{5.15}$$

Pit limits 431

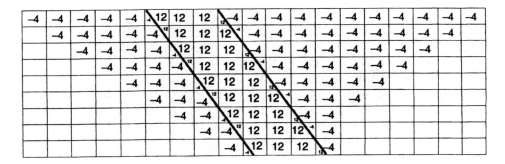

Figure 5.58. Initial economic block model.

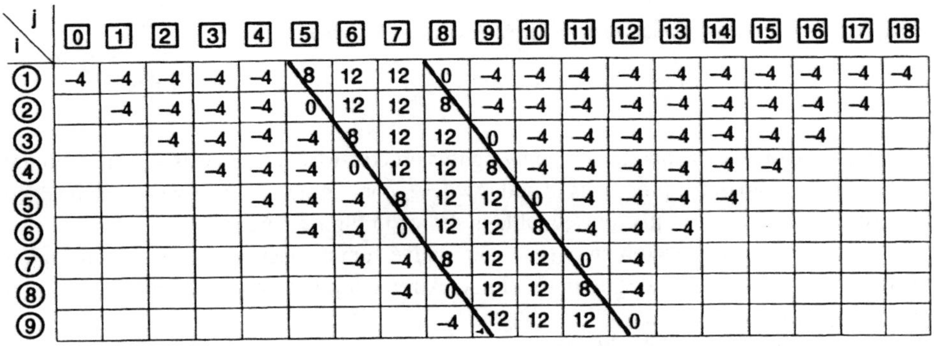

Figure 5.59. Final economic block model (Lerchs & Grossmann, 1965).

Row	Current Value	Revised Value
$i = 1$	12	12
$i = 2$	12	$24 = 12 + 12$
$i = 3$	8	$32 = 12 + 12 + 8$
$i = 4$	0	$32 = 12 + 12 + 8 + 0$
$i = 5$	-4	$28 = 12 + 12 + 8 + 0 - 4$
$i = 6$	-4	$24 = 12 + 12 + 8 + 0 - 4 - 4$
$i = 7$	-4	$20 = 12 + 12 + 8 + 0 - 4 - 4 - 4$
$i = 8$	-4	$16 = 12 + 12 + 8 + 0 - 4 - 4 - 4 - 4$

Figure 5.60. Calculation of the cumulative sums for column 6.

where M_{ij} is the profit realized in extracting a single column with block (i, j) at its base and m_{kj} is the net value of block (k, j). Applying the equation to find the value of the column for $j = 6, i = 3$

$$M_{36} = \sum_{k=1}^{3} m_{k6} = m_{16} + m_{26} + m_{36}$$
$$= 12 + 12 + 8 = 32$$

432 Open pit mine planning and design: Fundamentals

Columns

	[0]	[1]	[2]	[3]	[4]	[5]	[6]	[7]	[8]	[9]	[10]	[11]	[12]	[13]	[14]	[15]	[16]	[17]	[18]
⓪	0	0	0	0	0	0	0	0	0	0	0	0	0	0	0	0	0	0	0
①	-4	-4	-4	-4	-4	8	12	12	0	-4	-4	-4	-4	-4	-4	-4	-4	-4	-4
②	-8	-8	-8	-8	-8	8	24	24	8	-8	-8	-8	-8	-8	-8	-8	-8	-8	
③	-12		-12	-12	-12	4	32	36	20	8	-12	-12	-12	-12	-12	-12	-12		
④			-16	-16	0	32	48	32	0	-16	-16	-16	-16	-16	-16				
⑤				-20	-4	28	56	44	12	16	-20	-20	-20	-20					
⑥					-8	24	56	56	24	-8	-24	-24	-24						
⑦						20	52	64	36	4	24	-28							
⑧						16	48	64	48	16	-16	-32							
⑨							60	56	28	-4	32								

Figure 5.61. Completed cumulative sums (Lerchs & Grossmann, 1965).

The new table of values obtained by applying this process to all columns is shown in Figure 5.61. The next step in the process is to add row $i = 0$ containing zero's. A zero is also added at position $(i = 0, j = 0)$. This revised table is shown in Figure 5.61.

It is now desired to develop an overall cumulative sum as one moves laterally from left to right across the section. Beginning with the extreme top left hand real block, the values of three blocks are examined:

1. One directly above and to the left.
2. One on the left.
3. One directly below and to the left.

Of the three, that block which when its value is added to the block in question yields the most positive sum is selected. An arrow is drawn from the original block to that block. This sum is substituted for that originally assigned and becomes the value used for subsequent calculations.

Figure 5.62 shows the process for block (1, 1). This process is continued, working down the first column, then down the next column to the right, until all blocks have been treated. It should be pointed out that the reason some of the blocks on the section have not been filled in is that they fall outside of the bounds of the ultimate pit. Figure 5.63 shows the results when the process has been completed through column 7.

One can examine the relationship between the values in the current table to the initial blocks (Fig. 5.59). If one follows the arrows beginning at the value 32, the pit which results is indicated by the shaded line (Fig. 5.64).

Superimposing the same pit on the block model one finds that the cumulative value of the blocks is 32 (Fig. 5.65). Moving up to the block containing the value 60 in Figure 5.64 one follows the arrows to outline the pit. The value obtained by summing the blocks is also 60 (Fig. 5.65). Therefore the technique provides a running total of the value of the pits defined by following the arrows. At this point in the calculation, the optimum pit has a value of 84. Figure 5.66 shows the results when the summing process is completed. The optimum pit is that which has the maximum cumulative value. To determine this, one moves from right to left along row 1 until the largest value is encountered. The arrows are then followed around to give the optimum pit outline on the section.

1. Beginning with Block 1,1 (the upper left hand corner of the section) the neighboring blocks in the Row above (Row 0), the same row (Row 1) and the Row below (Row 2) of the column to the left of column [0] are examined.

2. Of these three, the largest positive block is chosen.

3. In this case it is Block 0,0.

4. The sum of Block 1,1 and Block 0,0 is found and substituted for the original value of Block 1,1 in the Figure. In this case it is -4 + 0 or -4. An arrow is drawn between Block 1,1 and the selected block. This process is continued moving down column 1. When completed, column 2 is begun working from the top down.

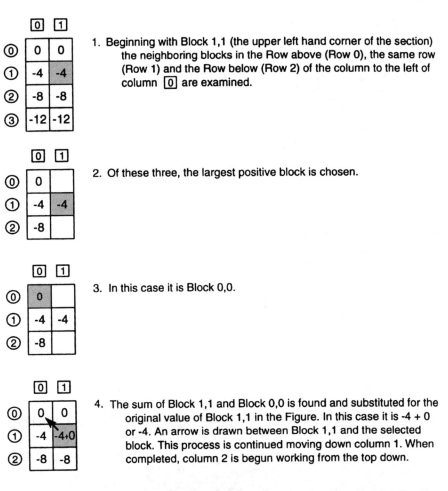

Figure 5.62. Procedure for determining the cumulative maximum value and maximizing direction.

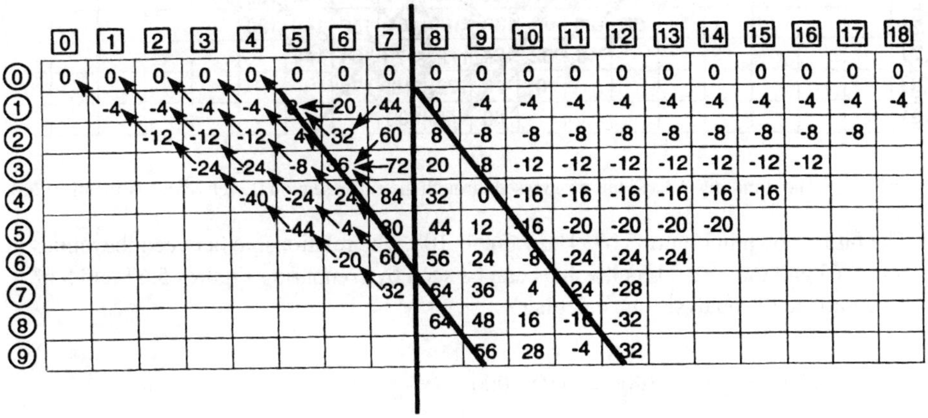

Figure 5.63. Progression of summing process through column 7.

434 Open pit mine planning and design: Fundamentals

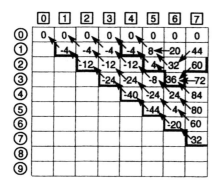

Figure 5.64. Pit determination and total value by following the arrows.

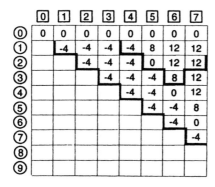

Figure 5.65. Individual block values for the two partial pits.

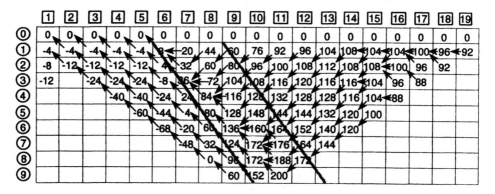

Figure 5.66. The summing process has progressed through the entire section (Lerchs & Grossmann, 1965).

The pit is shown in Figure 5.67. The value is 108. The relationship between the values in Figure 5.67 and the actual block values can be seen by comparing Figures 5.67 and 5.68.

To complete the analysis one calculates:

Net value = 108 × $1000 = $108,000

Total tons = 36 blocks × 10,000 tons/block = 360,000 tons

Tons ore = 20 blocks × 10,000 = 200,000 tons

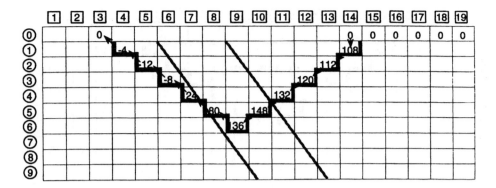

Figure 5.67. Optimum pit determination.

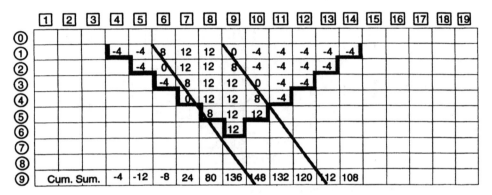

Figure 5.68. Optimum pit limits superimposed on the block model.

Tons waste = 16 blocks × 10,000 = 160,000 tons

Stripping ratio = $\dfrac{16}{20}$ = 0.8

Average profitability/ton = $\dfrac{\$108,000}{360,000}$ = \$0.30/ton

The expressions used (Lerchs & Grossmann, 1965) to calculate the 'derived' profit P_{ij} (as given in Fig. 5.66) are

$$P_{ij} = \begin{cases} 0, & i = 0 \\ M_{ij} + \max_{k=-1,0,1} P_{i+k,j-1} & i \neq 0 \end{cases} \quad (5.16)$$

The maximum is indicated by a arrow going from (i, j) to $(i + k, j - 1)$. P_{ij} is the maximum possible contribution of columns 1 to j to any feasible pit that contains (i, j) on its contour. If the element (i, j) is part of the optimum contour, then this contour to the left of element (i, j) can be traced by following the arrows starting from element (i, j). Any feasible pit must contain at least one element of the first row. If the maximum value of P in the first row is positive then the optimum contour is obtained by following the arrows from and to the left of this element. If all elements of the first row are negative, then no contour with positive profit exists.

436 Open pit mine planning and design: Fundamentals

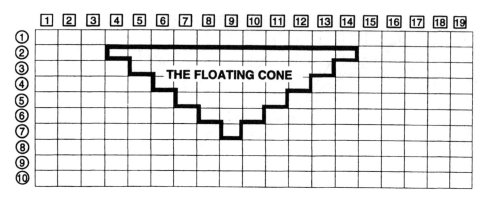

Figure 5.69. The floating cone used to evaluate the final pit limits.

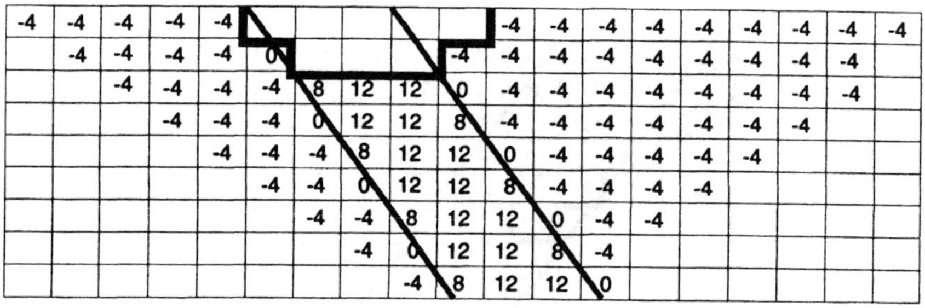

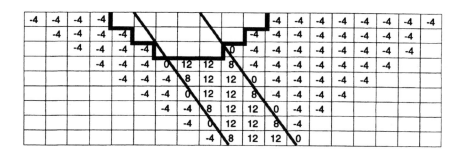

Figure 5.70. Situation after floating down two rows.

Figure 5.71. Situation after floating down three rows.

This same section (Fig. 5.59) can be evaluated using the floating cone technique. The 'stepped' cone of Figure 5.69 has been used to float over the section. Figures 5.70 through 5.73 show the intermediate pits with the final pit given in Figure 5.74. As can be seen the result is the same as when using the Lerchs-Grossmann approach. An advantage of the floating cone procedure is that a variety of slope angles can be modelled.

As shown in Figure 5.75, one can also examine the pit limits by considering the lengths along the perimeter. Consider the left-hand side (LHS) of the pit. The horizontal segments

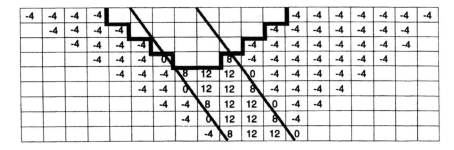

Figure 5.72. Situation after floating down four rows.

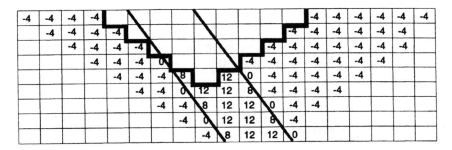

Figure 5.73. Situation after floating down five rows.

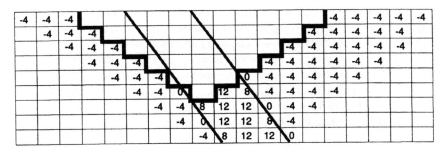

Figure 5.74. Situation after floating down six rows.

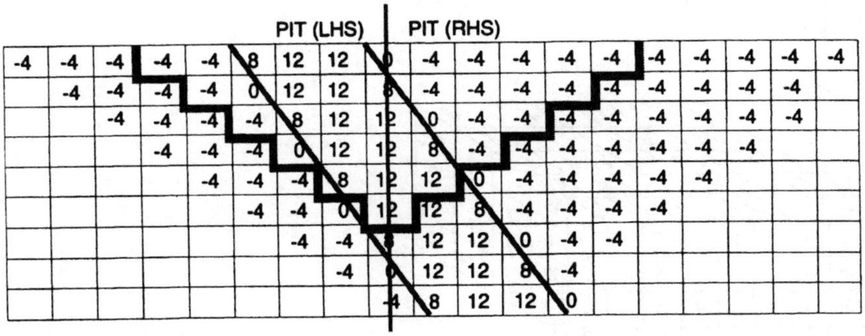

Figure 5.75. Pit resulting from the manual approach using the stripping curves.

438 *Open pit mine planning and design: Fundamentals*

are considered first to determine the vertical position. When the pit is slightly above the known true position, the sum of the horizontal segments is

$$-4 - 4 - 4 + 0 + 8 + 6 = +2$$

When it is slightly below than the sum is

$$-4 - 4 - 4 - 4 - 4 + 0 + 4 = -16$$

Examining the vertical segments to find the correct horizontal position is done next. When the contour is slightly inside the correct position the sum is

$$-4 - 4 - 4 + 0 + 8 + 12 = +8$$

When it is slightly outside, the sum is

$$-4 - 4 - 4 - 4 - 4 + 0 = -20$$

Therefore the correct position is as given by the floating cone and Lerchs-Grossmann procedures. The same procedure could be followed on the right hand side.

5.6 MODIFICATION OF THE LERCHS-GROSSMANN 2-D ALGORITHM TO A 2½-D ALGORITHM

The two-dimensional dynamic programming algorithm for determining the optimal configuration of blocks to be mined in cross-section was presented in the previous section. The technique is elegant yet simple. Like all other two-dimensional techniques it has the drawback that extensive effort is usually required to smooth out the pit bottom and the pit ends as well as to make sure that the sections fit with one another. As indicated by Johnson and Sharp (1971), this smoothing seldom results in an optimal three-dimensional pit. The modification which they proposed to the basic algorithm, as described in this section has been referred to as the 2½-D algorithm (Barnes, 1982). An overall view of the block model to be used is shown in Figure 5.76.

The column of blocks representing in the longitudinal section ($i = 4, j = 1$)

```
 2   Level 1
-3   Level 2
 1   Level 3
-7   Level 4
```

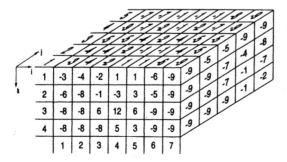

Figure 5.76. Block model used for the 2½-D projection bound (Johnson & Sharp, 1971).

Pit limits 439

This process is repeated for each of the other sections. The resulting longitudinal section is shown in Figure 5.79. The 2-D algorithm is now applied and the optimal contour becomes that given in Figure 5.80.

Figure 5.77 shows the 5 sections of a block model. Beginning with section 1, the optimum pit outlines are determined, given that one must mine to a given level k. In this case there are four levels to be evaluated. The results are shown in Figure 5.78. Each outline

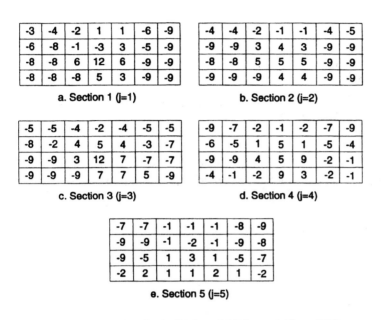

Figure 5.77. Block sections for the 2½-D model (Johnson & Sharp, 1971).

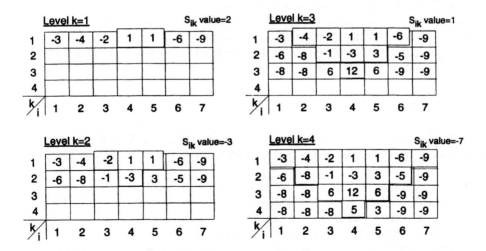

Figure 5.78. Analysis of section 1.

440 Open pit mine planning and design: Fundamentals

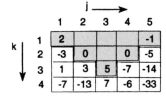

	j →				
	1	2	3	4	5
1	2	-1	-2	-1	-1
2	-3	0	-5	0	-5
3	1	3	5	-7	-14
4	-7	-13	7	-6	-33

Figure 5.79. Longitudinal section (Johnson & Sharp, 1971).

	j →				
	1	2	3	4	5
1	2				-1
2	-3	0		0	-5
3	1	3	5	-7	-14
4	-7	-13	7	-6	-33

Figure 5.80. Optimal longitudinal contour (Johnson & Sharp, 1971).

is found using a slightly modified version of the Lerchs-Grossmann 2-D algorithm. The net values for each of the four pits on this section are then determined. As can be seen these are:

Level 1: $S_{11} = 2$

Level 2: $S_{12} = -3$

Level 3: $S_{13} = 1$

Level 4: $S_{14} = -7$

To combine all of the transverse (ik) sections, a longitudinal section will now be taken. In viewing Figure 5.78, it is seen that the deepest mining on section 1 ($j = 1$) occurs at column $i = 4$. A column [$i = 4, j = 1$ (section 1)] of blocks will now be formed representing the net value of the cross-section as mined down to the various levels.

The net value for the material contained within the pit is found by summing the block values along the final contour. Here it becomes

Pit value $= 2 + 0 + 5 + 0 - 1 = 6$

Now that the bottom levels for each of the transverse sections have been determined, one goes back to these sections and selects the appropriate one. These are summarized in Figure 5.81.

The data and solution for a somewhat more complicated problem are given in Figures 5.82 and 5.83, respectively. The block height to length ratio along the longitudinal section (the axis of mineralization) is 1 to 2 while for the cross-sections the block height and length are equal (1 to 1). Hence to maintain a maximum pit slope of 45 degrees, the pit slope can change by two blocks per section along the length but by only block per column across the width.

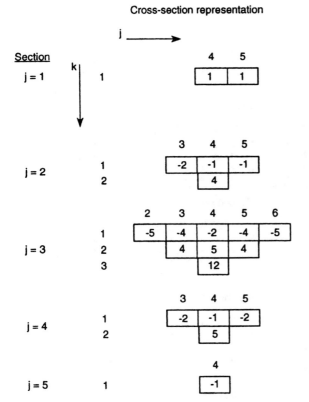

Figure 5.81. Optimal longitudinal pit sections (Johnson & Sharp, 1971).

5.7 THE LERCHS-GROSSMANN 3-D ALGORITHM

5.7.1 Introduction

When evaluating final pit dimensions based upon reserves expressed in the form of a grade block model, the overall objective is to find a grouping of blocks such that a selected parameter, for example:
- profit,
- metal content, or
- marginal value

is maximized. In preceding sections, some 2-dimensional approaches have been discussed. This is however a 3-dimensional problem and to obtain a true optimum such an approach is required.

For an orthogonal set of blocks there exist two basic geometries of interest for approximating an open-pit. They are:
- the 1–5 pattern, where 5 blocks are removed to gain access to one block on the level below and
- the 1–9 pattern, where 9 blocks are removed to gain access to one block on the level below.

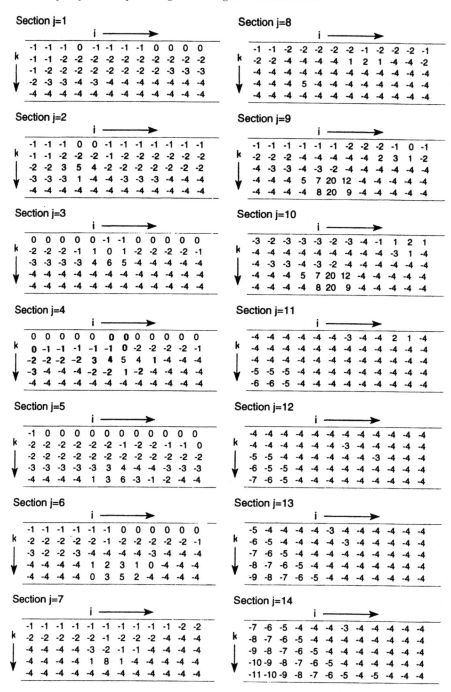

Figure 5.82. Block sections for a more complicated example (Johnson & Sharp, 1971).

Pit limits 443

	Optimal boundary blocks (level, k)												
Column (i)	1	2	3	4	5	6	7	8	9	10	11	12	Yields
Sections (j)													
1	0	0	0	0	0	0	0	0	0	0	0	0	0
2	0	0	0	0	1	1	1	0	0	0	0	0	-2
3	0	0	1	2	3	3	3	2	1	0	0	0	12
4	0	0	1	2	3	3	3	3	2	1	0	0	9
5	0	0	0	0	1	1	1	1	0	0	0	0	0
6	0	0	0	0	0	0	1	0	0	0	0	0	0
7	0	0	0	0	0	0	0	0	0	0	0	0	0
8	0	0	0	0	0	0	0	0	0	0	0	0	0
9	0	0	0	0	0	0	0	0	0	0	1	0	0
10	0	0	0	0	0	0	0	0	0	1	2	1	5
11	0	0	0	0	0	0	0	0	0	1	1	0	3
12	0	0	0	0	0	0	0	0	0	0	0	0	0
13	0	0	0	0	0	0	0	0	0	0	0	0	0
14	0	0	0	0	0	0	0	0	0	0	0	0	0

Optimal pit yields 27

Figure 5.83. The solution for the Figure 5.82 block model (Johnson & Sharp, 1971).

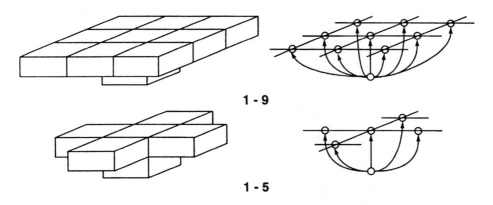

Figure 5.84. Representation of the 1-5 and 1-9 block constraints (Laurent et al., 1977).

The geometric configurations and equivalent graphic representations for these two geometries are shown in Figure 5.84. The nodes represent the physical blocks. The arrows (directed arcs) point toward those blocks immediately above which must first be removed before the underlying block can be mined. Each block has a weight associated with it. In general, the weight assigned is equal to the value of the parameter being maximized. Often this is the net economic value. The weight may be positive or negative.

Lerchs & Grossmann (1965) published the basic algorithm which when applied to a 3-D directed graph (block model) would yield the optimum final pit outline. This section presents in a simplified way, the basic concepts. They are illustrated by examples.

5.7.2 Definition of some important terms and concepts

There are a number of terms and concepts which are taken from graph theory (Lerchs & Grossmann, 1965; Laurent et al., 1977). The authors have tried to simplify them and yet retain their basic meaning.

Figure 5.85a is a section through a simple 2-D block model. As can be seen it consists of 6 blocks.

Each block is assigned a number (x_i) which indicates its location within the block model. In the case shown in Figure 5.85a, the block location are x_1, x_2, x_3, x_4, x_5, and x_6. One would know from the construction of the block model that the block designated as x_1 would actually have center coordinates of (2000, 3500, 6800). If there were 100,000 blocks in the block model, then x_i would go from x_1 to $x_{100,000}$.

The file of node locations may be expressed by

$$X = (x_i)$$

Figure 5.85b shows the 6 blocks simply redrawn as circles while maintaining their positions in 2-D space. For the graph theory application each of these circles is now called a 'node'. Straight line elements (called 'edges' in graph theory) are now added connecting the lower nodes to the nearest overlying neighbors. For the 3-D representations shown in Figure 5.84 there would be 9 edges for each underlying block in the 1-9 model and 5 for the 1-5 model. In this 2-D model there are 3 edges for each underlying block. The nearest overlying neighbors for node x_5 as shown in Figure 5.85c are nodes x_1, x_2 and x_3. The physical connection of node x_5 to node x_1 can be expressed either as (x_5, x_1) or (x_1, x_5). In Figure 5.85c, the nodes have been connected by 6 edges:

$$(x_1, x_5) = (x_5, x_1)$$
$$(x_2, x_5) = (x_5, x_2)$$

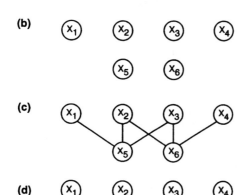

Figure 5.85. Steps in the model building process.

$$(x_3, x_5) = (x_5, x_3)$$
$$(x_2, x_6) = (x_6, x_2)$$
$$(x_3, x_6) = (x_6, x_3)$$
$$(x_4, x_6) = (x_6, x_4)$$

These edges (e_{ij}) can be described by

$$e_{ij} = (x_i, x_j)$$

The set containing all edges is given the symbol E:

$$E = (e_{ij})$$

A graph is defined as:

Graph: A graph $G = (X, E)$ is defined by a set of nodes x_i connected by ordered pairs of elements called edges $e_{ij} = (x_i, x_j)$.

The next step is to indicate which overlying blocks (nodes) must be removed prior to removing any given underlying block (node). This adds the required sequencing (flow) from the lowest most to the upper most blocks (nodes). To accomplish this, arrows are attached to the edges (lines) connecting the nodes (blocks) pointing in the direction of removal. This has been done in Figure 5.85d. By adding an arrowhead (direction) to an edge, the edge becomes an arc,

$$a_{kl} = (x_k, x_l)$$

An arc denoted by (x_k, x_l) means that the 'flow' is from node x_k to node x_l (the arrowhead is on the x_l end). The set containing all arcs in given the symbol A

$$A = (a_{kl})$$

The graph (G) consisting of nodes $\{X\}$ and arcs $\{A\}$ is called a directed graph.

Directed graph: A directed graph $G = (X, A)$ is defined by a set of nodes x_l connected by ordered pairs of elements $a_{kl} = (x_k, x_l)$ called the arcs of G.

One can consider the entire set of nodes (blocks) and the arcs connecting them (the directed graphs $G(X, A)$) or only a portion of it. A subset (Y) is referred to as a directed subgraph and represented by $G(Y, A_Y)$. An example of a directed subgraphs is shown in Figure 5.86. There are a great number of these subgraphs in the overall graph.

Subgraph: A directed subgraph $G(Y)$ is a subset of the directed graph $G(X, A)$. It is made up of a set Y of nodes and all of the arcs A_Y which connect them.

To this point in the discussion we have considered (1) the physical location of the blocks in space, (2) the connection of the blocks with one another and (3) the fact that overlying

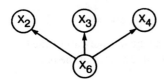

Figure 5.86. Example of a subgraph.

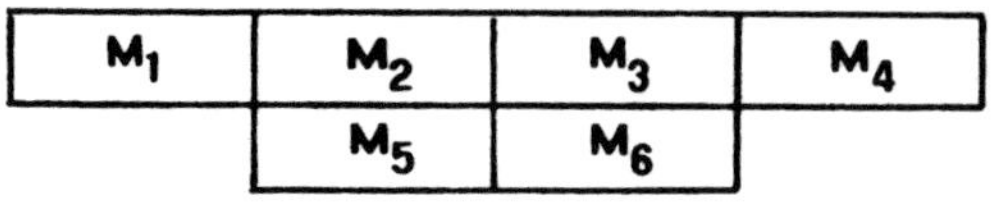

M_1	M_2	M_3	M_4
	M_5	M_6	

Figure 5.87. The weights assigned to the blocks of Figure 5.85.

blocks must be removed prior to mining lower blocks. Nothing has been said regarding the values of the individual blocks. Each block (x_i) has an associated weight (m_i). This is shown in Figure 5.87 for the blocks in Figure 5.85a. Although we have used net value as the assigned weight (m_i) to this point in the book other measures of worth or content can also be applied (profit, mineral content, etc.).

From a mining view point, the subgraph consisting of the four blocks. x_1, x_2, x_3 and x_5 could form a physically feasible pit. The subgraph consisting of blocks x_2, x_3, x_4 and x_6 could form another. A third possibility of a feasible pit is formed by the six blocks x_1, x_2, x_3, x_4, x_5 and x_6. There are many other feasible combinations. The subgraph x_2, x_3 and x_6 is not feasible since one of the overlying blocks x_4 has not been included. The term 'closure' is used to indicate a feasible subgraph.

Closure: Closure from the viewpoint of a mining engineer is simply a subgraph Y yielding a feasible pit.

Each one of these feasible pits (subgraphs) has an associated total weight (value). The challenge for the mining engineer is to find that one pit (subgraph) out of the great many possible which yields the maximum value. In graph theory this is referred to as finding the directed subgraph of 'maximum closure'.

Maximum closure: Again from the viewpoint of the mining engineer, maximum closure is that closure set, out of all those possible, which yields the maximum sum of block weights, i.e. where $M_Y = \sum m_i$ is a maximum.

A procedure based upon the application of graph theory is used to identify and sort through the various feasible pits in a structured way to find that yielding the maximum value. This corresponds to the optimum pit. To better follow the discussion the following definitions are introduced.

Circuit: A circuit is a path in which the initial node is the same as the find (terminal) node.

Chain: A chain is a sequence of edges in which each edge has one node in common with the succeeding edge.

Cycle: A cycle is a chain in which the initial and final nodes coincide.

Path: A path is a sequence of arcs such that the terminal node of each arcs is the initial node of the succeeding arc.

To illustrate the process, a discussion based upon a tree analogy is used. The terms 'tree', 'root', 'branch' and 'twig' are defined below:

Tree: A tree T is a connected and directed graph containing no cycles. A tree contains one more node than it does arcs. A rooted tree is a tree with a special node, the root.

Root: A root is one node selected from a tree. A tree may have only one root.

Branch: If a tree is cut into two parts by the elimination of one arc a_{kl}, the part of the tree not containing the root is called a branch. A branch is a tree itself. The root of the branch is the node of the branch adjacent to the arc a_{kl}.

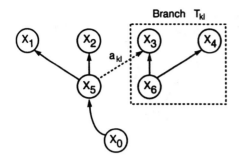

Figure 5.88. Example of a branch.

General Form					Assigned Weights			
x_1	x_2	x_3	x_4		-4	-4	-4	-4
	x_5	x_6				10	10	

Network Equivalent

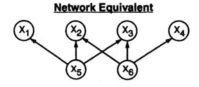

Figure 5.89. Example to be worked using the Lerchs-Grossmann 3-D algorithm.

Each arc a_{kl} of a tree T defines a branch T_{kl}. The weight W_{kl} of a branch T_{kl} is the sum of all weights associated with nodes of T_{kl}. An example of a branch is shown in Figure 5.88.

Twig: A twig is a branch of a branch.

As 'twigs' and 'branches' are added or cut from the 'tree', the value of the tree changes.

The Lerchs-Grossman algorithm is based upon a normalizing procedure in which a number of rules are followed. These will be demonstrated in detail in the next section.

5.7.3 *Two approaches to tree construction*

The algorithm starts with the construction of an initial tree T^0. This tree is then transformed into successive trees $T^1, T^2, \ldots, T^n$ following given rules until no further transformation is possible. The maximum closure is then given by summing the nodes of a set of well identified branches of the final tree. There are two approaches which may be used for generating the initial tree.

Approach 1. Construct an arbitrary tree having one connection to the root.

Approach 2. Construct a tree with each of the nodes connected directly to the root.

The simplest of these is Approach 2. Both approaches will however be applied to the simple example shown in Figure 5.89.

Although elementary, this is an interesting problem since the floating cone approach would suggest no mining at all. By inspection however one would strip the four waste

448 Open pit mine planning and design: Fundamentals

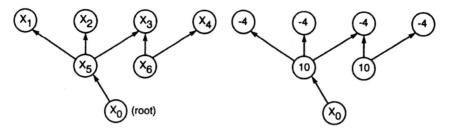

Figure 5.90. Addition of a root.

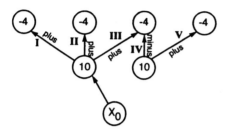

Figure 5.91. Labelling of plus and minus arcs.

blocks to uncover the two ore blocks. The net value for the resulting pit is expected to be +4.

The step-by-step approach used to demonstrate the techniques involved has been adapted from that presented by Laurent et al. (1977).

5.7.4 *The arbitrary tree approach (Approach 1)*

This approach (including the common normalizing procedure used with both Approaches 1 and 2) will be presented in step-by-step fashion.

Step 1. Begin by adding a root node x_0 to the directed graph and construct a tree of your choice keeping in mind the connection possibilities

From node x_5: (x_5, x_1)

(x_5, x_2)

(x_5, x_3)

From node x_6: (x_6, x_2)

(x_6, x_3)

(x_6, x_4)

The one chosen is shown in Figure 5.90. Each of the nodes (blocks) is connected to one of the others by a directed arc (arrow). One node is attached to the root.

Step 2. Each of the arcs is labelled with respect to whether it is directed away from the root (plus) or towards the root (minus). This is done in Figure 5.91.

Table 5.15. Labelling guide for arcs.

Case	Direction	Cumulative weight	Label
1	Plus	Positive	Strong
2	Plus	Null or negative	Weak
3	Minus	Positive	Weak
4	Minus	Null or negative	Strong

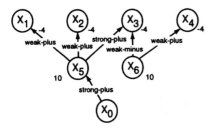

Figure 5.92. Labelling of weak and strong arcs.

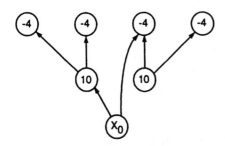

Figure 5.93. Creation of tree T^1.

Step 3. Beginning at the extremities each branch is worked back toward the trunk, summing the weights supported by the individual arcs. The objective is to add the words 'strong' or 'weak' to each of the arcs. Table 5.15 summarizes the labelling criteria.

We will begin on the left hand side of the tree. Roman numerals have been used to denote the segments for the discussion. For arc I, the direction is plus and the weight is negative. Thus the label to be attached is 'weak' (Case 2). For arc II, the same is true. For arc V, the direction is plus and the weight negative. The label is weak. It is also Case 2. When examining arc IV, the direction is minus and the cumulative weight is positive ($10 - 4 = 6$). Thus the label is weak (Case 3). Continuing to arc III, the direction is plus and the cumulative weight is positive ($10 - 4 - 4 = 2$). The label is 'strong' (Case 1). Figure 5.92 presents the resulting direction and label for each arc of this initial tree.

Step 4. The figure is now examined to identify strong arcs. Two actions are possible:

Action 1. A strong-minus arc: The arc (x_q, x_r) is replaced by a dummy arc (x_0, x_q). The node x_q is connected to the root.

Action 2. A strong-plus arc: The arc (x_k, x_l) is replaced by the dummy arc (x_0, x_l). The node x_l is connected to the root.

In this example there is one strong arc III. Since it is a 'strong-plus' arc, action 2 is taken. The arc connecting node x_5 (10) to node x_3 (−4) is removed. An arc connecting the root x_0 to node x_3 (−4) is drawn instead (Fig. 5.93). This becomes tree T^1.

450 *Open pit mine planning and design: Fundamentals*

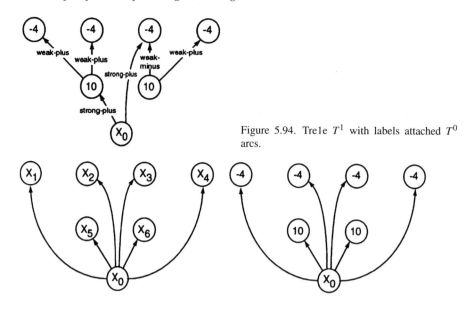

Figure 5.94. Trele T^1 with labels attached T^0 arcs.

Figure 5.95. Initial tree with all root connection.

Step 5. Tree T^1 is examined in the same way as before labelling the arcs as to whether they are 'plus' or 'minus' and 'weak' or 'strong'. This has been done in Figure 5.94.

Step 6. Any strong branches of the new tree not directly connected to the root are identified and the procedure discussed in Step 4 is followed. If there are no strong branches not connected to the root, the tree is said to be normalized and the process is over.

Step 7. The maximum closure consists of those nodes connected by strong arcs to the root. In this case the closure is

$$-4 - 4 + 10 - 4 - 4 + 10 = +4$$

5.7.5 *The all root connection approach (Approach 2)*

The step-by-step procedure is outlined below.

Step 1. Begin by adding a root node and connecting arcs between the root and each of the other nodes. For the example problem, this initial tree T^0 is shown in Figure 5.95. Note that all of the arcs are 'plus'.

Step 2. The set (graph) of directed arcs is now split into two groups. Those connected to the root by strong-plus arcs are included in group Y^0. The others are in group X-Y^0. In this case nodes x_5 and x_6 are in group Y^0. Their sum is 20.

Step 3. One must now look at the possible connections between the two groups. Following the sequencing constraints there are 6 directed arcs which can be drawn:

For x_5: (x_5, x_1)
                (x_5, x_2)
                (x_5, x_3)

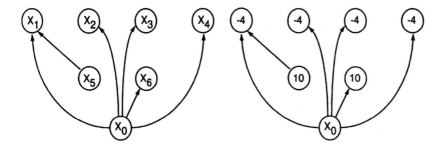

Figure 5.96. Selection of directed arc (x_5, x_1).

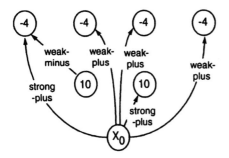

Figure 5.97. Labelling of the resulting arcs.

For x_6: (x_6, x_2)
(x_6, x_3)
(x_6, x_4)

One of these is selected. In this case it will be the connection (x_5, x_1). The directed arc (x_0, x_5) is removed and the directed arc (x_5, x_1) drawn. This is shown in Figure 5.96.

Step 4. The normalizing process is now followed. Each arc is labelled with respect to 'plus' or 'minus' and 'strong' or 'weak'. The result is shown in Figure 5.97. The arc(connection) between x_0 and x_1 is still strong-plus. Hence the members of the Y group are x_1, x_5 and x_6. The value of Y (closure) is 16.

Step 5. One now returns to step 3 to seek additional connections between the Y and X-Y (X without Y) groups. There are 5 feasible arcs

(x_5, x_2)
(x_5, x_3)
(x_6, x_2)
(x_6, x_3)
(x_6, x_4)

The arc (x_5, x_2) will be added to the tree and arc (x_0, x_2) dropped. This is shown in Figure 5.98.

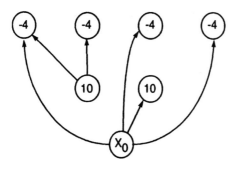

Figure 5.98. Addition of arc (x_5, x_2) and dropping arc (x_0, x_2) from the tree of Figure 5.97.

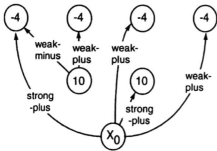

Figure 5.99. Labelling of the resulting arcs.

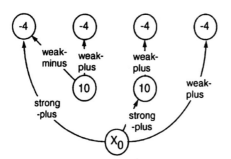

Figure 5.100. Addition of arc (x_6, x_3) and dropping arc (x_0, x_3) from the tree of Figure 5.97.

Step 6. The new tree is now normalized. The result is shown in Figure 5.99. The nodes included in Y are x_1, x_2, x_5 and x_6. The closure Y is 12.

Step 7. Returning to Step 5, there are now 3 possible connections remaining:

(x_5, x_3)

(x_6, x_3)

(x_6, x_4)

Here we will choose to add arc (x_6, x_3) and drop arc (x_0, x_3). (The choice (x_5, x_3) is an interesting one and will be evaluated later.) The resulting normalized tree is shown in Figure 5.100. The arc (x_0, x_6) remains strong-plus and hence the nodes included within Y are x_1, x_2, x_3, x_5, and x_6. The overall closure is 8.

Step 8. Returning to Step 5 there is one possible connection remaining: arc (x_6, x_4). Arc (x_0, x_4) is dropped and arc (x_6, x_4) added. The tree is normalized as before with the result

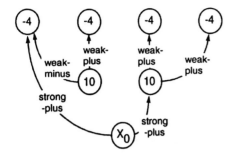

Figure 5.101. Addition of arc (x_6, x_4) and dropping arc (x_0, x_4).

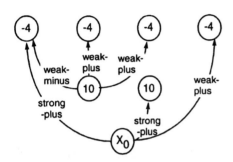

Figure 5.102. The resulting normalized tree for the alternative when arc (x_5, x_3) is added and arc (x_0, x_3) is dropped.

given in Figure 5.101. All of the nodes are now attached directly to the root by chains having one strong edge. There are no more connections to be tried.

Step 9. The maximum closure now is the cumulative sum of the nodes involved. In this case it is $+4$.

As was indicated, at the stage shown in Figure 5.98, it is possible to select connection (x_5, x_3). We will now return to this point and consider this option.

Step 7*. Arc (x_5, x_3) will be added and arc (x_0, x_3) dropped. The normalized tree is shown in Figure 5.102. As can be seen, the arc (x_0, x_1) has now become weak-plus.

The only member of the Y group is now x_6. The closure is 10.

Step 8*. One now considers the possible connections between the X-Y and Y groups. There are two possibilities.

(x_6, x_4)

(x_6, x_3)

Selecting (x_6, x_3) one obtains the normalized tree in Figure 5.103. The members of the Y group are x_6 and x_4. The closure is 6.

Step 9*. There is one possible connection between the two groups, that one being through arc (x_6, x_3). For the tree in Figure 5.104, arc (x_0, x_1) has been dropped and arc (x_6, x_3) added. The normalized tree is shown. As can be seen all of the nodes are connected to the root through a chain containing one strong edge. There are no more possible connections. The closure is the sum of the nodes which is $+4$.

454 Open pit mine planning and design: Fundamentals

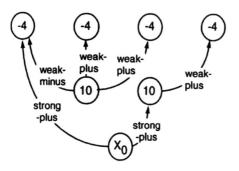

Figure 5.103. Normalized tree for connecting arc (x_6, x_3).

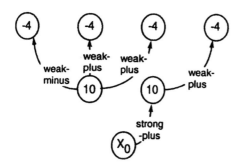

Figure 5.104. Normalized tree when adding arc (x_6, x_3) and dropping arc (x_0, x_1).

General Form			
X_1	X_2	X_3	X_4
	X_5	X_6	

Assigned Weights			
-10	-2	-2	-10
	10	20	

Figure 5.105. The example used to demonstrate tree cutting.

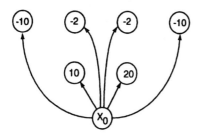

Figure 5.106. The arcs added to form the initial tree.

5.7.6 *The tree 'cutting' process*

The preceding problem was a simple case intended to familiarize the reader with the algorithm. In this example, the process of 'cutting' the tree during normalization will be demonstrated. The problem is as shown in Figure 5.105. The initial tree is as shown in Figure 5.106.

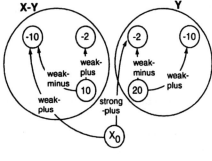

Figure 5.107. The normalized tree after 4 iterations.

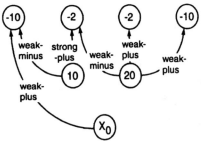

Figure 5.108. The arc (x_0, x_3) is dropped and arc (x_6, x_2) added.

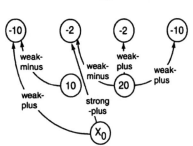

Figure 5.109. The arc (x_5, x_3) is cut and the arc (x_0, x_2) added.

The initial choices for connection are the same as before. They will be done in the following order:

For x_5: (x_5, x_1)

            (x_5, x_2)

For x_6: (x_6, x_3)

            (x_6, x_4)

After these 4 iterations the normalized tree would appear as in Figure 5.107.

The Y group is shown to consist of nodes x_3, x_4 and x_6. There is one remaining connection to be made between the groups (x_6, x_2). In Figure 5.108, arc (x_0, x_3) is dropped and arc (x_6, x_2) added. The tree is then normalized. As can be seen the arc (x_5, x_2) is strong-plus and not connected directly to the root. It must be cut in order to normalize the tree.

As discussed earlier the arc (x_k, x_l) is replaced by the dummy arc (x_0, x_l). In this case $x_k = x_5$, $x_l = x_2$. Thus the new arc is (x_0, x_2). This is shown in Figure 5.109. The resulting tree is in normalized form. All of the connections have been tried. The final closure is $20 - 10 - 2 - 2 = +6$. It can be seen that if Case 1 in Table 5.15 had been written as

456 Open pit mine planning and design: Fundamentals

'cumulative weight = null or positive' rather than just as 'cumulative weight = positive', then both nodes x_1 and x_5 would have been included in the maximum closure as well.

5.7.7 A more complicated example

Figure 5.110 shows a more complicated directed network to be analyzed using the Lerchs-Grossmann algorithm (Laurent et al., 1977). The results are shown in Figure 5.111. The interested reader is encouraged to select the upper two layers to work as an exercise. As can be seen, the two blocks at the pit bottom $+10$ and -10 would contribute a net of 0 to the section value and hence have not been mined.

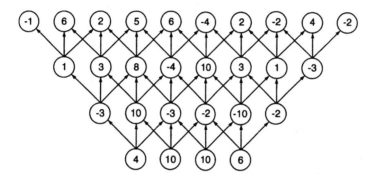

Figure 5.110. A more complicated example on which to practice the algorithm (Laurent et al., 1977).

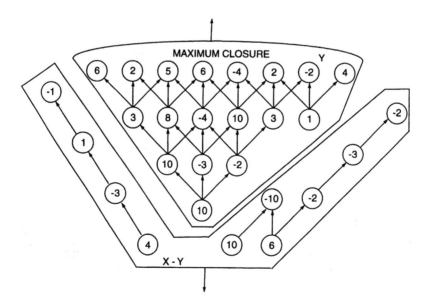

Figure 5.111. Maximum closure for the more complicated example (Laurent et al., 1977).

5.8 COMPUTER ASSISTED METHODS

5.8.1 *The RTZ open-pit generator*

Fairfield & Leigh (1969) presented a paper 'A computer program for the design of open-pits', which outlines a pit planning procedure in use by many mines today. This section will describe, using the material of Fairfield & Leigh (1969), the basic logic involved. The computer techniques using sections do not solve the problem of end sections or the smoothing of perimeters and consequently have definite limitations. The projection of volumes in the form of cones also has some of the same limitations. The approach described here is the projection of plan areas, specifically the projection of perimeter outlines.

The process begins with the development of a block model. Each block as a minimum would be assigned location coordinates and an index character dependent on the rock type or grade type.

The second step in the process is the selection of one or more base perimeters from which to generate the pit. These define the final horizontal extent of the pit at, or close to, the elevation of the pit's final base. Each perimeter so drawn becomes the trial base from which an overall pit is generated.

The information that is required to carry out these calculations is:
1. Size and shape of ore body.
2. Ore and rock types present.
3. Grades of ore.
4. Rock slope stability.
5. Size and shape of base perimeter.
6. Cutoff grades.
7. Depth of pit.
8. Unit working costs.

The trial base perimeter is defined by a series of short chords (Fig. 5.112).

The coordinates of the chord end points are read manually and then entered into the computer or are read and entered directly using a digitizer. Such a file of clockwise read perimeter points is given in Table 5.16.

At every horizon, proceeding in a clockwise order, each defining perimeter chord is considered in turn. For each, the defining end coordinates are located.

To illustrate the basic method, consider the portion of the pit base defined by chords H_1I_1 and I_1J_1 shown in Figure 5.113.

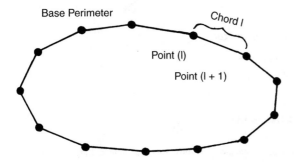

Figure 5.112. Trial based perimeter (Fairfield & Leigh, 1969).

458 Open pit mine planning and design: Fundamentals

The coordinate pairs H_1, I_1 and J_1 are projected outwards and upwards to the next upper horizon in a direction normal to the perimeter at that point. The next upper horizon occurs at a specified distance above the current horizon. The actual interval is determined by the requirements of the job at hand and is some fractional part (usually one-half) of the bench height. The slope angle used is that permitted in the particular rock type or at that pit position.

Table 5.16. Perimeter coordinates* for trial base (Fairfield & Leigh, 1969).

Perimeter coordinates

Y	X	Y	X	Y	X	Y	X
445.2	601.8	420.5	619.0	404.3	641.7	391.7	659.4
394.3	681.2	397.2	705.8	401.2	726.7	408.6	746.3
410.0	768.1	415.6	796.4	424.6	817.9	428.5	843.3
428.7	869.6	424.5	888.0	432.1	916.6	428.4	945.3
437.0	969.0	415.6	984.7	393.5	1001.5	378.6	1021.8
369.7	1050.1	371.4	1077.4	378.7	1106.8	382.8	1130.0
382.8	1150.3	382.8	1170.8	382.8	1195.7	395.4	1212.5
408.1	1229.5	420.5	1246.0	434.0	1264.0	447.3	1281.7
468.7	1287.3	490.6	1292.9	516.1	1298.5	546.1	1305.2
569.3	1299.1	502.9	1292.0	616.3	1280.2	640.4	1265.8
668.2	1242.7	681.3	1228.7	695.1	1214.9	716.2	1193.8
727.9	1166.9	746.5	1147.0	763.6	1125.1	776.1	1101.4
783.7	1075.3	783.0	1048.0	770.0	1027.6	754.6	1008.0
757.4	984.5	751.0	958.6	748.8	933.5	744.7	912.9
742.4	884.9	744.9	864.6	755.4	841.3	760.6	810.1
756.9	789.2	752.9	766.9	741.7	749.0	723.7	726.2
706.5	711.6	682.8	693.9	661.7	679.5	647.0	667.9
631.1	652.4	619.9	631.5	610.7	610.1	594.9	597.2
571.7	575.8	547.3	559.6	512.2	562.4	488.3	578.3
464.3	598.4						

* Coordinate pairs are connected reading from left to right and top to bottom in the table.

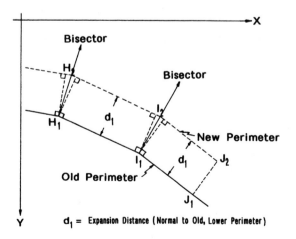

Figure 5.113. Projection of chords and perimeter points along the bisectors (Fairfield & Leigh, 1969).

If the slopes at the chord ends are different then the mean value is used.

$$\text{Slope of chord I} = \frac{1}{2}\,(\text{permitted slope at point (I)}$$
$$+ \text{permitted slope at point (I}+1)) \qquad (5.17)$$

Sufficient information is now available:
  (i) the coordinates of the chords defining the current perimeter,
  (ii) the average permitted rock slopes of these chord locations,
  (iii) the elevation difference to the next upper horizon,
to generate the pit perimeter on the next upper horizon. With this information the program works clockwise around the perimeter, considering pairs of adjacent chords in turn.

Consider the case where adjacent chords HI and IJ have the same permitted rock slopes θ and I is to be projected from Horizon 1 to Horizon 2 through a vertical height V. The situation is shown in plan in Figure 5.113 and in isometric view in Figure 5.114.

The calculation of the new coordinates of I on level 2 (I_2) is given below:

$$\text{Bisector of angle } H_1I_1J_1 = \overrightarrow{I_1A}$$
$$\text{Length of } I_1A = V \cot\theta$$
$$\text{Bearing angle of } I_1A = \alpha$$
$$\text{New coordinates of } I_2 = (x_1 + dx, y_1 - dy)$$
$$= (x_1 + I_1A \sin\alpha, y_1 - I_1A \cos\alpha)$$

As an example assume that
$$\theta = 45°$$
$$V = 10\,\text{m}$$
$$x_1 = 510.36$$
$$y_1 = 840.98$$
$$\overrightarrow{I_1A} = N50°E$$
$$\alpha = 50°$$

The horizontal distance I_1A is given by
$$I_1A = V \cot\theta = 10 \cot 45° = 10\,\text{m}$$

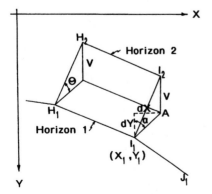

Figure 5.114. Isometric projection of chord HI (Fairfield & Leigh, 1969).

460 Open pit mine planning and design: Fundamentals

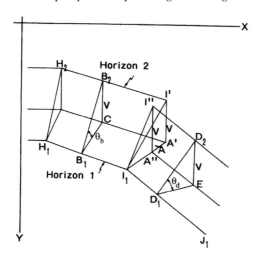

Figure 5.115. Isometric projection of adjacent chords HI and IJ (Fairfield & Leigh, 1969).

Using the bearing angle, one can calculate the distances dx and dy.

$$dx = I_1 A \sin \alpha = 10 \sin 50° = 7.66$$
$$dy = I_1 A \cos \alpha = 10 \cos 50° = 6.43$$

Hence the coordinates (x_2, y_2) of I_2 become

$$x_2 = x_1 + dx = 510.36 + 7.66 = 518.02$$
$$y_2 = y_1 - dy = 840.98 - 6.43 = 834.55$$

When adjacent chord slopes are not the same, (Fig. 5.115) an 'averaging' method is used in order to smooth out sharp changes in slope.

(i) The rock slope of any chord is the mean of the slopes as obtained at the chord end points.

$$\text{For chord } H_1 I_1, \text{ slope} = \theta_b = \frac{1}{2}(\theta_h + \theta_i)$$

$$\text{For chord } I_1 J_1, \text{ slope} = \theta_d = \frac{1}{2}(\theta_i + \theta_j)$$

(ii) The horizontal distance IA (where A is the midpoint of $A'A''$) is the mean of the distances BC and DE.

Since

$$B_1 C = V \cot \theta_b$$
$$D_1 E = V \cot \theta_d$$

Then

$$I_1 A = \frac{1}{2}(B_1 C + D_1 E)$$

or

$$I_1 A = \frac{1}{2} V \left(\cot \frac{\theta_h + \theta_i}{2} + \cot \frac{\theta_i + \theta_j}{2} \right)$$

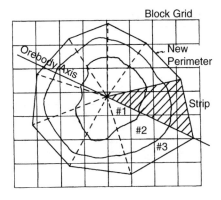

Figure 5.116. Superposition of the bench perimeters on the block model.

The new coordinates of I_2 are given by

New coordinates of $I_2 = (x + I_1 A \sin \alpha, y - I_1 A \cos \alpha)$

The program then continues around the perimeter and considers each pair of chords in turn and calculates the coordinates of the projected chord end points on the next upper horizon.

When the perimeter on the next upper horizon has thus been generated, the area lying within the new perimeter is scanned and, by reference to the rock type matrix and appropriate vertical interval, the volumes and tonnages of the various rock types between the lower and upper horizons are calculated. The principle adopted is analogous to that of obtaining areas by graph paper rather than by planimeter. Volumes are calculated down to the previous horizon on which a perimeter was generated.

The program works around the perimeter and considers each chord in turn. For each chord, the elemental area between the chord and the axis of the ore body is evaluated in terms of the types of rocks falling within the strip (see Fig. 5.116).

By referring to the current rock matrix as stored in the 'memory' of the computer, the zone between the chord and axis is scanned. Each rock type present is identified and counted in terms of blocks. Then, by reference to the cell dimensions and the level interval, the number of blocks is converted into a volume. In this way all the elements within the pit perimeter are evaluated and progressive totals of each rock type are built up. Table 5.17 shows the type of output which is obtained.

Before continuing to project the perimeter up to the next horizon two checks are made and adjustments carried out if necessary. First the perimeter chords are checked to ensure that the chord length is maintained within a specific range. If the chord length exceeds a certain permitted maximum, an intermediate coordinate point is introduced at the chord's mid point. On the other hand, in certain circumstances, the chord length may become very small and on further projection the end points may actually reverse their order. This will give rise to ambiguous calculations resulting in a badly distorted perimeter, so in order to avoid this situation points are dropped out as chord lengths become less than a certain permitted minimum. The program then scans the perimeter and checks for the development of sharp angles between adjacent perimeter chords, which are compared with a given minimum angle (set with the input data). If the acute chord angle is too sharp, perimeter points are adjusted until the angle becomes larger than the permitted minimum. This part of the program obviates the problem of the development of sharp pit-perimeter curves.

462 *Open pit mine planning and design: Fundamentals*

Table 5.17. Typical bench output using the RTZ pit generator (Fairfield & Leigh, 1969).

Test case 3
Base perimeter : 720 ft
Bench No: 11 (Elevation = 440 ft)
Footwall slope angle = 45°
Hangingwall slope angle = 55°

Rock type	1	2	3	4	5	6	7	8
Area (ft^2)	4423.4	29,537.0	59,059.9	45,353.7	14,685.8	1440.0	73,175.6	0.0
Volume (ft^3)	146,627.0	998,383.0	1,793,687.0	1,674,718.0	358,986.0	29,038.0	2,327,491.0	0.0

Total area = 227,675.8 ft^2
Total volume = 7,328,930 ft^3

Table 5.18. Typical total pit output using the RTZ pit generator (Fairfield & Leigh, 1969).

Test case 3: Final summary
Base perimeter : 720 ft
Footwall slope angle: 45°
Hangingwall slope angle: 55°

Rock type	1	2	3	4	5	6	7	8
Volume (ft^3)	3,348,438	5,979,359	17,496,752	13,867,850	9,031,000	1,029,359	51,243,824	7,470,351
Tonnage	304,403	519,944	1,458,062	1,109,428	694,692	76,249	3,660,273	
Cutoff grade (%)					0.6	0.5	0.4	0.3
Volumetric stripping ratio W : O					10.0	2.8	1.5	1.1
Total tonnage of ore (tons)					824,347	2,282,409	3,391,837	4,086,529
Total tonnage of waste (tons)					6,998,704	5,540,642	4,431,214	3,726,522
Average mill feed grade (%)					0.64	0.55	0.50	0.47
Total working costs (units)					27,143,601	32,642,173	34,924,076	36,024,585

Having carried out these checks the program continues to the next horizon where the complete cycle is repeated. Then so on until the final pit perimeter is reached.

Table 5.18 is a typical final summary output from the program. The following are given:
 (i) Pit perimeter coordinates.
 (ii) Volume and tonnage of ore.
 (iii) Volume and tonnage of waste.
 (iv) Stripping ratio.
 (v) Average grade of mill feed.
 (vi) Total working costs.

In cases where working costs and revenues are required additional input to the computer is necessary. Costs are fed in under headings: 'fixed costs per rock type unit' (that is, drilling, blasting, loading), 'depth variable costs' (that is, transport from various working levels), and 'fixed cost per unit of rock' (that is, supervision, services).

The basic pit design method described so far develops a single pit from a single base. Having completed a design for a single pit, the user can continue to add on additional pits either as incremental expansions to the first pit by incorporating a new base which joins the

Pit limits 463

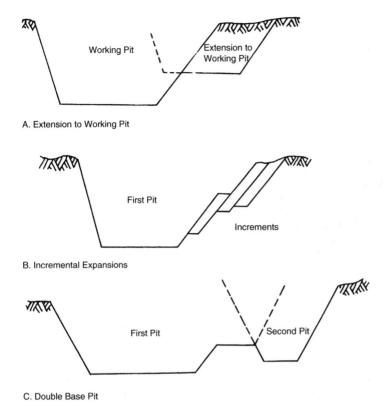

Figure 5.117. More complicated pit expansions (Golder Associates, 1981).

first pit, or by specifying a second base lying outside the first pit thereby creating a double base pit. The same procedure is adopted. A trial base perimeter for the second pit is input to the computer in the same form, and the program repeats the cycle of projection from level to level until the surface is reached.

The perimeter of this second pit will probably encroach on that of the first pit. However, with the aid of its memory the computer will take into consideration the fact that one pit has already been generated. As soon as encroachment occurs the complete volume bounded by the first pit perimeter will be treated as being mined out and hence air. This prevents any duplication that would otherwise occur. Third and further bases can be added by similar means. A second or further base can represent:

1. A working extension to the first pit (Fig. 5.117A).
2. An incremental expansion of the pit to test the sensitivity of the first pit to profitability (Fig. 5.117B).
3. A second pit which may merge with the first at the upper benches (Fig. 5.117C). A typical computer assisted pit design is shown in Figure 5.118.

5.8.2 *Computer assisted pit design based upon sections*

Open-pit design computerization which began in the early 1960's was driven largely by those involved in the mining of large, low grade deposits. The choice of a block model to

464 *Open pit mine planning and design: Fundamentals*

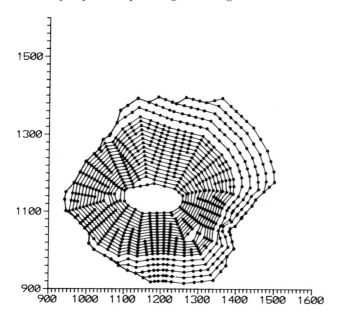

Figure 5.118. A typical pit design (Motta, 1988).

represent porphyry copper deposits, for example, was appropriate. The block size could be made still quite large and yet be small with respect to the size of the deposit and the scale over which the major grade changes take place. The distinction between ore and waste was made on the basis of cut off grade rather than on a sharp boundary.

There are a number of applications where block models are less satisfactory. Steeply dipping, relatively narrow vein/strata form deposits being one example. Here the ore is contained within definite, well defined boundaries. Grade variation within the ore zone is such that an average ore grade may be assigned. In such cases, the use of vertical sections still plays an important role. One computer assisted technique for dealing with such situations has been described by Luke (1972). The example which he presented forms the basis for this section.

Figure 5.119 illustrates the topography and geologic interpretation of a single cross-section through a steeply dipping iron orebody. A thin layer of overburden (alluvium, sand, gravel) overlies the bedrock. There is a sharp transition between the ore zone and the adjacent hanging- and footwall waste zones.

In the manual method of pit design using sections described in an earlier section, it will be recalled that the final pit limits were obtained through an iterative process by which the pit shape was 'floated' around on the section until the actual stripping ratios and average ore grades matched a point on the SR-grade curve.

Such a trial pit for this section is shown in Figure 5.120. The pit bottom has a width of 120 ft, the slope of the right hand slope (RHS) is 60° whereas the left hand slope (LHS) is reduced to 57° to include the presence of a 60 ft wide ramp segment. The straight line approximations for the slopes are used to simplify the process. Once the 'final' best position is determined then the functional mining parameters such as:
– the ramp(s),
– working bench heights,

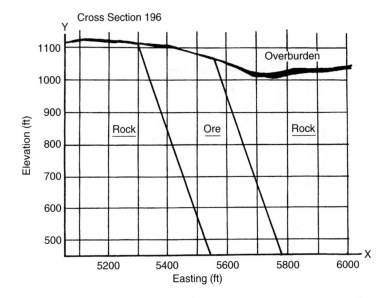

Figure 5.119. Topography and geologic interpretation of a single cross-section for a steeply inclined orebody (Luke, 1972).

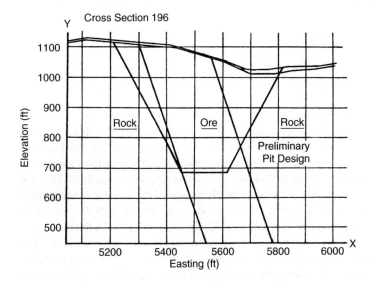

Figure 5.120. Preliminary pit design for example cross-section (Luke, 1972).

– berm widths, and
– bench face angles
are added (Fig. 5.121).

The design is then reexamined. Often significant changes occur between the simplified and actual pit designs. When the 'final' design on this section has been located, the areas of

466 Open pit mine planning and design: Fundamentals

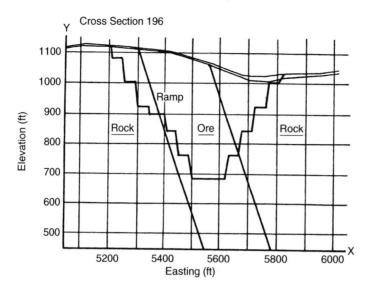

Figure 5.121. Preliminary pit design for example cross-section (Luke, 1972).

overburden, ore and waste are obtained by planimetering. Overall stripping ratios, etc. can then be calculated.

Design changes such as varying
- minimum pit width,
- slope angles,
- position/width of the ramp, and
- limiting stripping ratio

would require the entire process to be repeated.

Computer techniques have been developed to assist in the preparation of the sections and in making the necessary calculations. In this way a large number of potential designs can be evaluated quickly and inexpensively. Since the procedure largely follows the manual process, it is easy for the mine planner to understand what the program is accomplishing and to actively participate in computer-assisted design. The output of the process is a functional mine plan together with a working set of cross-sections. This section will discuss via an iron-ore example, originally presented by Luke (1972) the logic involved in the process.

The process begins with a description of the topography and the geologic data. The topographic relief and the overburden-bedrock contact are defined by a series of straight line segments. The end points of each segment making up the individual contour line (string) are defined either by hand or using a digitizer. In this case the surface relief will be denoted as the surface contour and the top of bedrock as the overburden contour. The values for cross-section 196 are given in Table 5.19.

From the cross-section (Fig. 5.119) it is seen that there are four material types present. Since they adjoin one another, they can be defined by four zones. If, for example, a pod of hangingwall waste exists in the ore then five zones would be needed. These zones must be described. In this simple case the ore-footwall and ore-hangingwall contacts can be described using end point coordinates (Table 5.20).

Table 5.19. Surface and overburden contour strings for cross-section 196 (Luke, 1972).

Surface contour		Overburden contour	
X	Y	X	Y
4870	1100	4870	1097
5000	1106	5000	1103
5114	1125	5111	1122
5251	1114	5222	1113
5298	1114	5305	1110
5416	1102	5406	1100
5539	1073	5419	1100
5613	1050	5523	1073
5673	1024	5590	1052
5723	1021	5710	1010
5743	1020	5770	1008
5838	1035	5842	1023
5940	1034	5940	1028
6000	1043	6057	1043
6100	1050	6121	1043
6121	1049	6200	1030
6200	1035		

Table 5.20. End points for zone lines (Luke, 1972).

	X_1	Y_1	X_2	Y_2
Ore-footwall contact	5298	1112	5550	440
Ore-hangingwall contact	5558	1065	5787	440

Table 5.21. Material descriptors (Luke, 1972).

Zone	Material type	Material description	Divisor
1	50	Rock waste	27.0
2	1	Ore	11.5
3	50	Rock waste	27.0
4	51	Overburden	64.8

The overburden zone lies between the surface and the overburden contour lines. The footwall (hangingwall) waste zone lies to the left (right) of the ore-footwall (hangingwall) zone line and below the overburden contour line. The ore zone lies between the contact lines and beneath the overburden contour line. Each zone corresponds to a particular material type which in turn has certain properties. In this case the area of each material included in the pit will first be determined. This will be changed to a volume by multiplying by the given section thickness (i.e. 100 ft). To convert this volume into the desired units of tons, cubic yards, cubic yards of equivalent rock, etc., certain factors are required. A table of such factors is given in Table 5.21. In this case the factor of 11.5 is used to convert volume (ft^3)

Table 5.22. Pit wall specifications (Luke, 1972).

	Bench height	Berm width	Wall angle	Valid berm elevation	Overburden angle	Toe coordinates	
						X	Y
Left	80	35	82	1000	45°	5500	680
Right	80	35	82	1000	45°	5620	680

into long tons of crude ore. The factor of 27 converts volume (ft^3) into cubic yards of rock waste. Since the overburden is much easier to remove than the rock waste, a factor of 2.4 is first introduced to convert volume of overburden (ft^3) into an equivalent volume (ft^3) of rock waste. The factor of 27 is then used to convert ft^3 into yd^3. The overall factor is the product of these two (64.8).

The pit design can now be superimposed upon the basic material-geometry model. As done earlier the pit is defined by a series of connected straight segments. The width and position of the pit bottom is first decided. For Figure 5.120 one can see that the width is 120 ft and the bottom elevation is 680. The end points of this segment form the toe positions of the left and right hand slopes. The walls are defined by:

1. Bench height. The bench elevation differential on the ultimate wall.
2. Bench width.
3. Wall angle. The bedrock wall angle from bench to bench, not the overall slope angle.
4. Valid bench elevation. The elevation of any existing berm if the pit is under current development, or the elevation of the first proposed bench. From this specified bench, the elevation of successive benches is determined from the bench height.
5. Overburden angle. The wall angle that can be maintained in overburden.
6. Toe position. The (X, Y) coordinates for the indicated intersection of the pit wall and pit floor.

The wall and pit bottom specifications for the trial pit are given in Table 5.22. The design specifies the location of the ramp as to
- left or right wall,
- ramp elevation when it crosses the section, and
- ramp width,

For the section shown it has been decided that the ramp should have

location = left wall,

elevation = 896 ft, and

width = 60 ft.

In manually superimposing the pit onto the section, the designer would locate the pit bottom and then using scale and protractor construct the pit walls. The ramp would be positioned at the proper location. The same procedure is followed with the computer except that end point coordinates of the segments are determined. In addition, the coordinates of the intersections between the string making up the pit and the zone/contour lines are determined as well. Once this has been done, the areas involved in each zone can be obtained. For the ore zone (Zone 2) using a planimeter one might start in the lower left corner (Fig. 5.122) and proceed around the loop in a clockwise motion eventually returning to the starting corner.

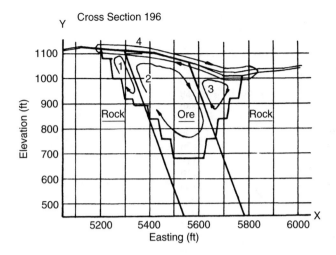

Figure 5.122. Summary of developed data for the example cross-section (Luke, 1972).

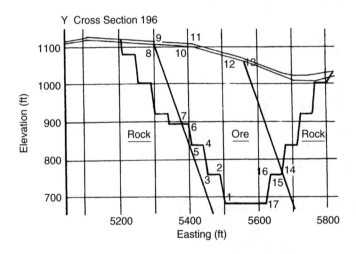

Figure 5.123. Ore area calculation based on the segment end points.

The same basic process is followed knowing the end point coordinates of the segments. Figure 5.123 shows the end point locations and Table 5.23 gives the coordinates.

To demonstrate the process by which areas are found consider the area excavated (A) from the bottom bench in Figure 5.124. The formula used is

$$A = \frac{1}{2}(Y_2 + Y_1)(X_2 - X_1)$$

where Y_1, X_1 are coordinates of initial point of segment and Y_2, X_2 the coordinates of final point of segment.

470 Open pit mine planning and design: Fundamentals

Table 5.23. Coordinates of the segment end points (Luke, 1972).

Point	X	Y
1	5500	680
2	5488.5	760
3	5453.5	760
4	5442.0	840
5	5407	840
6	5399.1	896
7	5379	896
8	5298	1112
9	5305	1110
10	5406	1100
11	5419	1100
12	5523	1073
13	5558	1065
14	5667.7	766.8
15	5666.6	760
16	5631.5	760
17	5620	680
1	5500	680

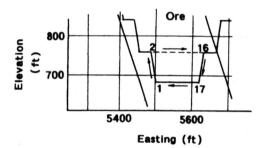

Figure 5.124. Figure used for the example calculation.

In this case, going clockwise around the figure are finds that

Segment 1: 1 to 2

$$A_1 = \frac{1}{2}(760 + 680)(5488.5 - 5500)$$
$$= -8280$$

Segment 2: 2 to 16

$$A_2 = \frac{1}{2}(760 + 760)(5631.5 - 5488.5)$$
$$= 108,680$$

Segment 3: 16 to 17

$$A_3 = \frac{1}{2}(680 + 760)(5620 - 5631.5)$$
$$= -8280$$

Pit limits 471

Segment 4: 17 to 1

$$A_4 = \frac{1}{2}(680 + 680)(5500 - 5620)$$
$$= -81,600$$

The total area is

$$A = A_1 + A_2 + A_3 + A_4 = -8280 + 108,680 - 8280 - 81,600 = 10,520 \text{ ft}^2$$

Using the formula for the area of the trapezoid one finds that

$$A_{tz} = \frac{1}{2}(b_1 + b_2)h$$

where

b_1 = length of lower base = 120 ft
b_2 = length of upper base = 143 ft
h = height = 80 ft

Substitution yields:

$$A_{tz} = \frac{1}{2}(120 + 143) \times 80 = 10,520 \text{ ft}^2$$

The results from applying this line integration method to zones 1, 2, 3 and 4 in Figure 5.122 are given below

Zone	Material type	Compound area (ft²)	Section thickness (ft)	Volume (ft³)
1	50	14,785	100	1,478,500
2	1	87,500	100	8,750,000
3	50	27,931	100	2,793,100
4	51	3264	100	326,400

Using the scaling factors these values are converted into those desired:

Footwall rock waste = 54,759 yd³
Ore = 760,870 long tons (lt)
Hangingwall rock waste = 103,448 yd³
Overburden (equivalent waste rock) = 5,037 yd³

The material totals are:

Ore = 760,870 lt
Waste = 163,224 yd³
SR = 0.215 yd³/lt

Through the use of this pit generator, a variety of pit locations can be tested. In the manual procedure described in Section 5.2, the lengths of the lines in ore and waste were compared. Here the ratio of waste to ore lying between two successive pit positions will be calculated. Two such positions are shown in Figure 5.125.

472 Open pit mine planning and design: Fundamentals

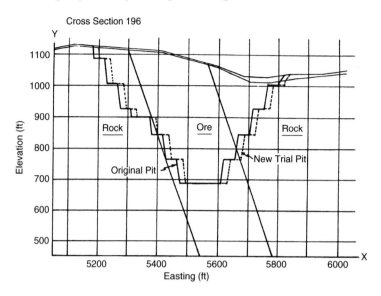

Figure 5.125. Results of the optimization as compared to the preliminary design of Figure 5.121 (Luke, 1972).

The left hand wall was moved 25 ft to the left and the right hand wall 15 ft to the left. For this new position the (ore)$_n$ and (waste)$_n$ totals are:

$$(\text{ore})_n = 787,366 \text{ lt}$$

$$(\text{waste})_n = 172,687 \text{ yd}^3$$

The change in the amount of ore between these two pits is

$$\Delta\text{ore} = (\text{ore})_n - (\text{ore})_o = 787,366 - 760,870 = 26,496 \text{ lt}$$

Similarly, the change in the amount of waste is

$$\Delta\text{waste} = (\text{waste})_n - (\text{waste})_o = 172,687 - 163,244 = 9443 \text{ yd}^3$$

For this change in geometry the incremental (or differential) stripping ratio (DSR) of the increment is equal to

$$\text{DSR} = \frac{\Delta\text{waste}}{\Delta\text{ore}} = \frac{9443}{26,496} = 0.356$$

This value must be compared to the break even (or limiting) stripping ratio (SRL) as applied at the pit periphery. Suppose for example that

$$\text{SRL} = 0.8$$

Since

$$\text{DSR} < \text{SRL}$$

this modification of the pit is desirable. The formula presented by Luke (1972) for use in guiding the changes is

$$\frac{\Delta T - \frac{\Delta W}{\text{SLR}}}{|\Delta T|} \geq 0 \qquad (5.18)$$

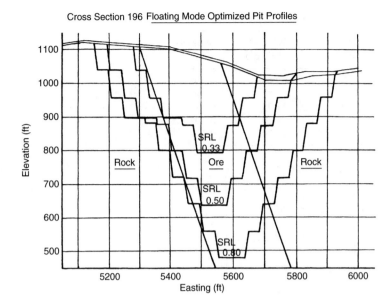

Figure 5.126. Pit profiles for three SRL values (Luke, 1972).

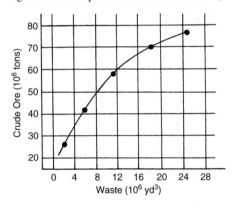

Figure 5.127. Curve of reserve level versus minimum stripping (Luke, 1972).

where

$$\Delta T = T_t - T_b,$$

$$\Delta W = W_t - W_b,$$

$$T = \text{units of ore},$$

$W =$ units of waste,
SRL $=$ incremental stripping ratio,
subscript b refers to the current best position,
subscript t refers to the new trial position.

The optimization can be done under a variety of constraints. One constraint, for example, might be that the pit floor level must remain at a given elevation. When SRL is the only constraint, then the pit can float both vertically and horizontally around the section. Figure 5.126 shows pits for section 196 under different SRL constraints. Figure 5.127 shows the tons of crude ore as a function of stripping for the different SRL scenarios.

Table 5.24. Summary for the optimum pits (Luke, 1972).

Incremental stripping ratio (SRL)	Crude ore (10^6 tons)	Waste (10^6 yd^3)	Stripping ratio
0.33	26.0	2.08	0.080
0.50	42.0	5.87	0.140
0.67	58.0	11.30	0.195
0.80	70.0	18.20	0.260
1.00	76.0	24.70	0.325

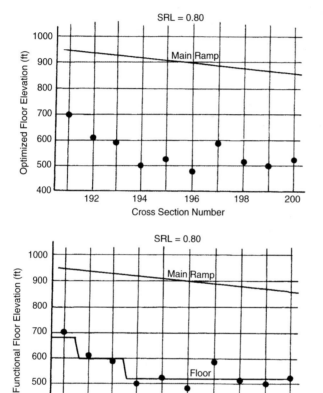

Figure 5.128. Plot of optimum pit floor elevation versus cross-section position in the pit (Luke, 1972).

Table 5.24 presents an overall stripping ratio summary. Such results are very useful for management in examining production decisions.

Figure 5.128 presents the final floor elevations for a series of adjacent sections in which a limit of SRL = 0.8 was imposed. Using this plot, decisions regarding bottom bench location at various locations in the pit can be made. Having decided this elevation the sections can then be rerun to obtain an optimum location. With regard to section 196, the pit bottom should be at 520 ft rather than 480 ft. As can be seen, this use of computer assist, greatly enhances the information base from which the mine planner can make decisions.

REFERENCES

Alford, C.G. & J. Whittle 1986. Application of Lerchs-Grossmann pit optimization to the design of open-pit mines. In: *Large Open-Pit Mining Conference* (J.R. Davidson, editor) 10: 201–208. Australasian Inst. of Min. and Met.

Axelson, A.H. 1964. A practical approach to computer utilization in mine planning. In: *4th APCOM, Golden, Colorado*: 593–622. Quarterly of the Colorado School of Mines.

Baafi, E.Y., Milawarma, E., and C. Cusack 1995. Computer aided design and analysis of Sapan Dalam open pit mine. 25th International Symposium on Application of Computers and Operations Research in the Mineral Industry: 1283–1288. Brisbane:AusIMM.

Barnes, R.J. 1982. Optimizing the ultimate pit. MS Thesis T-2611. Colorado School of Mines.

Barnes, R.J. & T.B. Johnson 1982. Bounding techniques for the ultimate pit limit problem. In: *Proceedings of the 17th APCOM Symposium* (R.J. Barnes and T.B. Johnson, editors). 263–273. New york: AIME.

Basu, A.J., Fonseca, M., & A.J. Richmond 2001. Open pit mining risk measured by stochastic simulation. 29th International Symposium on Application of Computers and Operations Research in the Mineral Industry: 87–92. Beijing:Balkema.

Berlanga, J.M., Cardona R. & M.A. Ibarra 1988. Recursive formulae for the floating cone algorithm. In: *Mine Planning and Equipment Selection* (R.K. Singhal, editor): 15–28. Rotterdam: A. A. Balkema.

Brackebusch, F.W. 1970. Mine planning applications at the Hecla Mining Company. Presented at the SME Fall Meeting, St. Louis, Missouri, Oct. 21–23, 1970: preprint 70-AR-312.

Brackebusch, F.W. 1971a. Hecla open-pit mine planning programs. Personal communication.

Brackebusch, F.W. 1971b. To design Lakeshore open-pit Hecla thinks big with small computer. *E/MJ* 172(7): 72–76.

Braticevic, D.B. 1984. Open-pit optimization method. In: *Proceedings 18th APCOM*: 133–138. London: Inst. Min. Met.

Caccetta, L. & L.M. Giannini 1986. Optimization techniques for the open-pit limit problem. *Bull. Proc. Australas. Inst. Min. Metall.* (12)291(8): 57–63.

Caccetta, L. & L.M. Giannini 1988. The generation of minimum search patterns. *The AusIMM Bulletin and Proceedings* (5)293(7): 57–61.

Cai, W.L. 1992. *Sensitivity Analysis of 3-D Model Block Dimensions in the Economic Open-Pit Limit Design*: 475–486. Littleton, CO: SME.

Cai, W.L. 1992. *Sensitivity Analysis of 3-D Model Block Dimensions in the Economic Open Pit Limit Design*. 23rd International Symposium on Application of Computers and Operations Research in the Mineral Industry: 475–486. Tucson: SME.

Caldwell, T. 1961. Use of computers in determining pit expansions and pit limits. In: *Transactions of Short Course on Computers and Compute Applications in the Mineral Industry*. Tucson, AZ, April 1961, Vol. II, Section M, Paper M: 1-M10 plus Appendix I.

Camus, J.P. 1992. Open pit optimization considering an underground alternative. 23rd International Symposium on Application of Computers and Operations Research in the Mineral Industry: 435–442. Tucson: SME.

Carlson, T.R., Erickson, J.D. O'Brian D.T. & M.T. Pana 1966. Computer techniques in mine planning. *Mining Engineering* 18(5): 53–56, 80.

Cavender, B. 1992. Determination of the optimum lifetime of a mining project using discounted cash flow and option pricing techniques. *Mining Engineering*. 44(10):1262–1268.

Celin, E. & P.A. Dowd 2002. The use of genetic algorithms for multiple cut-off grade optimization. 30th International Symposium on Application of Computers and Operations Research in the Mineral Industry: 769–780. Alaska: SME.

Chanda, E.K. 1999. Maximizing resource utilization in open pit design – A case study. 28th International Symposium on Application of Computers and Operations Research in the Mineral Industry: 359–366. Colorado School of Mines: CSM.

Chen, J., Li, J., Luo, Z., & D. Guo 2001. Development and application of open pit limits software for the combined mining of surface and underground. 29th International Symposium on Application of Computers and Operations Research in the Mineral Industry: 303–306. Beijing: Balkema.

Chen, T. 1976. 3-D pit design with variable wall slope capabilities. In: *14th APCOM, Penn State, October 1976*: 615–625.

Cherrier, T.E. 1968. A report on the Ceresco Ridge Extension of the Climax molybdenite deposit. MSc Thesis. University of Minnesota.

Crawford, J.T. 1979. Open-pit limit analysis – some observations on its use. In: *16th APCOM* (T.J. O'Neil, editor): 625–634.

Crawford, J.T. & R.K. Davey 1979. Case study in open-pit limit analysis. In: *Computer Methods for the 80's* (A. Weiss, editor): 310–318.

Dagdelen, K. 1992. Cutoff grade optimization. 23rd International Symposium on Application of Computers and Operations Research in the Mineral Industry: 157–168. Tucson: SME.

Dagdelen, K. 1993. An NPV maximization algorithm for open pit mine design. 24th International Symposium on Application of Computers and Operations Research in the Mineral Industry: 2(257–266). Montreal: CIMM.

Davey, R.K. 1979. Mineral block evaluation criteria. In: *Open-Pit Mine Planning and Design* (J.T. Crawford and W.A. Hustrulid, editors): 83–96. SME-AIME.

Diering, J.A.C. 1984. Algorithm for variable slope pit generation. In: *Proceedings 18th APCOM*: 113–121. London: Inst. Min. Met.

Dindiwe, C. & K. Sasaki 2001. Optimization of pit limit designs by the newly developed BPITC approach. 29th International Symposium on Application of Computers and Operations Research in the Mineral Industry: 249–256. Beijing:Balkema.

Dowd, P.A., Khalokakaie, R. & R.J. Fowell 2001. A Windows program for incorporating slope design in open pit optimization. 29th International Symposium on Application of Computers and Operations Research in the Mineral Industry: 189–196. Beijing:Balkema.

Drew, D. & E. Baafi 2002. Ultimate pit limit design using Microsoft Excel Spreadsheet. 30th International Symposium on Application of Computers and Operations Research in the Mineral Industry: 113–121. Alaska:SME.

Duval Corporation 1976. MINPAC – computerized aid to pit design and extraction scheduling (March).

Erarslan, K. & N. Celebi 2001. A simulative model for optimum open pit design. CIM Bulletin 94(1055): 59–68.

Erickson, J.D. 1968. Long-range open-pit planning. *Mining Engineering* 20(4): 75–78.

Fairfield, J.D. & R.W. Leigh 1969. A computer program for the design of open-pits. *Operations Research and Computer Applications in the Mineral Industries* 64(3): 329–340. Quarterly of the Colorado School of Mines.

Ferreira, F.M. & G. De Tomi 2001. Optimizing ultimate open pit limits with quality constraints. 29th International Symposium on Application of Computers and Operations Research in the Mineral Industry: 261–266. Beijing:Balkema.

Gamache, M., and G. Auger 2001. Analysis of maximum flow algorithms for ultimate pit contour problems. 29th International Symposium on Application of Computers and Operations Research in the Mineral Industry: 267–272. Beijing:Balkema.

Gauthier, F.J. & R.G. Gray 1971. Pit design by computer at Gaspe Copper Mines, Limited. *CIM Bulletin* (715)64(11): 95–102.

Grosz, R.W. 1969. The changing economics of surface mining: A case study: In: *A Decade of Digital Computing in the Mineral Industry* (A. Wiess, editor): 401–420. New York: AIME.

Hufford, G.A. 1986. Selection of a cutoff grade strategy for an open-pit mine. Society of Mining Engineers of AIME, SME Annual Meeting, New Orleans, March 2–6, 1986: preprint No. 86-49.

Huttagosol, P. 1988. Modified tree graph algorithm for ultimate pit limit analysis. MS Thesis T-3412. Colorado School of Mines.

Huttagosol, P. 1990. Transportation algorithm for ultimate pit limit. PhD Thesis T-3974. Colorado School of Mines.

Huttagosol, P. 1992. Optimizing open-pit limit by transportation algorithm. PhD Thesis. Colorado School of Mines.

Huttagosol, P. & R.E. Cameron 1992. A computer design of ultimate pit limit by using transportation algorithm. In: *23rd APCOM* (Y.C. Kim, editor): 443–460. Littleton, CO: SME.

Johnson, T.B. 1973. A comparative study of methods for determining ultimate open-pit mining limits. In: *Proceedings of the 11th APCOM Symposium, Univ. of Arizona, Tucson*.

Johnson, T.B. & D.G. Mickle 1971. Optimum design of an open-pit – an application in uranium. *Decision Making in the Mineral Industry*. CIM Special Volume 12: 331–326. Discussion: 337–338.

Johnson, T.B. & W.R. Sharp 1971. A three-dimensional dynamic programming method for optimal ultimate open-pit design. U.S. Bureau of Mines Report of Investigations 7553.

Kelsey, R.D. 1979. Cutoff grade economics. In: *Proceedings 16th APCOM* (T.J. O'Neil, editor): 286–292. SME-AIME.

Kim, Y.C. 1978. Ultimate pit design methodologies using computer models – the state of the art. *Mining Engineering* 30(10): 1454–1459.

Kim, Y.C., Cai, W. & W.L. Meyer 1988. A comparison of microcomputer based optimum pit limit design algorithms. SME Annual Meeting, Phoenix, AZ, Jan. 25–28, 1988: preprint 88-23.

Koenigsberg, E. 1982. The optimum contours of an open-pit mine: An application of dynamic programming. In: *Proceedings 17th APCOM* (T.B. Johnson and R.J. Barnes, editors): 274–287. SME-AIME.

Korobov, S.D. 1993. Parametric analysis of open pit mines. 24th International Symposium on Application of Computers and Operations Research in the Mineral Industry: 2(57–66). Montreal:CIMM.

Korobov, S. 1974. Method for determining optimal open-pit limits. Rapport Technique EP74-R-4. Department of Mineral Engineering, Ecole Polytechnique, Montreal, February 1974.

Koskiniemi, B.C. 1979. Hand methods. In: *Open-Pit Mine Planning and Design* (J.T. Crawford and W.A. Hustrulid, editors). SME/AIME: 187–195.

Laurent, M., Placet J. & W. Sharp 1977. Optimum design of open-pit mines. Gecamines Rapport No. 04/77. Lubumbashi.

Le Bel, G. 1993. Discussion of the paper. Determination of the optimum lifetime of a mining project using discounted cash flow and option pricing techniques. *Mining Engineering*. 45(11):1409–1412.

Leigh, R.W. & R.L. Blake 1971. An iterative approach to the optimal design of open-pits. *Decision-Making in the Mineral Industry*. CIM Special Volume 12: 254–260.

Lemieux, M. 1979. Moving cone optimizing algorithm. In: *Computer Methods for the 80's in the Mineral Industry* (A. Weiss, editor): 329–345. SME-AIME.

Lerchs, H. & I.F. Grossmann 1965. Optimum design of open-pit mines. *CIM Bulletin* 58: 47–54.

Linder, D.E. 1993. Software determines limestone ore body parameters. *Mining Engineering* 45(10): 1275.

Lipkewich, M.P. & L. Borgman. 1969. Two- and three-dimensional pit design optimization techniques. In: *A Decade of Digital Computing in the Mineral Industry* (A. Weiss, editor): 505–524. New york: AIME.

Luke, K.W. 1972. Functional optimization of open-pit mine design utilizing geologic cross-section data. *Trans. of the AIME* 252: 125–131. SME/AIME.

Mason, P.M. & R.F. Lea 1978. Practical approach to the use of computers in planning the Nchanga open-pit complex. In: *Proceedings of the Eleventh Commonwealth Mining and Metallurgical Congress* (M.J. Jones, editor): 535–541.

1989 Minerals Yearbook (Volume I) Metals and Minerals. 1991. U.S. Department of the Interior, Bureau of Mines. Washington, DC: U.S. Government Printing Office.

Motta, R. 1988. Personal communication.

Nilsson, D. 1997. Optimal final pit depth: Once again. *Mining Engineering.* 49(1): 71–72.

O'Brien, N. & F.J. Nowak 1966. An application of a computer to open-pit mine design. *Pit and Quarry* 58(2): 128–135.

Osborne, W.G. 1976. MINPAC – computerized aid to pit design and extraction scheduling. Spring Meeting Arizona Section. AIME.

Pana, M.T. 1965. The simulation approach to open-pit design. In: *5th APCOM, Tucson Arizona, March 16–19, 1965*: ZZ-1 to ZZ-24.

Pasieka, A.R. & G.V. Sotirow 1985. Planning and operational cutoff grades based on computerized net present value and net cash flow. *CIM Bulletin* (878)78(6): 47–54.

Phillips, D.A. 1972. Optimum design of an open-pit. In: *10th APCOM* (4): 145–147. Johannesburg: South African Institute of Mining & Metallurgy (SAIMM).

Picard, J-C. & B.T. Smith 1993. Optimal rate of return in open pit mine design. 24th International Symposium on Application of Computers and Operations Research in the Mineral Industry: 2(111–118). Montreal:CIMM.

The Planning and Operation of Open-Pit and Strip Mines – Proceedings of an International Conference, Univ. of Pretoria, April 9–13, 1984 (J.P. Deetlefs, editor). Symposium Series S7. Johannesburg: (SAIMM).

Pronk Van Hoogeveen, L.A.J., Cutland J.R. & M. Weir 1972. An open-pit design system for stratiform ore-bodies. In: *10th Application of Computers in the Minerals Industries Symposium*. Johannesburg: SAIMM.

Quang, N.C. & D.N. Trung 2001. Optimization of grouped open pits. 29th International Symposium on Application of Computers and Operations Research in the Mineral Industry: 257–260. Beijing: Balkema.

Reibell, H.V. 1969. Deep open-pit optimization. In: *A Decade of Digital Computing in the Mineral Industry* (A. Weiss, editor): 359–372. New York: AIME.

Robinson, R.H. 1975. Programming the Lerchs-Grossmann algorithm for open-pit design. In: *Proceedings of the 13th APCOM Symposium, Clausthal, Germany, 1975*: B-IV 1–17.

Robinson, R.H. & N.B. Prenn 1973. An open-pit design model. In: *Proceedings of the 10th APCOM Symposium*: 155–163. Johannesburg: SAIMM.

Sainsbury, G.M. 1970. Computer-based design of open cut mines. *Proc. Aust. Inst. Min. Met.* (6)(234): 49–57.

Saydam, S. & E. Yalcin 2002. Reserve and ultimate pit limit design analysis of Caldagi Nickel Deposit, Turkey. 30th International Symposium on Application of Computers and Operations Research in the Mineral Industry: 121–132. Alaska:SME.

Soderberg, A. & D.O. Rausch 1968. Chapter 4.1. Pit planning and layout. In: *Surface Mining* (E.P. Pfleider, editor): 141–165. New York: SME-AIME.

Strangler, R.L., Costa, J.F. & J.C. Koppe 2001. Risk in stripping ratio estimation. 29th International Symposium on Application of Computers and Operations Research in the Mineral Industry: 81–86. Beijing:Balkema.

Tanays, E., Cojean, R. & D. Hantz 1992. DEGRES: A software to design open-pit geometry and to draw open-pit plans. *Int. J. of Surface Mining and Reclamation* (2)6: 91–98.

Underwood, R. & B. Tolwinski. 1996. Lerchs Grossman algorithm from a dual simplex viewpoint. 26th International Symposium on Application of Computers and Operations Research in the Mineral Industry: 229–236. Penn. State University:SME.

Wang, Q. & H. Sevim 1992. Enhanced production planning in open pit mining through intelligent dynamic search. 23rd International Symposium on Application of Computers and Operations Research in the Mineral Industry: 461–474. Tucson:SME.

West, F.J. 1966. Open-pit planning at the Adams Mine. *CIM Bulletin* (657)59(3): 392–331.

Wharton, C.L. 1996. What they don't teach you in mining school – Tips and tricks with pit optimizers. In Surface Mining 1996 (H.W. Glen, editor): 17–22. SAIMM.

Whittle, J. 1988. Beyond optimization in open-pit design. 1st Canadian Conference on Computer Applications in the Mineral Industry, Laval University, Quebec City, March 7–9, 1988.

Whittle, J. 1989. The facts and fallacies of open-pit optimization. Whittle Programming Pty. Ltd. (1).

Wick, D.E. & E.M. Sunmoo 1974. Computer use in determining ore reserves and stripping ratios at Eagle Mountain. In: *Proceedings of the 12th APCOM* (T.B. Johnson and D.W. Gentry, editors), Volume II: H59-H71. Colorado School of Mines.

Williams, C.E. 1974. Computerized year-by-year open-pit mine scheduling. *Trans* 256. SME-AIME.

Williamson, D.R. & E.R. Mueller 1976. Ore estimation at Cyprus Pima Mine. Paper presented at the AIME Annual Meeting, Las Vegas, Feb. 22–26, 1976: preprint 76-AO-13.

Wright, E.A. MOVING CONE II – A simple algorithm for optimum pit limits design. 28th International Symposium on Application of Computers and Operations Research in the Mineral Industry: 367–374. Colorado School of Mines:CSM.

Xu, C. & P.A. Dowd 2002. Optimal orebody reconstruction and visualization. 30th International Symposium on Application of Computers and Operations Research in the Mineral Industry: 743–756. Alaska:SME.

Yamatomi, J., Mogi, G., Akaike, A. & U. Yamaguchi 1995. Selective extraction dynamic cone algorithm for three-dimensional open pit designs. 25th International Symposium on Application of Computers and Operations Research in the Mineral Industry: 267–274. Brisbane:AusIMM.

Yegulalp, T.M. & J.A. Arias 1992. A fast algorithm to solve the ultimate pit limit problem. 23rd International Symposium on Application of Computers and Operations Research in the Mineral Industry: 391–398. Tucson:SME.

Yegulalp, T.M. & J.A. Arias 1992. A fast algorithm to solve the ultimate pit limit problem. In: *Proceedings 23rd APCOM* (Y.C. Kim, editor): 391–397. Littleton, CO:SME.

Zhao, Y. & Y.C. Kim 1990. A new graph theory algorithm for optimal ultimate pit design. SME Annual Meeting, Salt Lake City, UT, Feb. 26–Mar. 1, 1990: preprint No.90-9.

Zhao, Y. & Y.C. Kim 1992. A new optimum pit limit design algorithm. In: *23rd APCOM* (Y.C. Kim, editor): 423–434. Baltimore: Port City Press/SME.

Zhao, Y. & Y.C. Kim 1992. A new optimum pit limit design algorithm. 23rd International Symposium on Application of Computers and Operations Research in the Mineral Industry: 423–434. Tucson:SME.

REVIEW QUESTIONS AND EXERCISES

1. What is the difference between a mineral inventory and an ore reserve?
2. Repeat the example in section 5.2.1 assuming that the wall on the left side of the pit is at 53° rather than 45°. Determine the:
   a. Final pit position
   b. Breakeven stripping ratio
   c. Overall stripping ratio
   d. Value of the section
   Use a slice of zero thickness. Compare the results to those for a slice of thickness 1.4u.
3. Repeat the example in section 5.2.1 assuming that the density of the ore is 1.2 times that of the waste.
4. Summarize the pit limit determination procedure in words.
5. What is meant by the term 'cutoff grade?'
6. What is the practical use of the cutoff grade? Can there be more than one cutoff grade?
7. What is meant by the breakeven cutoff grade?
8. Repeat the net value calculation example in section 5.2.2 assuming a mill recovery of 85% and a mill concentrate grade of 28%. All other factors remain constant. Complete all of the steps including development of the final curve (Figure 5.9).
9. What is meant by the concentration ratio?
10. Summarize the steps required for making a net value calculation.
11. Redo the example of section 5.2.3 assuming that the right hand slope is 45°.
12. The stripping ratio – ore grade curve in Figure 5.9 has been used to determine the final limits for the section shown in Figure 5.10. Repeat the example using the curve obtained in problem 8.
13. Summarize the process of locating the pit limits when the pit bottom is in waste.
14. How is the procedure modified if the ore and waste densities are different? Rework the example in section 5.2.3 if the ore density is $3.0 \, g/cm^3$ and the waste density is $2.5 \, g/cm^3$.
15. How is the procedure modified if there is a difference in the waste and ore mining costs? Rework the example in section 5.2.3 assuming the ore mining cost is $1.10/ton and the cost of mining waste is $0.85/ton.
16. Apply the curve developed in Problem 12 to the determination of the final limits in Figure 5.12.
17. Apply the curve developed in Problem 12 to Figure 5.13.
18. Show the development of equation (5.11) which relates the true stripping ratio for a radial section to the measured.
19. Summarize the steps outlined by Koskiniemi for developing a composite mine plan map from the sections.

480 *Open pit mine planning and design: Fundamentals*

20. Assuming that the width of the safety bench in Figure 5.27 is 35 ft, what is the road grade?
21. Using Figure 5.28, check the values given in Table 5.9 for the reserves in Table 5.9. What must have been the block size, block height and material density used?
22. Once the final pit outline has been determined, the material within the pit limits is re-evaluated regarding destination. What is the reason for this? How is the decision made?
23. With the advent of the computer, many companies now use block to model their deposits. To develop the mineable reserves, economic values must be applied to each block. Should the G&A costs be assigned to the ore only? To the ore and waste? Discuss. How can this decision be included in CSMine?
24. In a block model can you include depreciation costs? Minimum profit? To which blocks should such values be assigned?
25. As pits get deeper, the haulage costs increase. Assuming that the truck operating cost is $150/hour. The bench height is 15 m and the road grade is 10%. If the average truck speed is 10 mph uphill loaded and 20 mph down hill unloaded, what should be the assigned incremental haulage cost/level.
26. Discuss the floating cone process described in section 5.4.
27. In the running of a floating cone model to a particular level, the cone returns to the surface and repeats the process. Why are these scavenging runs performed?
28. Describe three problems with regard to the application of the floating cone technique.
29. List the positive aspects of the floating cone technique.
30. What is meant by the 'optimal' pit?
31. Summarize the steps in the Lerchs-Grossmann 2D algorithm.
32. Redo the Lerchs-Grossmann 2D assuming ore value $= +16$ and the waste value $= -4$.
33. Redo problem 32 using the floating cone technique.
34. Summarize the 2½ -D technique described in section 5.6.
35. Apply the 2½ -D process to sections 2 through 5. Compare your results to those given in Figure 5.81.
36. On section 4 the block at position $i = 5$, $k = 4$ has a value of 9 but is not being mined. Apply the 1–9 constraint model of Figure 5.84 to show why this result is correct.
37. Apply the 2½ -D procedure to the block sections in Figure 5.82. Check your results versus those given in Figure 5.83.
38. Define the following terms used with respect to the Lerchs-Grossmann 3D algorithm:
    a. Directed arc
    b. Edge
    c. Weight
    d. Node
    e. Graph
    f. Directed graph
    g. Sub graph
    h. Closure
    i. Maximum closure
    j. Circuit
    k. Chain
    l. Cycle
    m. Path

    n. Tree
    o. Root
    p. Branch
    q. Twig
39. Following the steps, apply the 'arbitrary tree approach (Approach 1)' to the example shown in Figure 5.89.
40. Following the steps, apply the 'all root connection approach (Approach 2)' to the example shown in Figure 5.89.
41. Apply the tree 'cutting' process to the problem shown in Figure 5.105.
42. Apply your skills to the more complicated example shown in Figure 5.110.
    a. Start with doing the first two layers.
    b. For more practice, choose the first three layers.
    c. Apply your skills to the full four layer problem whose solution is given in Figure 5.111.
43. Summarize the steps used in the RTZ open pit generator.
44. Redo one of the computer-assisted design examples considered by Luke. Check your answers with those given.

CHAPTER 6

Production planning

6.1 INTRODUCTION

In this chapter some of the production planning activities involved in an open pit mine will be discussed. Specifically, attention will be devoted to mine life – production rate determinations, push back design and sequencing, as well as providing some general guidance regarding both long and short range planning activities.

The basic objectives or goals of extraction planning have been well stated by Mathieson (1982):

– To mine the orebody in such a way that for each year the cost to produce a kilogram of metal is a minimum, i.e., a philosophy of mining the 'next best' ore in sequence.

– To maintain operation viability within the plan through the incorporation of adequate equipment operating room, haulage access to each active bench, etc.

– To incorporate sufficient exposed ore 'insurance' so as to counter the possibility of mis-estimation of ore tonnages and grades in the reserve model. This is particularly true in the early years which are so critical to economic success.

– To defer waste stripping requirements, as much as possible, and yet provide a relatively smooth equipment and manpower build-up.

– To develop a logical and easily achievable start-up schedule with due recognition to manpower training, pioneering activities, equipment deployment, infrastructure and logistical support, thus minimizing the risk of delaying the initiation of positive cash flow from the venture.

– To maximize design pit slope angles in response to adequate geotechnical investigations, and yet through careful planning minimize the adverse impacts of any slope instability, should it occur.

– To properly examine the economic merits of alternative ore production rate and cutoff grade scenarios.

– To thoroughly subject the proposed mining strategy, equipment selection, and mine development plan to 'what if' contingency planning, before a commitment to proceed is made.

Planning is obviously an ongoing activity throughout the life of the mine. Plans are made which apply to different time spans.

There are two kinds of production planning which correspond to different time spans (Couzens, 1979):
– Operational or short-range production planning is necessary for the function of an operating mine.
– Long-range production planning is usually done for feasibility or budget studies. It supplements pit design and reserve estimation work and is an important element in the decision making process.

Couzens (1979) has provided some very useful advice which should be firmly kept in mind by those involved in the longer term planning activities:

How would I plan this if I had to be the mine superintendent and actually make it work?

In guiding the planner, Couzens (1979) has proposed the following five planning 'commandments' or rules:

1. We must keep our objectives clearly defined while realizing that we are dealing with estimates of grade, projections of geology, and guesses about economics. We must be open to change.

2. We must communicate. If planning is not clear to those who must make decisions and to those who must execute plans, then the planning will be either misunderstood or ignored.

3. We must remember that we are dealing with volumes of earth that must be moved in sequence. Geometry is as important to a planner as is arithmetic.

4. We must remember that we are dealing with time. Volumes must be moved in time to realize our production goals. The productive use of time will determine efficiency and cost effectiveness.

5. We must seek acceptance of our plans such that they become the company's goals and not just the planner's ideas.

This chapter will focus on the longer term planning aspects both during feasibility studies and later production.

6.2 SOME BASIC MINE LIFE – PLANT SIZE CONCEPTS

To introduce this very important topic, an example problem will be considered. Assume that a copper orebody has been thoroughly drilled out and a grade block model constructed. Table 6.1 presents an initial estimate for the costs and recoveries.

Since they will be refined later, this step will be called Assumption 1.

The best estimate price (in this case $1/lb) is also selected (Assumption 2). From these values an economic block model is constructed. The break-even grade for final pit limit

Table 6.1. Costs used to generate the economic block model.

Mining cost (ore)	= $1.00/ton
Mining cost (waste)	= $1.00/ton
Milling cost	= $2.80/ton
G&A cost (mining)	= $0.17/ton
G&A cost (milling)	= $0.40/ton ore
Smelting, refining and sales	= $0.30/lb Cu
Overall metal recovery	= 78%

Table 6.2. Mineral inventory as a function of grade class interval.

Grade class interval (% Cu)	Tons (10^3)	Grade class interval (% Cu)	Tons (10^3)
>3.2 (Ave = 5.0)	25	1.5–1.6	205
3.1–3.2	7	1.4–1.5	130
3.0–3.1	15	1.3–1.4	270
2.9–3.0	5	1.2–1.3	320
2.8–2.9	5	1.1–1.2	570
2.7–2.8	10	1.0–1.1	460
2.6–2.7	33	0.9–1.0	550
2.5–2.6	40	0.8–0.9	420
2.4–2.5	15	0.7–0.8	950
2.3–2.4	25	0.6–0.7	980
2.2–2.3	30	0.5–0.6	830
2.1–2.2	30	0.4–0.5	1200
2.0–2.1	50	0.3–0.4	1050
1.9–2.0	75	0.2–0.3	1300
1.8–1.9	60	0.1–0.2	2700
1.7–1.8	150	<0.1	18,020
1.6–1.7	170		

determination is

$$g(\% \text{ Cu}) = \frac{\$1.00 + \$2.80 + \$0.40 + \$0.17}{0.78(1.00 - 0.30)\frac{2000}{100}}$$

$$= \frac{\$4.37}{10.92} \cong 0.40\%(\text{Cu})$$

It is noted that the costs chosen at this stage might be considered as 'typical'. No attempt has been made to include capital related costs nor a 'profit'. Using the floating cone, Lerchs-Grossmann or another technique, a final pit outline is generated. The intersections are transferred onto the corresponding benches of the grade block model. A mineral inventory using appropriate grade class intervals is created bench by bench. These are later combined to form a mineral inventory for the material within the pit. Table 6.2 is the result for the example orebody (Hewlett, 1961, 1962, 1963). The number of tons in each grade class interval (0.1%) are plotted versus grade. This is shown in Figure 6.1. From this curve a plot of total (cumulative) tonnage above a given cutoff grade can be constructed. The result is given in Figure 6.2. As can be seen, the curve is of the shape typical for a low grade copper orebody. Figure 6.3 shows the straight line expected when a lognormal plot is made of the data.

One can also determine the average grade of the material lying above a given cutoff grade. Figure 6.4 presents the results. Table 6.3 contains the data used in constructing Figures 6.2 and 6.3. These two curves are very useful when considering various plant size – mine life options.

At this point two destination options will be considered for the material contained within the pit:
– Destination A: Mill.
– Destination B: Waste dump.

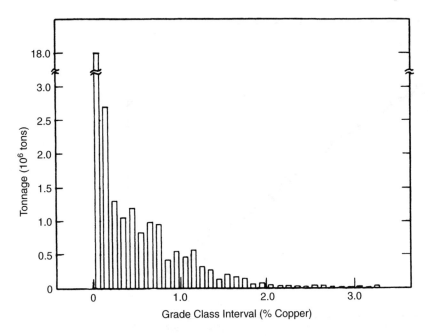

Figure 6.1. Tonnage versus grade class interval for the Silver Bell oxide pit (Hewlett, 1961, 1962).

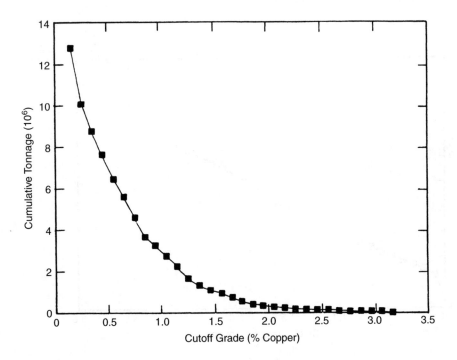

Figure 6.2. Cumulative tonnage versus cutoff grade for the Silver Bell oxide pit.

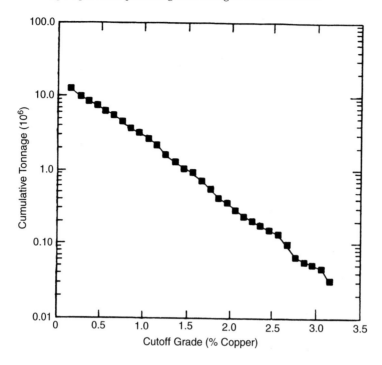

Figure 6.3. Logarithm of cumulative tonnage versus cutoff grade.

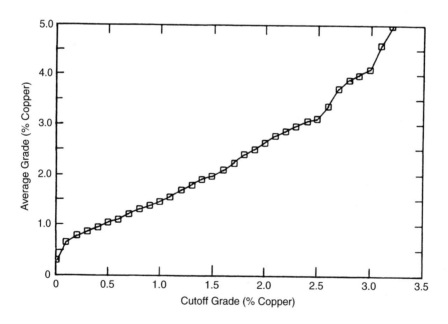

Figure 6.4. Average grade versus cutoff grade.

Table 6.3. Cumulative tons and average grade as a function of cutoff grade.

Grade (% Cu)		Tons above cutoff (10^3)
Cutoff	Average	
0.0	0.30	30,800
0.1	0.65	12,780
0.2	0.78	10,080
0.3	0.86	8780
0.4	0.94	7630
0.5	1.04	6430
0.6	1.11	5600
0.7	1.21	4620
0.8	1.32	3670
0.9	1.38	3250
1.0	1.47	2700
1.1	1.56	2240
1.2	1.70	1670
1.3	1.81	1350
1.4	1.92	1080
1.5	1.99	950
1.6	2.11	745
1.7	2.24	575
1.8	2.41	425
1.9	2.51	365
2.0	2.65	290
2.1	2.78	240
2.2	2.87	210
2.3	2.97	180
2.4	3.07	155
2.5	3.13	140
2.6	3.37	100
2.7	3.72	67
2.8	3.89	57
2.9	3.99	52
3.0	4.10	47
3.1	4.60	32
3.2	5.00	25

There are other possible destinations which can be considered later. This destination stipulation becomes Assumption 3. Since all of the material will eventually have to be removed from the pit there is no question concerning mining or not mining. For the sake of this example it will be assumed that the mineral distribution is uniform throughout the pit. This is generally not true and there will be high and low grade areas of various extent.

A cutoff grade must be selected differentiating ore (that going to the mill) and waste (that going to the dump). Assumption 4 will be that the mill cutoff grade is 0.40% (the same value used in the pit limit determination). From Figure 6.5 one finds that there are 7.8×10^6 tons of ore with an average grade (Fig. 6.6) of 0.92%.

The next question is with regard to the size of plant to be constructed. An equivalent question is 'What is the expected life of the property?' Although there are several ways of approaching this, the one chosen here is market based. It will be assumed (Assumption 5)

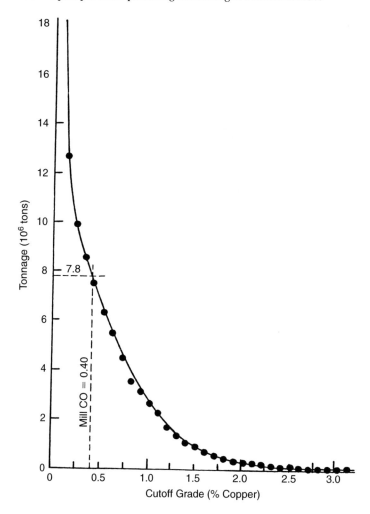

Figure 6.5. Mill tonnage for a 0.4 cutoff grade.

that a market survey has suggested that 5000 tons of copper metal can be sold every year. The yearly and daily production rates as well as the mine life can now be computed.

Assuming that:

Mill recovery = 80%

Combined smelter/refinery recovery = 97%

Operating days = 250 days/yr

we obtain the following milling rates (R_{mill})

$$R_{mill} = \frac{5000 \text{ tpy} \times 2000 \text{ lbs/ton}}{\frac{0.92}{100} \times 0.80 \times 0.97 \times 2000 \text{ lbs/ton}}$$

$$= 700{,}360 \text{ tpy}$$

$R_{mill} = 2801$ tpd

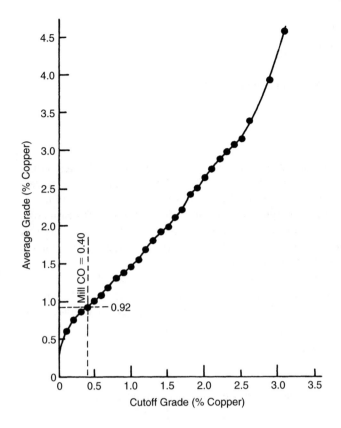

Figure 6.6. Average mill feed grade for a 0.4 cutoff grade.

Knowing the mill production rate and the ore reserves the mill/mine life can be calculated:

$$\text{Mine life (yrs)} = \frac{\text{Ore reserves (tons)}}{\text{Ore production rate (tpy)}}$$

$$= \frac{7{,}800{,}000}{700{,}360}$$

$$= 11.1 \text{ years}$$

The required mine production rate (R_{mine}) is:

$$R_{mine} = \frac{\text{Mineral reserve (tons)}}{\text{Mine life (yrs)}}$$

$$= \frac{30{,}800{,}000}{11.1} = 2{,}775{,}000 \text{ tpy}$$

$$R_{mine} = 11{,}100 \text{ tpd}$$

The total amount of copper recovered is

Copper recovered = 55,500 tons

and the overall stripping ratio is

$$\text{Overall SR} = \frac{23{,}000{,}000}{7{,}800{,}000} = 2.95$$

Knowing the mining and milling rates and the mine life

> Milling = 2801 tpd
> Mining = 11,000 tpd
> Mine life = 11.1 years

one can now go back to Assumption 1 and improve the operating cost estimates. Using these values one can recalculate the economic block values, the final pit limits, the grades-tonnages, etc. Eventually a solution will be found which changes little from run to run.

Knowing the plant size, the required capital investment can be determined. The cash flows are calculated as is the net present value. Obviously other economic indicators such as total profit, internal rate of return, etc., could be calculated as well. One might then return to Assumption 2 and examine the sensitivity with price.

It will be recalled that Assumption 4 dealt with the mill cutoff grade which was chosen as 0.4. If a mill cutoff grade of 0.2 is assumed instead the process can be repeated maintaining the 5000 ton copper output. As Figures 6.7 and 6.8 show, the ore tonnage and average grade become

> Ore tonnage = 10.8×10^6 tons
> Average ore grade = 0.77% Cu

With the same assumptions as before the milling rate, mill/mine life and mining rate become:

Mill rate:

$$R_{mill} = \frac{5000 \times 2000}{\frac{0.77}{100} \times 0.80 \times 0.97 \times 2000}$$

$$= 836{,}800 \text{ tpy}$$

$$R_{mill} = 3350 \text{ tpd}$$

Mill/mine life:

$$\text{Mill life} = \frac{10{,}800{,}000}{836{,}800} = 12.9 \text{ yrs}$$

Mining rate:

$$R_{mine} = \frac{30{,}800{,}000}{12.9} = 2{,}387{,}600 \text{ tpy}$$

$$R_{mine} = 9550 \text{ tpd}$$

The amount of copper recovered increases to

> Copper recovered = 64,500 tons

and the overall stripping ratio would drop to

$$\text{Overall SR} = \frac{20{,}000{,}000}{10{,}800{,}000} = 1.85 : 1$$

A summary of the results for the two mill cutoff grades is given in Table 6.4.

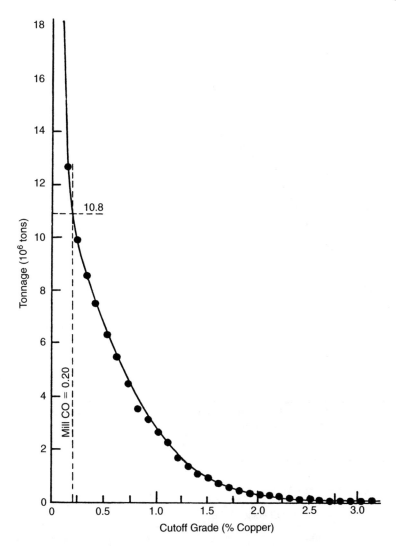

Figure 6.7. Mill tonnage for a 0.3 cutoff grade.

These would be expected to yield different economic results. The incremental financial analysis approach to this type of evaluation will be discussed in Section 6.6.

There are a number of iterations which must be performed as the various assumptions are examined. In this simple example, there were five assumptions made in order to proceed:
- Assumption 1. Cost-recovery values.
- Assumption 2. Commodity price.
- Assumption 3. Destination options.
- Assumption 4. Mill cutoff grade.
- Assumption 5. Yearly product mix.

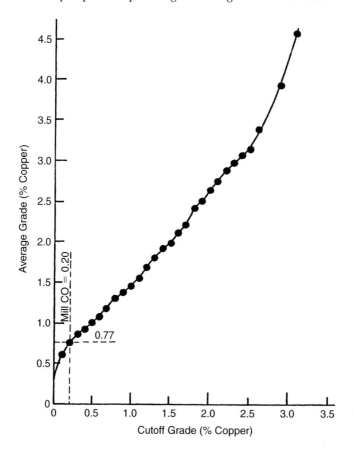

Figure 6.8. Average mill feed grade for a 0.3 cutoff grade.

Table 6.4. Summary for the two mill cutoff grades.

Quantity	Mill cutoff grade (% Cu)	
	0.2	0.4
Tons mined	30.8×10^6	30.8×10^6
Tons milled	10.8×10^6	7.8×10^6
Tons waste	20.0×10^6	23.0×10^6
Avg. ore grade	0.77%	0.92%
Milling rate (tpd)	3350	2800
Mining rate (tpd)	9550	11,100
Mine life (yrs)	12.9	11.1
Overall SR	1.85	2.95
Recovered copper (tons)	64,500	55,500

Clearly, a large number of combinations must be considered, many of which could provide satisfactory results. Because of the many uncertainties associated with the grade, tonnage, price, and cost data, the meaning of some sort of 'optimum' solution is tenuous at best.

6.3 TAYLOR'S MINE LIFE RULE

Taylor (1977, 1986, 1991) has, over the years, provided some very practical and useful advice regarding mine life. This section is based on extractions from his writings. In theory, it is possible to calculate an 'optimum' rate of extraction from an orebody. To do this, however, knowledge or precise assumption of the total tonnage and its sequential grades (including the effects of varying the cutoff grade), and of all costs and product prices throughout the project life is required. This information is unavailable for early studies and may indeed never reach high certainty or even be necessary.

Even with certain knowledge of everything, optimizing theory yields different answers depending on what quantity is selected to be maximized. The maximized quantity might be total profit, total cash flow, the net present value or the internal rate of return. Furthermore, the peaks of such curves are rather flat. Thus when allowing for the practical inaccuracies of data, the calculated results cannot be considered critical. Hence, although valid, a highly mathematical approach to mine life determination is seldom of practical use. Other ways must be found to provide a reasonable first approximation for mine life.

Too low a production rate sacrifices possible economies of scale and defers possible profits too far into the future. Conversely, too high a rate may drive up the project's capital cost beyond any ability to repay within the shortened life. Too high an output may be unsalable, while too short a life for a large enterprise may be wholly undesirable on social grounds. One hazard of short life mines merits special mention. Since base metal prices seem to move in cycles of four to seven years' duration, an operation of under four years' life may find itself depleting all its ore in a trough of the price cycle, and be left with neither ore nor time to recover.

In real life, rates of output are strongly limited or influenced by practical problems. One of the most important of these is working space. A mine may be able to increase output as it gets older solely because its ever expanding workings offer more points of attack.

In an open pit the working space for equipment and hence maximum production rate tends to vary with the area (ft^2) exposed while tonnage varies with volume (ft^3). Thus one might expect the production rate for groups of more-or-less similarly shaped orebodies to be proportional to the two-thirds power of the orebody tonnage. The life would then be proportional to the cube root of that tonnage.

Taylor (1977) studied many actual projects (some operating and others only planned) involving a wide range of orebody sizes, and shapes (other than thin deposits of very large lateral extent), for which the total ore reserves were reasonably well known before major design commenced. He found that the extraction rates seemed proportional to the three-quarters power of the ore tonnage rather than the two-thirds power. The designed lives were proportional to the fourth root of the tonnage.

This lead to the formulation of Taylor's rule, a simple and useful guide that states:

$$\text{Life (years)} \cong 0.2 \times \sqrt[4]{\text{Expected ore tonnage}} \tag{6.1}$$

In this equation, it is immaterial whether short or metric tons are used. It is more convenient to use quantities expressed in millions and except for special conditions, the practical range of variation seems to lie within a factor of 1.2 above and below. The rule can thus be restated as:

$$\text{Life (years)} \cong (1 \pm 0.2) \times 6.5 \times \sqrt[4]{\text{Ore Tonnage in millions}} \tag{6.2}$$

494 Open pit mine planning and design: Fundamentals

Table 6.5. Mine life as a function of ore tonnage (Taylor, 1977).

Expected ore (10^6 tons)	Median life (years)	Range of lives (years)	Median output (tpd)	Range of outputs (tpd)
0.5	3.5	4.5–3	80	65–100
1.0	6.5	7.5–5.5	450	400–500
5	9.5	11.5–8	1500	1250–1800
10	11.5	14.9–5	2500	2100–3000
25	14	17–12	5000	4200–6000
50	17	21–14	8400	7000–10,000
100	21	25–17	14,000	11,500–17,000
250	26	31–22	27,500	23,000–32,500
350	28	33–24	35,000	30,000–42,000
500	31	37–26	46,000	39,000–55,000
700	33	48–28	60,000	50,000–72,000
1000	36	44–30	80,000	65,000–95,000

At a preliminary stage, 'ore tonnage' could represent a reasonable though not optimistic estimate of the ore potential. Later, it could comprise the total of measured and indicated ore, including probable ore, but excluding possible or conjectural ore.

This empirical formula generates the values presented in Table 6.5.

The rule provides an appropriate provisional output rate for preliminary economic appraisals and will define a range of rates for comparative valuation at the intermediate stage after which a preferred single rate can be selected for use in the feasibility study.

6.4 SEQUENCING BY NESTED PITS

Of the various techniques used to develop mining sequences, the most common is to produce a nest of pits corresponding to various cutoff grades. From a practical point of view this is accomplished by varying the price of the metal (commodity) being extracted. The final pit limit is generally determined using the most likely price. For prices lower than this value, successively smaller pits will be produced. The pits will migrate toward the area of highest grade and/or lowest amount of stripping. This will be illustrated by way of an example.

The topographic map for a molybdenum prospect (Suriel, 1984) is shown in Figure 6.9. As can be seen, the majority of the deposit area has moderate surface relief with the exception of some steeper topography in the north-central project area. The relative position of the deposit is southwest of the hill. It is shown by the dashed lines on the figure.

A grade block model has been prepared using blocks 50 ft × 50 ft × 50 ft. The tonnage factor is 12.5 ft^3/st hence each block contains 10,000 tons. The following data were used in preparing the grade block model and in running the floating cone.

Mining cost = $0.74/st

Processing cost = $1.89/st ore

General and administrative (G&A) cost = $0.67/st ore

Mill recovery = 90%

Selling price = $6/lb contained molybdenum (F.O.B. mill site)

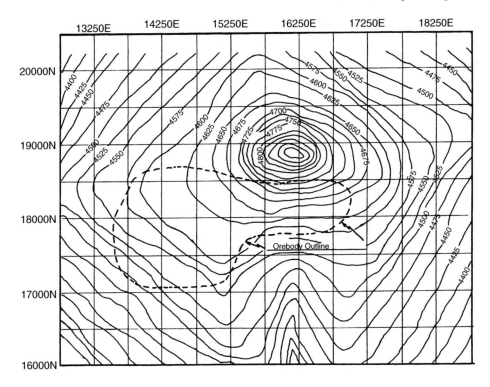

Figure 6.9. Topographic map of the project area (Suriel, 1984).

The ore grade is expressed in terms of % MoS_2. Since one pound of MoS_2 contains 0.60 lbs Mo, the equivalent price is $3.24/lb MoS_2.

In determining the final pit limits, the costs and revenues involved in mining and processing 1 ton of material containing X% MoS_2 are first determined.

$$\text{Cost (\$/st)} = 0.74 + 0.67 + 1.89 = \$3.30/\text{st}$$

$$\text{Revenue (\$/st)} = \frac{X}{100} \times 2000 \times 0.90 \times \$6.00 \times 0.60$$

$$= 64.8X$$

Equating costs and revenues

$$64.8X = 3.30$$

one finds that the breakeven grade (X%) is

$$X = 0.05\%\,MoS_2$$

This cutoff grade corresponds to a commodity price of $6/lb contained molybdenum. Using these costs and the $6/lb price one generates the pit shown in plan in Figure 6.10. It can be referred to as the $6 pit or equivalently the 0.05% cutoff pit.

For this example two smaller pits will be created. If a price of $2/lb Mo is selected instead, then by redoing the breakeven analysis one finds that

$$X = 0.15\%$$

496 Open pit mine planning and design: Fundamentals

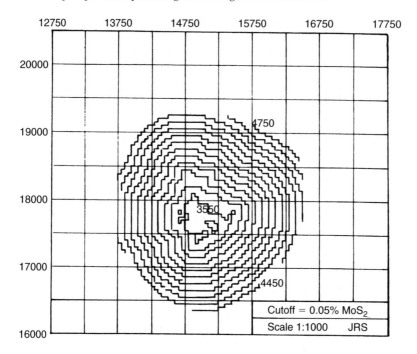

Figure 6.10. Pit outline for a 0.05 cutoff (Suriel, 1984).

The $2 pit is shown in Figure 6.11. For a price of $1.50/lb Mo, the breakeven grade is

$$X = 0.20\%$$

Figure 6.12 is a plan view of the resulting $1.50 pit. The three nested pits are shown on section 18000 N in Figure 6.13.

With this approach one would begin (Phase I) with the mining of the 0.20% MoS_2 ($1.50) cutoff pit. Phase II involves the material in the 0.15% MoS_2 ($2) cutoff pit and finally, Phase III, the 0.05% MoS_2 ($6) cutoff pit. Pits intermediate to these can be found by selecting the appropriate price. As can be seen in Figure 6.13, barren material overlies the orebody. All material down to the 4400 level will be stripped and sent to a waste dump. Its removal requires drilling and blasting. There is also some low grade material running 0–0.05% MoS_2 below this level. The grade-tonnage distribution for the overall pit is shown in Table 6.6. There are 102,970,000 tons above the 0.05% cutoff. The average grade is 0.186%. The grade-tonnage distributions for the 3 mining phases are given in Table 6.7.

The average grade of the material above 0.05% for each of the 3 phases is:

Phase I: $g = 0.225\%$

Phase II: $g = 0.182\%$

Phase III: $g = 0.176\%$

An initial decision has been made to mine the orebody over a period of 15 years. Thus the average milling rate will be of the order of 6.9×10^6 tons/year. Assuming that the mill operates 250 days/year the daily milling rate is 27,500 tpd. In reviewing the level plans,

Production planning 497

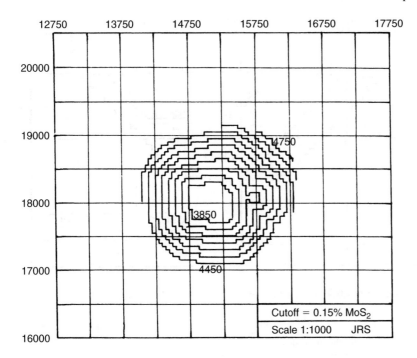

Figure 6.11. Pit outline for a 0.15 cutoff (Suriel, 1984).

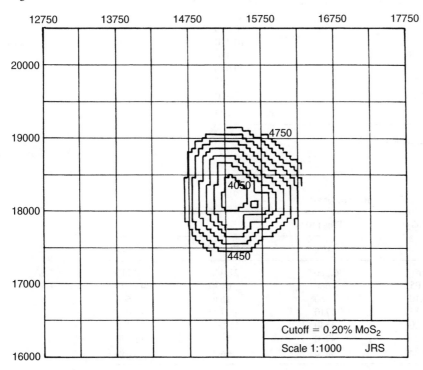

Figure 6.12. Pit outline for a 0.20 cutoff (Suriel, 1984).

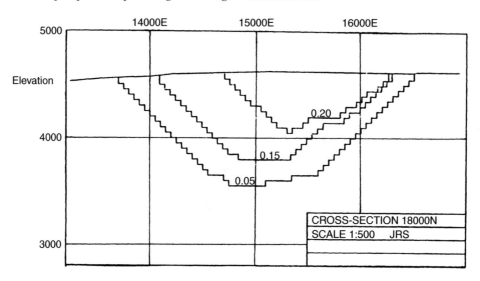

Figure 6.13. Cross-section 18000 N showing the three pit outlines (Suriel, 1984).

Table 6.6. Mineral inventory for the molybdenum pit.

Grade (% MoS_2)	Tons (10^3)
0 (overburden)	86,410
$0 < g < 0.05$	50,720
$0.05 \leq g < 0.10$	10,420
$0.10 \leq g < 0.15$	23,430
$0.15 \leq g < 0.20$	29,010
$0.20 \leq g < 0.25$	22,300
$g^* \geq 0.25$	17,810
	Total = 240,100

*The average grade is 0.30%.

Table 6.7. Mineral inventory by phase.

Grade (% MoS_2)	Tons (10^3)		
	Phase I	Phase II	Phase III
0 (overburden)	36,260	21,440	28,710
$0 < g < 0.05$	680	6170	43,870
$0.05 \leq g < 0.10$	180	1640	8600
$0.10 \leq g < 0.15$	1710	9700	12,020
$0.15 \leq g < 0.20$	4530	11,060	13,420
$0.20 \leq g < 0.25$	4260	5670	12,370
$g^* \geq 0.25$	5670	4860	7280

* The average grade is 0.30%.

it is seen that 57,700,000 tons of rock overburden (the Phase I and Phase II overburden) must be removed in a pre-production period in order to have 6 months ore supply available at the time of production. This leaves 182,400,000 tons to be mined (ore and waste) during the production period. Assuming that the mine also works 5 days/week, the daily mining rate is

$$\text{Mining rate} = 48{,}600 \text{ tpd}$$

The same equipment fleet will be used for the prestripping as the production mining. During the first year of stripping the average daily production rate will be assumed to be $\frac{1}{3}$ that maintained in later years due to equipment delivery, personnel training, and limited working place. Thus the time required for the prestripping is

$$t \text{ (yrs)} = \frac{57{,}700{,}000}{(48{,}600(t-1) + 16{,}200 \times 1)250}$$

$$\cong 5.4 \text{ yrs}$$

To complete the stripping in 5 years the actual waste mining rate will be

Year 1 = 18,000 tpd

Years 2 to 5 = 53,200 tpd

With this milling rate, there are approximately 2.4 years of reserves in Phase I and 4.8 years of reserves in Phase II. Phase III represents the reserves for years 7.2 through 15. The following data will be used:

Total material to be mined = 240,100,000 tons

Ore grade > 0.05% MoS_2 material = 102,970,000 tons

Average ore grade = 0.186% MoS_2

Mill recovery = 90%

Price = $3.24/lb MoS_2

Mining rate = 48,600 tpd

Milling rate = 27,500 tpd

Operating time = 250 days per year

Total material to be pre-stripped = 57,700,000 tons

Pre-stripping period = 5 years

Stripping rate year 1 = 16,200 tpd

Stripping rate years 2–5 = 48,600 tpd

This example is continued in the cash flow calculations of Section 6.5.

6.5 CASH FLOW CALCULATIONS

In Section 6.4 the concept of nested pits was discussed with respect to a molybdenum orebody. In this section the example will be continued with consideration of the resulting cash

Table 6.8. Assumed schedule for the pre-production period (Suriel, 1984).

	Year						
	1	2	3	4	5	6	7
Exploration	✓	✓					
Property acquisition		✓					
Development			✓	✓	✓	✓	✓

flows. The basic ideas involved with cash flow calculations were introduced in Chapter 2. The specific application, however, has been left to this section. Each state, province, and country has specific rules and regulations regarding taxes, depreciation, depletion, royalties, etc. Even for a given state these rules change with time. Therefore, the authors have selected a somewhat simplified example originally presented by Suriel (1984) to illustrate the calculations involved. Hopefully the reader can adapt the procedures to fit the application at hand. The example applies roughly to an orebody located in the State of Colorado in the year 1983. Certain aspects of the laws in effect at the time such as minimum tax and investment tax credit have been left out since they add unnecessary complication to the example. It has been assumed that the mining company is a division of a profitable corporation. Costs involved in the development of this new mine will be expensed whenever possible.

In the mine life two periods will be considered:
– pre-production period, and
– production period.

Today there is generally a third period, the post-production or closure period in which final reclamation takes place. This is not covered here. The pre-production period which is assumed to require 7 years, can be broken down into 3 different and distinct expenditure categories:
– detailed exploration,
– property acquisition, and
– infrastructure and mine development.

A representative schedule of activities is illustrated in Table 6.8. The first two years are used for detailed exploration work. Property acquisition takes place in year 2. The following 5 years are required for infrastructure and mine development. Hence, the pre-production period requires a total of 7 years. Production is expected to take place over a period of 15 years.

The basic line items in the pre-production cash flow calculation are given in Table 6.9. A brief discussion of each line item will be presented below. The numerical values for this example have been inserted in Table 6.10.

Lines 1 and 2: Project year and Calendar year. The first calendar year in which major expenditures occur is 1984. For the cash flow calculations this is selected as project year 1. Discounting will be done back to the beginning of year 1984 (project year 1).

Line 3: Capital expenditures. The 'capital' expenditures include a variety of different types of items ranging from hardware (mine and mill equipment) to royalties and property taxes. They occur at varying times in the cash flow table.

Table 6.9. Typical pre-production cash flow categories (Suriel, 1984).

1. Project year	1	2	Last year of
2. Calendar year	1984	1985...	pre-production
3. Capital expenditures:			
4. Property acquisition			
5. Royalties			
6. Exploration			
7. Development			
8. Mine and mill buildings			
9. Mine and mill equipment			
10. Property tax			
11. Working capital			
12. Total capital expenditures			
13. Cash generated due to tax savings:			
14. Exploration			
15. Development			
16. Depreciation			
17. Property tax			
18. Total cash generated			
19. Net cash flow			

Line 4: Property acquisition costs $2,000,000. It takes place in project year 2. This expenditure is a primary component of the depletion account which controls unit depletion allowance.

Line 5: Royalties. In addition to receiving $2,000,000 for the property, the original owners will receive a royalty. Normally these royalties are a certain percentage of the net smelter return (NSR). In this case the royalty is 5% of the mill return. No royalty is, therefore, paid during the pre-production period.

Line 6: Exploration. As defined in the U.S. tax law, exploration costs are those incurred prior to any development of the deposit. Exploration refers to the activities performed in order to determine the location, size, extent, quality, and quantity of a mineral occurrence. The reader is encouraged to compare this interpretation with that appropriate at the specific location and time. The U.S. Internal Revenue Service (IRS) allows the firm to choose between two separate methods in accounting for exploration expenditures: capitalize exploration costs into the depletion account and allocate them over time as production occurs, or treat exploration costs as annual expenses. For this example, the exploration costs are broken down as follows:

Year 1 = $1,500,000

Year 2 = $1,000,000

To simplify the example, these have been fully expensed in the years in which they occurred.

Line 7: Development. In this example site development consists of three individual line items:
  (a) Site preparation.
  (b) Pre-production stripping.
  (c) Plant site cleaning and mass excavation.

502 Open pit mine planning and design: Fundamentals

Table 6.10. Pre-production cash flow table ($1,000).

	1	2	3	4	5	6	7	
1. Project year	1	2	3	4	5	6	7	
2. Calendar year	1984	1985	1986	1987	1988	1989	1990	Total
3. Capital expenditures:								
4. Property acquisition	0	2000	0	0	0	0	0	2000
5. Royalties	0	0	0	0	0	0	0	
6. Exploration	1500	1000	0	0	0	0	0	2500
7. Development	0	0	3330	9842	9842	9842	9842	42,698
8. Mine/mill buildings	0	0	0	4836	13,057	1711	9948	29,552
9. Mine/mill equipment	0	0	10,000	0	9981	25,295	2999	48,275
10. Property tax	0	0	255	378	965	1654	1985	5237
11. Working capital	0	0	0	0	0	0	8353	8353
12. Total capital expenditures	1500	3000	13,585	15,005	33,744	38,349	33,127	138,310
13. Cash generated due to tax savings								
14. Exploration	690	460	0	0	0	0	0	1150
15. Development	0	0	1532	4527	4527	4527	4527	19,640
16. Depreciation	0	0	0	920	920	920	920	3680
17. Property tax	0	0	117	150	397	690	819	2173
18. Total cash generated	690	460	1649	5597	5844	6137	6266	26,643
19. Net cash flow	−810	−2540	−11,936	−9408	−27,900	−32,212	−26,861	−111,667

General site preparation (cost = $1,729,000) as well as the plant site cleaning (cost = $859,000) will occur in project year 3. The total amount of pre-production stripping is 57,700,000 tons. Since the expected cost per ton is $0.74, the total cost is $42,698,000. The stripping schedule is as follows:

Project year	Amount stripped (tons)	Cost ($)
3	4,500,000 (18,000 tpd)	3,330,000
4	13,300,000 (53,200 tpd)	9,842,000
5	13,300,000	9,842,000
6	13,300,000	9,842,000
7	13,300,000	9,842,000

These costs are expensed in the year incurred.

Line 8: Mine and mill buildings. There are 7 items which fall into this category. They are indicated below together with their cost and the project year in which the expenditure is made.

Item	Project year	Cost ($)
1. Concrete foundation and detailed excavation	4	$4,836,000
2. Open-pit maintenance facilities	7	$6,623,000
3. Concentrator building	5	$13,057,000
4. Concentrate storage and loading	7	$139,000
5. Capital cost of general plant service	6	$1,711,000
6. Tailings storage	7	$1,775,000
7. Water supply	7	$1,411,000

These items, considered as improvements to the property, will be depreciated (straight line) over a 20 year period beginning in production year 1. Capital gains and losses will be ignored.

Line 9: Mine and mill equipment. There are 6 items which fall into this category. They are indicated below together with their cost and the project year in which the expenditure is made.

Item	Project year	Cost ($)
1. Open-pit equipment	3	$10,000,000
2. Crushing plant, coarse ore storage and conveyors	5	$9,981,000
3. Grinding section and fine ore storage	6	$16,535,000
4. Flotation section	6	$6,985,000
5. Thickening and filtering	6	$1,775,000
6. Electric power supply and distribution	7	$2,999,000

The open-pit equipment is broken down as follows:

(a) Shovels (3-15 yd^3) = $3,000,000
(b) Trucks (12-150 ton) = $5,000,000
(c) Drills, graders, dozers = $2,000,000

The equipment in categories (b) and (c) will be replaced every 5 years. Since this fleet will be used for the stripping as well as the production mining, there will be a capital expense of $7,000,000 in project years 1, 6 and 11. The shovels are expected to last the life of the mine. Straight line depreciation over a period of 5 years is used. There is no salvage. The equipment will be put into use in the year purchased, hence, depreciation will begin then.

Items 2 through 6 will be depreciated over a 7 year period (straight line, no salvage) beginning in production year 1.

Line 10: Property tax. This tax, sometimes called ad valorem tax is one of the most common types of state tax. Property tax in Colorado is assessed on: (a) personal and real property and (b) ore sales.

Colorado taxes personal and real property at 30% of the actual value of the property. In this case personal property is that listed in line 9 and real property in line 8. The taxable value for real and personal property is determined by its base year value. The tax is then determined by multiplying the total assessed value by a rate called the 'pro mille' levy or in the U.S. simply the 'mill' levy. In Colorado the 'pro mille' levy varies from county to county. In this case study that used is 85 (corresponding to that applied by Jefferson County). This means that the appropriate factor is $85/1000$ or 0.085. The assessed value for the calculation of the property tax on ore sales is: (a) 25 percent of gross proceeds or (b) 100 percent of net proceeds, whichever is greater.

Gross proceeds are defined as the gross value of the ore produced minus treatment, reduction, transportation and sales cost at the mine mouth. Net proceeds is equal to the gross proceeds minus all costs associated with extraction of the ore. In Colorado, royalties are not deductible for either case. The pro mille levy is applied to the assessed value.

Line 11: Working capital. This represents the amount of money necessary to cover operating costs during a portion of the project's life. Working capital consists of cash, inventories (parts, supplies and concentrate) and accounts receivable. Working capital was estimated at four months of operating costs. It was allocated to the last year of the pre-production period before production start up (project year 7). The account is maintained during the production period and recovered at the end of project life. In this case, the amount of required working capital is $12,383,000.

Line 12: Total capital expenditures. This is the sum year by year of line items 4 through 11.

Line 13: Cash generated due to tax savings. As indicated earlier, the mining company is one division of a large profitable corporation. The corporate structure enables the company to expense most costs whenever possible and generate tax savings during the preproduction period. The federal tax rate is 46 percent.

To illustrate this assume that the corporation has an income subject to federal tax, prior to including this mining venture, of $10,000,000. At a tax rate of 46%, the corporation would pay $4,600,000 in federal tax. However, in project year 1, the mining division incurs an exploration expense of $1,500,000. If this is considered as an expense of the corporation, then the taxable income of the corporation would drop from $10,000,000 to $8,500,000. The tax on this amount is $3,910,000. Hence, there is a tax savings of $690,000 for the corporation. This has been included on the cash flow table for the mining company (Table 6.10) under 'Cash generated due to tax savings – Exploration'. Similar 'tax savings' occur with regard to development, depreciation and property tax when they are applied against other income.

Line 14: Exploration. The exploration expenses result in a tax savings of $690,000 (0.46 × $1,500,000) in project year 1 and $460,000 in project year 2.

Line 15: Development. Stripping, site preparation, etc., costs are multiplied by the tax rate and included here.

Line 16: Depreciation. Straight line depreciation of the mining equipment begins in project year 3. The yearly amount is multiplied by the tax rate.

Line 17: Property tax. The property tax is the assessed value multiplied by the tax rate.

Line 18: Total cash generated. The items in lines 14 through 17 are summed and entered here.

Line 19: Net cash flow. The total capital expenditures (line 12) are subtracted from the total cash generated (line 18). This is the net cash flow.

The basic line items in the production period cash flow calculation are given in Table 6.11. A brief discussion of these will be presented below.

Line 1, 2, and 3: Production, project and calendar year. The first year of production is 1991 which is the eighth year of the project. In doing the discounted cash flows and NPV calculations, they will be brought back to the beginning of the project.

Line 4: Revenue. Revenue is calculated by multiplying concentrate tonnage by the price of molybdenite concentrate at the mill site. The molybdenite concentrate price is estimated

Table 6.11. Typical production period cash flow categories (Suriel, 1984).

1. Production year
2. Project year
3. Calendar year
4. Revenue
5. Royalty
6. Net revenue
7. Mining cost
8. Processing cost
9. General cost
10. Property tax
11. Severance tax
12. Depreciation
13. State income tax
14. Net income after costs
15. Depletion
16. Taxable income
17. Federal income tax
18. Profit
19. Depreciation
20. Depletion
21. Cash flow
22. Capital expenditures
23. Working capital
24. Net cash flow

506 *Open pit mine planning and design: Fundamentals*

at $6 per pound of contained molybdenum. Concentrate tonnage is equal to ore tonnage multiplied by average grade, the mill recovery, and by the concentrate percent of Mo. (MoS_2 contains about 60% Mo.)

Line 5: Royalty. The royalty is 5 percent of the revenue.

Line 6: Net revenue. This is the difference between revenue and the royalty payment.

Line 7: Mining cost. The mining cost depends somewhat on the production rate as can be seen in Figure 6.14 and in Table 6.12. Because 57,700,000 tons have been mined in the pre-production period, 182,400,000 tons remain. The length of the production period has been selected as 15 years, and the mine will work 5 days per week (250 days per year). Thus the mining rate is 48,640 tons per day. From Figure 6.14, the mining cost of $0.77/ton has been selected.

Line 8: Processing cost. The mill will also run 5 days per week, 250 days per year. The mill cutoff grade has been selected as 0.05% MoS_2 hence there are 102,970,000 tons to be processed. The milling rate is therefore 27,460 tons/day and yearly ore production is 6,865,000 tons. The milling cost selected from Figure 6.15 and Table 6.12 is $1.93/ton.

Line 9: General and administrative cost. This cost has been computed per ton of ore processed. Its value as can be seen in Figure 6.16 and Table 6.12 is $0.67/ton. In some operations there is a G&A cost attached to waste removal as well. This has not been done here.

Line 10: Property tax. The property tax is computed in the same way as was discussed for the pre-production period. The assessed value is 30% of the initial cost and the pro mille rate is 85 (0.085). The value at the end of the pre-production period is $77,827,000. Applying the 30% assessed valuation and the 85 pro mille levy, the annual property tax is $1,985,000.

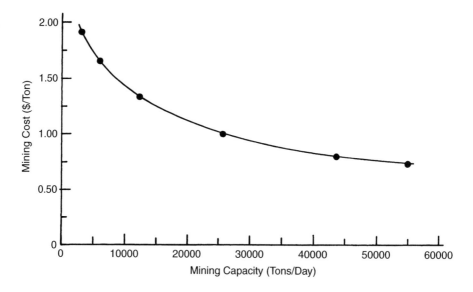

Figure 6.14. Mining cost versus mining capacity (Suriel, 1984).

This annual value will be carried throughout the production period. The book value has not been used and the effect of replacement capital has not been included.

The gross revenue from annual ore sales is $82,742,000. The net revenue after subtracting mining, processing, and general costs is $57,683,000. Note that royalties have not

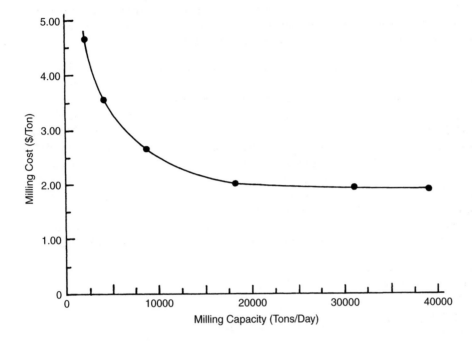

Figure 6.15. Milling cost versus milling capacity (Suriel, 1984).

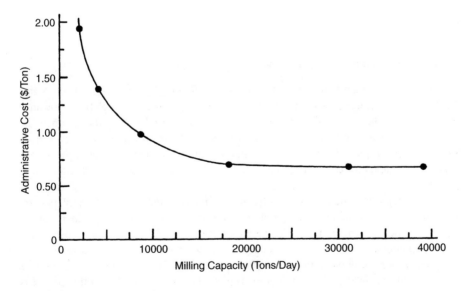

Figure 6.16. Administrative cost versus milling capacity (Suriel, 1984).

been subtracted. Applying the two rules (25 percent of gross proceeds or 100 percent of net proceeds) one finds that the greatest amount (which becomes the assessed valuation) is $57,683,000. Applying the 85 pro mille levy, this contribution to the property tax is $4,903,055. Thus the total annual property tax is $6,888,055.

Line 11: Severance tax. A severance tax is levied for removing or 'severing' the mineral from the earth and the state/country. Colorado imposes a such severance tax on molybdenum. It was imposed on production at a rate of $0.15 per ton of ore.

Line 12: Depreciation. The depreciation allowance is a deduction over a period of years for use, deterioration, wear and tear of depreciable assets used in generating income for the project. Depreciable assets for a mining project are grouped into either personal or real property. Mine and mill equipment are considered personal property. The total value is $48,275,000. The open pit mining equipment is depreciated over 5 years. The initial investment is $10,000,000 in project year 3 while replacement equipment (trucks, drills, etc.) at a cost of $7,000,000 is purchased in project years 8, 13, and 18. The remainder of the mine and mill equipment (initial cost of $38,275,000) is depreciated over a seven year period beginning at the start of production. Real property, including mine and mill buildings and improvements, amount to $29,552,000. A twenty year life is used for real property with depreciation beginning at the start of production. Thus the depreciation per year is $1,477,600. The depreciation by class and year is shown in Table 6.13.

Line 13: State income tax. The state tax is 5% of the net income. In computing the net income the following deductions are allowed:
– royalties,
– operating costs,
– depreciation,
– property tax,
– severance tax, and
– depreciation.

For production years 1 through 7 the state income tax is $1,864,000. For years 8 through 15 it is $2,138,000.

Line 14: Net income after costs. This is the net revenue (line 6) minus lines 7 through 13.

Line 15: Depletion. In order to recognize that minerals, oil, and gas are non-renewable assets which are depleted through production, the U.S. government permits a depletion allowance to be deducted prior to the calculation of federal taxable income. For mineral resources the depletion allowance is calculated and claimed by either of two methods, whichever gives the largest amount of pre-tax deduction. It is permissible to change methods from year to year. The two methods for calculating the depletion allowance are statutory depletion and cost depletion.

For statutory depletion, the amount which can be deducted is the smaller of: (a) 50 percent of net income or (b) a certain percentage of net revenue (revenue minus royalties). For the (a) calculation, net income is defined as revenues minus royalties, operating costs, state taxes, and depreciation. For (b) the percentage depletion for molybdenum is 22 percent. Note that the amount of depreciation varies throughout the life of the mine. For the first 7 years of production it is $8,346,000 per year. The annual net income during this period is $35,418,000 and hence, the allowable depletion on this basis is $17,709,000.

Table 6.12. Assumed operating cost data (Suriel, 1984).

Cutoff grade % MoS$_2$	Geologic reserves (× 10^6 st)	90% geologic reserves (× 10^6 st)	Average grade % MoS$_2$	Tonnage milled (tpd)	Tonnage mined (tpd)	Total mining cost ($/ton)	Total milling cost ($/ton)	General & administrative cost ($/ton)	Total operating cost ($/ton)
0.30	12.680	11.412	0.40	2173	3043	1.91	4.64	1.93	8.48
0.25	24.390	21.950	0.34	4181	5854	1.65	3.55	1.39	6.35
0.20	51.300	46.170	0.28	8794	12,312	1.33	2.65	0.97	4.95
0.15	107.010	96.309	0.22	18,345	25,682	1.00	1.98	0.68	3.66
0.10	181.490	163.341	0.18	31,113	43,558	0.80	1.91	0.67	3.38
0.05	228.280	205.452	0.16	39,134	54,787	0.74	1.89	0.67	3.30

Table 6.13. Depreciation by class and project year.

Depreciation class (yrs)	Project year depreciation ($1000)																					
	1	2	3	4	5	6	7	8	9	10	11	12	13	14	15	16	17	18	19	20	21	22
5	0	0	2000	2000	2000	2000	2000	1400	1400	1400	1400	1400	1400	1400	1400	1400	1400	1400	1400	1400	1400	1400
7	0	0	0	0	0	0	0	5468	5468	5468	5468	5468	5468	5468	0	0	0	0	0	0	0	0
20	0	0	0	0	0	0	0	0	1478	1478	1478	1478	1478	1478	1478	1478	1478	1478	1478	1478	1478	1478
Total	0	0	2000	2000	2000	2000	2000	8346	8346	8346	8346	8346	8346	8346	2878	2878	2878	2878	2878	2878	2878	2878

510   Open pit mine planning and design: Fundamentals

The annual net revenue is $78,605,000. Taking 22% of this number yields $17,293,000. The lesser of these two numbers is $17,293,000.

Cost depletion is based on the cost of the property, non-expensed exploration costs, number of units of mineral sold during the year and reserves available in the deposit at the end of the year. Cost depletion is calculated by multiplying the adjusted basis, (equivalent to all property acquisition costs plus capitalized exploration costs) by the tons mined during the year divided by remaining reserves. The adjusted basis is reduced each year by the depletion allowance claimed.

$$\text{Cost depletion} = (\text{Adjusted basis}) \times \frac{\text{Mineral units removed during the year}}{\text{Mineral units recoverable at start of the year}}$$

$$\text{Adjusted basis} = \text{Cost basis} \pm \text{Adjustments} - \text{Cumulative depletion allowance claimed}$$

In this case the cost of the property was $2,000,000. For the first year, 6,865,000 tons of ore are mined. The initial ore tonnage is 102,970,000. Hence, for year 1 the allowable cost depletion would be

$$\$2,000,000 \times \frac{6,865,000}{102,970,000} \times \$133,340$$

The greater of the cost and percentage depletion values is chosen. Thus, one chooses $17,293,000 to be deducted. However, according to the tax rules one must recapture the exploration expense of $2,500,000 by reducing the amount of depletion earned. Hence, in year 1, the allowed depletion deduction is $17,293,000 − $2,500,000 = $14,793,000.

In year 2 the statutory depletion allowance is $17,293,000. For the cost depletion calculation the adjusted basis is

$$\text{Adjusted basis} = \text{Cost basis} - \text{Cumulative depletion}$$

The cost basis is $2,000,000. The cumulative depletion already taken (through project year 8) is $14,793,000. Hence the adjusted basis is

$$\text{Adjusted basis} = \$2,000,000 - \$14,793,000 = -\$12,793,000$$

The cost of the property was fully recovered through depletion in production year 1 and hence only statutory depletion is used for the remaining mine life. In production year 2, a full depletion of $17,293,000 is taken since all previously expensed exploration costs have been recaptured.

Line 16: Taxable income. This is the net income after cost (line 14) minus depletion (line 15).

Line 17: Federal income tax. In this case the federal income tax is 46% of the taxable income.

Line 18: Profit. The profit is the taxable income minus the federal income tax.

Line 19 and 20: Depreciation and depletion. These two items which had been previously considered as expenses when computing taxes are now added to the profit to arrive at a cash flow.

Line 21: Cash flow. This is the sum of lines 18 through 20.

Line 22: Capital expenditures. In years 1, 6, and 11 new trucks, drills, dozers, etc., are purchased. The capital cost involved in each of these years is $7,000,000.

Line 23: Working capital. The working capital was allocated to the last year of the pre-production period before production start-up. The account is maintained during the production period and recovered at the end of project lift (project year 22).

Line 24: Net cash flow. The net cash flow for any particular year is the sum of lines 21, 22, and 23.

The cash flows during the production period are summarized in Table 6.14.

6.6 MINE AND MILL PLANT SIZING

6.6.1 *Ore reserves supporting the plant size decision*

In the preceding sections, the final pit limits have been determined using the following steps:

1. The metal content of a block together with forecasted sales prices and estimated full scale plant metallurgical recoveries are used for determining the revenue.

2. Production costs through to sales and shipping costs are estimated.

3. A net value is calculated for each block. A block with a net value equal to zero is one where the revenues equal the production costs. The grade corresponds to a mining cutoff.

4. A computer technique (such as the floating cone) is used to identify all blocks capable of paying for the costs of uncovering them. These are included within the final pit.

5. A mineral inventory – ore reserve is constructed.

6. Milling and mining rates are assumed and the associated capital costs determined. For example a series of different milling cutoff grades might be applied to the total pool of material. Mill ore quantities would be calculated. These reserves would be used to size the mill plant.

7. A financial analysis would be conducted for each alternative to determine the NPV, internal ROR, etc. The best option would be selected.

As can be seen, the introduction of the capital cost takes place at a very late stage. The profit is an output rather than an input. After having gone through this entire process, it may be that the rate of return is too low. The process would then have to be repeated focussing on the higher grade ore, using a different (lower) production rate, reducing the investment in mine and mill plant, etc. In the process just described a variety of materials of different economic value are being mined at any one time. Those with the highest values contribute more to paying off the investment and profit than those of lower value. The summing or integration of these values occurs for a given production year. This overall value is used as an entry in the cash flow table. To illustrate this concept, assume that there are 60 ore blocks to be mined in a given period. The net value distribution (at step 3, prior to introducing the capital cost and profit) is as follows:

Grade	Net value/block	No. of blocks	Net value
g_1	$0	10	$0
g_2	$10	10	$100
g_3	$20	10	$200
g_4	$30	10	$300
g_5	$40	10	$400
g_6	$50	10	$500

512 Open pit mine planning and design: Fundamentals

Table 6.14. Cash flow during mine production life ($1000).

	1	2	3	4	5	6	7	8	9	10	11	12	13	14	15	
1. Production year																
2. Project year	8	9	10	11	12	13	14	15	16	17	18	19	20	21	22	
3. Calendar year	1991	1992	1993	1994	1995	1996	1997	1998	1999	2000	2001	2002	2003	2004	2005	Total
4. Revenue	82,742	82,742	82,742	82,742	82,742	82,742	82,742	82,742	82,742	82,742	82,742	82,742	82,742	82,742	82,742	1,241,130
5. Royalty	4137	4137	4137	4137	4137	4137	4137	4137	4137	4137	4137	4137	4137	4137	4137	62,055
6. Net revenue	78,605	78,605	78,605	78,605	78,605	78,605	78,605	78,605	78,605	78,605	78,605	78,605	78,605	78,605	78,605	1,179,075
7. Mining cost	9363	9363	9363	9363	9363	9363	9363	9363	9363	9363	9363	9363	9363	9363	9363	140,445
8. Processing cost	13,249	13,249	13,249	13,249	13,249	13,249	13,249	13,249	13,249	13,249	13,249	13,249	13,249	13,249	13,249	198,735
9. General cost	4600	4600	4600	4600	4600	4600	4600	4600	4600	4600	4600	4600	4600	4600	4600	69,000
10. Property tax	6888	6888	6888	6888	6888	6888	6888	6888	6888	6888	6888	6888	6888	6888	6888	103,320
11. Severance tax	1030	1030	1030	1030	1030	1030	1030	1030	1030	1030	1030	1030	1030	1030	1030	15,450
12. Depreciation	8346	8346	8346	8346	8346	8346	8346	2878	2878	2878	2878	2878	2878	2878	2878	81,446
13. State income tax	1756	1756	1756	1756	1756	1756	1756	2030	2030	2030	2030	2030	2030	2030	2030	28,532
14. Net income after costs	33,373	33,373	33,373	33,373	33,373	33,373	33,373	38,567	38,567	38,567	38,567	38,567	38,567	38,567	38,567	542,147
15. Depletion	14,793	17,293	17,293	17,293	17,293	17,293	17,293	17,293	17,293	17,293	17,293	17,293	17,293	17,293	17,293	256,895
16. Taxable income	18,580	16,080	16,080	16,080	16,080	16,080	16,080	21,274	21,274	21,274	21,274	21,274	21,274	21,274	21,274	285,252
17. Federal income tax	8547	7397	7397	7397	7397	7397	7397	9786	9786	9786	9786	9786	9786	9786	9786	131,217
18. Profit	10,033	8683	8683	8683	8683	8683	8683	11,488	11,488	11,488	11,488	11,488	11,488	11,488	11,488	154,035
19. Depreciation	8346	8346	8346	8346	8346	8346	8346	2878	2878	2878	2878	2878	2878	2878	2878	81,446
20. Depletion	14,793	17,293	17,293	17,293	17,293	17,293	17,293	17,293	17,293	17,293	17,293	17,293	17,293	17,293	17,293	256,895
21. Cash flow	33,172	34,322	34,322	34,322	34,322	34,322	34,322	31,659	31,659	31,659	31,659	31,659	31,659	31,659	31,659	492,376
22. Capital expenditures	7000	0	0	0	0	7000	0	0	0	0	7000	0	0	0	0	21,000
23. Working capital	0	0	0	0	0	0	0	0	0	0	0	0	0	0	8353	8353
24. Net cash flow	26,172	34,322	34,322	34,322	34,322	27,322	34,322	31,659	31,659	31,659	24,659	31,659	31,659	31,659	23,306	463,023

There are 50 associated waste blocks each carrying a value of −$20. Thus the apparent total net value for this period would be $500. At first glance this might appear good. However when a cash flow calculation (now including the capital cost and profit) is made, it is found that the desired rate of return is too low. The required net value at step 3 is $1,100 rather than $500. Since there are 60 ore blocks involved, this would mean an average 'assessment' of $10/block. If this is done then the modified net value distribution becomes

Grade	Net value/block	No. of blocks	Net value
g_1	−$10	10	−$100
g_2	$0	10	$0
g_3	$10	10	$100
g_4	$20	10	$200
g_5	$30	10	$300
g_6	$40	10	$400

Clearly those blocks with grade g_1 are not contributing their fair share of the capital investment and profit. Thus the cutoff grade should be g_2 rather than g_1. The number of ore blocks would drop from 60 to 50. If the plant size is not changed, this translates into a shorter property life which changes the cash flow which in turn would affect the $10 average block assessment. If the mine life is maintained constant, then the size of the mill plant should be reduced. This also affects the cash flow. For this situation, perhaps the average assessment per block would drop to $5/block. The number of blocks would increase again and the process would be repeated.

One way of handling this problem early on is to include a cost item for capital and profit amongst those used to generate the economic block model. These costs would then be covered up front.

The alternative sequence of steps is listed below (Halls et al., 1969):

1. The metal content of a block, together with forecasted sales prices and estimated full-scale plant metallurgical recoveries is used for determining the revenue.

2. Production costs through to sales and shipping costs are estimated.

3. The amount of depreciation to be added to the operating costs is calculated on a straight line basis by dividing the estimated capital expenditure by the estimated ore reserve tonnage.

4. A minimum profit after tax, related to minimum acceptable return on investment is assigned to each block.

5. The net value for each block is calculated. The 'cutoff grade' used to separate ore from waste is that grade of material which produces a revenue equal to the production costs through to sales together with depreciation and minimum profit.

6. Using the above data the computer will include all blocks of material in the ore reserve which are capable of producing more than the minimum pre-determined profit (after taxes) after paying for the stripping costs necessary to uncover them.

There are no real difficulties in assessing the revenues to be produced from a block or the production costs involved in producing its metal for sale, including such considerations as cost variations due to varying haulage distances and rock types. It is, however, extremely difficult to estimate unit depreciation and required minimum profit per ton accurately before the ore reserve tonnage is known because both
– depreciation, and
– required minimum profit

514 *Open pit mine planning and design: Fundamentals*

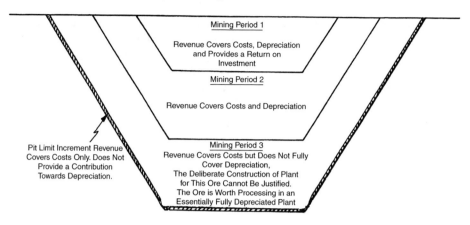

Figure 6.17. Economic pit limits as a function of included costs (Halls et al., 1970).

will vary with plant size and total ore tonnage. In order to estimate depreciation and profit
- an ore reserve tonnage,
- a production rate, and
- an effective tax rate

must be assumed. An iteration process is used until convergence between the assumed and final values is achieved.

The resulting pit is shown as the 'Mining Period 1' pit in Figure 6.17. A second pit outline can be produced by eliminating the required profit (profcost) assessed each block. The depreciation cost would remain, however. By eliminating the profit cost element and keeping the other costs and revenues the same, the positive block values would increase. Each positive block would be able to carry more stripping and the pit would increase in size. This is shown by the 'Mining Period 2' pit. The 'no-profit' condition only applies when determining the pit limit. The average ore grade of the material between the Mining Period 1 and 2 limits would be higher than the pit limit cutoff. Hence a profit is realized. To increase the size of the pit further, the depreciation cost would have to be dropped from the economic block model calculation. This further lowers the grade required to produce positive valued blocks. The pit expands to the limit shown by the heavy shaded line (Mining Period 3 limit). As indicated by Halls (1970), plant capacity cannot be deliberately constructed for this ore. The ore however is worth processing in an essentially fully depreciated plant. Even though a specific profit cost has not been attached to each block in Mining Period 3, there will still be an overall profit since the average ore grade in the period is higher than the cutoff. The Mining Period 2 pit encloses those reserves which should be considered when sizing the plant. The final pit outline would be the same as produced by the first procedure described in this section. Here however the reserves contained within specific mineable pits are used rather than a certain portion of the overall reserves. There are a number of objections (Halls, 1970) which have been raised concerning the requirement that each ore block irrespective of grade contribute an equal portion to profits and capital payback:

(a) The only practical way of determining unit depreciation for use in individual block evaluation is the straight-line method. In the financial evaluation of the ore reserve, the depreciation method employed will be the one which reduces the impact of taxation to a minimum. Also it is virtually impossible to incorporate in the cutoff calculations the unit depreciation which would apply to replacement capital needed throughout the life of the mine.

(b) The use of a minimum after-tax profit for each block does not ensure that the summation of the profits from each block will meet the corporate investment goal (except possibly in the case of an ideal mine, where the annual cash flows remain the same throughout its life).

(c) In the case of properties which can commence mining in a high-grade portion of the pit, or where taxation benefits are allowed in the initial stages of production, the accurate establishment of a profit factor to ensure that such increment of investment yields at least a minimum corporate return is an impossibility.

An alternative procedure based upon incremental financial analysis will now be discussed.

6.6.2 Incremental financial analysis principles

The incremental financial analysis approach described in this section is based upon material originally presented by Halls et al. (1969). The rate of return on the total investment is important for assessing the potential of a property as a whole. It does not however indicate the profitability of each capital increment. Only by considering a series of potential pit expansions and evaluating the yields and returns from each is it possible to ascertain that every increment of capital can pay its way. This evaluation process is termed incremental financial analysis. For a new property, the financial returns of alternative ore inventories and associated plant sizes for a given mine life (for example) are compared. One begins from the smallest tonnage ore inventory and plant size. Each progressively larger tonnage ore inventory is considered as a possible expansion. The additional cash flow developed from each expansion is determined as a rate of return on the additional (incremental) capital required for such an expansion. The largest tonnage and hence the lowest grade ore inventory in which every increment of invested capital yields at least the minimum desired corporate return is the optimum ore reserve for the assumed life. The process is repeated using other realistic lives for the property. These studies provide one basis for the selection of a depreciation life (depreciation) for the property. As discussed earlier, the actual mine life may be longer than this since once the invested capital has been fully depreciated, the cutoff grade drops (and the reserves increase). This method of optimizing ore reserves and plant size for a new property involves three steps:

Step 1. Calculate ore inventories using arbitrary cutoff grades.

Step 2. Prepare a financial analysis for each ore inventory.

Step 3. Select that ore inventory which is the optimum ore reserve by incremental financial analysis.

Each of these steps is outlined below.

In step 1, a series of arbitrarily chosen, but reasonable, cutoff grades are chosen. For an open pit copper mine cutoff grades in the range 0.3 to 0.7 percent equivalent copper might be chosen. Using the expected production costs and recoveries, the price needed for breakeven with the selected cutoff (0.7, for example) would be calculated. Note that no capital costs and profit is included in these calculations. The price together with the costs are then used to obtain an economic block model. A technique to provide a pit, such as the floating cone,

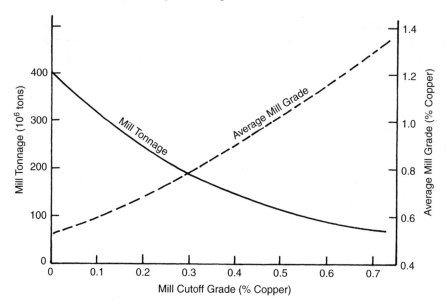

Figure 6.18. Pit material inventory for a 0.7 cutoff grade pit.

Table 6.15. Mill tonnage and average grade as a function of mill cutoff grade.

Mill cutoff grade (% Cu)	Mill tons ($\times 10^6$ st)	Avg. grade (% Cu)
0.70	72	1.29
0.60	88	1.17
0.50	116	1.02
0.40	148	0.90
0.30	188	0.78
0	400	0.52

is applied. This would be the 0.7% cutoff grade pit. The inventory of material contained would be created. One such example is shown in Figure 6.18.

The total tonnage contained is

$$T = 400 \times 10^6 \text{ tons}$$

and the average grade is

$$g = 0.52\% \text{ Cu}$$

If the mill cutoff were 0% Cu then this entire amount would be processed. For various mill cutoff grades, the total tons and grades are as given in Table 6.15.

All material has to be removed from the pit and the costs of depositing it on surface, whether waste or ore are virtually the same. Therefore some material which is below the mining cutoff grade, but within the confines of the pit may be processed at a profit.

Another mining cutoff grade would now be selected, for example 0.6%. The breakeven price would be calculated and the floating cone run on the deposit again using this price.

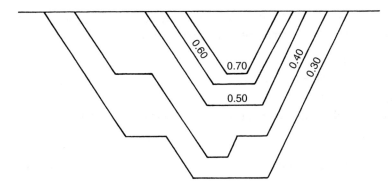

Figure 6.19. Pit outlines corresponding to different mining cutoff grades.

A table similar to Table 6.15 would be constructed for this condition. An idealized cross section showing the series of pits which might be produced is shown in Figure 6.19.

In step 2 a financial evaluation of each ore inventory is prepared. To do this the tonnage and grade to be mined from the inventory each year must be decided. Lane (1964) has discussed one way of doing this. His approach will be the topic of Section 6.7. Here the same life, 25 years, will be used to evaluate all inventories. In practice several realistic lives might be tried. Furthermore the production rate will be held constant. Obviously other variations can be applied. The following data are required to complete the financial evaluations:

(1) the estimated capital costs, broken down into the years in which the money will be spent (including replacement capital),

(2) the estimated operating costs for each production rate,

(3) the metal recoveries, based on metallurgical laboratory testing after discounting for commercial operating conditions,

(4) the long- and short-term estimates of metal sales prices,

(5) the legally applicable depletion and depreciation allowances which will minimize taxation, and

(6) the current and forecasted taxation rates.

The cash flows for each ore inventory are calculated on a year-by-year basis. Any one of a number of financial techniques which consider the time value of money can be utilized in determining the rates of return for both total investment and incremental investment. The discounted cash flow (DCF) method will be used here.

In step 3, the different pit expansions are compared. To illustrate the process, consider the pit generated with the 0.7% Cu cutoff. Two mill cutoffs are considered: 0.7% and 0.6%. The capital investments for mine and concentrator are:

Mill cutoff	Capital cost
0.7	$43,500,000
0.6	$46,832,000

The incremental capital cost is $3,332,000. The annual cash flow for each after taxes is:

Mill cutoff	Annual cash flow
0.7	$7,550,000
0.6	$8,323,000

The incremental annual cash flow is $773,000. These cash flows occur over a period of 25 years. The net present value of these incremental cash flows is given by

$$\text{NPV}_{\text{CF}} = \$773,000 \left[\frac{(1+i)^{25} - 1}{i(1+i)^{25}} \right]$$

where i is the interest rate.

These are achieved through the incremental capital investment of $3,332,000.

$$\text{NPV}_{\text{CI}} = -\$3,332,000$$

The interest rate (i) which makes the sum of these equal to zero is called the rate of return. In this case

$$-3,332,000 + 773,000 \left[\frac{(1+i)^{25} - 1}{i(1+i)^{25}} \right] = 0$$

$$i = 0.231$$

Thus there is a 23.1% effective rate of return (ROR) on the incremental investment. Since this exceeds the company's desired rate of 12%, the mill cutoff grade of 0.6% would become that used in future comparisons (0.5%, 0.4%, etc.) for the given mining cutoff (0.7%). This process would continue until ROR values are less than the base.

6.6.3 Plant sizing example

The incremental financial analysis process described in the previous section will now be demonstrated step-by-step using an example adapted from Halls et al. (1969).

1. Prices and costs are estimated. These are entered into the computer for a floating cone analysis or are used to generate a stripping ratio – grade curve for a hand analysis.

This is a breakeven analysis such that

Revenue = Costs

The revenue is calculated by

$$\text{Revenue (\$/ton)} = 2000 \text{ lbs/st} \times \frac{\text{Recovery (\%)}}{100} \times \frac{\text{Grade (\%)}}{100} \times \text{Price(\$/lb)} \quad (6.3)$$

The costs are given by

$$\text{Costs (\$/ton)} = \text{Mining (\$/st)} + \text{Milling (\$/st)} + \text{G\&A (\$/st)}$$
$$+ 2000 \text{ lbs/st} \times \frac{\text{Recovery (\%)}}{100} \times \frac{\text{Grade (\%)}}{100} \times \text{SRS (\$/lb)} \quad (6.4)$$

where SRS is refining, smelting, selling cost ($/lb).
Setting Equation (6.3) equal to (6.4) yields

$$\text{Mining (\$/ton)} + \text{Milling (\$/ton)} + \text{G\&A (\$/ton)}$$
$$= (\text{Price} - \text{SRS}) \times \frac{2000 \text{ lbs}}{\text{st}} \times \frac{\text{Recovery (\%)}}{100} \times \frac{\text{Grade (\%)}}{100} \quad (6.5)$$

Solving for the cutoff grade one finds that

$$\text{Grade (\%)} = \frac{[\text{Mining (\$/st)} + \text{Milling (\$/st)} + \text{G\&A (\$/st)}]10^4}{[\text{Price (\$/lb)} - \text{SRS (\$/lb)}]2000 \times \text{Recovery (\%)}} \quad (6.6)$$

If the expected price is used then the maximum (ultimate) pit will result.

2. The grades and tonnage in the pit are determined. This becomes the reserve of material to be mined.
   Assume the following values:

   Mining cost = $0.45/st
   Milling + G&A = $1.25/st
   Price = $0.40/lb
   SRS = $0.059/lb
   Recovery = 80%

(Note that for simplicity the G&A has been applied only to the mill (i.e. the ore).)
Then the mining cutoff grade becomes

   Grade (%) = 0.31%

One would generate a pit with this as the cutoff.

Table 6.16 shows that for a pit based upon a 0.3% Cu pit limit cutoff, the tonnage would be 1,410,000,000. The average grade would be 0.31% Cu (all material). If the mill cutoff grade were 0% then the entire amount would be sent to the mill. If only those blocks having a grade of +0.3% were sent to the mill, then the mill tonnage would be 390,000,000 having an average grade of 0.73% Cu. This would be the largest pit.

3. One could consider smaller pits as well since the biggest pit would not necessarily lead to the desired financial result. This result could be measured in NPV, total profit, return on investment, etc. Other pits can be generated corresponding to different pit limit cutoff grades. As can be seen from Equation (6.6), this is a simple matter of just changing the price. For a lower price, the cutoff grade must go up. In Table 6.16 the following results were obtained.

Pit limit cutoff grade (% Cu)	Total tons in pit ($\times 10^6$ st)	Avg. grade (% Cu)
0.3	1410	0.31
0.4	1206	0.33
0.5	875	0.36
0.6	650	0.38
0.7	400	0.40

For each of these ultimate pits all the material would be mined. One must decide whether it should go to the dump or to the mill.

4. Each of the potential pits is examined with respect to mill and dump material. Consider the pit developed assuming a final pit cutoff grade of 0.7%. The mill cutoff could be set at

Table 6.16. Summary of ore inventory for each mining plan (Halls et al., 1969).

Milling cutoff grade		Pit limit cutoff grade				
		0.7% Cu	0.6% Cu	0.5% Cu	0.4% Cu	0.3% Cu
0.7% Cu	Accum. tons Average grade Stripping ratio	80,000,000 1.20% Cu 4.00 : 1				
0.6% Cu	Accum. tons Average grade Stripping ratio	92,000,000 1.13% Cu 3.35 : 1	135,000,000 1.12% Cu 3.81 : 1			
0.5% Cu	Accum. tons Average grade Stripping ratio	105,000,000 1.06% Cu 2.81 : 1	160,000,000 1.03% Cu 3.06 : 1	195,000,000 0.99% Cu 3.49 : 1		
0.4% Cu	Accum. tons Average grade Stripping ratio	115,000,000 1.01% Cu 2.48 : 1	185,000,000 0.95% Cu 2.51 : 1	235,000,000 0.90% Cu 2.73 : 1	290,000,000 0.85% Cu 3.06 : 1	
0.3% Cu	Accum. tons Average grade Stripping ratio	130,000,000 0.93% Cu 2.08 : 1	205,000,000 0.89% Cu 2.17 : 1	265,000,000 0.84% Cu 2.30 : 1	330,000,000 0.79% Cu 2.51 : 1	390,000,000 0.73% Cu 2.62 : 1
0.0% Cu	Accum. tons Average grade Stripping ratio	400,000,000 0.40% Cu 0.0 : 1	650,000,000 0.38% Cu 0.0 : 1	875,000,000 0.36% Cu 0.0 : 1	1,206,000,000 0.33% Cu 0.0 : 1	1,410,000,000 0.31% Cu 0.0 : 1

0.3%, 0.4%, 0.5%, 0.6% or 0.7%. The amount of material to be milled and the average mill grade for each of these scenarios is given below.

Scenario	Milled tons (10^6)	Avg. grade (% Cu)	SR
7-7	80	1.20	4:1
7-6	92	1.13	3.35:1
7-5	105	1.06	2.81:1
7-4	115	1.01	2.48:1
7-3	130	0.93	2.08:1

The total amount of material to be moved is 400×10^6 tons and hence an overall stripping ratio SR can be calculated.

5. A mine life can be assumed. In this case 25 years is chosen. The mine/mill will work 300 days/year. Therefore the daily mining and milling rates can be calculated. For the scenario given above (scenario 7-7) one finds that:

$$\text{Milling rate} = \frac{80 \times 10^6}{25 \times 300} = 10{,}667 \text{ tpd}$$

$$\text{Ore mining rate} = \frac{80 \times 10^6}{25 \times 300} = 10{,}667 \text{ tpd}$$

$$\text{Stripping rate} = \frac{320 \times 10^6}{25 \times 300} = 42{,}667 \text{ tpd}$$

Overall mining rate = 53,334 tpd

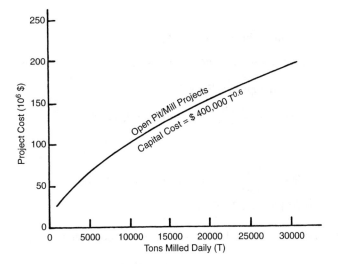

Figure 6.20. Open pit/mill project costs as a function of milling rate (O'Hara, 1980).

The results for pit 7 are given in the table below.

Scenario	Milling rate (tpd)	Mining rate (tpd)
7-7	10,667	53,334
7-6	12,267	53,334
7-5	14,000	53,334
7-4	15,333	53,334
7-3	17,333	53,334

6. Each of the different mining and milling rates has a certain associated capital cost. O'Hara (1980) has presented a curve, Figure 6.20, which could be used for estimating. In the present case (Halls et al., 1969), the costs which were used are:

Scenario	Capital cost (10^3 $)
7-7	$43,500
7-6	46,832
7-5	50,166
7-4	52,832
7-3	56,832

7. The mining, milling, and G&A operating costs can also be estimated using the curves developed by O'Hara (1980). In this case the following apply:

Ore mining cost = $0.45/st
Stripping cost = 0.45/st
Milling + G&A = $1.25/st
Mill recovery = 80%

The smelting, refining and sales cost are estimated at 5.9 ¢/lb.

8. One can calculate the gross revenues for the various scenario's. A price of 40 ¢/lb is used.

$$\text{Gross revenue (\$/yr)} = \text{Recovered copper} \times \text{Price}$$
$$= \text{Milling rate (t/day)} \times \text{days/yr} \times \frac{\text{Grade}(\%)}{100} \times 2000 \times \text{Price (\$/lb)}$$

For scenario 7-4:

$$\text{Gross revenue} = 15,333 \times 300 \times \frac{1.01}{100} \times 2000 \times 0.40 = \$29,734,000$$

Recovered copper = 74,334,384 lbs = 37,167 tons

9. The operating costs can be similarly computed. Scenario 7-4 will be continued.

Mining cost (ore + waste) = 0.45 × 53,334 × 300 = $7,200,000
Milling + G&A = 1.25 × 15,333 × 300 = $5,750,000
SRS = 0.059 × 74,334,384 = $4,386,000

10. Straight line depreciation will be used.

$$\text{Depreciation (\$/yr) for scenario 7-4} = \frac{\$52,832,000}{25 \text{ yrs}}$$

Depreciation ($/yr) = $2,113,280

11. Percentage depletion is computed for copper as 15% of the gross revenues. It cannot exceed 50% of the taxable income.

12. Taxes are computed at a rate of 52%.

13. A cash flow analysis can now be made of the various scenarios. This will be illustrated with scenario 7-4.

Cash flow analysis for scenario 7-4

1. Gross revenues   = $29,734,000
2. Operating expenses   = $17,336,000
   Mining   = 7,200,000
   Milling = 5,750,000
   SRS   = 4,386,000
           = 17,336,000
3. Net revenue   = $12,398,000
4. Depreciation   = $2,113,380
5. Taxable income before depletion   = $10,284,620
6. Depletion   = $4,460,100
   15% of gross revenues = $4,460,100
   50% of taxable income = $5,142,310
   (choose 15% of gross revenue since this is smaller)

7. Taxable income after depletion = $5,824,520
8. Taxes @52% = $3,028,750
9. Annual net income after taxes (line 5 minus line 8) = $7,255,870
10. Cash flow after taxes (line 9 plus line 4) = $9,369,250

(This cash flow analysis is similar to that done by Halls in Table 6.17.) This cash flow analysis has been done for all of the scenarios.

14. Calculate the internal ROR. It is assumed that there are 25 uniform positive cash flows (in years 1 through 25) which will be compared with the negative cash flow in year zero (capital expense). The interest rate which makes the sum (NPV) zero is to be determined. This is the internal ROR. For scenario 7-4 the situation is

$$-\$52,832,000 + \$9,353,000 \left[\frac{(1+i)^{25} - 1}{i(1+i)^{25}}\right] = 0$$

In this case the value of Halls ($9,353,000) rather than $9,369,250 has been used.
Solving for i, one finds that

$$i = 0.174$$

Therefore the rate of return on the invested capital is 17.4%.
The results of this calculation for all of the pit 7 scenarios are summarized below:

Scenario	Capital investment ($10³)	Annual cash flow after taxes ($10³)	Internal ROR (%)
7-7	43,500	7550	17.0
7-6	46,832	8323	17.5
7-5	50,166	8951	17.6
7-4	52,832	9353	17.4
7-3	56,832	9727	16.8

If the company's minimum desired ROR is 12% then all of these meet the criterion. The one chosen is scenario 7-3 due to the fact that the total profits would be highest.

15. An incremental financial analysis is now performed on these data. As can be seen the increase in the annual cash flows in going from scenario 7-7 to 7-3 is due to an increase in the capital investment. The question is 'what is the rate of return on the increment of capital needed to go from one scenario to the next?' These increments in capital and annual cash flows are summarized in Table 6.18. One now goes through the calculation of determining the rate of return on the invested incremental capital.

$$\Delta \text{Capital} = \Delta \text{CF} \left[\frac{(1+i)^{25} - 1}{i(1+i)^{25}}\right]$$

Examining the change between scenarios 7-7 and 7-6 one finds that

$$3,332,000 = 773,000 \left[\frac{(1+i)^{25} - 1}{i(1+i)^{25}}\right]$$

$$i = 23.1\%$$

524 *Open pit mine planning and design: Fundamentals*

Table 6.17. Summary of evaluations for ABC mine (Halls et al., 1969).

	Scenario designation								
	7-7	7-6	7-5	7-4	7-3	6-6	6-5	6-4	6-3
Ore inventories									
Tons (10^6)	80	92	105	115	130	135	180	185	205
Grade (% Cu)	1.20	1.13	1.06	1.0	0.83	1.12	1.03	0.95	0.89
Stripping ratio	4.00:1	3.35:1	2.81:1	2.48:1	2.08:1	3.81:1	3.06:1	2.51:1	2.17:1
Concentrator capacity for 25-year life (tons/day)	10,700	12,300	14,000	15,300	17,300	18,000	21,300	24,700	27,300
Capital investment for mine and concentrator (10^3 $)	$43,600	$48,832	$50,166	$52,832	$56,832	$62,500	$69,166	$75,834	$81,166
Copper production									
Annual tons (refined)	30,720	33,268	35,616	37,168	38,688	48,384	52,736	56,240	58,384
Total tons (refined)	768,000	831,700	890,400	929,200	967,200	1,209,600	1,318,400	1,408,000	1,544,400
Annual net income after taxes	$5810	$6462	$8945	$7239	$7455	$9397	$10,320	$10,960	$11,320
Annual cash flow after taxes	$7530	$8323	$8951	$9363	$9727	$11,897	$13,086	$13,993	$14,567
Financial analysis of investment									
Return on total investment-DCF	17.0%	17.5%	17.6%	17.4%	16.8%	18.8%	18.8%	18.2%	17.7%
Return on incremental investment of this									
Scenario compared to	—	7-7	7-6	7-6	7-4	7-4	6-6	6-5	6-4
Incremental investment required (10^3 $)	—	$3332	$3334	$2666	$4000	$9666	$8888	$6668	$5232
Incremental cash flow (10^3 $)	—	$773	$628	$4092	$374	$2,544	$1,189	$907	$574
DCF return on incremental investment	—	23.1%	18.6%	14.6%	8.9%	26.4%	17.6%	13.0%	9.4%
Extrapolated cutoff grade to provide 12% return on incremental investment	Economic cutoff for 0.7% Cu pit = 0.36% Cu					Economic cutoff for 0.6% Cu pit = 0.37% Cu			

Table 6.17. (Continued).

	Scenario designation					
	5-5	5-4	5-3	4-4	4-3	3-3
Ore reserves						
Tons (10^6)	195	235	265	290	330	390
Grade (% Cu)	0.99	0.9	0.84	0.84	0.79	0.73
Stripping ratio	3.49:1	2.73:1	2.30:1	3.16:1	2.65:1	2.62:1
Concentrator capacity for 25-year life (tons/day)	26,000	31,000	35,300	38,700	44,000	52,000
Capital investment for mine and concentrator (10^3\$)	\$80,000	\$90,667	\$98,667	\$103,000	\$112,666	\$120,000
Copper production						
Annual tons (refined)	61,776	67,680	71,200	78,880	83,424	91,104
Total tons (refined)	1,544,400	1,692,000	1,780,000	1,972,000	2,085,600	2,277,600
Annual next income after taxes (10^3\$)	\$11,752	\$12,763	\$13,779	\$13,728	\$14,431	\$14,812
Annual cash flow after taxes (10^3\$)	\$14,752	\$16,390	\$17,327	\$17,808	\$18,938	\$19,612
Financial analysis of investment						
Return on total investment-DCF	17.9%	17.8%	17.3%	17.0%	16.4%	15.9%
Return on incremental investment of this						
Scenario compared to	6-4	5-5	5-4	5-4	5-4	5-4
Incremental investment required (10^3\$)	\$4166	\$10,667	\$8,000	\$12,333	\$21,999	\$29,333
Incremental cash flow (10^3\$)	\$759	\$1638	\$937	\$1418	\$2548	\$3222
DCF return on incremental investment	17.9%	15.3%	10.2%	10.6%	10.7%	10.0%
Extrapolated cutoff grade to provide 12% return on incremental investment	Economic cutoff for 0.5% Cu pit = 0.34% Cu					

The scenario designations refer to the cutoff grades for the pit limits and the concentrator (i.e. scenario 5-4 means an ultimate pit cutoff grade of 0.5% Cu and concentrator cutoff grade of 0.4% Cu). All dollar values are recorded in thousands.

Table 6.18. Incremental capital investments and annual cash flows.

Scenario	ΔCapital investment ($10³)	ΔCash flow ($10³)
7-7		
	3332	773
7-6		
	3334	628
7-5		
	2666	402
7-4		
	4000	374
7-3		

Repeating this process for the others one finds

Scenario	i
7-7 → 7-6	23.1%
7-6 → 7-5	18.6%
7-5 → 7-4	14.6%
7-4 → 7-3	8.9%

Now it can be seen that the optimum scenario is 7-4 rather than 7-3. The milling rate would be 15,300 tpd.

16. Examine all of the other scenarios with respect to the best current one. For pit 7, the best scenario is 7-4. For pit 6, scenario 6-6 has a higher total ore tonnage than 7-4 with an acceptable rate of return on the additional capital expenditure. It thus becomes the best current alternative. Plan 6-5 is better than 6-6 and 6-4 is better than 6-5. Plan 6-3 is not acceptable. Pit 5 alternatives are compared against 6-4. As can be seen from the table, the overall best plan is between Plan 5-4 and Plan 5-3. The economic cutoff (mill cutoff) for the pit is 0.34% Cu.

This leads to the following operation

Ore reserve = 253,000,000 tons
Mill rate = 33,580 tpd
Mining rate = 116,600 tpd
Average ore grade = 0.864% Cu
Mine life = 25 years
Capital investment = $95,500,000

6.7 LANE'S ALGORITHM

6.7.1 *Introduction*

In 1964, K.F. Lane (Lane, 1964) presented what has become a classic paper entitled 'Choosing the Optimum Cut-off Grade'. This section will describe his approach and illustrate it with an example. As has been discussed earlier cutoff grade is the criterion normally used in mining to discriminate between ore and waste in the body of a deposit. Waste may either be left in place or sent to waste dumps. Ore is sent to the treatment plant for further processing and eventual sale.

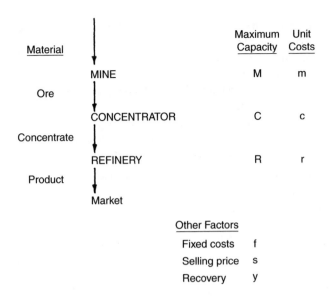

Figure 6.21. The model described by the Lane algorithm (Lane, 1964).

The choice of cutoff grade can directly affect profits. This chapter will examine the principles which determine the best choice of a cutoff grade under different circumstances.

A mining operation is considered to consist of three stages:
– mining,
– concentrating, and
– refining.

Each stage has its own associated costs and a limiting capacity. The operation as a whole will incur continuing fixed costs. The three most important economic criteria which can be applied are:

Case I: Maximum present value.
Case II: Maximum total profits.
Case III: Maximum immediate profit.

The maximum present value gives the economic optimum and is that generally applied lacking special circumstances. It is the one which will be used in this book. As has been shown by Lane (1964) the second and third correspond to the application of special discount rates in the first. Case II, maximum total profits, corresponds to a discount rate of zero percent whereas Case III is for a high value.

In this chapter, attention is focussed on choosing a cutoff grade to maximize the present value of the cash flow from the operation.

6.7.2 Model definition

Figure 6.21 is a diagrammatic representation of the elements and symbols used in the model.

Definitions of the maximum capacities, unit costs and quantities involved in the evaluation are presented below.

528 Open pit mine planning and design: Fundamentals

1. Maximum capacity: M is the maximum amount of material (ore and waste) that the mine can produce in a given time period (for example 1000 tons/year). It is therefore a restriction on the maximum rate of progress through the orebody.

C is the maximum amount of ore which can be put through the concentrator in a given time period (for example 500 tons/year), assuming unrestricted availability of input ore from the mine. A concentrate of fixed grade is produced.

R is the maximum amount of final product produced in the time period (for example 500 lbs/year), assuming unrestricted availability of concentrate from the concentrator. The maximum can be due to a restriction on refinery throughput or a market limitation.

2. Costs: m are the mining costs expressed in $/ton of material moved. These are assumed to be the same irrespective as to whether the material is classified as ore or waste. The unit mining costs include drilling, blasting, loading, hauling, etc.

c are the concentrating costs expressed in $/ton of material milled. The unit cost c includes crushing, grinding, floating, leaching, etc. It also includes some haulage if ore is trucked farther than waste (if not, this can become a credit item in calculating c).

r includes all costs incurred at the product and selling stages such as smelting, refining, packaging, freight, insurance, etc. These are expressed in terms of $ per unit of product. For copper it would be $/lb.

f, the fixed cost, includes all costs such as rent, administration, maintenance of roads and buildings, etc. which are independent of production levels (within normal limits of variation) but which would cease were the mine to be closed. It is expressed in terms of a fixed cost over the production period considered (for example 1 year). Other costs such as head office charges, depreciation, etc. are not included.

s, the selling price, is expressed in terms of selling price per unit of product. It is a gross figure provided all selling charges are included in r. If not they must be subtracted from s.

y, the recovery, is an overall figure for the concentrator and the refinery. It is that proportion of the mineral contained in the original ore feed retained in the final product.

3. Quantities: T is the length of the production period being considered (for example 1 year); Q_m is the quantity of material to be mined, Q_c is the quantity of ore sent to the concentrator and Q_r is the amount of product actually produced over this production period.

6.7.3 The basic equations

Using the definitions given in the preceding section, the basic equations can be developed. The total costs T_c are

$$T_c = mQ_m + cQ_c + rQ_r + fT \qquad (6.7)$$

Since the revenue R is

$$R = sQ_r \qquad (6.8)$$

the profit P is given by

$$P = R - T_c = sQ_r - (mQ_m + cQ_c + rQ_r + fT) \qquad (6.9)$$

Combining terms, yields

$$P = (s - r)Q_r - cQ_c - mQ_m - fT \qquad (6.10)$$

This is the basic profit expression. It can be used to calculate the profit from the next Q_m of material mined.

Table 6.19. Initial mineral inventory for the Lane example.

Grade (lbs/ton)	Quantity (tons)
0.0–0.1	100
0.1–0.2	100
0.2–0.3	100
0.3–0.4	100
0.4–0.5	100
0.5–0.6	100
0.6–0.7	100
0.7–0.8	100
0.8–0.9	100
0.9–1.0	100
	$Q_m = 1000$

6.7.4 An illustrative example

To introduce the reader in a soft way to the problem being explored in detail in this section consider the following example. A final pit has been superimposed on a mineral inventory. Within the pit outline are contained 1000 tons of material. The grade distribution is shown in Table 6.19. The associated costs, price, capacities, quantities and recovery are:

Costs

m = mining = $1/ton
c = concentrating = $2/ton
r = refining = $5/lb
f = fixed cost = $300/year

Price

s = $25/lb

Capacities

M = 100 tons/year
C = 50 tons/year
R = 40 lbs/year

Quantities

Q_m = amount to be mined (tons)
Q_c = amount sent to the concentrator (tons)
Q_r = amount of concentrator product sent for refining (lbs)

Recovery(Yield)

$y = 1.0$ (100% recovery is assumed).

530 Open pit mine planning and design: Fundamentals

There are a great number of possible mine, concentrator and refinery operating combinations. Which is the optimum? In this section the basic equations will be developed in addition to demonstrating the process. However, prior to beginning the theoretical treatment, it is considered useful to briefly consider just one of these operating combinations. The total amount of material to be mined Q_m is 1000 tons. If the mine is operated at capacity (100 tons/year) then the pit would be mined out over a time period of 10 years. Assuming that the grades (Table 6.19) are equally distributed throughout the pit and a concentrator cutoff grade of 0.50 lbs/ton is used, then 50 tons of material having an average grade of 0.75 lbs/ton would be sent to the concentrator every year. The other 50 tons would be sent to the waste dump. Since the concentrator capacity C is 50 tons/year, this is an acceptable situation. The concentrator product becomes the refinery feed. In this case it would be 37.5 lbs/year (0.75 lbs/ton × 50 tons/year). Since the refinery can handle 40 lbs/year, it would be operating at below its rated capacity R. This combination of mining rate and cutoff grade would yield a yearly profit P_y of

$$P_y = (25 - 5)37.5 - 2 \times 50 - 1 \times 100 - 300 = \$250$$

These profits would continue for 10 years and hence the total profit would be $2500. The NPV assuming an interest rate of 15% would be

$$\text{NPV} = \frac{250[(1.15)^{10} - 1]}{0.15(1.15)^{10}} = \$1,254.69$$

The first question to be asked is whether some other combination of mine production rate and concentrator cutoff grade would yield a better profit from this deposit? The larger question is whether the various plant capacities (with their associated costs) are optimum? The procedure described in this section is a way of determining the combination yielding the maximum profit for a given set of operating constraints. The constraints may then be changed (mine, concentrator and refining capacities, for example) and the profit corresponding to this new combination determined as well as how the various capacities should be utilized over the life of the pit.

6.7.5 Cutoff grade for maximum profit

Step 1. Determination of the economic cutoff grade – one operation constraining the total capacity

As indicated, the basic profit expression (6.10) is

$$P = (s - r)Q_r - cQ_c - mQ_m - fT$$

Calculate cutoff grade assuming that the mining rate is the governing constraint. If the mining capacity M is the applicable constraint, then the time needed to mine material Q_m is

$$T_m = \frac{Q_m}{M} \tag{6.11}$$

Equation (6.10) becomes

$$P = (s - r)Q_r - cQ_c - \left(m + \frac{f}{M}\right)Q_m \tag{6.12}$$

To find the grade which maximizes the profit under this constraint one first takes the derivative of (6.12) with respect to g.

$$\frac{dP}{dg} = (s - r)\frac{dQ_r}{dg} - c\frac{dQ_c}{dg} - \left(m + \frac{f}{M}\right)\frac{dQ_m}{dg} \tag{6.13}$$

However the quantity to be mined is independent of the grade, hence

$$\frac{dQ_m}{dg} = 0 \tag{6.14}$$

Equation (6.13) becomes

$$\frac{dP}{dg} = (s-r)\frac{dQ_r}{dg} - c\frac{dQ_c}{dg} \tag{6.15}$$

The quantity refined Q_r is related to that sent by the mine for concentration Q_c by

$$Q_r = \bar{g} y Q_c \tag{6.16}$$

where $\bar{g}$ is the average grade sent for concentration, and y is the recovery.
Taking the derivative of Q_r with respect to grade one finds that

$$\frac{dQ_r}{dg} = \bar{g} y \frac{dQ_c}{dg} \tag{6.17}$$

Substituting Equation (6.17) into (6.15) yields

$$\frac{dP}{dg} = [(s-r)\bar{g}y - c]\frac{dQ_c}{dg} \tag{6.18}$$

The lowest acceptable value of $\bar{g}$ is that which makes

$$\frac{dP}{dg} = 0$$

Thus the cutoff grade g_m based upon mining constraints is the value of $\bar{g}$ which makes

$$(s-r)\bar{g}y - c = 0$$

Thus

$$g_m = \bar{g} = \frac{c}{y(s-r)} \tag{6.19}$$

Substituting the values from the example yields

$$g_m = \frac{\$2}{1.0(\$25 - \$5)} = 0.10 \, \text{lbs/ton}$$

Calculate cutoff grade assuming that the concentrating rate is the governing constraint. If the concentrator capacity C is the controlling factor in the system, then the time required to mine and process a Q_c block of material (considering that mining continues simultaneously with processing) is

$$T_c = \frac{Q_c}{C} \tag{6.20}$$

Substituting Equation (6.20) into (6.10) gives

$$P = (s-r)Q_r - cQ_c - mQ_m - f\frac{Q_c}{C} \tag{6.21}$$

Rearranging terms one finds that

$$P = (s-r)Q_r - \left(c + \frac{f}{C}\right)Q_c - mQ_m$$

532 Open pit mine planning and design: Fundamentals

Differentiating with respect to g and setting the result equal to zero yields

$$\frac{dP}{dg} = (s-r)\frac{dQ_r}{dg} - \left(c + \frac{f}{c}\right)\frac{dQ_c}{dg} - m\frac{dQ_m}{dg} = 0$$

As before

$$\frac{dQ_m}{dg} = 0$$

$$\frac{dQ_r}{dQ_c} = gy$$

Thus

$$(s-r)\frac{dQ_r}{dg} = \left(c + \frac{f}{C}\right)\frac{dQ_c}{dg}$$

$$\frac{dQ_r}{dQ_c} = \frac{c + \frac{f}{C}}{s - r} = gy$$

The cutoff grade when the concentrator is the constraint is

$$g_c = \frac{c + \frac{f}{C}}{y(s-r)} \tag{6.22}$$

For the example, this becomes

$$g_c = \frac{\$2 + \frac{\$300}{50}}{1.0(\$25 - \$5)} = 0.40$$

Calculate cutoff grade assuming that the refining rate is the governing constraint. If the capacity of the refinery (or the ability to sell the product) is the controlling factor then the time is given by

$$T_r = \frac{Q_r}{R} \tag{6.23}$$

Substituting (6.23) into Equation (6.10) yields

$$P = (s-r)Q_r - \frac{fQ_r}{R} - cQ_c - mQ_m$$

or

$$P = \left(s - r - \frac{f}{R}\right)Q_r - cQ_c - mQ_m \tag{6.24}$$

Differentiating with respect to g and setting the result equal to zero gives

$$\frac{dP}{dg} = \left(s - r - \frac{f}{R}\right)\frac{dQ_r}{dg} - c\frac{dQ_c}{dg} - m\frac{dQ_m}{dg} = 0$$

Simplifying and rearranging gives

$$g_r = \frac{c}{\left(s - r - \frac{f}{R}\right)y} \tag{6.25}$$

For this example

$$g_r = \frac{\$2}{\left(\$25 - \$5 - \frac{\$300}{40}\right)1.0} = 0.16 \tag{6.26}$$

One can now calculate the amount of material which would be concentrated and refined under the various constraints as well as the time required. When the mining rate of 100 tons/year is the constraint

$$T_m = \frac{1000 \text{ tons}}{100 \text{ tons/year}} = 10 \text{ years}$$

Since the cutoff grade g_m is 0.10 lbs/ton, a quantity Q_c of 900 tons having an average grade of 0.55 lbs/ton would be sent to the concentrator. The total amount of product refined and sold Q_r is

$$Q_r = 900 \times 0.55 = 495 \text{ lbs}$$

Substituting these values into the profit equation gives

$$P_m = (\$25 - \$5)495 - \$2 \times 900 - \$1 \times 1000 - \$300 \times 10 = \$4100$$

The same procedure can be followed with the other two limiting situations. The results are given below:

Concentrator limit:

$$g_c = 0.40 \text{ lbs/ton}$$
$$Q_c = 0.60 \times 1000 = 600 \text{ tons}$$
$$C = 50 \text{ tons/year}$$
$$T_c = \frac{600}{50} = 12 \text{ years}$$
$$Q_m = \frac{1000}{12} = 83.3 \text{ tons/year}$$
$$\bar{g} = 0.7 \text{ lbs/tons}$$
$$Q_r = 600 \times 0.7 \times 1.0 = 420 \text{ lbs}$$
$$P_c = (25 - 5)420 - 2 \times 600 - 1 \times 1000 - 300 \times 12$$
$$= \$2600$$

Refinery limit:

$$g_r = 0.16 \text{ lbs/ton}$$
$$\bar{g} = 0.58 \text{ lbs/tons}$$
$$Q_c = 0.84 \times 1000 = 840 \text{ tons}$$
$$Q_r = 840 \times 0.58 \times 1.0 = 487.2 \text{ lbs}$$
$$T_r = \frac{487.2}{40} = 12.18 \text{ years}$$
$$M = \frac{1000}{12.18} = 82.1 \text{ tons/year}$$
$$C = \frac{840}{12.18} = 69 \text{ tons/year}$$
$$P_r = (25 - 5)487.2 - 2 \times 840 - 1 \times 1000 - 300 \times 12.18$$
$$= \$3410$$

Table 6.20. Total profits as a function of concentrator cutoff with mine operating at capacity.

g	Profits ($)		
	P_m	P_c	P_r
0.0	4000	1000	3250
0.1	4100	1700	3386
0.16	4064	2024	3410
0.2	4000	2200	3400
0.3	3700	2500	3287.50
0.4	3200	2600	3050
0.5	2500	2500	2687.50
0.6	1600	2200	2200
0.7	500	1700	1587.50
0.8	−800	1000	850
0.9	−2300	100	−12.50
0.95	−3125	−425	−511.25

In summary, for each operation taken as a single constraint, the optimum cutoff grades are:

$$g_m = 0.10$$
$$g_c = 0.40$$
$$g_r = 0.16$$

The total profits assuming the single constraint of mining, concentrating or refining are given as a function of cutoff grade in Table 6.20. The values have been plotted in Figure 6.22.

Step 2. Determination of the economic cutoff grade by balancing the operations

In the first step, it was assumed that only one of the operations was the limiting factor to production capacity.

A second type of cutoff is based simply on material balance. To be able to calculate this one needs to know the distribution of grades of the mined material. The average grade of the treated material can be found as a function of the chosen cutoff. The average grade, and the number of units involved are given as a function of cutoff grade in Table 6.21.

For both the mine and mill to be at their respective capacities, then

$$Q_m = 100 \text{ tons}$$
$$Q_c = 50 \text{ tons}$$

As can be seen from Table 6.21, the cutoff grade should be 0.5 lbs/ton. This balancing cutoff between mine and concentrator is expressed as g_{mc}. For the concentrator and the refinery to be at full capacity

$$Q_c = 50 \text{ tons}$$
$$Q_r = 40 \text{ tons}$$

The relationship between Q_c and Q_r is shown in Table 6.22.

In examining Table 6.22, it can be seen that the required balancing cutoff grade g_{cr} is

$$g_{cr} = 0.60$$

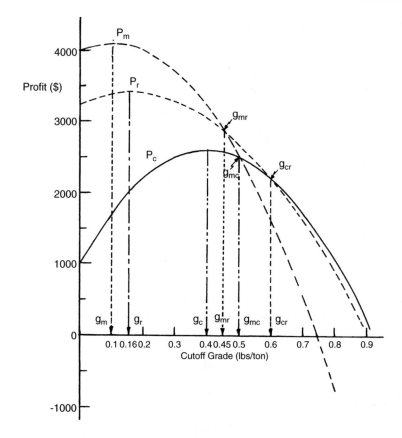

Figure 6.22. Total profit as a function of cutoff grade under different constraints.

Table 6.21. Concentrator feed as a function of concentrator cutoff with mine operating at capacity.

Mined amount (Q_m) (tons)	Concentrator cutoff grade (g_c) (lbs/ton)	Feed going to the concentrator (Q_c) (tons)
100	0	100
100	0.1	90
100	0.2	80
100	0.3	70
100	0.4	60
100	0.5	50
100	0.6	40
100	0.7	30
100	0.8	20
100	0.9	10

The final balancing cutoff is between the mine and the refinery. As seen in Table 6.23 (assuming 100% concentrator recovery) a cutoff grade of 0.4 yields 42 lbs of product whereas 0.5 yields 37.5 lbs. The desired level of 40 lbs lies between these two. Interpolating one finds that

$$g_{mr} \cong 0.456 \text{ lbs/ton}$$

536 Open pit mine planning and design: Fundamentals

Table 6.22. Refinery product as a function of concentrator cutoff with concentrator operating at capacity.

Amount to be concentrated (Q_c) (tons)	Concentrator cutoff grade (g_c) (lbs/ton)	Avg. conc. feed grade ($\bar{g}_c$) (lb/ton)	Refinery product (Q_r) (lbs)
50	0	0.5	25
50	0.1	0.55	27.5
50	0.2	0.5	30
50	0.3	0.65	32.5
50	0.4	0.7	35
50	0.5	0.75	37.5
50	0.6	0.8	40
50	0.7	0.85	42.5
50	0.8	0.9	45
50	0.9	0.95	47.5

Table 6.23. Refinery feed as a function of mine cutoff with the mine operating at capacity (assuming 100% concentratory recovery).

Mined amount (Q_m) (tons)	Mine cutoff grade (g_m) (lbs/ton)	Refinery product (Q_r) (lbs)
100	0	50
100	0.1	49.5
100	0.2	48
100	0.3	45.5
100	0.4	42
100	0.5	37.5
100	0.6	32
100	0.7	25.5
100	0.8	18
100	0.9	9.5

In summary, when the operations are taken in combination, the optimum cutoff grades are:

$$g_{mc} = 0.50$$

$$g_{cr} = 0.60$$

$$g_{mr} = 0.456$$

Step 3. Determining the overall optimum of the six cutoff grades

There are six possible cutoff grades. Three (g_{mc}, g_{cr}, and g_{mr}) are based simply upon the grade distribution of the mined material and capacities. The other three (g_m, g_c, and g_r) are based upon capacities, costs and the price. The objective is to find the cutoff grade which produces the overall maximum profit in light of the mining, concentrating and refining constraints. The local optimums for each pair of operations are first considered. The corresponding optimum grades for each pair (G_{mc}, G_{rc}, and G_{mr}) are selected using the following rules:

$$G_{mc} = \begin{cases} g_m & \text{if } g_{mc} \leq g_m \\ g_c & \text{if } g_{mc} \geq g_c \\ g_{mc} & \text{otherwise} \end{cases} \qquad (6.26a)$$

$$G_{rc} = \begin{cases} g_r & \text{if } g_{rc} \le g_r \\ g_c & \text{if } g_{rc} \ge g_c \\ g_{rc} & \text{otherwise} \end{cases} \quad (6.26b)$$

$$G_{mr} = \begin{cases} g_m & \text{if } g_{mr} \le g_m \\ g_r & \text{if } g_{mr} \ge g_r \\ g_{mr} & \text{otherwise} \end{cases} \quad (6.26c)$$

The overall optimum cutoff grade G is just the middle value of G_{mc}, G_{mr}, and G_{rc}. In our example the six possible cutoff grades are:

$g_m = 0.10$
$g_c = 0.40$
$g_r = 0.16$
$g_{mc} = 0.50$
$g_{mr} = 0.456$
$g_{cr} = 0.60$

Consider them in groups of three:

Group1 $\quad g_m = 0.10$
$\qquad\quad g_c = 0.40 \qquad$ Choose the middle value $G_{mc} = 0.40$
$\qquad\quad g_{mc} = 0.50$

Group2 $\quad g_m = 0.10$
$\qquad\quad g_r = 0.16 \qquad$ Choose the middle value $G_{mr} = 0.16$
$\qquad\quad g_{mr} = 0.456$

Group3 $\quad g_r = 0.10$
$\qquad\quad g_c = 0.40 \qquad$ Choose the middle value $G_{cr} = 0.40$
$\qquad\quad g_{cr} = 0.60$

Considering the three middle values

$G_{mc} = 0.40$
$G_{mr} = 0.16$
$G_{cr} = 0.40$

one chooses one numerically in the middle

$G = 0.40 \text{ lbs/ton}$

From Table 6.24, the average grade $\bar{g}_c$ of the material sent to the concentrator for a cutoff of 0.40 lbs/ton would be

$\bar{g}_c = 0.70 \text{ lbs/ton}$

For 100% recovery the quantities are

$Q_m = 1000 \text{ tons}$
$Q_c = 600 \text{ tons}$
$Q_r = 420 \text{ lbs}$

Table 6.24. Grade distribution for the first 100 ton parcel mined.

Grade (lb/ton)	Quantity (tons)
0.0–0.1	10
0.1–0.2	10
0.2–0.3	10
0.3–0.4	10
0.4–0.5	10
0.5–0.6	10
0.6–0.7	10
0.7–0.8	10
0.8–0.9	10
0.9–1.0	10
	$Q_m = 100$

Applying the respective capacities to these quantities one finds that

$$T_c = \frac{600}{50} = 12 \text{ years}$$

$$T_r = \frac{420}{40} = 10.5 \text{ years}$$

$$T_m = \frac{1000}{100} = 10 \text{ years}$$

Since the concentrator requires the longest time, it controls the production capacity. The total profit is

$$P = \$2600$$

and the profit per year P_y is

$$P_y = \frac{\$2600}{12 \text{years}} = \$216.70/\text{year}$$

The net present value of these yearly profits assuming an interest rate of 15% is

$$\text{NPV} = P_y \frac{(1+i)^{12} - 1}{i(1+i)^{12}} = \$216.70 \frac{1.15^{12} - 1}{0.15(1.15)^{12}} = \$1174.60$$

In summary: the concentrator is the controlling production limiter; concentrator feed = 50 tons/year; optimum mining cutoff grade = 0.40 lbs/ton (constant); total concentrator feed = 600 tons; average concentrator feed grade = 0.70 lbs/ton; years of production = 12 years; copper production/year = 35 lbs; total copper produced = 420 lbs; total profits = $2600.40; net present value = $1174.60.

6.7.6 *Net present value maximization*

The previous section considered the selection of the cutoff grade with the objective being to maximize profits. In most mining operations today, the objective is to maximize the net present value NPV. In this section the Lane approach to selecting cutoff grades maximizing NPV subject to mining, concentrating and refining constraints will be discussed.

For the example of the previous section a fixed mining cutoff grade of 0.40 lbs/ton was used. One found that

$$Q_m = 83.3 \text{ tons/year}$$
$$Q_c = 50 \text{ tons/year}$$
$$Q_r = 35 \text{ lbs/year}$$
$$f = \$300/\text{year}$$
$$\text{Lifetime} = 12 \text{ years}$$

The yearly profit would be

$$P_j = 35.0 \times \$20 - 50 \times \$2 - 83.3 \times \$1 - \$300$$
$$= \$216.70$$

when finding the maximum total profit, the profits realized in the various years are simply summed.

The total profit is, therefore

$$P_T = \sum_{j=1}^{12} P_j = 12 \times 216.7 = \$2600.40$$

The net present value for this uniform series of profits (Chapter 2) is calculated using

$$\text{NPV} = P_j \frac{(1+i)^n - 1}{i(1+i)^n}$$

Assuming an interest (discount) rate of 15% one finds that

$$\text{NPV} = 216.7 \frac{(1.15)^{12} - 1}{0.15 \times (1.15)^{12}} = \$1174.6$$

The question to be raised is 'Could the NPV be increased using a cutoff grade which, instead of being fixed, varies throughout the life of the mine?' If so, then 'What should be the cutoff grades as a function of mine life?' These questions have been addressed by Lane and are the subject of this section.

Assume that just prior to mining increment Q_m (shown to commence at time $t = 0$ in Figure 6.23 for simplicity), the present value of all remaining profits is V. This is composed of two parts. The first, PV_p, is from the profit P realized at time T by mining the quantity Q_m. The second, PV_w is obtained from profits realized by mining the material remaining after time T. These profits are indicated as P_1 occurring at time T_1, P_2 occurring at time T_2, etc., in Figure 6.23. The value of all these remaining profits for mining conducted after $t = T$ as expressed at time T is W. The present values of W and P, respectively, discounted to time $t = 0$ are given by

$$\text{PV}_w(t=0) = \frac{W}{(1+d)^T} \tag{6.27}$$

$$\text{PV}_p(t=0) = \frac{P}{(1+d)^T} \tag{6.28}$$

where d is the discount rate.

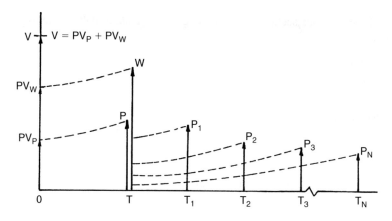

Figure 6.23. A diagrammatic representation of the NPV calculation.

The present value at time $t = 0$ is therefore

$$V = \frac{W}{(1+d)^T} + \frac{P}{(1+d)^T} \tag{6.29}$$

Since the present value at time $t = T$ of the remaining reserves is W, the difference v between the present values of the remaining reserves at times $t = 0$ and $t = T$ is

$$v = V - W \tag{6.30}$$

Equation (6.29) can be rewritten as

$$W + P = V(1+d)^T \tag{6.31}$$

Applying the binomial expansion to the term $(1+d)^T$ one finds that

$$(1+d)^T = 1 + Td + T\frac{(T-1)d^2}{2!} + \frac{T(T-1)(T-2)d^3}{3!} + \cdots \tag{6.32}$$

For d small, $(1+d)^T$ can be approximated by

$$(1+d)^T \approx 1 + Td \tag{6.33}$$

Combining Equations (6.31) and (6.33) results in

$$W + P = V(1 + Td) = V + VTd$$

or

$$V - W = P - VTd \tag{6.34}$$

Comparing Equations (6.30) and (6.34) one finds that

$$v = P - VTd \tag{6.35}$$

The profit P obtained through the mining of Q_m in time T is given as before by

$$P = (s-r)Q_r - cQ_c - mQ_m - fT \tag{6.36}$$

Combining Equations (6.35) and (6.36) yields

$$v = (s-r)Q_r - cQ_c - mQ_m - T(f + Vd) \tag{6.37}$$

Production planning 541

One would now like to schedule the mining in such a way that the decline in remaining present value takes place as rapidly as possible. This is because later profits get discounted more than those captured earlier. In examining Equation (6.37), this means that v should be maximized. As in the previous section one first takes the derivative of v with respect to grade. Setting the derivative equal to zero, one can solve for the appropriate cutoff grades subject to mining, concentrating and refining constraints.

Step 1. Determination of the economic cutoff grades – one operation constraining the total capacity

(a) *Calculate cutoff grade assuming that the mining rate is the governing constraint.*
The time T_m is given by

$$T_m = \frac{Q_m}{M} \qquad (6.38)$$

Substituting this into Equation (6.37) yields

$$v_m = (s-r)Q_r - cQ_c - \left[m + \frac{f+Vd}{M}\right]Q_m \qquad (6.39)$$

Differentiating (6.39) with respect to grade g gives

$$\frac{dv_m}{dg} = (s-r)\frac{dQ_r}{dg} - c\frac{dQ_c}{dg} - \left[m + \frac{f+Vd}{M}\right]\frac{dQ_m}{dg} \qquad (6.40)$$

However the quantity mined Q_m does not depend upon the grade:

$$\frac{dQ_m}{dg} = 0 \qquad (6.41)$$

Hence

$$\frac{dv_m}{dg} = (s-r)\frac{dQ_r}{dg} - c\frac{dQ_c}{dg} \qquad (6.42)$$

The relationship between the quantities refined Q_r and those sent for concentration Q_c is

$$Q_r = Q_c \bar{g}_c y \qquad (6.43)$$

where $\bar{g}_c$ is the average grade of ore sent for concentration and y is the recovery in concentration.
Thus

$$\frac{dQ_r}{dg} = \bar{g}_c y \frac{dQ_c}{dg} \qquad (6.44)$$

Substituting Equation (6.44) into (6.42) yields

$$\frac{dv_m}{dg} = [(s-r)\bar{g}_c y - c]\frac{dQ_c}{dg} \qquad (6.45)$$

The average grade $\bar{g}_c$ is defined as the mining cutoff (breakeven) grade g_m when

$$\frac{dv_m}{dg} = 0 \qquad (6.46)$$

Setting Equation (6.45) equal to zero and solving for $\bar{g}_c = g_m$ one finds that

$$g_m = \frac{c}{(s-r)y} \qquad (6.47)$$

Substituting the values from the example yields

$$g_m = \frac{\$2}{(\$25 - \$5)1.0} = 0.10 \text{ lbs/ton}$$

(b) *Calculate cutoff grade assuming that the concentrating rate is the governing constraint.* If the concentrator throughput rate is the limiting factor then the time T is controlled by the concentrator.

$$T = \frac{Q_c}{C} \tag{6.48}$$

where Q_c is the total number of tons which will be sent to the concentrator, and C is the tons/year capacity.

Equation (6.37) becomes

$$v_c = (s-r)Q_r - cQ_c - mQ_m - (f+dV)\frac{Q_c}{C} \tag{6.49}$$

$$v_c = (s-r)Q_r - \frac{c+f+dV}{C}Q_c - mQ_m - m \tag{6.50}$$

Since the total amount of material Q_m is fixed,

$$mQ_m = \text{const}$$

Thus the cutoff grade affects only Q_r and Q_c.

Substituting as before

$$Q_r = Q_c \bar{g}_c y$$

one finds that

$$v_c = \left[(s-r)\bar{g}_c y - \left(c + \frac{f+dV}{C}\right)\right] Q_c - mQ_m \tag{6.51}$$

To make v_c as large as possible the term

$$(s-r)\bar{g}_c y - \left(c + \frac{f+dV}{C}\right)$$

should be as large as possible. At breakeven (the cutoff grade), the term is zero. Thus

$$g_c = \frac{c + \frac{f+dV}{C}}{y(s-r)} \tag{6.52}$$

(c) *Calculate cutoff grade assuming the refining rate is the governing constraint.* If the refinery output is the limiting factor then the time T is controlled by the refinery,

$$T = \frac{Q_r}{R} \tag{6.53}$$

where Q_r is output of the refinery and R is the refining/sales capacity per year.

Substituting into Equation (6.37) yields

$$v_r = (s-r)Q_r - cQ_c - mQ_m - (f+dV)\frac{Q_r}{R} \tag{6.54}$$

$$v_r = \left(s - r - \frac{f+dV}{R}\right) Q_r - cQ_c - mQ_m \tag{6.55}$$

Production planning

As before

$$Q_r = g_r y Q_c$$

Thus

$$V_r = \left(s - r - \frac{f+dV}{R}\right) g_r y Q_c - cQ_c - mQ_m$$

The total amount of material in the pit is fixed, therefore

$$mQ_m = \text{const}$$

Maximizing the expression for V_r one finds that

$$\left(s - r - \frac{f+dV}{R}\right) g_r y Q_c = cQ_c$$

Solving yields

$$g_r = \frac{c}{\left(s - r - \frac{f+dV}{R}\right) y} \qquad (6.56)$$

In summary, this first type of cutoff grade determination is based upon finding the grade for which the net increase in overall present value is zero. The expressions are as in formulas (6.47), (6.52) and (6.56):

$$g_m = \frac{c}{(s-r)y}$$

$$g_c = \frac{c + \frac{f+dV}{C}}{y(s-r)}$$

$$g_r = \frac{c}{\left(s - r - \frac{f+dV}{R}\right) y}$$

As can be seen, the expressions for g_c and g_r contain the unknown value of V.

Step 2. Determination of the economic cutoff grade by balancing the operations
This step is exactly the same as that discussed in the previous section. Hence only the results will be presented here.

$$g_{mc} = 0.50$$
$$g_{cr} = 0.60$$
$$g_{mr} = 0.456$$

Step 3. Determining the optimum of the six cutoff grades
There are six possible cutoff grades. Three are based simply upon the grade distribution of the mined material and capacities.

$$g_{mc} = 0.50$$
$$g_{cr} = 0.60$$
$$g_{mr} = 0.456$$

The other three are based upon capacities and cost/price. Substituting in the known values one finds that

$$g_m = 0.10$$

$$g_c = \frac{c + \frac{f+dV}{c}}{y(s-r)} = \frac{2 + \frac{300+0.15V}{50}}{1.0(25-5)}$$

$$= \frac{8 + 0.003V}{20}$$

$$g_r = \frac{c}{\left(s - r - \frac{f+dV}{R}\right)y}$$

$$= \frac{2}{(25 - 5 - \frac{300+0.15V}{40})1.0} = \frac{2}{12.5 - 0.00375V}$$

Of these, two of the limiting economic cut-off grades are not known initially since they depend upon knowing the overall present value. This in turn depends upon the cutoff grade. Since the unknown V appears in the equations an iterative process must be used.

An optimum grade will be determined for each of the three pairs of operations. This will be followed by finding the optimum of the three final candidates. For the mine and the concentrator considered as a pair, there are three possible candidates for the optimum cutoff grade G_{mc}. These are g_m, g_c, and g_{mc}. The following rules are used to select G_{mc}.

$$G_{mc} = \begin{cases} g_m & \text{if } g_{mc} \leq g_m \\ g_c & \text{if } g_{mc} \geq g_c \\ g_{mc} & \text{otherwise} \end{cases}$$

This simple sorting algorithm yields the middle value. Treating the concentrator and refinery as a pair, the optimum G_{rc} is found from

$$G_{rc} = \begin{cases} g_r & \text{if } g_{rc} \leq g_r \\ g_c & \text{if } g_{rc} \geq g_c \\ g_{rc} & \text{otherwise} \end{cases}$$

Finally the optimum cutoff grade G_{mr} when the mine and refinery are treated as a pair is

$$G_{mr} = \begin{cases} g_m & \text{if } g_{mr} \leq g_m \\ g_r & \text{if } g_{mr} \geq g_r \\ g_{mr} & \text{otherwise} \end{cases}$$

As the first step in the iteration process, it will be assumed that $V = 0$.
Applying these rules to the example values

$$g_m = 0.10$$
$$g_c = 0.40$$
$$g_r = 0.16$$
$$g_{mc} = 0.50$$
$$g_{rc} = 0.60$$
$$g_{mr} = 0.456$$

Table 6.25. Concentrator product as a function of concentrator cutoff grade with mine operating at capacity.

Concentrator cutoff (lbs/ton)	Concentrator feed (tons)	Average feed grade (lbs/ton)	Concentrator/refinery product (lbs)
0.0	100	0.50	50
0.1	90	0.55	49.5
0.2	80	0.60	48
0.3	70	0.65	45.5
0.4	60	0.70	42
0.5	50	0.75	37.5
0.6	40	0.80	32
0.7	30	0.85	25.5
0.8	20	0.90	18
0.9	10	0.95	9.5
1.0	0	1.00	0

one finds that

$$G_{mc} = 0.40$$
$$G_{rc} = 0.40$$
$$G_{mr} = 0.16$$

The overall optimum cutoff grade G is the middle value of G_{mc}, G_{mr}, and G_{rc}.

$$G = \text{middle value } (G_{mc}, G_{mr}, G_{rc})$$

In this case

$$G = 0.40$$

Step 4. Calculation of quantities

The next step in the procedure is to determine the maximum quantities Q_m, Q_c and Q_r which could be produced and not violate the capacities. Assume that 100 tons are mined ($Q_m = 100$). The grade distribution of this material is as shown in Table 6.24.

From Table 6.25, a cutoff grade of 0.40 would yield an average feed grade of 0.70. Each of the capacities must, however, be considered. A mining capacity ($Q_m = M$) of 100 tons would mean 60 tons to the concentrator and 42 product units. Both the concentrator and refinery capacities are exceeded. In meeting the concentrator capacity ($Q_c = 50$), the required mining and refinery capacities are:

$$Q_m = \frac{5}{6} \times 100 = 83.3$$
$$Q_r = 50 \times 0.70 = 35$$

These are both less than the maximum values. In meeting the refinery capacity of $Q_r = 40$, the required concentrating and mining capacities are:

$$Q_c = \frac{40}{0.7} = 57.1$$
$$Q_m = \frac{57.1}{60} 100 = 95.2$$

546 Open pit mine planning and design: Fundamentals

Thus the concentrating capacity is violated. The result is that the concentrator is the bottleneck.

In the further calculations

$$Q_m = 83.3$$
$$Q_c = 50$$
$$Q_r = 35$$

The profit from time period T is expressed as

$$P = (s-r)Q_r - cQ_c - mQ_m - fT$$

For $T = 1$ year one finds that

$$P = (25-5) \times 35 - 2 \times 50 - 1 \times 83.3 - 300 \times 1 = \$216.7$$

Since the total amount of material to be mined from the pit is $Q = 1000$ units, the number n of years required is

$$n = \frac{1000}{83.3} = 12 \text{ years}$$

The present value V corresponding to 12 equally spaced payments of $P = \$216.7$ using an interest rate of 15 percent is

$$V = \frac{216.7[1.15^{12} - 1]}{0.15(1.15)^{12}} = \$1174.6$$

This value of V becomes the second approximation of V (the first was $V = 0$) for use in the formulas to calculate g_c and g_r.

$$g_c = \frac{c + \frac{f+dV}{C}}{y(s-r)} = \frac{2 + \frac{300 + 0.15 \times 1174.6}{50}}{25 - 5} = 0.576$$

$$g_r = \frac{c}{s - r - \frac{f+dV}{R}y} = \frac{2}{20 - \frac{300 + 0.15 \times 1174.6}{40}} = 0.247$$

The new six choices become

$$g_m = 0.10$$
$$g_c = 0.576$$
$$g_r = 0.247$$
$$g_{mc} = 0.50$$
$$g_{rc} = 0.60$$
$$g_{mr} = 0.456$$

Applying the rules to select the overall pair optimum yields

$$G_{mc} = 0.50$$
$$G_{rc} = 0.576$$
$$G_{mr} = 0.247$$

The overall optimum is the middle value

$$G = 0.50$$

Returning to the grade distribution Table 6.24 one finds that the average grade is 0.75. If the mining rate $Q_m = 100$, then $Q_c = 50$ and $Q_r = 37.5$. Both the mine and the concentrator are at their rated capacities.

The profit in a given year is

$$P = (s - r)Q_r - cQ_c - mQ_m - fT$$
$$= 20 \times 37.5 - 2 \times 50 - 1 \times 100 - 300$$
$$= \$250$$

The number of years is

$$n = \frac{Q}{Q_m} = \frac{1000}{100} = 10 \text{ years}$$

The present value becomes

$$V = 250 \frac{1.15^{10} - 1}{0.15 \times 1.15^{10}} = \$1254.7$$

This becomes the third estimate for V to be used in calculating g_c and g_r.

$$g_c = \frac{2 + \frac{300 + 0.15 \times 1254.7}{50}}{20} = 0.588$$

$$g_r = \frac{2}{20 - \frac{300 + 0.15 \times 1254.7}{40}} = 0.257$$

The six possible values are:

$$g_m = 0.10$$
$$g_c = 0.588$$
$$g_r = 0.257$$

$$g_{mc} = 0.50$$
$$g_{rc} = 0.60$$
$$g_{mr} = 0.456$$

The optimum pairs are:

$$G_{mc} = 0.50$$
$$G_{rc} = 0.588$$
$$G_{mr} = 0.257$$

and the overall optimum ($G = 0.50$) is the same as found with the previous estimate. Hence in year 1

Optimum cutoff grade = 0.50 lbs/ton
Quantity mined = 100 tons
Quantity concentrated = 50 tons
Quantity refined = 37.5 lbs
Profit = $250

Table 6.26. Reserve distribution at the end of year 1.

Grade (lbs/ton)	Quantity (tons)
0.0–0.1	90
0.1–0.2	90
0.2–0.3	90
0.3–0.4	90
0.4–0.5	90
0.5–0.6	90
0.6–0.7	90
0.7–0.8	90
0.8–0.9	90
0.9–1.0	90
	Total = 900

Table 6.27. Reserve distribution at the start of year 8.

Grade (lbs/ton)	Quantity (tons)
0.0–0.1	30
0.1–0.2	30
0.2–0.3	30
0.3–0.4	30
0.4–0.5	30
0.5–0.6	30
0.6–0.7	30
0.7–0.8	30
0.8–0.9	30
0.9–1.0	30
	Total = 300

The reserves must now be adjusted to those given in Table 6.25 and the process is repeated assuming $V = 0$, calculating g_c and g_r, etc.

Through year 7, it will be found that the optimum cutoff grade remains at 0.50 with the quantities mined, concentrated and refined being 100, 50 and 37.5, respectively. The annual profit is $250. The reserves going into year 8 are those given in Table 6.27. The balancing grades remain at

$$g_{mc} = 0.50$$
$$g_{rc} = 0.60$$
$$g_{mr} = 0.456$$

The first approximation for the economic cutoff grades ($V = 0$) is

$$g_m = 0.10$$
$$g_c = 0.40$$
$$g_r = 0.16$$

The optimum values of the pairs are

$$G_{mc} = G_{rc} = 0.40$$
$$G_{mr} = 0.16$$

The overall optimum is 0.40, and the quantities are

$$Q_m = 83.3$$
$$Q_c = 50$$
$$Q_r = 35$$

The profit is $216.7 as before. The number of years becomes

$$n = \frac{300}{83.3} = 3.6 \text{ years}$$

The present value V becomes

$$V = 216.7 \frac{(1.15)^{3.6} - 1}{0.15(1.15)^{3.6}} = 571.2$$

Substituting this into the formulas for g_c and g_r yields

$$g_c = 0.486$$
$$g_r = 0.193$$

Combining them with the others

$$g_m = 0.10$$
$$g_{mc} = 0.50$$
$$g_{rc} = 0.60$$
$$g_{mr} = 0.456$$

yields

$$G_{mc} = 0.486$$
$$G_{rc} = 0.486$$
$$G_{mr} = 0.193$$

The overall optimum cutoff is $G = 0.486$ and the average grade above cutoff drops to 0.743:

Tons	Grade	Tons × Grade
4.2	0.493	2.07
30	0.55	16.50
30	0.65	19.50
30	0.75	22.50
30	0.85	25.50
30	0.95	28.50
Total = 154.2	Avg = 0.743	Sum = 114.57

There are 154.2 ore tons out of the 300 tons remaining to be mined. Since the concentrator capacity is 50 tons/year, the mine life would be

$$n = \frac{154.2}{50} = 3.08 \text{ years}$$

550 Open pit mine planning and design: Fundamentals

Table 6.28. Reserve distribution at the start of year 9.

Grade (lbs/ton)	Quantity (tons)
0.0–0.1	20.27
0.1–0.2	20.27
0.2–0.3	20.27
0.3–0.4	20.27
0.4–0.5	20.27
0.5–0.6	20.27
0.6–0.7	20.27
0.7–0.8	20.27
0.8–0.9	20.27
0.9–1.0	20.27
	Total = 202.27

The yearly mine production becomes

$$Q_m = \frac{300}{3.08} = 97.3$$

and $Q_r = 37.15$. Calculating the profit one finds that

$$P = 20 \times 37.15 - 2 \times 50 - 1 \times 97.3 - 300 = 245.7$$

The corresponding present value is

$$V = 245.7 \frac{(1.15)^{3.08} - 1}{0.15 \times (1.15)^{3.08}} = \$572.96$$

Repeating the process with this new estimate of V yields

$$g_c = 0.486$$
$$g_r = 0.193$$

These are the same as before. Hence the values for year 8 are

$$G = 0.486$$
$$Q_m = 97$$
$$Q_c = 50$$
$$Q_r = 37.1$$
$$\text{Profit} = \$245.7$$

The reserves are those given in Table 6.28. In year 9, the initial values for a cutoff of 0.4 are

$$Q_m = 83.3$$
$$Q_c = 50$$
$$Q_r = 35$$

The profit would be $216.7.
Based upon this mining rate, the reserves would last

$$n = \frac{202.7}{83.3} = 2.43 \text{ years}$$

and the present value is

$$V = 216.7 \frac{(1.15)^{2.43} - 1}{0.15 \times (1.15)^{2.43}} = \$416.0$$

Recomputing g_r and g_c we obtain

$$g_c = 0.462$$
$$g_r = 0.183$$

The other possible values are:

$$g_m = 0.10$$
$$g_{mc} = 0.50$$
$$g_{rc} = 0.60$$
$$g_{mr} = 0.456$$

The optimum pair values are:

$$G_{mc} = 0.462$$
$$G_{rc} = 0.462$$
$$G_{mr} = 0.183$$

The middle value of these is

$$G = 0.462$$

Examining the reserve distribution suggests that there are 109.05 tons out of the total 202.7 tons which are above cutoff. The average grade of this remaining ore is 0.731:

Tons	Grade	Tons × Grade
7.70	0.48	3.70
20.27	0.55	11.15
20.27	0.65	13.18
20.27	0.75	15.20
20.27	0.85	17.23
20.27	0.95	19.26
Total = 109.05	Avg = 0.731	Sum = 79.72

Since the maximum concentrating rate is 50 tons/year the life is

$$\frac{109.05}{50} = 2.18 \text{ years}$$

The amount of product is

$$Q_r = 50 \times 0.731 = 36.55$$
$$Q_m = \frac{202.7}{2.18} = 93$$

The profit becomes

$$P = 20 \times 36.55 - 2 \times 50 - 1 \times 93 - 300 = \$238$$

Table 6.29. Reserve distribution at the start of year 10.

Grade (lbs/ton)	Quantity (tons)
0.0–0.1	11
0.1–0.2	11
0.2–0.3	11
0.3–0.4	11
0.4–0.5	11
0.5–0.6	11
0.6–0.7	11
0.7–0.8	11
0.8–0.9	11
0.9–1.0	11
	Total = 110

The present value is

$$V = 238 \frac{(1.15)^{2.18} - 1}{0.15 \times 1.15^{2.18}} \times \$417$$

Iterating again does not change the values. The new distribution is shown in Table 6.29. In year 10, the initial values for a cutoff of 0.4 yields:

$$Q_m = 83.3$$
$$Q_c = 50$$
$$Q_r = 35$$
$$\text{Profit} = \$216.7$$

Based upon this mining rate, the reserves would last

$$n = \frac{110}{83.3} = 1.32 \text{ years}$$

The present value is

$$V = 216.7 \frac{(1.15)^{1.32} - 1}{0.15 \times (1.15)^{1.32}} = \$243.4$$

Calculating g_c and g_r and using this approximation for V yields

$$g_c = 0.437$$
$$g_r = 0.172$$

The other possible values are:

$$g_m = 0.10$$
$$g_{mc} = 0.50$$
$$g_{rc} = 0.60$$
$$g_{mr} = 0.456$$

The optimum pair values are:

$G_{mc} = 0.437$
$G_{rc} = 0.437$
$G_{mr} = 0.172$

The optimum value is $G = 0.437$. Examining the reserve distribution suggests that there are 62 tons of the 110 tons remaining which are above this cutoff.

Tons	Grade	Tons × Grade
7	0.469	3.28
11	0.55	6.05
11	0.65	7.15
11	0.75	8.25
11	0.85	9.35
11	0.95	10.45
Total = 62	Avg = 0.718	Sum = 44.53

The average grade is 0.718. The number of years would be

$$n = \frac{62}{50} = 1.24 \text{ years}$$

The rate of mining and refining would be

$$Q_m = \frac{110}{1.24} = 89$$

$Q_r = 35.9$

and the profit would become

$$P = 20 \times 35.9 - 2 \times 50 - 1 \times 89 - 300$$
$$= \$229$$

The present value is

$$V = \$229 \frac{(1.15)^{1.24} - 1}{0.15 \times (1.15)^{1.24}} = \$243$$

Further iteration yields no change. In year 11, the grade distribution is shown in Table 6.30. The initial values ($V = 0$), yield a cutoff of 0.4 and

$Q_m = 21$
$Q_c = 12.6$
$Q_r = 8.8$

The time would be the largest of

$$T_m = \frac{21}{100} = 0.21$$

$$T_c = \frac{12.6}{50} = 0.25$$

$$T_t = \frac{8.8}{40} = 0.22$$

Table 6.30. Reserve distribution at the start of year 11.

Grade (lbs/ton)	Quantity (tons)
0.0–0.1	2.1
0.1–0.2	2.1
0.2–0.3	2.1
0.3–0.4	2.1
0.4–0.5	2.1
0.5–0.6	2.1
0.6–0.7	2.1
0.7–0.8	2.1
0.8–0.9	2.1
0.9–1.0	2.1
	Total = 21.0

which is again controlled by the concentrator. The profit is

$$P = 20 \times 8.8 - 12.6 \times 2 - 21 \times 1 - \frac{300}{4}$$
$$= \$54.8$$

The present value is

$$V = \$54.8 \frac{(1.15)^{0.25} - 1}{0.15 \times (1.15)^{0.25}} = \$12.5$$

Solving for g_c and g_r yields

$$g_c = 0.402$$
$$g_r = 0.161$$

Combining with the others

$$g_m = 0.16$$
$$g_{mc} = 0.50$$
$$g_{rc} = 0.60$$
$$g_{mr} = 0.456$$

one finds that

$$G_{mc} = 0.402$$
$$G_{rc} = 0.402$$
$$G_{mr} = 0.161$$

The cutoff grade is

$$G = 0.402$$

The distribution is only slightly changed and further iteration is not warranted.

Table 6.31. The production schedule determined by the first pass.

Year	Optimum cutoff grade (lbs/ton)	Quantity mined (tons)	Quantity concentrated (tons)	Quantity refined (lbs)	Profit ($)	Net present value
1	0.5	100	50	37.5	250	1255
2	0.5	100	50	37.5	250	1193
3	0.5	100	50	37.5	250	1122
4	0.5	100	50	37.5	250	1040
5	0.5	100	50	37.5	250	946
6	0.5	100	50	37.5	250	838
7	0.5	100	50	37.5	250	714
8	0.486	97	50	37.1	245.7	574
9	0.462	93	50	36.55	238	417
10	0.437	89	50	35.9	229	243
11	0.40	21	12.6	8.8	55	53

Table 6.32. The final schedule for the manual example.

Year	Optimum cutoff grade (lbs/ton)	Quantity mined (tons)	Quantity concentrated (tons)	Quantity refined (lbs)	Profit ($)	Net present value
1	0.50	100.0	50.0	37.5	250.0	1257.8
2	0.50	100.0	50.0	37.5	250.0	1196.5
3	0.50	100.0	50.0	37.5	250.0	1126.0
4	0.50	100.0	50.0	37.5	250.0	1044.9
5	0.50	100.0	50.0	37.5	250.0	951.7
6	0.50	100.0	50.0	37.5	250.0	844.5
7	0.50	100.0	50.0	37.5	250.0	721.1
8	0.49	97.2	50.0	37.1	245.6	579.3
9	0.46	93.0	50.0	36.6	238.2	420.6
10	0.44	88.7	50.0	35.9	229.5	245.5
11	0.41	21.0	12.5	8.8	54.7	52.8

The net present value is calculated using the yearly profits.

$$NPV = 250 \frac{(1.15)^7 - 1}{0.15 \times (1.15)^7} + \frac{245.7}{(1.15)^8} + \frac{238}{(1.15)^9} + \frac{229}{(1.15)^{10}} + \frac{55}{(1.15)^{10.25}}$$
$$= 1040 + 80.32 + 67.70 + 56.61 + 13.12$$
$$\cong \$1258$$

Step 5. Repetition of the iteration process
In Table 6.31, the present value column reflects the current approximation to V as each years cutoff grade was calculated. The present value of $1258 obtained using the yearly profits should be the same as that shown in the table for year 1. Since the values are not the same ($1258 versus $1255), the process is repeated from the beginning using $V = \$1258$ as the initial estimate for V. Using a computer this iterative procedure is completed in fractions of a second. The final results are shown in Table 6.32. The NPV is slightly higher than the $1255 which would have been obtained by maintaining a constant cutoff grade of 0.5.

In summary:
- Initially the mine and the concentrator are in balance, both operating at capacity. In the last few years, the concentrator is the limiter.
  - The cutoff grade begins at 0.50 lbs/ton and drops to 0.41 lbs/ton at the end of mine life.
  - Mine life is slightly more than 10 years.
  - Total copper produced = 380.9 lbs.
  - Total profits = $2518.
  - Net present value = $1257.80.

This net present value should be compared to that of $1174.60 obtained with the fixed cutoff grade.

6.8 MATERIAL DESTINATION CONSIDERATIONS

6.8.1 Introduction

The term 'cutoff grade' is a rather poorly defined term in the mining literature. A major reason for this is that there are many different cutoff grades. Furthermore the values change with time, mining progress, etc. A cutoff grade is simply a grade used to assign a destination label to a parcel of material.

The destination can change. During the evaluation of final pit limits, the destinations to be assigned are:
- to the surface, and
- left in the ground.

Once the destination 'to the surface' has been assigned, then the destination label 'where on the surface' must be assigned as well. In the distant past there were really only two surface destinations:
- to the mill, and
- to the waste dump.

A grade was used to assign the location. The distinction between destinations was called the mill cutoff grade. In more recent times, the potential future value of material carrying values has been recognized. Hence the lean (low grade) ore dump has become a destination. Thus the 3 destinations require 2 distinguishing grades:

Destination	Assignment
– to the mill	} mill cutoff grade
– to the lean ore dump	
	} waste cutoff grade
– to the waste dump	

Today there are many more possible destinations as our ability to handle and treat materials have improved. Leach dumps/leach pads are a common destination. An active stockpile is a less common destination.

This section will deal with alternate destinations to the mill and waste dump. These will be discussed with respect to cutoff grade. However the reader should remember that these simply are a way of assigning material destinations.

6.8.2 The leach dump alternative

For copper minerals and some others as well, the leach dump (or pad) is an alternative destination to the mill or waste dump. Although this section will focus on copper, the approach used is typical. Two cutoff grades must be determined:
1. Waste dump – leach cutoff.
2. Leach dump – mill cutoff.

In one copper leaching process, sulfuric acid is percolated through the dump. The copper is taken into solution as cupric sulfate or cuprous sulfate. The collected solution is then run through a pachuca tank containing shredded iron where the iron goes into solution and the copper is precipitated. The reactions are given below:

Dump

$$\text{Leach material} + H_2SO_4 \rightarrow CuSO_4 \text{ or } Cu_2SO_4$$

Pachuca tank

$$CuSO_4 + Fe \rightarrow Cu + FeSO_4 \cdot H_2O$$

The precipitated copper is then sent to the smelter. A full discussion of copper recovery from dump material is far beyond the scope of this book. The rate of recovery and the overall percent recovery depend upon both the ore and waste minerals present, the type of sprinkling and collection system, the size distribution of the particles involved, the way the material was placed, etc. Here the results presented by Davey (1979c) will be used for illustrative purposes.

For copper mineralization type 1, no recovery is expected when the copper content is below 0.2%. Using mineral dressing terminology, this would be the 'Fixed Tails' limit. Thus, when considering material destinations, the waste dump – leach dump cutoff criteria would be:

Waste dump destination: copper content < 0.2%

Leach dump destination: copper content $\geq$ 0.2%

There will be another cutoff grade above which all material should be sent to the mill. To determine this value one must look both at the total copper recovery through leaching and the recovery as a function of time. For copper mineralization type 1, a fairly uniform distribution of the various grades may be assumed for the low grade material having a head grade of 0.4% copper or less. In this case

Head grade $\leq$ 0.4%

the percent recovery is expressed by

$$\text{Recovery} = \frac{\text{Head grade} - \text{Fixed tails limit}}{\text{Head grade}} \times 100\% \tag{6.57a}$$

If the head grade is better than 0.4%,

Head grade > 0.4%

then the copper recovery by leaching is 50%

$$\text{Recovery} = 50\% \tag{6.57b}$$

Four examples will be used to demonstrate the use of the recovery formula.

558 *Open pit mine planning and design: Fundamentals*

Table 6.33. Recovery factor as a function of year for the leach example (Davey, 1979c).

Year	Recovery factor
1	0.40
2	0.30
3	0.14
4	0.10
5	0.06

1. Head grade $= 0.45\%$ Cu

   Recovery $= 50\%$

   $$\text{Recovered copper} = \frac{0.45}{100} \times 0.5 \times 2000 = 4.5 \text{ lbs/ton}$$

2. Head grade $= 0.40\%$ Cu

   $$\text{Recovery} = \frac{0.40 - 0.20}{0.40} \times 100 = 50\%$$

   $$\text{Recovered copper} = \frac{0.40}{100} \times 0.5 \times 2000 = 4.0 \text{ lbs/ton}$$

3. Head grade $= 0.30\%$ Cu

   $$\text{Recovery} = \frac{0.30 - 0.20}{0.30} \times 100 = 33.3\%$$

   $$\text{Recovered copper} = \frac{0.30}{100} \times 0.333 \times 2000 = 2 \text{ lbs/ton}$$

4. Head grade $= 0.15\%$ Cu

   Recovery $= 0\%$

   Recovered copper $= 0$ lbs/ton

The copper is recovered from the ton of material over some period of time, normally several years. The amount recovered per year from the ton normally decreases with time as shown below.

Yearly recovery (lbs) $=$ Recovery factor $\times$ Total lbs to be recovered   (6.58b)

The factors are given in Table 6.33.

For a copper ore running 0.55% (head grade), the amount of copper in situ is 11 lbs/ton. Of this 5.5 lbs is expected to be recovered over the 5-year period. For year 1, one would expect to recover

   Year 1 copper recovered $= 0.40 \times 5.5$ lbs $= 2.20$ lbs

For year 2 the result would be

   Year 2 copper recovered $= 0.30 \times 5.5$ lbs $= 1.65$ lbs

Table 6.34 summarizes the overall recovery.

Table 6.34. Estimated recovery as a function of year (Davey, 1979c).

Year	Estimated recovery (lbs/ton)
1	2.20
2	1.65
3	0.77
4	0.55
5	0.33
	Total = 5.50

To demonstrate the application, consider the following example for a leach ore running 0.55% Cu. The costs of mining and transporting the mineral are assumed to be the same for both the milling and leaching alternatives. Since the material will be removed from the pit in any case, these are considered as sunk costs. An interesting exercise for the reader is to calculate the value of potential leach material when deciding the ultimate pit limits. As can be seen in Table 6.35, the net value per ton of material when considered for dump leaching is $2.07.

The same process is repeated for the milling alternative. This is done in Table 6.36.

As can be seen, the net value is now $2.75/ton of material treated. Obviously, the destination of the material should be the mill.

The process illustrated in Table 6.35 and 6.36 is repeated until a breakeven grade between leaching and milling is found (see Figure 6.24). In this particular case it is 0.45% Cu. Material carrying grades above this level should be sent to the mill. Material grading between 0.2 and 0.45% copper should go to the leach dump.

In this simplified calculation, two important considerations have been left out. The first is the cost of capital associated with the mill. Inclusion of these costs will increase the breakeven grade. However, possible by-product credits from gold, molybdenum, etc., which would be realized by milling and not leaching have not been included. Obviously, a more detailed analysis would take into account both of these factors.

It may also be that the mill capacity is taken up with higher grade material. In such a case another destination, a stockpile, might be considered. Cutoff grades are involved in the decision as well.

If these cutoff grades are applied to the tonnage-grade curve given in Figure 6.25, one finds the following.

Class of material	Grade %	Tons	Average grade* (%)
Mill ore	≥ 0.45	7.4×10^6	0.96
Leach ore	$0.20 < g < 0.45$	3.4×10^6	0.33
Waste	≤ 0.20	20.0×10^6	–

*See Figure 6.26.

In this particular case, the actual production results were

Class of material	Tons	Average grade % Cu
Mill ore	7.5×10^6	0.93
Leach	6.4×10^6	0.35
Waste	16.9×10^6	–

Table 6.35. Economic analysis regarding the leaching of material containing 0.55% copper (Davey, 1979c).

	$/ton of material
A. Production cost breakdown	
1. Precipitation cost	
($0.20/lb of recovered copper)	
– Recovery = 5.5 lbs/ton	
– Precipitation cost = 5.5 × $0.20 = $1.10	$1.10
– The precipitate runs 85% copper	
2. Treatment cost	$1.08
Smelting	
($50/ton of precipitate)	
– Ratio of concentration = $\dfrac{2000 \times 0.85}{5.5}$ = 309.09	
– Smelting cost/ton of leach material = $\dfrac{\$50}{309.09}$ = $0.16	
– Recovery = 97.5%	
– Freight (smelter → refinery)	
($50/ton of blister copper)	
– Freight cost = $\dfrac{5.5 \times 0.98}{2000}$ × $50 = $0.1351	
Refining	
($130/ton of blister copper)	
– Refining cost = $\dfrac{5.404}{2000}$ × $130 = $0.351	
– Recovery = 99.7%	
Selling and delivery	
($0.01/lb Cu)	
cost = 0.01 × 5.39 = $0.054	
General plant	
($0.07/lb Cu)	
cost = 0.07 × 5.39 = $0.38	
3. Total production cost	$2.18
B. Revenue ($/ton of leach material)	
1. Total value of in situ saleable copper	
(price $1.00/lb)	
5.39 lbs × $1/lb = $5.39	
(discount rate $i = 0\%$)	
2. Discounted value	
(discount rate = 12.5%)	

Year	Recovered copper (lbs)	Discounted value ($)
1	2.15	1.92
2	1.62	1.28
3	0.75	0.53
4	0.54	0.34
5	0.32	0.18
		$4.25

  3. In general no by-products are recovered.
  4. It has been assumed that costs and prices have inflated at the same rate.

C. Net value ($/ton) of leach material
    Net value = Revenue − Costs
            = $4.25 − $2.18 = $2.07

Table 6.36. Milling evaluation of 0.55% copper feed (Davey, 1979c).

	$/ton of material
A. Production cost breakdown	
1. Milling cost	
– Operating cost	= $2.80
– G&A cost (15% of operating)	= 0.42

    Mill recovery = 80%
    Mill concentrate = 20% Cu

2. Shipment of mill concentrate to smelter
   ($1.40/ton of concentrate)
   Rate of concentration

$$r = \frac{2000 \times 0.20}{\frac{0.55}{100} \times 2000 \times 0.80} = 45.45$$

   Freight cost $= \dfrac{\$1.40}{45.45} = \$0.03$ = 0.03

3. Treatment cost
   Smelting
   ($50/ton of concentrate)

   – Cost $= \dfrac{\$50}{45.45} = \$1.10$ = 1.10

   – Recovery = 97.5%
   Freight (smelter → refinery)
   ($50/ton of blister copper)

   Blister transport $= \$50 \times \dfrac{0.975 \times 8.8}{2000} = \$0.21$ = $0.21

   Refining cost
   ($130.00/ton of blister copper)
   – Recovery = 99.8%
   – Cost $= \$130 \times \dfrac{8.58}{2000} = \$0.56$ = $0.56

   Selling and delivery
   ($0.01/lb of copper)
     Cost $= 8.56 \times 0.01 = \$0.09$ = $0.09

   General plant
   ($0.07/lb of copper)
     Cost $= 8.56 \times 0.07 = \$0.60$ = $0.60

4. Total production cost   = $5.81

B. Revenues ($/ton)
   There are a total of 8.56 lbs of copper recovered. A price of $1/lb will be used.
   Hence
     Revenue $= 8.56 \times \$1/\text{lb} = \$8.56$

C. Net value ($/ton)
   The net value is given as the revenue minus the costs. It has been assumed here
   that there is no appreciable time required in the processing chain so that
   discounting is not used.
   Net value $= \$8.56 - \$5.81 = \$2.75$

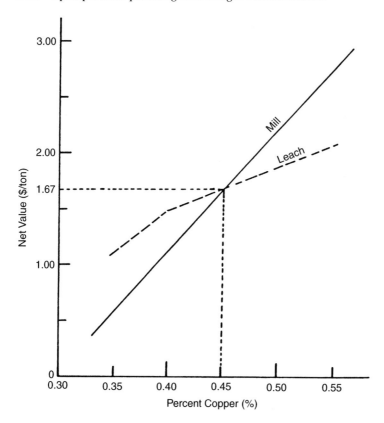

Figure 6.24. Breakeven between milling and leaching.

6.8.3 The stockpile alternative

As has been discussed there are a series of cutoff grades which are applied during the life of a mine-mill complex. Early in the life, there is a desire to recover the capital investment as early as possible. Since the mill capacity, in terms of tons per day, is normally fixed, it is better to use this capacity with higher rather than lower grade material. Hence the initial mill cutoff grade may be fairly high. As the mine-mill complex matures, the mill cutoff will normally decrease. Assume that for a copper operation the set mill cutoff is 0.7% Cu. The mining cutoff (based upon covering the production costs, smelting, refining, sales, etc.) could be 0.3%. If the material has to be mined to reach deeper higher grade ore, then the breakeven cutoff for the mill destination could be even lower than 0.3%. As indicated, because of mill capacity restrictions and the requirement for rapid capital payback, the mill cutoff is 0.7%. Therefore a large amount of material which, if produced later in the life of the mine would go to the mill, is now below cutoff. One possibility is to stockpile the material lying between 0.3% and 0.7% for later treatment by the mill. An alternative would be to send the material to leach dumps for a quicker return of the costs expended in the mining. No rehandling (with its added cost) would be required. Also there would not be a cost associated with the storage areas themselves.

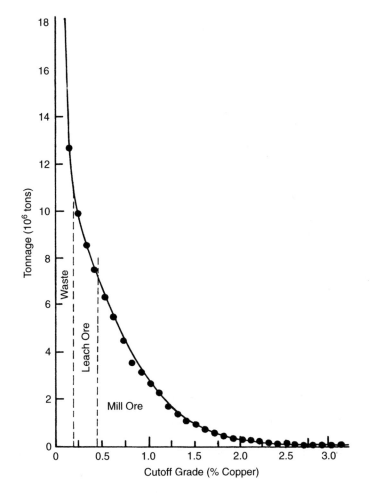

Figure 6.25. Superposition of the mill-leach and leach-waste cutoff grades on the Silver Bell oxide pit cumulative tonnage curve.

There are, in summary, a number of disadvantages with ore stockpiles (Schellman, 1989):

1. A rehandling step is necessary. The costs involved may be significant.

2. Space is required to accommodate the various qualities. Often such space is scarce near the operations.

3. Over time some materials become more difficult to treat.

4. Additional expense is incurred in tracking the various qualities of material (sampling, production control).

For these and other reasons, stockpiles have not been popular with most mining operations. There are some exceptions however.

Taylor (1985) justified the use of the stockpile system through the use of some actual examples from mines. At the Craigmont copper mine, a substantial sub-grade stockpile that averaged about 0.6% Cu was accumulated from the initial high grade open pit. The operation then moved underground, where sublevel caving yielded a daily output of about

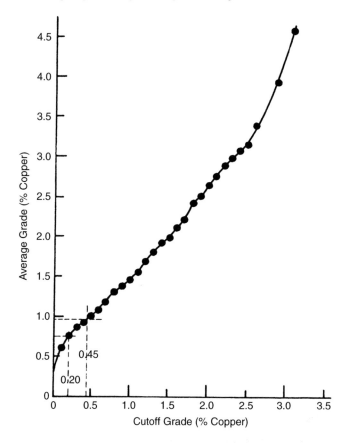

Figure 6.26. Superposition of the mill-leach and leach-waste cutoff grades on the Silver Bell oxide pit average grade curve.

4500 tons at 1.8% Cu. For many years the stockpile provided the remaining 1500 tons per day that were needed to supply the mill. Another practical example is the Gibraltar copper mine. A cutoff grade of 0.3% Cu was used in the early years and more than 100,000,000 short tons, grading down to 0.25% Cu, were stockpiled. In 1982–83 this surface ore enjoyed a cost advantage of at least $2.50 Canadian/ton over ore that had yet to be exposed in the pit. At the then low copper price, the 0.27% grade had a higher net value than the new 0.37% Cu ore.

A leach dump is a form of stockpile since cutoff grades have been used to select the material for placement. Generally no rehandle is involved. Today a number of such 'waste' dumps are being retreated with improved technology and there is reason to believe that this process will occur even for the 'waste' dumps being created today.

There are a number of advantages to having stockpiles:

1. They can be used for blending to ensure a constant head grade to the mill. Normally recoveries are higher when such fluctuations are low.

2. They can be used as a buffer to make up for production short falls due to various reasons.

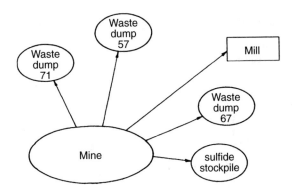

Figure 6.27. Schematic representation of various material destination (Schellman, 1989).

3. The metal recovery is generally far higher in the mill than in the leach dump. Thus resource utilization is better. Since the overall recovery is higher the overall costs get spread over a bigger base.

4. The presence of a buffer offers more flexibility in examining design options. For example one haulage road might be used instead of two due to the production security provided by the buffer. Steeper pit slopes might be considered.

In the future it is expected that the use of stockpiles will play a much more important part in the life of mines.

Computerized truck dispatching has simplified the changing of destinations. As can be seen in Figure 6.27, the dumping points are the concentrator plant, the three waste dumps (71, 57, and 67) and the sulfide stockpile. The addition of new dumps would not introduce major changes in the loading-haulage operation. Strategies regarding the number and types of stockpiles as well as how to best incorporate the stockpile into the overall operating plan (when and how much to draw as compared to primary production) must be developed.

Lane (1988) has provided a rather complete discussion regarding use of the stockpile alternative.

Schellman (1989) recently modified Lane's algorithm to include the stockpiling option with the grades below current mill cutoff being segregated. The optimum time to draw upon the stockpile as well as the tons and grades moved have been determined for three options. His approach will be illustrated using the data given in Table 6.37.

There are a total of 600×10^6 tons to be mined (100×10^6 tons in each phase). The capacities and costs used in applying the standard Lane approach described in Section 6.7 are given in Table 6.38. The results without stockpiling are given in Table 6.39. As can be seen the total profits (P_T) are $\$484.2 \times 10^6$ and the net present value (NPV) is $\$150.33 \times 10^6$. Of the total tonnage mined 319×10^6 goes to the mill and the remainder 281×10^6 tons to the waste dump. The cutoff grade in the final year is 0.23% Cu.

In the proposed system, a series of stockpiles are generated with material which is below the cutoff grade for each period but above the lowest cutoff grade for the whole project. Considering only increment 1 in the preceding example (Table 6.39), all material below 0.5% and above 0.23% will be sent to the stockpile. For the second increment (year 6), material below 0.53% and above 0.23% will be sent to the stockpile.

Three different alternatives for handling the stockpiled material will be considered.

Table 6.37. Input data for the demonstration example (Lane, 1964) (The material amounts given are in 10^6 tons).

Grade categories (%Cu)	Pit increments in mining sequence					
	1	2	3	4	5	6
0.00–0.15	14.4	15.9	17.9	20.3	23.4	27.7
0.15–0.20	4.6	5.1	5.5	6.3	7.2	8.3
0.20–0.25	4.4	4.9	5.4	6.0	6.7	7.7
0.25–0.30	4.3	4.7	5.3	5.6	6.4	7.3
0.30–0.35	4.2	4.5	4.9	5.5	6.2	6.7
0.35–0.40	4.1	4.4	4.7	5.3	5.6	6.3
0.40–0.45	3.9	4.3	4.6	4.9	5.4	5.7
0.45–0.50	3.8	4.1	4.5	4.8	5.1	5.3
0.50–0.55	3.7	3.9	4.2	4.5	4.6	4.7
0.55–0.60	3.6	3.8	3.9	4.2	4.4	4.3
0.60–0.65	3.4	3.6	3.8	3.9	4.0	3.7
0.65–0.70	3.3	3.5	3.7	3.7	3.6	3.3
+0.70	42.3	37.5	31.6	25.0	17.4	9.0
Average % Cu	1.13	1.07	1.00	0.93	0.87	0.80

Table 6.38. Capacities and costs used in the demonstration example (Lane, 1964).

Capacities:		
	Mine	20 million tons material/year
	Mill	10 million tons ore/year
	Refinery	90,000 tons product/year
Costs:		
	Mining	$0.50/ton of material from pit
	Mill	$0.60/ton of ore to mill
	Refining	$50.0/ton of product
	Fixed charges	$4 million/year
	Selling price	$550/ton of product
	Recovery	90%

Alternative A. Mill feed coming simultaneously from the mine and stockpile

In this alternative, the analysis of whether or not to send material from the stockpile to the mill is done taking into account the profit generated by the material in the stockpile and the profit generated by the material in the push back (pit increment). If the profitability of the ore in the stockpile is greater than that of the ore in the mine, ore from both the stockpile and the mine will feed the mill. The comparison is established between the same grade categories for both alternatives. If, for example, at the mine the mineral grades between 0.4% and 0.7% and at the stockpile it grades between 0.4% and 0.55%, only the mineral between 0.4% and 0.55% will be examined. If the profitability of the stockpiled ore is better, then ore between 0.4% and 0.55% will be sent from the stockpile. Ore with grades between 0.55% and 0.7% will be sent from the mine. Mined material between 0.23% and 0.55% would be sent to the stockpile.

This stockpile system is dynamic, in the sense that, for some determined year, the stock pile is receiving ore from the mine, and sending ore to the mill. When the mining of the

Table 6.39. Final production schedule for the demonstration example (Schellman, 1989).

Year	Pit increment	Cutoff grade (% Cu)	Quantity mined (10⁶ tons)	Quantity concentrated (10⁶ tons)	Quantity refined (10⁶ tons)	Profit (10⁶ $)	Total profit (10⁶ $)	Net present value (10⁶ $)
1	1	0.50	17.9	10.0	90.0	26.1	484.2	150.33
2	1	0.50	17.9	10.0	90.0	26.1	458.1	146.80
3	1	0.50	17.9	10.0	90.0	26.1	432.0	142.75
4	1	0.50	17.9	10.0	90.0	26.1	405.9	138.08
5	1	0.50	17.9	10.0	90.0	26.1	379.8	132.72
6	1	0.50	10.7	6.0	54.2	15.7	353.7	126.55
6	2	0.53	8.0	4.0	34.2	9.1	338.0	126.55
7	2	0.5	20.0	10.0	85.9	23.0	328.9	120.69
8	2	0.53	20.0	10.0	85.9	23.0	305.9	115.83
9	2	0.53	20.0	10.0	85.9	23.0	282.9	110.23
10	2	0.53	20.0	10.0	85.9	23.0	259.9	103.80
11	2	0.53	20.0	10.0	85.9	23.0	259.9	103.80
11	3	0.47	7.8	3.9	29.5	7.0	222.8	96.40
12	3	0.47	20.0	10.0	76.1	18.0	215.8	89.80
13	3	0.47	20.0	10.0	76.1	18.0	197.8	85.22
14	3	0.47	20.0	10.0	76.1	18.0	178.8	79.95
15	3	0.47	20.0	10.0	76.4	18.0	161.8	73.90
16	3	0.45	12.3	6.4	47.7	11.3	143.8	66.94
16	4	0.41	7.2	3.6	24.0	4.8	132.5	66.94
17	4	0.41	20.0	10.0	66.5	13.2	127.7	60.87
18	4	0.41	20.0	10.0	66.5	13.2	114.5	56.77
19	4	0.40	19.4	10.0	65.6	13.1	101.3	52.06
20	4	0.38	18.8	10.0	64.5	12.9	88.2	46.78
21	4	0.33	14.6	8.1	51.2	10.2	75.3	40.91
21	5	0.35	3.8	1.9	11.0	1.7	65.3	40.91
22	5	0.34	19.5	10.0	56.5	8.5	63.5	35.18
23	5	0.33	19.0	10.0	55.8	8.4	55.0	31.92
24	5	0.32	18.4	10.0	55.1	8.3	46.5	28.27
25	5	0.30	17.9	10.0	54.3	8.2	38.1	24.17
26	5	0.29	17.3	10.0	53.3	8.0	29.9	19.6
27	5	0.27	4.2	2.5	13.1	2.0	21.8	14.50
27	6	0.27	14.1	7.5	34.7	2.8	20.3	14.50
28	6	0.26	18.3	10.0	45.7	3.7	17.3	11.91
29	6	0.26	18.0	10.0	45.4	3.7	13.6	10.00
30	6	0.25	17.7	10.0	44.9	3.6	9.9	7.83
31	6	0.24	17.3	10.0	44.4	3.6	6.3	5.39
32	6	0.23	14.6	8.7	37.9	3.0	2.7	2.63

particular push back is completed, then the process continues. The stockpile would be considered as a new push back.

Alternative B
In this alternative if the profit from the ore in the stockpile is greater than that for ore in the push back, the mill is fed with material exclusively coming from the stockpile.

The material to feed the mill is sent either from the stockpile or from the push back, but not from both places simultaneously.

568 *Open pit mine planning and design: Fundamentals*

Alternative C
In this alternative, material is not sent to the mill from the stockpile during the mining of the pit. Material is stockpiled until all the material in the pit is exhausted. Only at that moment does the stockpile start to work as if it were a new push back.

The three stockpile alternatives will now be applied to the earlier example. The cost involved with the stockpile is assumed to be $0.225/ton. This is 45% of the original mining cost. This incremental cost is $0.20/ton to cover material re-handling and $0.025/ton for increased pit supervision, sampling, etc. The results from alternative A are given in Table 6.40. As can be seen the total profit is significantly higher than the no stockpile alternative. The NPV is only slightly higher because the major contribution occurs late in the life of the property. Table 6.41 is a year-by-year breakdown of the stockpile content. This is theoretical since one clearly would not keep so many separate grade categories.

The results from alternative B are shown in Table 6.42. As can be seen the NPV and the undiscounted profits are better than without stockpiling, but not as good as for alternative A. This is because the stockpile only starts to send ore to the mill in year 28. In the previous alternative, the first time that the stockpile sends ore to the mill is in year 15. Moreover, using material from the stockpile in years 28 through 35 means that almost all of the material in the stockpile is used. At the end of year 40, mining the remaining material in the stockpile is not economically profitable even though there is material in the stockpile between the grades of 0.23% and 0.26%.

Alternative C (Table 6.43) has a better net present value and a better undiscounted profit than without stockpiling. This alternative also has a slightly higher net present value than alternative B, but smaller than alternative A. From an operational point of view, this alternative has several advantages. It is not necessary to blend material as in the production schedule generated by alternative A or to stop the mine production for 8 years and then start to produce again, as in alternative B.

The stockpile option should always be considered in open pit mine planning. The potential contribution to the NPV depends upon each particular case. It is particularly dependent upon the spread between the highest and the lowest grades in the project cutoff grade strategy. If that spread is significant, the stockpile alternative should be considered for the project.

6.9 PRODUCTION SCHEDULING

6.9.1 *Introduction*

Production scheduling is a very important part of the mining process. This will be demonstrated through a simple example. Consider the property shown in Figure 6.28 in which 10 ore blocks are overlain by 10 waste blocks.

A production rate of 5 blocks per year (irrespective of whether the blocks are ore or waste) will be assumed. The net value for an ore block is $2 and the cost of removing the waste is $1/block. The total cost involved in waste removal would be $10 and the ore value is $20.

If both ore and waste could be mined instantaneously, the net present value would be $10. However due to practical constraints, they cannot. Therefore a number of scheduling scenarios must be considered.

Table 6.40. Final schedule for the alternative A stockpile (Schellman, 1989).

Year	Pit increment	Cutoff grade (% Cu)	Quantity mined (10^6 tons)	Quantity concentrated (10^6 tons)	Quantity refined (10^6 tons)	Profit (10^6 \$)	Total profit (10^6 \$)	Net present value (10^6 \$)
1	1	0.50	17.8	10.0	89.8	26.0	528.4	150.56
2	1	0.50	17.8	10.0	89.8	26.0	502.4	147.13
3	1	0.50	17.8	10.0	89.8	26.0	476.4	143.19
4	1	0.50	17.8	10.0	89.8	26.0	450.4	138.66
5	1	0.50	17.8	10.0	89.8	26.0	424.4	133.45
6	1	0.50	11.2	6.3	56.6	16.4	398.4	127.46
6	2	0.53	7.4	3.7	31.8	8.5	382.0	127.46
7	2	0.53	20.0	10.0	85.9	23.0	373.5	121.69
8	2	0.53	20.0	10.0	85.9	23.0	350.5	116.97
9	2	0.53	20.0	10.0	85.9	23.0	327.5	111.55
10	2	0.53	20.0	10.0	85.9	23.0	304.5	105.31
11	2	0.53	12.8	6.4	55.0	14.7	281.5	98.14
11	3	0.47	7.2	3.6	27.4	6.5	266.8	98.14
12	3	0.47	20.0	10.0	76.1	18.0	260.3	91.66
13	3	0.47	20.0	10.0	76.1	18.0	242.3	87.36
14	3	0.47	20.0	10.0	76.1	18.0	224.3	82.42
15	***	0.47	1.0	10.0	4.6	1.2	206.3	76.73
15	3	0.47	18.7	9.0	71.5	17.2	205.1	76.73
16	3	0.47	13.1	6.4	48.8	11.4	187.9	69.82
16	4	0.41	6.9	3.5	23.1	4.6	176.5	69.82
17	4	0.41	20.0	10.0	66.5	13.2	171.9	64.28
18	4	0.41	20.0	10.0	66.5	13.2	158.7	60.70
19	4	0.41	20.0	10.0	66.5	13.2	145.5	56.57
20	***	0.41	2.2	2.2	9.2	2.2	132.3	51.83
20	4	0.41	17.3	7.8	57.0	11.7	130.1	51.83
21	4	0.41	13.6	6.6	43.8	8.4	118.4	45.64
21	5	0.35	6.4	3.2	18.3	2.8	110.0	45.64
22	5	0.35	20.0	10.0	57.2	8.6	107.2	41.35
23	5	0.35	20.0	10.0	57.2	8.6	98.6	38.94
24	5	0.35	20.0	10.0	57.2	8.6	90.0	36.17
25	***	0.35	3.7	3.7	15.4	3.7	81.4	32.98
25	5	0.35	15.6	6.3	42.5	6.6	77.7	32.98
26	5	0.35	14.7	7.2	40.5	5.6	71.1	27.60
26	6	0.29	5.3	2.6	12.6	1.0	65.5	27.60
27	6	0.29	20.0	10.0	47.6	3.8	64.5	25.08
28	6	0.29	20.0	10.0	47.6	3.8	60.7	25.03
29	6	0.29	20.0	10.0	47.6	3.8	56.9	24.98
30	***	0.29	5.5	5.5	18.8	3.6	53.1	24.91
30	6	0.29	13.6	4.5	27.6	1.6	49.5	24.91
31	6	0.29	16.0	7.8	36.7	2.5	47.9	23.50
31	***	0.31	2.2	2.2	6.8	0.8	45.6	23.50
32	***	0.31	10.0	10.0	36.5	6.0	44.6	23.79
33	***	0.30	10.0	10.0	36.5	6.0	38.6	21.36
34	***	0.29	10.0	10.0	35.1	5.3	32.6	18.57
35	***	0.28	10.0	10.0	34.5	5.0	27.3	16.05
36	***	0.27	10.0	10.0	33.7	4.6	22.3	13.47
37	***	0.27	10.0	10.0	32.9	4.2	17.7	10.90
38	***	0.26	10.0	10.0	30.8	3.1	13.5	8.35
39	***	0.25	10.0	10.0	29.0	2.3	10.4	6.45
40	***	0.24	10.0	10.0	27.2	1.4	8.1	5.17
41	***	0.23	10.0	10.0	27.2	1.4	6.7	4.58
42	***	0.23	10.0	10.0	27.2	1.4	5.3	3.91
43	***	0.23	10.0	10.0	28.2	1.9	3.9	3.13
44	***	0.23	8.2	8.2	24.2	2.0	2.0	1.7

*** Mineral from stockpile.

Table 6.41. Stockpile grade and tonnage (expressed in 10^6 tons) distribution by year for alternative A (Schellman, 1989).

STOCKPILE GRADE DISTRIBUTION YEAR 1

GRADE CATEGORIES %		AVRG	TONNAGE
.230	.250	.240	.5
.250	.300	.275	.8
.300	.350	.325	.7
.350	.400	.375	.7
.400	.450	.425	.7
.450	.500	.475	.7

TOTAL: 4.1

STOCKPILE GRADE DISTRIBUTION YEAR 2

GRADE CATEGORIES %		AVRG	TONNAGE
.230	.250	.240	.9
.250	.300	.275	1.5
.300	.350	.325	1.5
.350	.400	.375	1.5
.400	.450	.425	1.4
.450	.500	.475	1.3

TOTAL: 8.1

STOCKPILE GRADE DISTRIBUTION YEAR 3

GRADE CATEGORIES %		AVRG	TONNAGE
.230	.250	.240	1.4
.250	.300	.275	2.3
.300	.350	.325	2.2
.350	.400	.375	2.2
.400	.450	.425	2.1
.450	.500	.475	2.0

TOTAL: 12.2

STOCKPILE GRADE DISTRIBUTION YEAR 4

GRADE CATEGORIES %		AVRG	TONNAGE
.230	.250	.240	1.9
.250	.300	.275	3.1
.300	.350	.325	3.0
.350	.400	.375	2.9
.400	.450	.425	2.8
.450	.500	.475	2.7

TOTAL: 16.3

STOCKPILE GRADE DISTRIBUTION YEAR 5

GRADE CATEGORIES %		AVRG	TONNAGE
.230	.250	.240	2.3
.250	.300	.275	3.8
.300	.350	.325	3.7
.350	.400	.375	3.6
.400	.450	.425	3.5
.450	.500	.475	3.4

TOTAL: 20.4

STOCKPILE GRADE DISTRIBUTION YEAR 6

GRADE CATEGORIES %		AVRG	TONNAGE
.230	.250	.240	2.6
.250	.300	.275	4.3
.300	.350	.325	4.2
.350	.400	.375	4.1
.400	.450	.425	3.9
.450	.500	.475	3.8

TOTAL: 22.9

STOCKPILE GRADE DISTRIBUTION YEAR 7

GRADE CATEGORIES %		AVRG	TONNAGE
.230	.250	.240	3.2
.250	.300	.275	5.2
.300	.350	.325	5.1
.350	.400	.375	5.0
.400	.450	.425	4.8
.450	.500	.475	4.6
.500	.528	.514	.4

TOTAL: 28.4

STOCKPILE GRADE DISTRIBUTION YEAR 8

GRADE CATEGORIES %		AVRG	TONNAGE
.230	.250	.240	3.8
.250	.300	.275	6.2
.300	.350	.325	6.0
.350	.400	.375	5.9
.400	.450	.425	5.6
.450	.500	.475	5.4
.500	.528	.514	.9

TOTAL: 33.8

STOCKPILE GRADE DISTRIBUTION YEAR 9

GRADE CATEGORIES %		AVRG	TONNAGE
.230	.250	.240	4.4
.250	.300	.275	7.1
.300	.350	.325	6.9
.350	.400	.375	6.7
.400	.450	.425	6.5
.450	.500	.475	6.3
.500	.528	.514	1.3

TOTAL: 39.2

STOCKPILE GRADE DISTRIBUTION YEAR 10

GRADE CATEGORIES %		AVRG	TONNAGE
.230	.250	.240	5.0
.250	.300	.275	8.1
.300	.350	.325	7.8
.350	.400	.375	7.6
.400	.450	.425	7.3
.450	.500	.475	7.1
.500	.528	.514	1.8

TOTAL: 44.6

STOCKPILE GRADE DISTRIBUTION YEAR 11

GRADE CATEGORIES %		AVRG	TONNAGE
.230	.250	.240	5.4
.250	.300	.275	8.7
.300	.350	.325	8.4
.350	.400	.375	8.2
.400	.450	.425	7.9
.450	.469	.459	2.9
.469	.500	.484	4.7
.500	.528	.514	2.0

TOTAL: 48.1

STOCKPILE GRADE DISTRIBUTION YEAR 12

GRADE CATEGORIES %		AVRG	TONNAGE
.230	.250	.240	6.0
.250	.300	.275	9.7
.300	.350	.325	9.3
.350	.400	.375	9.1
.400	.450	.425	8.8
.450	.469	.459	3.2
.469	.500	.484	4.7
.500	.528	.514	2.0

TOTAL: 53.0

Table 6.41. (Continued).

STOCKPILE GRADE DISTRIBUTION YEAR 13			STOCKPILE GRADE DISTRIBUTION YEAR 16			STOCKPILE GRADE DISTRIBUTION YEAR 19			STOCKPILE GRADE DISTRIBUTION YEAR 22		
GRADE CATEGORIES %	AVRG	TONNAGE	GRADE CATEGORIES %	AVRG	TONNAGE	GRADE CATEGORIES %	AVRG	TONNAGE	GRADE CATEGORIES %	AVRG	TONNAGE
.230 .250	.240	6.7	.230 .250	.240	8.4	.230 .250	.240	10.5	.230 .250	.240	12.5
.250 .300	.275	10.8	.250 .300	.275	13.6	.250 .300	.275	16.9	.250 .300	.275	20.1
.300 .350	.325	10.3	.300 .350	.325	12.9	.300 .350	.325	16.2	.300 .350	.325	19.3
.350 .400	.375	10.1	.350 .400	.375	12.5	.350 .400	.375	15.7	.350 .400	.375	17.5
.400 .450	.425	9.7	.400 .410	.405	2.5	.400 .410	.405	3.1	.400 .410	.405	3.4
.450 .469	.459	3.6	.410 .450	.430	9.7	.410 .450	.430	9.7	.410 .450	.430	9.4
.469 .500	.484	4.7	.450 .469	.459	4.4	.450 .469	.459	4.4	.450 .469	.459	4.3
.500 .528	.514	2.0	.469 .500	.484	4.6	.469 .500	.484	4.6	.469 .500	.484	5.1
			.500 .528	.514	2.2	.500 .528	.514	2.2	.500 .528	.514	2.0
	TOTAL:	57.9		TOTAL:	70.7		TOTAL:	83.3		TOTAL:	93.5

STOCKPILE GRADE DISTRIBUTION YEAR 14			STOCKPILE GRADE DISTRIBUTION YEAR 17			STOCKPILE GRADE DISTRIBUTION YEAR 20			STOCKPILE GRADE DISTRIBUTION YEAR 23		
GRADE CATEGORIES %	AVRG	TONNAGE	GRADE CATEGORIES %	AVRG	TONNAGE	GRADE CATEGORIES %	AVRG	TONNAGE	GRADE CATEGORIES %	AVRG	TONNAGE
.230 .250	.240	7.3	.230 .250	.240	9.1	.230 .250	.240	11.2	.230 .250	.240	13.3
.250 .300	.275	11.8	.250 .300	.275	14.7	.250 .300	.275	18.0	.250 .300	.275	21.3
.300 .350	.325	11.3	.300 .350	.325	14.0	.300 .350	.325	17.3	.300 .350	.325	20.5
.350 .400	.375	11.0	.350 .400	.375	13.6	.350 .400	.375	16.7	.350 .400	.375	17.5
.400 .450	.425	10.6	.400 .410	.405	2.7	.400 .410	.405	3.3	.400 .410	.405	3.4
.450 .469	.459	3.9	.410 .450	.430	9.7	.410 .450	.430	9.4	.410 .450	.430	9.4
.469 .500	.484	4.7	.450 .469	.459	4.4	.450 .469	.459	4.3	.450 .469	.459	4.3
.500 .528	.514	2.0	.469 .500	.484	4.6	.469 .500	.484	5.1	.469 .500	.484	5.1
			.500 .528	.514	2.2	.500 .528	.514	2.0	.500 .528	.514	2.0
	TOTAL:	62.7		TOTAL:	74.9		TOTAL:	87.2		TOTAL:	96.8

STOCKPILE GRADE DISTRIBUTION YEAR 15			STOCKPILE GRADE DISTRIBUTION YEAR 18			STOCKPILE GRADE DISTRIBUTION YEAR 21			STOCKPILE GRADE DISTRIBUTION YEAR 24		
GRADE CATEGORIES %	AVRG	TONNAGE	GRADE CATEGORIES %	AVRG	TONNAGE	GRADE CATEGORIES %	AVRG	TONNAGE	GRADE CATEGORIES %	AVRG	TONNAGE
.230 .250	.240	7.9	.230 .250	.240	9.8	.230 .250	.240	11.7	.230 .250	.240	14.1
.250 .300	.275	12.9	.250 .300	.275	15.8	.250 .300	.275	18.8	.250 .300	.275	22.6
.300 .350	.325	12.3	.300 .350	.325	15.1	.300 .350	.325	18.0	.300 .350	.325	21.7
.350 .400	.375	11.9	.350 .400	.375	14.7	.350 .400	.375	17.5	.350 .400	.375	17.5
.400 .450	.425	11.5	.400 .410	.405	2.9	.400 .410	.405	3.4	.400 .410	.405	3.4
.450 .469	.459	4.2	.410 .450	.430	9.7	.410 .450	.430	9.4	.410 .450	.430	9.4
.469 .500	.484	4.6	.450 .469	.459	4.4	.450 .469	.459	4.3	.450 .469	.459	4.3
.500 .528	.514	2.2	.469 .500	.484	4.6	.469 .500	.484	5.1	.469 .500	.484	5.1
			.500 .528	.514	2.2	.500 .528	.514	2.0	.500 .528	.514	2.0
	TOTAL:	67.5		TOTAL:	79.1		TOTAL:	90.2		TOTAL:	100.1

Table 6.41. (Continued).

STOCKPILE GRADE DISTRIBUTION YEAR 25				STOCKPILE GRADE DISTRIBUTION YEAR 27				STOCKPILE GRADE DISTRIBUTION YEAR 29				STOCKPILE GRADE DISTRIBUTION YEAR 31			
GRADE CATEGORIES %		AVRG	TONNAGE	GRADE CATEGORIES %		AVRG	TONNAGE	GRADE CATEGORIES %		AVRG	TONNAGE	GRADE CATEGORIES %		AVRG	TONNAGE
.230	.250	.240	14.9	.230	.250	.240	16.4	.230	.250	.240	18.3	.230	.250	.240	19.9
.250	.300	.275	23.8	.250	.293	.272	22.6	.250	.293	.272	25.2	.250	.293	.272	22.6
.300	.350	.325	22.9	.293	.300	.297	3.4	.293	.300	.297	3.4	.293	.300	.297	3.3
.350	.400	.375	18.5	.300	.350	.325	23.8	.300	.350	.325	23.8	.300	.350	.325	23.1
.400	.410	.405	3.6	.350	.400	.375	18.5	.350	.400	.375	18.5	.350	.400	.375	18.2
.410	.450	.430	8.5	.400	.410	.405	3.6	.400	.410	.405	3.6	.400	.410	.405	3.5
.450	.469	.459	4.4	.410	.450	.430	8.5	.410	.450	.430	8.5	.410	.450	.430	8.6
.469	.500	.484	4.8	.450	.469	.459	4.4	.450	.469	.459	4.4	.450	.469	.459	4.9
.500	.528	.514	1.6	.469	.500	.484	4.8	.469	.500	.484	4.8	.469	.500	.484	5.0
				.500	.528	.514	1.6	.500	.528	.514	1.6	.500	.528	.514	1.5
												.528	.550	.539	4.4
															4.8
		TOTAL:	103.0			TOTAL:	107.7			TOTAL:	112.0			TOTAL:	124.5

STOCKPILE GRADE DISTRIBUTION YEAR 26				STOCKPILE GRADE DISTRIBUTION YEAR 28				STOCKPILE GRADE DISTRIBUTION YEAR 30				STOCKPILE GRADE DISTRIBUTION YEAR 32			
GRADE CATEGORIES %		AVRG	TONNAGE	GRADE CATEGORIES %		AVRG	TONNAGE	GRADE CATEGORIES %		AVRG	TONNAGE	GRADE CATEGORIES %		AVRG	TONNAGE
.230	.250	.240	15.5	.230	.250	.240	17.3	.230	.250	.240	19.1	.230	.250	.240	19.5
.250	.293	.272	21.4	.250	.293	.272	23.9	.250	.293	.272	26.3	.250	.284	.267	21.3
.293	.300	.297	3.4	.293	.300	.297	3.4	.293	.300	.297	3.3	.284	.293	.289	5.5
.300	.350	.325	23.8	.300	.350	.325	23.8	.300	.350	.325	23.1	.293	.300	.297	3.2
.350	.400	.375	18.5	.350	.400	.375	18.5	.350	.400	.375	18.2	.300	.350	.325	22.7
.400	.410	.405	3.6	.400	.410	.405	3.6	.400	.410	.405	3.5	.350	.400	.375	17.9
.410	.450	.430	8.5	.410	.450	.430	8.5	.410	.450	.430	8.6	.400	.410	.405	3.5
.450	.469	.459	4.4	.450	.469	.459	4.4	.450	.469	.459	4.9	.410	.450	.430	8.5
.469	.500	.484	4.8	.469	.500	.484	4.8	.469	.500	.484	5.0	.450	.469	.459	4.8
.500	.528	.514	1.6	.500	.528	.514	1.6	.500	.528	.514	1.5	.469	.500	.484	4.9
												.500	.528	.514	1.4
												.500	.528	.514	4.3
												.528	.550	.539	4.7
												.550	.600	.575	1.6
		TOTAL:	105.5			TOTAL:	109.8			TOTAL:	113.6			TOTAL:	123.9

Table 6.41. (Continued).

STOCKPILE GRADE DISTRIBUTION YEAR 33

GRADE CATEGORIES %	AVRG	TONNAGE
.230 – .250	.240	19.5
.250 – .278	.264	17.6
.278 – .284	.281	3.8
.284 – .293	.289	5.5
.293 – .300	.297	3.2
.300 – .350	.325	19.7
.350 – .400	.375	15.5
.400 – .410	.405	3.0
.410 – .450	.430	7.3
.450 – .469	.459	4.1
.469 – .500	.484	4.3
.500 – .528	.514	1.3
.528 – .550	.539	3.7
.550 – .600	.575	1.4
	TOTAL:	113.9

STOCKPILE GRADE DISTRIBUTION YEAR 34

GRADE CATEGORIES %	AVRG	TONNAGE
.230 – .250	.240	19.5
.250 – .271	.261	13.3
.271 – .278	.275	4.2
.278 – .284	.281	3.8
.284 – .293	.289	5.5
.293 – .300	.297	3.2
.300 – .350	.325	16.6
.350 – .400	.375	13.0
.400 – .410	.405	2.5
.410 – .450	.430	6.2
.450 – .469	.459	3.5
.469 – .500	.484	3.6
.500 – .528	.514	1.1
.528 – .550	.539	3.4
.550 – .600	.575	1.2
	TOTAL:	103.9

STOCKPILE GRADE DISTRIBUTION YEAR 35

GRADE CATEGORIES %	AVRG	TONNAGE
.230 – .250	.240	19.5
.250 – .265	.258	9.4
.265 – .271	.268	4.0
.271 – .278	.275	4.2
.278 – .284	.281	3.8
.284 – .293	.289	4.7
.293 – .300	.297	2.7
.300 – .350	.325	14.0
.350 – .400	.375	11.0
.400 – .410	.405	2.1
.410 – .450	.430	5.2
.450 – .469	.459	2.9
.469 – .500	.484	3.0
.500 – .528	.514	.9
.500 – .528	.514	2.6
.528 – .550	.539	2.9
.550 – .600	.575	1.0
	TOTAL:	93.9

STOCKPILE GRADE DISTRIBUTION YEAR 36

GRADE CATEGORIES %	AVRG	TONNAGE
.230 – .250	.240	19.5
.250 – .265	.258	9.4
.265 – .271	.268	4.0
.271 – .278	.275	4.2
.278 – .284	.281	3.1
.284 – .293	.289	3.8
.293 – .300	.297	2.3
.300 – .350	.325	11.5
.350 – .400	.375	9.1
.400 – .410	.405	1.8
.410 – .450	.430	4.3
.450 – .469	.459	2.4
.469 – .500	.484	2.5
.500 – .528	.514	.7
.500 – .528	.514	2.2
.528 – .550	.539	2.4
.550 – .600	.575	.8
	TOTAL:	83.9

STOCKPILE GRADE DISTRIBUTION YEAR 37

GRADE CATEGORIES %	AVRG	TONNAGE
.230 – .250	.240	19.5
.250 – .261	.256	7.2
.261 – .265	.263	2.2
.265 – .271	.268	4.0
.271 – .278	.275	3.4
.278 – .284	.281	2.5
.284 – .293	.289	3.1
.293 – .300	.297	1.8
.300 – .350	.325	9.3
.350 – .400	.375	7.3
.400 – .410	.405	1.4
.410 – .450	.430	3.5
.450 – .469	.459	2.0
.469 – .500	.484	2.0
.500 – .528	.514	.6
.500 – .528	.514	1.8
.528 – .550	.539	1.9
.550 – .600	.575	.7
	TOTAL:	73.9

STOCKPILE GRADE DISTRIBUTION YEAR 38

GRADE CATEGORIES %	AVRG	TONNAGE
.230 – .245	.238	15.0
.245 – .250	.248	4.5
.250 – .261	.256	7.2
.261 – .265	.263	1.7
.265 – .271	.268	3.1
.271 – .278	.275	2.7
.278 – .284	.281	2.0
.284 – .293	.289	2.4
.293 – .300	.297	1.4
.300 – .350	.325	7.3
.350 – .400	.375	5.7
.400 – .410	.405	1.1
.410 – .450	.430	2.7
.450 – .469	.459	1.5
.469 – .500	.484	1.6
.500 – .528	.514	.5
.500 – .528	.514	1.4
.528 – .550	.539	1.5
.550 – .600	.575	.5
	TOTAL:	63.9

Table 6.41. (Continued).

STOCKPILE GRADE DISTRIBUTION YEAR 39

GRADE CATEGORIES %	AVRG	TONNAGE
.230 – .245	.238	15.0
.245 – .250	.248	4.5
.250 – .261	.256	5.5
.261 – .265	.263	1.3
.265 – .271	.268	2.4
.271 – .278	.275	2.1
.278 – .284	.281	1.5
.284 – .293	.289	1.9
.293 – .300	.297	1.1
.300 – .350	.325	5.7
.350 – .400	.375	4.4
.400 – .410	.405	.9
.410 – .450	.430	2.1
.450 – .469	.459	1.2
.469 – .500	.484	1.2
.500 – .528	.514	.4
.500 – .528	.514	1.1
.528 – .550	.539	1.2
.550 – .600	.575	.4
	TOTAL:	53.9

STOCKPILE GRADE DISTRIBUTION YEAR 40

GRADE CATEGORIES %	AVRG	TONNAGE
.230 – .245	.238	12.3
.245 – .250	.248	3.6
.250 – .261	.256	4.5
.261 – .265	.263	1.1
.265 – .271	.268	2.0
.271 – .278	.275	1.7
.278 – .284	.281	1.2
.284 – .293	.289	1.5
.293 – .300	.297	.9
.300 – .350	.325	4.6
.350 – .400	.375	3.6
.400 – .410	.405	.7
.410 – .450	.430	1.7
.450 – .469	.459	1.0
.469 – .500	.484	1.0
.500 – .528	.514	.3
.500 – .528	.514	.9
.528 – .550	.539	1.0
.550 – .600	.575	.3
	TOTAL:	43.9

STOCKPILE GRADE DISTRIBUTION YEAR 41

GRADE CATEGORIES %	AVRG	TONNAGE
.230 – .245	.238	9.5
.245 – .250	.248	2.8
.250 – .261	.256	3.5
.261 – .265	.263	.8
.265 – .271	.268	1.5
.271 – .278	.275	1.3
.278 – .284	.281	1.0
.284 – .293	.289	1.2
.293 – .300	.297	.7
.300 – .350	.325	3.6
.350 – .400	.375	2.8
.400 – .410	.405	.5
.410 – .450	.430	1.3
.450 – .469	.459	.8
.469 – .500	.484	.8
.500 – .528	.514	.2
.500 – .528	.514	.7
.528 – .550	.539	.7
.550 – .600	.575	.3
	TOTAL:	33.9

STOCKPILE GRADE DISTRIBUTION YEAR 42

GRADE CATEGORIES %	AVRG	TONNAGE
.230 – .245	.238	6.7
.245 – .250	.248	2.0
.250 – .261	.256	2.5
.261 – .265	.263	.6
.265 – .271	.268	1.1
.271 – .278	.275	.9
.278 – .284	.281	.7
.284 – .293	.289	.8
.293 – .300	.297	.5
.300 – .350	.325	2.5
.350 – .400	.375	2.0
.400 – .410	.405	.4
.410 – .450	.430	.9
.450 – .469	.459	.5
.469 – .500	.484	.5
.500 – .528	.514	.2
.500 – .528	.514	.5
.528 – .550	.539	.5
.550 – .600	.575	.2
	TOTAL:	24.0

STOCKPILE GRADE DISTRIBUTION YEAR 43

GRADE CATEGORIES %	AVRG	TONNAGE
.230 – .245	.238	3.9
.245 – .250	.248	1.2
.250 – .261	.256	1.4
.261 – .265	.263	.3
.265 – .271	.268	.6
.271 – .278	.275	.5
.278 – .284	.281	.4
.284 – .293	.289	.5
.293 – .300	.297	.3
.300 – .350	.325	1.5
.350 – .400	.375	1.1
.400 – .410	.405	.2
.410 – .450	.430	.5
.450 – .469	.459	.3
.469 – .500	.484	.3
.500 – .528	.514	.1
.500 – .528	.514	.3
.528 – .550	.539	.3
.550 – .600	.575	.1
.431	.438	2.9
	TOTAL:	16.8

STOCKPILE GRADE DISTRIBUTION YEAR 44

GRADE CATEGORIES %	AVRG	TONNAGE
.230 – .245	.238	1.6
.245 – .250	.248	.5
.250 – .261	.256	.6
.261 – .265	.263	.1
.265 – .271	.268	.3
.271 – .278	.275	.2
.278 – .284	.281	.2
.284 – .293	.289	.2
.293 – .300	.297	.1
.300 – .350	.325	.6
.350 – .400	.375	.5
.400 – .410	.405	.1
.410 – .450	.430	.2
.450 – .469	.459	.1
.469 – .500	.484	.1
.500 – .528	.514	.0
.500 – .528	.514	.1
.528 – .550	.539	.1
.550 – .600	.575	.0
.431	.438	2.9
	TOTAL:	8.5

Table 6.42. Final schedule for the alternative B stockpile (Shellman, 1989).

Year	Pit increment	Cutoff grade (% Cu)	Quantity mined (10^6 tons)	Quantity concentrated (10^6 tons)	Quantity refined (10^6 tons)	Profit (10^6 $)	Total profit (10^6 $)	Net present value (10^6 $)
1	1	0.50	17.8	10.0	89.8	26.0	515.5	150.41
2	1	0.50	17.8	10.0	89.8	26.0	489.5	146.96
3	1	0.50	17.8	10.0	89.8	26.0	463.5	142.99
4	1	0.50	17.8	10.0	89.8	26.0	437.5	138.42
5	1	0.50	17.8	10.0	89.8	26.0	411.5	133.18
6	1	0.50	11.2	6.3	56.6	16.4	385.5	127.15
6	2	0.53	7.4	3.7	31.8	8.5	369.1	127.15
7	2	0.53	20.0	10.0	85.9	23.0	360.6	121.34
8	2	0.53	20.0	10.0	85.9	23.0	337.6	116.57
9	2	0.53	20.0	10.0	85.9	23.0	314.6	116.57
10	2	0.53	20.0	10.0	85.9	23.0	291.6	104.78
11	2	0.53	12.8	6.4	55.0	14.7	268.6	97.53
11	3	0.47	7.2	3.6	27.4	6.5	253.9	97.53
12	3	0.47	20.0	10.0	76.1	18.0	247.4	90.96
13	3	0.47	20.0	10.0	76.1	18.0	229.4	86.56
14	3	0.47	20.0	10.0	76.1	18.0	211.4	81.49
15	3	0.47	20.0	10.0	76.1	18.0	193.0	75.67
16	3	0.45	12.8	6.6	49.6	11.8	175.4	68.97
16	4	0.45	6.0	3.4	22.5	4.5	163.6	68.97
17	4	0.41	20.0	10.0	66.5	13.2	159.1	63.06
18	4	0.41	20.0	10.0	66.5	13.2	145.9	59.29
19	4	0.41	19.6	10.0	65.9	13.1	132.7	54.95
20	4	0.39	19.2	10.0	65.2	13.0	119.6	50.07
21	4	0.37	14.5	7.8	50.1	10.0	106.6	44.57
21	5	0.35	4.4	2.2	12.5	1.9	69.6	44.57
22	5	0.35	20.0	10.0	57.2	8.6	94.7	39.37
23	5	0.34	19.7	10.0	56.8	8.6	86.1	36.66
24	5	0.33	19.2	10.0	56.2	8.5	77.5	33.59
25	5	0.32	18.7	10.0	55.5	8.4	69.0	30.14
26	5	0.31	17.8	10.0	54.1	8.2	60.6	26.27
27	5	0.3	0.3	0.2	1.00	0.2	52.4	22.04
27	6	0.27	19.6	9.8	46.7	3.7	52.2	22.04
28	***	0.26	10.0	10.0	32.4	4.0	48.5	21.46
29	***	0.26	10.0	10.0	32.4	4.0	44.5	20.72
30	***	0.26	10.0	10.0	32.4	4.0	40.5	19.87
31	***	0.26	10.0	10.0	32.4	4.0	40.5	19.87
32	***	0.26	10.0	10.0	32.4	4.0	32.5	17.78
33	***	0.26	10.0	10.0	32.4	4.0	28.5	16.5
34	***	0.26	10.0	10.0	32.4	4.0	24.5	15.02
35	***	0.26	10.0	10.0	32.4	4.0	20.5	13.31
36	6	0.26	18.2	10.0	45.6	3.7	16.5	11.36
37	6	0.25	17.8	10.0	45.6	3.6	12.8	9.37
38	6	0.25	17.5	10.0	44.7	3.6	9.2	7.14
39	6	0.24	16.4	10.0	43.3	3.4	5.6	4.62
40	6	0.23	10.5	6.4	27.7	2.2	2.2	1.89

*** Mineral from stockpile.

Table 6.43. Final schedule for the alternative B stockpile (Shellman, 1989).

Year	Pit increment	Cutoff grade (% Cu)	Quantity mined (10^6 tons)	Quantity concentrated (10^6 tons)	Quantity refined (10^6 tons)	Profit (10^6 \$)	Total profit (10^6 \$)	Net present value (10^6 \$)
1	1	0.5	17.8	10	89.8	26	529.6	150.46
2	1	0.5	17.8	10	89.8	26	503.6	147.01
3	1	0.5	17.8	10	89.8	26	477.6	143.06
4	1	0.5	17.8	10	89.8	26	451.6	138.5
5	1	0.5	17.8	10	89.8	26	425.6	133.27
6	1	0.5	11.2	6.3	56.6	16.4	399.6	127.25
6	2	0.53	7.4	3.7	31.8	8.5	383.2	127.25
7	2	0.53	20	10	85.9	23	374.7	121.45
8	2	0.53	20	10	85.9	23	351.7	116.7
9	2	0.53	20	10	85.9	23	328.7	111.24
10	2	0.53	20	10	85.9	23	305.7	104.95
11	2	0.53	12.8	6.4	55	14.7	282.7	97.73
11	3	0.47	7.2	3.6	27.4	6.5	268	97.73
12	3	0.47	20	10	76.1	18	261.5	91.19
13	3	0.47	20	10	76.1	18	243.5	86.82
14	3	0.47	20	10	76.1	18	225.5	81.79
15	3	0.47	20	10	76.1	18	207.5	76.01
16	3	0.45	12.8	6.7	49.8	11.8	189.5	69.36
16	4	0.41	6.7	3.3	22.2	4.4	177.7	69.36
17	4	0.41	20	10	66.5	13.2	173.2	63.52
18	4	0.41	20	10	66.5	13.2	160.1	59.82
19	4	0.4	19.5	10	65.6	13.1	146.9	55.57
20	4	0.38	18.8	10	64.6	12.9	133.8	50.82
21	4	0.36	15.1	8.5	53.4	10.7	120.9	45.55
21	5	0.35	3.00	1.5	8.6	1.3	110.2	45.55
22	5	0.34	19.5	10	56.5	8.8	108.9	40.43
23	5	0.33	19	10	55.9	8.4	100.4	37.95
24	5	0.32	18.5	10	55.1	8.3	92	35.2
25	5	0.3	17.8	10	54.1	8.2	83.7	32.14
26	5	0.29	17.3	10	53.4	8.0	75.5	28.79
27	5	0.27	5.0	3.0	15.8	2.4	67.5	25.07
27	6	0.27	31.1	7	32.4	2.6	65.1	25.07
28	6	0.26	18.3	10	45.8	3.7	62.5	23.85
29	6	0.26	17.8	10	45	3.6	58.8	23.72
30	6	0.26	17.7	10	44.9	3.6	58.8	23.72
31	6	0.24	17.1	10.0	44.2	3.5	51.6	23.57
32	6	0.23	15.9	9.7	42.0	3.3	48.1	23.57
32	***	0.30	0.3	0.3	0.9	0.1	44.8	23.57
33	***	0.30	10.0	10.0	36.2	5.8	44.7	23.69
34	***	0.30	10.0	10.0	35.4	5.4	38.9	21.41
35	***	0.29	10.0	10.0	34.9	5.2	33.5	19.18
36	***	0.28	10.0	10.0	33.9	4.7	28.3	16.87
37	***	0.27	10.0	10.0	34.4	4.9	23.6	14.70
38	***	0.27	10.0	10.0	33.3	4.4	18.7	11.97
39	***	0.26	10.0	10.0	31.7	3.6	14.3	9.35
40	***	0.25	10.0	10.0	29.8	2.6	10.7	7.14
41	***	0.24	10.0	10.0	27.6	1.6	8.1	5.58
42	***	0.23	10.0	10.0	27.9	1.7	6.5	4.86
43	***	0.23	10.0	10.0	28.5	2.0	4.8	3.87
44	***	0.23	9.7	9.7	29.3	2.8	2.8	2.42

*** Mineral from stockpile.

Production planning 577

Figure 6.28. Simple sequencing example.

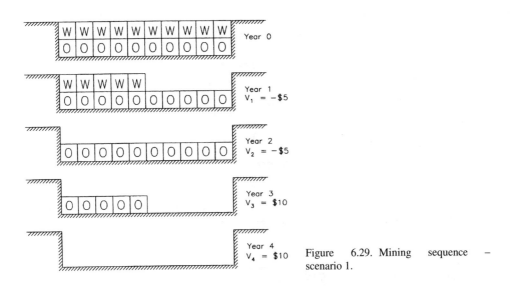

Figure 6.29. Mining sequence – scenario 1.

Scenario 1. Removal of waste followed by ore mining
For operational simplicity it would be best if all of the waste could be stripped (removed) first followed later by ore mining. This is shown in Figure 6.29.

The net present value for this sequence assuming an interest rate of 10% is

$$\text{NPV} = \frac{-\$5}{(1.10)^1} + \frac{-\$5}{(1.10)^2} + \frac{\$10}{(1.10)^3} + \frac{\$10}{(1.10)^4}$$
$$= -\$4.55 - \$4.13 + \$7.51 + \$6.83 = \$5.66$$

Scenario 2. One year of pre-stripping followed by both ore (3 blocks/year) and waste (2 blocks/year) mining
This alternative would require mining 5 blocks of waste in year 1. In years 2 and 3 three blocks of ore would be mined for every two blocks of waste. The final year would have 1 block of waste and 4 blocks of ore. The sequencing is shown in Figure 6.30.

The net present value is now

$$\text{NPV} = \frac{-\$5}{(1.10)^1} + \frac{\$4}{(1.10)^2} + \frac{\$4}{(1.10)^3} + \frac{\$7}{(1.10)^4}$$
$$= -\$4.54 + \$3.31 + \$3.01 + \$4.78 = \$6.56$$

Scenario 3. Mining of waste is maintained one block ahead of ore
Comparing scenarios 1 and 2, there was an improvement in the net present value when the time lag between stripping and mining was shortened. In this third scenario, the stripping

578 Open pit mine planning and design: Fundamentals

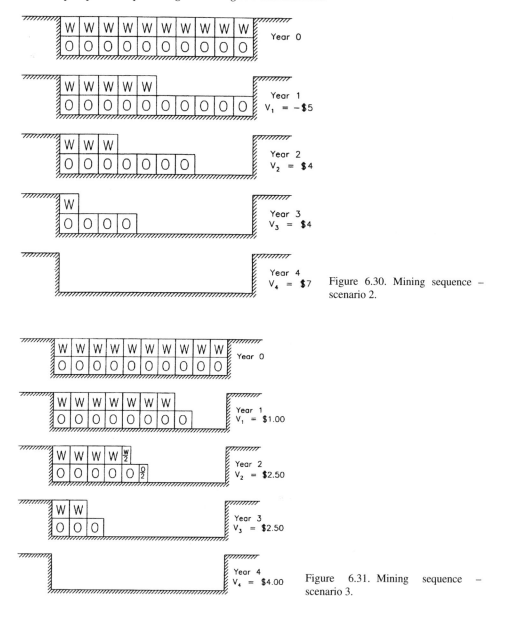

Figure 6.30. Mining sequence – scenario 2.

Figure 6.31. Mining sequence – scenario 3.

will be kept only one block ahead of ore mining, to make the time lag even shorter. This is shown in Figure 6.31.

The net present value is

$$\text{NPV} = \frac{\$1}{(1.10)^1} + \frac{\$2.50}{(1.10)^2} + \frac{\$2.50}{(1.10)^3} + \frac{\$4}{(1.10)^4}$$
$$= \$0.91 + \$2.07 + \$1.88 + \$2.73 = \$7.59$$

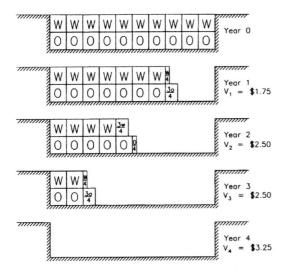
Figure 6.32. Mining sequence –scenario 4.

Scenario 4. Stripping is maintained one half block ahead of ore mining
A major improvement in the NPV was observed between scenarios 2 and 3. To explore this further, consider the situation when the stripping lead is cut to one-half block (Figure 6.32).
The NPV is

$$NPV = \frac{\$1.75}{(1.10)^1} + \frac{\$2.50}{(1.10)^2} + \frac{\$2.50}{(1.10)^3} + \frac{\$3.25}{(1.10)^4}$$
$$= \$1.59 + \$2.07 + \$1.88 + \$2.22 = \$7.76$$

This would appear to be the most favorable alternative of the four scenarios. However, suppose that in reducing the stripping lead it is found that the operating costs for both ore and waste increase by $0.05/block, perhaps through lack of sufficient working space or through neglect of drilling precision because of time pressures. Hence the cost for waste removal increases to $1.05/block and the net ore revenue drops to $1.95/block.
The actual NPV is

$$NPV = \frac{\$1.50}{(1.10)^1} + \frac{\$2.25}{(1.10)^2} + \frac{\$2.25}{(1.10)^3} + \frac{\$3.00}{(1.10)^4}$$
$$= \$1.36 + \$1.86 + \$1.69 + \$2.05 + \$6.96$$

In this case scenario 3 remains the most attractive.

Scenario 5. Mining rate doubled
It has been suggested that with the purchase of more equipment, the mining rate could be increased to 10 blocks per year. There would be an increase in the equipment ownership costs to be charged against both ore and waste however. The resulting values are:

Waste cost = $1.10/block

Ore revenue = $1.90/block

Stripping will be kept one block ahead of ore mining as in scenario 3.

580 Open pit mine planning and design: Fundamentals

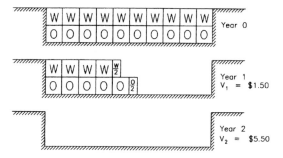

Figure 6.33. Mining sequence – scenario 5.

The scheduling is shown in Figure 6.33.
The NPV is

$$\text{NPV} = \frac{\$2.50}{(1.10)^1} + \frac{\$5.50}{(1.10)^2} = \$2.27 + \$4.55 = \$6.82$$

As can be seen scenario 3 remains the most favorable.
If there had been no additional cost then

$$\text{NPV} = \frac{\$3.50}{(1.10)^1} + \frac{\$6.50}{(1.10)^2} = \$3.18 + \$5.37 = \$8.55$$

and scenario 5 would have been the most favorable.

In summary, this very simple example has demonstrated some important aspects of production scheduling. The NPV is dependent upon

1. The time interval between stripping and ore mining. It is highest when the lead time is short. With added costs associated with shortening the lead time, however, there may or may not be an improvement in NPV.

2. The production rate. For the same unit cost, the highest NPV is achieved with the highest production rate. With added costs with increasing production rate, there may or may not be an improvement in NPV.

6.9.2 Phase scheduling

Several mining areas or mine phases, typically three or more, are active at any given time during the life of a mine. Of these, one or two would be in the process of being stripped, another being mined for ore and the last nearing exhaustion. This section will describe a procedure which can be used to help sequence the phases so that the desired ore stream is produced. The procedure and the illustrative example have been adapted from Mathieson (1982).

The hypothetical deposit is shown in section in Figure 6.34. The orebody, located in rock type 2, is overlain by waste (rock type 1). The planning scheduling will be described in a step-by-step manner.

1. The phases are first designed. The slope angles used are selected based upon initial geotechnical investigations. For this example it is assumed that the orebody is of uniform grade. Hence the phases A through F (shown in Figure 6.35) have been designed, subject to access constraints, to progressively mine the 'next best' ore in terms of annual stripping ratio.

Production planning 581

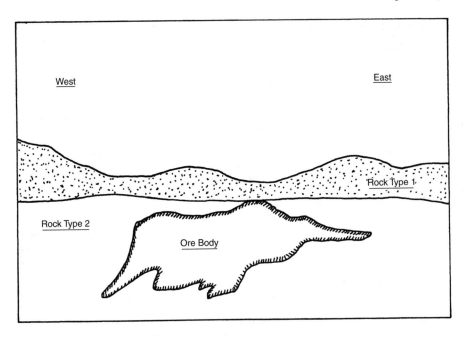

Figure 6.34. Hypothetical deposit for the sequencing study (Mathieson, 1982).

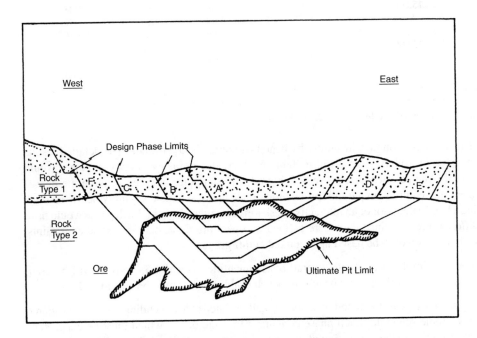

Figure 6.35. Development scheme for the hypothetical deposit (Mathieson, 1982).

582 Open pit mine planning and design: Fundamentals

Table 6.44. Tonnage-grade inventory by phase and bench (Mathieson, 1982).

Bench	Thousands of tons					
	'A' phase		'B' phase		'C' phase	
	Waste	Ore	Waste	Ore	Waste	Ore
5100	1500					
5050	3200					
5000	5000		200		400	
4950	3800		1800		1500	
4900	1500		2000		1800	
4850	400	1000	1500		2200	
4800	300	900	400	900	1600	
4750	200	800	300	900	300	1000
4700			200	700	500	2000
4650			100	600	800	2200
4600					300	1700
4550					100	700
	15,900	2700	6500	3100	9500	7600

Table 6.45. Summary of phase quantities (Mathieson, 1982).

Phase	Thousands of tons			Ore life* (yrs)	Cumulative ore life* (yrs)
	Waste above first ore bench	Waste on ore benches	Ore		
A	15,000	900	2700	1.08	1.08
B	5500	1000	3100	1.24	2.32
C	7500	2000	7600	3.04	5.36
D	12,800	3800	12,500	5.00	10.36
E	18,200	4900	15,100	6.04	16.40
F	22,000	4500	13,000	5.20	21.60
Total	81,000	17,100	54,000	21.60	

*Assuming an annual milling rate of 2,500,000 tons.

2. The ore-waste tonnage inventory by bench and phase is determined. The detailed results from the first 3 phases are given in Table 6.44. At this point the mining engineer would 'mine' the ore on each successive bench gradually stepping, in order, through the phases to meet the required annual mill production. Such an ore schedule would typically be done using a hand calculator or an interactive desk top computer program. Each phase would have a different ore life since they were defined on operating rather than schedule constraints. The first trials would be based upon a fixed cutoff grade.

3. For this simple example the overall phase quantities given in Table 6.45 will be used. Normally those broken down bench-by-bench (Table 6.44) would be considered.

4. These phases will be mined in sequence and no inter-phase blending will be considered in this simple schedule. Each phase contains a tonnage of ore which must be exposed or developed prior to the exhaustion of ore from the previous phase. A 2.5 million ton per year milling rate is assumed. This is divided into the ore tons available in each phase to determine

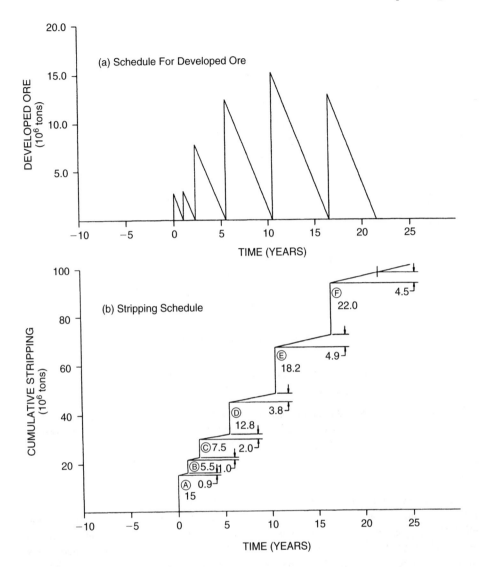

Figure 6.36. Developed ore and stripping schedules.

phase life. A saw-tooth plot of available ore versus time is made such as is shown in Figure 6.36a. This illustrates the availability and depletion of ore from the various phases.

5. The planner is now able to define the points in time at which the waste stripping must be completed for any given phase in order to sustain the ore supply. For phase A ore production to commence, the 15 million tons of waste lying above the first bench must have first been removed. During the mining of Phase A there is an additional 900,000 tons of internal waste assumed to be evenly distributed and 'locked-up' with the ore. In order for the phase B ore to be available, 21,400,000 tons of waste must be removed. The cumulative waste tons versus time plot is shown in Figure 6.36b. The vertical steps in the plot correspond to the respective

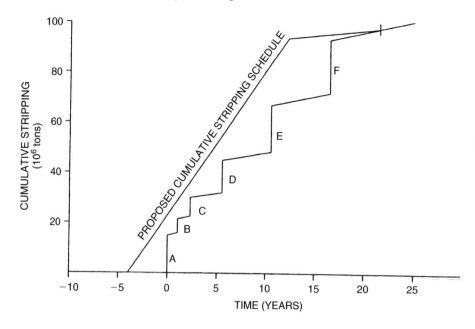

Figure 6.37. Proposed cumulative stripping schedule for the example orebody.

overburden stripping quantities above the first ore bench of each phase. The latter segments represent the progressive mining of internal waste.

6. The next step is to arrive at a 'smoothed' stripping schedule which exceeds the minimum. A possible schedule is shown as the straight line superimposed on Figure 6.37. It consists of a 4 year pre-production period totalling 20 million tons followed by a constant stripping rate of 5 million tons per year through year 15. Beyond this only the internal waste of phase F remains.

The detailed preproduction stripping schedule is shown in Figure 6.38. It consists of 2.5 million tons during the first year when the crews are being trained and equipment is being delivered. During years 2 and 3 the stripping rate is 5 million tons/year. Finally in year 4 the rate is increased to the total material rate (ore plus waste) which will be sustained nearly throughout the remaining life. It can be seen from the figure that the required stripping is completed prior to the required time by various amounts.

Phase	Early completion (months)
A	8
B	9
C	4
D	5
E	9
F	16

7. The initial curve of developed ore versus time is now adjusted to reflect the early completion of the stripping. This construction is shown in Figure 6.39. The cumulative stripping and developed ore curves are used. For phase B one moves horizontally at the required

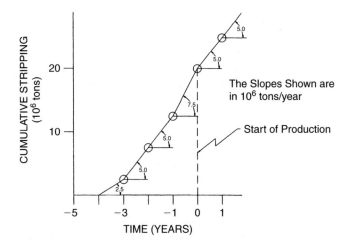

Figure 6.38. Preproduction stripping schedule.

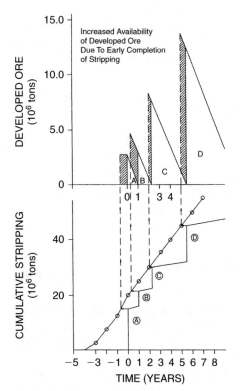

Figure 6.39. Construction showing modification of ore availability.

cumulative stripping level to the actual stripping curve. Then one proceeds vertically to the developed ore curve. As can be seen there are approximately 2 million tons of ore from phase A remaining. Due to the early development of phase B, an additional 3.1 million tons of ore become available. The construction is simply to extend the slanted portion of

586 *Open pit mine planning and design: Fundamentals*

the phase B ore reserve upward until it meets the vertical line from the stripping graph. This process is repeated for phases C through F. For phase A, the ore in the phase simply becomes available earlier. The cross hatched areas indicate contingency ore available in the case that the stripping schedule falls behind. The modified developed ore curve represents the predicted inventory balance with time. A further acceleration of stripping is sometimes done in the trial scheduling process to guard against possible surprises in mineable reserves. This can help to avoid:
- an unexpected crash stripping program,
- a forced reduction in mill feed,
- a temporary lowering of the cutoff grade to sustain planned concentrate production.

8. With this thorough understanding of the orebody and its development options, the pit planner presents his/her findings to management.

9. Final mine plan period maps are drawn up to test the viability of the plan. Some refinements in the ramping and phasing strategy, etc., may be needed but major changes are unlikely.

10. To this point, a series of logical pit development phases have been defined based on the 'next best' profitable ore and a fixed cutoff. The plan can now be fine tuned. Alternative ore and waste schedules based on variable production rate and cutoff grade strategies can be developed using the computed tonnage – grade inventories within each successive phase. Such schedules can then be compared economically through standard internal ROR analysis. A visual comparison can be achieved by plotting cumulative operating cash flow with time.

11. Once a 'final' production schedule has been decided upon, the planner generates a series of period end plans. These might be for example:
  (a) end of preproduction,
  (b) years 1–5 in yearly increments,
  (c) years 10, 15, etc.

These plans would be based on both the phase designs and the paper schedules. They would constitute a vital test on the mineability of the proposed plan. Shovel and drill deployment and any internal temporary ramping would also be considered in detail.

6.9.3 *Block sequencing using set dynamic programming*

Introduction

In 1974, Roman (1974) described an algorithm for determining the optimum mining sequence and pit limits patterned after one originally presented by Lerchs & Grossmann (1965). The process will be demonstrated through the use of a 2-dimensional example. Figure 6.40 is a schematic representation of a slice through a block model.

An index number representing the column and row position for each block is assigned. The first step in the process is to convert the grade block model into an economic block model. To assign the appropriate costs and revenues a decision must be made at this point regarding the destination of each block. Three possibilities might be:
- mill,
- leach dump, and
- waste dump.

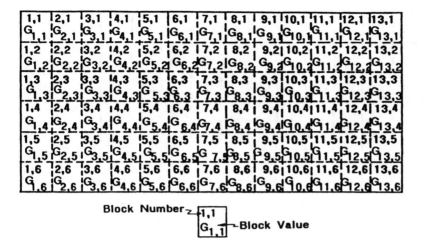

Figure 6.40. Schematic of the ore deposit showing block numbers and grades (Roman, 1974).

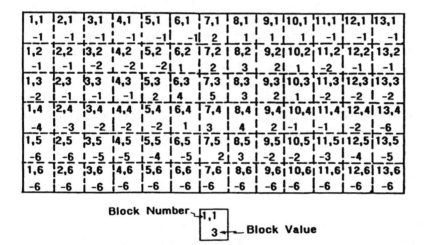

Figure 6.41. Schematic of the ore deposit showing block numbers and block values (Roman, 1974).

The net block value is determined by subtracting the mining and processing costs from the revenues. The mining costs are for the block alone and do not include stripping costs. Figure 6.41 shows the resulting economic block model. At this point a constraint relating to the final pit slope is introduced.

Constraint 1: The pit wall slope may not exceed 1:1 at any point.

Applying the floating cone procedure introduced in the previous chapter one would arrive at the final pit shown in Figure 6.48. This same result will be achieved using the technique described in this section. The problem is to determine the sequence in which the blocks should be mined so that the net present value for the section is a maximum.

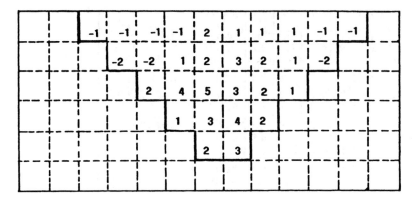

Figure 6.42. Schematic of the ore deposit with the final pit limits as determined using the floating cone superimposed.

The optimum sequence. To begin the sequence optimizing process, the economic block model is scanned to determine the maximum outline which the future pit could assume. Obviously all of the positive blocks on the section must be included and the pit limit slopes obeyed. The objective is to identify the location of the last block which might be mined. In this example, the last block has been selected as the bottom vertice of the inverted triangle containing all blocks of positive value. This triangle shown in Figure 6.43 has been constructed in accordance with constraint 1. An alternative procedure would be to select a hypothetical block (a block that does not exist) lying on or below the lowest level of positive blocks. To simplify the discussion, the triangle approach has been used. There are 36 blocks included within the triangle. If each block corresponds to a unit time period (of unspecified length), 36 time periods are required to mine all of the blocks. Block (7,6) as seen in Figure 6.44 is the last one to be mined. It is mined in period 36. In order to mine block (7,6) one must first mine blocks (6,5), (7,5) and (8,5). At this point a second constraint, one regarding sequencing, will be introduced.

Constraint 2: *Each mining level may be entered at only one point.*
With this constraint in place there are only two sequencing options for the mining of blocks in time periods 35 and 36:
Option 1. Mine block (6,5) followed by block (7,6) (Figure 6.45a).
Option 2. Mine block (8,5) followed by block (7,6) (Figure 6.45b).
The third option (shown in Figure 6.45c) of mining block (7,5) in period 35 means that both blocks (8,5) and (6,5) had been mined earlier. This requires that two separate entries be made on level 5 thus violating constraint 2. Hence this is not an option.
 In time period 34 there are several choices for the block to be mined depending upon the block mined in period 35. If the last two blocks mined are (6,5) and (7,6) then the possible sequences for the last 3 periods are

(1) $(7,5) \rightarrow (6,5) \rightarrow (7,6)$
(2) $(8,5) \rightarrow (6,5) \rightarrow (7,6)$
(3) $(5,4) \rightarrow (6,5) \rightarrow (7,6)$

Production planning 589

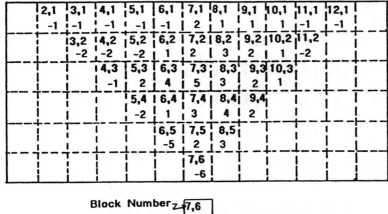

Figure 6.43. Schematic of the ore deposit showing the triangle containing the maximum pit superimposed (Roman, 1974).

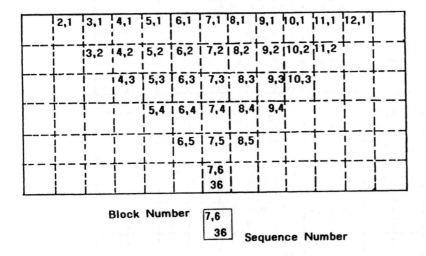

Figure 6.44. Blocks which must be removed prior to mining block 36.

On the other hand if blocks (8,5) and (7,6) are mined last then the possible sequences are:

(4) $(6,5) \rightarrow (8,5) \rightarrow (7,6)$

(5) $(7,5) \rightarrow (8,5) \rightarrow (7,6)$

(6) $(9,4) \rightarrow (8,5) \rightarrow (7,6)$

These 6 possibilities are shown in Figure 6.46.

Sequences 2 and 4 however involve mining the same 3 blocks in just a different order. An economic evaluation is performed to determine the most attractive of the two alternatives.

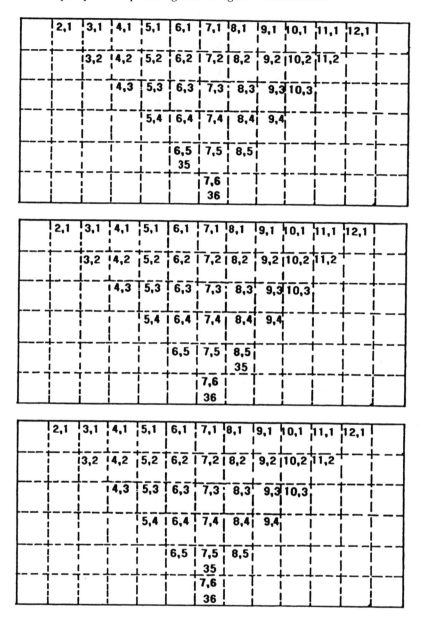

Figure 6.45. Possible sequences for mining blocks 35 and 36.

The least attractive is dropped from further consideration. Choosing an interest rate of 10% and discounting to the beginning of time period 34 one finds:

Sequence 2: $(8, 5) \rightarrow (6, 5) \rightarrow (7, 6)$

$$\text{NPV}_2 = \frac{\$3}{(1.1)^1} - \frac{\$5}{(1.1)^2} - \frac{\$6}{(1.1)^3} = -\$5.91$$

1.

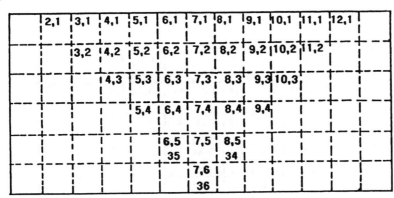

2.

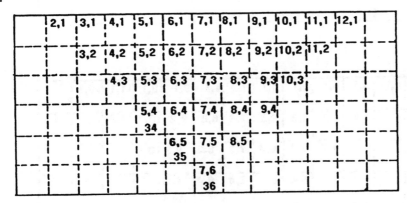

3.

Figure 6.46. Possible sequences for mining blocks 34, 35, and 36.

Sequence 4: $(6,5) \to (8,5) \to (7,6)$

$$NPV_2 = \frac{-\$5}{(1.1)^1} + \frac{\$3}{(1.1)^2} - \frac{\$6}{(1.1)^3} = -\$6.57$$

4.

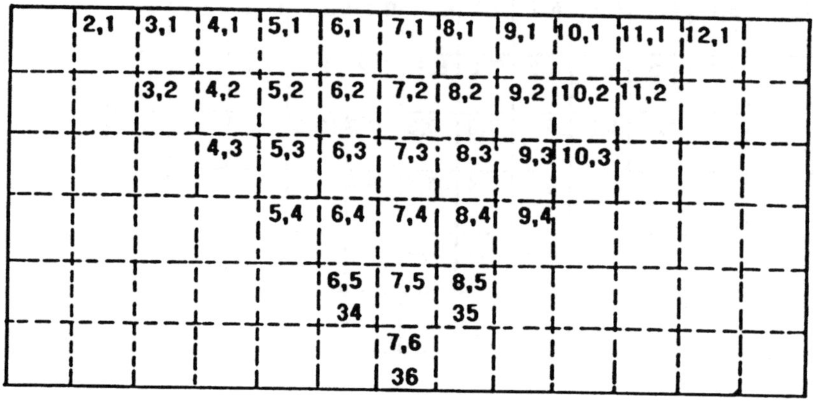

5.

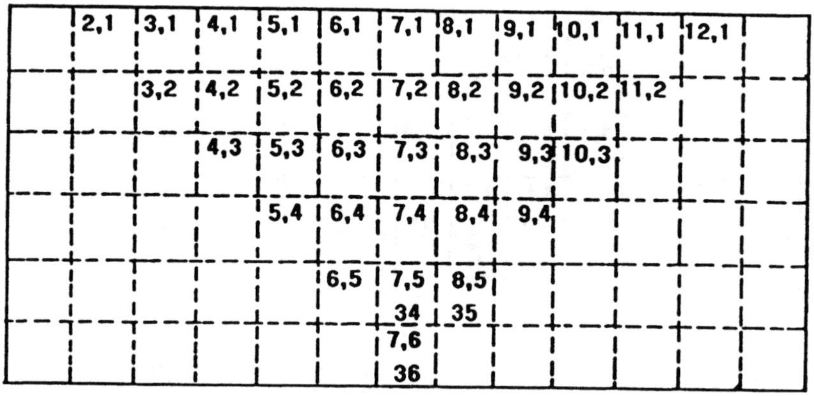

6.

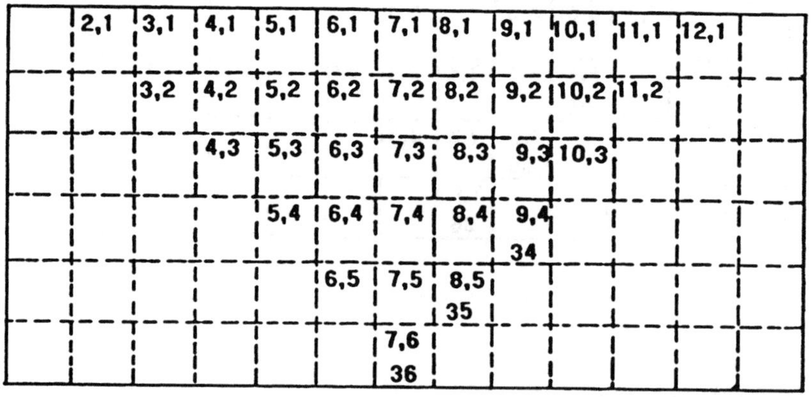

Figure 6.46. (Continued).

Sequence 4 is the least attractive of the two and is dropped. The five block combinations for periods 34 through 36 which must be included when sequencing the remaining 33 periods are:

Sequence	Mining order
1	(7,5) → (6,5) → (7,6)
2	(8,5) → (6,5) → (7,6)
3	(5,4) → (6,5) → (7,6)
4	(7,5) → (8,5) → (7,6)
5	(9,4) → (8,5) → (7,6)

The remaining choices for sequencing the final four blocks after eliminating duplicate combinations of blocks by the present value analysis are:

Sequence	Mining order
1	(8,5) → (7,5) → (6,5) → (7,6)
2	(9,4) → (8,5) → (6,5) → (7,6)
3	(7,5) → (5,4) → (6,5) → (7,6)
4	(8,5) → (5,4) → (6,5) → (7,6)
5	(4,3) → (5,4) → (6,5) → (7,6)
6	(9,4) → (7,5) → (8,5) → (7,6)
7	(10,3) → (9,4) → (8,5) → (7,6)

This process is continued until all 36 blocks have been included. In this final stage the various sequences will just be permutations of the same combination. Consequently the optimum sequence can be determined through a present value calculation. Figure 6.47 shows the section with the 36 blocks numbered in the order that they are to be removed.

Determination of the optimum pit
Once the optimum sequence has been found, the pit outline can be developed. The procedure for determining which of the 36 blocks in the optimum sequence are actually to be mined is as follows:

1) Identify the last block in the optimum sequence with a positive value. Drop all blocks after this one.

2) Examine the remaining sequence of blocks and identify the latest negative block scheduled for mining. If there are a number of negative blocks in a row select the earliest in the row. Determine the present worth for the sequence extending from identified negative block to the end.

3) If the present value is negative, drop all these blocks from the optimum sequence and repeat step 2. If the subsequence has a positive present value, replace the subsequence by an equivalent block value at the end of the sequence.

4) Repeat steps 2 and 3 until the first mined block is included in the subsequence. The final present value is that of the optimum pit.

Applying these rules to the example problem, it is seen that blocks 31 through 36, which are all negative, are dropped by inspection. Continuing along the sequence it is seen that two adjacent blocks (13,1) and (4,2) corresponding to mining periods 26 and 27 respectively, are negative. The net present value for this subsequence (blocks 26 to 30) discounted to the

594 *Open pit mine planning and design: Fundamentals*

a. **Block Number and Sequence**

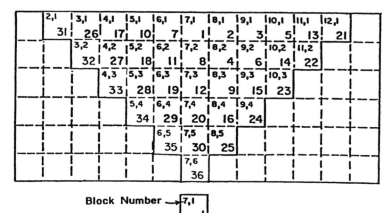

b. **Block Value and Sequence**

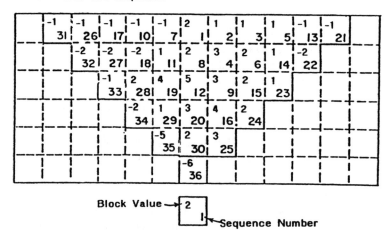

Figure 6.47. Schematic of the deposit showing the optimum block mining sequence (Roman, 1974).

beginning of period 26 is

$$\text{NPV} = \frac{-\$1}{(1.1)^1} - \frac{\$2}{(1.1)^2} + \frac{\$2}{(1.1)^3} + \frac{\$1}{(1.1)^4} + \frac{\$2}{(1.1)^5}$$

$$= \$0.87$$

Since it is positive, this subsequence, referred to as subsequence SS1 is retained. The next switch between negative and positive blocks occurs after block 21. The net present value for the subsequence SS2 through to the beginning of period 21 is

$$\text{NPV} = \frac{-\$1}{(1.1)^1} - \frac{\$2}{(1.1)^2} + \frac{\$1}{(1.1)^3} + \frac{\$2}{(1.1)^4} + \frac{\$3}{(1.1)^5} + \frac{\$0.87}{(1.1)^6}$$

$$= \$1.91$$

a. Block Number and Sequence

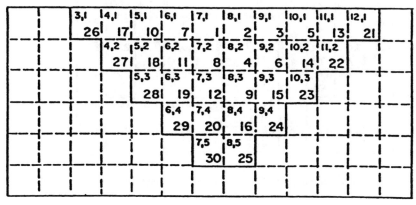

b. Block Value and Sequence

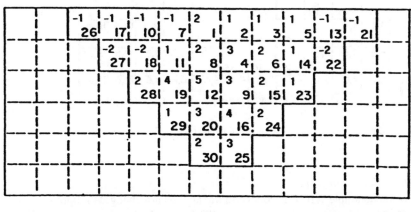

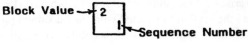

Figure 6.48. The final pit outline and optimum mining sequence (Roman, 1974).

It is positive and hence retained. This process of calculating net present value is continued until the first block to be mined is included. The final pit is shown in Figure 6.48. The overall NPV for the pit on this section is

NPV (overall) = $12.60

Simply summing the values of the blocks included within the outlined yield

Sum = $34

This would be the result if the pit could be mined instantaneously at time zero.

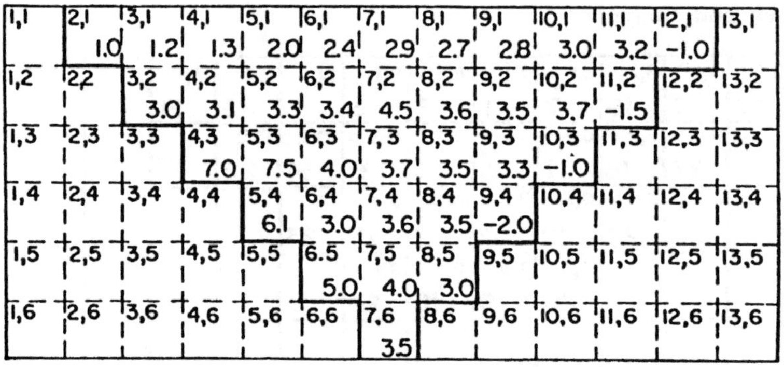

596 *Open pit mine planning and design: Fundamentals*

[Figure 6.49 grid with block values and maximum pit limits]

Figure 6.49. Hypothetical deposit showing block values and maximum pit limits (Roman, 1974).

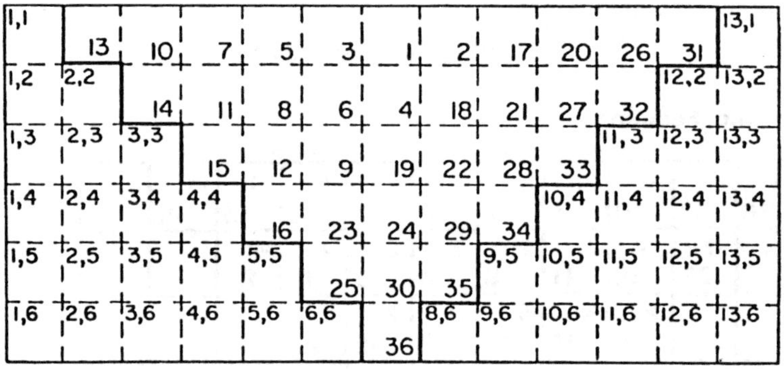

[Figure 6.50 grid with mining sequence numbers]

Figure 6.50. Mining sequence and pit limit for an interest rate of 5% (Roman, 1974).

Time value of money influence

In the previous example, a 10% discount rate was applied in arriving at the optimum sequence and pit. The question arises as to the influence this rate has both on the sequence and the final pit. Figure 6.49 shows the section to be used with the maximum pit superimposed. Using the techniques and constraints previously described the pits corresponding to the use of 5%, 10%, 20%, and 50% discount rates are shown in Figures 6.50 through 6.53 respectively. As can be seen, a definite effect is observed. At first glance it would appear that the reduction in pit size between discount rates of 5% and 10% is incorrect since the strip of blocks

$$(12, 1) \rightarrow (11, 2) \rightarrow (10, 3) \rightarrow (9, 4) \rightarrow (8, 5) \rightarrow (7, 6)$$

when summed has a positive value (+1). However when taking sequencing into account the discounted value of this strip (at a rate of 10%) is

$$\text{NPV} = \frac{\$3.5}{(1.1)^{36}} - \frac{\$3.0}{(1.1)^{35}} - \frac{\$2.0}{(1.1)^{34}} - \frac{\$1}{(1.1)^{33}} - \frac{\$1.5}{(1.1)^{32}} - \frac{\$1.0}{(1.1)^{31}}$$
$$= -\$0.025$$

For higher discount rates this sub-sequence is even more negative.

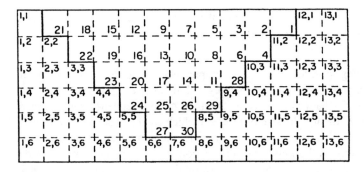

Figure 6.51. Mining sequence and pit limit for an interest rate of 10% (Roman, 1974).

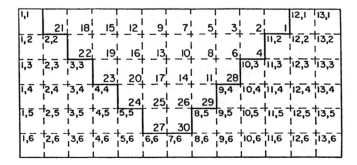

Figure 6.52. Mining sequence and pit limit for an interest rate of 20% (Roman, 1974).

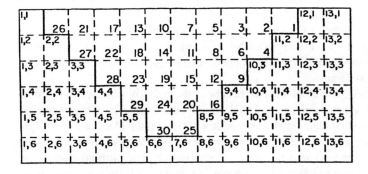

Figure 6.53. Mining sequence and pit limit for an interest rate of 50% (Roman, 1974).

The rate to be used depends upon a number of factors, the principal one being that of company policy. In the early stages of a property, a high rate of return is generally desired in order to repay the investment as quickly as possible. Later when the investment has been repaid, another value may apply. Thus the procedure outlined is not intended to be used just once in the life of a property and the derived sequence to be followed without change.

598 *Open pit mine planning and design: Fundamentals*

Rather with changing conditions, it can be rerun on the blocks remaining at any given time and a new optimum sequence developed.

Summary

The described sequencing procedure is quite simple. Although the examples were 2-dimensional in nature, programs containing the algorithm have been written for the true 3-dimensional pit problem. Changes in constraints on the mining sequence can be incorporated relatively easily. In theory, the size of problem which can be handled is unlimited. However as was seen in the 2-dimensional hand example, the number of sequences to be considered increases rapidly as the number of blocks contained in the deposit increases. A relatively large and fast computer is required to perform the calculations for even a small deposit. A deposit for which ore characteristics fluctuate radically and continually along any direction will require more computer time and storage to evaluate. On the other hand, adding constraints on the mining sequence tends to simplify the problem. If in the problem of sequencing the 36 blocks constraint 2 is dropped, the required computer time increases by a factor of 30 and the storage by 20 times.

Rather than attempting to provide an overall optimum sequence for an entire deposit consisting of many thousands of blocks, the mine plan may be first broken down into a series of phases. The sequencing procedure can then be applied to each of the phases in turn. In this way a series of sub-optimizations is realized. Being sub-optimizations, due care should be taken in putting these phases together, but it is possible in practice to build up a good mine plan in this way.

6.9.4 *Some scheduling examples*

Dagdelen (1985) has applied the Lagrangian parametrization technique for optimum open pit mine production scheduling. A full discussion of the procedure is beyond the scope of this textbook. An ultimate pit limit contour is first obtained using Lerchs-Grossmann's 3-D algorithm. A series of constraints such as: (a) mining capacity, (b) milling capacity, (c) mill feed grade, (d) geometric (i.e. slope limitation), are then introduced.

The program searches through the set of candidate blocks and selects those yielding a maximum net present value for the deposit while meeting the constraints. The results of applying this procedure to a small high grade copper deposit are included here to demonstrate the effect of changing constraints.

The grade block model (25 × 20 blocks) for benches 1 through 6 of this deposit is shown in Figure 6.54. The pit limits obtained using a 45 degree slope constraint have been superimposed. Each of the blocks has plan dimensions 100 ft × 100 ft and the block (bench) height is 45 ft. The assumptions made in the development of the economic block model are given in Table 6.46. The level by level statistics of reserves in the ultimate pit limit contour are given in Table 6.47. The tonnage factor for both ore and waste is 12.0 ft^3/ton, hence there are 37,500 tons of material per block. The overall material in the pit consists of

Material	Blocks	Tons	Grade
Waste	505	18,937,500	0
Ore	63	2,362,500	3.68% Cu
Total	568	21,300,300	NA

If all the material in the pit could be mined instantaneously then the net value (revenues-costs) would be about $57,632,000. It will be assumed to be mined over a 3 year period

Bench 1

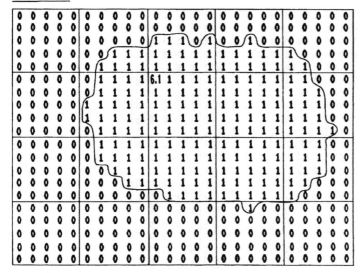

Bench 2

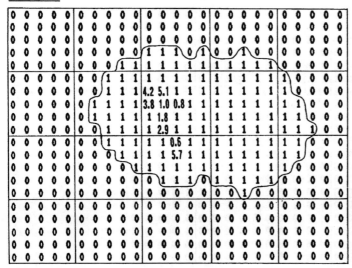

Figure 6.54. Grade block model with final pit limits superimposed. The number 1 denotes waste with grade 0 (Dagdalen, 1985).

instead. A discounting rate of 12.5% will be assumed. The precedence (order of mining) and slope constraints which apply are respectively

– In order to remove a given block, overlying blocks in the cone of influence must be removed first.
 – Pit slopes may not exceed the given maximum values (45° in this case).
 – The bench faces are considered to be vertical.

Bench 3

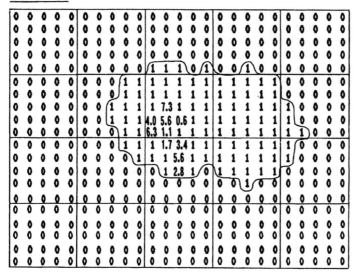

Bench 4

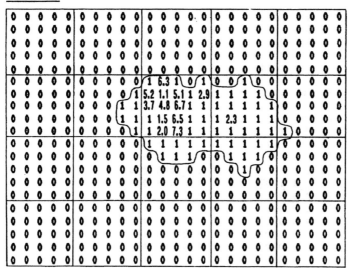

Figure 6.54. (Continued).

Constraint example 1
The first schedule will be made subject to the following ore mining (milling) capacity constraint:
1. Mine 19 ore blocks in period 1.
2. Mine 21 ore blocks in period 2.
3. Mine 23 ore blocks in period 3.

The problem is to schedule the operation such that the above restrictions are not violated and, at the same time, the total discounted profits before tax are maximized. The resulting

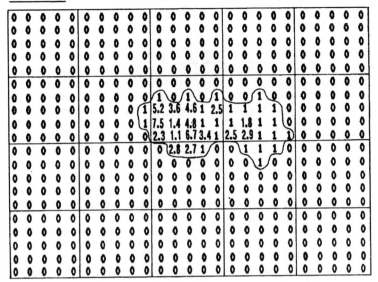
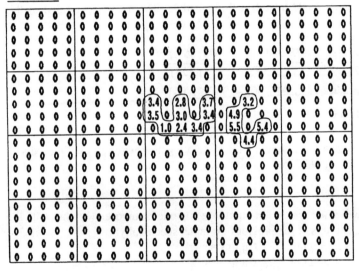

Figure 6.54. (Continued).

Table 6.46. Values used in the economic block model calculation (Dagdalen, 1985).

Price of copper	$0.80 per lb
Mining costs	$0.85 per ton of material
Milling costs	$2.20 per ton of ore
Smelting, refining and marketing	$0.25 per lb of Cu
G&A	$0.15 per ton of ore
Overall metal recovery	85%

Table 6.47. Summary of the reserves within the ultimate pit limits by level (Dagdalen, 1985).

Bench	Number of ore blocks	Ave. grade	Number of waste blocks
1	1	6.10	196
2	9	2.88	134
3	10	3.84	95
4	13	4.26	55
5	16	3.49	25
6	14	3.57	0
Total	63	3.68	505

schedules are shown in Figure 6.55 and in Table 6.48. In Figure 6.55, the numbers indicate the period in which each block is scheduled to be mined.

The average grade of the ore mined is 3.7% Cu in the first year, goes up to 4.4% Cu in the second year and decreases to 3.0% Cu in year 3 (Table 6.49).

According to the schedule, very little stripping is required in year 1; the amount of stripping is more than doubled in year 2; and most of the stripping is done in year 3. No cost adjustments have been made to reflect this schedule. The net present value of this schedule is

$$\text{NPV} = \frac{\$20,352,206}{1.125} + \frac{\$25,341,619}{(1.125)^2} + \frac{\$11,857,556}{(1.125)^3}$$
$$= \$18,090,850 + \$20,023,008 + \$8,327,941$$
$$= \$46,441,800$$

Imposing yearly ore tonnage constraints on the system together with discounting reduced the net present value by $11,190,225.

Constraint example 2
For this example, an additional restriction on mill feed grade is imposed together with the ore mining capacity constraints of the previous example; specifically:
  1. Mine 19 ore blocks averaging 3.7% Cu in year 1.
  2. Mine 21 ore blocks averaging 3.7% Cu in year 2.
  3. Mine 23 ore blocks averaging 3.7% Cu in year 3.

As indicated in the discussion of the ultimate pit reserves, the average grade of the total reserve within the ultimate pit contour was 3.68% Cu. Hence, this new restriction is to force the operation to mine as close to this average grade as possible.

The optimum solution to the above mining system is depicted in Figure 6.56. The numbers on the different benches indicate the years in which the block will be mined.

The summary statistics by bench and year are given in Table 6.50.

All of the constraints are satisfied except for the average grade requirement in year 2.

In year 2 the average grade is 3.6% as compared to the required 3.7%. The reason for this is a lack of available blocks. Table 6.51 summarizes the results. The NPV for this schedule, $45,929,335, is slightly less than that obtained in the previous example.

Production planning 603

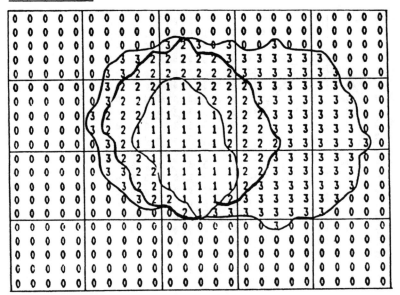

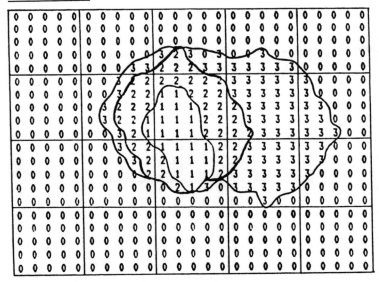

Figure 6.55. Proposed mining sequence for years 1, 2 and 3 under constraint set 1 (Dagdalen, 1985).

This reduction is caused by blending some of the low grade material with high grade in period 2.

These example studies can be expanded to include other conditions to determine the effects and costs of different constraints on the system.

604 *Open pit mine planning and design: Fundamentals*

Bench 3

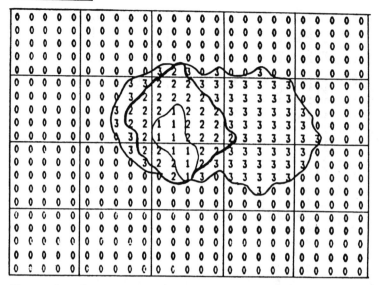

Bench 4

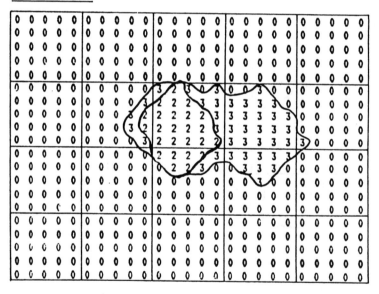

Figure 6.55. (Continued).

6.10 PUSH BACK DESIGN

6.10.1 *Introduction*

The first step in the practical planning process is to break the overall pit reserve into more manageable planning units. These units are commonly called sequences, expansions, phases,

Bench 5

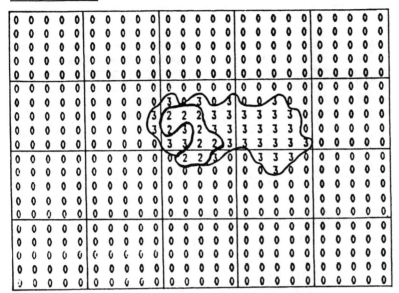

Bench 6

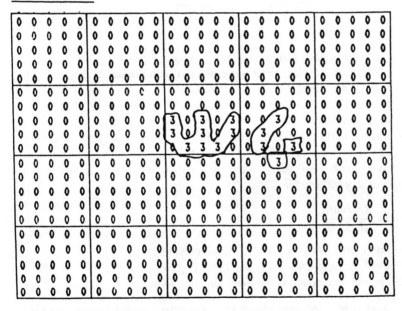

Figure 6.55. (Continued).

working pits, slices or push backs. In the beginning quite coarse divisions covering periods of several years may be used. They are a preliminary attempt to relate

    the geometry of mining

                to

    the geometry of the ore distribution.

Table 6.48. Level by level statistics for the optimum schedule under constraint set 1 (Dagdalen, 1985).

Bench	Year 1			Year 2			Year 3		
	Ore blocks	Ave. Grade	Waste blocks	Ore blocks	Ave. grade	Waste blocks	Ore blocks	Ave. grade	Waste blocks
1	1	6.10	42	0	0	50	0	0	104
2	9	2.88	15	0	0	44	0	0	75
3	9	4.20	0	1	0.6	32	0	0	63
4	0	0	0	11	4.56	14	2	2.60	41
5	0	0	0	9	4.60	0	7	2.07	25
6	0	0	0	0	0	0	14	3.57	0
Total	19	3.67	57	21	4.39	140	23	3.03	308

Table 6.49. Production summary under constraint set 1 (Dagdalen, 1985).

Year	Ore grade (% Cu)	Ore	Tonnage waste	Total	SR
1	3.67	712,500	2,137,500	2,850,000	3.0
2	4.39	787,500	5,250,000	6,037,500	6.7
3	3.03	862,500	11,550,000	12,412,500	13.4
Total		2,362,500	18,937,500	21,300,000	8.0

Phase planning should commence with mining that portion of the orebody which will yield the maximum cash flow. Succeeding phases are ordered with respect to their cash flow contribution. Eventually the ultimate pit limits are reached.

By studying the ore grade distribution (particularly as depicted on bench plans) and the topography the mining engineer can, in most cases, arrive at a logical pit development strategy in a relatively short time. Figure 6.57 shows a two dimensional representation of the phases used to extract an ore reserve (Mathieson, 1982). The extraction sequence proceeds from that phase having the highest average profit ratio (APR) to the lowest. In this case they proceed in alphabetical order A to G.

$$\text{Profit ratio} = \frac{\text{Revenue}}{\text{All costs}}$$

The incremental profit ratio, computed at the final pit boundary is 1.

A basic overview of the steps involved in phase planning (Mathieson, 1982) are indicated below. These will be expanded upon in the succeeding sections.

1. Before design work is initiated some preliminary judgements must be made regarding
   – the probable maximum ore and waste mining rate required in a given phase,
   – the size and type of equipment to be used. This determines the required minimum operating bench width necessary,
   – appropriate working, interramp, and final slope angles.

2. Using the constraints given in step 1, the mining engineer then proceeds to design a series of phases in some detail, complete with haul roads. He ensures that ramp access

Bench 1

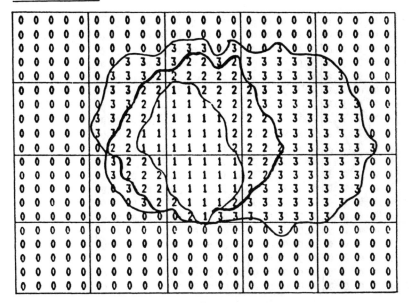

Bench 2

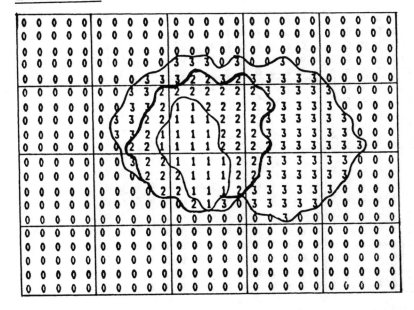

Figure 6.56. Proposed mining sequence for years 1, 2 and 3 under constraint set 2 (Dagdalen, 1985).

to each active bench is provided. The transition between phases is carefully planned. The designer is not constrained by having to include a certain quantity or number of years of ore supply in each phase. The variable ore and waste quantities by phase will be subsequently scheduled.

Bench 3

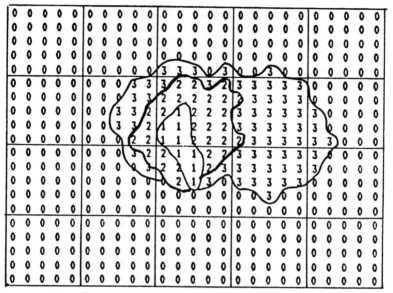

Bench 4

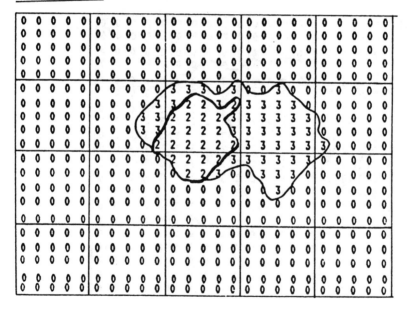

Figure 6.56. (Continued).

3. Once complete, the phase designs are transferred onto the bench plans. Bench by bench tonnages and grades are evaluated. A tonnage-grade inventory for each phase of the logically defined pit development sequence is then available.

4. The last task is to determine annual mining schedules based on mill feed or product requirements.

Production planning 609

Bench 5

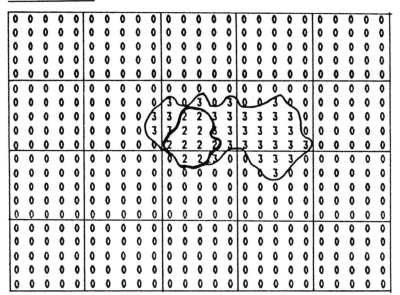

Bench 6

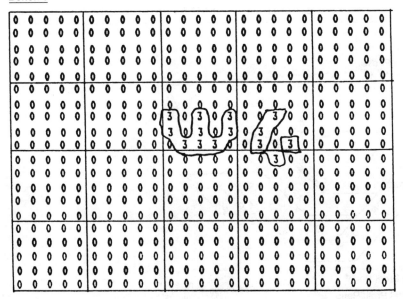

Figure 6.56. (Continued).

It may be necessary to repeat this process several times before an adequate plan is developed.

The development of phases makes it possible to determine the amount of pre-production stripping required. Although the natural tendency is to keep the amount of such stripping

Table 6.50. Level by level statistics for the optimum schedule under constraint set 2 (Dagdalen, 1985).

Bench	Year 1			Year 2			Year 3		
	Ore blocks	Ave. Grade	Waste blocks	Ore blocks	Ave. grade	Waste blocks	Ore blocks	Ave. grade	Waste blocks
1	1	6.1	42	0	0	50	0	0	104
2	9	2.88	15	0	0	44	0	0	75
3	9	4.20	0	1	0.6	31	0	0	64
4	0	0	0	10	4.16	13	3	4.60	42
5	0	0	0	10	3.34	0	6	3.73	25
6	0	0	0	0	0	0	14	3.57	0
Total	19	3.67	57	21	3.60	138	23	3.75	310

Table 6.51. Production summary under constraint set 2 (Dagdalen, 1985).

Year	Ore grade (% Cu)	Ore	Tonnage waste	Total	SR
1	3.67	712,500	2,137,500	2,850,000	3.0
2	3.60	787,500	5,175,000	5,962,500	6.6
3	3.75	862,500	11,625,000	12,487,500	13.5
Total		2,362,500	18,937,500	21,300,000	8.0

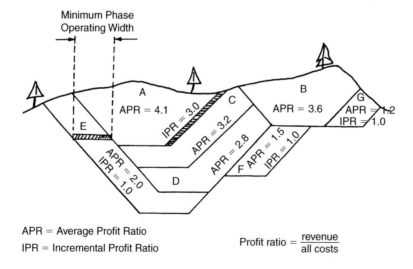

APR = Average Profit Ratio
IPR = Incremental Profit Ratio

Profit ratio = $\dfrac{\text{revenue}}{\text{all costs}}$

Figure 6.57. Pit sequencing in order of decreasing value (Mathieson, 1982).

to a minimum since it appears as a major negative cash flow early in the mining process, it is important that enough work be done to (1) expose a sustaining ore supply, and (2) keep the mine in a condition that allows it to be operated efficiently at all times. Sometimes the preproduction work is contracted. However, in new large tonnage truck-shovel mines, the pre-production period provides an opportunity to build an organization and to gain operating experience.

6.10.2 *The basic manual steps*

This series of steps in manual push back design has been provided by Crawford (1989a).

1. Start with the ultimate pit limit design.
   – Develop detailed data of ore grade and stripping distributions for various cutoff grades in zones around the designed pit circumference and in pit shell progression between the beginning surface topography (or pit surface) and the design pit limit. These data should include locations of ore zones (these vary with cutoff) and the impact of the differences between operating and ultimate pit slopes. Of particular interest should be locating high ore grade and low stripping zones on level plan maps and cross-sections.
   – Manual planning methods are essentially trial and error approaches.

2. Planning goals typically comprise one or more of the following:
   – Maximize NPV economics.
   – Provide stable cash flow patterns.
   – Uniform ore grade, grade decline curve, or high grading. Frequently high grading is used during the initial investment pay back period. The level(s) of ore grade will be related to the cutoff criteria.
     – Uniform stripping ratio or classic stripping ratio curve.
     – Uniform total tonnage rate (ore + waste).
     – Uniform ore tonnage rate.
     – Uniform or variable rate of product output.

3. Operating design criteria for push back design:
   – Operating and remnant bench widths.
   – Slopes between operating benches and roads.
   – Overall operating slope.
   – Road widths and grades.
   – Bench height.
   Some typical values are as follows:
     – Remnant bench width equal to bench height.
     – Push back widths normally 200–500 ft depending on size of pit and orebody characteristics.
     – Minimum push back widths (single cut passes) about 80 ft for small equipment and 135–150 ft for 25–30 yd shovels and 150–200 ton trucks.
     – Road widths 50–80 ft depending on width of equipment. Maximum road grade 8–12 percent.

4. Laying out of one to several push backs. Evaluating whether they satisfy the goals, individually and collectively, is a more or less cut and try process. Normally a push back is laid out according to the operating criteria in plan and cross-section views. The plan view approach is used here. The operating criteria are expressed in geometric parameters (feet and degrees). A useful tool is to make a scaled bench crest and toe pattern including operating bench and road widths.

612 *Open pit mine planning and design: Fundamentals*

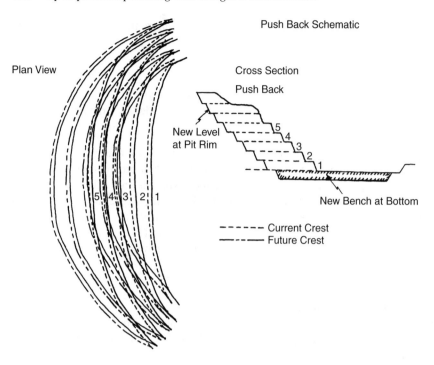

Figure 6.58. Push back schematic in plan and section views (Crawford, 1989a).

5. The push backs are shown on plan view maps as a progression of movements of bench level toes and crests from initial topography to ultimate pit limit (see Fig. 6.58). New levels are created as the push back progresses at the pit rim and at the bottom. New bottom levels are established on the basis of minimum level size and ore grades. Normally, new bottom levels are encouraged by the need to hold stripping at reasonable levels. In addition to pit geometry, ore/waste interface lines for the selected ore cutoff must be plotted.

6. Calculations of volumes of ore and waste are done using a planimeter to measure areas, and the average grades within push backs are determined. The volumes of material to be removed from each bench are based on the average of the areas encompassed by the movements of the bench crest and toe from initial to new position by a push back, and the average bench height for the zone covered. Ore and waste volumes are calculated separately. The average ore grade is determined by averaging the block values within the ore zones. In multiple push backs a push back serves as the initial location for a subsequent push back. The calculated values are evaluated against the various goals for acceptability of the individual push back or series of push backs.

7. For plan view calculations, the planimeter is used to determine the areas of crest and toe movements; commonly called crest to crest and toe to toe calculations. If the pit geometry is sufficiently regular only the toe to toe calculation may be necessary. To achieve accurate results, the calculation of volumes for new levels at the pit rim or at the bottom, along irregular pit rim elevations, for roads, at the ultimate pit intercept and for irregular bench heights require special planimeter techniques. The key is to divide up the volumes to be

Production planning 613

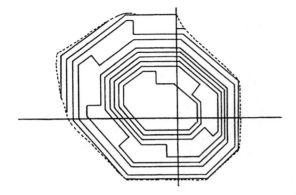

Figure 6.59. The Phase I pit with haulage road (Mathieson, 1982).

calculated into rectangular, parallelogram type solids with flat tops and bottoms and reliable average heights. The areas to be calculated must be closed polygons. The geometric layout prior to calculation is critical for accurate results. All the benches and roads must be described in the form of crests, toes, and average heights. Each bench and its related parts are calculated separately. Frequently, the drawing of a few cross-sections is helpful to keep the plan view drawing from becoming too confusing.

6.10.3 Manual push back design example

This section describes the construction procedure involved in designing a push back together with the layout of the main haulage road. An initial Phase I pit (Figure 6.59) already exists (Mathieson, 1982). The following information applies:

- The bench toe lines are shown
- Bench height = 45 ft
- Toe-toe distance = 40 ft
- Road width = 120 ft
- Road grade = 10%

Careful examination of the bench plans and the sections have indicated that the push back should involve the south and east portion of the pit. The north and northeast sections of the pit together with their portion of the haulroad will remain unchanged. The final pit design for Phase II must conform with this current geometry. It has been decided that the width of the push back should be 320 ft. This is a multiple of the basic 40 ft dimension and provides the desired
- tonnage for the phase, and
- operating space.

Figure 6.60 shows the basic push back area and the region where no changes will occur. The toe of the bottom bench has been described by a series of straight line segments. It is not necessary that they be straight lines but this facilitates the construction for this example. In Figure 6.61, that part of the Phase I pit involved directly in the push back has been removed. Sectors A and B will be modified for a smooth transition into the new pit. In Figure 6.62 the initial construction lines are shown. Lines a–b and c–d are drawn at the desired push back

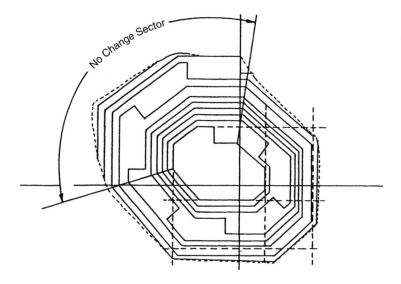

Figure 6.60. Construction lines for the push back superimposed.

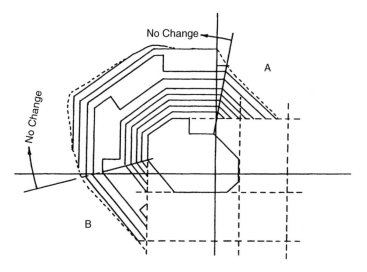

Figure 6.61. The push back section removed. Sectors A and B are transition regions.

distance of 320 ft parallel to the existing pit bottom lines. Lines f–a and b–c which form the pit corners are drawn at 45°. They are parallel to the corresponding lines of the current pit and at a distance of 320 ft. Line d–e has been drawn as shown recognizing that the pit will become more elongated in the east-west direction on that side. In Figure 6.63 the initial attempt at a pit bottom design is shown. The transitional sectors A and B have been removed

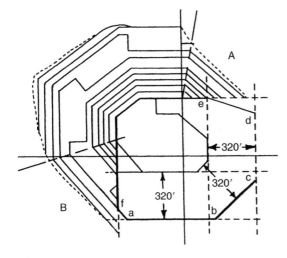

Figure 6.62. The dimensions used for constructing the new pit bottom.

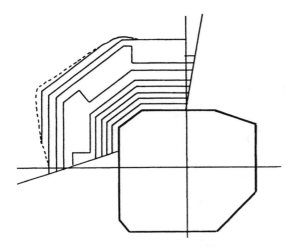

Figure 6.63. The initial design for the bottom of the Phase II pit.

from the drawing. To assist in constructing the upper bench from this base pit, lines have been drawn at the ends of the line segments perpendicular to the segment. This is shown in Figure 6.64. From each corner a line bisecting the angle described by the two radiating lines is drawn. This is shown by the heavy lines in Figure 6.65. A series of lines spaced at 40 ft (the toe-toe distance) are drawn parallel to those of the pit bottom. They are extended to join those of the existing pit (if appropriate) or to the corner lines. The drawing of the lines for the right hand side of the pit has been delayed pending the extension of the haulroad. In Figure 6.66, haulroad extension is shown. Due to the 10% grade and the 45 ft bench height, the distance between the adjacent road elevation lines shown is 450 ft. Figures 6.67 and 6.68 show the extension of the road down to the pit floor. In Figure 6.68 it can be seen that material in sector

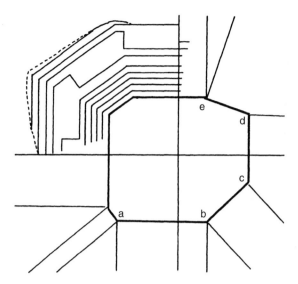

Figure 6.64. Construction of normals at the segment end points.

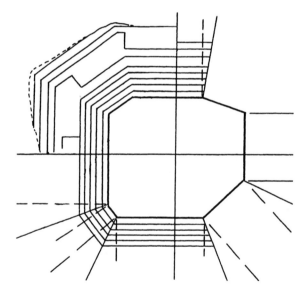

Figure 6.65. Construction of toe lines representing higher benches.

C must be mined to allow the road to daylight. A straight stretch having a length of one road width (120 ft) is added (Fig. 6.69). From this point (g), three segments g–h, h–i, and i–d have been drawn connecting to segment d–e. This provides the trucks with smooth access to the pit bottom and facilitates future road extension as the pit is deepened. As before, the bisecting lines are drawn from the segment corners (Fig. 6.70). In Figure 6.71, the parallel bench lines are added and the lower design is completed. As can be seen in Figure 6.71, there is no access

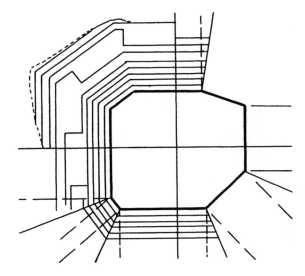

Figure 6.66. Extension of the road into the new phase.

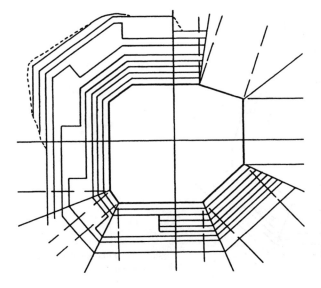

Figure 6.67. Extension of the road along the south side of the pit.

to the upper benches on the southeast side of the pit. This is corrected in Figure 6.72 by the addition of a road. The intersection of the Phase II pit with the surface topography is also shown.

With a workable design completed, one can now examine the tonnages involved. This is done on a bench by bench basis. Figure 6.73 shows the resulting bench geometry changes between the Phase I and Phase II pits. Figure 6.74 is the grade block model for bench 3835 with the toe lines for the Phase I and II pits superimposed. The next step in the process is to determine the grade-tonnage distribution for each level. Each of the complete blocks in the block model has plan dimensions 100 ft × 100 ft. For these, since the scale used is

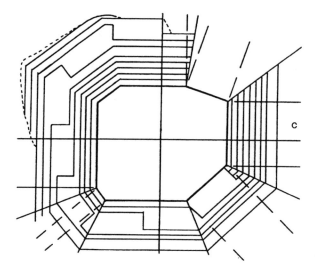

Figure 6.68. The road reaches the pit bottom elevation.

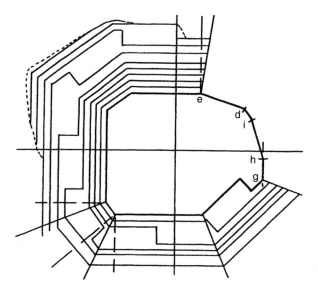

Figure 6.69. Final segments added to the pit bottom.

$1'' = 200$ ft, the block plan area is 0.25 in². However, in viewing Figure 6.74 it can be seen that
- the toe lines create many partial blocks,
- there are often several adjacent block with the same grade.

To save time, the contiguous blocks of the same grade are first identified. The area involved is determined using a planimeter. The number of square inches obtained is written on the bench map (see Fig. 6.75). Next the areas of the individual whole or partial blocks are found. These values are also entered on to the map. A summary sheet such as shown in Table 6.52 is prepared for the bench. The planimetered areas (in²) are converted into square feet (ft²)

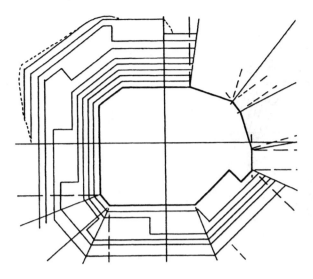

Figure 6.70. Construction at the segment corners.

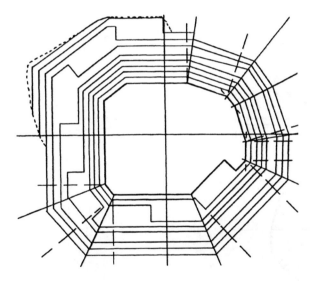

Figure 6.71. Road to pit bottom completed.

using the scale factor and then into tons via the bench height (45 ft) and the tonnage factor (12.5 ft³/ton). Hence

$$1 \text{ in}^2 \text{ area} = \frac{200 \text{ ft} \times 200 \text{ ft} \times 45 \text{ ft}}{12.5 \text{ ft}^3/\text{ton}} = 144{,}000 \text{ tons}$$

This conversion factor is applied to the areas in Table 6.52.

The total area on this bench involved in phase II mining is 17.57 in² hence the tonnage is

$$\text{Tons} = 17.57 \text{ in}^2 \times 144{,}000 \text{ tons/in}^2$$
$$= 2{,}530{,}080 \text{ tons}$$

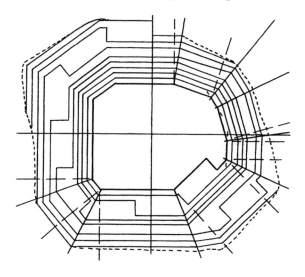

Figure 6.72. Road access to upper benches added.

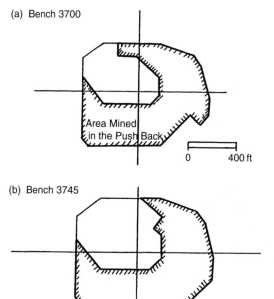

Figure 6.73. Area mine by bench between Phases I and II.

Sometimes grade classes are used. In Table 6.53 an interval of 0.05% Cu has been employed.

Expressed in this form it is quite easy to summarize the overall results from the entire push back. Applying one or more cutoff grades, the amount of material falling in each category is found.

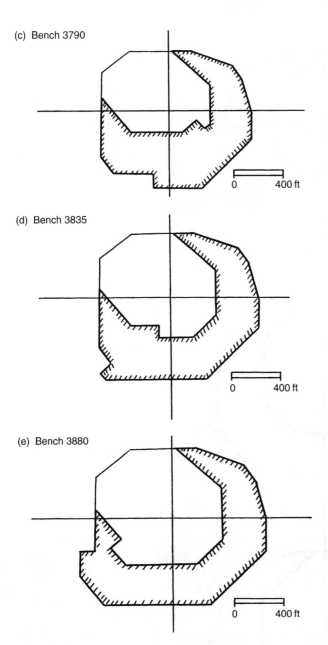

(c) Bench 3790

(d) Bench 3835

(e) Bench 3880

Figure 6.73. (Continued).

These results unfortunately may be unacceptable. For example, the overall stripping ratio might be too high, the average grade too low, the tonnage too low, etc. Another push back design must then be done in the same fashion. Eventually an acceptable plan will emerge.

622 *Open pit mine planning and design: Fundamentals*

(f) Bench 3925

(g) Bench 3970

(h) Bench 4015

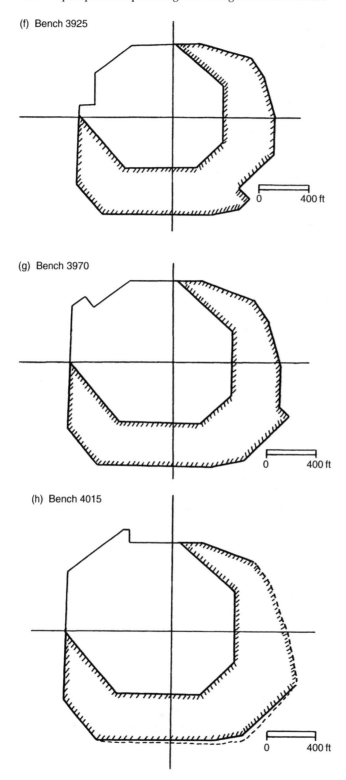

Figure 6.73. (Continued).

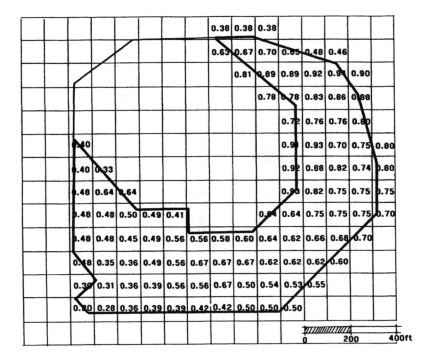

Figure 6.74. The final pit outline has been superimposed on the grade block model for bench 3835.

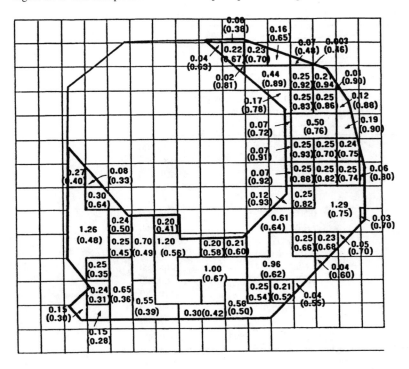

Figure 6.75. The areas are calculated.

Table 6.52. Summary tonnage-grade for the Phase II mining of bench 3835.

Grade (% Cu)	Plan area (in²)	Tons (10³)	Grade (% Cu)	Plan area (in²)	Tons (10³)
0.28	0.15	21.6	0.63	0.04	5.8
0.30	0.15	21.6	0.64	0.91	131.0
0.31	0.24	34.6	0.65	0.16	23.0
0.33	0.08	11.5	0.66	0.25	36.0
0.35	0.25	36.0	0.67	1.22	175.7
0.36	0.65	93.6	0.68	0.23	33.1
0.38	0.06	8.6	0.70	0.56	80.6
0.39	0.55	79.2	0.71	0.07	10.1
0.40	0.27	38.9	0.74	0.25	36.0
0.41	0.20	28.8	0.75	1.53	220.3
0.42	0.30	43.2	0.76	0.53	76.3
0.45	0.25	36.0	0.78	0.17	24.5
0.46	0.01	1.4	0.80	0.25	36.0
0.48	1.33	191.5	0.81	0.02	2.9
0.49	0.70	100.8	0.82	0.50	72.0
0.50	0.82	118.1	0.83	0.25	36.0
0.53	0.21	30.2	0.86	0.25	36.0
0.54	0.25	36.0	0.88	0.37	53.3
0.55	0.04	5.8	0.89	0.44	63.4
0.56	1.20	172.8	0.90	0.01	1.4
0.58	0.20	28.8	0.91	0.07	10.1
0.60	0.04	5.8	0.92	0.32	46.1
0.61	0.21	30.2	0.93	0.37	53.3
0.62	0.96	138.2	0.94	0.21	30.2

Table 6.53. The data from Table 6.51 reorganized into grade intervals.

Grade interval	Area (in²)	Tons
0.26–0.30	0.30	43,200
0.31–0.35	0.57	82,080
0.36–0.40	1.53	220,320
0.41–0.45	0.75	108,000
0.46–0.50	2.86	411,840
0.51–0.55	0.50	72,000
0.56–0.60	1.44	207,360
0.61–0.65	2.28	328,320
0.66–0.70	2.26	325,440
0.71–0.75	1.85	266,400
0.76–0.80	0.42	60,480
0.81–0.85	0.77	110,880
0.86–0.90	1.07	154,080
0.91–0.95	0.97	139,680
Total	17.57	2,530,080

It is obviously possible to evaluate various push back options without going to the trouble of including the haulroads. However, a considerable amount of material is involved in later adding such a road to the design and the results can change markedly.

6.10.4 *Time period plans*

Most of the pit planner's work (Couzens, 1979) is done on plan or bench maps. These show:
– topography or surface contour,
– location of ore,
– geologic boundaries, and
– design limits.

Pit composite maps showing the shape of the mine at the end of each planning period should be kept up. These enable the planner to:
– avoid conflicts between features of the plan,
– provide a picture of the access at each stage of development,
– illustrate the actual working-slopes, operating room and spacial relationships between ore and waste.

The transition from phase plans to time period plans should be made as soon as the phase designs are complete enough to set the overall pattern. The yearly plans:
– Enable definite production goals to be set in space as well as in quantities of material to be moved.
– Allow better economic evaluations than the phase average provide.
– Give a better definition of the relationship of the phases to each other as they overlap in the complete mine operation. They show actual operating slopes and haulage routes.

Figures 6.76, 6.77, and 6.78 show such a 3-year progression of benches in an open pit mine (Couzens, 1979).

The midbench contours have been plotted. The haulroads, stripping areas and part of the waste dumps are also shown. In this system the labelling of elevations is as follows (Couzens, 1979):

(a) Outside of the pit the contours are labelled with their true elevations.

(b) Inside of the pit:
– The labelled elevations refer to the bench toe elevations.
– The elevations of bench centerlines are one-half the bench height above the bench toe elevation. Thus it is the flat areas between center lines that are labelled.
– On ramps, the bench centerlines cross the ramp halfway between benches. The labels are positioned at would be the actual bench elevation on the road.
– For more explanation of the labelling, (see Section 4.7).

At an operating mine there will be a number of different plans covering different periods. The engineering staff is generally responsible for:
– annual ore reserve estimation,
– yearly or multi-yearly plans regarding the progression of the pit, changes in haul roads, etc.,
– quarterly plans, and
– monthly plans.

626 *Open pit mine planning and design: Fundamentals*

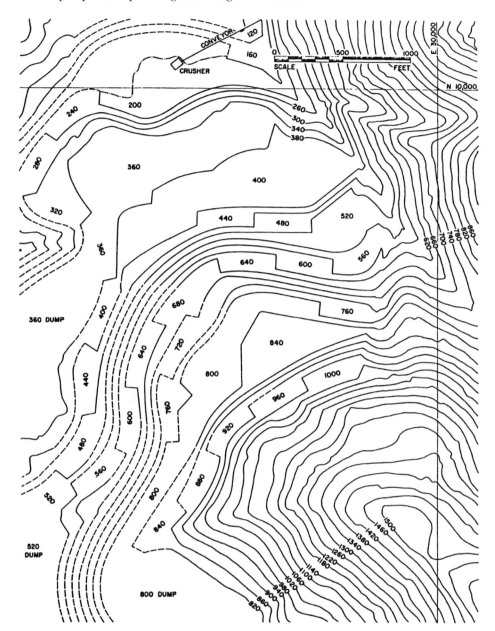

Figure 6.76. Bench composite for year 1 (Couzens, 1979).

The operating staff develop:
- weekly plans, and
- daily plans

within the longer range framework.

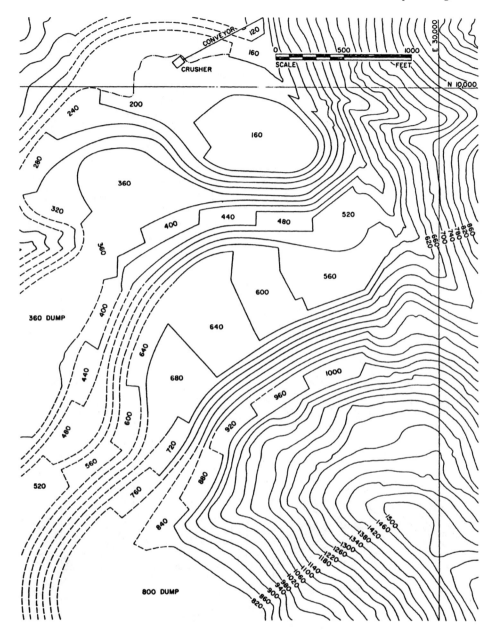

Figure 6.77. Bench composite for year 2 (Couzens, 1979).

6.10.5 *Equipment fleet requirements*

Once phase plans have been developed, equipment fleet requirements can be examined (Couzens, 1979). A graph showing the total tonnage movement and waste/ore ratios can be prepared. On such a graph, the planner can see what must be done to adjust or smooth out

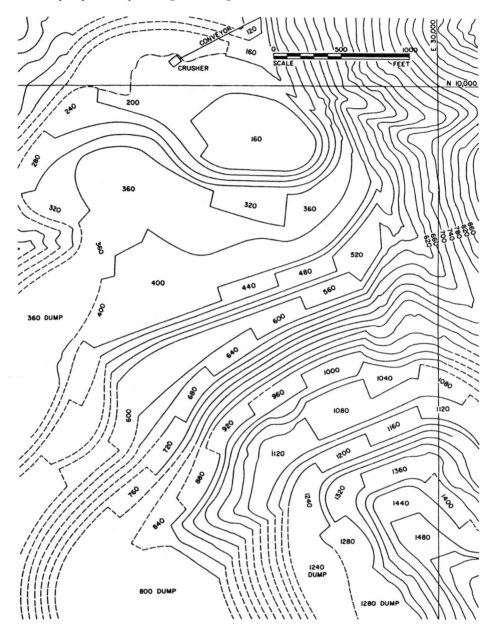

Figure 6.78. Bench composite for year 3 (Couzens, 1979).

the production. Figure 6.79 shows such a graph for a trial mining plan before smoothing has occurred. In this case the milling rate was constant and the plan was worked out to achieve:
- a good blend of ore,
- good ore exposure, and
- good operating conditions.

Production planning 629

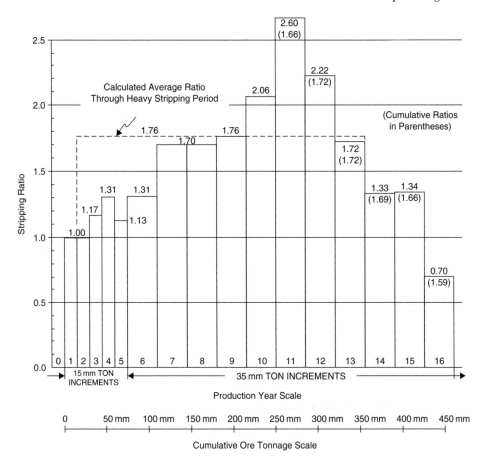

Figure 6.79. Waste/ore ratio versus time in a trial mining plan before smoothing (Couzens, 1979).

The amount of waste to be removed was determined by these conditions and as a result the waste/ore ratios vary. For the heavy stripping period, years 2–13, the average waste/ore ratio is 1.76. One attempt at replanning has been done in Figure 6.80. The peaks have been redistributed both earlier and later from their original positions.

Figure 6.81 illustrates a type of production scheduling graph. The various relationships between total tonnage movement, ore requirements, waste ratio, and shovel shifts are shown. A presentation to management of this type makes it possible to communicate the mining schedule better than just bare statements of tonnage and waste/ore ratios.

6.10.6 *Other planning considerations*

Dump planning is also a part of the mine planners job (Couzens, 1979). There are a number of factors which enter the scene at this stage:
– length of hauls,
– the required lifts,

630 *Open pit mine planning and design: Fundamentals*

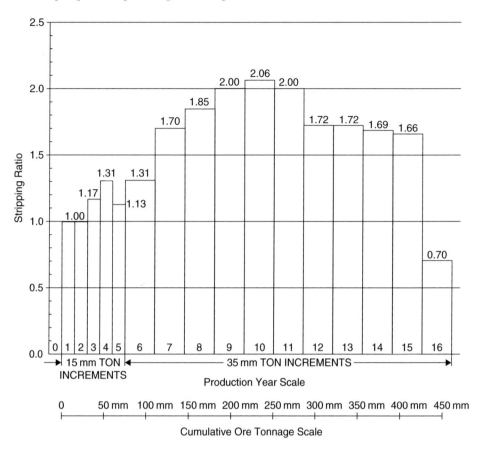

Figure 6.80. Waste/ore ratios versus time in a trial mining plan after smoothing (Couzens, 1979).

– the relationship between dumps and property constraints, other installations, drainage, etc., and
– reclamation and environmental requirements.

Pit planning should include an estimate of where the dumps are going to be at each stage. This should be done in connection with the haulage study. The planner can look at the trade-off between an additional lift and a longer haul. Dump planning can have an important bearing on pit planning, particularly in the haulage layout, scheduling and equipment estimating areas.

Water management planning forms a part of overall mine planning. In new pit design it is important to estimate the amount of expected water. This can be done by looking at rainfall records, drill logs and hydrologic reports.

Streams or even dry arroyos coming into the pit area may have to be diverted in order to avoid bringing surface water into the pit.

For environmental reasons the planner may have to consider what happens to water that is removed from the pit. Can it be discharged into natural drainage or must it be

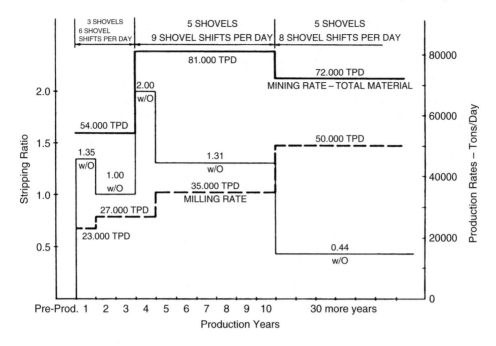

Figure 6.81. Example of a final production schedule graph based upon a balanced mining plan (Couzens, 1979).

impounded, treated, or recycled? Water affects blasting, equipment operation and maintenance, road construction and sometimes even ore quality. It is difficult to operate with much water in the pit. Water also has an adverse effect on slope stability. To manage the water, ditches, drains and/or dewatering systems may have to be included in the planning.

In summary, mine planning includes a rather wide spectrum of activities. It must be done carefully and thoroughly if the mine operation is to be successful.

6.11 THE MINE PLANNING AND DESIGN PROCESS – SUMMARY AND CLOSING REMARKS

In closing this chapter, it is appropriate and necessary to try and provide a perspective or helicopter view of the mine planning and design process. The individual component parts described in the preceding chapters and sections must be knit together if they are to function as a whole and produce the desired mine design.

Table 6.54 developed by Crawford (1989b) is a very useful summary of the steps that one goes through with a new orebody. In examining an expansion of a current mine, some simplifications arise due to the availability of better initial data. The steps are, however, basically the same. The iterative nature of the planning and design process has been very aptly termed circular analysis by Dohm (1979). The process and the included components are simply and rather elegantly presented in Figure 6.82.

Table 6.54. Logical sequenced approach for evaluating greenfield open pit ore deposits (after Crawford, 1989b).

1. Establish a long term price forecast and maximum practical marketable volume per year.
2. Make a geologic reserve assessment. Develop a grade versus tonnage curve for the contiguous portion of the orebody including overburden. Select a typical cut-off grade for the commodity (gold, copper, iron, etc.), type of deposit (deep, shallow, etc.), and likely mining/processing system. Determine an average grade and strip ratio and then determine average mill and mining rates for the maximum marketable volume per year or 10 year life whichever is lower.
3. Develop block net values and costs for both ore and waste using mining and milling rates ranging above and below those determined in No. 2. Incorporate capital costs and return considerations in the ore and waste block evaluations as feasible. Because timing and various inter-block interactions cannot be directly addressed, these will be only approximate.
4. Develop sets ultimate of pit designs flexing (varying) prices and/or milling/mining rates and costs to get logical concentric nests of pits. Use a block value ≥ 0 as cutoff.
5. Develop short range plans, using average operating slopes, within each ultimate pit. Holding costs and rates constant, flex the price starting above the price of ultimate pit and work down in increments. This approach should mine the best ore/waste combinations first. A set of short-range mining segments will be generated in this process and thereby fix the feasible mine geometry options for a given ultimate pit.
6. Develop production, revenue, operating and capital cost schedules over time for each short-range segment and over full mine life from development through closure. These data will not normally be derived from the block data directly but they must be related for consistency.
7. Optimize the value of each short range segment by flexing cutoff criteria and mining/ore processing rate being careful not to exceed the marketable maximum volume, using NPV methods. The production and economic schedules in No. 6 will be iterated to accomplish this. The choice of NPV discount rate will impact the results of the total evaluation process significantly.
8. Evaluate the NPV of the total pit plan segment sequence to confirm optimum.
9. Presumably the combination of ultimate pit, short-range pit plan segment sequence, and mining/ore processing rate yielding the maximum NPV regardless of mine life would be the preferred option for developing the deposit under evaluation.

Except for the simplest cases, there are often a large number of different combinations which are both possible and realistic. Of these, there is a small subset which would be acceptable. Eventually one must arrive at a 'best' few. Based upon technical, political, company cultural, or other reasons, one of these is eventually selected and the development phase is entered.

The rapid advances which have been made in the power, speed and friendliness of computerized design tools over the past few years have produced major changes in the activities performed by mining engineers in modern mining companies.

Largely relieved of routine number crunching and drafting, the design/planning engineer can focus on the task – problem solving – for which he/she has been educated/trained. Creative alternative solutions must be identified, developed and evaluated and recommendations made regarding an action path, if companies are to survive in a very competitive world. There should and must be some time for dreaming, perhaps not the 'Impossible Dream' but of innovative solutions to the tough problems facing mining today, and the even tougher ones coming. The continuing challenge of supplying the world's mineral requirements at

Figure 6.82. Circular analysis (Dohm, 1979).

reasonable cost under ever more stringent regulations will require the best minds and mines. Glückauf!

REFERENCES

Akaike, A. and K. Dagdelen 1999. A strategic production scheduling method for an open pit mine. 28th International Symposium on Application of Computers and Operations Research in the Mineral Industry: 729–738. Colorado School of Mines:CSM.

Anonymous 1990. Orebody modeling and mine planning software. *Mining Engineering* 42(8): 976–981.

Appiah, P.B. & J.R. Sturgul 1992. Optimal mine operating strategy for a fluctuating price. *Int. J. Surface Mining and Reclamation* 6(3): 121–128.

Arioglu, E. 1988. Examination of empirical formulae for predicting optimum mine output. *Trans. Instn. Min. Metall. (Sect. A: Min. Industry)* 97(1): A54–A55.

Baird, B.K., and P.C. Satchwell 2001. Application of economic parameters and cutoffs during and after pit optimization. *Mining Engineering*. 53(2): 33–40.

Barnes, R.J. & L. Bertrand 1990. Operational ore/waste classification at a surface gold mine. *Int. Journal of Surface Mining and Reclamation* 4: 26–29.

Belanger, M. 1993. Application of ORK to a secondary enriched porphyry copper deposit (Chile). 24th International Symposium on Application of Computers and Operations Research in the Mineral Industry: 2(241–248). Montreal:CIMM.

Belobgraidich, W., R.G. Trumbly & R.K. Davey 1979. Computer-assisted long-range mine planning practices at Ray Mines Division – Kennecott Copper Corporation. *Computer Methods for the 80's*. (A. Weiss, editor): 349–358. New York: SME.

Bernabe, D., and K. Dagdelen 2002. Comparative analysis of open pit mine scheduling techniques for strategic mine planning of the Tintaya Copper mine in Peru. Pre-print Paper 02–125. 2002 SME Annual Meeting, Feb 26–28. Phoenix, AZ.

Blackwell, G., Anderson, M., and K. Ronson 1999. Simulated grades and open pit mine planning – Resolving opposed positions. 28th International Symposium on Application of Computers and Operations Research in the Mineral Industry: 205–216. Colorado School of Mines:CSM.

Blackwell, G.H. 2000. Open pit mine planning with simulated gold grades. CIM Bulletin 93(1039): 31–37.

Blackwell, M.R.L. 1971. Some aspects of the evaluation and planning of the Bougainville copper project. *Decision-Making in the Mineral Industry.* CIM Special Volume No. 12. Montreal: 261–269.

Cai, W., and A.F. Banfield 1993. Long range open pit sequencing – a comprehensive approach. 24th International Symposium on Application of Computers and Operations Research in the Mineral Industry: 2(11–18). Montreal:CIMM.

Cai, W-L., and A.F. Banfield 1996. Some practical aspects of open pit mine planning. 26th International Symposium on Application of Computers and Operations Research in the Mineral Industry: 277–284. Penn. State University:SME.

Cai, W.L. 2001. Design of open pit phases with consideration of schedule constraints. 29th International Symposium on Application of Computers and Operations Research in the Mineral Industry: 217–222. Beijing:Balkema.

Camus, J.P. 1993. Optimization of the production schedule in the case of several orebodies feeding a stand alone plant. 24th International Symposium on Application of Computers and Operations Research in the Mineral Industry: 2(19–26). Montreal:CIMM.

Camus, J.P., and S.G. Jarpa 1996. Long range planning at Chuquicamata mine. 26th International Symposium on Application of Computers and Operations Research in the Mineral Industry: 237–242. Penn. State University:SME.

Cavender, B. 1992. Determination of the optimum lifetime of a mining project using discounted cash flow and option pricing techniques. *Mining Engineering* 44(10): 1262–1268.

Chuanqing, C. & Q. Wu 1989. Chapter 50. An approach to the mining design system based on a deposit model. *Proceedings APCOM 21.* (A. Weiss, editor): 521–528. SME.

Couzens, T.R. 1979. Aspects of production planning: Operating layout and phase plans. *Open Pit Mine Planning and Design.* (J.T. Crawford and W.A. Hustrulid, editors): 219–231. New York: SME-AIME.

Couzens, T.R. 1990. Personal communication. Cyprus Miami Mining, October 1990.

Crawford, J.T. 1989a. Push back design steps. Personal communication.

Crawford, J.T. 1989b. Logical sequenced approach for evaluating greenfield open pit ore deposits. Personal communication.

Crawford, J.T., III & R.K. Davey 1979. Case study in open-pit limit analysis. *Computer Methods for the 80's.* (A. Weiss, editor): 310–318. New York: SME-AIME.

Dagdalen, K. 1985. Optimum multi-period open pit mine production scheduling. PhD Thesis T-3073. Colorado School of Mines.

Dagdalen, K. 1992. Cutoff grade optimization. In: *Proceedings 23rd APCOM Proceedings* (Y.C. Kim, editor): 157–165. Littleton, CO: SME.

Davey, R.K. 1979a. Short-range mine planning. In: *Computer Methods for the 80's* (A. Weiss, editor): 600–609. New York: SME-AIME.

Davey, R.K. 1979b. Mine production reporting. In: *Computer Methods for the 80's* (A. Weiss, editor): 653–661. New York: AIME.

Davey, R.K. 1979c. Mineral block evaluation criteria. In: *Open Pit Mine Planning and Design* (J.T. Crawford and W.A. Hustrulid, editors): 83–96. SME-AIME.

David, P.A. & A.H. Onur 1992. Optimizing open pit design and sequencing. In: *Proceedings 23rd APCOM* (Y.C. Kim, editor): 411–422. Littleton, CO: SME.

Davis, R.E. & C.E. Williams 1973. Optimization procedures for open pit mine scheduling. In: *11th International Symposium on Computer Applications in the Minerals Industry* (J.R. Sturgul, editor) 2(4): C1-C18. University of Arizona.

Denby, B., Schofield, D., and T. Surme 1998. Genetic algorithms for flexible scheduling of open pit operations. 27th International Symposium on Application of Computers and Operations Research in the Mineral Industry: 605–616. London:IMM.

De Souza e Silva, K., Moura, M. Mourão, Lanna, O., Saliby, E., and J. Fleurisson 1999. Short term mine planning: Selection of working sites in iron ore mines. 28th International Symposium on Application of Computers and Operations Research in the Mineral Industry: 763–770. Colorado School of Mines:CSM.

Desbarats, A. & M. David 1984. Influence of selective mining on optimum pit design. *CIM Bulletin* (867)77(7): 49–56.

Discussion to Halls, Bellum, Lewis paper 1970. (Bulletin. No. 758)79: A26–A30. Reply. 1970. (Bulletin No. 764) 79(7): A107–108.

Dohm, G.C., Jr. 1979. Circular analysis – open pit optimization. In: *Open Pit Mine Planning and Design* (J.T. Crawford and W.A. Hustrulid, editors). New York: Society of Mining Engineers of the AIME.

Dohm, G.C., Jr. 1992. Open pit feasibility studies. In: *SME Mining Engineering Handbook*. 2nd Edition (H.L. Hartman, editor) 2: 1278–1297. Littleton, CO: SME.

Dohmen, A., and M. Dohmen 2001. Visualization of opencast mine planning. 29th International Symposium on Application of Computers and Operations Research in the Mineral Industry: 319–322. Beijing:Balkema.

Dowd, P.A., and A.H. Onur 1992. Optimizing open pit design and sequencing. 23rd International Symposium on Application of Computers and Operations Research in the Mineral Industry: 411–422. Tucson:SME.

Dowd, P.A., and C. Xu 1995. Financial evaluation of mining projects. 25th International Symposium on Application of Computers and Operations Research in the Mineral Industry: 1247–254. Brisbane:AusIMM.

Dowd, P.A., and C. Xu 1999. The financial evaluation of polymetallic mining projects. 28th International Symposium on Application of Computers and Operations Research in the Mineral Industry: 385–392. Colorado School of Mines:CSM.

Drebenstedt, C., and R. Grafe 2001. Stereoscopic 3D-display for mine planning and geological modeling. 29th International Symposium on Application of Computers and Operations Research in the Mineral Industry: 323–328. Beijing:Balkema.

Dubnie, A. 1972. *Surface Mining Practice in Canada*. Mines Branch Information Circular 292(10): 109 pp.

Elbrond, J. & P.A. Dowd 1975. The sequence of decisions on cutoff grades and rates of production. *13th APCOM Symposium*: 13–113. Germany, Verlag Gluckauf.

Elevli, B. 1988. Open pit mine production scheduling. MS Thesis T-3531. Colorado School of Mines.

Elevli, B. 1992. Open pit mine production scheduling using operations research and artificial intelligence. 189 pp. PhD Thesis T-4125. Colorado School of Mines.

Elevli, B., K. Dagdelen, & M.D.G. Salamon 1989. Single time period production scheduling of open pit mines. For Presentation at the SME Annual Meeting, Las Vegas, Nevada, Feb. 27–Mar. 2, 1989. Preprint No. 89–157.

Erickson, J.D. 1968. Long range open-pit planning. *Mining Engineering* 20(4): 75–77.

Ewanchuk, H.G. 1970. Open-pit mining at Bethlehem Copper Corporation, Ltd., Highland Valley, British Columbia, Canada. *Proceedings of the 9th Commonwealth Mining and Metallurgical Congress*. (M.J. Jones, editor): 415–441.

Faust, W.A. 1989. Mine scheduling at the Chimney Creek Mine. Presented at the SME-AIME Annual Meeting, Las Vegas, Feb. 27–Mar. 2, 1989. Preprint No. 89–194.

Franklin, P.J. 1985. Computer-aided short-term mine planning. *CIM Bulletin* (879)78: 49–52.

Fraser, S. 1971. Computer-aided scheduling of open-pit mining operations. *Decision-Making in the Mineral Industry* CIM Special Volume No. 12: 379–382.

Fytas, K. & P.N. Calder 1986. Chapter 11 – A computerized model of open pit short and long range production scheduling. In: *19th APCOM Symposium* (R.V. Ramani, editor): 109–119. SME.

Gershon, M. 1986. A blending-based approach to mine planning and production scheduling. In: *Proceedings 19th APCOM* (R.V. Ramani, editor): 120–126. Littleton, CO: SME.

Gershon, M.E. 1987. An open-pit production scheduler: algorithm and implementation. *Mining Engineering* 39(8): 793–795.

Gibbs, B. & S.A. Krajewski 1991. Workshop attendees compare ore modeling and mine planning software systems. *Mining Engineering* 43(7): 732–737.

Gill, D.K. 1966. Open pit mine planning. *Mining Congress Journal* 52(7): 48–51.

Golder Associates 1981. Open pit design program – user guide. Personal communication.

Gurtler, D.J. & L.R. Grobl 1990. Development and implementation of economic material splits at Chino Mines Company. Presentation at the SME Annual Meeting, Salt Lake City, Utah, Feb. 26–Mar. 1, 1990. Preprint No. 90–124.

Gurtler, D.J. 1990. Development and implementation of economic material splits at Chino Mines. *Mining Engineering* 42(4): 343–344.

Hack, D.R. 2002. Comparison of commercial mine planning packages available to the aggregate industry. 30th International Symposium on Application of Computers and Operations Research in the Mineral Industry: 311–320. Alaska:SME.

Halatchev, R.A. 1996. The factor of compromise in the assessment of open pit long-term production plans. In Surface Mining 1996 (H.W. Glen, editor):31–36. SAIMM.

Halatchev, R. 2002. The time aspect of the optimum long-term open pit production sequencing. 30th International Symposium on Application of Computers and Operations Research in the Mineral Industry: 133–146. Alaska:SME.

Halatchev, R.A. 2003. Open pit production sequencing: Pseudo-economic discretization versus technological discretization. Pre-print Paper 03–52. 2003 SME Annual Meeting, Feb 24–26. Cincinnati, OH.

Hall, C.J. 1987. Discussion of 'An open-pit production scheduler: algorithm and implementation'. *Mining Engineering* (12): 1107–1108.

Halls, J.L. 1970. The basic economics of open pit mining. In: *Planning Open Pit Mines* (P.W.J. Van Rensburg, editor). Johannesburg: SAIMM.

Halls, J.L., D.P. Bellum & C.K. Lewis 1969. Determination of optimum ore reserves and plant size by incremental financial analysis. *Instn Min Metall (Min. Sect A)* (Bulletin No. 746) 78(1): A20–A26.

Heath, K.C.G. 1980. Relationship between ore reserve, mill feed and real tonnages and grades. *Instn. of Mining & Met.* 89(7): A128–130.

Herfst, U. 1989. Evaluation of stacking alternatives in gold heap leach operations. MS Thesis T-3711. Colorado School of Mines.

Hewlett, R.F. 1961. Application of computers to open pit ore reserve estimation. In: *Short Course on Computers and Computer Applications in the Mineral Industry* (J.C. Dotson, editor) 1: I-1-I-19. Tucson: The University of Arizona.

Hewlett, R.F. 1962. Use of high-speed data reduction and processing in the mineral industry. IC 8099: 82 pp.

Hewlett, R.F. 1963. *A Basic Computer Program for Computing Grade and Tonnage of Ore Using Statistical and Polygonal Methods.* USBM RI 6292: 20.

Hickson, R.J. 1991. Grasberg open-pit: Ertsberg's big brother comes onstream in Indonesia. *Mining Engineering* 43(4): 385–391.

Hodson, D. & C. Rich 1986. *Open-pit production control at Rössing. The Planning and Operation of Open-Pit and Strip Mines* (J.P. Deetlefs, editor): 225–233. Johannesburg: SAIMM.

Hoerger, S., Bachmann, J., Criss, K., and E. Shortridge 1999. Long term mine and process scheduling at newmont's Nevada operations. 28th International Symposium on Application of Computers and Operations Research in the Mineral Industry: 739–748. Colorado School of Mines:CSM.

Hoerger, S., Hoffman, L., and F. Seymour 1999. Mine planning at Newmont's Nevada operations. *Mining Engineering.* 51(10): 26–30.

Hoerger, S., Hoffman, L., and F. Seymour 1999. Nevada planning at Newmont Gold Company. Pre-print Paper 99-036. 1999 SME Annual Meeting, Mar 1–3. Denver, CO.

Iles, C.D. & G.A. Perry 1983. Sierrita incremental pit design system. *Mining Engineering* 35(2): 152–155.

Jerez, R., and S.V. Tivy 1992. Developments in linear programming applications at Cyprus Minerals Company. 23rd International Symposium on Application of Computers and Operations Research in the Mineral Industry: 519–534. Tucson:SME.

Ji, C.S., and Y.D. Zhang 2001. A criteria to optimize mine production rate. 29th International Symposium on Application of Computers and Operations Research in the Mineral Industry: 299–302. Beijing:Balkema.

Johnson, T.B. 1968. Optimum pit mine production scheduling. PhD Thesis. University of California, Berkeley.

Johnson, T.B. 1969. *Optimum open pit mine production scheduling. A Decade of Digital Computing in the Mineral Industry* (A. Weiss, editor): 539–562. New York: Society of Mining Engineers of AIME.

Johnson, T.B., Dagdelen, K., and S. Ramazan 2002. Open pit mine scheduling based on the fundamental tree algorithm. 30th International Symposium on Application of Computers and Operations Research in the Mineral Industry: 147–162. Alaska:SME.

Kelsey, R.D. 1979. Cutoff grade economics. Chapter 28. In: *Proceedings 16th APCOM* (T.J. O'Neil, editor): 286–292. Littleton, CO: SME.
Kim, Y.C. 1979. Production scheduling – technical overview. *Computer Methods for the 80's* (A. Weiss, editor): 610–614. SME of AIME: New York.
King, B. 1999. Schedule optimization of large complex mining operations. 28th International Symposium on Application of Computers and Operations Research in the Mineral Industry: 749–762. Colorado School of Mines:CSM.
King, B.M. 1998. Impact of rehabilitation and closure costs on production rate and cutoff grade strategy. 27th International Symposium on Application of Computers and Operations Research in the Mineral Industry: 617–630. London:IMM.
Knowles, A. 1996. Computer-aided mine design, production and cost management. In Surface Mining 1996 (H.W. Glen, editor):263–272. SAIMM.
Koski, A.E. 1994. Computerized mine planning in Michigan iron ore. *Mining Engineering*. 46(12): 1338–1342.
Koski, A.E. 2000. Mine planning software models Empire Mine ore. *Mining Engineering*. 52(9): 47–52.
Kvapil, R., J. McMorran & M. Dotson 1982. Rapid open pit design system. *World Mining* 35(3): 66–70.
Lane, K.F. 1964. Choosing the optimum cutoff grade. *Colorado School of Mines Quarterly* 59: 811–829.
Lane, K.F. 1979. Chapter 27. Commercial aspects of choosing cutoff grades. In: *Proceedings 16th APCOM* (T.J. O'Neil, editor): 280–285. SME-AIME.
Lane, K.F. 1988. *The Economic Definition of Ore Cutoff Grades in Theory and Practice*. London: Mining Journal Books, Ltd.
Lane, K.F., D.J. Hamilton, & J.J.B. Parker 1984. Cutoff grades for two minerals. In: *18th APCOM*: 485–492. London: Inst. Min. Met.
Leigh, R.W. & R.L. Blake 1971. An iterative approach to the optimal design of open pits. *Decision-Making in the Minerals Industry*, CIM Special Volume No. 12: 254–260.
Lerchs, H. & I.F. Grossman 1965. Optimum design of open-pit mines. *CIM Bulletin*: 47–54.
Lestage, P., L. Mottola, R. Scherrer, and F. Soumis. 1993. A computerized tool for short range production planning in use at the Mont Wright operation. 24th International Symposium on Application of Computers and Operations Research in the Mineral Industry: 2(67–74). Montreal:CIMM.
Lewis, C.K. 1969. An 'economic life' for property evaluation. *E/MJ* 170(10): 78–79.
Li, Z. 1989. A theoretical approach to determination of mine life and design capacity. *Int. J. of Surface Mining* (1)3: 49–50.
Lill, J.W. 1986. *Operating features and control aspects of the Palabora copper open pit. The Planning and Operation of Open-Pit and Strip Mines* (J.P. Deetlefs, editor): 161–174. Johannesburg: SAIMM.
Lizotte, Y. & J. Elbrond 1982. Choice of mine-mill capacities and production schedules using open-ended dynamic programming. *CIM Bulletin* (832)75(3): 154–163.
Lizotte, Y. 1988. Economic and technical relations between open-pit design and equipment selection. In: *Mine Planning and Equipment Selection* (R.K. Singhal, editor): 3–13. Rotterdam: A.A. Balkema.
Mann, C. & F.L. Wilke 1992. Open pit short term mine planning for grade control – A combination of CAD-techniques and linear programming. In: *Proceedings 23rd APCOM* (Y.C. Kim, editor): 487–497. Littleton, CO: SME.
Mann, C., and F.L. Wilke 1992. Open pit short term mine planning for grade control - A combination of CAD techniques and linear programming. 23rd International Symposium on Application of Computers and Operations Research in the Mineral Industry: 487–498. Tucson:SME.
Mathieson, G.A. 1982. *Open pit sequencing and scheduling*. Presented at the First International SME-AIME Fall Meeting, Honolulu, Hawaii, Sept. 4–9 1982: Preprint No. 82–368.
Melamud, A., and D.S. Young 1993. A system for optimizing the interdependence of operating cost and three dimensional pit development. 24th International Symposium on Application of Computers and Operations Research in the Mineral Industry: 2(75–82). Montreal:CIMM.
Mellish, M., A.H. Preller & D.A. Tutton 1987. Integrated open-pit planning approach and systems. *Trans. Instn. Min. Metall (Sect A: Mine Industry)* 96(10): A162–A170.
Milner, T.E. 1977. Long range mine planning at Gibraltar Mines Limited. *CIM Bulletin* (786)70: 89–94.
Morey, P.G. & K.J. Ashley 1984. Open pit mine design. 1984 NWMA Short Course. Mine Feasibility – Concept to Completion, Dec. 3–5, 1984, Spokane, WA.
Mutmansky, J.M. 1979. Computing and operations research techniques for production scheduling. In: *Computer Methods for the 80's* (A. Weiss, editor): 615–625. New York: SME-AIME.

Nelson, L.R. & M.R. Goldstein 1979. Interactive short-range mine planning – A case study. *Computer methods for the 80's* (A. Weiss, editor): 592–599. New York: Society of Mining Engineers.

Nilsson, D. & K. Burgher 1982. Determining the optimal depth of an open pit. *E&MJ* 183(7): 73–80.

O'Hara, T.A. 1980. Quick guides to the evaluation of orebodies. *CIM Bulletin* (814)73: 87–99.

Osborne, W.G. MINPAC – computerized aid to pit design and extraction scheduling. Paper presented at the Spring Meeting, Arizona Section, AIME, May 20, 1976.

Pana, M.T. & R.K. Davey 1973. *Open-pit mine design. SME Mining Engineering Handbook* (I.A. Given, editor). New York: Society of Mining Engineers of the AIME.

Peralta, A., and P.N. Calder 1993. Analysis of the blending problem in open pit production scheduling. 24th International Symposium on Application of Computers and Operations Research in the Mineral Industry: 2(98–105). Montreal:CIMM.

Perez, A.S., Oreja, E.C., and J.M. Perez 1993. Short term planning and production schedule of open pit mines. 24th International Symposium on Application of Computers and Operations Research in the Mineral Industry: 2(106–110). Montreal:CIMM.

Pincomb, A. 2003. Aggregate reserves planning & value. Pre-print Paper 03–7. 2003 SME Annual Meeting, Feb 24–26. Cincinnati, OH.

Recny, C.J. 1982. *The influence of geologic characteristics on production capacity and their relation to costs and profitability of mining projects. Mineral Industry Costs* (J.R. Hoskins, compiler): 45–62. Northwest Mining Association.

Richmond, A.J. 2001. Maximum profitability with minimum risk and effort. 29th International Symposium on Application of Computers and Operations Research in the Mineral Industry: 45–50. Beijing:Balkema.

Roman, R.J. 1973. *The use of dynamic programming for determining mine-mill production schedules. 10th APCOM Symposium*: 165–170. South African Inst. Mining Met., Johannesburg.

Roman, R.J. 1974. The role of time value of money in determining an open pit mining sequence and pit limits. In: *12th Int. Symp. on the Applications of Computers and Mathematics in the Minerals Industry* (T.B. Johnson and D.W. Gentry, editors) CSM, vol. I: C-72–C-85.

Rudenno, V. 1979. Chapter 24 – Determination of optimum cutoff grades. In: *Proceedings, 16th APCOM* (T.J. O'Neil, editor): 261–268. SME.

Schellman, M.G. 1989. Determination of an optimum cutoff grade policy considering the stockpile alternative. MS Thesis T-3741. Colorado School of Mines.

Sims, D.E. 1979. Open-pit long-range mine planning using O.R.E. In: *Computer Methods for the 80's* (A. Weiss, editor): 359–370. New York: SME.

Smith, M.L. 2001. Integrating conditional simulation and stochastic programming: An application in production scheduling. 29th International Symposium on Application of Computers and Operations Research in the Mineral Industry: 203–208. Beijing:Balkema.

Soderberg, A. 1959. Elements of long range open pit planning. *Mining Congress Journal* 45(4): 54–57, 58, 62.

Stuart, N.J. 1992. Pit optimization using solid modelling and the Lerchs-Grossmann algorithm. *Int. J. of Surface Mining and Reclamation* 6(1).

Sturgul, J.R. 1985. Optimum life of a mine: Declining production case. *Int. Journal of Mining Engineering* 3: 161–166.

Suriel, J.R. 1984. A procedure for preliminary economic evaluation of open pit mines. ME Thesis ER-2905. Colorado School of Mines.

Taylor, H.K. 1972. General background theory of cutoff grades. *Institution of mining and metallurgy transactions*, Section A 81: A160–170.

Taylor, H.K. 1977. Mine valuation and feasibility studies. *Mineral Industry Costs* (J.R. Hoskins and W.R. Green, editors). 1–17. Spokane, WA: Northwest Mining Association.

Taylor, H.K. 1985. Cutoff grades – some further reflections. *Trans. Instn. Min. Metall.* (Sect. A: Min. Industry) 94(10): A204–216.

Taylor, H.K. 1986. Rates of working of mines – a simple rule of thumb. *Trans. Instn. Min. Metall.* (Sect A: Min. Industry) 95: A203–204.

Taylor, H.K. 1988. Discussion of the paper 'Examination of Empirical formulae for predicting optimum mine output' by E. Arioglu. *Trans. Instn. Min. Metall.* (Sect A: Min. Industry) 97(7): A160.

Taylor, H.K. 1991. Ore-reserves – the mining aspects. *Trans. Instn. Min. Metall.* (Sect A: Min. Industry) 100(9–12): A146–A158.

Taylor, H.K. 1992. Personal communication.

Thomas, G.S. 1996. Pit optimization and mine production scheduling - The way ahead. 26th International Symposium on Application of Computers and Operations Research in the Mineral Industry: 221–228. Penn. State University:SME.

Tolwinski, B. & R. Underwood 1992. An algorithm to estimate the optimal evolution of an open pit mine. In: *Proceedings 23rd APCOM* (Y.C. Kim, editor): 399–409. Littleton, CO: SME.

Tolwinski, B., and R. Underwood 1992. An algorithm to estimate the optimum evolution of an open pit mine. 23rd International Symposium on Application of Computers and Operations Research in the Mineral Industry: 399–410. Tucson:SME.

Tolwinski, B., and R. Underwood 1993. Scheduling mine production while maintaining constant mill feeds. 24th International Symposium on Application of Computers and Operations Research in the Mineral Industry: 2(155–162). Montreal:CIMM.

Tolwinski, B. 1998. Scheduling production for open pit mines. 27th International Symposium on Application of Computers and Operations Research in the Mineral Industry: 651–662. London:IMM.

Topuz, E. 1979. An optimal determination of capital investment, capacity and cutoff grade under multiple objectives. In: *Proceedings Council on Economics,* AIME 108th Annual Meeting: 97–107.

Toutin, E. & J. Camus 1988. Aspectos economicos y de diseno en la planificacion de la mina Sur-Sur. *Minerales* (177)42: 53–60. (Translated by M. Schellman, 1988).

Vieira, J.L., F.D. Durão, F.H. Muge, and J.Q. Rogado 1995. The integration of short-term planning with mineral processing simulation. 25th International Symposium on Application of Computers and Operations Research in the Mineral Industry: 303–312. Brisbane:AusIMM.

Wang, Q. & H. Sevim 1992. Enhanced production planning in open pit mining through intelligent dynamic search. In: *Proceedings 23rd APCOM* (Y.C. Kim, editor): 461–471. Littleton, CO: SME.

Wang, Q., and H. Sevim 1993. An alternative to parameterization in finding a series of maximum-metal pits for production planning. 24th International Symposium on Application of Computers and Operations Research in the Mineral Industry: 2(168–175). Montreal:CIMM.

Wang, Q., and H. Sun 2001. A theorem on open pit planning optimization and its application. 29th International Symposium on Application of Computers and Operations Research in the Mineral Industry: 295–298. Beijing:Balkema.

Wells, H.M. 1978. Optimization of mining engineering design in mineral valuation. *SME Transactions* 30(12): 1676–1684. Discussion by R.C. Kirkman, p. 406, April 1979.

Wharton, C.L. 1996. Optimization of cutoff grades for increased profitability. In Surface Mining 1996 (H.W. Glen, editor):101–104. SAIMM.

Whittle, J., and C.L. Wharton 1995. Optimizing cut-offs over time. 25th International Symposium on Application of Computers and Operations Research in the Mineral Industry: 261–266. Brisbane:AusIMM.

Whittle. J. 1999. A decade of open pit mine planning and optimization – The craft of turning algorithms into packages. 28th International Symposium on Application of Computers and Operations Research in the Mineral Industry: 15–24. Colorado School of Mines:CSM.

Wilke, F.L. & T. Reimer 1979. Optimizing the short term production schedule for an open pit iron ore mining operation. In: *Computer Methods for the 80's* (A. Weiss, editor): 642–646.

Wilke, F.L., K. Mueller & E.A. Wright 1984. Ultimate pit and production scheduling optimization. In: *Proceedings of the 18th APCOM Symposium*: 29–38. London: IMM.

Wilkinson, W.A. 2002. Elements of graphical mine planning. Pre-print Paper 02–46. 2002 SME Annual Meeting, Feb 26–28. Phoenix, AZ.

Williams, C.E. 1974. Computerized year-by-year open-pit mine scheduling. *SME Trans.* 256(12): 309–317.

Winkelmann, S.P. 1990. Newmont Gold outlines its mine planning procedures. *Mining Engineering* 42(4): 333–338.

Winkler, B.M. 1998. Mine production scheduling using linear programming and virtual reality. 27th International Symposium on Application of Computers and Operations Research in the Mineral Industry: 663–672. London:IMM.

Winkler, B.M., and P. Griffin 1998. Mine production scheduling with linear programming- development of a practical tool. 27th International Symposium on Application of Computers and Operations Research in the Mineral Industry: 673–680. London:IMM.

Wright, E.A. 1986. Dynamic programming in open pit mining sequence planning – a case study. In: *21st APCOM Proceedings*: 415–422.

Wright, E.A. 1987. The use of dynamic programming for open pit mine design: Some practical implications. *Mining Science and Technology* 4(2): 97–104.

Yun, Q. & T. Yegalalp 1982. Optimum scheduling of overburden removal in open pit mines. *Transactions CIM Bulletin* (848)75(12): 80–83.

Yun, Q.X., G.Q. Huang, J.M. Xie & J.X. Liu 1988. A CAD optimization package for production scheduling in open-pit mines. In: *Mine Planning and Equipment Selection* (R.K. Singhal, editor): 77–85. Rotterdam: A.A. Balkema.

Zhang, Y., Q. Cai, L. Wu, and D. Zhang 1992. Combined approach for surface mine short term planning optimization. 23rd International Symposium on Application of Computers and Operations Research in the Mineral Industry: 499–506. Tucson:SME.

Zhao, Y., and Y.C. Kim 1993. Optimum mine production sequencing using Lagrangian parameterization approach. 24th International Symposium on Application of Computers and Operations Research in the Mineral Industry: 2(176–186). Montreal:CIMM.

REVIEW QUESTIONS AND EXERCISES

1. Summarize the basic objectives or goals of extraction planning.
2. What are the two types of production planning?
3. Summarize the five planning 'commandments.'
4. What are the assumptions that need to be made in the mine life – plant size determination?
5. In section 6.2, an example has been presented to illustrate the mine life – plant size determination. Redo the example assuming a copper price of $1.50/lb rather than $1.00.
6. Repeat the example assuming that the copper price is $1.00/lb but the overall recovery is 88%.
7. Taylor has provided some useful insights into deciding the appropriate mine life. Summarize the factors which should be considered.
8. Develop the reasoning as to why one might expect the life of a mine to be related to the cubic root of the ore tonnage.
9. Apply Taylor's rule to the example presented in section 6.2. Do the results agree? Discuss.
10. Discuss the concept behind 'nested' pits. In practice, how are they developed?
11. Apply Taylor's rule to the molybdenum example discussed in section 6.4. Comments?
12. What would be the affect on the operation using the molybdenum prices of today?
13. What are the basic line items in a pre-production cash flow table?
14. In the U.S., how are exploration costs handled?
15. What is an ad valorem tax?
16. What is meant by a pro mille levy?
17. What is meant by the term working capital?
18. What is meant by the terms
    a. Project year
    b. Production year
    c. Calendar year
19. Briefly discuss the basic line items in the production period cash flow calculations.
20. In Figure 6.14, the mining cost is seen to vary with the production rate. Based on these data, what would be the estimated cost if the rate was to be increased to 100,000 tpd? Is this reasonable? Why or why not?
21. Discuss the concept of depletion.
22. Discuss the two types of depletion.

Production planning   641

23. Redo the example assuming a 20 year mine life rather than 15 years. What is the effect on the NPV?
24. What is the difference between profit and cash flow?
25. Summarize the steps leading up to the determination of the final pit limits and the sizing of the mine/mill plant.
26. At what point in the process are the capital costs introduced? What effect does this have? How can this problem be minimized?
27. Discuss the significance of Figure 6.17. In practice how would this be generated?
28. In mining period 3, why is there still an overall profit produced?
29. Summarize the objections to the idea that each ore block irrespective of grade should contribute an equal portion to profits and capital payback?
30. Discuss the concept behind 'Incremental Financial Analysis'? What are the steps involved?
31. In section 6.6.2 an example is given for the pit generated with an 0.7% Cu mining cutoff. Two mill cutoffs, 0.7% and 0.6%, were considered and it was decided that the 0.6% cutoff was an improvement. Should the cutoff be further reduced to 0.5%? To 0.4%?
32. Repeat the calculations in the plant sizing example given in section 6.6.3.
33. Summarize the basic concepts behind Lane's algorithm.
34. In section 6.7.4 an illustrative example concerning Lane's algorithm has been presented. Redo the calculation assuming a concentrator cutoff of 0.4% Cu and no constraints are violated.
35. Show the development of equations (6.19), (6.22), and (6.25).
36. Repeat the calculations in section 6.7.5 regarding the cutoff grade for maximum profits.
37. Develop the three curves shown in Figure 6.22.
38. If instead of seeking the cutoff grade that maximizes profits, suppose that one was interested in maximizing the NPV. What approach would be taken?
39. In the example a fixed cutoff grade was assumed throughout the life of the mine. Could the NPV be improved by varying the cutoff grade throughout the life? How could this question be addressed?
40. Repeat the calculations for the example in section 6.7.6 for year 1.
41. What is the difference in the results obtained using the maximum profit and maximum NPV cutoff values?
42. What is the practical meaning of the cutoff grade? How many cutoff grades are there? Briefly discuss each one.
43. What is the difference between a leach dump and a leach pad? What is vat leaching?
44. For the leach dump alternative presented in section 6.8.2, assume that the grade is 0.35% Cu. What is the expected recovery in lbs/ton?
45. Repeat the example in Table 6.35 assuming leaching of material with a grade of 0.45% Cu.
46. Repeat the example in Table 6.36 assuming that material of grade 0.45% Cu is sent to the mill.
47. How should the value of the potential leach material be included when deciding final pit limits?
48. How should the differences in capital cost be taken into account when making leaching versus mill decisions?
49. Discuss the advantages and disadvantages of the stockpile alternative.
50. What were the three stockpile alternatives studied by Schellman?

51. How did the three alternatives compare with respect to the non-stockpiling option?
52. When should the stockpile alternative be considered in open pit mine planning?
53. Repeat the production scheduling example in section 6.9.1 but assume a production rate of 4 blocks per year.
54. Summarize the step-by-step approach to phase scheduling.
55. Phase scheduling is discussed in section 6.9.2. It has been done assuming an annual milling rate of 2,500,000 tons. Repeat the example assuming a milling rate of 2,000,000 tons.
56. What is meant by contingency ore? What are the advantages and disadvantages?
57. Apply the floating cone approach to the block model shown in Figure 6.41. Do you get the same result as shown in Figure 6.48?
58. What is the goal when applying the dynamic programming approach?
59. List the constraints that have been applied in the example in section 6.9.3?
60. Summarize in a simple way the procedure used in the dynamic programming approach as described in section 6.9.3.
61. In the example of section 6.9.3, how has the 'value' of the section been changed by including the fact that all of the blocks cannot be mined simultaneously?
62. In Figure 6.49, how does the choice of discount rate affect the pit limits? The mining sequence?
63. Consider the simple economic block model shown below:

    1 2 3 2 1
       2 2 4
          3

    Using the approach of Roman, in which order should the blocks be mined to yield the maximum NPV. Assume a discount rate of 10% and the same constraints that were used in his example.
64. Redo problem 63 with constraint 2 removed.
65. Discuss the approach being applied in section 6.9.4. What constraints were applied? How does this approach compare to that introduced in section 6.9.3?
66. What is the first step in the practical planning process?
67. What are some of the terms given to the practical pit planning units?
68. What is the simple objective of the planning process?
69. What is meant by the term 'average profit ratio'? How is it applied?
70. Summarize the steps involved in phase planning as indicated by Mathieson.
71. Summarize the steps in manual pushback design offered by Crawford?
72. In section 6.10.3, an example of a pushback design has been provided. The width of the pushback is 320 ft. Redo the example for a width of 200 ft. Assume all of the other values to be the same.
73. Summarize the concepts dealing with time period plans.
74. With regard to planning, what are the different responsibilities with regard to the engineering and operating groups?
75. In reviewing figures 6.76, 6.77 and 6.78, where did the mining take place over this 2 year period from the end of year 1 to the end of year 3? Provide an estimate of the amount of material removed.
76. At what point in the planning process do you evaluate the equipment fleet requirements? On what basis is it done?
77. What might be included on a production scheduling graph?

78. What are some of the other duties of the mine planner?
79. What is meant by the term 'green field open pit ore deposit'?
80. Summarize the steps in a logical sequenced approach for evaluating a 'green field open pit ore deposit.'
81. In practical terms, what is meant by 'circular analysis'? For an open pit mine evaluation, what are the logical components and sub-components?
82. To a miner, what is meant by the German expression 'Glückauf'?

CHAPTER 7

Reporting of mineral resources and ore reserves

7.1 INTRODUCTION

When describing and classifying mineral resources, it is important that the terms being used are precisely defined and accurately applied. Otherwise, there is the real possibility that potentially very costly misrepresentations and misunderstandings will arise. When considering a property for purchase, or lease, or simply considering the purchase of mining shares, the nature and potential of the assets being represented must be properly and clearly described. Today, the popular expression for this is 'transparency'. For the unscrupulous promoter of mineral properties (or mining shares), the lack of rules, precise definitions and specified procedures are essential for success. Smoke, mirrors, and sometimes a pinch of salt are the 'rules' of the game. Scams involving mineral properties have been around forever and the stories surrounding them provide interesting table discussions at mining meetings. There is, however, enough honest risk in mining without the active participation of scoundrels. The inclusion of this chapter is intended to acquaint the reader with some of the current guidelines regarding the public reporting of exploration results, mineral resources and ore reserves.

During the period of 1969–1970, Australia was home to a series of stock scams involving nickel, primarily, but other metals as well (Sykes (1978), Sykes (1988)). In response, the Australian Mining Industry Council established a committee to examine the issue. The effort was quickly joined by the Australasian Institute of Mining and Metallurgy (AusIMM) and in 1971 the Australasian Joint Ore Reserves Committee (JORC) was formed. Between 1972 and 1989, JORC issued a number of reports which made recommendations on public reporting and ore reserve classification. In 1989 the first version of the JORC Code was released. Since that time, it has gone through a number of revisions. With the gracious permission of the AusIMM (2005), the metal/nonmetal related portion of the most recent document (JORC 2004) has been included as section 7.2 of this chapter.

Over the years, the JORC Code has served as the model for the codes of a number of countries including those of Canada, the United States, South Africa, the United Kingdom/Europe, Chile and Peru. The authors have found the Canadian 'Estimation of Mineral Resources and Mineral Reserves – Best Practice Guidelines' to be quite comprehensive and it is included with the kind permission of the Canadian Institute of Mining, Metallurgy and Petroleum (CIM, 2004) as section 7.3.

It might be argued that even with a rigorous Code in place abuses will occur. A case in point is the recent scam/scandal involving the Canadian company Bre-X Minerals Ltd. This

Reporting of mineral resources and ore reserves 645

prospect located in the headwaters of the Busang River on the Island of Borneo in Indonesia was obtained by the company in 1993. A drilling program was begun and in February of 1997, the company geologist was talking of 200 million ounces of gold. The stock which was traded on the Alberta, Toronto and Montreal stock exchanges in Canada as well as the NASDAQ National Market in the United States went along for the ride rising from pennies to a high of $250 (Can). In late March 1997 after due-diligence drilling by another company revealed insignificant amounts of gold, the share price plunged to near zero. It is a fascinating story and the interested reader is referred to one of the several books that have been written about it (Francis (1997), Goold and Willis (1997)) or to the numerous articles appearing on the Internet (see for example, www.brexclass.com/graphics/docs/complaint.pdf).

Given the world wide nature of the minerals business, there is considerable interest in the development of a uniform reserve/resource reporting system. The interested reader is referred to the extensive reference list included at the end of the chapter.

7.2 THE JORC CODE – 2004 EDITION

7.2.1 Preamble

The authors wish to express their sincere thanks to the Australasian Institute of Mining and Metallurgy (AusIMM) for permission to include the JORC Code – 2004 Edition as part of this book. The original format has been somewhat modified to conform to that used in the rest of the book. The interested reader is encouraged to refer to the original which is available on the JORC website (www.jorc.org).

7.2.2 Foreword

The Australasian Code for Reporting of Mineral Resources and Ore Reserves (the 'JORC Code' or 'the Code') sets out minimum standards, recommendations and guidelines for Public Reporting of Exploration Results, Mineral Resources and Ore Reserves in Australasia. It has been drawn up by the Joint Ore Reserves Committee of The Australasian Institute of Mining and Metallurgy, Australian Institute of Geoscientists and the Minerals Council of Australia. The Joint Ore Reserves Committee was established in 1971 and published a number of reports which made recommendations on the classification and Public Reporting of Ore Reserves prior to the first release of the JORC Code in 1989. Revised and updated editions of the Code were issued in 1992, 1996 and 1999. The 2004 edition of the Code (effective December 2004) included in this section supercedes all previous editions.

7.2.3 Introduction

The Sections which belong to the Code are printed in ordinary text. The guidelines, placed after the respective Code clauses to provide improved assistance and guidance to readers, are printed in italics. They do not form part of the Code but should be considered persuasive when interpreting the Code. The same formatting has been applied to Table 7.1 – 'Check List of Assessment and Reporting Criteria' to emphasize that it is not a mandatory list of assessment and reporting criteria.

The Code has been adopted by The Australasian Institute of Mining and Metallurgy and the Australian Institute of Geoscientists and is therefore binding on members of those

Table 7.1. Checklist of assessment and reporting criteria (AusIMM, 2005).

CRITERIA	EXPLANATION
SAMPLING TECHNIQUES AND DATA (criteria in this group apply to all succeeding groups)	
Drilling techniques	Drill type (eg. core, reverse circulation, open-hole hammer, rotary air blast, auger, Bangka etc.) and details (eg. core diameter, triple or standard tube, depth of diamond tails, face-sampling bit or other type, etc.). Measures taken to maximise sample recovery and ensure representative nature of the samples.
Logging	Whether core and chip samples have been logged to a level of detail to support appropriate Mineral Resource estimation, mining studies and metallurgical studies. Whether logging is qualitative or quantitative in nature. Core (or costean, channel etc.) photography.
Drill sample recovery	Whether core and chip sample recoveries have been properly recorded and results assessed. In particular whether a relationship exists between sample recovery and grade and whether sample bias may have occurred due to preferential loss/gain of fine/coarse material.
Other sampling techniques	Nature and quality of sampling (eg. cut channels, random chips etc.) and measures taken to ensure sample representivity.
Sub-sampling techniques and sample preparation	If core, whether cut or sawn and whether quarter, half or all core taken. If non-core, whether riffled, tube sampled, rotary split etc. and whether sampled wet or dry. For all sample types, the nature, quality and appropriateness of the sample preparation technique. Quality control procedures adopted for all sub-sampling stages to maximise representivity of samples. Measures taken to ensure that the sampling is representative of the in situ material collected. Whether sample sizes are appropriate to the grainsize of the material being sampled.
Quality of assay data and laboratory tests	The nature, quality and appropriateness of the assaying and laboratory procedures used and whether the technique is considered partial or total. Nature of quality control procedures adopted (eg. standards, blanks, duplicates, external laboratory checks) and whether acceptable levels of accuracy (ie. lack of bias) and precision have been established.
Verification of sampling and assaying	The verification of significant intersections by either independent or alternative company personnel. The use of twinned holes.
Location of data points	Accuracy and quality of surveys used to locate drill holes (collar and down-hole surveys), trenches, mine workings and other locations used in Mineral Resource estimation. Quality and adequacy of topographic control.
Data density and distribution	Data density for reporting of exploration results. Whether the data density and distribution is sufficient to establish the degree of geological and grade continuity appropriate for the Mineral Resource and Ore Reserve estimation procedure(s) and classifications applied. Whether sample compositing has been applied.
Audits or reviews	The results of any audits or reviews of sampling techniques and data.
REPORTING OF EXPLORATION RESULTS (criteria listed in the preceding group apply also to this group)	
Mineral tenement and land tenure status	Type, reference name/number, location and ownership including agreements or material issues with third parties such as joint ventures, partnerships, overriding royalties, native title interests, historical sites, wilderness or national park and environmental settings. In particular the security of the tenure held at the time of reporting along with any known impediments to obtaining a licence to operate in the area.

(Continued)

Table 7.1. (Continued).

CRITERIA	EXPLANATION
REPORTING OF EXPLORATION RESULTS (criteria listed in the preceding group apply also to this group)	
Exploration done by other parties	Acknowledgement and appraisal of exploration by other parties.
Geology	Deposit type, geological setting and style of mineralisation.
Data aggregation methods	In reporting exploration results, weighting averaging techniques, maximum and/or minimum grade truncations (eg. cutting of high grades) and cut-off grades are usually material and should be stated. Where aggregate intercepts incorporate short lengths of high grade results and longer lengths of low grade results, the procedure used for such aggregation should be stated and some typical examples of such aggregations should be shown in detail. The assumptions used for any reporting of metal equivalent values should be clearly stated.
Relationship between mineralisation widths and intercept lengths	These relationships are particularly important in the reporting of exploration results. If the geometry of the mineralisation with respect to the drill hole angle is known, its nature should be reported. If it is not known and only the down-hole lengths are reported, there should be a clear statement to this effect (eg. 'downhole length, true width not known').
Diagrams	Where possible, maps and sections (with scales) and tabulations of intercepts should be included for any material discovery being reported if such diagrams significantly clarify the report.
Balanced reporting	Where comprehensive reporting of all exploration results is not practicable, representative reporting of both low and high grades and/or widths should be practised to avoid misleading reporting of exploration results.
Other substantive exploration data	Other exploration data, if meaningful and material, should be reported including (but not limited to): geological observations; geophysical survey results; geochemical survey results; bulk samples – size and method of treatment; metallurgical test results; bulk density, groundwater, geotechnical and rock characteristics; potential deleterious or contaminating substances.
Further work	The nature and scale of planned further work (eg. tests for lateral extensions or depth extensions or large-scale step-out drilling).
ESTIMATION AND REPORTING OF MINERAL RESOURCES (criteria listed in the first group, and where relevant in the second group, apply also to this group)	
Database integrity	Measures taken to ensure that data has not been corrupted by, for example, transcription or keying errors, between its initial collection and its use for Mineral Resource estimation purposes. Data validation procedures used.
Geological interpretation	Nature of the data used and of any assumptions made. The effect, if any, of alternative interpretations on Mineral Resource estimation. The use of geology in guiding and controlling Mineral Resource estimation. The factors affecting continuity both of grade and geology.
Estimation and modelling techniques	The nature and appropriateness of the estimation technique(s) applied and key assumptions, including treatment of extreme grade values, domaining, interpolation parameters, maximum distance of extrapolation from data points. The availability of check estimates, previous estimates and/or mine production records and whether the Mineral Resource estimate takes appropriate account of such data. The assumptions made regarding recovery of by-products. In the case of block model interpolation, the block size in relation to the average sample

(Continued)

Table 7.1. (Continued).

CRITERIA	EXPLANATION
ESTIMATION AND REPORTING OF MINERAL RESOURCES (criteria listed in the first group, and where relevant in the second group, apply also to this group)	
	spacing and the search employed. Any assumptions behind modelling of selective mining units (eg. non-linear kriging). The process of validation, the checking process used, the comparison of model data to drillhole data, and use of reconciliation data if available.
Cut-off grades or parameters	The basis of the cut-off grade(s) or quality parameters applied, including the basis, if appropriate, of equivalent metal formulae.
Mining factors or assumptions	Assumptions made regarding possible mining methods, minimum mining dimensions and internal (or, if applicable, external) mining dilution. It may not always be possible to make assumptions regarding mining methods and parameters when estimating Mineral Resources. Where no assumptions have been made, this should be reported.
Metallurgical factors or assumptions	The basis for assumptions or predictions regarding metallurgical amenability. It may not always be possible to make assumptions regarding metallurgical treatment processes and parameters when reporting Mineral Resources. Where no assumptions have been made, this should be reported.
Tonnage factors (in situ bulk densities)	Whether assumed or determined. If assumed, the basis for the assumptions. If determined, the method used, the frequency of the measurements, the nature, size and representativeness of the samples.
Classification	The basis for the classification of the Mineral Resources into varying confidence categories. Whether appropriate account has been taken of all relevant factors. ie. relative confidence in tonnage/grade computations, confidence in continuity of geology and metal values, quality, quantity and distribution of the data. Whether the result appropriately reflects the Competent Person(s)' view of the deposit.
Audits or reviews	The results of any audits or reviews of Mineral Resource estimates.
ESTIMATION AND REPORTING OF ORE RESERVES (criteria listed in the first group, and where relevant in other preceding group, apply also to this group)	
Mineral Resource estimate for conversion to Ore Reserves	Description of the Mineral Resource estimate used as a basis for the conversion to an Ore Reserve. Clear statement as to whether the Mineral Resources are reported additional to, or inclusive of, the Ore Reserves.
Cut-off grades or parameters	The basis of the cut-off grade(s) or quality parameters applied, including the basis, if appropriate, of equivalent metal formulae. The cut-off grade parameter may be economic value per block rather than metal grade.
Mining factors or assumptions	The method and assumptions used to convert the Mineral Resource to an Ore Reserve (ie either by application of appropriate factors by optimisation or by preliminary or detailed design). The choice of, the nature and the appropriateness of the selected mining method(s) and other mining parameters including associated design issues such as pre-strip, access, etc. The assumptions made regarding geotechnical parameters (eg. pit slopes, stope sizes, etc.), grade control and pre-production drilling. The major assumptions made and Mineral Resource model used for pit optimisation (if appropriate). The mining dilution factors, mining recovery factors, and minimum mining widths used and the infrastructure requirements of the selected mining methods.

(Continued)

Table 7.1. (Continued).

CRITERIA	EXPLANATION
ESTIMATION AND REPORTING OF ORE RESERVES (criteria listed in the first group, and where relevant in other preceding group, apply also to this group)	
Metallurgical factors or assumptions	The metallurgical process proposed and the appropriateness of that process to the style of mineralisation. Whether the metallurgical process is well-tested technology or novel in nature. The nature, amount and representativeness of metallurgical testwork undertaken and the metallurgical recovery factors applied. Any assumptions or allowances made for deleterious elements. The existence of any bulk sample or pilot scale testwork and the degree to which such samples are representative of the orebody as a whole.
Cost and revenue factors	The derivation of, or assumptions made, regarding projected capital and operating costs. The assumptions made regarding revenue including head grade, metal or commodity price(s), exchange rates, transportation and treatment charges, penalties, etc. The allowances made for royalties payable, both Government and private.
Market assessment	The demand, supply and stock situation for the particular commodity, consumption trends and factors likely to affect supply and demand into the future. A customer and competitor analysis along with the identification of likely market windows for the product. Price and volume forecasts and the basis for these forecasts. For industrial minerals the customer specification, testing and acceptance requirements prior to a supply contract.
Others	The effect, if any, of natural risk, infrastructure, environmental, legal, marketing, social or governmental factors on the likely viability of a project and/or on the estimation and classification of the Ore Reserves. The status of titles and approvals critical to the viability of the project, such as mining leases, discharge permits, government and statutory approvals.
Classification	The basis for the classification of the Ore Reserves into varying confidence categories. Whether the result appropriately reflects the Competent Person(s)' view of the deposit. The proportion of Probable Ore Reserves which have been derived from Measured Mineral Resources (if any).
Audits or reviews	The results of any audits or reviews of Ore Reserve estimates.

organizations. It is endorsed by the Minerals Council of Australia and the Securities Institute of Australia as a contribution to best practice.

The JORC Code requires the Competent Person(s) on whose work the Public Report of Exploration Results, Mineral Resources or Ore Resources is based, to be named in the report. The report or attached statement must say that the person consents to the inclusion in the report of the matters based on their information in the form and context in which it appears, and must include the name of the person's firm or employer.

7.2.4 Scope

The main principles governing the operation and application of the JORC Code are transparency, materiality and competence. 'Transparency' requires that the reader of a Public Report is provided with sufficient information, the presentation of which is clear and unambiguous, to understand the report and is not misled. 'Materiality' requires that a Public Report contains all the relevant information which investors and their professional advisers would reasonably require, and reasonably expect to find in the report, for the purpose of making a reasoned and balanced judgment regarding the Exploration Results, Mineral Resources or Ore Reserves being reported. 'Competence' requires that the Public Report is based on work which is the responsibility of suitably qualified and experienced persons who are subject to an enforceable professional code of ethics.

Reference in the Code to a Public Report or Public Reporting is to a report or reporting on Exploration Results, Mineral Resources or Ore Reserves, prepared for the purpose of informing investors or potential investors and their advisors. This includes a report or reporting to satisfy regulatory requirements.

The Code is a required minimum standard for Public Reporting. JORC also recommends its adoption as a minimum standard for other reporting. Companies are encouraged to provide information in their Public Reports which is as comprehensive as possible.

Public Reports include, but are not limited to: company annual reports, quarterly reports and other reports to the Australian or New Zealand Stock Exchanges or required by law. The Code applies to other publicly released company information in the form of postings on company web sites and briefings for shareholders, stockbrokers and investment analysts. The Code also applies to the following: environmental statements; Information Memoranda; Expert Reports, and technical papers referring to Exploration Results, Mineral Resources or Ore Reserves.

The term 'regulatory requirements' is not intended to cover reports provided to State and Government agencies for statuary purposes, where providing information to the investing public is not the primary intent. If such reports become available to the public, they would not normally be regarded as Public Reports under the JORC Code.

It is recognized that situations may arise where documentation prepared by Competent Persons for internal company purposes or similar non-public purposes does not comply with the JORC Code. In such situations, it is recommended that the documentation includes a prominent statement to this effect.

While every effort has been made within the Code and Guidelines to cover most situations likely to be encountered in the Public Reporting of exploration results, Mineral Resources and Ore Reserves, there may be occasions when doubt exists as to the appropriate form of disclosure. On such occasions, users of the Code and those compiling reports to comply with the Code should be guided by its intent, which is to provide a minimum standard for Public Reporting and to ensure that such reporting contains all information which investors and their professional advisers would reasonably require, and reasonably expect to find in the report, for the purpose of making a reasoned and balanced judgment regarding the Exploration Results, Mineral Resources or Ore Reserves being reported.

The Code is applicable to all solid minerals, including diamonds, other gemstones, industrial minerals and coal, for which Public Reporting of Exploration Results, Mineral Resources and Ore Reserves is required by the Australian and New Zealand Stock Exchanges.

JORC recognizes that further review of the Code and Guidelines will be required from time to time.

7.2.5 *Competence and responsibility*

A Public Report concerning a company's Exploration Results, Mineral Resources or Ore Reserves is the responsibility of the company acting through its Board of Directors. Any such report must be based on, and fairly reflect the information and supporting documentation prepared by a Competent Person or Persons. A company issuing a Public Report shall disclose the name(s) of the Competent Person or Persons, state whether the Competent Person is a full-time employee of the company, and, if not, name the Competent Person's employer. The report shall be issued with the written consent of the Competent Person or persons as to the form and context in which it appears. Documentation detailing Exploration Results, Mineral Resource and Ore Reserve estimates, on which a Public report on Exploration Results, Mineral Resources and Ore Reserves is based, must be prepared by, or under the direction of, and signed by, a Competent Person or Persons. The documentation must provide a fair representation of the Exploration Results, Mineral Resources or Ore reserves being reported.

A 'Competent Person' is a person who is a Member or Fellow of The Australasian Institute of Mining and Metallurgy or of the Australian Institute of Geoscientists, or of a 'Recognized Overseas Professional Organization' ('ROPO') included in a list promulgated from time to time. A 'Competent Person' must have a minimum of five years experience which is relevant to the style of mineralization and type of deposit under consideration and to the activity which that person is undertaking. If the Competent Person is preparing a report on Exploration results, the relevant experience must be in exploration. If the Competent Person is supervising the estimation of Mineral Resources, the relevant experience must be in the estimation, assessment and evaluation of Mineral Resources. If the Competent Person is estimating, or supervising the estimation of Ore Reserves, the relevant experience must be in the estimation, assessment, evaluation and economic extraction of Ore Reserves.

The key qualifier in the definition of a Competent Person is the word 'relevant'. Determination of what constitutes relevant experience can be a difficult area and common sense has to be exercised. For example, in estimating Mineral Resources for vein gold mineralization, experience in a high-nugget, vein-type mineralization such as tin, uranium etc. will probably be relevant whereas experience in (say) massive base metal deposits may not be. As a second example, for a person to qualify as a Competent Person in the estimation of Ore Reserves for alluvial gold deposits, considerable (probably at least five years) experience in the evaluation and economic extraction of this type of mineralization would be needed. This is due to the characteristics of gold in alluvial systems, the particle sizing of the host sediment, and the low grades involved. Experience with placer deposits containing minerals other than gold may not necessarily provide appropriate relevant experience.

The key word 'relevant' also means that it is not always necessary for a person to have five years experience in each and every type of deposit in order to act as a Competent Person if that person has relevant experience in other deposit types. For example, a person with (say) 20 years experience in estimating Mineral Resources for a variety of metalliferous hard-rock deposit types may not require five years specific experience in (say) porphyry copper deposits in order to act as a Competent Person. Relevant experience in the other deposit types could count towards the required experience in relation to porphyry copper deposits.

In addition to experience in the style of mineralization, a Competent Person taking responsibility for the compilation of Exploration Results or Mineral Resource estimates should have sufficient experience in the sampling and analytical techniques relevant to the deposit under consideration to be aware of problems which could affect the reliability of the data. Some appreciation of extraction and processing techniques applicable to that deposit type would also be important.

As a general guide, persons being called upon to act as Competent Persons should be clearly satisfied in their own minds that they could face their peers and demonstrate competence in the commodity, type of deposit and situation under consideration. If doubt exists, the person should either seek opinions from appropriately experienced colleagues or should decline to act as a Competent Person.

Estimation of Mineral Resources may be a team effort (for example, involving one person or team collecting the data and another person or team preparing the estimate). Estimation of Ore Reserves is very commonly a team effort involving several technical disciplines It is recommended that, where there is a clear division of responsibility within a team, each Competent Person and his or her contribution should be identified, and responsibility accepted for that particular contribution. If only one Competent Person signs the Mineral Resource or Ore Reserve documentation, that person is responsible and accountable for the whole of the documentation under the Code. It is important in this situation that the Competent Person accepting overall responsibility for a Mineral Resource or Ore Reserve estimate and supporting documentation prepared in whole or in part by others, is satisfied that the work of the other contributors is acceptable.

Complaints made in respect of the professional work of a Competent Person will be dealt with under the disciplinary procedures of the professional organization to which the Competent Person belongs.

When an Australian listed or New Zealand listed company with overseas interests wishes to report an overseas Exploration Results, Mineral Resource or Ore Reserve estimates prepared by a person who is not a member of The AusIMM, the AIG or a ROPO, it is necessary for the company to nominate a Competent Person or Persons to take responsibility for the Exploration Results, Mineral Resource or Ore Reserve estimate. The Competent Person or Persons undertaking this activity should appreciate that they are accepting full responsibility for the estimate and supporting documentation under Stock Exchange listing rules and should not treat the procedure merely as a 'rubber-stamping' exercise.

7.2.6 Reporting terminology

Public Reports dealing with Exploration Results, Mineral Resources and/or Ore Reserves must only use the terms set out in Figure 7.1.

The term 'Modifying Factors' is defined to include mining, metallurgical, economic, marketing, legal, environmental, social and governmental considerations.

Figure 7.1 sets out the framework for classifying tonnage and grade estimates to reflect different levels of geological confidence and different degrees of technical and economic evaluation. Mineral Resources can be estimated mainly by a geologist on the basis of geo-scientific information with some input from other disciplines. Ore Reserves, which are a modified sub-set of the Indicated and Measured Mineral Resources (shown within the dashed outline in Figure 7.1), require consideration of the Modifying Factors affecting extraction, and should in most instances be estimated with input from a range of disciplines.

Measured Mineral Resources may convert to either Proved Ore Reserves or Probable Ore Reserves. The Competent Person may convert Measured Mineral Resources to Probable Ore Reserves because of uncertainties associated with some or all of the Modifying Factors which are taken into account in the conversion from Mineral Resources to Ore Reserves. This relationship is shown by the broken arrow in Figure 7.1. Although the trend of the broken arrow includes a vertical component, it does not, in this instance, imply a reduction

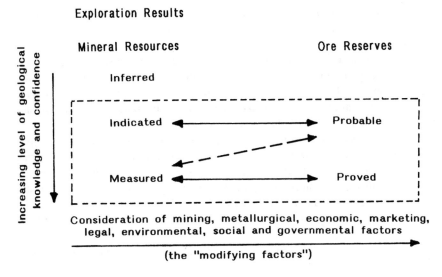

Figure 7.1. General relationship between Exploration Results, Mineral Resources and Ore Reserves. AusIMM. 2005.

in the level of geological knowledge or confidence. In such a situation these Modifying Factors should be fully explained.

7.2.7 *Reporting – General*

Public Reports concerning a company's Exploration Results, Mineral Resources or Ore Reserves should include a description of the style and nature of mineralization.

A company must disclose relevant information concerning a mineral deposit which could materially influence the economic value of that deposit to the company. A company must promptly report any material changes in its Mineral Resources or Ore Reserves.

Companies must review and publicly report on their Mineral Resources and Ore Reserves at least annually.

Throughout the Code, if appropriate, 'quality' may be substituted for 'grade' and 'volume' may be substituted for 'tonnage'.

7.2.8 *Reporting of Exploration Results*

Exploration Results include data and information generated by exploration programs that may be of use to investors. The Exploration Results may or may not be part of a formal declaration of Mineral Resources or Ore reserves.

The reporting of such information is common in the early stages of exploration when the quantity of data available is generally not sufficient to allow any reasonable estimates of Mineral resources.

If a company reports Exploration Results in relation to mineralization not classified as a Mineral Resource or an Ore Reserve, then estimates of tonnage and average grade must not be assigned to the mineralization except under very strict conditions.

Examples of Exploration Results include results of outcrop sampling, assays of drill hole intercepts, geochemical results and geophysical survey results.

Public Reports of Exploration Results must contain sufficient information to allow a considered and balanced judgment of their significance. Reports must include relevant information such as exploration context, type and method of sampling, sampling intervals and methods, relevant sample locations, distribution, dimensions and relative location of all relevant assay data, data aggregation methods, land tenure status plus information on any of the other criteria listed in Table 7.1 that are material to an assessment.

Public Reports of Exploration Results must not be presented so as to unreasonably imply that potentially economic mineralization has been discovered. If true widths of mineralization are not reported, an appropriate qualification must be included in the Public report.

Where assay and analytical results are reported, they must be reported using one of the following methods, selected as the most appropriate by the Competent Person:
- either by listing all results, along with sample intervals (or size, in the case of bulk samples), or
- by reporting weighted average grades of mineralized zones, indicating clearly how grades were calculated.

Reporting of selected information such as isolated assays, isolated drill holes, assays of panned concentrates or supergene enriched soils or surface samples, without placing them in perspective is unacceptable.

Table 7.1 is a checklist and guideline to which those preparing reports on Exploration Results, Mineral Resources and Ore Reserves should refer. The check list is not prescriptive and, as always, relevance and materiality are overriding principles which determine what information should be publicly reported.

It is recognized that it is common practice for a company to comment on and discuss its exploration in terms of target size and type. Any such information relating to exploration targets must be expressed so that it cannot be misrepresented or misconstrued as an estimate of Mineral Resources or Ore reserves. The terms Resource(s) or Reserve(s) must not be used in this context. Any statement referring to potential quantity and grade of the target must be expressed as ranges and must include (1) a detailed explanation of the basis for the statement, and (2) a proximate statement that the potential quantity and grade is conceptual in nature, that there has been insufficient exploration to define a Mineral Resource and that it is uncertain if further exploration will result in the determination of a Mineral Resource.

7.2.9 Reporting of Mineral Resources

A 'Mineral Resource' is a concentration or occurrence of material of intrinsic economic interest in or on the Earth's crust in such form and quantity that there are reasonable prospects for eventual economic extraction. The location, quantity, grade, geological characteristics and continuity of a Mineral Resource are known, estimated or interpreted from specific geological evidence and knowledge. Mineral Resources are sub-divided, in order of increasing geological confidence, into Inferred, Indicated and Measured categories.

Portions of a deposit that do not have reasonable prospects for eventual economic extraction must not be included in a Mineral Resource. If the judgment as to 'eventual economic extraction' relies on untested practices or assumptions, this is a material matter which must be disclosed in a public report.

The term 'Mineral Resource' covers mineralization, including dumps and tailings, which has been identified and estimated through exploration and sampling and within which Ore Reserves may be defined by the consideration and application of the Modifying Factors.

The term 'reasonable prospects for eventual economic extraction' implies a judgment (albeit preliminary) by the Competent Person in respect of the technical and economic factors likely to influence the prospect of economic extraction, including the approximate mining parameters. In other words, a Mineral Resource is not an inventory of all mineralization drilled or sampled, regardless of cut-off grade, likely mining dimensions, location or continuity. It is a realistic inventory of mineralization which, under assumed and justifiable technical and economic conditions, might, in whole or in part, become economically extractable.

Where considered appropriate by the Competent Person, Mineral Resource estimates may include material below selected cut-off grade to ensure that the Mineral resources comprise bodies of mineralization of adequate size and continuity to properly consider the most appropriate approach to mining. Documentation of Mineral Resource estimates should clearly identify any diluting material included, and Public reports should include commentary on the matter if considered material.

Any material assumptions made in determining the 'reasonable prospects for eventual economic extraction' should be clearly stated in the Public Report.

Interpretation of the word 'eventual' in this context may vary depending on the commodity or mineral involved. For example, for some coal, iron ore, bauxite and other bulk minerals or commodities, it may be reasonable to envisage 'eventual economic extraction' as covering time periods in excess of 50 years. However for the majority of gold deposits, application of the concept would normally be restricted to perhaps 10 to 15 years, and frequently to much shorter periods of time.

Any adjustment made to the data for the purpose of making the Mineral resource estimate, for example by cutting or factoring grades, should be clearly stated and described in the Public Report.

Certain reports (e.g. coal inventory reports, exploration reports to government and other similar reports not intended primarily for providing information for investment purposes) may require full disclosure of all mineralization, including some material that does not have reasonable prospects for eventual economic extraction. Such estimates of mineralization would not qualify as Mineral Resources or Ore Reserves in terms of the JORC.

An 'Inferred Mineral Resource' is that part of a Mineral Resource for which tonnage, grade and mineral content can be estimated with a low level of confidence. It is inferred from geological evidence and assumed but not verified geological and/or grade continuity. It is based on information gathered through appropriate techniques from locations such as outcrops, trenches, pits, workings and drill holes which may be limited or of uncertain quality and reliability.

An Inferred Mineral Resource has a lower level of confidence than that applying to an Indicated Mineral Resource.

The inferred category is intended to cover situations where a mineral concentration or occurrence has been identified and limited measurements and sampling completed, but where the data are insufficient to allow the geological and/or grade continuity to be confidently interpreted. Commonly, it would be reasonable to expect that the majority of Inferred Mineral Resources would upgrade to Indicated Mineral Resources with continued exploration. However, due to the uncertainty of Inferred Mineral Resources, should not be assumed that such upgrading will always occur.

Confidence in the estimate of Inferred Mineral resources is usually not sufficient to allow the results of the application of technical and economic parameters to be used for detailed planning. For this reason, there is not direct link from an Inferred Resource to any category of Ore Reserves (see Figure 1). Caution should be exercised if this category is considered in technical and economic studies.

An 'Indicated Mineral Resource' is that part of a Mineral Resource for which tonnage, densities, shape, physical characteristics, grade and mineral content can be estimated with a reasonable level of confidence. It is based on exploration, sampling and testing information gathered through appropriate techniques from locations such as outcrops, trenches, pits, workings and drill holes. The locations are too widely or inappropriately spaced to confirm geological and/or grade continuity but are spaced closely enough for continuity to be assumed.

An Indicated Mineral Resource has a lower level of confidence than that applying to a Measured Mineral Resource, but has a higher level of confidence than that applying to an Inferred Mineral Resource.

Mineralization may be classified as an Indicated Mineral Resource when the nature, quality, amount and distribution of data are such as to allow confident interpretation of the geological framework and to assume continuity of mineralization. Confidence in the estimate is sufficient to allow the appropriate application of technical and economic parameters and to enable an evaluation of economic viability.

A 'Measured Mineral Resource' is that part of a Mineral Resource for which tonnage, densities, shape, physical characteristics, grade and mineral content can be estimated with a high level of confidence. It is based on detailed and reliable exploration, sampling and testing information gathered through appropriate techniques from locations such as outcrops, trenches, pits, workings and drill holes. The locations are spaced closely enough to confirm geological and/or grade continuity.

Mineralization may be classified as a Measured Mineral Resource when the nature, quality, amount and distribution of data are such as to leave no reasonable doubt, in the opinion of the Competent Person determining the Mineral Resource, that the tonnage and grade of the mineralization can be estimated to within close limits and that any variation from the estimate would not significantly affect potential economic viability.

This category requires a high level of confidence in, and understanding of, the geology and controls of the mineral deposit.

Confidence in the estimate is sufficient to allow the application of technical and economic parameters and to enable an evaluation of economic viability that has a greater degree of certainty than an evaluation based on an Indicated Mineral Resource.

The choice of the appropriate category of Mineral Resource depends upon the quantity, distribution and quality of data available and the level of confidence that attaches to those data. The appropriate Mineral Resource category must be determined by a Competent Person or Persons.

Mineral Resource classification is a matter for skilled judgment and Competent Persons should take into account those items in Table 7.1 which relate to confidence in Mineral Resource estimation.

In deciding between Measured Mineral Resources and Indicated Mineral Resources, Competent Persons may find it useful to consider, in addition to the phrases in the two definitions relating to geological and grade continuity, the phrase in the guideline to the definition for Measured Mineral Resources: '.... any variation from the estimate would be unlikely to significantly affect potential economic viability'.

In deciding between Indicated Mineral Resources and Inferred Mineral Resources, Competent Persons may wish to take into account, in addition to the phrases in the two definitions relating to geological and grade continuity, the guideline to the definition for Indicated Mineral Resources: 'Confidence in the estimate is sufficient to allow the application of technical and economic parameters and to enable an evaluation of economic viability', which contrasts with the guideline to the definition for Inferred Mineral Resources: 'Confidence in the estimate of Inferred Resources is usually not sufficient to allow the results of the application of technical and economic parameters to be used for detailed planning' and 'Caution should be exercised if this category is considered in technical and economic studies'.

The Competent Person should take into consideration issues of the style of mineralization and cutoff grade when assessing geological and grade continuity.

Cutoff grades chosen for the estimation should be realistic in relation to the style of mineralization.

Mineral Resource estimates are not precise calculations, being dependent on the interpretation of limited information on the location, shape and continuity of the occurrence and on the available sampling results. Reporting of tonnage and grade figures should reflect the order of accuracy of the estimate by rounding off to appropriately significant figures and, in the case of Inferred Mineral Resources, by qualification with terms such as 'approximately'.

In most situations, rounding to the second significant figure should be sufficient. For example 10,863,000 tonnes at 8.23 per cent should be stated as 11 million tonnes at 8.2 per cent. There will be occasions, however, where rounding to the first significant figure may be necessary in order to convey properly the uncertainties in estimation. This would usually be the case with Inferred Mineral Resources.

To emphasize the imprecise nature of a Mineral Resource estimate, the final result should always be referred to as an estimate not a calculation.

Competent Persons are encouraged, where appropriate, to discuss the relative accuracy and/or confidence of the Mineral resource estimates. The statement should specify whether it relates to global or local estimates, and, if local, state the relevant tonnage or volume. Where a statement of the relative accuracy and/or confidence is not possible, a qualitative discussion of the uncertainties should be provided (refer to Table 7.1).

Public Reports of Mineral Resources must specify one or more of the categories of 'Inferred', 'Indicated' and 'Measured'. Categories must not be reported in a combined form unless details for the individual categories are also provided. Mineral Resources must not be reported in terms of contained metal or mineral content unless corresponding tonnages and grades are also presented. Mineral Resources must not be aggregated with Ore Reserves.

Public Reporting of tonnages and grades outside the categories covered by the Code is generally not permitted.

Estimates of tonnage and grade outside of the categories covered by the Code may be useful for a company in its internal calculations and evaluation processes, but their inclusion in Public Reports could cause confusion.

Table 7.1 provides, in a summary form, a list of the main criteria which should be considered when preparing reports on Exploration Results, Mineral Resources and Ore Reserves. These criteria need not be discussed in a Public Report unless they materially affect estimation or classification of the Mineral Resources.

It is not necessary, when publicly reporting, to comment on each item in Table 7.1, but it is essential to discuss any matters which might materially affect the reader's understanding or interpretation of the results or estimates being reported. This is particularly important where inadequate or uncertain data affect the reliability of, or confidence in, a statement of Exploration Results or an estimate of Mineral Resources or Ore Reserves; for example, poor sample recovery, poor repeatability of assay or laboratory results, limited information on bulk densities, etc.

If there is doubt about what should be reported, it is better to err on the side of providing too much information rather than too little.

Uncertainties in any of the criteria listed in Table 7.1 that could lead to under- or over-statement of resources should be disclosed.

Mineral Resource estimates are sometimes reported after adjustment from reconciliation with production data. Such adjustments should be clearly stated in a Public Report of Mineral Resources and the nature of the adjustment or modification described.

The words 'ore' and 'reserves' must not be used in describing Mineral Resource estimates as the terms imply technical feasibility and economic viability and are only appropriate when all relevant Modifying Factors have been considered. Reports and statements should continue to refer to the appropriate category or categories of Mineral Resources until technical feasibility and economic viability have been established. If re-evaluation indicates that the Ore Reserves are no longer viable, the Ore Reserves must be reclassified as Mineral Resources or removed from Mineral Resource/Ore Reserve statements.

It is not intended that re-classification from Ore Reserves to Mineral Resources or vice-versa should be applied as a result of changes expected to be of a short term or temporary nature, or where company management has made a deliberate decision to operate on a non-economic basis. Examples of such situations might be a commodity price fluctuations expected to be of short duration, mine emergency of a non-permanent nature, transport strike etc.

7.2.10 *Reporting of Ore Reserves*

An 'Ore Reserve' is the economically mineable part of a Measured and/or Indicated Mineral Resource. It includes diluting materials and allowances for losses, which may occur when the material is mined. Appropriate assessments and studies have been carried out, and include consideration of and modification by realistically assumed mining, metallurgical, economic, marketing, legal, environmental, social and governmental factors. These assessments demonstrate at the time of reporting that extraction could reasonably be justified. Ore Reserves are sub-divided in order of increasing confidence into Probable Ore Reserves and Proved Ore Reserves.

In reporting Ore Reserves, information on estimated mineral processing recovery factors is very important, and should always be included in Public Reports.

Ore Reserves are those portions of Mineral Resources which, after the application of all mining factors, result in an estimated tonnage and grade which, in the opinion of the Competent Person making the estimates, can be the basis of a viable project after taking account of all relevant Modifying Factors.

Ore Reserves are reported as inclusive of marginally economic material and diluting material delivered for treatment or dispatched from the mine without treatment.

The term 'economically mineable' implies that extraction of the Ore Reserve has been demonstrated to be viable under reasonable financial assumptions. What constitutes the term 'realistically assumed' will vary with the type of deposit, the level of study that has been carried out and the financial criteria of the individual company. For this reason, there can be no fixed definition for the term 'economically mineable'.

In order to achieve the required level of confidence in the Modifying Factors, appropriate studies will have been carried out prior to the determination of Ore Reserves. The studies will have determined a mine plan that is technically achievable and economically viable and from which the Ore Reserves can be derived. It may not be necessary for these studies to be at the level of a final feasibility study.

The term 'Ore Reserve' need not necessarily signify that extraction facilities are in place or operative, or that all necessary approvals have been received. It does signify that there are reasonable expectations of such approvals or contracts. The Competent Person should consider the materiality of any unresolved matter that is dependent on a third party on which extraction is contingent. If there is doubt about what should be reported, it is better to err on the side of providing too much information rather than too little.

Any adjustment made to the data for the purpose of making the Ore Reserve estimate, for example by cutting of factoring grades, should be clearly stated and described in the Public Report.

Where companies prefer to use the term 'Mineral Reserves' in their Public Reports, e.g. for reporting industrial minerals or for reporting outside Australasia, they should state clearly that this is being used with the same meaning as 'Ore Reserves', defined in this Code. If preferred by the reporting company, 'Ore Reserve' and 'Mineral Resource' estimates for coal may be reported as 'Coal Reserve' and 'Coal Resource' estimates.

JORC prefers the term 'Ore Reserve' because it assists in maintaining a clear distinction between a 'Mineral Resource' and an 'Ore Reserve'.

A 'Probable Ore Reserve' is the economically mineable part of an Indicated, and in some circumstances, a Measured Mineral Resource. It includes diluting materials and allowances for losses which may occur when the material is mined. Appropriate assessments and studies have been carried out, and include consideration of and modification by realistically assumed mining, metallurgical, economic, marketing, legal, environmental, social and governmental factors. These assessments demonstrate at the time of reporting that extraction could reasonably be justified.

A Probable Ore Reserve has a lower level of confidence than a Proved Ore Reserve but is of sufficient quality to serve as the basis for a decision on the development of the deposit.

A 'Proved Ore Reserve' is the economically mineable part of a Measured Mineral Resource. It includes diluting materials and allowances for losses which may occur when the material is mined. Appropriate assessments and studies have been carried out, and include consideration of and modification by realistically assumed mining, metallurgical, economic, marketing, legal, environmental, social and governmental factors. These assessments demonstrate at the time of reporting that extraction could reasonably be justified.

A Proved Ore Reserve represents the highest confidence category of reserve estimate. The style of mineralization or other factors could mean that Proved Ore Reserves are not achievable in some deposits.

The choice of the appropriate category of Ore Reserve is determined primarily by the relevant level of confidence in the Mineral Resource and after considering any uncertainties in the Modifying Factors. Allocation of the appropriate category must be made by a Competent Person or Persons.

The Code provides for a direct two-way relationship between Indicated Mineral Resources and Probable Ore Reserves and between Measured Mineral Resources and Proved Ore Reserves. In other words, the level of geological confidence for Probable Ore Reserves is similar to that required for the determination of Indicated Mineral Resources, and the level of geological confidence for Proved Ore Reserves is similar to that required for the determination of Measured Mineral Resources.

The Code also provides for a two-way relationship between Measured Mineral Resources and Probable Ore Reserves. This is to cover a situation where uncertainties associated with any of the Modifying Factors considered when converting Mineral Resources to Ore Reserves may result in there being a lower degree of confidence in the Ore Reserves than in the corresponding Mineral Resources. Such a conversion would not imply a reduction in the level of geological knowledge or confidence.

A Probable Ore Reserve derived from a Measured Mineral Resource may be converted to a Proved Ore Reserve if the uncertainties in the Modifying Factors are removed. No amount of confidence in the Modifying Factors for conversion of a Mineral Resource to an Ore Reserve can override the upper level of confidence that exists in the Mineral Resource. Under no circumstances can an Indicated Mineral Resource be converted directly to a Proved Ore Reserve (see Figure 7.1).

Application of the category of Proved Ore Reserve implies the highest degree of confidence in the estimate, with consequent expectations in the minds of readers of the report. These expectations should be borne in mind when categorizing a Mineral Resource as Measured.

Ore Reserve estimates are not precise calculations. Reporting of tonnage and grade figures should reflect the relative uncertainty of the estimate by rounding off to appropriately significant figures.

To emphasize the imprecise nature of an Ore Reserve, the final result should always be referred to as an estimate not a calculation.

Competent Persons are encouraged, where appropriate, to discuss the relative accuracy and/or confidence of the Ore Reserve estimates. The statement should specify whether it relates to global or local estimates, and, if local, state the relevant tonnage or volume. Where a statement of the relative accuracy and/or confidence is not possible, a qualitative discussion of the uncertainties should be provided (refer to Table 7.1).

Public Reports of Ore Reserves must specify one or other or both of the categories of 'Proved' and 'Probable'. Reports must not contain combined Proved and Probable Ore Reserve figures unless the relevant figures for each of the individual categories are also provided. Reports must not present metal or mineral content figures unless corresponding tonnage and grade figures are also given.

Public Reporting of tonnage and grade outside the categories covered by the Code is generally not permitted.

Estimates of tonnage and grade outside of the categories covered by the Code may be useful for a company in its internal calculations and evaluation processes, but their inclusion in Public Reports could cause confusion.

Ore Reserves may incorporate material (dilution) which is not part of the original Mineral Resource. It is essential that this fundamental difference between Mineral Resources and Ore Reserves is borne in mind and caution exercised if attempting to draw conclusions from a comparison of the two.

When revised Ore Reserve and Mineral Resource statements are publicly reported they should be accompanied by reconciliation with previous statements. A detailed account of differences between the figures is not essential, but sufficient comment should be made to enable significant changes to be understood by the reader.

In situations where figures for both Mineral Resources and Ore Reserves are reported, a statement must be included in the report which clearly indicates whether the Mineral Resources are inclusive of, or additional to the Ore Reserves.

Ore Reserve estimates must not be aggregated with Mineral Resource estimates to report a single figure.

In some situations there are reasons for reporting Mineral Resources inclusive of Ore Reserves and in other situations for reporting Mineral Resources additional to Ore Reserves. It must be made clear which form of reporting has been adopted. Appropriate forms of clarifying statements may be:

'The Measured and Indicated Mineral Resources are inclusive of those Mineral Resources modified to produce the Ore Reserves.' or 'The Measured and Indicated Mineral Resources are additional to the Ore Reserves'.

In the former case, if any Measured and Indicated Mineral Resources have not been modified to produce Ore Reserves for economic or other reasons, the relevant details of these unmodified Mineral Resources should be included in the report. This is to assist the reader of the report in making a judgment of the likelihood of the unmodified Measured and Indicated Mineral Resources eventually being converted to Ore Reserves.

Inferred Mineral Resources are by definition always additional to Ore Reserves.

The reported Ore Reserve figures must not be aggregated with the reported Mineral Resource figures. The resulting total is misleading and is capable of being misunderstood or of being misused to give a false impression of a company's prospects.

Table 7.1 provides, in a summary form, a list of the main criteria which should be considered when preparing reports on Exploration Results, Mineral Resources and Ore Reserves. These criteria need not be discussed in a Public Report unless they materially affect estimation or classification of the Ore Reserves. Changes in economic or political factors alone may be the basis for significant changes in Ore Reserves and should be reported accordingly.

Ore Reserve estimates are sometimes reported after adjustment from reconciliation with production data. Such adjustments should be clearly stated in a Public Report of Ore Reserves and the nature of the adjustment or modification described.

7.2.11 Reporting of mineralized stope fill, stockpiles, remnants, pillars, low grade mineralization and tailings

The Code applies to the reporting of all potentially economic mineralized material. This can include mineralized fill, remnants, pillars, low grade mineralization, stockpiles, dumps and tailings (remnant materials) where there are reasonable prospects for eventual economic extraction in the case of Mineral Resources, and where extraction is reasonably justifiable in the case of Ore Reserves. Unless otherwise stated, all other parts of the Code (including Figure 7.1) apply.

Any mineralized material as described here can be considered to be similar to in situ mineralization for the purposes of reporting Mineral Resources and Ore Reserves. Judgments about the mineability of such mineralized material should be made by professionals with relevant experience.

662 *Open pit mine planning and design: Fundamentals*

If there are no reasonable prospects for the eventual economic extraction of all or part of the mineralized material as described here, then this material cannot be classified as either Mineral Resources or Ore Reserves. If some portion of the mineralized material is sub-economic, but there is a reasonable expectation that it will become economic, then this material may be classified as a Mineral Resource. If technical and economic studies have demonstrated that economic extraction could reasonably be justified under realistically assumed conditions, then the material may be classified as an Ore Reserve.

The above guidelines apply equally to low grade in situ mineralization, sometimes referred to as 'mineralized waste' or 'marginal grade material', and often intended for stockpiling and treatment towards the end of mine life. For clarity of understanding, it is recommended that tonnage and grade estimates of such material be itemized separately in Public Reports, although they may be aggregated with total Mineral Resource and Ore Reserve figures.

Stockpiles are defined to include both surface and underground stockpiles, including broken ore in stopes, and can include ore currently in the ore storage system. Mineralized material in the course of being processed (including leaching), if reported, should be reported separately.

7.3 THE CIM BEST PRACTICE GUIDELINES FOR THE ESTIMATION OF MINERAL RESOURCES AND MINERAL RESERVES – GENERAL GUIDELINES

7.3.1 *Preamble*

The authors wish to express their sincere thanks to the Canadian Institute of Mining, Metallurgy and Petroleum (CIM) for permission to include the 'Estimation of Mineral Resources and Mineral Reserves Best Practices Guidelines' as part of this book. These Guidelines were adopted by the CIM Council on November 23, 2003. The format has been somewhat modified to conform to that used in the rest of the book. The interested reader is encouraged to refer to the original which is available on the CIM website (www.cim.org/committees/estimation.cfm).

7.3.2 *Foreword*

These guidelines have been prepared by the Canadian Institute of Mining and Metallurgy and Petroleum (CIM) led 'Estimation Best Practices Committee'. They are intended to assist the Qualified Person(s) (QP) in the planning, supervision, preparation and reporting of Mineral Resource and Mineral Reserve (MRMR) estimates. All MRMR estimation work from which public reporting will ensue must be designed and carried out under the direction of a QP. A QP is defined as 'an individual who is an engineer or geoscientist with at least five (5) years of experience in mineral exploration, mine development, mine operation, project assessment or any combination of these; has experience relevant to the subject matter of the mineral project and technical report; and is a member in good standing of a professional association'.

The 'General Guidelines' section of the document (that included in this section) deals primarily with the description of best practice as it applies to metalliferous deposits.

In planning, implementing and directing any estimation work, the QP should ensure that practices followed are based on methodology that is generally accepted in the industry and that the provisions of the Exploration Best Practices Guidelines have been adhered to during the exploration phase that led to the delineation of the resource.

In addition to assisting the QP in the preparation of MRMR estimates, these 'Best Practice Guidelines' are intended to ensure a consistently high quality of work and foster greater standardization of reporting in publicly disclosed documents.

Qualified Person

The Qualified Person will base the MRMR estimation work on geological premises, interpretation and other technical information as the QP deems appropriate. In addition, the QP will select an estimation method, parameters and criteria appropriate for the deposit under consideration. In planning, implementing and supervising any estimation work, the QP will ensure that the methods employed and the practices followed can be justified on technical merit and/or are generally accepted in the industry.

Because a MRMR model is based fundamentally on accurate geological interpretation and economic understanding, the persons responsible for the Mineral Resource and subsequent Mineral Reserve estimation should have a firm understanding of geology, mining, and other issues affecting the estimate. This level of understanding would normally be developed through acquiring appropriate geological, mining and Mineral Reserve preparation experience in a relevant operating mine.

While the reporting QP ultimately will have responsibility for the resulting estimate, he or she should have access to other QP, in the compilation of the estimate, who have suitable training or experience in disciplines that may fall outside the expertise of the reporting QP. This will allow appropriate consideration of all factors affecting the estimate including, for example, geology and geological interpretation, metallurgy, mining and social, legal and environmental matters.

Definitions

- Mineralization: 'material of potential interest. Mineral Resources and Mineral Reserves are economic subsets of such mineralization'.

- Quality Assurance/Quality Control (QA/QC): for the purpose of this document;
Quality Assurance means:
'All of those planned or systematic actions necessary to provide adequate confidence in the data collection and estimation process', and
Quality Control means:
'The systems and mechanisms put in place to provide the Quality Assurance. The four steps of quality control include; setting standards; appraising conformance; acting when necessary and planning for improvements'.

- Mineral Resource: a concentration or occurrence of natural, solid, inorganic, or fossilized organic material in or on the Earth's crust in such form and quantity and of such a grade or quality that it has reasonable prospects for economic extraction. The location, quantity, grade, geological characteristics, and continuity of a Mineral Resource are known, estimated or interpreted from specific geological evidence and knowledge'.

- Mineral Reserve: 'the economically mineable part of a Measured or Indicated Mineral Resource demonstrated by at least a Preliminary Feasibility Study. This study must include adequate information on mining, processing, metallurgical, economic, and other relevant factors that demonstrate, at the time of reporting, that economic extraction can be justified. A Mineral Reserve includes diluting materials and allowances for losses that may occur when the material is mined'.

• Estimate: (verb) 'to judge or approximate the value, worth, or significance of; to determine the size, extent, or nature of'. (noun) 'an approximate calculation; a numerical value obtained from a statistical sample and assigned to a population parameter'.

• Preliminary Feasibility Study: 'a comprehensive study of the viability of a mineral project that has advanced to the stage where the mining method, in the case of underground mining, or the open pit configuration, in the case of an open pit, has been established and which, if an effective method of mineral processing has been determined includes a financial analysis based on reasonable assumptions of technical, engineering, operating, and economic factors and evaluation of other relevant factors which are sufficient for a QP, acting reasonably, to determine if all or part of the Mineral Resource may be classified as a Mineral Reserve'.

• Deposit: 'a natural occurrence of mineral or mineral aggregate, in such quantity and quality to invite exploitation'.

• Classification and Categorization: 'a mineral deposit may be subdivided into two Classes, Mineral Resources and Mineral Reserves. Each of these Classes may be subdivided into Categories: Measured, Indicated and Inferred in the case of Mineral Resources and Proven and Probable in the case of Mineral Reserves'.

7.3.3 *The Resource Database*

The Resource Database is established by the collection, verification, recording, storing and processing of the data and forms the foundation necessary for the estimation of MRMR. The establishment of a QA/QC program of all data is essential during this process.

Components of the Resource Database typically will include geological data (e.g. lithology, mineralization, alteration, and structure), survey data, geophysical data, geochemical data, assay data, rock quality and bulk density information and activity dates.

As stated in the CIM Standards and as noted above, a Mineral Resource must have reasonable prospects of economic extraction. Consequently, preliminary data and information concerning a number of factors (e.g. mining, metallurgy, economics and social and environmental sensitivity) will be collected and assessed during the estimation of a Mineral Resource.

General comments

• A database consists of two types of data, primary data and interpreted data. Primary data are parameters amenable to direct physical measurement. Examples include assays, survey data, and geological observations. Interpreted data sets are derivations or interpretations of primary information. Examples are geological projections and block models.

• Bulk density is an important parameter that should be measured and recorded at appropriate intervals, and in an appropriate manner, for the deposit. The choice of methods for determining the bulk density of a particular deposit will depend on the physical characteristics of the mineralization and the available sampling medium.

- The QP should be diligent in ensuring that the final database fairly represents the primary information. Data verification is an essential part of finalizing the resource database.

- The Resource Database provides a permanent record of all the data collected from the work carried out, the date of the work, observations and comments from the results obtained. It should be readily available for future reference. The database provides all of the information necessary to enable current and future geological interpretations and modeling.

- Although most databases are generally maintained in an electronically-stored digital format, hand-printed tables with well-organized information may also form a database. It is recommended that data be stored digitally, using a documented, standard format and a reliable medium that allows for easy and complete future retrieval of the data.

Primary data visualization

- It is essential that the systematic recording of geological observations from mapping and drill hole logging be entered into an organized database.

- Data collection and display must foster a good geological understanding of a deposit as a prerequisite for the Mineral Resource estimation process (see Section 7.3.4).

- The important primary data must be identified and accurately presented in three dimensions, typically on a set of plans and sections. Examples are lithologies, structural measurements, assays, etc.

- Where local mine coordinates are used on geological maps and sections, a mechanism for conversion to universal coordinates must be provided. Maps and sections must include appropriate coordinates, elevation, scale, date, author(s) and appropriate directional information.

- Data positioning information should be relative to a common property co-ordinate system and should include the methodology and accuracy used to obtain that information. Accurate location of data points is essential. If data points are referred to a particular map or grid, those reference data should be included, the map properly identified and the coordinate system clearly stated.

- If primary data have been intentionally omitted from the presentation, they should be identified with an explanatory note for their exclusion.

Interpreted data visualization

- The geological interpretation including mineralization and its controls (e.g. structure, alteration, and lithology) is essential for MRMR estimation. The primary data (i.e. from outcrops, trenches and drill holes) should be clearly identifiable and be distinct from the interpreted data so that it may be utilized in subsequent interpretations and Mineral Resource estimates.

- The relevant geophysical/geochemical/topographic data used to support the interpretation of faults or boundaries must be included or referenced appropriately.

- Since the mineralizing episode(s) and related features of the geology are critical aspects in the MRMR estimations, they must be clearly represented. Examples are controlling features, style(s) and age(s) of mineralization, boundaries of the mineralization, and zonation of the mineralization.

Data collection, recording, storing and processing

- Primary data collected must be recorded even if not used in the MRMR estimation.

- Original assay data should be stored in the units of measure as received by the laboratory (e.g. large ppm values should not be reported as percentages). The analytical method used must be described.

- Analytical data should be converted into common units of measure provided the analytical technique supports the conversion. The original and converted assay should be reported, including the conversion factor(s).

- Data that have been acquired over multiple periods and by various workers should be verified and checked prior to entry into the database. In addition, data records should possess unique identifiers (e.g. unique drill hole, zone and sample numbers, etc.). A distinction must be made between data collected by different methodologies (e.g. reverse circulation holes versus diamond drill holes, etc.) and an explanation of how these data sets are integrated, should be provided.

- Upon the reporting of MRMR estimates, all the tabulations and defining parameters become part of the database. Summations, tabulations, maps, assumptions and related parameters, for example cut-off grade(s), commodity price(s), dilution, losses, plant recovery(ies) become interpreted data and must be enumerated.

- Mine production data are primary and must be incorporated into the MRMR database. Best practice includes routine reporting of reconciliations and monitoring systems implemented during the operational phases of the project. These results will be used for revisions in the MRMR estimation.

- Periodic review of data to ensure its integrity is recommended.

- Duplicate, secure off site storage of data is recommended.

QA/QC

- QA/QC must be addressed during the collection, recording and storage of any of the data ultimately used in the MRMR estimation. This program should be concerned with, but not limited to: data verification, drill sample recovery, sample size, sample preparation, analytical methods, the use of duplicates/blanks/standards, effects of multiple periods of data acquisition and consistency of interpretation in three dimensions. The sample preparation description should include aliquot weight used in the laboratory. The results of the QA/QC program form part of the database and must be recorded.

7.3.4 Geological interpretation and modeling

The purpose of this section is to give guidance to the QP responsible for estimating MRMR. These Guidelines outline requirements for interpretation of geological data, the consideration of economic and mining criteria and the linkage of that information to the grade distribution of the MRMR model as described in Section 7.3.5.

Geological data

• Comprehensive geology and reliable sample information remain the foundation of MRMR estimates.

• Information used for MRMR estimation should include surface geology at suitable scales (lithologies, mineralogical zones, structural regimes, alteration, etc.), topographical data, density information, a complete set of all available sample results and surveyed locations of all sample sites (chips, drill samples, etc.).

• All geological information within the deposit should be transposed from plan onto sections (or vice versa) to confirm reliability and continuity using all available data (drill holes, mine workings, etc.). Two directions of vertical sections (usually orthogonal) and plans should be used to ensure manual interpretations are internally consistent.

• Geological interpretation is frequently completed in a three dimensional (3-D) computer environment. Computer assisted interpretations should be validated on plan and orthogonal section to evaluate the reliability of the geological interpretation.

• Understanding the relationship between the mineralization and the geological processes that govern its geometry is essential. Mineralized limits (whether sharp or gradational) within which the MRMR are to be determined must be interpreted and depicted on maps, plans and sections.

Geological interpretations

• MRMR modeling should be developed within a regional context. Accordingly, the regional geology and property geology are important parts of the geological database.

• The interpretation of geological field data (lithology, structure, alteration and mineralized zones, etc.) should include direct input from individuals with mapping or core logging experience on the deposit.

• Field data should be presented in their entirety, in an unmodified form. Every effort must be made to analyze these data in an unbiased, scientific fashion to develop a 'Geological Concept' which forms the underlying premise on which the geologic interpretation is developed. The concept should include, among others, geological setting, deposit type, styles of mineralization, mineralogical characteristics and genesis.

• The styles of the mineralization under investigation must be identified to allow the modeler to establish geological controls for mineralization and permit more accurate interpolation of grades within the model.

• The geological interpretation and ideas regarding genesis of the deposit should be reviewed in the context of the resultant MRMR model. Aspects and assumptions, for which field data are incomplete, should be clearly identified.

Controls of mineralization

• Once the geological framework of the deposit has been reasonably established, geological controls for mineralization and the limits of those controls are determined. Attention to

668   *Open pit mine planning and design: Fundamentals*

detail is vital for early recognition of important features that control the spatial distribution, variability and continuity of economic mineralization.

• Mineralization may be defined or limited by some combination of features such as structure, lithology and the alteration envelope. These limits or boundaries should be used to constrain the interpolation of grade or quality within the MRMR model.

• When determining limits of mineralization, the estimator must recognize that many mineral deposits comprise more than one type of mineralization. The characteristics of each type will likely require different modeling techniques and/or parameters.

Mining and economic requirements

• By definition, a Mineral Resource must have 'reasonable prospects of economic extraction'.

• Factors significant to project economics must be considered for both Mineral Resource and Mineral Reserve estimates. These will include the extraction characteristics for both the mining and processing method selected as affected by geotechnical, grade control, and metallurgical, environmental and economic attributes.

• For a Mineral Resource, factors significant to project economics should be current, reasonably developed and based on generally accepted industry practice and experience. Assumptions should be clearly defined. For Mineral Reserves, parameters must be detailed with engineering complete to Preliminary Feasibility standards as defined in the CIM Standards.

• Mining assumptions for a Mineral Reserve include: continuity of mineralization, methods of extraction, geotechnical considerations, selectivity, minimum mining width, dilution and percent mine extraction.

• Cut-off grade or cut-off net smelter return (NSR) used for MRMR reporting are largely determined by reasonable long term metal price(s), mill recovery and capital and operating costs relating to mining, processing, administration and smelter terms, among others. All assumptions and sensitivities must be clearly identified.

• Cut-off grade must be relevant to the grade distribution. The mineralization must exhibit sufficient continuity for economic extraction under the cut-off applied.

Three dimensional computer modeling

• MRMR models can be generated with or without the use of 3-D computer software. However, it is likely that any MRMR estimate that is included in a feasibility study will be in the form of a 3-D computer model. This section refers only to those MRMR models generated using such techniques.
• The modeling technique(s) adopted for a project should be appropriate for the size, distribution and geometry of the mineralized zones. The technique should also be compatible with the anticipated mining method(s) and size and type of equipment.
• The QP must analyze the grade distribution to determine if grade compositing is required. Where necessary, assay data should be composited to normalize the grade distribution and to adequately reflect the block size and production units.
• The size of the blocks in the model will be chosen to best match mining selectivity, drill hole and sample density, sample statistics and anticipated grade control method. A change in

cut-off grade or economic limit and selectivity of the mining method(s) frequently requires the development of new models and perhaps increased drill definition to properly evaluate the mineral deposit in question.
- An aspect of block modeling is the loss of critical geological and assay information through smoothing of details inherent in the modeling technique. General validation of the block model against raw data is required to ensure reliability.

Selection of software

- This section refers only to those MRMR models generated using software. It is recognized that MRMR can be estimated using other methods without the use of computers and software.

- The software and the version used should be clearly stated.

- There is a number of adequate, commercially available data handling and modeling software packages currently in use. The person responsible for the development of the MRMR model should have appropriate knowledge of the software, methodology, limitations and underlying assumptions utilized during the modeling process.

7.3.5 Mineral Resource estimation

This section considers important factors in estimating a Mineral Resource and documenting the estimation process. It provides guidelines to the QP responsible for the Mineral Resource estimate with respect to data analysis, sample support, model setup and interpolation. Critical elements to the Mineral Resource estimate are the consideration of the appropriate geological interpretation and the application of reasonably developed economic parameters, based on generally accepted industry practice and experience. While innovation is encouraged, comparisons with other tested methods are essential, prior to publicizing or reporting estimates. Optimization of the Mineral Resource interpretation in consideration of economic parameters is an iterative process.

Data density

- A key initial step prior to the commencement of estimating a Mineral Resource is the assessment of data adequacy and representativeness of the mineralization to be modeled. If the number and spatial distribution of data are inadequate, an estimation is required of how much additional data are needed before a Mineral Resource calculation can meaningfully be done. The QP responsible for modeling must ensure that the available information and sample density allow a reliable estimate to be made of the size, tonnage and grade of the mineralization in accordance with the level of confidence established by the Mineral Resource categories in the CIM Standards.

Integration of geological information

- The deposit geology forms the fundamental basis of the Mineral Resource estimation. The data must be integrated into, and reconciled with, the geological interpretation as part of the estimation process. The interpretation should include the consideration and use of reasonable assumptions on the limits and geometry of the mineralization, mineralization controls and internal unmineralized or 'waste' areas (i.e. dikes or sills). Interpretive information should be continuously reassessed as knowledge of the geological characteristics of a deposit improves.

Listing/recording the data set

- All data and information used in the Mineral Resource estimation must be identified, catalogued and stored for future reference and audits. Any portion of the pertinent data acquired during the exploration and development of the property that is not used in the Mineral Resource estimation must be identified and an explanation provided for its exclusion.

- Sampling, sample preparation, assaying practice and methodologies must be clearly described and an explanation given for the choice of the particular methods used. A comment as to their effectiveness should also be provided.

- Particular care should be taken in recording, analyzing and storing data and results from QA/QC programs related to the Mineral Resource estimation.

Data analysis

- The principal purpose of data analysis is to improve the quality of estimation through a comprehensive understanding of the statistical and spatial character of variables on which the estimate depends. This would include establishment of any interrelationships among the variables of interest, recognition of any systematic spatial variation of the variables (e.g. grade, thickness), definition of distinctive domains that must be evaluated independently for the estimate, and identification and understanding of outliers. In particular, it will be necessary to understand the extent to which 'nugget effect' affects the mineralized sample population. This is often a major concern for precious metal deposits and may be important in other types of deposits.

- Data analysis should be comprehensive and be conducted using appropriate univariate, bivariate, and/or multivariate procedures. Univariate procedures include statistical summaries (mean, standard deviation, etc.), histograms and probability plots. Bivariate procedures include correlation studies, evaluation of scatter plots and regression analysis whereas multivariate analysis might involve procedures such as multiple regression (e.g. bulk density – metal relationships) and multiple variable plots (e.g. triangular diagrams).

- Variography is an aspect of data analysis that assists in defining the correlation and range of influence of a grade variable in various directions in three dimensions.

- Outlier recognition and treatment of outliers is an important part of the data analysis. An outlier is an observation that appears to be inconsistent with the majority of the data and attention for the purposes of Mineral Resource estimation usually is directed to those that are high relative to most data. The modeler must state how an outlier is defined and how it is treated during the resource estimation process (i.e. grade cutting strategy, restricted search philosophy).

Sample support

- Sample or data support (size, shape and orientation of samples) must be considered. Data for the Mineral Resource estimate generally are obtained from a variety of supports and statistical parameters can vary substantially from one support to another. If composites are used as a basis for estimation, the data must be combined in a manner to produce composites of approximately uniform support prior to grade estimation.

- Selection of a composite length should be appropriate for the data and deposit (e.g. bench or half bench height, dominant assay interval length, vein thickness). Commonly compositing is specific to a geological domain.

Economic parameters

- The cut-off grade or economic limit used to define a Mineral Resource must provide 'reasonable prospects for economic extraction'. In establishing the cutoff grade, it must realistically reflect the location, deposit scale, continuity, assumed mining method, metallurgical processes, costs and reasonable long-term metal prices appropriate for the deposit. Assumptions should be clearly defined.

- Variations within the resource model (rock characteristics, metallurgy, mining methods, etc.) that may necessitate more than one cut-off grade or economic limit in different parts of the deposit model must be an ongoing consideration.

Mineral Resource model

- The Mineral Resource estimation techniques employed are dependent to a degree on the size and geometry of the deposit and the quantity of available data. Currently, most resource models are computer models constructed using one of several specialized commercially available software packages. Simple geometric methods may be acceptable in some cases (e.g. early stage deposit definition) but three-dimensional modeling techniques may be more appropriate for advanced projects.

- Model parameters (e.g. block size, model orientation) should be developed based on mining method (e.g. open pit versus underground, blast hole versus cut and fill mining), deposit geometry and grade distribution (e.g. polymetallic zoning in a sulphide deposit).

Estimation techniques

- The QP responsible for the Mineral Resource model must select a technique to estimate grades for the model. Methods range from polygonal or nearest neighbor estimates, inverse distance to a power, various kriging approaches (e.g. ordinary kriging, multiple indicator kriging) through to more complex conditional simulations. The choice of method will be based on the geology and complexity of grade distribution within the deposit and the degree to which high-grade outliers are present.

- In some complex models, it may be necessary to use different estimation techniques for different parts of the deposit.

- The QP should ensure that the selected estimation method is adequately documented and should not rely solely on the computer software to produce a comprehensive document or report 'trail' of the interpolation process.

Mineral Resource model validation

- The QP must ensure the Mineral Resource model is consistent with the primary data. The validation steps should include visual inspection of interpolated results on suitable plans and sections and compared with the composited data, checks for global and local bias (comparison of interpolated and nearest neighbor or declustered composite statistics), and a change of support check (degree of grade smoothing in the interpolation). It is recommended

672 Open pit mine planning and design: Fundamentals

that manual validation of all or part of a computer-based Mineral Resource estimate be completed.

• For Mineral Resource models of deposits that have had mine production or are currently being mined, the validation mist include a reconciliation of production to the Mineral Resource model.

• A final step, best practice includes the re-evaluation of the economic parameters to confirm their suitability.

• As per the CIM Standards, Mineral Resource estimation involves the classification of resources into three classes. The criteria used for classification should be described in sufficient detail so that the classification is reproducible by others.

7.3.6 Quantifying elements to convert a Mineral Resource to a Mineral Reserve

This section forms the logical extension of the topics discussed in Section 7.3.5, 'Mineral Resource Estimation', and addresses factors required for the conversion of a Mineral Resource into a Mineral Reserve. These factors are provided, in the form of a checklist, for assembling information that should be considered prior to the process of estimating Mineral Reserves. This checklist referred to as quantifying elements or modifying factors, is not intended to be exhaustive. The QP should ensure that these elements/factors have been considered in adequate detail to demonstrate that economic extraction can be justified. The appropriate level of detail for each of these elements/factors is left to the discretion of the QP. However, in aggregate, the levels of detail and engineering must meet or exceed the criteria contained in the definition of a Preliminary Feasibility Study.

Quantifying Element or Modifying Factor Check List:

a) Mining:
- data to determine appropriate mine parameters, (e.g. test mining, RQD)
- open pit and/or underground
- production rate scenarios
- cut-off grade (single element, multiple element, dollar item)
- dilution: included in the Mineral Resource model or external factor(s)
  recovery with respect to the Mineral Resource model
- waste rock handling
- fill management (underground mining)
- grade control method
- operating cost
- capital cost
- sustaining capital cost

b) Processing
- sample and sizing selection: representative of planned mill feed, measurement of variability, is a bulk sample appropriate
- product recoveries
- hardness (grindability)
- bulk density

- presence and distribution of deleterious elements
- process selection
- operating cost
- capital cost
- sustaining capital cost

c) Geotechnical/Hydrological
- slope stability (open pit)
- ground support strategy (underground), test mining
- water balance
- area hydrology
- seismic risk

d) Environmental
- baseline studies
- tailings management
- waste rock management
- acid rock drainage issues
- closure and reclamation plan
- permitting schedule

e) Location and infrastructure
- climate
- supply logistics
- power source(s)
- existing infrastructure
- labor supply and skill level

f) Marketing elements or factors
- product specification and demand
- off-site treatment terms and costs
- transportation costs

g) Legal elements or factors
- security of tenure
- ownership rights and interests
- environmental liability
- political risk (e.g. land claims, sovereign risk)
- negotiated fiscal regime

h) General costs and revenue elements or factors
- general and administrative costs
- commodity price forecasts
- foreign exchange forecasts
- inflation
- royalty commitments
- taxes
- corporate investment criteria

i) Social issues
- sustainable development strategy
- impact assessment and mitigation
- negotiated cost/benefit agreement
- cultural and social influences

7.3.7 Mineral Reserve estimation

This section considers important factors in estimating a Mineral Reserve and documenting the estimation process. As a Mineral Reserve estimate represents the collation of work carried out by numerous professional disciplines, the QP producing the Mineral Reserve estimate must understand the significance of each discipline's work in order to assess economic viability. In addition, the QP should recognize that the time from discovery, to production, through to closure, of a mine is often measured in years and this timeframe makes good documentation an important aspect of the estimation process.

Preparation

The QP should document and use a methodology in estimating Mineral Reserves to ensure no significant factor is ignored. Pre-planning is important to identify the factors affecting the Mineral Reserve estimate. Utilizing a checklist to ensure all aspects are considered is good practice.

Mineral Reserve definition and classification is covered by the CIM Standards. Definitions do change from time-to-time and in the compilation of a Mineral Reserve estimate the QP should ensure the current definitions are being used. Of significance are the requirements that the material forms the basis of an economically viable project.

The test of economic viability should be well documented as part of the Mineral Reserve estimation process. The requirement for economic viability implies determination of annual cash flows and inclusion of all the parameters that have an economic impact.

Classification

The CIM Standards provide two categories for the definition of the Mineral Reserve, Proven Mineral Reserve and Probable Mineral Reserve and the QP must ensure that the minimum criteria are met prior to assigning these categories. The QP should be mindful of all the inputs used in establishing the Mineral Reserve that affect the confidence in the categories. The methodology of establishing the classification should be well documented and easily understood. Best practice includes providing a narrative description of the qualitative reasons behind the classification selection.

Where practical, empirical evidence (e.g. production data) should be used to calibrate and justify the classification.

Verification of inputs

It is the responsibility of the QP to ensure the verification of all inputs to the Mineral Reserve estimate. As the Mineral Reserve estimate is based on many data inputs, including the Mineral Resource model, it is important that the inputs and the consistency of the inputs be validated as part of the Mineral Reserve estimation process. A defined methodology to achieve this is considered best practice and the use of a protocol such as the checklist contained in Section 7.3.6 is recommended. Identification of critical aspects of the Mineral Reserve estimate is an important part of the input verification.

Application of Cut-off grade

Cut-off grade is a unit of measure that represents a fixed reference point for the differentiation of two or more types of material. Owing to the complexity of Mineral Reserve estimates, numerous cut-off grades may be required to estimate a Mineral Reserve, (e.g. the set point defining waste from heap leach ore and the set point defining heap leach ore from milled ore).

The cut-off grade(s) (the economic limit or pay limit) should be clearly stated, unambiguous and easily understood. Complex ores may require complicated procedures to determine cut-off grades and to define the Mineral Reserve. The procedures used to establish the cut-off strategies should be well documented, easily available for review, and clearly stated in disclosure statements.

Cut-off grade must be relevant to the grade distribution modeled for the Mineral Resource. If cut-off grades are outside the specified range, the QP must review model-reliability and a new model might be necessary.

A key objective of Mineral Reserve estimation is the successful extraction and delivery of a Mineral Resource for processing at the grade estimated. Due consideration should be given to the problems associated with selective mining where the cut-off grade is set high relative to the average grade of the Mineral Resource.

Practicality of mining

The practicalities of the mining/processing rates and methods for a deposit are important considerations in the estimation of a Mineral Reserve. The QP must assess the various proposals when estimating a Mineral Reserve. Care should also be taken to ensure that the mining equipment selected is appropriate for the deposit. Inappropriate equipment selection may have an effect on both dilution and extraction. The QP must have a high level of confidence in the viability of the mining and processing methods considered in determining the Mineral Reserves.

A QP should, when appropriate, consider of alternative mine/plant configurations. Selecting the appropriate mining and processing methods and rates may involve several iterations and will involve input from members of other disciplines. Trial evaluations, referred to as 'trade-off' or 'scoping' studies, may be required as a prelude to the completion of a Preliminary Feasibility Study.

Project risk assessment

While the classification of the Mineral Reserve allows the QP to identify technical risk in broad terms, best practice includes the establishment of a methodology to identify and rank risks associated with each input of the Mineral Reserve estimate. This will assist the QP in establishing the Mineral Reserve categorization, thus providing an understanding of the technical risk associated with the Mineral Reserve estimate. This methodology, ranking and analysis should be well documented.

Peer reviews

Best practice includes the use of an internal peer review of the Mineral Reserve estimate including inputs, methodology, underlying assumptions, the results of the estimate itself, and test for economic viability.

Audits/Governance

Upon completion of a Preliminary Feasibility Study, or in the case of significant changes to a Mineral Reserve estimate, best practice includes completion of a properly scoped audit carried out by an impartial QP. The audit should consider the methodology used, test the reasonableness of underlying assumptions, and review conformity to Mineral Reserve definitions and classification. The methodology for Mineral Reserve risk identification, assessment and management should also be included in the Mineral Reserve audit. The audit should be documented, distributed and responded to in a manner that recognizes good corporate governance.

Documentation

There are often several iterations of evaluations carried out over a protracted period of time prior to completion of a Preliminary Feasibility Study. Best practice includes appropriate documentation of the inputs/methodology/risks/assumptions used in these valuations so these will be available for future Mineral Reserve estimates. Information should be easily retrievable, readily available and catalogued in a manner that allows easy assessment of the history of the evaluations carried out and records the location of all relevant information/reports/etc. It is important to ensure that the information used in an evaluation, and understanding gained of a mineral deposit, is available for future work. Care should be taken in storage and consideration given to the continuous evolution of computer file formats and the impact this may have on previous work. File conversion of historic work into formats that allow continued access is recommended.

Mineral Reserve statements

Mineral Reserve statements should be unambiguous and sufficiently detailed for a knowledgeable person to understand the significance of, for example, cut-off grade and its relationship to the Mineral Resource. In the case of open pit Mineral Reserve estimates, the waste:ore ratio (the strip ratio) should be unambiguously stated. There should be an obvious linkage of the Mineral Reserve estimate to the Mineral Resource estimate provided in disclosure documents. Best practice includes documentation of those linkages (e.g. dilution and mining recovery) that were used in preparing the Mineral Reserve estimation.

7.3.8 Reporting

This section is primarily a compilation of references regarding reporting standards that should be considered when preparing reports on MRMR estimates. Although these standards are intended for public disclosure, they also represent the minimum requirement for best practice for all reporting.

National Instrument 43-101, Form 43-l01Fl and Companion Policy 43-101CP, establish standards for all oral and written disclosure made by an issuer concerning mineral projects that are reasonably likely to be made available to the public. All disclosure concerning mineral projects including oral statements and written disclosure in, for example, news releases, prospectuses and annual reports is to be based on information supplied by or under the supervision of a QP. Disclosure of information pertaining to MRMR estimation is to be made in accordance with industry standard definitions contained in the CIM Standards which have been incorporated by reference into the NI 43-101.

One of the objectives of the Estimation Best Practice Guidelines is to foster greater standardization of reporting in publicly disclosed documents. The recommendations included below represent further guidance and attempt to develop a reporting template, which should help reporting Canadian companies achieve greater standardization.

The QP should familiarize themselves with current disclosure regulations as part of preparing a MRMR estimate.

Reporting units

In the preparation of all reports and press releases, either metric or imperial units may be used. However, the following provisos apply:
- Reports must maintain internal consistency – metric and imperial units should not be used in different parts of the same report.
- The mixing of metric and imperial units (e.g. oz/tonne) is never acceptable.

The Committee considers that reporting in metric units is preferable.

Technical reports

A technical report shall be in accordance with Form 43-101 Fl, NI 43-101. The obligation to file a technical report arises in a number of different situations. These are set out in NI 43-101 in Part 4.

The CIM Standards on Mineral Resources and Mineral Reserves referenced in NI 43-101, provide additional guidance for reporting of MRMR estimates. A listing of the main recommendations and requirements is as follows:

(a) The QP is encouraged to provide information that is as comprehensive as possible in Technical Reports on MRMR.

(b) Fundamental data such as commodity price used and cut-off grade applied must be disclosed.

(c) Problems encountered in the collection of data or with the sufficiency of data must be clearly disclosed.

(d) Modifying factors applied to MRMR estimates such as cutting of high grades or resulting from reconciliation to mill data must be identified and their derivation explained.

(e) MRMR estimates are not precise calculations and, as a result should be referred to as estimates.

(f) Tonnage and grade figures should reflect the order of accuracy of the estimate by rounding off to the appropriate number of significant figures.

(g) Technical Reports of a Mineral Resource must identify one or more categories of 'Inferred', 'Indicated' and 'Measured' and Technical Reports of Mineral Reserves must specify one or both categories of 'Proven' and 'Probable'. Categories must not be reported in combined form unless details of the individual categories are also provided. Inferred Mineral Resources cannot be combined with other categories and must always be reported separately.

(h) Mineral Resources must never be added to Mineral Reserves and reported as total Resources and Reserves. MRMR must not be reported in terms of contained metal or mineral content unless corresponding tonnages, grades and mining, processing and metallurgical recoveries are also presented.

(i) In cases where estimates for both Mineral Resources and Mineral Reserves are reported, a clarifying statement must be included that clearly indicates whether Mineral Resources are inclusive or exclusive of Mineral Reserves.

The Estimation Best Practice Committee recommends that Mineral Resources should be reported separately and exclusive of Mineral Reserves.

(j) Mineral Reserves may incorporate material (dilution) which is not part of the original Mineral Resource and exclude material (mining losses) that is included in the original Mineral Resource. It is essential that the fundamental differences between these estimates be understood and duly noted.

(k) In preparing a Mineral Reserve report, the relevant Mineral Resource report on which it is based should be developed first. The application of mining and other criteria to the Mineral Resource can then be made to develop a Mineral Reserve statement that can also be reconciled with the previous comparable report. A detailed account of the differences between current and previous estimates is not required, but sufficient commentary should be provided to enable significant differences to be understood by the reader. Reconciliation of estimates with production whenever possible is required.

(l) Where Mineral Reserve estimates are reported, commodity price projections, operating costs and mineral processing/metallurgical recovery factors are important and must be included in Technical Reports.

The Committee considers that when reporting a Mineral Reserve mineable by open pit methods, the waste-to-ore ration must be disclosed.

(m) Reports must continue to refer to the appropriate categories of Mineral Resources until technical feasibility and economic viability have been established by the completion of at least a Preliminary Feasibility Study.

(n) Reporting of mineral or metal equivalence should be avoided unless appropriate correlation formulae including assumed metal prices, metallurgical recoveries, comparative smelter charges, likely losses, payable metals, etc. are included.

(o) Broken mineralized inventories, as an example, surface and underground stockpiles, must use the same basis of classification outlined in the CIM Standards. Mineralized material being processed (including leaching), if reported, should be reported separately.

(p) Reports of MRMR estimates for coal deposits should conform to the definitions and guidelines on Paper 88-21 of the Geological Survey of Canada. 'A Standardized Coal Resource/Reserve Reporting System for Canada'.

(q) When reporting MRMR estimates relating to an industrial mineral site, QP must make the reader aware of certain special properties of these commodities and relevant standard industry specifications.

(r) Reports of MRMR estimates of diamonds or gemstones must conform to the definitions and guidelines found in 'Reporting of Diamond Exploration Results, Identified Mineral Resources and Ore Reserves' published by the Association of Professional Engineers, Geologists and Geophysicists of the Northwest Territories. These definitions and guidelines remain in force until/if they are replaced by guidelines of the Diamond Exploration Best Practice Committee, the relevant sections of these guidelines, or other guidelines which may be accepted by the CSA or CIM.

Annual reports

Written disclosure (including annual reports) of MRMR is covered by Part 3.4 of NI 43-101. Further reference is made in Parts 1.3, 1.4, 2.1, and 2.2 of NI 43-101.

An issuer shall ensure that all written disclosure of MRMR on a property material to an issuer includes:

(a) the effective date of each estimate of MRMR;

(b) details of quantity and grade or quality of each category of MRMR;

(c) details of key assumptions, parameters and methods used to estimate the MRMR;

(d) a general discussion of the extent to which the estimate of MRMR may be materially affected by any known environmental, permitting, legal, title, taxation, socio-political, marketing, or other relevant issues; and

(e) a statement that Mineral Resources which are not Mineral Reserves do not have demonstrated economic viability.

Further to these requirements, the Committee recommends that:

(a) MRMR estimates should be reported as a tabulation.

(b) The name of the appropriate QP must be included with the estimate. The relationship of the QP to the reporting company should be stated. Note that in the estimation of Mineral Reserves, the services of a number of different QP are likely to have been employed. Under CSA guidelines a corporation may designate a reporting QP with overall responsibility for the estimates and, if so, the name must be included. In some Canadian Provinces, it may not be appropriate to designate a reporting QP.

(c) These data remain 'estimates' and should be reported as such.

(d) NI 43-101 Part 3.4 (c) requires those details of key assumptions, parameters and methods used to estimate MRMR must be included. These details could be included as a footnote in the MRMR section:
- Metal prices assumptions.
- Cut-off grades.
- Ore losses and dilution.
- Mill recoveries.
- Estimation methodologies.
- It is essential that the estimates conform to the CIM Standards on Mineral Resources and Mineral Reserves, Definitions and Guidelines, or equivalent foreign code as described in Part 7 of NI 43-101. A note stating the standard being used must be included.
- Year-to-year changes in MRMR must be included, together with the reasons for the changes.
- A statement whether the Mineral Resources are inclusive or exclusive of Mineral Reserves. In the interests of standardization, the Committee recommends that Mineral Resources should be reported exclusive of Mineral Reserves in Annual Reports.
- Date of the estimate of MRMR.

Press releases

The content of press releases discussing MRMR is covered in Section 3.0 of Appendix B of Disclosure Standard No. 1450-025 of the Toronto Stock Exchange (TSX), and Corporate Finance Manual Appendix 3F-Mining Standards Guidelines-Policy 3.3-Timely Disclosure of the TSX Venture Exchange.

Section 3.1 (Definitions) states that estimates 'must conform to the definitions contained in NI 43-101'. Section 3.2 (Use) covers a number of points regarding reporting:

• All MRMR estimations must disclose the name of the QP responsible for the estimate and the relationship of the QP the reporting company. The company must also state whether, and how, any independent verification of the data has been published.

• The statement must make a clear distinction between Mineral Resources and Mineral Reserves.

• MRMR should, wherever possible, be published in such a manner so as not to confuse the reader as to the potential of the deposit. Inferred Resources must not be aggregated with Indicated and Measured Resources. Any categories of MRMR that are aggregated must also be disclosed separately.

• When Mineral Reserves are first reported, the key economic parameters of the analysis must be provided. These will include:
  • Operating and capital cost assumptions.
  • Commodity prices (If commodity prices used differ from current prices of the commodities which could be produced, an explanation should be given, including the effect on the economics of the project if current prices were used. Sensitivity analysis may be used in this section).
  • All reported quantities of MRMR must be expressed in terms of tonnage and grade or characteristics. Contained ounces must not be disclosed out of the context of the tonnage and grade of a deposit.
  • MRMR for polymetallic deposits may not be disclosed in terms of 'metal equivalents' except in limited circumstances as set out in NI 43-101 Fl, 19(k) and in the CIM Standards on MRMR. It is also inappropriate to refer to the gross value or in situ value of MRMR.

The Committee recommends that any press release that reports initial estimates of MRMR include all of the information listed under 'Annual Reports' above. Subsequent press releases may refer back to the initial press release. It should be noted that the CSA is reviewing this requirement.

Annual Information Filing

The requirements for reporting for an AIF (Annual Information Filing) are set out in the CSA document National Instrument 44-101 and Form 44-101 Fl.

7.3.9 Reconciliation of Mineral Reserves

Production monitoring and reconciliation of Mineral Reserves are the ultimate activities by which the QP can continuously calibrate and refine the Mineral Reserve estimate. While this section is primarily concerned with Mineral Reserves, the only valid confirmation of both the Mineral Resource and Mineral Reserve estimate is through appropriate production monitoring and reconciliation of the estimates with mine and mill production.

The QP must take into consideration the results of any grade-tonnage reconciliation in any public disclosure of MRMR estimates.

Production monitoring is the grade control and tonnage accounting function performed at an operating mine. This function provides the information required to minimize dilution, maximize mineral recovery and supply a consistent and balanced feed to the process plant as required. Grade and tonnage control comprises representative sampling of production sources, establishing ore/waste boundaries and accurately recording production tonnes and grade.

Reconciliation is required to validate the Mineral Reserve estimate and allows a check on the effectiveness of both estimation and operating practices. Since the MRMR estimates are based on much wider spaced sampling than used for production, reconciliation identifies anomalies, the resolution of which may prompt changes to the mine/processing operating practices and/or to the estimation procedure.

Production monitoring

The following should be given consideration in effective production monitoring and forms part of an ongoing quality control program, which takes into account the closer spaced production sampling and mapping:

Minimize Dilution/Maximize Mineral Recovery

- When ore/waste contacts are visual, mine operators can classify ore and waste easily and send material to the correct destination, however, production monitoring is still required.

- Where assay boundaries are used to delineate ore, appropriately spaced representative samples are required to estimate the locations of economic margins.

- Given the negative economic consequences of misclassification, diligence is necessary to ensure that mined ore and waste types are delivered to the appropriate destinations.

Characterize the ore to ensure the requested metallurgical balance is achieved.

- This may require geological mapping and logging of blast hole chips if ore types are visually distinguishable.

- Where characterization is non-visual, other testing is required to achieve appropriate blending.

Characterize waste to allow for potential mixing (blending) or separation based on environmental requirements (e.g. acid rock drainage control).

Monitor deleterious or by-product constituents that might compromise or enhance mill recoveries and concentrate quality.

Record accurately production tonnes and grade to permit reconciliation of mine production to the processing facility and ultimately to the Mineral Reserve estimate. Mine production needs to be reconciled to mine surveys on a regular basis, commonly monthly.

Ensure that accurate measurements of in-situ bulk densities for various ore and waste types have been determined so that volumes can be converted appropriately to tonnes. Periodic checks are required of in situ bulk densities, truck and bucket factors and weightometers.

Develop appropriate sampling protocols and continuously evaluate them to ensure representative sampling in both the mine and plant.

Ensure that acceptable QA/QC procedures are being followed at all laboratories being utilized.

Maps of workings/benches at appropriate scales must be kept current to provide:

• Geological information to compare to the MRMR model and update where appropriate.

• Ore type classification information to compare to the MRMR model and provide data for blending requirements of the process plant.

• Structural information that may impact MRMR continuity or provide valuable geotechnical information.

• Current, detailed, grade distribution from production sampling which, when combined with geological information, may be used to improve the grade interpolation in the MRMR estimation process.

• Information that will assist in quantifying dilution and mining losses, which can be used for future MRMR estimations.

Volumetric surveys should be retained so they can be used for future reconciliations.

Production reconciliation

The following should be considered in undertaking production reconciliation:

Reconciliation of mine and mill

• Reconciliation between the mine and mill production should be done on a regular basis but monitored on a daily basis to ensure accuracy of sampling and record keeping. Current best practice is considered to be reconciliation on a monthly basis.

• A reconciliation provides checks for discrepancies, which may require changes to operational procedures or the MRMR model.

• Mine production is usually reconciled to the plant since measurement in the plant is generally accepted to be more accurate. Significant discrepancies and resulting adjustment factors should be explained and reported.

• On a yearly basis, mill production should be reconciled with the final concentrate, bullion or mineral shipped and resulting adjustment factors should be explained and reported.

Reconciliation of production and MRMR estimates:

• Reconciliation should be done at least annually to coincide with the corresponding MRMR statement.

• Reconcile production to estimated depletion of Mineral Reserves; any discrepancies in grade and/or tonnes should be explained and appropriate changes should be made to operating practice or the MRMR estimation process.

Annual review of remaining MRMR

• Remaining MRMR at operating mines should be reviewed at least annually and should reflect changes in the underlying criteria, including long term commodity price forecasts, increases or decrease in costs, changes in metallurgical processing performance, and changes in mining methods, dilution or mining recovery.

- Cut-off grades or economic limits should be reassessed and updated at least annually.

- Remaining MRMR should be adjusted for improved geological interpretation due to drilling or mapping.

- The rationale for any changes to operating practice or to MRMR estimation procedures must be documented.

End-of-year MRMR estimates should be reconciled with previous year's MRMR estimates by showing a balance sheet detailing the changes due to mining extraction, commodity price change, cost changes, additions or deletions due to drilling or mining losses/gains, among others.

7.3.10 Selected References

The Committee considers that there are several documents and publications which are essential or useful in dealing with best practice requirements for the estimation of MRMR.
CIM Standards on Mineral Resources and Reserves – Definitions and Guidelines. Prepared by the CIM Standing Committee on Reserve Definitions, October 2000. CIM Bulletin Vol. 93, No. 1044, pp 53–61
Vallée, M. and Sinclair, A.J. (eds.) (1998), 'Quality Assurance, Continuous Quality Improvement and Standards in Mineral Resource Estimation' Exploration and Mining Geology (Volume 7, nos. I and 2, 1998).
Mineral Resource and Ore Reserve Estimation – The AusIMM Guide to Good Practice, Monograph 23. Editor: A.C. Edwards, The Australasian Institute of Mining and Metallurgy: Melbourne.
National Instrument 43-101 – Standards of Disclosure for Mineral Projects. Ontario Securities Commission Bulletin 7815, November 17, 2000.
Exploration Best Practice Guidelines: included in 'CIM Standards on Mineral Resources and Reserves – Definitions and Guidelines'. Prepared by the CIM Standing Committee on Reserve Definitions, October, 2000. CJM Bulletin Vol. 93, No. 1044, pp 53–61.
Guidelines for the Reporting of Diamond Exploration Results. Available at www.cim.org and in press.

REFERENCES

Anonymous. 1999. Guide for reporting exploration information, mineral resources and mineral reserves – an abstract. Mining Engineering. 51(6):82–84.
Arik, A. 1999. An alternative approach to resource classification. 28th International Symposium on Application of Computers and Operations Research in the Mineral Industry: 45–54. Colorado School of Mines.
Arik, A. 2002. Comparison of resource methodologies with a new approach. 30th International Symposium on Application of Computers and Operations Research in the Mineral Industry: 57–64. Alaska:SME.
AusIMM. 2004. JORC information. www.jorc.org/main.
AusIMM. 2005. Australasian Code for the reporting of exploration results, mineral resources and ore reserves: The JORC Code – 2004 Edition. www.jorc.org/pdf/jorc2004print.pdf.
AusIMM. 1995. Code and guidelines for technical assessment and/or valuation of mineral and petroleum assets and mineral and petroleum securities for independent expert reports (The VALMIN Code). http://www.ausimm.com.au/codes/valmin/valcode0.asp.
AusIMM. 1998a. The revised and updated VALMIN Code: Code and guidelines for technical assessment and/or valuation of mineral and petroleum assets and mineral and petroleum securities for independent expert reports (The VALMIN Code). http://www.ausimm.com.au/codes/valmin.
AusIMM. 1998b. The revised VALMIN Code and guidelines: An aide memoire to assist its interpretation. http://www.ausimm.com.au/codes/valmin/eexp11.asp.

Bre-X. 2005. Bre-X/Bresea shareholder class action information web site. www.brexclass.com
CIM. 2000. Exploration Best Practice Guidelines. Aug 20. www.cim.org/definitions/exploration BESTPRACTICE.pdf
CIM. 2003a. Standards and guidelines for valuation of mineral properties. February. www.cim.org/committees/CIMVal_final_standards.pdf
CIM. 2003b. Guidelines for the reporting of diamond exploration results. Mar 9. www.cim.org/committees/diamond_exploration_final.cfm.
CIM. 2003c. Estimation of mineral resources and mineral reserves: Best practice guidelines. Nov 23. www.cim.org/committees/estimation.cfm.
CIM. 2004. Definition Standards – On mineral resources and mineral reserves. www.cim.org/committees/StdsAppNovpdf
CIM. 2005. Standards and guidelines. www.cim.org/committees/guidelinesStandards_main.cfm
Camisani-Calzolari, F.A., and D.G. Krige. 2001. The SAMREC Code seen in a global context. 29th International Symposium on Application of Computers and Operations Research in the Mineral Industry: 39–44. Beijing:Balkema.
Camisani-Calzolari, F.A. 2004. National and international codes for reporting mineral resources and reserves: Their relevance, future and comparison. JSAIMM. 104(5):297–300.
Cawood, F.T. 2004. Towards a mineral property valuation code: Considerations for South Africa. JSAIMM. 104(1):35–44.
Cawood, F.T. 2004. Mineral property valuation in South Africa: A basket of assets and legal rights to consider. JSAIMM. 104(1):45–52.
Cawood, F.T. 2004. The Mineral and Petroleum Resources Development Act of 2002: A paradigm shift in mineral policy in South Africa. JSAIMM. 104(1):53–63.
Cawood, F.T., and A.S. Macfarlane. 2003. The Mineral and Petroleum Royalty Bill – Report to the National Treasury.
Diatchkov, S.A. 1994. Principles of classification of reserves and resources in the CIS countries. Mining Engineering. 46(3):214–217.
Dowd, P.A. 1999. Reserves and resources – Continuity and levels of confidence. 28th International Symposium on Application of Computers and Operations Research in the Mineral Industry: 93–100. Colorado School of Mines.
Ellis, T.R. 2002. The U.S. mineral property valuation patchwork of regulations and standards. CIM Bulletin 95(1059):110–118.
Ellis, T.R. 2000. The difference between a value estimate and an appraisal. Pre-print Paper 00-119. 2000 SME Annual Meeting, Feb 28–Mar 1. SLC, UT.
Ellis, T.R. 2002. Recent developments in international valuation standards. 2002. Pre-print Paper 02-168. 2002 SME Annual Meeting, Feb 26–28. Phoenix, AZ.
Ellis, T.R., Abbott, D.M., Jr., and H.J. Sandri. 1999. Trends in the regulation of mineral deposit valuation. Pre-print Paper 99-029. 1999 SME Annual Meeting, Mar 1–3. Denver, CO.
Ellis, T.R. 2001. U.S. and international valuation standards – The future. Pre-print Paper 01-161. 2001 SME Annual Meeting, Feb 26–28. Denver, CO.
Ellis, T.R. 2004. International perspective on U.S. minerals appraisal standards development. Mining Engineering. 56(3):44–48.
Ellis, T.R. 2005. Implementation of the international valuation standards. Pre-print Paper 05-093. 2005 SME Annual Meeting, Feb 28-Mar 2. SLC, UT.
Francis, D. 1997. Bre-X: The Inside Story. Key Porter Books, Toronto. 240 pp.
Goold, D., and A. Willis. 1997. Bre-X Fraud. McClelland & Stewart. 272 pp.
Heintz, J.H. 2002. Valuation of mineral interests in condemnation cases. Pre-print Paper 02-74. 2002 SME Annual Meeting, Feb 26–28. Phoenix, AZ.
Kirk, W.S. 1998. Iron ore reserves and mineral resource classification systems. Skillings Mining Review. June 6. Pp 4–11.
Kral, S. 2003. Experts discuss reserves reporting at SME meeting. Mining Engineering. 55(12):23–26.
Lawrence, R.D. 2002. Valuation of mineral properties without mineral resources: A review of market-based approaches. CIM Bulletin 95(1060):103–106.
Lawrence, M.J. 2002. The VALMIN Code – The Australian experience. CIM Bulletin 95(1058):76–81.
Lilford, E.V., and R.C.A. Minnitt. 2002. Methodologies in the valuation of mineral rights. JSAIMM. 102(7):369–384.

Limb, J.S. 2001. Mineral appraisal issues in court testimony. Pre-print Paper 01-90. 2001 SME Annual Meeting, Feb 26–28. Denver, CO.
Mcfarlane, A.S. 2002. A Code for the valuation of mineral properties and projects in South Africa. JSAIMM. 102(1):37–48.
Mwasinga, P.P. 2001. Approaching resource classification: General practices and the integration of geostatistics. 29th International Symposium on Application of Computers and Operations Research in the Mineral Industry: 97–107. Beijing:Balkema.
Rendu, J.M. 1999. International definitions of mineral resources and mineral reserves – A U.S. perspective. 28th International Symposium on Application of Computers and Operations Research in the Mineral Industry: 27–34. Colorado School of Mines:CSM.
Rendu, J.M. 2003. SME meets with the SEC – resources and reserves reporting discussed. Mining Engineering. 55(7):35.
Rendu, J.M. 2005. SME submits reserves, resources reporting recommendations to SEC. Mining Engineering. 57(8):8–9.
Roscoe, W.E. 2002. Valuation of mineral exploration properties using the cost approach. CIM Bulletin 95(1059):105–109.
SAIMM. 2000. South African Code for reporting of mineral resources and mineral reserves (the SAMREC Code). www.saimm.co.za/pages/comppages/samrec_version.pdf.
SEC. 2005. Description of property by users engaged or to be engaged in significant mining operations. Guide 7. www.sec.gov/divisions/corpfin/forms/industry.htm#secguide7/.
Sinclair, A. 2001. High-quality, axiomatic to high-quality resource/reserve estimates. CIM Bulletin 94(1049):37–41.
SME. 1991. A guide for reporting exploration information, resources and reserves. Mining Engineering. 53(4):379–384.
SME. 1999. A guide for reporting exploration information, mineral resources, and mineral reserves. www.smenet.org.digital_library/index.cfm.
SME. 2003. Reporting mineral resources and reserves: Regulatory, financial, legal, accounting, managerial and other aspects. Conference October 1–3, 2003. SME.
Spence, K.N., and W.E. Roscoe. 2001. Development of Canadian standards and guidelines for valuation of mineral properties. Pre-print Paper 01-138. 2001 SME Annual Meeting, Feb 26–28. Denver, CO.
Spence, K. 2002. An overview of valuation practices and the development of a Canadian code for the valuation of mineral properties. CIM Bulletin. 95(1057):77–82.
Stephenson, P.R., and P.T. Stoker. 1999. Classification of mineral resources and ore reserves. 28th International Symposium on Application of Computers and Operations Research in the Mineral Industry: 55–68. Colorado School of Mines:CSM.
Subelj, A. 2002. Why 'World Code' has to cover all mineral resources, not only 'Reserves'. 30th International Symposium on Application of Computers and Operations Research in the Mineral Industry: 3–12. Alaska:SME.
Sykes, T. 1978. The Money Miners: Australia's Mining Boom 1969–1970. Wildcat Press, Sydney. 388 pages.
Sykes, T. 1988. Two Centuries of Panic: A History of Corporate Collapses in Australia. Allen & Unwin. London. 593 pages.
Sykes, T. 1996. The Official History of Blue Sky Mines. Australian Financial Review. 165 pages.
USGS. 1976. Principles of the mineral resource classification system of the U.S. Bureau of Mines and U.S. Geological Survey. Bulletin 1450-A.
USGS. 1980. Principles of a resource/reserve classification for minerals. USGS Circular 831. http://pubs.er.usgs.gov/pubs/cir/cir831/ or http://imcg.wr.usgs.gov/usbmak/c831.html.
Valleé, M. 1999. Toward resource/reserve estimation, inventory and reporting standards. 28th International Symposium on Application of Computers and Operations Research in the Mineral Industry: 69–76. Colorado School of Mines.
Valleá, M. 1999. Resource/reserve inventories: What are the objectives? CIM Bulletin 92(1031): 151–155.

REVIEW QUESTIONS AND EXERCISES

1. Why is it important that the terms used to describe and classify mineral resources be precisely defined and accurately applied?

2. What is meant by the popularly-used term 'transparency'?
3. Who benefits from 'smoke-and-mirrors' in the mining business?
4. What does JORC denote? Why was it formed? What did it produce? Why was it important to the worldwide mining business?
5. Briefly describe the Bre-X scandal. Who were the main participants? Whose resource/reserve definitions were used? What was the outcome?
6. What are the main principles governing the operation and application of the JORC code? Briefly describe each.
7. What is meant by a Public Report or Public Reporting? What is the relationship to the JORC code?
8. What is meant by a 'Competent Person'?
9. What is the role of a 'Competent Person'?
10. What are the qualifications of a 'Competent Person'?
11. What is meant by relevant experience?
12. Are there penalties for unprofessional work? What is the enforcement mechanism?
13. Figure 7.1 sets out the framework for classifying tonnage and grade estimates to reflect different levels of geological confidence and different degrees of technical and economic evaluation. How is the figure to be applied in practice?
14. What are meant by modifying factors?
15. What are the rules regarding the reporting of exploration results?
16. How can Table 7.1 be profitably used?
17. What is meant by the term 'prescriptive'? What is meant by the term 'guideline'? How do the two terms differ?
18. What are the rules regarding the reporting of mineral resources?
19. What is covered by the term 'Mineral Resource'?
20. Why is the term 'reasonable prospects for eventual extraction' important?
21. What is meant by the term 'Inferred Mineral Resource'?
22. What is meant by the term 'Indicated Mineral Resource'?
23. What is meant by the term 'Measured Mineral Resource'?
24. How do the three categories listed in problems 21, 22 and 23 differ? On what basis do you choose the appropriate category?
25. In reporting resources, what is the guidance concerning significant figures?
26. What is covered by the term 'Ore Reserve'?
27. What is meant by the term 'economically mineable'?
28. What is meant by the term 'Probable Ore Reserve'?
29. What is meant by the term 'Proved Ore Reserve'?
30. How do the two categories listed in problems 28 and 29 differ? On what basis do you choose the appropriate category?
31. Discuss in detail the conversion process between resources and reserves and within categories.
32. How is the relative accuracy and/or confidence in the estimates expressed?
33. How should Ore Reserve and Mineral Resource data be reported?
34. What is meant by 'reconciliation with production data'?
35. How does the JORC Code deal with stockpiles, low grade mineralization and tailings?
36. What is the purpose of the 'CIM Best Practice Guidelines for the Estimation of Mineral Resources and Mineral Reserves – General Guidelines'?

37. What is meant by the term 'Qualified Person (QP)'? Is it the same as a 'Competent Person' as used in the JORC code?
38. Define the following terms based on the CIM code:
    – Quality Assurance
    – Quality Control
    – Mineral Resource
    – Mineral Reserve
    – Preliminary Feasibility Study
39. Describe the establishment, content and importance of the Resource Database.
40. What is the difference between primary data and interpreted data? Give an example of each.
41. Describe the guidelines for visualization of the primary data.
42. What are the guidelines concerning the visualization of interpreted data?
43. What are the guidelines for data collection, recording, storing and processing?
44. Discuss the required data QA/QC program.
45. In section 7.3.4 'Geological Interpretation and Modeling' the following appears:

    'These guidelines outline requirements for interpretation of geological data, the consideration of economic and mining criteria and the linkage of that information to the grade distribution of the MRMR model'.

    Are these 'guidelines' or 'requirements'? What is the difference? Are these prescriptive?
46. Summarize the material in section 7.3.4 'Geological Interpretation and Modeling' using a step-by-step approach. Does it make good sense? Is it a good checklist for the first part of a senior thesis project? Why or why not?
47. Summarize the material in section 7.3.5 'Mineral Resource Estimation' using a step-by-step approach. Does it make good sense? Is it a good checklist for the second part of a senior thesis project? Why or why not?
48. Review the checklist included in section 7.3.6 'Quantifying Elements to Convert a Mineral Resource to a Mineral Reserve.' How would this be used as part of a senior thesis project? Is it complete? Are there factors missing?
49. Summarize the material in section 7.3.7 'Mineral Reserve Estimation' using a step-by-step approach. Does it make good sense? Does this form a good checklist for a senior thesis project? Why or why not?
50. Summarize the discussion on the selection and application of cutoff grade.
51. How does the 'practicality' of mining enter into the process?
52. What is the value of performing a risk assessment? How should it be performed?
53. What is a Mineral Reserve audit?
54. How should mineral reserves be stated?
55. The senior thesis is presented as a comprehensive technical report. Which items under the heading Technical Report in section 7.3.8 'Reporting' would be useful in that regard? Summarize them.
56. Summarize the discussion of 'Mineral Reserve Reconciliation'.
57. Summarize the discussion on 'Production Monitoring'.
58. Summarize the discussion on 'Production Reconciliation'.

CHAPTER 8

Responsible mining

8.1 INTRODUCTION

At the beginning of Chapter 1, 'ore' was defined as:

Ore: A natural aggregation of one or more solid minerals that can be mined, processed and sold at a profit.

The meaning of 'profit' was defined very simply in Equation (1.1) as

$$\text{Profits} = \text{Revenues} - \text{Costs} \tag{1.1}$$

Replacing 'revenues' and 'costs' by their equivalent expressions in Equation (1.1), one obtains Equation (1.4).

$$\text{Profits} = \text{Material sold(units)} \times (\text{Price/unit} - \text{Cost/unit}) \tag{1.4}$$

The profit (or loss) incurred is often referred to as the 'bottom line' since it is the final calculation made in a financial analysis and appears at the bottom of a financial ledger sheet. In simple terms, it is the amount of money remaining after all of the bills have been paid. At the conclusion of a presentation on a potential mining project, the first focus is on the 'bottom line.' Other financial indicators such as the NPV, ROR, DCFROR and the payback period as discussed in Chapter 2 will also be closely examined. For a project to be considered a 'Go,' the ROR, for example, must be greater than the 'hurdle rate' (the minimum ROR), required by the company. If the projected rate of return is below the 'hurdle', then the project may be re-evaluated using other parameters/assumptions or put aside. The potential 'ore reserve' is no more but remains a mineralized occurrence or mineral concentration.

The requirement to make a profit commensurate with the risk being taken is a key aspect of any business but particularly in mining where the risks can be very high. Because of this, it is of value to consider the notion of 'profit' in today's world in greater detail.

In this respect, Equation (1.4) can be re-written as

$$\text{Profit/unit} = \text{Price/unit} - \text{Cost/unit} \tag{8.1}$$

For metals and many other minerals, the price received per unit is set on the international market and will be considered fixed for the purposes of this discussion. The Cost/unit is something that is project dependent. For reasons that will become obvious as the reader traverses this chapter, it will be expressed as

$$\text{Cost/unit} = \text{Extraction cost/unit} + \text{Environmental cost/unit} + \text{Social cost/unit} \tag{8.2}$$

For any project there will be environmental and social impacts. It is clear that neither the natural nor the societal setting of the orebody will be the same pre- and post-project. Measures will be taken to address these changes. Each level of mitigation will have an associated cost that will directly impact the bottom line. When simply considering profits, one should set the environmental and social cost components in Equation (8.2) to zero. This has, unfortunately, been the approach taken by many mining operations in the past and even by some today. However, positive changes have occurred. In much of the world today, this approach is unacceptable and the project would not go forth. Hence the profits would be zero. The other extreme would be to allocate a very high level of funding to mitigate all potential adverse environmental and social impacts. When these costs are entered into Equation (8.2), and subsequently into Equation (8.1), the resulting profits become unacceptably low and the project would not go forth. This forms the other extreme.

Assuming that the project on the whole is a potentially 'good' one for all of the stakeholders, significant efforts must be made to establish appropriate levels of environmental and social mitigation lying between these two extreme cases that will still lead to an acceptable economic result. This approach and the eventual results achieved are encompassed by the phrase 'Responsible Mining.' This phrase has a very clear meaning and directly includes the 'Economic Responsibility', the 'Environmental Responsibility', and the 'Social Responsibility' components that all responsible mining companies must satisfactorily address when embarking on a new project and when continuing with a current project.

Today, unfortunately, it is not the phrase 'Responsible Mining' which is on the tongues and in the minds of mining executives but rather the phrase 'Sustainable Development.' Although a connection between mining and sustainable development can and has been made, the connection must be described as rather weak. 'Responsible Mining' is one important *response* to 'Sustainable Development.' As such, the continuing world-wide focus on 'Sustainable Development' is important to maintaining and even accelerating the pace toward the universal adoption of 'Responsible Mining' practices.

This chapter will review some of the milestones in the 'Sustainable Development' movement and the actions being taken by the mining industry in moving toward the universal adoption of 'Responsible Mining' practices.

8.2 THE 1972 UNITED NATIONS CONFERENCE ON THE HUMAN ENVIRONMENT

In June of 1972, the United Nations Conference on the Human Environment (UNCHE) met in Stockholm to consider the need for a common outlook and for common principles to inspire and guide the peoples of the world in the preservation and enhancement of the human environment. At the conclusion of the conference it was proclaimed that (UNEP (1972a, 1972b)):

1. Man is both creature and molder of his environment, which gives him physical sustenance and affords him the opportunity for intellectual, moral, social and spiritual growth. In the long and tortuous evolution of the human race on this planet a stage has been reached when, through the rapid acceleration of science and technology, man has acquired the power to

transform his environment in countless ways and on an unprecedented scale. Both aspects of man's environment, the natural and the man-made, are essential to his well-being and to the enjoyment of basic human rights the right to life itself.

2. The protection and improvement of the human environment is a major issue which affects the well-being of peoples and economic development throughout the world; it is the urgent desire of the peoples of the whole world and the duty of all Governments.

3. Man has constantly to sum up experience and go on discovering, inventing, creating and advancing. In our time, man's capability to transform his surroundings, if used wisely, can bring to all peoples the benefits of development and the opportunity to enhance the quality of life. Wrongly or heedlessly applied, the same power can do incalculable harm to human beings and the human environment. We see around us growing evidence of man-made harm in many regions of the earth: dangerous levels of pollution in water, air, earth and living beings; major and undesirable disturbances to the ecological balance of the biosphere; destruction and depletion of irreplaceable resources; and gross deficiencies, harmful to the physical, mental and social health of man, in the man-made environment, particularly in the living and working environment.

4. In the developing countries most of the environmental problems are caused by under development. Millions continue to live far below the minimum levels required for a decent human existence, deprived of adequate food and clothing, shelter and education, health and sanitation. Therefore, the developing countries must direct their efforts to development, bearing in mind their priorities and the need to safeguard and improve the environment. For the same purpose, the industrialized countries should make efforts to reduce the gap between themselves and the developing countries. In the industrialized countries, environmental problems are generally related to industrialization and technological development.

5. The natural growth of population continuously presents problems for the preservation of the environment, and adequate policies and measures should be adopted, as appropriate, to face these problems. Of all things in the world, people are the most precious. It is the people that propel social progress, create social wealth, develop science and technology and, through their hard work, continuously transform the human environment. Along with social progress and the advance of production, science and technology, the capability of man to improve the environment increases with each passing day.

6. A point has been reached in history when we must shape our actions throughout the world with a more prudent care for their environmental consequences. Through ignorance or indifference we can do massive and irreversible harm to the earthly environment on which our life and well being depend. Conversely, through fuller knowledge and wiser action, we can achieve for ourselves and our posterity a better life in an environment more in keeping with human needs and hopes. There are broad vistas for the enhancement of environmental quality and the creation of a good life. What is needed is an enthusiastic but calm state of mind and intense but orderly work. For the purpose of attaining freedom in the world of nature, man must use knowledge to build, in collaboration with nature, a better environment. To defend and improve the human environment for present and future generations has become an imperative goal for mankind – a goal to be pursued together with, and in harmony with, the established and fundamental goals of peace and of worldwide economic and social development.

7. To achieve this environmental goal will demand the acceptance of responsibility by citizens and communities and by enterprises and institutions at every level, all sharing equitably in common efforts. Individuals in all walks of life as well as organizations in many fields, by their values and the sum of their actions, will shape the world environment of the future.

The twenty-six principles included in the Stockholm Declaration (UNEP (1972a, 1972b)) are listed below:

Principle 1: Man has the fundamental right to freedom, equality and adequate conditions of life, in an environment of a quality that permits a life of dignity and well-being, and he bears a solemn responsibility to protect and improve the environment for present and future generations. In this respect, policies promoting or perpetuating apartheid, racial segregation, discrimination, colonial and other forms of oppression and foreign domination stand condemned and must be eliminated.

Principle 2: The natural resources of the earth, including the air, water, land, flora and fauna and especially representative samples of natural ecosystems, must be safeguarded for the benefit of present and future generations through careful planning or management, as appropriate.

Principle 3: The capacity of the earth to produce vital renewable resources must be maintained and, wherever practicable, restored or improved.

Principle 4: Man has a special responsibility to safeguard and wisely manage the heritage of wildlife and its habitat, which are now gravely imperiled by a combination of adverse factors. Nature conservation, including wildlife, must therefore receive importance in planning for economic development.

Principle 5: The non-renewable resources of the earth must be employed in such a way as to guard against the danger of their future exhaustion and to ensure that benefits from such employment are shared by all mankind.

Principle 6: The discharge of toxic substances or of other substances and the release of heat, in such quantities or concentrations as to exceed the capacity of the environment to render them harmless, must be halted in order to ensure that serious or irreversible damage is not inflicted upon ecosystems. The just struggle of the peoples of all countries against pollution should be supported.

Principle 7: States shall take all possible steps to prevent pollution of the seas by substances that are liable to create hazards to human health, to harm living resources and marine life, to damage amenities or to interfere with other legitimate uses of the sea.

Principle 8: Economic and social development is essential for ensuring a favorable living and working environment for man and for creating conditions on earth that are necessary for the improvement of the quality of life.

Principle 9: Environmental deficiencies generated by the conditions of under-development and natural disasters pose grave problems and can best be remedied by accelerated development through the transfer of substantial quantities of financial and technological assistance as a supplement to the domestic effort of the developing countries and such timely assistance as may be required.

Principle 10: For the developing countries, stability of prices and adequate earnings for primary commodities and raw materials are essential to environmental management, since economic factors as well as ecological processes must be taken into account.

Principle 11: The environmental policies of all States should enhance and not adversely affect the present or future development potential of developing countries, nor should they hamper the attainment of better living conditions for all, and appropriate steps should be taken by States and international organizations with a view to reaching agreement on meeting the possible national and international economic consequences resulting from the application of environmental measures.

Principle 12: Resources should be made available to preserve and improve the environment, taking into account the circumstances and particular requirements of developing countries and any costs which may emanate from their incorporating environmental safeguards into their development planning and the need for making available to them, upon their request, additional international technical and financial assistance for this purpose.

Principle 13: In order to achieve a more rational management of resources and thus to improve the environment, States should adopt an integrated and coordinated approach to their development planning so as to ensure that development is compatible with the need to protect and improve environment for the benefit of their population.

Principle 14: Rational planning constitutes an essential tool for reconciling any conflict between the needs of development and the need to protect and improve the environment.

Principle 15: Planning must be applied to human settlements and urbanization with a view to avoiding adverse effects on the environment and obtaining maximum social, economic and environmental benefits for all. In this respect projects which are designed for colonialist and racist domination must be abandoned.

Principle 16: Demographic policies which are without prejudice to basic human rights and which are deemed appropriate by the Governments concerned should be applied in those regions where the rate of population growth or excessive population concentrations are likely to have adverse effects on the environment of the human environment and impede development.

Principle 17: Appropriate national institutions must be entrusted with the task of planning, managing or controlling the environmental resources of States with a view to enhancing environmental quality.

Principle 18: Science and technology, as part of their contribution to economic and social development, must be applied to the identification, avoidance and control of environmental risks and the solution of environmental problems and for the common good of mankind.

Principle 19: Education in environmental matters, for the younger generation as well as adults, giving due consideration to the underprivileged, is essential in order to broaden the basis for an enlightened opinion and responsible conduct by individuals, enterprises and communities in protecting and improving the environment in its full human dimension. It is also essential that mass media of communications avoid contributing to the deterioration of the environment, but, on the contrary, disseminates information of an educational nature on the need to protect and improve the environment in order to enable all to develop in every respect.

Principle 20: Scientific research and development in the context of environmental problems, both national and multinational, must be promoted in all countries, especially the developing countries. In this connection, the free flow of up-to-date scientific information and transfer of experience must be supported and assisted, to facilitate the solution of environmental problems; environmental technologies should be made available to developing countries on terms which would encourage their wide dissemination without constituting an economic burden on the developing countries.

Principle 21: States have, in accordance with the Charter of the United Nations and the principles of international law, the sovereign right to exploit their own resources pursuant to their own environmental policies, and the responsibility to ensure that activities within their jurisdiction or control do not cause damage to the environment of other States or of areas beyond the limits of national jurisdiction.

Principle 22: States shall cooperate to develop further the international law regarding liability and compensation for the victims of pollution and other environmental damage caused by activities within the jurisdiction or control of such States to areas beyond their jurisdiction.

Principle 23: Without prejudice to such criteria as may be agreed upon by the international community, or to standards which will have to be determined nationally, it will be essential in all cases to consider the systems of values prevailing in each country, and the extent of the applicability of standards which are valid for the most advanced countries but which may be inappropriate and of unwarranted social cost for the developing countries.

Principle 24: International matters concerning the protection and improvement of the environment should be handled in a cooperative spirit by all countries, big and small, on an equal footing. Cooperation through multilateral or bilateral arrangements or other appropriate means is essential to effectively control, prevent, reduce and eliminate adverse environmental effects resulting from activities conducted in all spheres, in such a way that due account is taken of the sovereignty and interests of all States.

Principle 25: States shall ensure that international organizations play a coordinated, efficient and dynamic role for the protection and improvement of the environment.

Principle 26: Man and his environment must be spared the effects of nuclear weapons and all other means of mass destruction. States must strive to reach prompt agreement, in the relevant international organs, on the elimination and complete destruction of such weapons.

8.3 THE WORLD CONSERVATION STRATEGY (WCS) – 1980

One of the products of the United Nations Conference on the Human Environment (UNCHE) was the establishment of the United Nations Environment Program (UNEP). In 1980, the document 'World Conservation Strategy: Living Resource Conservation for Sustainable Development' was produced. It was commissioned by UNEP, prepared by the International Union for Conservation of Nature and Natural Resources (IUCN), and jointly financed by UNEP and the World Wildlife Fund (WWF).

Although IUCN spear-headed the overall preparation of the 'Strategy,' many governments, non-government organizations and individuals from both developed and developing countries participated in its development. The draft reports were circulated for comment to a large number of specialists on ecology, threatened species, protected areas, environmental planning, environmental policy, law and administration, and environmental education (Munro, 1980).

In 1980, the 'World Conservation Strategy' report was published in pack format intended for decision-makers and, as such, had a very limited distribution. An unofficial version of the 'Strategy' is the publication 'How to Save the World: Strategy for World Conservation' prepared by Robert Allen (Allen, 1980). It was based on the material contained in the report but oriented toward the general reader. The material contained in this section is extracted directly from that publication (Allen, 1980).

The intention of the 'The World Conservation Strategy' was to stimulate a more focused approach to living resource conservation and to provide policy guidance on how this can be carried out. The 'Strategy,' according to Scott (1980), is the first time that:(1) it has been clearly shown how conservation can contribute to the development objectives of governments, industry, commerce, organized labor, and the professions, and (2) development has been suggested as a major means of achieving conservation instead of being viewed as an obstruction to it.

The present authors consider the following points from the book to be especially relevant to mining readers (Allen, 1980).

1. The biosphere is like a self-regenerating cake, and conservation is the conduct of our affairs so that we can have our cake and eat it too. As long as certain bits of cake are not consumed and consumption of the rest is kept within certain limits, the cake will renew itself and provide for continuing consumption. For people to gain a decent livelihood from the earth without undermining its capacity to go on supporting them, they must conserve the biosphere. This means doing three things:
   1. Maintaining essential ecological processes and life-support systems
   2. Preserving genetic diversity
   3. Utilizing species and ecosystems sustainably.

2. Although environmental modification is natural and a necessary part of development, this does not mean that all modification leads to development (nor that preservation impedes development). While it is inevitable that most of the planet will be modified by people and that much of it will be transformed, it is not at all inevitable that such alterations will achieve the social and economic objectives of development. *Unless it is based on conservation, much development will continue to have unacceptably harmful side effects, provide reduced benefits or even fail altogether; and it will become impossible to meet the needs of today without foreclosing the achievement of tomorrow* (emphasis added by current authors).

3. A world strategy for the conservation of the earth's living resources is needed for three reasons:
   a. The need for conservation is so pressing that it should be in the forefront of human endeavor. Yet for most people and their governments, conservation is an obscure peripheral activity perpetrated by birdwatchers. One consequence of this view is that development, which should be the main means of solving human problems, is so little affected by

conservation that too often it adds to human problems by destroying or degrading living resources essential for human welfare. A world strategy is needed to focus the attention of the world on conservation.

b. National and international organizations concerned with conservation, whether governmental or non-governmental, are ill-organized and fragmented, split up among different interests such as agriculture, forestry, fisheries and wildlife. As a result, there is duplication of effort, gaps in coverage, competition for money and influence, and conflict. A world strategy is needed to promote that effort and define the areas where cooperation is most needed.

c. The action required to cure the most serious current conservation problems and to prevent still worse ones takes time: time for planning, education, training, better organization and research. When such action is undertaken, it takes time for the biosphere to respond: reforestation, the restoration of degraded land, the recovery of depleted fisheries, and so on are not instantaneous processes. Because time is running out a world strategy is necessary: (1) to determine the priorities, (2) to indicate the main obstacles to achieving them and (3) to propose ways of overcoming the obstacles.

4. For development practitioners, the 'Strategy' proposes ways of improving the prospects of *sustainable development – development that is likely to achieve lasting satisfaction of human needs and improvement of the quality of human life – by integrating conservation into the development process* (emphasis added by current authors). It also attempts to identify those areas where the interests of conservation and of development are most likely to coincide, and therefore where a closer partnership between the two processes would be particularly advantageous.

Although the focus of the 'World Conservation Strategy' is on living things, the main obstacles cited have relevance to mining. The obstacles are (Allen, 1980):

1. The belief that the conservation of living resources is a specialized activity rather than a process that cuts across and must be considered by all sectors of activity.

2. The consequent of failure to integrate conservation with development.

3. A development process that is generally inflexible and needlessly destructive, because of inadequate environmental planning and a lack of rational allocation of land and water uses.

4. The lack of capacity, because of inadequate legislation, to conserve; poor organization (notably government agencies with insufficient mandates and a lack of coordination); lack of trained personnel; and a lack of basic information on priorities, on the productive and regenerative capacities of the living resources concerned, and on trade-offs between one management option and another.

5. The lack of support for conservation, because of a lack of awareness (other than at the most superficial level) of the need for conservation and of the responsibility to conserve amongst those who use or have an impact on living resources, including in many cases, governments.

6. Failure to deliver conservation–based development where it is most needed, notably the rural areas of developing countries.

The publication 'How to Save the World: Strategy for World Conservation' is still available today and its reading is strongly encouraged. The 'World Conservation Strategy' is credited as being the source of the phrase 'sustainable development' which is in such wide use today.

In this document it is very correctly applied since it concerns living, renewable objects and things.

8.4 WORLD COMMISSION ON ENVIRONMENT AND DEVELOPMENT (1987)

In 1983, the 'World Commission on Environment and Development' was established under the auspices of the United Nations Environment Program (UNEP). The Commission's directive was to re-examine the critical environment and development problems on the planet and to formulate realistic proposals to solve them. A global agenda for change was to be developed working within the principle of Environmentally Sustainable Development (Alcor, 2005).

Some of the specific goals were (Brundtland, 1987):

(1) to propose long-term environmental strategies for achieving sustainable development by the year 2000 and beyond;

(2) to recommend ways in which concern for the environment may be translated into greater co-operation among developing countries and between countries at different stages of economic and social development and lead to the achievement of common and mutually supportive objectives that take account of the inter-relationships between people, resources, environment, and development;

(3) to consider ways and means by which the international community can deal more effectively with environmental concerns;

(4) to help define shared perceptions of long-term environmental issues and the appropriate efforts needed to deal successfully with the problems of protecting and enhancing the environment, a long-term agenda for action during the coming decades, and aspirational goals for the world community.

Mrs. Gro Harlem Brundtland, then Prime Minister of Norway, chaired the commission.

The Commission's report 'Our Common Future: The World Commission on Environment and Development' (often referred to as simply 'Our Common Future' or 'The Brundtland Report') was delivered in 1987. In the report, population and human resources, food security, species and ecosystems, energy, and the 'urban challenge' of humans in their built environment are examined. Recommendations are made for institutional and legal changes in order to confront common global problems. These include the development and expansion of international institutions for cooperation and legal mechanisms to confront common concerns. It called for increased cooperation with industry (Alcor, 2005).

As part of their work, the Commission (Our Common Future, 1987) defined sustainable development as:

> 'development that meets the needs of the present without compromising the ability of future generations to meet their own needs.'

The words are very similar to the definition provided in 1980 as part of the 'World Conservation Strategy.' However, its meaning has been considerably broadened. Sustainable development is considered to be a *process of change* (emphasis added by current authors) in which

> 'the exploitation of resources, the direction of investments, the orientation of technological development, and institutional change are made consistent with future as well as present needs (Our Common Future, 1987).'

In Chapter 2, which is entitled 'Towards Sustainable Development,' there are some additional statements regarding what they mean by sustainable development, both in general, and also with respect to non-renewable resources. A few selected by the present authors are presented below (Our Common Future, 1987):

1. There are two key concepts in sustainable development:
   - the concept of 'needs', in particular the essential needs of the world's poor, to which overriding priority should be given.
   - the idea of limitations imposed by the state of technology and social organization on the environment's ability to meet present and future needs.

2. Sustainable development requires that societies meet human needs both by increasing productive potential and by ensuring equitable opportunities for all.

3. Settled agriculture, the diversion of water courses, the extraction of minerals, the emission of heat and noxious gases into the atmosphere, commercial forests, and genetic manipulation are all examples of human intervention in natural systems during the course of development. Until recently, such interventions were small in scale and their impact limited. Today's interventions are more drastic in scale and impact, and more threatening to life-support systems both locally and globally. This need not happen. At a minimum, sustainable development must not endanger the natural systems that support life on Earth: the atmosphere, the waters, the soils, and the living beings.

4. In general, renewable resources like forests and fish stocks need not be depleted provided the rate of use is within the limits of regeneration and natural growth.

5. For non-renewable resources, like fossil fuels and minerals, their use reduces the stock available for future generations. But this does not mean that such resources should not be used. In general, the rate of depletion should take into account the criticality of that resource, the availability of technologies for minimizing depletion, and the likelihood of substitutes being available. Thus land should not be degraded beyond reasonable recovery.

6. With minerals and fossil fuels, the rate of depletion and the emphasis on recycling and economy of use should be calibrated to ensure that the resource does not run out before acceptable substitutes are available.

7. Sustainable development requires that the rate of depletion of non-renewable resources should foreclose as few future options as possible.

8. So-called free goods like air and water are also resources. Sustainable development requires that the adverse impacts on the quality of air, water and other natural elements are minimized so as to sustain the ecosystem's overall integrity.

9. In essence, sustainable development is a process of change in which the exploitation of resources, the direction of investments, the orientation of technological development, and institutional change are all in harmony and enhance both current and future potential to meet human needs and aspirations.

10. Many problems of resource depletion and environmental stress arise from disparities in economic and political power. An industry may get away with unacceptable levels of air and water pollution because the people who bear the brunt of it are poor and unable to complain

effectively. A forest may be destroyed by excessive felling because the people living there have no alternatives or because timber contractors generally have more influence than forest dwellers.

For mining, the above should be kept firmly in mind in addition to the simple and often quoted 'sustainable development' definition statement.

This section will close with a listing of what the authors of the report felt to be required in the pursuit for sustainable development (Our Common Future, 1987):
- a political system that secures effective citizen participation in decision making,
- an economic system that is able to generate surpluses and technical knowledge on a self-reliant and sustained basis,
- a social system that provides for solutions for the tensions arising from disharmonious development,
- a production system that respects the obligation to preserve the ecological base for development,
- a technological system that can search continuously for new solutions,
- an international system that fosters sustainable patterns of trade and finance, and
- an administrative system that is flexible and has the capacity for self-correction.

As the authors of the document indicate, these requirements are more in the nature of goals that should underlie national and international action on development. They stress that what matters is the sincerity with which these goals are pursued and the effectiveness with which departures from them are corrected.

8.5 THE 'EARTH SUMMIT'

The United Nations Conference on Environment and Development (UNCED) was held in Rio de Janeiro, Brazil in 1992 (UNCED, 1992a). It is often referred to as the 'Earth Summit' and that name has been used here. It had the following goals: (1) building upon the United Nations Conference on the Human Environment held in Stockholm in 1972; (2) establishing a new and equitable global partnership through the creation of new levels of cooperation among States, key sectors of societies and people; and (3) working towards international agreements which respect the interests of all and protect the integrity of the global environmental and developmental system (UNCED, 1992b).

Several important documents were produced, two of which, 'The Rio Declaration' (UNCED, 1992b) and 'Agenda 21' (UNCED, 1992c) will be reproduced in this section.

8.5.1 *The Rio Declaration*

Recognizing the integral and interdependent nature of the Earth, our home, the following principles were proclaimed (UNCED, 1992b):

Principle 1: Human beings are at the centre of concerns for sustainable development. They are entitled to a healthy and productive life in harmony with nature.

Principle 2: States have, in accordance with the Charter of the United Nations and the principles of international law, the sovereign right to exploit their own resources pursuant to their own environmental and developmental policies, and the responsibility to ensure that activities within their jurisdiction or control do not cause damage to the environment of other States or of areas beyond the limits of national jurisdiction.

Principle 3: The right to development must be fulfilled so as to equitably meet developmental and environmental needs of present and future generations.

Principle 4: In order to achieve sustainable development, environmental protection shall constitute an integral part of the development process and cannot be considered in isolation from it.

Principle 5: All States and all people shall cooperate in the essential task of eradicating poverty as an indispensable requirement for sustainable development, in order to decrease the disparities in standards of living and better meet the needs of the majority of the people of the world.

Principle 6: The special situation and needs of developing countries, particularly the least developed and those most environmentally vulnerable, shall be given special priority. International actions in the field of environment and development should also address the interests and needs of all countries.

Principle 7: States shall cooperate in a spirit of global partnership to conserve, protect and restore the health and integrity of the Earth's ecosystem. In view of the different contributions to global environmental degradation, States have common but differentiated responsibilities. The developed countries acknowledge the responsibility that they bear in the international pursuit of sustainable development in view of the pressures their societies place on the global environment and of the technologies and financial resources they command.

Principle 8: To achieve sustainable development and a higher quality of life for all people, States should reduce and eliminate unsustainable patterns of production and consumption and promote appropriate demographic policies.

Principle 9: States should cooperate to strengthen endogenous capacity-building for sustainable development by improving scientific understanding through exchanges of scientific and technological knowledge, and by enhancing the development, adaptation, diffusion and transfer of technologies, including new and innovative technologies.

Principle 10: Environmental issues are best handled with the participation of all concerned citizens, at the relevant level. At the national level, each individual shall have appropriate access to information concerning the environment that is held by public authorities, including information on hazardous materials and activities in their communities, and the opportunity to participate in decision-making processes. States shall facilitate and encourage public awareness and participation by making information widely available. Effective access to judicial and administrative proceedings, including redress and remedy, shall be provided.

Principle 11: States shall enact effective environmental legislation. Environmental standards, management objectives and priorities should reflect the environmental and developmental context to which they apply. Standards applied by some countries may be inappropriate and of unwarranted economic and social cost to other countries, in particular developing countries.

Principle 12: States should cooperate to promote a supportive and open international economic system that would lead to economic growth and sustainable development in all countries, to better address the problems of environmental degradation. Trade policy measures for environmental purposes should not constitute a means of arbitrary or unjustifiable discrimination or a disguised restriction on international trade. Unilateral actions to deal with environmental challenges outside the jurisdiction of the importing country should be avoided. Environmental measures addressing trans-boundary or global environmental problems should, as far as possible, be based on an international consensus.

Principle 13: States shall develop national law regarding liability and compensation for the victims of pollution and other environmental damage. States shall also cooperate in an expeditious and more determined manner to develop further international law regarding liability and compensation for adverse effects of environmental damage caused by activities within their jurisdiction or control to areas beyond their jurisdiction.

Principle 14: States should effectively cooperate to discourage or prevent the relocation and transfer to other States of any activities and substances that cause severe environmental degradation or are found to be harmful to human health.

Principle 15: In order to protect the environment, the precautionary approach shall be widely applied by States according to their capabilities. Where there are threats of serious or irreversible damage, lack of full scientific certainty shall not be used as a reason for postponing cost-effective measures to prevent environmental degradation.

Principle 16: National authorities should endeavor to promote the internalization of environmental costs and the use of economic instruments, taking into account the approach that the polluter should, in principle, bear the cost of pollution, with due regard to the public interest and without distorting international trade and investment.

Principle 17: Environmental impact assessment, as a national instrument, shall be undertaken for proposed activities that are likely to have a significant adverse impact on the environment and are subject to a decision of a competent national authority.

Principle 18: States shall immediately notify other States of any natural disasters or other emergencies that are likely to produce sudden harmful effects on the environment of those States. Every effort shall be made by the international community to help States so afflicted.

Principle 19: States shall provide prior and timely notification and relevant information to potentially affected States on activities that may have a significant adverse trans-boundary environmental effect and shall consult with those States at an early stage and in good faith.

Principle 20: Women have a vital role in environmental management and development. Their full participation is therefore essential to achieve sustainable development.

Principle 21: The creativity, ideals and courage of the youth of the world should be mobilized to forge a global partnership in order to achieve sustainable development and ensure a better future for all.

Principle 22: Indigenous people and their communities and other local communities have a vital role in environmental management and development because of their knowledge and traditional practices. States should recognize and duly support their identity, culture and interests and enable their effective participation in the achievement of sustainable development.

Principle 23: The environment and natural resources of people under oppression, domination and occupation shall be protected.

Principle 24: Warfare is inherently destructive of sustainable development. States shall therefore respect international law providing protection for the environment in times of armed conflict and cooperate in its further development, as necessary.

Principle 25: Peace, development and environmental protection are interdependent and indivisible.

Principle 26: States shall resolve all their environmental disputes peacefully and by appropriate means in accordance with the Charter of the United Nations.

Principle 27: States and people shall cooperate in good faith and in a spirit of partnership in the fulfillment of the principles embodied in this Declaration and in the further development of international law in the field of sustainable development.

8.5.2 *Agenda 21*

The climax of the Earth Summit was the adoption of Agenda 21. Although weakened by compromise and negotiation during the generation process, it offered a wide-ranging blueprint for action to achieve sustainable development worldwide (UNCED, 1992a). The contents are indicated below (UNCED, 1992c).

Chapter 1. Preamble

SECTION I. SOCIAL AND ECONOMIC DIMENSIONS

Chapter 2. International cooperation to accelerate sustainable development in developing countries and related domestic policies
Chapter 3. Combating poverty
Chapter 4. Changing consumption patterns
Chapter 5. Demographic dynamics and sustainability
Chapter 6. Protecting and promoting human health conditions
Chapter 7. Promoting sustainable human settlement development
Chapter 8. Integrating environment and development in decision-making

SECTION II. CONSERVATION AND MANAGEMENT OF RESOURCES FOR DEVELOPMENT

Chapter 9. Protection of the atmosphere
Chapter 10. Integrated approach to the planning and management of land resources

Chapter 11. Combating deforestation
Chapter 12. Managing fragile ecosystems; combating desertification and drought
Chapter 13. Managing fragile ecosystems: sustainable mountain development
Chapter 14. Promoting sustainable agriculture and rural development
Chapter 15. Conservation of biological diversity
Chapter 16. Environmentally sound management of biotechnology
Chapter 17. Protection of the oceans, all kinds of seas, including enclosed and semi-enclosed seas, and coastal areas and the protection, rational use and development of their living resources
Chapter 18. Protection of the quality and supply of freshwater resources; application of integrated approaches to the development, management and use of water resources
Chapter 19. Environmentally sound management of toxic chemicals, including prevention of illegal international traffic in toxic and dangerous products
Chapter 20. Environmentally sound management of hazardous wastes, in hazardous wastes
Chapter 21. Environmentally sound management of solid wastes and sewage-related issues
Chapter 22. Safe and environmentally sound management of radioactive wastes

SECTION III. STRENGTHENING THE ROLE OF MAJOR GROUPS

Chapter 23. Preamble
Chapter 24. Global action for women towards sustainable and equitable development
Chapter 25. Children and youth in sustainable development
Chapter 26. Recognizing and strengthening the role of indigenous and their communities
Chapter 27. Strengthening the role of non-governmental organizations; partners for sustainable development
Chapter 28. Local authorities' initiatives in support of Agenda 21
Chapter 29. Strengthening the role of workers and their trade unions
Chapter 30. Strengthening the role of business and industry
Chapter 31. Scientific and technological community
Chapter 32. Strengthening the role of farmers

SECTION IV. MEANS OF IMPLEMENTATION

Chapter 33. Financial resources and mechanisms
Chapter 34. Transfer of environmentally sound technology, cooperation and capacity-building
Chapter 35. Science for sustainable development
Chapter 36. Promoting education, public awareness and training
Chapter 37. National mechanisms and international cooperation for capacity-building in developing countries

Chapter 38. International institutional arrangements
Chapter 39. International legal instruments and mechanisms
Chapter 40. Information for decision-making

8.6 WORLD SUMMIT ON SUSTAINABLE DEVELOPMENT (WSSD)

In 2002, the World Summit on Sustainable Development (WSSD) was held in Johannesburg, South Africa to review the progress made on Agenda 21 during the intervening ten-year period (WSSD, 2002a). Specific attention was placed on reviewing (1) the obstacles encountered, (2) the lessons learned during the implementation process, and (3) new factors which have emerged (WSSD, 2002b). The major output of the Summit was contained in two documents, 'Plan of Implementation for the World Summit on Sustainable Development' (WSSD, 2002c) and the 'Political Declaration' (WSSD, 2002d).

The 'Plan of Implementation' contains several direct references to the mining sector in addition to a wide range of proposals that will impact upon the activities of the sector.

The contents of the 'Plan of Implementation' document prepared at the World Summit on Sustainable Development are listed below (WSSD, 2002c).

   I. Introduction
  II. Poverty eradication
 III. Changing unsustainable patterns of consumption and production
  IV. Protecting and managing the natural resource base of economic and social development
   V. Sustainable development in a globalizing world
  VI. Health and sustainable development
 VII. Sustainable development of small-island developing States
VIII. Sustainable development for Africa
  IX. Other regional initiatives
       A. Sustainable development in Latin America and the Caribbean
       B. Sustainable development in Asia and the Pacific
       C. Sustainable development in the West Asia region
       D. Sustainable development in the Economic Commission for Europe region
   X. Means of implementation
  XI. Institutional framework for sustainable development
       A. Objectives
       B. Strengthening the institutional framework for sustainable development at the international level
       C. Role of the General Assembly
       D. Role of the Economic and Social Council
       E. Role and function of the Commission on Sustainable Development
       F. Role of international institutions
       G. Strengthening institutional arrangements for sustainable development at the regional level
       H. Strengthening institutional frameworks for sustainable development at the national level
       I. Participation of major groups

Paragraph 46 of this document (Section X – Means of implementation) which deals directly with mining, minerals and metals, is provided below (WSSD, 2002c):

Mining, minerals and metals are important to the economic and social development of many countries. Minerals are essential for modern living. Financing the contribution of mining, minerals and metals to sustainable development includes actions at all levels to:

a) Support efforts to address the environmental, economic, health and social impacts and benefits of mining, minerals and metals throughout their life cycle including workers health and safety, and use a range of partnerships, furthering existing activities at the national and international levels among interested Governments, intergovernmental organizations, mining companies and workers and other stakeholders to promote transparency and accountability for sustainable development;

b) Enhance the participation of stakeholders, including local and indigenous communities and women, to play an active role in minerals, metals and mining development throughout the life cycles of mining operations, including after closure for rehabilitation purposes in accordance with national regulations and taking into account significant trans-boundary impacts.

c) Foster sustainable mining practices through the provision of financial, technical and capacity-building support to developing countries and countries with economies in transition for the mining and processing of minerals, including small-scale mining, and where possible and appropriate, improve value-added processing, upgrade scientific and technological information and reclaim and rehabilitate degraded sites.

8.7 MINING INDUSTRY AND MINING INDUSTRY-RELATED INITIATIVES

8.7.1 *Introduction*

To this point in this chapter, the focus has been on some of the major sustainable development initiatives organized under the auspices of the United Nations. This section will deal with some of the sustainable development oriented activities involving various sectors of the world mining industry.

8.7.2 *The Global Mining Initiative (GMI)*

In the fall of 1998, discussions between the chairmen and chief executives of several of the world's largest mining companies were held concerning the social and environmental challenges facing the mining and metals industries. Their common concerns were: (1) the signs of increasing resistance to new mining projects from a wide range of stakeholders worldwide; (2) growing concerns over perceptions of health and environmental threats resulting from the use of metals; and (3) the effectiveness of industry associations. These discussions led to the birth of the Global Mining Initiative (GMI). This Initiative was aimed at ensuring that an industry essential to the well-being of a changing world was responsive to global needs and challenges (GMI (1998a, 1998b)).

The participating mining companies hoped to learn how they could manage these issues more effectively in order to maximize the contribution made to the wider transition to sustainable development. They hoped that non-industry participants in the initiative, such as governments, international organizations and other representatives of civil society, would share in the effort to define legitimate mutual expectations and to clarify where the boundaries lay for action by different actors (GMI, 1998c).

The GMI was concerned with the full range of issues in the mining, minerals and metals cycles, including (GMI, 1998c):
- access to land and resources
- exploration
- project development and secondary development impacts
- governance of mining projects, their place in social and economic development and issues of capacity building
- rent capture and distribution
- mining operations
- stewardship of resources such as water and biodiversity
- energy use
- management of waste
- social and environmental aspects of mine closure
- primary and subsequent stages of processing
- the trade in materials produced by mining
- how those materials are used – their consumption, recycling and disposal

It was noted that since the 'Earth Summit' at Rio de Janeiro in 1992, the relationship between the mining industry and the environmental movement could be characterized as being more positive. Both have realized that co-operation and open discussion are essential for real progress towards sustainable development. The mining industry has accepted that, alone, it cannot 'rewrite its reputation' and needs to work with a wide range of non-industry participants, such as governments, international organizations like the World Bank, and the NGO community, in order to achieve a common understanding of the challenges and priorities (GMI, 1998b).

The GMI planned three major activities the objective of which was to reach a clearer definition and understanding of the positive part the mining and minerals industry can play in making the transition to sustainable patterns of economic development. These activities were (GMI, 1998b):

1. The Mining, Minerals and Sustainable Development (MMSD) project. This activity, sponsored by 30 mining companies, had the aim of 'identifying how mining and minerals can best contribute to the global transition to sustainable development'. The project was commissioned by the mining working group of the World Business Council for Sustainable Development, managed by a committee with a majority of non-industry representatives, and coordinated by Richard Sandbrook, co-founder of Friends of the Earth. The work was undertaken with the respected think-tank, the International Institute for Environment and Development (GMI, 1998b).

2. A careful re-appraisal of all the major bodies representing mining interests around the world. The aim was to refine association activities to give a more concerted approach to sustainable development issues. In this way, the activities can be focused via a global structure which would serve as a pro-active advocate for the mineral industries. The result was the creation of a strong global organization, the International Council on Mining and Metals (ICMM), to represent and lead the industry in meeting the challenges of sustainable development (GMI, 1998b).

3. Organization of the 'Global Mining Initiative Conference.' This conference which was held in Toronto in May 2002 (just prior to the World Summit on Sustainable Development held in Johannesburg, September 2002), provided a platform for industry leaders to discuss key issues and recommendations of the MMSD report with leaders of government, international organizations and NGOs (GMI, 1998b).

Following publication of the final report of the Mining, Minerals and Sustainable Development (MMSD) project and the close of the Global Mining Initiative Conference, the Global Mining Initiative had served its function and ceased to exist as an entity.

8.7.3 International Council on Mining and Metals (ICMM)

As indicated in section 8.7.2, the International Council on Mining and Metals (ICMM) was created under the auspices of the Global Mining Initiative. It has as its vision 'A viable mining, minerals and metals industry that is widely recognized as essential for modern living and a key contributor to sustainable development.' The mission statement of the ICMM may be summarized as: (1) ensuring the continued access to land, capital and markets by the mining, minerals and metals industry (as well as building trust and respect) by demonstrating its ability to contribute successfully to sustainable development; (2) offering strategic industry leadership towards achieving continuous improvements in sustainable development performance; and (3) providing a common platform for the industry to share challenges and responsibilities as well as to engage with key constituencies on issues of common concern at the international level, based on science and principles of sustainable development. This mission, they believe, is best achieved by acting collectively (ICCM, 2002a).

For the mission to be accomplished it must be accompanied by an action plan with clearly stated goals and timely deliverables. The ICMM goals are (ICCM, 2002a) to:

1. Offer strategic leadership to achieve improved sustainable development performance in the mining, minerals and metals industry.

2. Represent the views and interests of its members and serve as a principal point of engagement with the industry's key constituencies in the international arena.

3. Promote science-based regulations and material-choice decisions that encourage market access and the safe production, use, reuse, and recycling of metals and minerals.

4. Identify and advocate the use of good practices to address sustainable development issues within the industry.

The ICMM consists of a Council, the Executive Working Group, and various Committees and Task Forces (ICCM, 2002b). The Council which is made up of the CEOs of all ICMM member companies and associations is ICMM'S principal governing body. The Executive Working Group comprises nominated representatives from each of the corporate and association members. This group facilitates input from members on cross-cutting issues and ensures the effective implementation of the ICMM work program. Task Force chairs are appointed from this group. Committees and Task Forces are appointed by the Council to develop policy and pursue programs as required. The current task forces (ICCM, 2002b) are:

- Integrated Materials Management
- Community and Social Development
- Health and Safety
- Environmental Stewardship

The ICMM Principles are as follows (ICCM, 2002c):
- Implement and maintain ethical business practices and sound systems of corporate governance
- Integrate sustainable development considerations within the corporate decision-making process
- Uphold fundamental human rights and respect cultures, customs and values in dealings with employees and others who are affected by our activities
- Implement risk management strategies based on valid data and sound science
- Seek continual improvement of our health and safety performance
- Seek continual improvement of our environmental performance
- Contribute to conservation of biodiversity and integrated approaches to land use planning
- Facilitate and encourage product design, use, re-use, recycling and disposal of our products
- Contribute to the social, economic and institutional development of the communities in which we operate
- Implement effective and transparent engagement, communication and independently verified reporting arrangements with our stakeholders.

At the conclusion of the Global Mining Initiative Conference held May 12–15, 2002, the ICMM Toronto Declaration was issued (ICMM, 2002d). Based on a shared desire to enhance the contribution that mining and metals can make to social and economic development, the following realizations were put forth (ICMM, 2002d):
- that successful mining and metals processing operations require the support of the communities in which they operate;
- that respect for these communities and a serious engagement with them is required to ensure that mining and metals processing are seen as beneficial for the community and the company;
- that successful companies will respect fundamental human rights, including workplace rights and the need for a healthy and safe workplace: and
- that successful companies will accept their environmental stewardship responsibilities for their facility locations.

To give expression to these values will take dedicated and focused action on our part. We cannot achieve this alone. Progress towards sustainable development will be the product of continuing engagement with government and civil society. This engagement, which will have to occur at all levels of our industry, will at times involve trade-offs and difficult choices (ICMM, 2002d).

ICMM recognizes that (ICMM, 2002d):
- The MMSD Report and the process on which it was based, including the regional programs, have elevated and informed the debate leading to a way forward for the sector.
- Decisive and principled leadership is required at this critical time.
- Accountability, transparency and credible reporting is essential.
- Its Members, in satisfying their obligations to shareholders, must do business in a manner that merits the trust and respect of key constituencies, including the communities in which they operate.
- Constructive and value-adding engagement among constituencies at the local, national, and global levels is essential.
- Its Members must move beyond a regulatory-compliance-based mind set to effectively manage the complex trade-offs of economic, environmental, and social issues.

- The industry requires additional capacity to be effective in advancing sustainable development.
- The rates and responsibilities of the diverse parties comprising governments, civil society, and business are different and must be respected.
- Artisanal, small-scale mining and orphan site legacy issues are important and complex. However, they are beyond the capacity of ICMM to resolve. Governments and international agencies should assume the lead role in addressing them.

ICMM will (ICMM, 2002d):
- Expand the current ICMM Sustainable Development Charter to include appropriate areas recommended in the MMSD Report.
- Develop best-practice protocols that encourage third-party verification and public reporting.
- Engage in constructive dialogue with key constituencies.
- Assist Members in understanding the concepts and application of sustainable development.
- Together with the World Bank and others, seek to enhance effective community development management tools and systems.
- Promote the concept of integrated materials management throughout the minerals value chain wherever relevant.
- Promote sound science-based regulations and material-choice decisions that encourage market access and the safe use, reuse and recycling of metals and minerals.
- Create an emergency response regional register for the global mining, metals and minerals industry.
- In partnership with IUCN – The World Conservation Union and others, seek to resolve the questions associated with protected areas and mining.

8.7.4 Mining, Minerals, and Sustainable Development (MMSD)

The Mining, Minerals and Sustainable Development (MMSD) project was commissioned as part of the Global Mining Initiative. It was an independent two-year process of consultation and research aimed at understanding how to maximize the contribution of the mining and minerals sector to sustainable development at the global, national, regional and local levels. Managed by the International Institute for Environment and Development (IIED) in London, under contract to the World Business Council for Sustainable Development, the project began in April 2000. It was designed to produce concrete results in the form of a final report and a series of working papers as well as to create a dialogue process that could be carried forward into the future (MMSD, 2000).

The general objectives of MMSD (MMSD, 2000) were:

1. To assess global mining and minerals use in terms of the transition to sustainable development. This was to cover the current contribution – both positive and negative – to economic prosperity, human well-being, ecosystem health and accountable decision-making, as well as the track record of past practice.

2. To identify how the services provided by the minerals system could be delivered in accordance with sustainable development in the future.

3. To propose key elements of an action plan for improving the minerals system.

4. To build platforms of analysis and engagement for ongoing cooperation and networking among all stakeholders.

A listing of the contents of the final report 'Breaking New Ground' released in May 2002 is presented below (MMSD, 2002a):
- Executive Summary
- Table of Contents
- Titles and Copyright
- Foreword and Statements by Assurance Group and Sponsors Group
- Introduction

Part I: A Framework for Change
   1. The Minerals Sector and Sustainable Development

Part II: Current Trends and Actors
   2. Producing and Selling Minerals
   3. A Profile of the Minerals Sector
   4. The Need for and Availability of Minerals
   5. Case Studies on Minerals

Part III: Challenges
   6. Viability of the Minerals Industry
   7. The Control, Use, and Management of Land
   8. Minerals and Economic Development
   9. Local Communities and Mines
   10. Mining, Minerals and the Environment
   11. An Integrated Approach to Using Minerals
   12. Access to Information
   13. Artisanal and Small-Scale Mining
   14. Sector Governance: Roles, Responsibilities and Instruments for Change

Part IV: Responses and Recommendations
   15. The Regional Perspectives
   16. Agenda for Change

Appendices
   Appendix 1. The MMSD Project
   Appendix 2. MMSD Consultation Activities
   Acronyms and Abbreviations
   Index
   Bibliography

As the authors of the report indicate (MMSD, 2002b), given the limited time and resources available, they could not even begin to address all of the issues that will ever be faced by the mining and minerals industries. The report does, however, provide a starting point for identifying different concerns and getting processes under way.

8.7.5 *The U.S. Government and federal land management*

The material included in this section has been extracted directly from the excellent publication 'Sustainable Development and Its Influence on Mining Operations On Federal Lands – A Conversation In Plain Language' prepared by representatives of the U.S. Bureau of Land Management and the U.S. Forest Service (Anderson, et al., 2002). These agencies administer 262 million acres and 194 million acres of federal lands, respectively, in the United States.

The remainder of this section has been taken directly from their publication.

Sustainable development is about ensuring human well-being while respecting ecosystem well-being and the earth's environmental limits and capacities. It encompasses environmental and social issues, as well as economic activity. These are interrelated and actions in any domain may, over time, impact all aspects of life in the region where we live.

A sustainable development perspective applied to resource management puts multiple use-including conservation, production, remediation, and land stewardship into a larger, integrated picture of resource management activities. Sustainable development gives us a checklist to work from, such as: What are the environmental and social impacts of an economic proposal? What are the economic and social implications of an environmental regulation?

The continuity of supply of resources obtainable through mining, and the sound management of these resources and the environment, are essential parts of sustainable development requiring a long-term view. We remain concerned not only about current results and impacts and the well being of the present generation, but also about cumulative impacts and the wellbeing of our children and grandchildren. This approach is not new to natural resource managers, but in some cases it is a welcome change for economic production and consumption to be managed with these broader and more long-term values in mind.

Simply put, sustainable development means thinking more broadly and longer-term about our national, corporate, and individual actions and how they relate to our environment and community It also means regularly checking our progress, as well as learning from experience and studying how we can better meet human needs. Perhaps more importantly for federal agencies, a sustainable development perspective in resource management gives us the opportunity to create better relationships with and among our stakeholders, including local, state, regional, tribal, corporate, and non-government communities of interest. Each of us can contribute knowledge, information, or resources to help us accomplish together what we cannot do alone.

The question arises as to why include minerals when they are a finite resource and not sustainable? It is true that individual mineral deposits are finite, but that does not mean minerals and metals have no place in sustainable development. Rather, sustainable development can provide the foundation for a policy framework that ensures minerals and metals are produced, used, reused, recycled and, if necessary, stored for the future (landfills) in a manner that respects the economic, social, and environmental needs of the local, national, and global communities. Within this framework, the benefits provided by minerals and mining are acknowledged, as is the reality that geology dictates the location of mineral deposits. Moreover, sustainable development makes good business sense because improving the efficiency of extracting and processing mineral resources creates both economic and environmental rewards.

The basic components of sustainable development (not in order of importance) are: social well-being, environmental health, and economic prosperity. Essentially, sustainable development requires that social, environmental, and economic issues be integrated in decision making. In all decisions, the long-term effects on resources and capital and the capacity for future creation of benefits should be considered. Decision making by natural resource managers should be broad, participatory, and also interdisciplinary.

Concern for economic and technological efficiency, for local environmental quality including planning for cleanup and reclamation at the closure of a mine, and concern for the social well-being of the local mine community and nearby population have long been mineral industry issues. Sustainable development provides a context within which to integrate these concerns.

Employees with the United States Departments of Interior and Agriculture have been working with their stakeholders to show how the social, environmental, and economic components of sustainable development could apply to mining operations. Following are examples of each.

Social: This component relates to community responsibilities. It is aimed at alerting companies, governments, and others to the need for enhancing the health of people and their communities, while maintaining profitable companies. Further, it raises the need for communities to understand and agree upon the distribution of cost, benefits, and risks of any proposed project or activity. It includes concepts such as:

- Respecting the cultures, customs, and values of individuals and groups whose livelihoods may be affected by exploration, mining, and processing
- Respecting the authority of national, regional, and tribal governments; taking into account their development objectives; contributing information related to mining and metal processing activities; and supporting the sharing of economic benefits generated by operations
- Recognizing local communities and other affected organizations and engaging with them in an open, honest, and effective process of consultation and communication from exploration through production to mine closure
- Assessing the social and cultural impacts of proposed activities
- Reducing to acceptable levels, as recommended by all stakeholders, the adverse social impacts on communities of activities related to exploration, extraction, and closure of mining and processing facilities
- Promoting health and safety both on and off the project site
- Developing one-on-one programs to support the wellbeing of employees' families in mining communities, such as activities and educational opportunities for spouses and children of mine employees

Environmental: This component relates to environmental stewardship. It is aimed at alerting companies, governments, and others to the need for enhancing environmental conditions over the long term. It includes concepts such as:

- Making environmental management a high priority
- Planning for mine closure beginning with exploration and mine approval
- Establishing environmental accountability in industry and government at the highest management and policymaking levels
- Adopting best practices to minimize environmental degradation and adapting them to local conditions as necessary
- Using energy and materials that conserve resources and avoid waste and expensive cleanup
- Conducting environmental impact assessments and collecting baseline data for flora and fauna, soil, and underground and surface waters
- Determining the capacity of the land for uses other than mining
- Minimizing noise and dust during operations
- Handling hazardous materials safely
- Minimizing pollution during operations

- Developing a mine waste management plan that includes tailings dam inspections, emergency checks, and hazard prevention
- Reclaiming the land to prevent erosion and planting native species targeting the same density and diversity of plants that were there before mining

Economic: This component relates to economic and financial actions, impacts, and policies. It is aimed at recognizing that the health of the economy has to be maintained as a principal means for achieving our quality of life. It includes concepts such as:

- Assessing the economic impacts of proposed and ongoing activities and developing management policies that maximize positive and minimize negative community and household impacts
- Working with local communities to develop strategies for sustaining their economies after mine closures and encouraging the establishment of other sustainable local and regional business activities looking for continuing improvements in design and efficiency that will help both profitability and competitiveness while reducing wastes released into the environment
- Investing to optimize long term returns to investment rather than immediate returns
- Investing in programs that improve the skills and thus productivity of the workforce with the goal of creating both economic and social benefits
- Encouraging suppliers to use energy efficient materials and technologies

In closing, the authors of the report indicate that these examples reflect current thinking about how sustainable development principles could apply to mining operations on federal land and do not constitute policy or regulation for the mining industry. The Forest Service and Bureau of Land Management have embraced sustainable development because the concept is complementary to and consistent with each agency's mission to provide for many uses of federal lands. This includes developing natural resources and working with stakeholders to achieve a sustainable future for our lands and for our communities.

8.7.6 *The position of the U.S. National Mining Association (NMA)*

The National Mining Association (NMA) includes more than 325 member companies involved in all aspects of the mining industry. In September 2002, the NMA adopted the Sustainable Development Pledge and the Sustainable Development Principles Statement reproduced below (NMA, 2002):

NMA Sustainable Development Pledge

The members of the National Mining Association pledge to conduct their activities in a manner that recognizes the needs of society and the needs for economic prosperity, national security and a healthy environment. Accordingly, we are committed to integrating social, environmental, and economic principles in our mining operations from exploration through development, operation, reclamation, closure and post closure activities, and in operations associated with preparing our products for further use.

NMA Sustainable Development Principles Statement

1. From an environmental standpoint this involves:
   - Recognizing and being responsive to possible environmental impacts of exploration activities;

- Developing approaches to mine planning and development that are responsive to possible environmental impacts through every stage of the mining cycle including closure and post closure activities;
- Planning in advance for the timely reclamation of sites in accordance with site specific criteria and recognizing community priorities, needs and interests as the mine approaches and reaches closure;
- Assisting in addressing legacy issues through existing mechanisms and laws and by working with appropriate government bodies to establish responsible, balanced and cost effective solutions;
- Being a leader in developing, establishing and implementing good environmental practices;
- Promoting the safe use, recycling and disposal of products through an understanding of their life cycles;
- Developing and promoting new technologies that continue to improve efficiencies and environmental performance in our mining and processing operations and in the use of our products;
- Recognizing that the potential for climate change is a special concern of global scope that requires significant attention and a responsible approach cutting across all three of the sustainable development pillars: environmental, social and economic;
- Encouraging climate policies that promote fuel diversity, development of technology and long term actions to address climate concerns in order to ensure that technological and financial resources are available to support the needs of the future;
- Supporting additional research to improve scientific understanding of the existence, causes and effects of climate change and to enhance our understanding of carbon absorbing sinks; advancements in technology to increase efficiencies in electrical generation and capture and sequester carbon dioxide; voluntary programs to improve efficiency and reduce greenhouse gas emission intensity; and, constructive participation in climate policy formulation on bath international and national levels.

2. From a social standpoint this involves:
   - Being committed to the safety, health, development and well being of our employees;
   - Respecting human rights;
   - Treating our employees with respect, promoting diversity, and providing competitive compensation programs consistent with performance and industry practice;
   - Being a progressive and constructive partner to advance the economic, educational, and social infrastructures of the communities in which we operate;
   - Respecting the cultures, customs and values of people wherever we operate, being responsive to- and respecting - community needs and priorities and encouraging and participating in an open and on-going dialogue with constituencies; and,
   - Adhering to the highest ethical business practices in all our operations and interacting with communities in a responsible manner.

3. From an economic standpoint this involves:
   - Creating wealth and products that contribute to economic prosperity;
   - Helping eliminate poverty and providing economic opportunities;
   - Contributing to national, regional and local economic well-being and security through creation of employment opportunities, wage payments, purchase of goods and materials and payment of fair and competitive taxes and usage fees; and,

- Allowing shareholders and investors to earn a fair and equitable return commensurate with the risk they take.

8.7.7 The view of one mining company executive

Yearley (2003) in his 2003 D.C. Jackling Award lecture entitled 'Sustainable Development for the Global Mining and Metals Industry' provided some insights as to how a large mining company views 'sustainable development. Some of the key points from his lecture are summarized below (Yearley, 2003).

1. Fifteen years ago, few in the mining industry had heard the phrase sustainable development, let alone knew what it meant. Today, it is a different story. Industry leaders have embraced the concept. They have taken action to incorporate the inherent values and principles of sustainable development into the policies and modus operandi of their companies.

2. What does sustainable development mean for the mining and metal producers in operational terms? For years, this question was extensively debated both within and outside the industry. Today, there is broad consensus that sustainable development requires three things:
   - Integrated approaches to decision making on a full, life-cycle basis that satisfy obligations to shareholders and that are balanced and supported by sound science and social, environmental and economic analysis within a framework of good governance.
   - Consideration of the needs of current and future generations.
   - Establishment of meaningful relationships with key constituencies based on mutual trust and a desire for mutually beneficial outcomes, including those inevitable situations that require informed trade-offs.

3. There is also a consensus view that sustainable development is a journey rather than a destination. That means the concept evolves in response to changing societal values, priorities and needs. To move forward against this backdrop, the mining and minerals sector needs to adapt, to create a new way of doing business.

4. Improving performance is in our interest. There is a clear business case. Low returns and a poor social-environmental record make mining a risky investment. When mining projects are restricted, delayed or halted, revenues and profits are lost. Companies succeed by minimizing cost and risks. Investors invest in sectors that can manage their business costs and risks. Sustainable development strategies can help minimize costs as follows:
   - Lower labor costs – providing clean, safe working conditions can improve productivity, result in fewer union disputes and increase retention rates.
   - Lower health costs – healthy communities and workers will again be more productive.
   - Lower production costs – waste reduction, recycling and energy efficiency can significantly lower production costs.
   - Lower regulatory burden – a trusted company can enjoy a smoother permitting and regulatory path.
   - Lower closure costs – terminal liabilities can be more accurately predicted, managed and controlled.
   - Lower cost borrowing – managing risk will improve relations with lenders.
   - Lower insurance costs.

- Improved investor relations.
- Ultimately, a company that implements sustainable development policies can become the mining company of choice.

5. Managing environmental and social impact is not just a cost to be minimized, but an opportunity to enhance business value and ensure the long-term success of the company is fully realized. The evidence for mining companies during the last few years is compelling. It shows that mining companies that embrace environmental issues are more efficient, have superior management and represent a lower risk profile. This improved reputation ensures a company's social license to operate. The ultimate prize is access to land and markets combined with competitive advantage through improved performance. That is why the leading companies will act on sustainable development – it is in the interest of their shareholders. The message is that change is not always easy. There will be cost implications but the rewards are also potentially high.

6. There is a strong business case for change. The evidence for mining and metal companies to engage in this process has been compelling over the past few years. It shows that those companies that embrace environmental and social issues are more efficient, have superior management and better relationships with local communities, and represent a lower risk profile. This improved reputation will ensure a company's social license to operate and makes the company a better investment prospect.

7. Change is not always easy. There will be cost implications but the rewards are also potentially high. Ultimately, those companies that implement sustainable development policies will reap the benefits and become mining companies of choice from the standpoint of the investment and financial community, governments and other key constituencies.

8. Enhancing the contribution of mining, minerals and metals to sustainable development will require a concerted effort by governments, multilateral organizations, civil society groups as well as industry. Key priorities include:
- Ensure regulatory and material choice decisions based on precise and explicit criteria, cost-effective/timely risk assessments taking into account special characteristics of minerals and metals.
- Ensure openness and transparency and views of all stakeholders incorporated in decision-making processes likely to affect them.
- Establish market incentives to encourage product design, re-cyclability and economic collection and recovery of metals.
- Ensure benefits from mineral development are more fully realized by addressing the capacity-building needs of developing countries.

9. All constituents need to work together and have fair rules of engagement. The mining and metals industry is trying to reform and put in place the transparency and accountability systems that have been demanded. It is and will remain important that governments and civil society groups are held to the same high standards as are expected by companies.

10. A new course has been set for the industry. Sustainable development offers an excellent policy framework that will enable industry to demonstrate why minerals and metals should be considered materials of choice and to highlight the contributions that mining can make to economic and social development. There is a good business case for sustainable development, one that offers the opportunity – based on improved economic, environmental and

social performance – to build trust and improve our reputation, while enhancing shareholder value.

8.8 'RESPONSIBLE MINING' – THE WAY FORWARD IS GOOD ENGINEERING

8.8.1 Introduction

It is hoped that the reader now has a much better idea of the world in which new mining projects will be born. The successful ones will enter this world in the form of a three-legged stool whose legs, economics, environmental, and social/society, have the same length. This is as shown in Figure 8.1a. Clearly, if one or more of the legs is somewhat short with respect to the others (or missing entirely), the stool (project) will not be stable, or at least not perform in the way intended (Figures 8.1b and 8.1c). Some stools will be short (small) while others will be tall (large) but the stability (success) requirement of equal length legs remains the same.

In all of the foregoing text, little was indicated in the way of a path forward. One of the key elements, if not the key element, in the opinion of the authors, is an emphasis on good engineering. Unfortunately, good engineering has been absent or in very short supply in many past mining operations. Today, the production/operations group is still generally 'king' of the mine and represents the 'profit' center. Engineering is relegated to one of the many 'cost' centers and therefore labeled as an item to be minimized.

Logically, one could argue that production should be labeled a cost center since this is the home to all unit operations and their easily calculated costs. The true 'profit' center lies in engineering since this is where the optimal extraction of the resource is designed and planned, while keeping firmly in mind the economic, environmental and the societal aspects of a responsible project.

The operative phrase for future mining must be (Moss, 2005)

'Plan the mine, mine the plan'

In this way, the engineering and production/operations team, working together, form a very strong profit center. Inadequacies exhibited by either partner can lead to very high, unexpected costs, in one or more of the three legs – economic, environmental, and social. This is not 'Responsible Mining' and the consequences may be severe.

Today, even while the demand for well-engineered mines is increasing world-wide, the supply of the required mining, minerals, metallurgical/process, and geological engineers is small and decreasing rather than growing. Furthermore, the teaching corps required to train the future engineers is also diminishing through retirements and the lack of new-hires at the mineral's universities. Increasing demand and diminishing supply is not a good recipe for meeting the 'Responsible Mining' challenges of the future.

8.8.2 The Milos Statement

The 'Milos Statement: Contribution of the Minerals Professional Community to Sustainable Development' is used to close this chapter. This document was formulated and endorsed by the minerals professional community during their meeting at Milos, Greece in May 2003 (Karmis, 2003). This group is comprised of engineers, scientists, technical experts, and academics who work in, consult for; study, or are in some other manner associated with the

Responsible mining 717

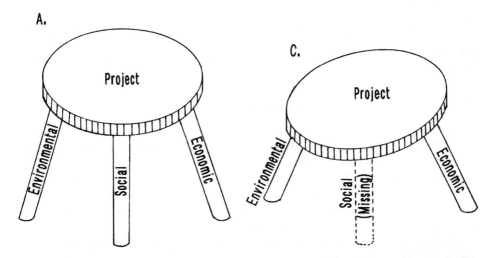

Figure 8.1. The three-legged stool analogy illustrating 'Responsible Mining' (A) and two undesirable alternatives (B and C).

minerals industry. The continuing efforts of this group will have a large impact as to how well we succeed toward the goal of 'Responsible Mining.'

The Milos Statement is:

Society's transition towards a sustainable future cannot be achieved without the application of the professional principles, scientific knowledge, technical skills, educational and research capabilities, and democratic processes practiced by our community. Our members share a mutual responsibility with all individuals to insure that our actions meet the needs of today without compromising the ability of future generations to satisfy their own needs.

What we believe:

We believe minerals are essential to meeting the needs of the present while contributing to a sustainable future.

The process of civilization is one of advancing intellectual, social and cultural development for all of humankind. Ah important part of the history of civilization is the history of scientific discoveries and technological advancements that have turned raw materials into resources, and in so doing provided the means for increased human well-being. The benefits and services derived from minerals, metals and fuels can contribute to the achievement of a sustainable future because it is the inherent characteristics of these resources that make productivity and consumption gains possible.

Achievement of a balance among economic prosperity, environmental health, and social equity will require significant changes in business strategies and operations, personal behaviors and public policies. Our minerals professional community can assist societies in balancing the need for mineral raw materials against the need to protect the environment and societies from unnecessary adverse impacts as they strive to improve quality of life.

Our vision for the future:

Our minerals community will support the transition to a sustainable future through integrated and experimental use of science, engineering and technologies as resources to people, catalysts for learning, providers of increased quality of life, protectors of the environmental and human health and safety.

What needs to be done to achieve our vision:

Professional Responsibility:
- Work to ensure that minerals and the capability to produce minerals are available to meet the needs of current and future generations.
- Encourage the development, transfer, and application of technologies that support sustainable actions throughout the product and mine life cycles.
- Give high priority to identifying solutions for pressing environmental and developmental challenges as related to sustainable development.
- Address social equity, poverty reduction and other societal needs as integral to minerals and mining related endeavors.
- Participate in the global dialog on sustainable development.
- Engage in all stages of the decision-making process, not only the project execution phase.

Education, Training, and Development:
- Attract the best people to the fields of mining and minerals by encouraging, facilitating and rewarding excellence.
- Build-up and maintain a critical mass of engineering, technical, scientific and academic capacity through improved education and training.
- Promote the teaching of sustainability principles in all engineering programs at all academic levels.
- Support and commit funding to the infrastructure that enables nations to provide mineral education, professional training, information, and research.
- Prevent the loss of core competencies.
- Transfer existing knowledge to the next generation of our community prior to and after retirement.

• Create a global exchange in academic training, as well as apprenticeship and internships programs.
• Support professional growth and interaction through books, articles, symposia, short courses and conferences on minerals and mining in sustainable development.

Communication:
• Share and disseminate sound information, knowledge, and technology.
• Share information on every aspect of minerals and mining to all appropriate audiences through journals, conferences, magazines, the Internet and other appropriate media.
• Disseminate technical information on sustainability, and the role of the minerals, metals and fuels in sustainable development, including information on the role of minerals in maintaining a high quality of life.
• Promote the achievements and capabilities of mineral community professionals to managers and executives, policy makers and the general public.

8.9 CONCLUDING REMARKS

'Responsible Mining' must be the message for now and especially in the future for everyone working in the mining industry. It must be practiced daily. Exceptions must be quickly identified and corrected. If we are to continue to have the privilege of producing the minerals that the world's people require, we must demonstrate that we are up to the task of extracting them in a responsible way. The three-legged stool image of economics, environment, and society should be kept clearly in mind.

REFERENCES

Alcor. 2005. Notes on Our Common Future. thttp://alcor.concordia.ca/~raojw/crd/reference/reference 001377.html.
Allen, R. 1980. How to Save the World: Strategy for World Conservation. Kogan Page Limited, London. 150 pp.
Alvarez, S.M., and B.A. Ridolfi. 1999. Moon Creek reclamation project: creative solutions for an historic mine site. Pre-print Paper 99-088. 1999 SME Annual Meeting, Mar 1–3. Denver, CO.
Anderson, B., Berry, D., and D. Shields. 2002. Sustainable development and its influence on mining operations on federal lands – A conversation in plain language. April. BLM/WO/GI-02/008+3000.
Anderson, K. 2004. Influence, rules and post-closure sustainability: Who will govern after the gates are closed? Pre-print Paper 04-167. 2004 SME Annual Meeting, Feb 23–25. Denver, CO.
Anonymous. 2002a. ICMM Toronto Declaration. May 15.
Anonymous. 2002b. 2002 Sustainable Development Report: Sustaining a healthy balance. Noranda Inc. /Falconbridge Limited.
Anonymous. 2003. Milos statement on the contribution of the minerals professional community to sustainable development: Draft for comment. Feb 23.
Anonymous. 2005a. World Commission on Environment and Development. http://www.wsu.edu:8080/~susdev/WCED87.html.
Arell, P., and D. Folkes. 2004. The superfund hazard ranking system and mining sites. Pre-print Paper 04-48. 2004 SME Annual Meeting, Feb 23–25. Denver, CO.
Barker, J.M., and V.T. McLemore. 2004. Sustainable development and industrial minerals. Pre-print Paper 04-171. 2004 SME Annual Meeting, Feb 23–25. Denver, CO.
Baird, B.K., and P.C. Satchwell. 1999. Application of economic parameters and cutoffs during and after pit optimization. Pre-print Paper 99-069. 1999 SME Annual Meeting, Mar 1–3. Denver, CO.
Basu, A.J., and V. Carabias-Hütter. 2004. Indicators of sustainability and reporting. Pre-print Paper 04-159. 2004 SME Annual Meeting, Feb 23–25. Denver, CO.

Berry, R.C. 1999. Mining and the environmental impact statement: Guidelines for baseline studies. Pre-print Paper 99-024. 1999 SME Annual Meeting, Mar 1–3. Denver, CO.

Bingham, N. 1994. Mining's image – what does the public really think. Mining Engineering. 46(3): 200–203.

Botts, S.D., and F. Cantuarias. 2003. Creating sustainability at Antimina. CIM Bulletin 96(1074): 84–86.

Botts, S.D. 2003. The Antimina Project: The challenge of sustainable development in Peru. Pre-print Paper 03-33. 2003 SME Annual Meeting, Feb 24–26. Cincinnati, OH.

Brice, W.C., and C.G. Buttleman. 2000. Reclaiming aggregate mining sites in Minnesota using private/public partnerships. Pre-print Paper 00-103. 2000 SME Annual Meeting, Feb 28-Mar 1. SLC, UT.

Brundtland, G.H. 1987. Foreword to Our Common Future: The World Commission on Environment and Development. 1987. Oxford University Press. Oxford. p.ix.

Cavender, B.W. 1994. Strategic ramifications of corporate environmental policy. Mining Engineering. 46(3): 204–207.

Cooney, J.P. 2000. People, participation, and partnership. CIM Bulletin 93(1037): 41–45.

Corn, R.M. 2000. Consequences of the Forest Service CERCLA action at Mansfield Canyon. Pre-print Paper 00-77. 2000 SME Annual Meeting, Feb 28-Mar 1. SLC, UT.

Cragg, A.W. 1998. Sustainable development and mining: Opportunity or threat to the industry? CIM Bulletin 91(1023): 45–50.

Czarnowsky, A.W. 2001. Controversial mining projects – Can you ever be prepared? Pre-print Paper 01-112. 2001 SME Annual Meeting, Feb 26–28. Denver, CO.

Daniel, S.E. 2003. Environmental due diligence for exploration managers. CIM Bulletin 96(1074): 87–90.

Daniels, W.L., Schroeder, P., Nagle, S., Zelazny, L., and M. Alley. 2002. Reclamation of prime farmland following mineral sands mining in Virginia. Pre-print Paper 02-132. 2002 SME Annual Meeting, Feb 26–28. Phoenix, AZ.

Davis, G.A. 2004. An empirical investigation of mining and sustainable development. Pre-print Paper 04-136. 2004 SME Annual Meeting, Feb 23–25. Denver, CO.

Dorey, R., Duckett, R., and D. Salisbury. 1999. Operational management for closure and reclamation – Kennecott Ridgeway mine. Pre-print Paper 99-134. 1999 SME Annual Meeting, Mar 1–3. Denver, CO.

Dunn, W.J. 2000. Beyond 'Bead'n Trinkets': A systematic approach to community relations for the next millennium. CIM Bulletin 93(1037): 41–45.

Ednie, H. 1999. The challenge of changing public perceptions of the industry. CIM Bulletin 92(1031): 9–10.

Eger, P., Melchert, G., and J. Wagner. 2000. Using passive treatment system for mine closure – a good approach or risky alternative. Mining Engineering. 52(9): 78–84.

Fitzgerald, A.D. 2004. Environmental and social responsibility practices: Moving to a global level playing field. CIM Bulletin 97(1078): 87–90.

Fox, F.D. 2003. Flambeau and Ridgeway Mines – Lessons learned. Pre-print Paper 03-141. 2003 SME Annual Meeting, Feb 24–26. Cincinnati, OH.

Gardner, J.S., and P. Sainato. 2005. Mountaintop mining and sustainable development in Appalachia. Pre-print Paper 05-067. 2005 SME Annual Meeting, Feb 28-Mar 2. SLC, UT.

Garisto, N.C., et al. 2000. Environmental monitoring at mine sites – Future directions. CIM Bulletin 93(1038): 53–61.

Gediga, J., Harsch, M., Saur, K., Schuckert, M., and P. Eyerer. 1996. Lifecycle assessment – An effective tool for environmental management. In Surface Mining 1996 (H.W. Glen, editor): 307–312. SAIMM.

GMI. 1998a. About ICMM. www.globalmining.com/about.php.

GMI. 1998b. Global Mining Initiative. www.angloamerican.co.uk/corporateresponsibility/sustainable development/gmi.

GMI. 1998c. GMI Issues. www.globalmining.com/gmi_issues_covered.php.

Gorton, W.T. 2004. Primer for engineering and technical professionals regarding reclamation and environmental surety bonds. Mining Engineering. 56(11): 36–38.

Gowens, J.K. 1998. A different role in designing new mining projects. CIM Bulletin 91(1026): 86–87.

Hamaguchi, R.A. 1999. Innovative reclamation research at Highland Valley Copper. CIM Bulletin 92(1033): 78–84.

Harris, L. 1995. Yanacocha – Incas + 450 years: A model for investment in Peruvian mining. E/MJ. 196(8): 32–36.

Heath, J.S. 2000. Environmental alternative dispute resolution for mining – An introduction. Pre-print Paper 00-86. 2000 SME Annual Meeting, Feb 28-Mar 1. SLC, UT.

Hodge, R.A. 2004. Tracking progress toward sustainability: Linking the power of measurement and story. Pre-print Paper 04-158. 2004 SME Annual Meeting, Feb 23–25. Denver, CO.

Hogeman, D.C. 2002. Growing greener: A watershed approach. Pre-print Paper 02-121. 2002 SME Annual Meeting, Feb 26–28. Phoenix, AZ.

Horswill, D.H., and H. Sandvik. 2000. Mining and sustainable development at Red Dog. Mining Engineering. 52(11): 25–31.

Horswill, D.H. 2000. Mining and sustainable development. Pre-print Paper 00-81. 2000 SME Annual Meeting, Feb 28-Mar 1. SLC, UT.

ICMM. 2002a. Mission. www.globalmining.com/about.php.

ICMM. 2002b. Organization. www.globalmining.com/about_organisat.php.

ICMM. 2002c. Principles. www.globalmining.com/icmm_principles.php.

ICMM. 2002d. ICMM Toronto Declaration. www.icmm.com/library_pub_detail.php?rcd=62.

James, P.M. 1999. 1999 Jackling Lecture: The miner and sustainable development. Mining Engineering. 51(6): 89–92.

Johnson, D., and K. Ward. 2001. EPA's EPCRA assistance visits: Getting prepared (EPCRA actually consists of sections other than 313?). Pre-print Paper 01-6. 2001 SME Annual Meeting, Feb 26–28. Denver, CO.

Johnson, D.C., and H.F. Letient. 2004. Environmental planning in mine waste management – The Huckleberry Mines experience. CIM Bulletin 97(1077): 62–65.

Journeaux, D. 2005. Rock remediation techniques for mining and quarry operations. Pre-print Paper 05-112. 2005 SME Annual Meeting, Feb 28-Mar 2. SLC, UT.

Joyce, S., and I. Thomsom. 2000. Earning a social license to operate: Social acceptability and resource development in Latin America. CIM Bulletin 93(1037): 49–53.

Kaldany, R. 2004. Sustainable mining development: listening to stakeholders. Mining Journal Supplement. London. January. Pp 8–11.

Kaliampakos, D.C., and A.G. Benardos. 2000. Quarrying and sustainable development in large urban centres: A contradiction in terms. CIM Bulletin 93(1040): 86–89.

Karmis, M. 2003. The Milos Statement: Contribution of the Minerals Professional Community to Sustainable Development. May. SME, Inc.

Kempton, J.H. 2000. Cost-saving strategies for environmental compliance. Pre-print Paper 00-125. 2000 SME Annual Meeting, Feb 28-Mar 1. SLC, UT.

Kempton, J.H., and D. Adkins. 2000. Strategies for permitting mines with delayed environmental impacts. Pre-print Paper 00-126. 2000 SME Annual Meeting, Feb 28-Mar 1. SLC, UT.

Kent, A., Ringwald, J., and M. Scoble. 2004. The evolution of sustainable mining practice: CIM's potential role. CIM Bulletin. 97(1083): 38–45.

Ketellapper, V.L. 2004. Conducting sustainable development concurrently with the reclamation of abandoned mines at the French Gulch site near Breckenridge, Colorado. Pre-print Paper 04-160. 2004 SME Annual Meeting, Feb 23–25. Denver, CO.

Kral, S. 2004. Sustainable development keys SME annual meeting. Mining Engineering. 56(6): 58:67.

Langer, W.H., Giusti, C., and G. Barelli. 2003. Sustainable development of natural aggregate with examples from Modena Province, Italy. Pre-print Paper 03-45. 2003 SME Annual Meeting, Feb 24–26. Cincinnati, OH.

Langer, W.H., and M.L. Tucker. 2004. Quarry expansion – Sustainable management of natural aggregate. Mining Engineering. 56(1): 23–28.

Lima, H.M., and P. Wathern. 1999. Mine closure: a conceptual review. Mining Engineering. 51(11): 41–45.

Livo, K.E., and D.H. Knepper, Jr. 2004. Non-evasive exploration in an environmentally sensitive world. Pre-print Paper 04-172. 2004 SME Annual Meeting, Feb 23–25. Denver, CO.

Macedo, J.M., Campos, D., and S. Chadwick. 2004. Alumar: Improving local communities capacities for implementation: A sustainable development model, São Luis-Maranhão, Brazil. Pre-print Paper 04-173. 2004 SME Annual Meeting, Feb 23–25. Denver, CO.

McAllister, M.L., Scoble, M., and M. Veiga. 1999. Sustainability and the Canadian mining industry at home and abroad. CIM Bulletin 92(1033): 85–92.

McLemore, V.T., and D. Turner. 2004. Sustainable development and exploration. Pre-print Paper 04-170. 2004 SME Annual Meeting, Feb 23–25. Denver, CO.

Meyers, K.L., Espell, R., and C. Deringer. 2003. AA Pad reclamation performance review at Barrick's Goldstrike mine. Pre-print Paper 03-97. 2003 SME Annual Meeting, Feb 24–26. Cincinnati, OH.
MMSD. 2000. MMSD project. www.globalmining.com/gmi_mmsd_project.php.
MMSD. 2002a. Breaking New Ground. Earthspan Publications.
MMSD. 2002b. Executive Summary for Breaking New Ground. www.iied.org/mmsd.
Moellenberg, D.L. 2005. Trends in hard rock mined land reclamation and financial assurance requirements. Pre-print Paper 05-027. 2005 SME Annual Meeting, Feb 28-Mar 2. SLC, UT.
Mudder, T., and K. Harvey. 1999. The state of mine closure: Concepts, commitments, and cooperation. Pre-print Paper 99-047. 1999 SME Annual Meeting, Mar 1–3. Denver, CO.
Munro, D.A. 1980. Preface to 'How to Save the World: Strategy for World Conservation' by Robert Allen. Kogan Page Limited, London. 150 pp.
Our Common Future: The World Commission on Environment and Development. 1987. Oxford University Press. Oxford. 383 pp.
Nalven, J.G., and J.A. Sturgess. 1999. Toxic release inventory reporting and consequences for a typical gold mining facility. Pre-print Paper 99-048. 1999 SME Annual Meeting, Mar 1–3. Denver, CO.
Nash, D. 2001. The importance of environmental compliance programs. Pre-print Paper 01-10. 2001 SME Annual Meeting, Feb 26–28. Denver, CO.
NMA. 2002. The National Mining Association Sustainable Development Principles. http://www.nma.org/policy/sustainable_dev.asp.
Pinter, A. Burger, and A. Spitz. 2000. NEMA, mining and metallurgy and the social environment: Implications from the melting pot. JSAIMM. 100(3): 153–154.
Robbins, D.A. 1999. Mining industry benefits and concerns on environmental remote sensing. Pre-print Paper 99-050. 1999 SME Annual Meeting, Mar 1–3. Denver, CO.
Robertson, A., and S. Shaw. 2004. Use of the multiple accounts analysis process for sustainability optimization. Pre-print Paper 04-163. 2004 SME Annual Meeting, Feb 23–25. Denver, CO.
Scott, P. 1980. Foreword to 'How to Save the World: Strategy for World Conservation' by Robert Allen. Kogan Page Limited, London. 150 pp.
Semenoff, M.N. 2001. Talking trash? – The RCRA cleanup reforms. Pre-print Paper 01-50. 2001 SME Annual Meeting, Feb 26–28. Denver, CO.
Smithen, A.A. 1999. Environmental considerations in the preparation of bankable feasibility documents. JSAIMM. 99(6): 317–320.
Stephenson, H.G., Van Den Bussche, B., and P. Curry. 1996. Reclamation, rehabilitation and development of abandoned mine land at Canmore, Alberta. CIM Bulletin 89(??):???.
Swarbrick, B. 2000. Mining in a multi-stakeholder environment: Falconbridge's Raglan project. Pre-print Paper 00-23. 2000 SME Annual Meeting, Feb 28-Mar 1. SLC, UT.
Tilton, J.E. 2003. Creating wealth and competitiveness in mining. Mining Engineering. 55(9): 15–21.
Tuttle, S., and R. Sisson. 1999. Closure plan for the proposed millennium project. CIM Bulletin 92(1026): 95–100.
UNCED. 1992a. UN Conference on Environment and Development (1992). http://www.un.org/geninfo/bp/enviro.html.
UNCED. 1992b. Report of the United Nations Conference on Environment and Development. Annex I - Rio Declaration on Environment and Development. http://www.un.org/documents/ga/conf151/aconf15126-1annex1.htm.
UNCED. 1992c. Agenda 21 – Table of Contents. http://www.un.org/essa/sustdev/documents/agenda21/english/agenda21toc.htm.
UNCED. 1992d. Agenda 21 – Full text. http://www.un.org/esa/sustdev/agenda21.htm.
UNEP. 1972a. Report of the UN Conference on the Human Environment, Stockholm.
UN Doc.: A/CONF.48/14/Rev1 at 3(1973), 11 ILM 1416 (1972).
UNEP. 1972b. Report of the UN Conference on the Human Environment. http://www.unep.org/Documents/Default.asp?DocumentID=97.
Upton, B., Harrington, T., and S. Mendenhall. 2004. Sustainable development and mine closure planning. A case study: Golden Sunlight Mine, Jefferson County, Montana. Pre-print Paper 04-162. 2004 SME Annual Meeting, Feb 23–25. Denver, CO.
Wagner, D.M., Osmundson, M., Biggs, G., and C.J. Twaroski. 2001. Class 1 Areas Impact Analyses for a proposed expansion of a taconite processing facility in northeast Minnesota. Pre-print Paper 01-24. 2001 SME Annual Meeting, Feb 26–28. Denver, CO.

Wahrer, R.J. 2003. So you have a T/E species on your permit: How to survive. Pre-print Paper 03-102. 2003 SME Annual Meeting, Feb 24–26. Cincinnati, OH.

Weber, I. 2005. Actualizing sustainable mining: 'Whole mine, whole community, whole planet' through 'industrial ecology' and community-based strategies. Pre-print Paper 05-102. 2005 SME Annual Meeting, Feb 28-Mar 2. SLC, UT.

Wilson, T.E., and T.M. Dyhr. 2004. Cost trends – Environmental management of mine operations. Pre-print Paper 04-125. 2004 SME Annual Meeting, Feb 23–25. Denver, CO.

Neuman, L. 2004. The ten most common mistakes that delay the review process. Pre-print Paper 04-20. 2004 SME Annual Meeting, Feb 23–25. Denver, CO.

Werniak, J. 2002. Working with nature and humanity. Canadian Mining Journal. August. Pp 16–18.

Wright, D., and J. Samborski. 2004. Sustainability: More than just reclamation. Pre-print Paper 04-174. 2004 SME Annual Meeting, Feb 23–25. Denver, CO.

WSSD. 2002a. Johannesburg Summit 2002 feature story – The Johannesburg Summit Test: What will change? http://www.johannesburgsummit.org/html/whats_new/feature_story41.html.

WSSD. 2002b. The World Summit on Sustainable Development. www.natural-resources.org/minerals/CD/sustdev.htm.

WSSD. 2002c. Plan of Implementation of the World Summit on Sustainable Development. www.un.org/esa/sustdev/documents/WSSD_POI_PD/English/WSSD_PlanImpl.pdf.

WSSD. 2002d. Johannesburg Declaration on Sustainable Development. www.un.org/esa/sustdev/documents/WSSD_POI_PD/English/POI_PD.htm.

Yearley, D.C. 2003. 2003 Jackling Lecture: Sustainable development for the global mining and metals industry. Mining Engineering. 55(8): 45–48.

Yernberg, W.R. 2001. Antamina – Peruvian mine of the future. Mining Engineering. 53(12): 19–21.

REVIEW QUESTIONS AND EXERCISES

1. What is meant by 'the bottom line'?
2. What is meant by 'the hurdle rate'?
3. Discuss the significance of equation (8.2).
4. In practice, how would you produce the 'Environmental cost/unit' and 'Social cost/unit' cost components?
5. Who are the 'stakeholders' in a mining project?
6. Discuss the meaning of 'Economic responsibility,' 'Environmental responsibility,' and 'Social responsibility' with regard to a mining project.
7. What is meant by the phrase 'Responsible mining'?
8. What is meant by the term 'Sustainable development'?
9. What is meant by 'Responsible mining practices'?
10. Summarize the five most important statements, in your opinion, contained in the proclamation made at the conclusion of the UNCHE.
11. From a mining viewpoint, list the ten most important principles included in the Stockholm Declaration. Be prepared to discuss the reasons for your selection.
12. What was the intention of the World Conservation Strategy?
13. How does the 'Strategy' relate to mining?
14. How does the 'Strategy' define sustainable development?
15. What are the obstacles cited? How should mining respond?
16. What were the goals of the 'World Commission on Environment and Development'? How do they affect the mining business?
17. How did the World Commission define 'sustainable development'?

18. Discuss the impact of the following statement on mining:
    'Sustainable development is considered to be a process of change in which the exploitation of resources, the direction of investments, the orientation of technological development, and institutional change are made consistent with future as well as present needs.'
19. What were identified by the World Commission as the two key concepts in sustainable development?
20. What is a non-renewable resource?
21. In the report 'Our Common Future', what were some of the important concepts regarding non-renewable resources?
22. What are some of the requirements to achieve sustainable development? Are they reasonable? Are they achievable?
23. What were the goals of the 'Earth Summit'?
24. From a mining viewpoint, summarize the ten most important principles, in your opinion, proclaimed? Be prepared to discuss your reasoning.
25. If you were in the mining business, which of the chapters in Agenda 21 would you be most interested in reading? Be prepared to discuss your reasoning.
26. Summarize the main things related to mining that came out of the World Summit on Sustainable Development. If you were a mining executive, how would you respond?
27. What was the Global Mining Initiative(GMI)? Who was the driving force behind it?
28. What were the main mining issues in question?
29. What is meant by the statement 'The mining industry has accepted that, alone, it cannot re-write its reputation'? What is the suggested way forward?
30. What were the three main activities identified by the GMI? What was the result?
31. What is the International Council on Mining and Metals (ICMM)?
32. What are the goals of the ICMM?
33. Who are the current members of the ICMM?
34. What are the principles of the ICMM? How do they relate to the principles put forth in the UN – sponsored conferences?
35. How is it expected that the ICMM-stated principles will get put into practice?
36. What was the Mining, Minerals and Sustainable Development (MMSD) project?
37. What were the objectives of the MMSD?
38. In what order would you read the MMSD report? Why?
39. The U.S. Bureau of Land Management (BLM) and the U.S. Forest Service are two major land administrators in the U.S. Why is it important to understand their view on sustainable development?
40. Summarize the view of these two government agencies with respect to sustainable development as expressed in section 8.7.5. As a citizen, are you comfortable with this view? As a miner, are you comfortable with this view? Be prepared to discuss your reasoning.
41. Summarize the point of view of the National Mining Association (NMA) with respect to environmental aspects as contained in their Sustainable Development Principles statement.
42. Summarize the point of view of the National Mining Association (NMA) with respect to social aspects as contained in their Sustainable Development Principles statement.
43. Summarize the point of view of the National Mining Association (NMA) with respect to economic aspects as contained in their Sustainable Development Principles statement.

44. In the opinion of Yearly, sustainable development, from the mining viewpoint, requires three things. What are they?
45. Discuss the concept that 'sustainable development is a journey rather than a destination.'
46. Discuss the need to make a 'business case' for sustainable development. Why is this important?
47. In what ways can sustainable development strategies help to minimize mining costs?
48. Comment on the statement 'Managing environmental and social impacts is not just a cost to be minimized.'
49. What are the key priorities for enhancing the contribution of mining, minerals and metals to sustainable development?
50. What role must mine engineering play in 'Responsible Mining'?
51. Is mine engineering a 'cost' center or a 'profit' center? Be prepared to discuss this.
52. In your own words, what is the practical meaning of the phrase 'Plan the mine, mine the plan'?
53. What is the way forward for mining in light of the substantial emphasis on sustainable development worldwide?
54. What is the significance of the Milos statement?
55. What is the Milos vision?
56. To achieve the Milos vision, what needs to be done? Summarize this in your own words.
57. How do you fit into the Milos vision and into its achievement?
58. Clearly state your view on the practical meaning of 'Responsible Mining.'

Index

access
  associated volume 285, 350–352
  pit 290
Agenda 21, 701
angle of repose 274, 299, 929, 954
angular exclusion 235
anisotropy 890–892
  direction 763, 820, 829
  ratio 763, 820–821, 829
arcs, labelling 448–449
Arizona Copper property
  description 740
  drill hole data 909–910
  geologic description 905–906
  historical background 904
  location 904
  mineralization 906–909
  mining considerations 910–911
  topography 905
assays, undefined 759
assessment and reporting
  criteria, checklist 645, 646–649
at-mine-revenue (AMR) 88
Australasian Joint Ore
  Reserves Committee
  (JORC) 644, 645
average annual investment
  (AAI) 144–145
average annual investment
  cost (AAIC) 145
average profit ratio (APR) 606
avoirdupois weight 61

basic smelter return (BSR) 87, 91
bearing capacity 355–356
bench composites 191, 758, 806, 811

bench representation
  changing from one type to another 330
  median lines 328–330
  toes and crests 326–330
bench, geometry
  alternative designs 273–275
  back break 271
  bank width 271
  bench floor 271
  bench width 271
  crest 271, 326–330
  cut 271
  definitions 270–277
  face angle 271, 272, 303
  height 271–275
  median lines 328–330
  plan representation 326–330
  safety/catch bench 272–274, 296–302
  selection 275–276
  toe 271, 326–330
  working bench 271–272, 280, 296
benefits, worker 156
berms
  median 274, 364–365
  parallel 364
  safety 274, 299, 314
Bingham Canyon Mine 115–118, 180
block
  coordinate system 792–793
  dimensions 227, 741, 820
  grade assignment 235
  grid 229
  grid definition 748–750
  kriging 258
  mode 761
  model creation, steps 740–743

models 227–235
sequencing 586
size 227
two- and three-dimensional modeling 821–822
value assignment 839
versus point representation 230
block model mode 815
  assigning block values 820–826
  creating block model 826–831
  defining block model grid 816
  file description 832–834
  reading a block file 832
  saving a block file 831–832
  surface topography 816–820
block modelling
  assigning values in 418–420
  creation, steps 740–743, 826–831
  data file 743–744
  economic 418–420
block plots
  configure command 848–851
  control keys 840
  controlling the blocks plotted 851–853
  creating 765–769
block reports 862–865
block sequencing
  Lagrangian parametrization 598–604
  set dynamic programming 586–598
branch 446

727

728 *Index*

breakeven grade
  between milling and leaching 559–562
  cutoff grade definition 396
Bureau of Land Management (BLM) 25

California bearing ratio (CBR) 356–358
Canadian Institute of Mining, Metallurgy and Petroleum (CIM) 662
capital cost, inclusion in block model 513–515
cash flow
  calculations 55, 499
  components 56
  definition 49
  economic concept 46
  example 49–51, 54–55, 522–523
  factors considered 56
cash flow example, molybdenum mine
  pre-production 500–505
  production 505–511
catch/safety bench 272–274
cell, size 243, 865
centrifugal force 362, 363
chaotic variance 246, 247
checklist, planning data 7–11
CIF, definition 61
CIM, guidelines
  geological interpretation and modeling 666–669
  mineral reserve estimation 674–676
  mineral reserves reconciliation 680–681
  mineral resources estimation 669–673
  reporting 676–680
  resource database 664–666
circular failure 314, 320
closure, graph theory 446
CMJ (Southam), Reference Manual and Buyers Guide 96
Codelco Andina Copper property
  background information 941
  geology 941–942
  geotechnical slope analysis and design 943–946
  structural geology 942–943

  unit operations and initial costs, for pit generation 946–947
Codelco Norte Copper property 947
  economic analysis 958
  geology 949–950
  geotechnical information 950–952
  leach pad design and operation 954–955
  location and access 949
  material handling systems 952–953
  metallurgical testing/process development 953–954
  mine design and plan 955
  open pit geometry 952
  unit operations and manpower 955–957
cohesion, slope 315
composite files
  creating 741
  description 813–815
  reading 813
  saving 812
  storing and loading 759
composite mode 745, 758, 788, 806
  calculation 758–759, 807–810
composites
  bench/collar 811
  calculation 758–759
  dialog command 812
  highest bench 811
  interval 812
  maximum number 812
  mode 806–815
  ore-zone composite 190
  reasons and benefits 191
  types 806
compositing
  bench 190–191
  benefits 191
  definition 187
  ore-zone 190, 193
cone, volume of 209
contour plots
  configure command 853–859
  contour filling 850–853
  creating 770–772, 810–816
  menu control keys 840

  menu value descriptions 840
coordinate system 749
  description 792–793
copper mine
  assumptions 483–492
  examples 563–564
cost
  actual operations 96
  calculations, detailed 137
  current equipment, supplies and labor 153–158
  depletion basis 52–53
  escalation 88, 96
  estimating 95
  hourly earnings 112, 113, 114
  indexes 88, 96, 111, 114, 117
  types 95–96
cost estimation 95
  actual operations 96
  detailed example 137
  escalation 96
  hourly earnings 112, 113, 114
  O'Hora cost estimators 116
  quick and dirty mining 152–153
  types 95
Cost Reference Guide 153
covariance 252–254
critical path representation 24
cross slope 361
CSMine, basics
  changing modes 790–791
  coordinate system description 792–793
  data window 793
  file types 788–789
  formatting the data screen 791–792
  printing data 792
  project file 789–790
  sorting data 792
CSMine, data
  printing 792
  sorting 792
  window 793
CSMine, installing 738–739
CSMine, program
  design overview 744–745
  hardware requirements 738
  installing 738
  running 739
  warranty 737
CSMine, section
  creation 801

CSMine, slope command
    783–786
CSMine, software features
    788
CSMine, tutorial 737
    Arizona Copper property
        description 740
    block mode 743–744, 761
    composite mode 758–761
    creation of block model
        740–744
    drill hole mode 747
    executing commands 745
    program design overview
        744–745
    starting 738–739, 745–747
CSMine, user's guide
    basics 788–793
    block model mode 815
    block plots 848–853
    block reports 862–865
    composite mode 806–815
    contour plots 853–859
    drill hole mode 793
    economic block values
        834–840
    pit modeling 840–848
    plotting pit profiles
        859–861
    statistics 865
    variogram modeling 882
CSMine, warranty 737
cut
    drive-by 293
    drop 278–287
    frontal 290–293
    parallel 293
    sequencing 302–303
cutoff grade
    definition 392, 393
    dump versus mill 392–394
    incremental financial
        analysis 515–518
    Lane approach 538
    material destinations 556
    stockpile 562–568
cyanide solutions 33–34

data file
    creating a block model
        743–744
De Wijsian model 250
density
    metals 196
    minerals 195
    rock types 196

techniques for determining
    194–198
depletion
    cost 52, 510
    example 54–55, 508–510
    percentage 52
    rates 53
    unit 52
development
    definition 2
differential stripping ratio
    (DSR) 472
discounted cash flow (DCF)
    50
discounted cash flow rate of
    return (DCFROR)
    calculation 51
    definition 50
    example 51–52
distributions
    grade 238
    logarithmic 239
    normal 239
double benches 273, 274
drill hole
    construction of
        cross-section 199–203
    examples 188
    iron ore 199–203
    log 185–186
drill hole mode
    block grid 748–750
    data file description
        794–795
    creating plan map 750–754
    plotting plan map 796–801
    plotting section map
        801–806
    reading a file 747, 795
    section map, creation
        755–757
drive-by cuts 293
drop cut 278–287
dump leaching
    concept 557
    recovery 558–559
    time value of money effect
        596
    versus milling 559–561
dump planning 629–631

earnings, hourly 112–114
Earth Summit
    Agenda 21 701–703
    Rio Declaration 698–701
econometric models 86

economic block model
    description 418–420
    capital and profit costs
        513–515
economic block values
    assigning 773–774
    calculation 834–836
    creating economic block
        model 839–840
    example 837–839
    formula evaluation
        836–839
economics
    cash flow 49
    DCF 50
    DCFROR 50
    depletion 52–55
    depreciation 51
    future worth 46
    payback period 48
    present value, equal
        uniform payments
        47–48
    present worth 47
    rate of return (ROR) 48–49
Engineering News Record
    (ENR)
    cost indices 111
environmental
    Environmental Impact
        Statement (EIS) 25,
        35–36
    Environmental Protection
        Agency (EPA) 32
    National Environmental
        Policy Act (NEPA) 25
    recommended
        Environmental
        Assessment (ERA) 25
environmental planning
    procedures
    initial evaluation 35–37
    permits and approvals 40
    strategy 37–38
    team 38–40
equipment fleet requirements
    627–629
equipment ownership cost
    AAI 144
    AAIC 145
    depreciation 144
erosion protection 30
escalation
    cost 96
estimation variance 253, 823,
    826

730 *Index*

exploration
  definition 2
  information 3, 4
  IRS 50
exploratory drilling
  drill hole log 185, 192
  structural data 186
  techniques 184

F.O.B. (free-on-board)
  definition 61
feasibility study
  activity network 26
  content 12–14
  definition 12
  functions 14
  preparation 19–24
Federal Land Policy and
    Management Act
    (FLPMA) 25
final pit outline
  cross-sections 398
  generation 411–416
  longitudinal sections 206,
    405–411
  radial sections 405–411
  smoothing 412
final pit, sections
  end completion 207
  outline 204
  side completion 203
  tons and grade 207
floating cone
  3-D 840
  advantages 436
  bounds 780–781
  description 420
  example 421–429
  missing combinations
    425–427
  problem combinations
    428–429
  unprofitable extensions 427
forecasting, price 76
formatting, data screen
  791–792
freight rates 61–63, 90
friction angles 315
fringe benefits 136
front, mining 280–283
frontal cuts 290–292
future worth, cash flow 46

geologic planning information
  183–187
geometric pit limits 777–778

geometrical considerations
  bench geometry 270–277,
    326–330
  final pit slope angles 312
  geometric sequencing
    374–377
  ore access 277
  pit expansion process 290
  pit slope geometry
    303–312
  road construction 352
  addition of road 330
  stripping ratios 369–374
geostatistics
  cells 243, 244
  example 243–248
  lag distance 241
  log normal distribution 241
  nugget effect 246
  range of sample influence
    242
  semi-variance 243
  semi-variogram 243
  sill 246
  variance 243
Global Mining Initiative
  (GMI) 704–706
grade
  anisotropy 234, 247,
    250
  assignment 235–251
  contours 212–218
  statistical assignment
    235–251
graph theory 444–447
greenfield deposit, evaluation
    steps 632–633

highest bench elevation 759

IDS technique 232–235
  block model grid, value
    822–823
  CSMine 816, 820
income tax
  federal 510
  state 508
incremental financial analysis
  plant sizing example
    518–526
  principles 515–518
indexes, tables of
  cost 111
  ENR 111
  producer price 112
  productivity 112–113

instantaneous stripping ratio
    (ISR)
  calculation 372–374
  example 372–374
intermediate valuation report
  content 12, 13–14
Internal Revenue Service (IRS)
  development definition 55
  exploration definition 55
  exploration versus
    development 55
International Council on
    Mining and Metals
    (ICMM) 706–708
interpolation rules
  inverse distance 232
  inverse distance squared
    (IDS) 232–235, 236
  linear 232
  polygon 226
  rule of nearest points
    230–231
interramp angles 304–311
inverse distance squared (IDS)
  816, 820, 822–823, 827

Joint Ore Reserves Committee
  (JORC) 644
JORC Code –2004 edition
  competence and
    responsibility 650–652
  economic mineralized
    material 661–662
  exploration results,
    reporting 653–654
  general reporting 653
  mineral resources, reporting
    654–658
  ore reserves, reporting
    658–661
  reporting terminology
    652–653
  scope 649–650

Kennecott Barneys Canyon
    Gold property
  climate 934
  geologic setting 932–933
  geotechnical data 933
  mining data 935–936
  ore processing 934–935
  resource definition 933
  topography and surface
    conditions 934
key block
  coordinates 750, 816

kriging
  block value 258, 823–826
  block versus point 258
  common problems 259–260
  concept 252–254
  example 254–258
  lognormal 825–826
  ordinary block 825
  ordinary point 823, 824, 825
  universal 826
kriging/IDS parameters 829–831

lag distance 242
Lagrange multipliers 254, 824
landform reclamation 34–35
Lane's algorithm
  equation development 528–529
  illustrative example 529–530
  model definition 527–528
  modified Lane's algorithm 565–568
  net present value maximization 538
leaching
  copper example 557–562
  tailings and slime ponds 33
  types 33–34
Lerchs–Grossmann 2-D algorithm 429–438
Lerchs–Grossmann 2½-D algorithm 438–441
Lerchs–Grossmann 3-D algorithm, terms and concepts
  arcs 443–446
  branch 446–447
  chain 446
  circuit 446
  closure 446
  cycle 446
  description 441–447
  directed graph 445
  example 447–456
  graph 445
  maximum closure 446
  path 446
  root 446
  subgraph 445
  tree 446
  tree construction 447–454

  tree cutting 454–456
  twig 447
level plans 411, 414
linear interpolation, grade 231
logging, drill core 168
ltu, definition 57

map scale factor 798
maps
  coordinates 177–178
  general area 174
  general mine 174–178
  isometric view 180, 183
  preparation guidelines 177–180
  purpose 168
  quadrangle 170–171
  scale 168–171
  types 170, 173
material destinations
  considerations 556
  cutoff grade(s) 556
  in pit materials 416–418
  leach dump alternative 557–562
  stockpile alternative 562–568
Matheron (spherical) model 250
maximum search distance 821
median lines, bench 330, 413–415
mill(e) levy 504, 508
Milos Statement 716–719
mine development phases 5–7
mine life
  concepts 483, 493
  example, small copper deposit 483–492
  molybdenum mine example 494–499
  Taylor's rule 493–494
mine planning
  accuracy, estimation 17–18
  data collection checklist 7–11
  development phases 5–7
  phases 11–16
  planning costs 17
  studies 11–12
mine reclamation 24
mine waste
  dumps 31
  management 32–33
  tailings and slime ponds 33
  tailings dams 33

mineral deposit, stages 2
mineral inventory
  average grade versus cutoff grade 486
  definition 168
  grade class interval 484
  tonnage versus cutoff grade 485, 486
  tonnage versus grade class interval 485
mineral reserve
  reconciliation 680–683
  estimation 674–676
mineral resource
  estimation 669–672
  indicated 4
  inferred 4
  measured 3
mineral resource to mineral reserve
  quantifying elements 672–674
mineral resources and ore reserves
  CIM best practice guidelines 662
  JORC Code –2004 edition 645
  reporting 644, 652–662
Mining Cost Service 154
mining department, planning activities 28
mining industry and related initiatives 704
  GMI 704–706
  ICMM 705, 706–708
  MMSD 705, 708–709
  U.S. government and federal land management 709–712
  U.S. NMA 712–714
Mining Minerals and Sustainable Development (MMSD) 705, 706, 708–709
mining/milling rates
  copper example 483–492
  molybdenum example 494–499
Minnesota Natural Iron property
  access 913–914
  climatic condition 914–915
  economic basis 918–920
  general geologic setting 916–917

Index

Minnesota Natural Iron
property (*contd*)
historical background
915–916
initial hand design 918
mine-specific geology
917–918
topography 916
Minnesota Taconite property
general geology
927–928
history 926–927
location 926
mining data 929–931
ore processing 931–932
structural data 928–929
topography and surface
condition 927
mode command 748, 758,
761, 791
Mohr-Coulomb failure
criterion 314
molybdenum mine, example
cash flow, pre-production
500–505
cash flow, production
505–511
mining/milling rate
494–499
nested pits 494–499
MSHA rules 364
mtu, definition 57
multiple use management 25

nested pits, sequencing
494–499
net smelter return (NSR)
definition 86–88
example 90–92
net value
block 418–420
calculation for copper
392–398
versus grade curve
395–396
net value versus grade curve
construction 395–398
use 398
Newmont Gold property
deposit mineralization
938–939
general geologic setting
937–938
initial pit modeling
parameters 940

local climatic conditions
940
property location 936–937
topography and surface
conditions 939–940
NMA
principle 712–714
sustainable development
pledge 712–714
nugget effect 246, 250, 885,
886
number of blocks
block mode 763
maximum number, point
830
plan maximum 828

O'Hara cost estimator
concentrator operating costs
135
daily operating cost
134–136
general plant cost 130–132
mill capital cost 116–118
mine capital cost 116–118
original 116–118
other operating costs
136–137
personnel requirements 140
pit operating cost 134–135
updated 118
ore
definition 1
jobs 2
meaning 1–2
reserves 4, 5
ore reserve
definition 4, 5
probable 4, 5
proven 4, 5
ore reserve, calculation
block models 227–235
by level 416
final 416
grade contours 212–219
horizontal sections
219–227
pit 203–212, 216–222
polygons 223–227
triangular prisms 219–223
vertical sections 198–212
ore-zone composites
190–191
orebody
important books 903–904
journals 903

orebody description
block models 227
composite calculation
187–193
geologic information
183–187
kriging 251
method of horizontal
sections 219
method of vertical sections
198, 212
mine map 168
statistical basis for grade
assignment 235
tonnage factor calculation
193–198
Orebody, case examples
Codelco Andina Copper
property 941–947
Codelco Norte Copper
property 947–958
CSMine Arizona Copper
property 904–911
Kennecott Barneys Canyon
Gold property 932–936
Minnesota Natural Iron
property 911–920
Minnesota Taconite
property 925–932
Newmont Gold property
936–940
Utah Iron property
920–925
outer economic bound
778–780, 784, 847, 861

parallel cuts 293–302
payback period 48
percent payment 90
permits, list 40
phase planning, steps
606–608
phase scheduling
pre-production stripping
584–585
saw-tooth ore-time plots
583
steps 580–586
pit
plots 781–783, 840
slopes 841, 844, 924–925
pit extensions 463
pit generator, RTZ 457–463
pit limit determination
computer assisted 463–474
floating cone 420–429

hand methods 389
Lerchs–Grossmann 2-D 429–438
Lerchs–Grossmann 2½-D 438–441
Lerchs–Grossmann 3-D 441
RTZ pit generator 457–463
pit limits, hand methods
 comparison of sections 406
 cutoff grade 392–393
 example, pit bottom and one side in ore 404–405
 example, pit bottom in ore 404–405
 example, pit bottom in waste 398–403
 final pit outline 411–416
 net value vs grade curve 395–396
 radial sections 405–411
 smoothing 412
 stripping cost 396
 stripping ratio vs grade curve 396–398
pit modeling
 changing pit limit restrictions 847–848
 entering pit slopes 844–847
 geometric pit limit restrictions 841
 pit slopes 841
 positive apexed cone limits 841–844
 surface topography restrictions 841
 three-dimensional floating cone 844
pit slopes
 bench 270, 302
 final 312–326
 interramp 304–311
 overall 304–311
plan map, creation 750–754
Plan of Implementation 703–704
planar failure 314–320
planning studies
 accuracy 17–18
 conceptual 11–12
 costs 17
 feasibility 12
 phases 11–16
 preliminary/pre-feasibility 11–12

plant size
 ore reserves supporting 511–515
plotting pit profiles 859–861
polygons
 computer generated 231
 interpolation rules 226, 227
 ore reserve calculation 249
pre-production stripping 121, 584, 585, 609
present value, calculation
 series of equal uniform payments 47–48
 single payment 47
price
 current mineral 55–61
 econometric models 86
 forecasting 76–86
 historical data 61–76
 trend analysis 76–86
 trends 63
price–cost relationship 92–93
prism, volume of 208–209
Producer Price Index 112, 114
production planning
 cash flow calculations 499
 design process 631–633
 goals 482
 Lane's algorithm 526
 material destination considerations 556
 mine and mill plant sizing 511
 mine planning 631–633
 plant size concept 483–492
 production scheduling 568
 push back design 604
 rules/commandments 483
 sequencing, nested pits 494–499
 Taylor's mine rule 493–494
 types 483
production scheduling, simplified
 example 568
production, definition 2
productivity
 calculation 150
 definition 112
 index 114, 115
profit cost, inclusion in block model 514
profits 2
project phases 5
property evaluation, greenfield deposits 632

property ownership 912
pushback design, manual
 example 613
 steps 611–613
pushback, definition 375
quadrangle maps 170, 171
radial sections 405–411
range of influence
 definition 243, 246
 example 243–248
rate of return (ROR)
 definition 49
 example 49, 50
ratio of concentration 393
realization cost 88
reclamation management
 cyanide leach systems 33–34
 mine waste 32–33
 soil 30
 surface and ground water 31–32
 tailings and slime ponds 33
 water 31
reclamation plan
 content 28–29
 purpose 28
reclamation standards 29–31
reporting mineral resources and ore reserves
 exploration results 653–654
 general 653
 mineral resources 654–658
 ore reserves 658–661
 terminology 652–653
reserve
 definition 4
 probable 4, 5
 proven 4, 5
resource
 definition 3
 indicated 4, 5
 inferred 4, 5
 measured 4, 5
responsible mining 688, 716–719
 1972 UNCHE Conference 689–693
 Earth Summit 698–703
 good engineering 716–719
 mining industry and related initiatives 704–716
 WCS 693–696

734 Index

responsible mining (contd)
  World Commission on Environment and Development 696–698
  WSSD 703–704
restrictions command
  floating cone bound 780–781
  geometric pit limits 777–778
  outer economic bound 778–779
  Rio Declaration 698–701
  Stockholm Declaration 691–693
  surface constraint 775–777
revegetation 30–31, 34
revenues, estimation of 55
Reverse Polish Notation (RPN) 774, 835–840
Rio Declaration 698–701
road, addition to a pit
  factors 333
  inside the wall 336–341
  length 341–342, 346
  outside the wall 341–344
  switchbacks 344–347
  volume 280–282, 347–352
road, construction
  base 356, 358
  California Bearing Ratio (CBR) 356–358
  equivalent wheel loads 355
  particle gradation 357–358
  section design 353–358
  soil bearing capacity 355–356
  subbase 353
  subgrade 353
  tips 368–369
  tire contact pressure 354–355
  wearing surface 353, 355–357
road, geometry
  cross slope 361
  gradients 365–367
  median berm 364–365
  parallel berm 364
  superelevation 362
  width 359–361
road, maintenance tips 368–369
rock quality designation (RQD) 189

rolling radius, tire 299, 364
root, graph theory 444, 446
royalties 501, 508
RTZ open pit generator 457–463
rule-of-nearest points 230–231

safety
  bench 299–302
  berm 299–300, 364–365
  factor, slope 315, 317–318, 322, 324
save and print commands 786–787
search ellipse 763, 765, 820–821
section
  development 178–183
  manual example from iron ore 197–203, 204
  types 406
semi-variance 243
sequencing
  bench cut 301, 302
  geometric 374–377
  Lagrangian parametrization 598–602, 606, 610
  nested pits 494–499
  set dynamic programming 586
shovel
  cutting height 275
  digging radius 275, 276, 279
  specifications 276
  swing angle 290–291, 293
shovel cuts
  drive-by 293
  example 287, 289
  frontal 290–292
  parallel 293–302
  sequencing 301–303
  sidehill 289
  width 279, 301
sill 246, 250, 885
Similkameen mine 96, 97
skewness 238, 870
slime ponds 33
slope angle
  bench 271–272, 303–304
  final 312
  interramp 304–308
  overall 304–309, 311

slope, design charts
  circular 314, 320, 321, 322, 323
  planar 314–320
slope, failure modes
  3-D effects 324
  circular 320, 321, 322, 323
  planar 314–320
  toppling 314
  wedge 314
slope, problem
  classification 312
  data presentation 322–323
  economic aspects 324–325
  safety factor 317, 318, 320, 324
slope, properties
  cohesion 315
  friction angle 315
  ground water 322, 324
  stress condition 313
slopes command 783–786
sludge analysis 185
smelter, contracts
  content 89–90
  escalation provisions 88
  model 90
  percent deduction 87
  unit deduction 87
smelter, deductions
  percent 87
  unit 87
smelter, return 86–92
spherical model 246–248, 250, 888
spotting, shovel
  double 294, 297–299
  single 294, 295–296
standard deviation 84, 243, 869–870
statistics
  description 865–871
  distribution 871–878
  EX1.CMP data set 865
  EX2.CMP data set 865
  ore body 238–242
statistics command 881–882
transform command 878–881
statistics, orebody
  additive constant, β 241
  average grade 241
  basis for grade assignment 235–251

cumulative frequency 239, 241
cumulative probability 240, 241
grade distribution 238, 241
histogram 238
logarithmic transformation 239
normal distribution 239, 241
normalization 241
range 241
skewness 238, 239
standard deviation 242
variance 241
Stockholm Declaration 691–693
stockpiling 562–568
straight line depreciation 51, 144, 505
stripping ratio
  breakeven 391, 374, 397
  definition 372
  differential (DSR) 472
  example 372
  generalized formula 402–403
  instantaneous 372, 373–374, 389–390, 392
  overall 372, 373–374, 392
  versus grade curve (parallel sections) 396–397
  versus grade curve (radial sections) 405–411
structural domains 323–324
stu, definition 57
subdrill 275
superelevation 362–364, 368
surface
  constraints 775–777
  topography file 741, 787, 816–820
sustainable development, mining
  requirement 698
switchbacks 344–347, 368

tailings ponds 9, 33
tax, inclusion in cash flow calculations 49–51, 54–55
taxes
  calculation 504
  federal 504, 510
  property 504, 506
  severance 508
  state 508

Taylor's mine life rule 493–494
tension cracks, slopes 315, 319–320, 323
time period plans 625–627
tonnage factor
  calculation 187
  definition 193–194
  example 197–198
toppling failure 314
tree
  construction 447–454
  cutting 454–456
  definition 446
trend analysis 76–86
triangular prisms
  area 223
  average grade 220
  concept 219–223
  thickness 223
troy weight 61
twig 447

U.S. Government and federal land management 709–712
ultimate pit, definition 388
United Nations Conference on the Human Environment (UNCHE) 689, 693, 698
Stockholm Declaration 691–693
Utah Iron property
  background 920–921
  climate 922–923
  general geology 923–924
  initial cost estimates 925
  mineral processing 924
  mineralization 924
  mining history of the district 921–922
  other considerations 925
  pit slopes 924–925
  topography and surface vegetation 922

variance
  chaotic 246
  estimation 253–254, 257, 823, 826
  geostatistical 243, 882
  statistical 241–242
  structured 246
variogram
  construction 242–243
  definition 243
  information from 250

minimum number of pairs 246
parameters 828
variogram modeling
  anisotropy 890–892
  experimental 887–890
  variogram command 892
variograms, models
  de Wijsian 249–250
  deposit quantification 250–251
  linear 248–249
  Matheron/spherical 250
  vertical sections, method of 198
volumes, calculation of
  frustum of a cone 209
  prism 208
  right circular cone 209

warranty, software 737
waste dumps, design 32–33, 369
water management
  Clean Water Act (NPDES) 31
  ground water 31–32
  planning 630–631
  surface 31–32
  tailings and slime ponds 33
wedge failure 314
weighted average grades
  areas 206–207
  holes 190–193, 208
  volumes 207
weights and measures, units
  avoirdupois 61
  tons 57
  troy 61
Work Breakdown Structure (WBS) 19–22
Work Classification Structure 20
working bench 270, 271–272, 296–297
working capital 133, 504, 511
World Commission on Environment and Development
  specific goals 696
World Conservation Strategy (WCS) 693
  obstacles 695–696
World Summit Sustainable Development (WSSD) 703–704, 706